100 Years of Chronogeometrodynamics: The Status of the Einstein's Theory of Gravitation in Its Centennial Year

Special Issue Editors

Lorenzo Iorio
Elias C. Vagenas

MDPI • Basel • Beijing • Wuhan • Barcelona • Belgrade

MDPI

Special Issue Editors

Lorenzo Iorio
Ministero dell'Istruzione
dell' Università e della Ricerca
Italy

Elias C. Vagenas
Kuwait University
Kuwait

Editorial Office
MDPI AG
St. Alban-Anlage 66
Basel, Switzerland

This edition is a reprint of the Special Issue published online in the open access journal *Universe* (ISSN 2218-1997) from 2015–2017 (available at: http://www.mdpi.com/journal/universe/special_issues/100years).

For citation purposes, cite each article independently as indicated on the article page online and as indicated below:

Author 1, Author 2. Article title. *Journal Name*. **Year**. Article number/page range.

First Edition 2017

ISBN 978-3-03842-482-6 (Pbk)
ISBN 978-3-03842-483-3 (PDF)

Cover figure: iStockphoto.com – ID 509689300

Table of Contents

About the Special Issue Editors

Lorenzo Iorio was born in Bari, Italy, in 1971. He received his precollege and college education there, obtaining a degree in physics from the University of Bari in 1997. In 2002, he earned the Ph.D. Degree in physics from the Department of Physics "Michelangelo Merlin", the University of Bari, where he also completed his postdoctoral studies. Later, he go qualified by the Italian Ministry for Education, University and Research (MIUR) as an Associate Professor for Astronomy, Astrophysics and Earth and Planetary Physics, and for Theoretical Physics of the Fundamental Interactions, respectively. His research activity is currently in the field of gravitational physics, in particular, experimental/observational tests of general relativity and modified models of gravity in several astronomical and astrophysical scenarios, and their theoretical interpretation. He is author of more than 200 peer-reviewed articles, with over 3000 citations and an h index of 37 (source: The SAO/NASA Astrophysics Data System). He is Editor-in-Chief of the international peer-reviewed journal Universe, and serves as an associated editor of the international peer-reviewed journals Galaxies and Frontiers in Astronomy and Space Sciences. http://digilander.libero.it/lorri/homepage_of_lorenzo_iorio.htm.

Elias C. Vagenas was born in 1971, in Athens, Greece. He studied at the University of Athens, where he was awarded his Ph.D. degree in 2002 under the supervision of Professor Theodosios Christodoulakis. Then he moved to Barcelona, Spain, for a postdoctoral position at the University of Barcelona from Jan. 2003 till Dec. 2004. Later he moved back to Athens for a second postdoctoral position at the University of Athens from Jan. 2005 to Jun. 2007. In 2008, he was elected as Researcher C at the Research Center for Astronomy and Applied Mathematics (RCAAM) of the Academy of Athens. In 2011, he got his tenure and was promoted to a Researcher B. In Sept. 2013, he moved to Kuwait University as Associate Professor and joined the Theoretical Physics Group in the Department of Physics. He serves as Editorial Board Member for the international peer-review journals: Universe, Open Physics, Open Physics Journal, International Journal of Modern Physics D, ISRN Mathematical Physics, Gravitation and Cosmology, and Advances in High Energy Physics. He acts as referee for more than 15 journals. In 2007, he co-authored with Professor Saurya Das (University of Lethbridge, Canada) and Sean P. Robinson (MIT, USA) a paper published in the International Journal of Modern Physics D which received an "Honorable Mention" in the Gravity Research Foundation Annual Essay Competition. In 2013, he co-authored with D. Singleton and T. Zhu a paper which received the Fourth Prize in 2013 FQXi's Essay Contest.

Preface to "100 Years of Chronogeometrodynamics: The Status of the Einstein's Theory of Gravitation in Its Centennial Year"

Theoretical cosmology is a vastly developing science, bringing a fundamental link between particle physics, theoretical physics and astrophysics. Its recent success and huge increase of publications are related to the fact that we are living in the era of the so-called precision cosmology. Hence, more and more accurate bounds are obtained on the cosmological parameters through the whole cosmic history, starting from the Big Bang up to the currently observed Universe. Observational confirmations for the early and late accelerations come from data related to the cosmic microwave background, the large scale structure, the baryonic acoustic oscillations, the supernovae, the Hubble flow, the gravitational lensing, among others. The overall picture of the Universe is that of an accelerating dynamical system dominated by some cosmic fluids giving rise to the current 'speed up' (dark energy or cosmological constant), a dark matter component accounting for the clustering and the stability of structures, and some percentage of luminous matter, radiation and neutrinos.

Several questions arise about the basic gravitational theory which governs the cosmic evolution. The above observations point out that either most of the Universe content is unknown (and then the problem of dark energy and dark matter should be solved from the particle physics side) or General Relativity should be reviewed in view of some extension, retaining the good results at local scales and addressing the ultraviolet and infrared problems at cosmic scales. At best, General Relativity is a reasonable approximation for the description of the Universe evolution at local scales. However, it should be qualitatively modified, especially at the very early (ultraviolet) scales and, eventually, at the current and even at the future epochs (infrared scales). This revision can be related to the three not yet understood components of the universe, specifically inflation, dark matter and dark energy. This is precisely the Modified Gravity Approach (for a review, see S. Nojiri and S.D. Odintsov, "Unified Cosmic History in Modified Gravity: from F(R) Theory to Lorentz Non-Invariant Models," Phys. Rept. 505, (2011) 59 and S. Capozziello and M. De Laurentis, "`Extended Theories of Gravity," Phys. Rept. 509 (2011) 167) which suggests the way to consider all the three dark components of the universe history under the standard of a single gravity theory. In such an approach, modified gravity acts in different ways at high curvature regime and at low curvature regime, so that it is capable of causing early-time inflation, formation of structures at intermediate epochs, as well as late-time acceleration. As some extra benefit, the modification of gravity could describe also dark matter (see S. Capozziello and M. De Laurentis "The Dark Matter Problem from f(R) Gravity Viewpoint" Annalen Phys. 524 (2012) 545). It is not surprising then that this current volume is devoted to these questions. The authors are specialists considering the deep issues of General Relativity, its validity, applications, and modifications. Specifically, in this volume, topics related to torsion gravity, dark energy, inflation and singularities are taken into account. The role of topology and non-zero cosmological constant are also discussed. The volume can represent a useful overview on the current status of General Relativity and the fundamental questions which are still open in theoretical cosmology and in the foundations of gravitational physics. It aims to be an overall and self-contained work that, without any claim of completeness, could be a useful research instrument.

Salvatore Capozziello
Università di Napoli "Federico II", Italy
Sergei D. Odintsov
ICREA and Space Science Inst., CSIC, Barcelona, Spain
July 2017

universe

MDPI

Article

Editorial for the Special Issue 100 Years of Chronogeometrodynamics: The Status of the Einstein's Theory of Gravitation in Its Centennial Year

Lorenzo Iorio

Ministero dell' Istruzione dell' Università e della Ricerca (M.I.U.R), Fellow of the Royal Astronomical Society (F.R.A.S.), Viale Unità di Italia, 68 70125 Bari, Italy; lorenzo.iorio@libero.it; Tel.: +39-3292399167

Received: 6 March 2015; Accepted: 17 April 2015; Published: 27 April 2015

Abstract: The present Editorial introduces the Special Issue dedicated by the journal *Universe* to the General Theory of Relativity, the beautiful theory of gravitation of Einstein, a century after its birth. It reviews some of its key features in a historical perspective, and, in welcoming distinguished researchers from all over the world to contribute it, some of the main topics at the forefront of the current research are outlined.

Keywords: general relativity and gravitation; classical general relativity; gravitational waves; quantum gravity; cosmology; experimental studies of gravity

PACS: 04.; 04.20.-q; 04.30.-w; 04.60.-m; 98.80.-k; 04.80.-y

1. Introduction

This year marks the centenary of the publication of the seminal papers [1–3] in which Albert Einstein laid down the foundations of his theory of gravitation, one of the grandest achievements of the human thought which is the best description currently at our disposal of such a fundamental interaction shaping the fabric of the natural world. It is usually termed "General Theory of Relativity" (GTR, from *Allgemeine Relativitätstheorie*), often abbreviated as "General Relativity" (GR). It replaced the Newtonian concept of "gravitational force" with the notion of "deformation of the chronogeometric structure of spacetime" [4] due to all forms of energy weighing it; as such, it can be defined as a chronogeometrodynamic theory of gravitation [5].

GTR is connected, in a well specific sense, to another creature of Einstein himself, with Lorentz [6] and Poincaré [7,8] as notable predecessors, published in 1905 [9]: the so-called Special (or Restricted) Theory of Relativity (STR). The latter is a physical theory whose cornerstone is the requirement of covariance of the differential equations expressing the laws of physics (originally only mechanics and electromagnetism) under Lorentz transformations of the spacetime coordinates connecting different inertial reference frames, in each of which they must retain the same mathematical form. More precisely, if

$$A(x, y, z, t), B(x, y, z, t), C(x, y, z, t), \ldots \qquad (1)$$

represent the state variables of a given theory depending on spacetime coordinates x, y, z, t and are mutually connected by some mathematical relations

$$f(A, B, C, \ldots) = 0 \qquad (2)$$

representing the theory's fundamental equations, the latter ones can always be mathematically written in a covariant form under a generic transformation from the old coordinates to the new ones as

$$f'\left(A'\left(x',y',z',t'\right),B'\left(x',y',z',t'\right),C'\left(x',y',z',t'\right),\ldots\right)=0 \tag{3}$$

In general, the new functional relations f' connecting the transformed state variables $A',B',C',\ldots$ are different from the ones of f. If, as for the Lorentz transformations, it turns out

$$f'=f \tag{4}$$

which does not necessarily implies that also the state variables $A,B,C,\ldots$ remain unchanged, then it is said that the equations of the theory retain the same form. It is just the case of the Maxwell equations, in which the electric and magnetic fields E,B transform in a given way under a Lorentz transformation in order to keep the form of the equations connecting them identical, which, instead, is not retained under Galilean transformations [10]. In the limiting case of the Galilean transformations applied to the Newtonian mechanics, it turns out that the theory's equations are even invariant in the sense that also the state variables remain unchanged, *i.e.*, it is

$$F'-ma'=0 \tag{5}$$

with

$$F'=F \tag{6}$$

$$a'=a$$

As such, strictly speaking, the key message of STR is actually far from being: "everything is relative", as it might be seemingly suggested by its rather unfortunate name which, proposed for the first time by Planck [11] (*Relativtheorie*) and Bucherer [12] (*Relativitätstheorie*), became soon overwhelmingly popular (see also [13]). Suffice it to say that, in informal correspondence, Einstein himself would have preferred that its creature was named as *Invariantentheorie* (Theory of invariants) [14], as also explicitly proposed-unsuccessfully-by Klein [15]. Note that, here, the adjective "invariant" is used, in a looser sense, to indicate the identity of the mathematical functional form connecting the transformed state variables.

Notably, if the term "relativity" is, instead, meant as the identity of all physical processes in reference frames in reciprocal translational uniform motion connected by Lorentz transformations, then, as remarked by Fock [16], a name such as "Theory of Relativity" can, to some extent, be justified. In *this specific sense*, relativity geometrically corresponds to the maximal uniformity of the pseudo-Euclidean spacetime of Poincaré and Minkowski in which it is formulated. Indeed, given a $N-$dimensional manifold, which can have constant curvature, or, if with zero curvature, can be Euclidean or pseudo-Euclidean, the group of transformations which leave identical the expression for the squared distance between two nearby points can contain at most $(1/2)N(N+1)$ parameters. If there is a group involving all the $(1/2)N(N+1)$ parameters, then the manifold is said to have maximal uniformity. The most general Lorentz transformations, which leave unchanged the coefficients of the expression of the 4-dimensional distance between two nearby spacetime events, involve just 10 parameters. Now, in the pseudo-Riemannian spacetime of GTR the situation is different because, in general, it is not uniform at all in the geometric sense previously discussed. Following Fock [16], it can be effectively illustrated by a simple example whose conclusion remains valid also for the geometry of the 4-dimensional spacetime manifold. Let us think about the surface of a sphere, which is a 2-dimensional manifold of a very particular form. It is maximally uniform since it can be transformed into itself by means of rotations by any angle about an arbitrary axis passing through the centre, so that the associated group of transformations has just three parameters. As a result, on a surface of a sphere there are neither preferred points nor preferred directions. A more general non-spherical surface of revolution has only partial uniformity since it can be transformed into itself by rotation about an axis which is now fixed, so that the rotation angle is the only arbitrary parameter left. There are privileged points and lines: the poles through which the axis passes, meridians, and latitude circles. Finally, if we consider a surface of general form, there will be no transformations taking it into itself, and it will possess no uniformity whatsoever. Thus, it should be clear that the generality of the form of the surface is a concept antagonistic to the concept of uniformity. Returning now to the concept of relativity *in the aforementioned specified sense*, it is related to uniformity in all those cases in which the spacetime metric can be considered fixed. This occurs not only in the Minkowskian spacetime, but also in the Einsteinian one, provided only that the physical processes one considers have no practical influence on the metric. Otherwise, it turns out that relativity can, to a certain extent, still be retained only if the non-uniformity generated by heavy masses may be treated as a local perturbation in infinite Minkowskian spacetime. To this aim, let us think about a laboratory on the Earth's surface [16]. If it was turned upside down, relativity would be lost since the physical processes in it would be altered. But, if the upset down laboratory was also parallel transported to the antipodes, relativity would be restored since the course of all the processes would be the same as at the beginning. In this example, a certain degree of relativity was preserved, even in a non-uniform spacetime, because the transformed gravitational field g' in the new coordinate system $\{x'\}$ has the same form as the old field g in the old coordinates $\{x\}$, *i.e.*,

$$\{x\} \mapsto \{x'\} \tag{7}$$

$$g(x) \mapsto g'\left(x'\right) = g(x) \tag{8}$$

Such considerations should have clarified that relativity, in the previously specified sense, either does no exist at all in a non-uniform spacetime like the Einsteinian one, or else it does exist, but does *not* go *beyond* the relativity of the Minkowskian spacetime. *In this sense*, the gravitational theory of Einstein *cannot* be a *generalization of his theory of space and time of 1905*, and its notion of relativity along with its related concept of maximal uniformity was *not* among the concepts subjected to *generalization*. Since the greatest possible uniformity is expressed by Lorentz transformations, there *cannot* be a *more general* principle of relativity than that discussed in the theory of 1905. All the more, there cannot be a general principle of relativity having physical meaning which would hold with respect to arbitrary frames of references. As such, both the denominations of "General Relativity" and "General Theory of Relativity" are confusing and lead to misunderstandings. Furthermore, such adjectives reflect also an incorrect understanding of the theory itself since they were adopted referring to the covariance of the equations with respect to arbitrary transformations of coordinates accompanied by the transformations of the coefficients of the distance between two events in the 4-dimensional spacetime. But it turned out that such kind of covariance has actually nothing to do with the uniformity or non-uniformity of spacetime [16,17]. Covariance of equations *per se* is just a merely mathematical property which in no way is expression of any kind of physical law. Suffice it to think about the Newtonian mechanics and the physically equivalent Lagrange equations of second kind which are covariant with respect to arbitrary transformations of the coordinates. Certainly, nobody would state that Newtonian mechanics contains in itself "general" relativity. A principle of relativity-Galilean or Einsteinian-implies a covariance of equations, but the converse is not true: covariance of differential equations is possible also when no principle of relativity is satisfied. Incidentally, also the the adjective "Special" attached to the theory of 1905 seems improper in that it purports to indicate that it is a special case of "General" Relativity.

In the following, for the sake of readability, we will adhere to the time-honored conventions by using STR and GTR (or GR) for the Einsteinian theory of space and time of 1905 and for his gravitational theory of 1915, respectively.

Of course, the previous somewhat "philosophical" considerations are, by no means, intended to undermine the credibility and the reliability of the majestic theory of gravitation by Einstein, whose concordance with experiments and observations has been growing more and more over the latest decades [18].

Below, some key features of GTR, to which the present Special Issue is meritoriously and timely dedicated, are resumed in a historical perspective [19–21], without any pretence of completeness. It is hoped that the distinguished researchers who will kindly want to contribute it will provide the community of interested readers with the latest developments at the forefront of the research in this fascinating and never stagnant field.

In the following, Greek letters $\mu, \nu, \varrho \ldots$ denote 4-dimensional spacetime indexes running over $0, 1, 2, 3$, while Latin ones $i, j, k, \ldots$, taking the values $1, 2, 3$, are for the 3-dimensional space.

2. The Incompatibility of the Newtonian Theory of Gravitation with STR

In the framework of the Newtonian theory of universal gravitation [22], the venerable force-law yielding the acceleration a imparted on a test particle by a mass distribution of density ρ could be formally reformulated in the language of the differential equations governing a field-type state variable Φ, known as potential, through the Poisson equation [23]

$$\nabla^2 \Phi = 4\pi G \rho \tag{9}$$

where G is the Newtonian constant of gravitation, so that

$$a = -\nabla \Phi \tag{10}$$

Nonetheless, although useful from a mathematical point of view, such a field was just a non-dynamical entity, deprived of any physical autonomous meaning: it was just a different, mathematical way of telling the same thing as the force law actually did [20]. It is so because, retrospectively, in the light of STR, it was as if, in the Newtonian picture, the gravitational interaction among bodies would take place *de facto* instantaneously, irrespectively of the actual distance separating them, or as if gravity would be some sort of occult, intrinsic property of matter itself. Remarkably, such a conception was opposed by Newton himself who, in the fourth letter to R. Bentley in 1692, explicitly wrote [24]: "[…] Tis inconceivable that inanimate brute matter should (without the mediation of something else which is not material) operate upon & affect other matter without mutual contact; as it must if gravitation in the sense of Epicurus be essential & inherent in it. And this is one reason why I desired you would not ascribe {innate} gravity to me. That gravity should be innate inherent & {essential} to matter so that one body may act upon another at a distance through a vacuum without the mediation of any thing else by & through which their action or force {may} be conveyed from one to another is to me so great an absurdity that I believe no man who has in philosophical matters any competent faculty of thinking can ever fall into it. Gravity must be caused by an agent {acting} consta{ntl}y according to certain laws, but whether this agent be material or immaterial is a question I have left to the consideration of my readers.". In the previous quotation, the text in curly brackets { … } is unclear in the manuscript, but the editor of the original document is highly confident of the reading.

In the second half of the nineteenth century, with the advent of the Maxwellian field theory of electromagnetism [25] scientists had at disposal a mathematically coherent and empirically well tested model of a physical interaction among truly dynamical fields which propagate as waves even *in vacuo* at the finite speed of light c transferring energy, momentum and angular momentum from a point in space to another. Now, STR is based on two postulates: The Principle of Relativity, extended by Einstein to all physical interactions, and another principle that states that the speed of light is independent of the velocity of the source. In this form, it retains its validity also in GTR. The latter is an immediate consequence of the law of propagation of an electromagnetic wave front which is straightforwardly obtained from the Maxwell equations obeying, by construction, the Principle of Relativity itself since they turned out to be covariant under Lorentz transformations. It necessarily follows [16] that there exists a maximum speed for the propagation of any kind of physical action.

This is numerically equal just to the speed of light *in vacuo*. If there was no single limiting velocity but instead different agents, e.g., light and gravitation, propagated *in vacuo* with different speeds, then the Principle of Relativity would necessarily be violated as regards at least one of the the agents. Indeed, it would be possible to choose an inertial frame traveling just at the speed of the slower agent in which the differential equations governing its course would take a particular form with respect to that assumed in all the other frames, thus predicting spurious, unphysical phenomena. It is reminiscent of the famous first *gedankenexperiment* made by Einstein about STR around 1895–1896 described by himself as follows [26]: "[...] Wenn man einer Lichtwelle mit Lichtgeschwindigkeit nachläuft, so würde man ein zeitunabhängiges Wellenfeld vor sich haben. So etwas scheint es aber doch nicht zu geben!" ["If one goes after a light wave with light velocity, then one would have a time-independent wavefield in front of him. However, something like that does not seem to exist!"] Indeed, the Maxwell equations *in vacuo*, in their known form, do not predict stationary solutions. That posed severe challenges to the Newtonian gravitational theory [8], which necessarily would have had to abandon its strict force-law aspect in favor of a genuine field-type framework making the Poisson equation covariant under Lorentz transformations [19,27].

Furthermore, as pointed out by Einstein himself [28], Newtonian universal gravitation did not fit into the framework of the maximally uniform spacetime of SRT for the deepest reason that [16], while in SRT the inertial mass m_i of a material system had turned out to be dependent on its total energy, in the Newtonian picture the gravitational mass m_g, did not. At high speeds, when the change in the inertia of a body becomes notable, this would imply a breakdown of the law of free fall, whose validity was actually well tested, although only at non-relativistic regimes (see Section 3).

Finally, it can be remarked also that the required Lorentz covariance would have imposed, in principle, also the existence of a new, magnetic-type component of the gravitational field so to yield some sort of gravitational inductive phenomena and travelling waves propagating at the finite speed of light *in vacuo*. Unfortunately, at the dawn of the twentieth century, there were neither experimental nor observational evidence of such postulated manifestations of a somehow relativistic theory of gravitation.

3. The Equivalence Principle and Its Consequences

3.1. The Equality of the Inertial and Gravitational Masses Raised to the Status of a Fundamental Principle of Nature

Luckily, at that time, Einstein was pressed also by another need: The quest for a coherent framework to consistently write down the laws of physics in arbitrary frames of references moving according to more complicated kinematical laws than the simple uniform translation. In 1907 [29], Einstein realized that the bridge across such two apparently distinct aspects could have been represented by the equality of the inertial and gravitational masses, known at that time to a 5×10^{-8} accuracy level thanks to the Eötvös experiment [30].

That was an empirical fact well known since the times of Galilei thanks to the (likely) fictional [31–33] tales of his evocative free fall experiments [34] allegedly performed from the leaning tower of Pisa around 1590. Newton himself was aware of the results by Galilei, and made his own experiments with pendulums of various materials obtaining an equality of inertial and gravitating masses to a 10^{-3} level of relative accuracy. Indeed, in the Proposition VI, Theorem VI, Book III of his *Principia* [22] Newton wrote [35]: "It has been, now for a long time, observed by others, that all sorts of heavy bodies [...] descend to the Earth from equal heights in equal times; and that equality of times we may distinguish to a great accuracy, by the help of pendulums. I tried experiments with gold, silver, lead, glass, sand, common salt, wood, water, and wheat. I provided two wooden boxes, round and equal: I filled the one with wood, and suspended an equal weight of gold (as exactly as I could) in the centre of oscillation of the other. The boxes, hanging by equal threads of 11 feet, made a couple of pendulums perfectly equal in weight and figure, and equally receiving the resistance of the air. And, placing the one by the other, I observed them to play together forwards and backwards, for a long time, with equal vibrations. And therefore the quantity of matter in the gold (by Cors. I and VI, Prop. XXIV, Book II) was to the quantity of matter in the wood as the action of the motive force (or *vis motrix*) upon all the gold to the action of the same upon all the wood; that is, as the weight of the one to the weight of the other: and the like happened in the other bodies. By these experiments, in bodies of the same weight, I could manifestly have discovered a difference of matter less than the thousandth part of the whole, had any such been." Interestingly, in the Proposition VI, Theorem VI, Book III of his *Principia* [22], Newton looked also the known motions of the natural satellites of Jupiter to make-from a phenomenological point of view -a further convincing case for the equality of the inertial and gravitational masses. Indeed, if the ratios of the gravitational to the inertial mass of Jupiter and of its satellites were different, the orbits of the Jovian moons about their parent planet would be unstable because of an imperfect balancing of the centrifugal acceleration and the Jupiter centripetal attraction caused by a residual, uncancelled force due to the Sun's attractions on either Jupiter and its moons themselves. Indeed, Newton wrote [36]: "[...] that the weights of Jupiter and of his satellites towards the Sun are proportional to the several quantities of their matter, appears from the exceedingly regular motions of the satellites (by Cor. III, Prop. LXV, Book I). For if some of those bodies were more strongly attracted to the Sun in proportion to their quantity of matter than others, the motions of the satellites would be disturbed by that inequality of attraction (by Cor. II, Prop. LXV, Book I). If, at equal distances from the Sun, any satellite, in proportion to the quantity of its matter, did gravitate towards the Sun with a force greater than Jupiter in proportion to his, according to any given proportion, suppose of *d* to *e*; then the distance between the centres of the Sun and of the satellite's orbit would be always greater than the distance between the centres of the Sun and of Jupiter, nearly as the square root of that proportion: as by some computations I have found. [...]" In principle, the Newtonian gravitational theory would have not lost its formal consistency even if experiments-all conducted at low speeds with respect to *c*-would have returned a different verdict about m_i/m_g. Nonetheless, one cannot help but notice as the very same name chosen by Newton for the universally attractive force regulating the courses of the heavens, *i.e.*, gravitation, may point, somehow, towards a not so accidental nature of the equality of inertial and gravitating masses. Indeed, it comes from the Latin word *gravis* ('heavy') with several Indoeuropean cognates [37], all with approximately the same meaning related to the weight of common objects on the Earth's surface: Sanskrit, *guruḥ* ('heavy, weighty, venerable'); Greek, βάροζ ('weight') and βαρύζ ('heavy in weight'); Gothic, *kaurus* ('heavy'); Lettish, *gruts* ('heavy'). It is tempting to speculate that, perhaps, Newton had some sort of awareness of the fundamental nature of that otherwise merely accidental fact. It seems not far from the position by Chandrasekhar who wrote [38]: "There can be no doubt that Newton held the *accurate proportionality of the weight 'to the masses of matter which they contain'* as inviolable".

Whatever the case, Einstein promoted it to a truly *fundamental* cornerstone on which he erected his beautiful theoretical building: the Equivalence Principle (EP). Indeed, the postulated *exact* equality of the inertial and gravitational mass implies that, in a given constant and uniform gravitational field, all bodies move with the same acceleration in exactly the same way as they do in an uniformly accelerated reference frame removed from any external gravitational influence. In this sense, an uniformly accelerated frame in absence of gravity is equivalent to an inertial frame in which a constant and uniform gravitational field is present. It is important to stress that the need of making the universality of the free fall, upon which the EP relies, compatible with the dictates of the SRT was not at all a trivial matter [21] (cfr. Section 1), and the merit of keeping the law of free fall as a *fundamental* principle of a viable relativistic theory of gravitation which could not reduce to a mere extension of the Newtonian theory to the SRT must be fully ascribed to Einstein. To better grasp the difficulties posed by such a delicate conceptual operation, let us think about an inertial reference frame *K* in which two stones, differing by shape and composition, move under the action of a uniform gravitational field starting from the same height but with different initial velocities; for the sake of simplicity, let us assume that, while one of the two stones is thrown horizontally with an initial velocity with respect to *K*, the other one falls vertically starting at rest [21]. Due to the universality of free fall, both the stones reach the ground simultaneously. Let us, now, consider an inertial frame K' moving uniformly at a speed equal to the horizontal component of the velocity of the projectile; in this frame, the kinematics of the two objects gets interchanged: the projectile has no horizontal velocity so that now it falls vertically, while the stone at rest acquires an horizontal velocity making it move parabolically in the opposite direction with respect to K'. According to the universality of the free fall, also in this case they should come to the rest at the same time. But this is in disagreement with the relativity of the simultaneity of the SRT. Moreover, another source of potential tension between the universality of the free fall and the SRT is as follows [21]. According to the latter one, a change in the energy of a body corresponds to a change also in its inertial mass, which acts as a "brake". On the other hand, since the inertial mass is equivalent to the gravitational mass, which, instead, plays the role of "accelerator", the correct relativistic theory of gravitation necessarily implies that also the gravitational mass should depend in an exactly known way from the total energy of the body. Actually, other scientists like, e.g., Abraham [27] and Mie [39] were willing to discard the Galileo's law of universality of free fall to obtain a relativistic theory of gravitation.

The heuristic significance of the original form of the EP unfolded in the findings by Einstein that identical clocks ticks at different rates if placed at different points in a gravitational potential, a feature which was measured in a laboratory on the Earth's surface in 1960 [40] by means of the Mössbauer effect which has recently received a general relativistic interpretation [41], and the gravitational redshift of the spectral lines emitted at the Sun's surface with respect to those on the Earth, which was measured only in the sixties of the last century [42] following the 1925 measurement with the spectral lines in the companion of Sirius [43]. Furthermore, it turned out that the speed of light in a gravitational field is variable, and thus light rays are deflected, as if not only an inertial mass but also a gravitational mass would correspond to any form of energy. Einstein [28] was also able to calculate the deflection of the apparent position of background stars due to the Sun's gravitational potential, although the value he found at that time was only half of the correct one later predicted with the final form of his GTR [44] and measured in 1919 [45,46] (see Section 4). In 1912, he [47,48] explored the possibility of gravitational lensing deriving the basic features of the lensing effect, which will be measured for the first time not until 1979 [49]. It must be noted [19,21] that this theory of the constant and uniform gravitational field went already *beyond* STR. Indeed, because of the dependence of the speed of light and the clock rates on the gravitational potential, STR definition of simultaneity and the Lorentz transformation themselves lost their significance (cfr. Section 1). *In this specific sense*, it can be said that STR can hold only in absence of a gravitational field.

The existence of non-uniformly accelerated reference frames like, e.g., those rotating with a time-dependent angular velocity $\Omega(t)$, naturally posed the quest for a further generalization of the EP able to account for spatially and temporally varying gravitational fields as well. The extension of the EP to arbitrarily accelerating frames necessarily implies, in principle, the existence of further, non-uniform, non-static (either stationary and non-stationary) and velocity-dependent gravitational effects, as guessed by Einstein [50–52]. They were later fully calculated by Einstein himself [53] and others [54–60] with the final form of the GTR (see Section 4 and [61–63] for critical analyses of the seminal works), which could not be encompassed by the gravito-static Newtonian framework. Indeed, it must be recalled that the inertial acceleration experienced by a body (slowly) moving with velocity v' with respect to a rotating frame K' is

$$a'_\Omega = 2\Omega \times v' + \dot{\Omega} \times r' + \Omega \times \left(\Omega \times r'\right) \tag{11}$$

At least to a certain extent, such new gravitational effects, some of which have been measured only a few years ago [64–67], might be considered as reminiscent of the Machian relational conceptions of mechanics [68–71].

Such a generalization of the EP to arbitrary gravitational fields lead Einstein to reformulate it as follows: in any infinitesimal spacetime region (*i.e.*, sufficiently small to neglect either spatial and temporal variations of gravity throughout it), it is always possible to find a suitable non-rotating coordinate system K_0 in which any effect of gravity on either test particle motions and any other physical phenomena is absent. Such a local coordinate system can ideally be realized by a sufficiently small box moving in the gravitational field freely of any external force of non-gravitational nature. Obviously, it appeared natural to assume the validity of STR in K_0 in such a way that all the reference frames connected to it by a Lorentz transformation are physically equivalent. *In this specific sense*, it could be said that the Lorentz covariance of all physical laws is still valid in the infinitely small.

At this point, still relying upon the EP, it remained to construct a theory valid also for arbitrarily varying gravitational fields by writing down the differential equations connecting the gravitational potential, assumed as state variable, with the matter-energy sources and requiring their covariance with respect to a fully general group of transformations of the spacetime coordinates.

3.2. Predictions of the Equivalence Principle

A step forward was done in 1914 when, in collaboration with Grossmann, Einstein [72], on the basis of the Riemannian theory of curved manifolds, was able to introduce the ten coefficients $g_{\mu\nu}$ of the symmetric metric tensor g by writing down the square of the spacetime line element $(ds)^2$ between two infinitely near events in arbitrary curvilinear coordinates x^μ as

$$(ds)^2 = g_{\mu\nu} dx^\mu dx^\nu \tag{12}$$

As a consequence, the equations of motion of a test particle, the energy-momentum theorem and the equations of the electromagnetism *in vacuo* were simultaneously written in their generally covariant ultimate form. In particular, from the right-hand-side of the geodesic equation of motion of a test particle

$$\frac{d^2 x^\alpha}{ds^2} = -\Gamma^\alpha_{\beta\varrho} \frac{dx^\beta}{ds} \frac{dx^\varrho}{ds} \tag{13}$$

where the Christoffel symbols

$$\Gamma^\alpha_{\beta\varrho} \doteq \frac{1}{2} g^{\alpha\sigma} \left(\frac{\partial g_{\sigma\beta}}{\partial x^\varrho} + \frac{\partial g_{\sigma\varrho}}{\partial x^\beta} - \frac{\partial g_{\beta\varrho}}{\partial x^\sigma} \right) \tag{14}$$

are constructed with the first derivatives of $g_{\mu\nu}$, it was possible to straightforwardly identify the components of g as the correct state variables playing the role of the Newtonian scalar potential Φ. Indeed, to a first-order level of approximation characterized by neglecting terms quadratic in v/c and the squares of the deviations of the $g_{\mu\nu}$ from their STR values

$$\eta_{00} = +1 \tag{15}$$

$$\eta_{ij} = -\delta_{ij}$$

the geodesic equations of motion for the spatial coordinates become

$$\frac{d^2 x^i}{dt^2} = -c^2 \Gamma^i_{00} \tag{16}$$

Furthermore, if the gravitational field is assumed static or quasi-static and the time derivatives can be neglected, the previous equations reduce to

$$\frac{d^2 x^i}{dt^2} = \frac{c^2}{2} \frac{\partial g_{00}}{\partial x^i} \tag{17}$$

By posing

$$\Phi \doteq -\frac{1}{2} c^2 (g_{00} - 1) \tag{18}$$

so that

$$g_{00} = 1 - \frac{2\Phi}{c^2} \qquad (19)$$

the Newtonian acceleration is obtained. The additive constant up to which the potential is defined is fixed in such a way that Φ vanishes when g_{00} assumes its STR value η_{00}. It is worthwhile remarking that, to the level of approximation adopted, only g_{00} enters the equations of motion, although the deviations of the other metric coefficients from their STR values may be of the same order of magnitude. It is this circumstance that allows to describe, to a first order approximation, the gravitational field by means of a single scalar potential.

In analogy with the geodesic equations of motion for a test particle, also those for the propagation of electromagnetic waves followed. Indeed, the worldlines of light rays are, thus, geodesics curves of null length

$$(ds)^2 = 0 \qquad (20)$$

$$\frac{d^2 x^\alpha}{d\lambda^2} = -\Gamma^\alpha_{\beta\varrho} \frac{dx^\beta}{d\lambda} \frac{dx^\varrho}{d\lambda}$$

where λ is some affine parameter.

The components of the metric tensor g are not assigned independently of the matter-energy distributions, being determined by field equations.

A further consequence of EP and the fact that, to the lowest order of approximation, g_{00} is proportional to the Newtonian potential Φ is that, in general, it is possible to predict the influence of the gravitational field on clocks even without knowing all the coefficients $g_{\mu\nu}$; such an influence is actually determined by g_{00} through

$$d\tau = \sqrt{g_{00}}dt \qquad (21)$$

where τ is the reading of a clock at rest. Instead, it is possible to predict the behaviour of measuring rods only knowing all the other coefficients g_{0i}, g_{ik}. Indeed, it turns out that the square of the distance dl between two nearby points in the 3-dimensional space is given by [73]

$$(dl)^2 = \left(-g_{jh} + \frac{g_{0j} g_{0h}}{g_{00}} \right) dx^j dk^h \qquad (22)$$

Thus, the field g determines not only the gravitational field, but also the behaviour of clocks and measuring rods, *i.e.*, the chronogeometry of the 4-dimensional spacetime which contains the geometry of the ordinary 3-dimensional space as a particular case. Such a fusion of two fields until then completely separated-metric and gravitation-should be regarded as a major result of GTR, allowing, in principle, to determine the gravitational field just from local measurements of distances and time intervals.

4. The Field Equations for the Metric Tensor and Their Physical Consequences

4.1. The Field Equations

The differential equations for the g tensor itself followed in 1915 [1–3].

The tortuous path [21] which lead to them can be sketchily summarized as follows [19]. According to the EP, the gravitational mass of a body is exactly equal to its inertial mass and, as such, it is proportional to the total energy content of the body. The same must, then, hold also in a given gravitational field for the force experienced by a body which is proportional to its (passive) gravitational mass. It is, thus, natural to assume that, conversely, only the energy possessed by a material system does matter, through its (active) gravitational mass, as for as its gravitational field is concerned. Nonetheless, in STR the energy density is not characterized by a scalar quantity, being, instead, the 00 component of the so-called stress-energy tensor T. It follows that also momentum and stresses intervene on the same footing as energy itself. These considerations lead to the assumption that no other material state variables than the components $T_{\mu\nu}$ of T must enter the gravitational field equations. Moreover, in analogy with the Poisson equation, T must be proportional to a differential expression G of the second order containing only the state variables of the gravitational field, *i.e.*, the components of the metric tensor *g*; because of the required general covariance, G must be a tensor as well. The most general expression for it turned out to be

$$G_{\mu\nu} = c_1 R_{\mu\nu} + c_2 g_{\mu\nu} R + c_3 g_{\mu\nu} \qquad (23)$$

where R is the contracted curvature tensor whose components are

$$R_{\mu\nu} = \frac{\partial \Gamma^{\alpha}_{\mu\alpha}}{\partial x^{\nu}} - \frac{\partial \Gamma^{\alpha}_{\mu\nu}}{\partial x^{\alpha}} + \Gamma^{\beta}_{\mu\alpha}\Gamma^{\alpha}_{\nu\beta} - \Gamma^{\alpha}_{\mu\nu}\Gamma^{\beta}_{\alpha\beta} \qquad (24)$$

and R is its invariant trace. The coefficients c_1, c_2, c_3 were determined by imposing that the stress-energy tensor satisfies the energy-momentum conservation theorem. By neglecting the third term in G, which usually plays a negligible role in the effects which will be discussed in this Section (see Section 5 for phenomena in which it may become relevant), the Einstein field equations became [1,2]

$$G = -T \qquad (25)$$

with

$$G_{\mu\nu} = R_{\mu\nu} - \frac{1}{2} g_{\mu\nu} R \qquad (26)$$

and $\varkappa$ is a constant which is determined by comparison with the Newtonian Poisson equation. By contraction, one gets

$$R = T \qquad (27)$$

where T is the trace of T, so that

$$R_{\mu\nu} = -\left(T_{\mu\nu} - \frac{1}{2} g_{\mu\nu} T\right) \qquad (28)$$

This is the generally covariant form of the gravitational field equations to which, after many attempts, Einstein came in 1915 [3].

The same field equations were obtained elegantly by Hilbert through a variational principle [74]. On the reciprocal influences between Einstein and Hilbert in the process of obtaining the GTR field equations and an alleged priority dispute about their publication, see [75].

It should be noted [19] that GTR, *per se*, yields neither the magnitude nor the sign (attraction or repulsion of the gravitational interaction) of $\varkappa$ which are, instead, retrieved from the observations. For weak and quasi-static fields generated by pressureless, extremely slowly moving matter of density ρ, the right-hand-side of the field equation for the 00 component becomes

$$-\frac{1}{2} c^2 \rho \qquad (29)$$

indeed, the only non-vanishing component of the matter stress-energy tensor is

$$T_{00} = \rho c^2 \qquad (30)$$

so that

$$T = -\rho c^2 \qquad (31)$$

Since the time derivatives and the products of the Christoffel symbols can be neglected, the 00 component of the Ricci tensor reduces to

$$R_{00} = \frac{1}{2} \nabla^2 g_{00} = -\frac{\nabla^2 \Phi}{c^2} \qquad (32)$$

Thus, it is

$$\nabla^2 \Phi = \frac{1}{2} c^4 \rho \tag{33}$$

the Poisson equation really holds. A comparison with the Newtonian equation tells that $\varkappa$ is positive, being equal to

$$\varkappa = \frac{8\pi G}{c^4} = 2 \times 10^{-43} \mathrm{kg}^{-1} \mathrm{m}^{-1} \mathrm{s}^2 \tag{34}$$

the spacetime can, thus, be assimilated to an extremely rigid elsatic medium.

4.2. First Predictions of the Theory and Confrontation with Observations

In the same year [44], Einstein readily employed his newborn theory to successfully explain the long-standing issue of the anomalous perihelion precession of Mercury [76]. To this aim, and also in order to derive the correct value of the deflection of a light ray grazing the Sun's limb [44] through the Fermat principle, it was necessary to know not only the coefficient g_{00} of the gravitational field of a point mass, as in the Newtonian approximation, but also the other metric coefficients g_{ij}. Since the spacetime outside a spherical body is isotropic, the off-diagonal metric coefficients g_{0i} are identically zero: otherwise, they would induce observable effects capable of distinguishing between, e.g., two opposite spatial directions (see Section 3.2). Moreover, it was also required to approximate g_{00} itself to a higher order. Einstein [44] solved that problem by successive approximations. The *exact* vacuum solution was obtained one year later by Schwarzschild [77] and, independently, Droste [78]; their results are virtually indistinguishable from those of Einstein. Relevant simplifications were introduced one year later by Weyl [79], who used cartesian coordinates instead of spherical ones, and worked on the basis of the action principle instead of recurring to the differential equations for the field *g*. Schwarzschild [80] extended the validity of his solution also to the interior of a material body modelled as a sphere of incompressible fluid. Having in hand this exact solution of the Einstein field equations revolutionized the successive development of GTR. Indeed, instead of dealing only with small weak-field corrections to Newtonian gravity, as Einstein had initially imagined would be the case, fully nonlinear features of the theory such as gravitational collapse and singularity formation could be studied, as it became clear decades later. About the Schwarzschild solution, the Birkhoff's Theorem [81] was proved in 1923. According to it, even without the assumption of staticity, the Schwarzschild metric is the *unique* vacuum solution endowed with spherically symmetry. As a consequence, the external field of a spherical body radially pulsating or radially imploding/exploding is not influenced at all by such modifications of its source.

The successful explanation of the anomalous perihelion precession of Mercury was a landmark for the validity of GTR since, as remarked in [82,83], it was a successful *retrodiction* of an effect which was known for decades. In particular, Weinberg wrote [83]: "It is widely supposed that the true test of a theory is in the comparison of its predictions with the results of experiment. Yet, with the benefit of hindsight, one can say today that Einstein's successful explanation in 1915 of the previously measured anomaly in Mercury's orbit was a far more solid test of general relativity than the verification of his calculation of the deflection of light by the sun in observations of the eclipse of 1919 or in later eclipses. That is, in the case of general relativity a *retrodiction*, the calculation of the already-known anomalous motion of Mercury, in fact provided a more reliable test of the theory than a true *prediction* of a new effect, the deflection of light by gravitational fields.

I think that people emphasize prediction in validating scientific theories because the classic attitude of commentators on science is not to trust the theorist. The fear is that the theorist adjusts his or her theory to fit whatever experimental facts are already known, so that for the theory to fit to these facts is not a reliable test of the theory.

But [. . .] no one who knows anything about how general relativity was developed by Einstein, who at all follows Einstein's logic, could possibly think that Einstein developed general relativity in order to explain this precession. [. . .] Often it is a successful *prediction* that one should really distrust.

In the case of a true prediction, like Einstein's prediction of the bending of light by the sun, it is true that the theorist does not know the experimental result when she develops the theory, but on the other hand the experimentalist does know about the theoretical result when he does the experiment. And that can lead, and historically has led, to as many wrong turns as overreliance on successful retrodictions. I repeat: it is not that experimentalists falsify their data. [...] But experimentalists who know the result that they are theoretically supposed to get naturally find it difficult to stop looking for observational errors when they do not get that result or to go on looking for errors when they do. It is a testimonial to the strength of character of experimentalists that they do not always get the results they expect".

The final work of Einstein on the foundations of GTR appeared in 1916 [84].

In the same year, de Sitter [85] was able to derive a further consequence of the static, spherically symmetric spacetime of the Schwarzschild solution: the precession of the orbital angular momentum of a binary system, thought as a giant gyroscope, orbiting a non-rotating, spherical body such as in the case of the Earth-Moon system in the Sun's field. Some years later, Schouten [86] and Fokker [87] independently obtained the same effect by extending it also to spin angular momenta of rotating bodies. Such an effect is mainly known as de Sitter or geodetic precession. It was measured decades later in the field of the Sun by accurately tracking the orbit of the Earth-Moon system with the Lunar Laser Ranging technique [88,89], and in the field of the Earth itself with the dedicated Gravity Probe B (GP-B) space-based experiment [64] and its spaceborne gyroscopes.

In 1964 [90], Shapiro calculated a further prediction of the static Schwarzschild spacetime: The temporal delay, which since then bears his name, experienced by travelling electromagnetic waves which graze the limb of a massive body as the Sun in a back-and-forth path to and from a terrestrial station after having been sent back by a natural or artificial body at the superior conjunction with our planet. In its first successful test performed with radar signals [91], Mercury and Venus were used as reflectors. Latest accurate results [92] relied upon the Cassini spacecraft en route to Saturn.

4.3. The General Approximate Solution by Einstein

In 1916 [93], Einstein, with a suitable approximation method, was able to derive the field generated by bodies moving with arbitrary speeds, provided that their masses are small enough. In this case, the $g_{\mu\nu}$ differ slightly from the STR values $\eta_{\mu\nu}$, so that the squares of their deviations $h_{\mu\nu}$ with respect to the latter ones can be neglected, and it is possible to keep just the linear part of the field equations. Starting from their form [1,2]

$$R_{\mu\nu} - \frac{1}{2}g_{\mu\nu}R = -T_{\mu\nu} \qquad (35)$$

working in the desired approximation, they can be cast into a linearized form in terms of the auxiliary state variables

$$\bar{h}_{\mu\nu} \doteq h_{\mu\nu} - \frac{1}{2}\delta_\mu^\nu h \qquad (36)$$

where δ_μ^ν is the Kronecker delta, and h is the trace of h which is a tensor only with respect to the Lorentz transformations. A further simplification can be obtained if suitable spacetime coordinates, satisfying the gauge condition

$$\frac{\partial \bar{h}_{\alpha\beta}}{\partial x^\beta} = 0 \qquad (37)$$

known as Lorentz gauge (or Einstein gauge or Hilbert gauge or de Donder gauge or Fock gauge), are adopted. The resulting differential equations for the state variables $\bar{h}_{\mu\nu}$ are

$$\Box \bar{h}_{\mu\nu} = -2T_{\mu\nu} \qquad (38)$$

which is the inhomogeneous wave equation; $\Box$ is the STR form of the d'Alembertian operator. The usual method of the retarded potentials allows to obtain

$$\bar{h}_{\mu\nu} = \frac{1}{2\pi} \int \frac{T_{\mu\nu}\left(x',y',z',t-r/c\right)}{r} dx' dy' dz' \tag{39}$$

Among other things, it implies that the action of gravity propagates to the speed of light: a quite important results which, some years ago, was the subject of dispute [94–96] boosted by the interpretation of certain VLBI measurements of the time delay suffered at the limb of Jupiter by electromagnetic waves from distant quasars [97,98].

4.3.1. Gravitational Waves

The form of the gravitational waves in empty regions follows from the Lorentz gauge condition and the inhomogeneous wave equation by posing T = 0: it was studied by Einstein in [99], where he also calculated the emission and the absorption of gravitational waves. It turned out that, when oscillations or other movements take place in a material system, it emits gravitational radiation in such a way that the total power emitted along all spatial directions is determined by the third temporal derivatives of the system's moment of inertia

$$I_{ij} = \int \rho x^i x^j dx^1 dx^2 dx^3 \tag{40}$$

Instead, when a gravitational wave impinges on a material system whose size is smaller than the wave's wavelength, the total power absorbed is determined by the second temporal derivatives of its moment of inertia [99].

Gravitational waves were *indirectly* revealed for the first time [100–102] in the celebrated Hulse-Taylor binary pulsar PSR B1913+16 [103,104]. *Direct* detection (some of) their predicted effects in both terrestrial [105–110] and space-based laboratories [111–117] from a variety of different astronomical and astrophysical sources [118], relentlessly chased by at least fifty years since the first proposals by Gertsenshtein and Pustovoit [119] of using interferometers and the pioneering attempts by J. Weber [120] with its resonant bars [121], is one of the major challenges of the current research in relativistic physics [122,123].

Conversely, by assuming their existence, they could be used, in principle, to determine key parameters of several extreme astrophysical and cosmological scenarios which, otherwise, would remain unaccessible to us because of lack of electromagnetic waves from them [124] by establishing an entirely new "Gravitational Wave Astronomy" [122,125]. A recent example [126] is given by the possibility that the existence of primordial gravitational waves may affect the polarization of the electromagnetic radiation which constitutes the so-called Cosmic Microwave Background (CMB), discovered in 1965 [127]. In this case, the polarizing effect of gravity is indirect since the field of the gravitational waves does not directly impact the polarization of CMB, affecting, instead, the anisotropy of the spatial distribution of CMB itself. Indeed, the polarization of CMB is a direct consequence of the scattering of the photons of the radiation with the electrons and positrons which formed the primordial plasma, existing in the primordial Universe at the so-called decoupling era [128]. At later epochs, when the temperature fell below 3000 K°, the radiation decoupled from matter, photons and electrons started to interact negligibly, and the polarization got "frozen" to the values reached at the instant of decoupling. Thus, mapping the current CMB's polarization state has the potential of providing us with direct information of the primordial Universe, not contaminated by the dynamics of successive evolutionary stages. In particular, it turns out that the presence of metric fluctuations of tensorial type, *i.e.*, of gravitational waves, at the epoch in which the CMB radiation interacted with the electrons of the cosmic matter getting polarized, may have left traces in terms of polarization modes of B type [129,130]. They could be currently measurable, provided that the intensity of the cosmic background of gravitational waves is strong enough. An example of cosmic gravitational background able to produce, in principle, such an effect is represented by the relic gravitational radiation produced during the inflationary epochs. The gravitational waves produced in this way are distributed over a very wide frequency band $\Delta\omega(t)$ which is generally time-dependent. In order to characterize the intensity of such relic gravitational waves, it turns out convenient to adopt the spectral energy density

$$\varepsilon_h(\omega,t) \doteq \frac{d\varepsilon(t)}{d\ln\omega} \tag{41}$$

defined as the energy density $\varepsilon(t)$ per logarithmic interval of frequency, normalized to the critical energy density ε_{crit} (see Section 5.4), *i.e.*, the dimensionless variable

$$\Omega_h(\omega,t) \doteq \frac{1}{\varepsilon_{crit}}\frac{d\varepsilon}{d\ln\omega} \tag{42}$$

The simplest inflationary models yield power-law signatures for it. In 2014 [131], the BICEP2 experiment at the South Pole seemed to have successfully revealed the existence of the B modes; the measured values seemed approximately in agreement-at least in the frequency band explored by BICEP2-with a cosmic gravitational radiation background corresponding to the aforementioned power-law models. More recently [132], a joint analysis of data from ESA's Planck satellite and the ground-based BICEP2 and Keck Array experiments did not confirm such a finding.

4.3.2. The Effect of Rotating Masses

The previously mentioned solution $\bar{h}_{\mu\nu}$ of the inhomogenous wave equation in terms of the retarded potentials was used by Thirring [56,57,61] to investigate, to a certain extent, the relative nature of the centrifugal and Coriolis fictitious forces arising in a rotating coordinate system with respect to another one connected with the static background of the fixed stars. Indeed, according to a fully relativistic point of view, they should also be viewed as gravitational effects caused by the rotation of the distant stars with respect to a fixed coordinate system. At first sight, it may seem that such a possibility is already included in the theory itself in view of the covariance of the field equations. Actually, it is not so because the boundary conditions at infinite distance play an essential role in selecting, *de facto*, some privileged coordinate systems, in spite of a truly "relativistic" spirit with which the theory should be informed. In other words, although the equations of the theory are covariant, the choice of the boundary conditions at spatial infinity, which are distinct from and independent of the field equations themselves, would pick up certain coordinate systems with respect to others, which is a conceptual weakness of an alleged "generally relativistic" theory. Thus, Thirring [56] did not aim to check the full equivalence of the gravitational effects of the rotation of the whole of the distant stars of the Universe with those due to the rotation of the coordinate system with respect to them, assumed fixed. Indeed, he considered just a rotating hollow shell of finite radius D and mass M, so to circumvent the issue of the boundary conditions at infinite distance by setting the spacetime metric tensor there equal to the Minkowskian one. By assuming M small with respect to the whole of the fixed stars, so to consider the departures of the $g_{\mu\nu}$ coefficients from their STR values $\eta_{\mu\nu}$ small inside the shell, the application of the previously obtained Einsteinian expression for $\bar{h}_{\mu\nu}$ to the shell yielded that a test particle inside the hollow space inside it is affected by accelerations which are formally identical to the centrifugal and Coriolis ones, apart from a multiplicative scaling dimensionless factor as little as

$$\frac{GM}{c^2D} \tag{43}$$

This explains the failures by Newton [133] in attributing the centrifugal curvature of the free surface of water in his swirling bucket to the relative rotation of the bucket itself and the water, and the Friedländer brothers [134] who unsuccessfully attempted to detect centrifugal forces inside a heavy rotating flywheel.

Another application of the approximate solution $\bar{h}_{\mu\nu}$ of the inhomogeneous wave equation allowed to discover that, while in either GTR and the Newtonian theory the gravitational field of a static, spherical body is identical to that of a point mass [81], it is not so-in GTR-if the body rotates. Indeed, Einstein [53], Thirring and Lense [54] calculated the (tiny) precessions affecting the orbits of test particles as natural satellites and planets moving in the field of rotating astronomical bodies such as the Sun and some of its planets. Such a peculiarity of the motion about mass-energy currents, universally known as "Lense-Thirring effect" by historical tradition (cfr. [62] for a critical historical analysis of its genesis), was subjected to deep experimental scrutiny in the last decades [65–67].

In the sixties of the twentieth century, another consequence of the rotation of an astronomical body was calculated within GTR: the precession of an orbiting gyroscope [135,136], sometimes dubbed as "Pugh-Schiff effect". The GP-B experiment [137], aimed to directly measure also such an effect in the field of the Earth, was successfully completed a few years ago [64], although the final accuracy obtained ($\sim$19%) was worse than that expected ($\sim$1% or better) .

4.4. Black Holes and Other Physically Relevant Exact Solutions of the Field Equations

4.4.1. The Reissner-Nordström Metric

In 1916, Reissner [138] solved the coupled Einstein-Maxwell field equations and found the metric which describes the geometry of the spacetime surrounding a pointlike electric charge Q. One year later, Weyl [79] obtained the same metric from a variational action principle. In 1918, Nordström [139], generalized it to the case of a spherically symmetric charged body. The metric for a non-rotating charge distribution is nowadays known as the Reissner-Nordström metric; in the limit $Q \to 0$, it reduces to the Schwarzschild solution.

The physical relevance of the Reissner-Nordström metric in astronomical and astrophysical scenarios depends on the existence of macroscopic bodies stably endowed with net electric charges.

4.4.2. Black Holes

One of the consequences of the vacuum Schwarzschild solution was that it predicts the existence of a surface of infinite red-shift at

$$r = r_{\mathrm{g}} \doteq \frac{2GM}{c^2} \tag{44}$$

Thus, if, for some reasons, a body could shrink so much to reduce to such a size, it would disappear from the direct view of distant observers, who would not be anymore able to receive any electromagnetic radiation from such a surface, later interpreted as a spatial section of an "event horizon" [140–142]. A "frozen star", a name common among Soviet scientists from 1958 to 1968 [143,144], would have, then, formed, at least from the point of view of an external observer. In 1968 [143,145], Wheeler renamed such objects with their nowadays familiar appellative of "black holes" [146].

In fact, both Eddington in 1926 [147] and Einstein in 1939 [148], although with arguments at different levels of soundness, were firmly convinced that such bizarre objects could not form in the real world. Instead, in 1939 [149], Oppenheimer and Snyder demonstrated that, when all the thermonuclear sources of energy are exhausted, a sufficiently heavy star will unstoppably collapse beyond its Schwarzschild radius to end in a spacetime singularity. The latter one is not to be confused with the so-called "Schwarschild singularity" occurring in the Schwarzschild metric at $r = r_{\mathrm{g}}$, which was proven in 1924 [150] to be unphysical, being a mere coordinate artifact; nonetheless, it took until 1933 for Lemaître to realize it [151]. In 1965, Penrose [152], in his first black hole singularity theorem, demonstrated that the formation of a singularity at the end of a gravitational collapse was an inevitable result, and not just some special feature of spherical symmetry. A black hole is the 4-dimensional spacetime region which represents the future of an imploding star: it insists on the 2-dimensional spatial critical surface determined by the star's Schwarzschild radius. The 3-dimensional spacetime hypersurface delimiting the black hole, *i.e.*, its event horizon, is located in correspondence of the critical surface [142].

4.4.3. The Kerr Metric

In 1963, the third physically relevant *exact* vacuum solution of the Einstein field equations was found by Kerr [153]. It describes the spacetime metric outside a rotating source endowed with mass M and proper angular momentum J. It was later put in a very convenient form by Boyer and Lindquist [154]. At that time, it was generally accepted that a spherical star would collapse to a black hole described by the Schwarzschild metric. Nonetheless, people was wondering if such a dramatic fate of a star undergoing gravitational collapse was merely an artifact of the assumed perfect spherical symmetry. Perhaps, the slightest angular momentum would halt the collapse before the formation of an event horizon, or at least before the formation of a singularity. In this respect, finding a metric for a rotating star would have been quite valuable.

Contrary to the Schwarzschild solution [80], the Kerr one has not yet been satisfactorily extended to the interior of any realistic matter-energy distribution, despite several attempts over the years [155]. Notably, according to some researchers [156–158], this limit may have no real physical consequences since the exterior spacetime of a rotating physically likely source is *not* described by the Kerr metric whose higher multipoles, according to the so-called "no-hair" conjecture [159,160], can all be expressed in terms of M and J [161,162], which is not the case for a generic rotating star [163]. Moreover, the Kerr solution does not represent the metric during any realistic gravitational collapse; rather, it yields the asymptotic metric at late times as whatever dynamical process produced the black hole settles down, contrary to the case of a non-rotating collapsing star whose exterior metric is described by the Schwarzschild metric at all times. The Birkhoff's Theorem [81] does not hold for the Kerr metric.

The enormous impact that the discovery by Kerr has had in the subsequent fifty years on every subfield of GTR and astrophysics as well is examined in [158]; just as an example, it should be recalled that, at the time of the Kerr's discovery, the gravitational collapse to a Schwarzschild black hole had difficulty in explaining the impressive energy output of quasars, discovered and characterized just in those years [164,165], because of the "frozen star" behavior for distant observers. Instead, the properties of the event horizon were different with rotation taken into account. A comparison of the peculiar features of the Schwarzschild and the Kerr solutions can be found in [166].

4.4.4. The Kerr-Newman Metric

In 1965 [167], a new *exact* vacuum solution of the Einstein-Maxwell equations of GTR appeared: the Kerr-Newman metric [168]. It was obtained from the Reissner-Nordström metric by a complex transformation algorithm [169] without integrating the field equations, and is both the spinning generalization of Reissner-Nordström and the electrically charged version of the Kerr metric. Such solutions point towards the possibility that charged and rotating bodies can undergo gravitational collapse to form black holes just as in the uncharged, static case of the Schwarzschild metric.

Leaving the issue of its physical relevance for astrophysics applications out of consideration, the Kerr-Newman metric is the most general static/stationary black hole solution to the Einstein-Maxwell equations. Thus, it is of great importance for theoretical considerations within the mathematical framework of GTR and beyond. Furthermore, understanding this solution also provides valuable insights into the other black hole solutions, in particular the Kerr metric.

5. Application to Cosmology

5.1. Difficulties of Newtonian Cosmologies

The birth of modern cosmology might be dated back to the correspondence between Newton and Bentley in the last decade of the seventieth century [170], when the issue of the applicability of Newtonian gravitational theory to a static, spatially infinite (Euclidean) Universe uniformly filled with matter was tackled. In four letters to R. Bentley, Newton explored the possibility that matter might be spread uniformly throughout an infinite space. To the Bentley's suggestion that such an even distribution might be stable, Newton replied that, actually, matter would tend to collapse into large massive bodies. However, he apparently also thought that they could be stably spread throughout all the space. In particular, in his letter of 10 December 1692, Newton wrote [171]: "it seems to me that if [. . .] all the matter in the Vniverse was eavenly scattered throughout all the heavens, & every particle had an innate gravity towards all the rest & the whole space throughout which this matter was scattered, was but finite: the matter on the outside of this space would by its gravity tend towards all the matter on the inside & by consequence fall down to the middle of the whole space & there compose one great spherical mass But if the matter was eavenly diffused through an infinite space, it would never convene into one mass but some of it convene into one mass & some into another so as to make an infinite number of great masses scattered at great distances from one another throughout all that infinite space."

Connected with the possibility that matter would fill uniformly an infinite space, and, thus, indirectly with the application of Newtonian gravitation to cosmology, there was also the so-called Olbers paradox [172], some aspects of which had been previously studied also by Kepler [173], Halley [174,175] and de Chéseaux [176]. According to it, although the light from stars diminishes as the square of the distance to the star, the number of stars in spherical shells increases as the square of the shell's radius. As a result, the accumulated effect of the light intensity should make the night sky as bright as the surface of the Sun. In passing, the Olbers paradox touched also other topics which will become crucial in contemporary cosmology like the temporal infinity of the Universe and its material content, and its spatial infinity as well. At the end of the nineteenth century, Seeliger [177] showed that, in the framework of the standard Newtonian theory, matter cannot be distributed uniformly throughout an infinite Universe. Instead, its density should go to zero at spatial infinity faster than r^{-2}; otherwise, the force exerted on a point mass by all the other bodies of the Universe would be undeterminate because it would be given by a non-convergent, oscillating series. Later, Einstein [178] critically remarked that, if the potential was finite at large distances as envisaged by Seeliger [177] to save the Newtonian law, statistical considerations would imply a depopulation of the fixed stars ensemble, assumed initially in statistical equilibrium. The possibility of an infinite potential at large distances, corresponding to a finite or vanishing not sufficiently fast matter density, already ruled out by Seeliger himself, was excluded also by Einstein [178] because it would yield unrealistically fast speeds of the distant stars. Seeliger [179] demonstrated also that matter density could be different from zero at arbitrary distances if the standard Poisson equation was modified as

$$\nabla^2 \Phi - \Lambda \Phi = 4\pi G \rho \tag{45}$$

It admits

$$\Phi = -\frac{4\pi G \rho}{\Lambda} \tag{46}$$

as a viable solution for a uniform matter density, thus making an evenly filled Universe stable. For a discussion of the problems encountered by the Newtonian theory of gravitation to cosmology, see, e.g., [180].

The inadequacy of Newtonian gravitation to cosmological problems can be also inferred in view of the modern discoveries concerning the expansion of the Universe over the eons (see Section 5.2) which, in conjunction with the finite value of c, yielded to the notion of observable Universe. As previously recalled in Section 3.2, the gravitational interaction among macroscopic bodies can be adequately described, to the first approximation, by the non-relativistic Newtonian model. Such an approximation is applicable over spatial scales ranging from laboratory to planetary, stellar, and galactic systems.

On the other hand [181], the Newtonian model cannot be applied, not even to the first approximation, to correctly describe gravity over cosmological distances of the order the Hubble distance

$$D_H \doteq \frac{c}{H_0} \sim 10^{26}\text{m} \tag{47}$$

where [182]

$$H_0 = (67.3 \pm 1.2)\text{kms}^{-1}\text{Mpc}^{-1} \tag{48}$$

is the current value of the Hubble parameter (see Section 5.4), which fixes the maximum spatial distance accessible to current observations (the radius of the observable universe is proportional to D_H through a numerical coefficient which, according to the present-day cosmological parameters, is equal to 3.53). Indeed, the absolute value of the potential of the mass equivalent to the energy density ε enclosed in a spherical volume of radius $\sim D_H$ is

$$|\Phi_H| = \frac{4}{3}\frac{\pi G \varepsilon}{H_0^2} \tag{49}$$

The condition of validity of the Newtonian approximation is that, for any test particle of mass m, the gravitational potential energy $m|\Phi_H|$ resulting from the interaction with the cosmological mass of the observable Universe is much smaller than its rest energy mc^2. Instead, it turns out [181]

$$\frac{4}{3}\frac{\pi G \varepsilon}{H_0^2 c^2} \sim 1 \tag{50}$$

It follows that the Newtonian approximation is not valid at the Hubble scale, and a correct dynamical description of the Universe to cosmological scales must necessarily rely upon a relativistic theory of gravity.

5.2. Relativistic Cosmological Models

GTR, applied to cosmology for the first time in 1917 by Einstein himself [178], was able to put such a fundamental branch of our knowledge on the firm grounds of empirical science.

In the following, we will try to follow the following terminological stipulations [183]. We will generally use the word "Universe" to denote a model of the cosmological spacetime along with its overall matter-energy content; as we will see, the relativistic Universe is the space woven by time and weighed by all forms of energy (matter-either baryonic and non-baryonic-, radiation, cosmological constant). As such, the Universe has neither center nor borders, neither inside nor outside. Instead, by means of "universe" we will denote the observable portion of the cosmological spacetime delimited by a cosmological horizon unavoidably set by the fact that all the physical means (electromagnetic and gravitational radiation, neutrinos, cosmic rays) by which we collect information from objects around us travel at finite speeds. Its spatial section is a centered on the Earth-based observer with a radius equal to

$$3.53 D_H = 51.3\text{Gly} = 15.7\text{Gpc} \tag{51}$$

5.2.1. The Static Einstein Model

In 1917, Einstein [178] showed that, following his field equations in their original form, it would not be possible to choose the boundary conditions in such a way to overcome simultaneously the depopulation and the observed small stellar velocities issues. Instead, in principle, it is mathematically possible to modify them in as much as the same way as it was doable with the Poisson equation by introducing a Λ term which yielded

$$R_{\mu\nu} + \Lambda g_{\mu\nu} = -\left(T_{\mu\nu} - \frac{1}{2}g_{\mu\nu}T\right) \tag{52}$$

Some years later, Cartan [184] demonstrated that the most general form of the Einstein field equations necessarily implies the Λ term. It turned out that a Universe uniformly filled with constant matter density ρ and non-vanishing

$$\Lambda_E = \frac{4\pi G\rho}{c^2} \tag{53}$$

would rest in equilibrium. Moreover, since it would be spatially closed with

$$
\begin{aligned}
g_{00} &= 1 \\
g_{0i} &= 0 \\
g_{ij} &= -\left[\delta_{ij} + \frac{x_i x_j}{S^2 - (x_1^2 + x_2^2 + x_3^2)}\right]
\end{aligned}
\tag{54}
$$

and radius S connected with Λ by

$$\Lambda = \frac{1}{S^2} \tag{55}$$

there would not be the need of choosing suitable boundary conditions at infinity, thus removing the aforementioned "non-relativistic" drawback of the theory (see Section 4). It should be noted that if such a 4$-$dimensional cylindrical Universe did not contain matter, there would not be any gravitational field, *i.e.*,

$$T_{\mu\nu} = 0 \tag{56}$$

would imply

$$g_{\mu\nu} = 0 \tag{57}$$

Thus, the postulate of the complete relativity of inertia would be met. In the Einstein spatially hyperspherical model, the spacetime trajectories of moving bodies and light rays wind around spirals on the surface of a cylinder in such a way that if one watched a spaceship moving away from her/him, it first would diminish in size but then would come back beginning to magnify again. Thirteen years later, Eddington [185] showed that the static Einsteinian model is, actually, unstable.

It may be interesting to note [20] how the Einstein's Universe is, in fact, no less liable to the Olbers paradox than the Newtonian one; indeed, the light emitted by a star would endlessly circumnavigate the static spherical space until obstructed by another star.

5.2.2. The de Sitter Model

In 1917, de Sitter [186,187] found a solution for the modified Einstein field equations with $\Lambda \neq 0$ yielding a 4-dimensional hyperbolic Universe

$$g_{\mu\nu} = \frac{\eta_{\mu\nu}}{\left(1 - \frac{\Lambda}{12}\eta_{\alpha\beta}x^\alpha x^\beta\right)^2} \tag{58}$$

$$\Lambda = \frac{3}{S^2}$$

with non-zero gravitational field even in absence of matter, thus differing from the Einstein model. It allowed also a sort of spatial (and not material) origin of inertia, which would be relative to void space: a hypothetical single test particle existing in the otherwise empty de Sitter Universe would have inertia just because of Λ.

At the time of the Einstein and de Sitter models, there were not yet *compelling* means to observationally discriminate between them [188], although their physical consequences were remarkably different. Suffice it to say that the spacetime geometry of the de Sitter Universe implied that, although static, test particles would have escaped far away because of the presence of the Λ term, unless they were located at the origin. Such a recessional behaviour was known as "de Sitter effect".

As said by Eddington [189], "the de Sitter Universe contains motion without matter, while the Einstein Universe contains matter without motion".

After having lost appeal with the advent of the genuine non-stationary Fridman-Lemaître solutions (see Section 5.2.3), the de Sitter model was somewhat revamped in the framework of the inflationary phase characterized by an ultrafast expansion that it is believed to have occurred in the early stages of the universe [190–192].

5.2.3. The Fridman-Lemaître-Robertson-Walker Expanding Models

In the twenties of the last century, the first truly non-static theoretical models of the Universe were proposed by Fridman [193,194]. Indeed, he found new solutions of the Einstein field equations with Λ representing spatially homogeneous and isotropic cosmological spacetimes filled with matter-energy modeled as a perfect fluid generally characterized only by time-varying density $\rho(t)$, and endowed with an explicitly time-dependent universal scaling factor $S(t)$ for the spatial metric having constant curvature $k = 0, \pm 1$ throughout all space. If $k = +1$, the 3-dimensional space is *spherical* and necessarily *finite* (as the hypersphere); if $k = 0$, it is *Euclidean*; if $k = -1$, it is *hyperbolic*. Euclidean and hyperbolic spaces can be either finite or infinite, depending on their topology [195,196] which, actually, is *not* determined by the Einstein field equations governing only the dynamical evolution of $\rho(t), S(t)$. Importantly, viable solutions exist also in absence of the cosmological Λ term for all the three admissible values of the spatial curvature parameter k. The Einstein and de Sitter models turned out [193] to be merely limiting cases of an infinite family of solutions of the Einstein field equations for a positive, time-varying matter density, any one of which would imply, at least for a certain time span, a general recession-or oncoming, since the solutions are symmetric with respect to time reversal-of test particle. According to their dynamical behaviour, the Fridman's models are classified as *closed* if they recollapse, *critical* if they expand at an asymptotically zero rate, and *open* if they expand indefinitely. In this respect, a spherical universe can be open if Λ is positive and large enough, but it cannot be infinite. Conversely, Euclidean or hyperbolic universes, generally open, can be closed if $\Lambda < 0$; their finiteness or infiniteness depends on their topology, not on their material content. Fridman's simplifying assumptions were much weaker than those of either Einstein and de Sitter, so that they defined a much likelier idealization of the real world [20], as it turned out years later: indeed, the russian scientist was interested only in the mathematical aspects of the cosmological solutions of the Einstein equations.

Approximately in the same years, a body of observational evidence pointing towards mutual recessions of an increasingly growing number of extragalactic nebulae was steadily accumulating [197–201] from accurate red-shifts measurements, probably unbeknownst to Fridman. In 1929, Hubble [200] made his momentous claim that the line-of-sight speeds of the receding galaxies are proportional to their distances from the Earth. If, at first, the de Sitter model, notwithstanding its material emptiness, was regarded with more favor than the Einstein one as a possible explanation of the observed red-shifts of distant nebulæ, despite the cautiousness by de Sitter himself [187], it would have been certainly superseded by the more realistic Fridman ones, if only they had been widely known at that time (Fridman died in 1925). It may be that a role was played in that by the negative remark by Einstein about a claimed incompatibility of the non-stationary Fridman's models with his field equations [202], later retracted by the father of GTR because of an own mathematical error in his criticism [203].

At any rate, in 1927, Lemaître [204], who apparently did never hear of the Fridman's solutions, rederived them and applied them to the physical universe with the explicit aim of founding a viable explanation of the observed recessions of galaxies (the red-shifted nebulæ had been recognized as extra-galactic objects analogous to our own galaxy in 1925 by Hubble [205]). Lemaître [204] also showed that the static solution by Einstein is unstable with respect to a temporal variation of matter density. Enlightened by the Hubble's discovery [200], and, perhaps, also struck by the criticisms by Lemaître [204] and Eddington [185] to his own static model, Einstein fully acknowledged the merits of the non-static Fridman-type solutions rejecting outright his cosmological Λ term as unnecessary and unjustifiable [206].

Interestingly, in 1931, Lemaître [207] did not appreciate the disown by Einstein of his cosmological constant Λ, which, instead, was retained by the belgian cosmologist an essential ingredient of the physical Universe for a number of reasons, one of which connected also with quantum mechanics, which, however, convinced neither Einstein nor the scientific community, at least until the end of the nineties of the last century [208]. His "hesitating" model was characterized by positive spatial curvature ($k = +1$), and a positive cosmological constant so that its perpetual expansion is first decelerated, then it enters an almost stationary state, and finally it accelerates, thus resolving the problem of the age of the Universe and the time required for the formation of galaxies.

The formal aspects of the homogeneous and isotropic expanding models were clarified and treated in a systematic, general approach in the first half of the thirties of the last century by Robertson [209–211] and Walker [212]. Today, the spacetime tensor g of standard expanding cosmologies is commonly named as Fridman-Lemaître-Robertson-Walker (FLRW) metric (see Section 5.4).

5.3. The Einstein-de Sitter Model

In 1932, Einstein and de Sitter [213] published a brief note of two pages whose aim was to simplify the study of cosmology. About their work, as reported by Eddington [214], Einstein would have told him: "I did not think the paper very important myself, but de Sitter was keen on it", while de Sitter wrote to him: "You will have seen the paper by Einstein and myself. I do not myself consider the result of much importance, but Einstein seemed to think that it was". At any rate, such an exceedingly simplified solution, characterized by dust-like, pressureless matter, $k = 0, \Lambda = 0$ and perpetual, decelerating expansion, served as "standard model" over about six decades, to the point of curb researches on other models. In it, mutual distances among test particles grow as $t^{2/3}$. Such a behaviour is unstable in the sense that it can only occur if $k = 0$ exactly; for tiny deviations from such a value, the expansion would gradually depart from the trajectory of the Einstein-de Sitter model. Actually, it represented the best description of the cosmic expansion as it was known for the next sixty years. The fact that the observed behaviour of the physical universe was still so close to that particular expansion rate suggested that the instability had not yet had the time to manifest itself significantly. But, after all, the universe had been expanding for about several billions of years, as if it had started just from the very spacial initial conditions of the Einstein-de Sitter model. This peculiar situation, later known as "the flatness problem", motivated, among other things, the studies on the cosmic inflation in the eighties of the last century [190–192]. The Einstein-de Sitter model has now been abandoned, also because it would imply a too short age of the Universe given by

$$t_0 = \frac{2}{3}\frac{1}{H_0} = 9.6\text{Gyr} \tag{59}$$

For a recent popular account on the panoply of possible Universes allowed by GTR, see [215].

In passing, let us note that the expanding cosmological models by GTR, along with the associated finite age of the Universe, represent the framework to correctly solve the Olbers paradox [216,217].

5.4. Some Peculiar Characteristics of the FLRW Models

The assumptions of homogeneity *and* isotropy of the spatial sections of the FLRW models are of crucial importance. It must be stressed that they are, in general, distinct requirements. Homogeneity does not generally imply isotropy: for instance, think about a universe filled with galaxies whose axes of rotation are all aligned along some specific spatial direction, or a wheat field where the ears grow all in the same direction. Conversely, a space which is isotropic around a certain point, in the sense that the curvature is the same along all the directions departing from it, may well not be isotropic in other points, or, if some other points of isotropy exist, the curvature there can be different from each other: an ovoid surface is not homogenous since its curvature varies from point to point, but the space is isotropic around its two "vertices". Instead, the same vale of the curvature in all the directions, *i.e.*, the same amount of isotropy, around all points of space implies homogeneity [218]. As far as our location is concerned, it can be said phenomenologically that isotropy about us holds in several physical aspects to a high level of accuracy, as demonstrated, e.g., by the CMB which is isotropic at a 10^{-5} level. In view of the Copernican spirit, it is commonly postulated that every other observer located everywhere would see the same situation, thus assuring the homogeneity as well: it is the content of the so-called Cosmological Principle. The fact that the curvature of the spatial parts of the FLRW models is the same everywhere, and that they are expanding over time, according to the Weyl principle [219], admit a peculiar foliation of the spacetime which allows for an unambiguous identification of the spatial sections of simultaneity and of the bundle of time-like worldlines orthogonal to them as worldlines of fundamental observers at rest marking a common, cosmic time. Thus, it is possible to describe the spacetime of the Universe as the mathematical product of a $3-$dimensional Riemannian space with the temporal axis. In comoving dimensionless spatial coordinates r, θ, ϕ, the line element can be written as

$$(ds)^2 = c^2(dt)^2 - S(t)\left[\frac{(dr)^2}{1 - kr^2} + r^2(d\theta)^2 + r^2 \sin^2\theta(d\phi)^2 \right] \tag{60}$$

The Einstein field equations, applied to the FLRW metric with a pressureless cosmic fluid as standard source with matter and radiation densities ρ_m, ρ_r, respectively, yield the Fridman equation

$$\dot{S}^2 = \frac{8}{3}\pi G(\rho_m + \rho_r)S^2 - kc^2 + \frac{1}{3}\Lambda c^2 S^2 \tag{61}$$

By defining the Hubble parameter as

$$H \doteq \frac{\dot{S}}{S} \tag{62}$$

and the critical density as

$$\rho_{crit} \doteq \frac{3H^2}{8\pi G} \tag{63}$$

it is possible to recast the Fridman equation in the form

$$\Omega_m + \Omega_r + \Omega_\Lambda + \Omega_k = 1 \tag{64}$$

or also

$$\Omega_{tot} = 1 - \Omega_k \tag{65}$$

by posing

$$\Omega_{tot} \doteq \Omega_m + \Omega_r + \Omega_\Lambda \tag{66}$$

with the dimensionless parameters entering Equation (??) defined as

$$\Omega_m \doteq \frac{8}{3}\frac{\pi G\rho_m}{H^2} > 0 \tag{67}$$

$$\Omega_r \doteq \frac{8}{3}\frac{\pi G\rho_r}{H^2} > 0$$

$$\Omega_\Lambda \doteq \frac{1}{3}\frac{\Lambda c^2}{H^2} \lessgtr 0$$

$$\Omega_k \doteq -\frac{kc^2}{S^2 H^2} \lessgtr 0$$

At present epoch, $\Omega_{r,0} \sim 0$, so that the normalized Fridman equation reduces to

$$\Omega_{m,0} + \Omega_{\Lambda,0} + \Omega_{k,0} = 1 \tag{68}$$

or also

$$\Omega_{tot,0} = 1 - \Omega_{k,0} \tag{69}$$

Results collected in the last twenty years from a variety of observational techniques (e.g., SNe Ia [220–222], Baryon acoustic oscillations [223], WMAP [224], Planck [182]), interpreted within a FRLW framework, point towards an observable universe whose spatial geometry is compatible with an Euclidean one (such a possibility, in view of the unavoidable error bars, is impossible to be proved experimentally with certainty: on the contrary, it could be well excluded should the ranges of values for $\Omega_{tot,0}$ did not contain 1), and whose dynamical behaviour is characterized by a small positive cosmological constant Λ which makes it accelerating at late times. By assuming $\Omega_{tot,0} = 1$ *exactly*, as allowed by the experimental data and predicted by the inflationary paradigm, the values for the other normalized densities are inferred by finding [182,224]

$$\Omega_{m,0} \sim 0.3, \Omega_{\Lambda,0} \sim 0.7 \tag{70}$$

6. Summary

After its birth, GTR went to fertilize and seed, directly as well as indirectly, many branches of disparate sciences as mathematics [225–230], metrology [231–234], geodesy [236–238], geophysics [239–241], astronomy [242–247], astrophysics [248–252], cosmology [181,253–255], not to say about the exquisite technological spin-off [256–270] due to the long-lasting efforts required to put to the test various key predictions of the theory [18,271]. Moreover, once some of them have been or will be successfully tested, they have or will become precious tools for determine various parameters characterizing several natural systems, often in extreme regimes unaccessible with other means: gravitational microlensing for finding extrasolar planets, even of terrestrial size [272,273], weak and strong gravitational lensing to map otherwise undetectable matter distributions over galactic, extragalactic and cosmological scales [274–276], frame-dragging to measure angular momenta of spinning objects like stars and planets [277–280], gravitational waves to probe, e.g., quantum gravity effects [281], modified models of gravity [282,283] and cosmic inflationary scenarios [131,132], to characterize tight binary systems hosting compact astrophysical objects like white dwarves, neutron stars and black holes [284–289], and to investigate extremely energetic events like, e.g., supernovæ explosions [290].

However, GTR has its own limits of validity, and presents open problems [291]. At certain regimes, singularities plague it [225,292–295]. Connected to this issue, there is also a major drawback of the theory of gravitation of Einstein, *i.e.*, its lingering inability to merge with quantum mechanics yielding a consistent theory of quantum gravity [296–302]. Moreover, in view of the discoveries made in the second half of the last century about the seemingly missing matter to explain the rotation curves of galaxies [303–305] and the accelerated expansion of the Universe [220–222], it might be that GTR need to be modified [306–311] also at astrophysical and cosmological scales in order to cope with the issue of the so-called "dark" [312] components of the matter-energy content of the Universe known as Dark Matter and Dark Energy.

We consider it appropriate to stop here with our sketchy review. Now, we give the word to the distinguished researchers who will want to contribute to this Special Issue by bringing us towards the latest developments of the admirable and far-reaching theory of gravitation by Einstein. At a different level of coverage and completeness, the interested reader may also want to consult the recent two-volume book [313,314].

References

1. Einstein, A. Zur allgemeinen Relativitätstheorie. *Sitzungsber. Kön. Preuß. Akad. Wiss. zu Berlin* **1915**, 778–786. (In German)
2. Einstein, A. Zur allgemeinen Relativitätstheorie (Nachtrag). *Sitzungsber. Kön. Preuß. Akad. Wiss. zu Berlin* **1915**, 799–801. (In German)
3. Einstein, A. Die Feldgleichungen der Gravitation. *Sitzungsber. Kön. Preuß. Akad. Wiss. zu Berlin* **1915**, 844–847. (In German)
4. Damour, T. General Relativity Today. In *Gravitation and Experiment. Poincaré Seminar 2006. Progress in Mathematical Physics. Volume 52*; Damour, T., Duplantier, B., Rivasseau, V., Eds.; Birkhäuser: Basel, Switzerland, 2007; pp. 1–49.
5. Torretti, R. The Geometric Structure of the Universe. In *Philosophy and the Origin and Evolution of the Universe*; Agazzi, E., Cordero, A., Eds.; Kluwer Academic Publisher: Dordrecht, The Netherlands, 1991; pp. 53–73.
6. Lorentz, H.A. Electromagnetische verschijnselen in een stelsel dat zich met willekeurige snelheid, kleiner dan die van het licht, beweegt. *Koninklijke Akademie van Wetenschappen te Amsterdam. Wis-en Natuurkundige Afdeeling. Verslagen van de Gewone Vergaderingen* **1904**, *12*, 986–1009. (In Dutch)
7. Poincaré, H. Sur la dynamique de l' électron. *Comptes-rendus des séances de l'Académie des Sci.* **1905**, *140*, 1504–1508. (In French)
8. Poincaré, H. Sur la dynamique de l' électron. *Rendiconti del Circolo Matematico di Palermo* **1906**, *21*, 129–175. (In French) [CrossRef]
9. Einstein, A. Zur Elektrodynamik bewegter Körper. *Ann. Phys.* **1905**, *17*, 891–921. (In German) [CrossRef]
10. Preti, G.; de Felice, F.; Masiero, L. On the Galilean non-invariance of classical electromagnetism. *Eur. J. Phys.* **2009**, *30*, 381–391. [CrossRef]
11. Planck, M. Die Kaufmannschen Messungen der Ablenkbarkeit der β-Strahlen in ihrer Bedeutung für die Dynamik der Elektronen. *Phys. Z.* **1906**, *7*, 753–761. (In German)
12. Bucherer, A.H. Ein Versuch, den Elektromagnetismus auf Grund der Relativbewegung darzustellen. *Phys. Z.* **1906**, *7*, 553–557. (In German)
13. Stachel, J.; Cassidy, D.C.; Renn, J.; Schulmann, R. Editorial Note: Einstein on the Theory of Relativity. In *The Collected Papers of Albert Einstein. Volume 2: The Swiss Years: Writings, 1900–1909*; Princeton University Press: Princeton, NJ, USA, 1989; pp. 253–274.
14. Holton, G.; Elkana, Y. *Albert Einstein: Historical and Cultural Perspectives-The Centennial Symposium in Jerusalem*; Princeton University Press: Princeton, NJ, USA, 1982; p. xv.
15. Klein, F. Über der geometrischen Grundlagen der Lorentzgruppe. *Jahresbericht der Deutschen Mathematiker-Vereinigung* **1910**, *19*, 281–300. (In German)
16. Fock, V.A. *Theory of Space, Time, and Gravitation*, 2nd ed.; Pergamon Press: New York, NY, USA, 1964.
17. Cartan, E. La théorie des groupes et la géométrie. *L'enseignement Math.* **1927**, *26*, 200–225. (In French)
18. Will, C.M. The Confrontation between General Relativity and Experiment. *Living Rev. Relativ.* **2006**, *163*, 146–162. [CrossRef]
19. Pauli, W. Relativitätstheorie. In *Encyklopädie der matematischen Wissenschaften*; Teubner: Leipzig, Germany, 1921; volume 5, pt. 2, art. 19. (In German)
20. Torretti, R. *Relativity and Geometry*; Dover: Mineola, NY, USA, 1996; pp. 33–34.
21. Renn, J. *Auf den Schultern von Riesen und Zwergen. Einsteins unfollendete Revolution*; Wiley-VCH: Weinheim, Germany, 2006. (In German)
22. Newton, Is. *Philosophiæ Naturalis Principia Mathematica*; For the Royal Society by Joseph Streater: London, UK, 1687. (In Latin)
23. Poisson, S.D. Remarques sur une équation qui se présente dans la théorie des attractions des sphéroïdes. *Nouveau Bull. Soc. Philomath. Paris* **1813**, *3*, 388–392. (In French)
24. Newton, Is. *Original letter from Isaac Newton to Richard Bentley. 25 February 1692/3*; Trinity College Cambridge 189.R.4.47, ff. 7-8; Trinity College Library: Cambridge, UK, 1692; Available online: http://www.newtonproject.sussex.ac.uk/view/texts/normalized/THEM00258 (accessed on 20 April 2015).
25. Maxwell, J.C. *A Treatise on Electricity and Magnetism*; Clarendon Press: Oxford, UK, 1873.
26. Einstein, A. Autobiographische Skizze. In *Helle Zeit-Dunkle Zeit. In Memoriam Albert Einstein*; Seelig, C., Ed.; Europa Verlag: Zürich, Switzerland, 1956; pp. 9–17. (In German)

27. Abraham, M. Una nuova teoria della gravitazione. *Il Nuovo Cimento.* **1912**, *4*, 459–481. [CrossRef]
28. Einstein, A. Über den Einfluß der Schwerkraft auf die Ausbreitung des Lichtes. *Ann. Phys.* **1911**, *340*, 898–908. (In German) [CrossRef]
29. Einstein, A. Über das Relativitätsprinzip und die aus demselben gezogenen Folgerungen. *Jahrbuch der Radioaktivität* **1907**, *4*, 411–462. (In German)
30. von Eötvös, R. Über die Anziehung der Erde auf verschiedene Substanzen. *Mathematische und Naturwissenschaftliche Berichte aus Ungarn* **1890**, *8*, 65–68. (In German)
31. Adler, C.G.; Coulter, B.L. Galileo and the Tower of Pisa experiment. *Am. J. Phys.* **1978**, *46*, 199–201. [CrossRef]
32. Segre, M. Galileo, Viviani and the tower of Pisa. *Stud. History Philos. Sci. Part A.* **1989**, *20*, 435–451. [CrossRef]
33. Crease, R.P. The Legend of the Leaning Tower. In *The Laws of Motion. An Anthology of Current Thought*; Hall, L.E., Ed.; The Rosen Publishing Group: New York, NY, USA, 2006; pp. 10–14.
34. Drake, S. *Galileo at Work. His Scientific Biography*; University of Chicago Press: Chicago, IL, USA, 1978.
35. Chandrasekhar, S. *Newton's Principia for the Common Reader*; Clarendon Press: Oxford, UK, 1995; p. 362.
36. Chandrasekhar, S. *Newton's Principia for the Common Reader*; Clarendon Press: Oxford, UK, 1995; p. 364.
37. Peeters, C. Gothic *kaurus*, Sanskrit *guruḥ*, Greek βαρζ. *Indogermanische Forschungen* **2010**, *79*, 33–34.
38. Chandrasekhar, S. *Newton's Principia for the Common Reader*; Clarendon Press: Oxford, UK, 1995; p. 369.
39. Mie, G. Grundlagen einer Theorie der Materie. *Ann. Phys.* **1913**, *40*, 1–66. (In German) [CrossRef]
40. Pound, R.V.; Rebka, G.A. Apparent Weight of Photons. *Phys. Rev. Lett.* **1960**, *4*, 337–341. [CrossRef]
41. Corda, C. Interpretation of Mössbauer experiment in a rotating system: A new proof for general relativity. *Ann. Phys.-New York* **2015**, *355*, 360–366. [CrossRef]
42. Blamont, J.E.; Roddier, F. Precise Observation of the Profile of the Fraunhofer Strontium Resonance Line. Evidence for the Gravitational Red Shift on the Sun. *Phys. Rev. Lett.* **1961**, *7*, 437–439. [CrossRef]
43. Adams, W.S. The Relativity Displacement of the Spectral Lines in the Companion of Sirius. *Proc. Natl. Acad. Sci. USA* **1925**, *11*, 382–387. [CrossRef] [PubMed]
44. Einstein, A. Erklärung der Perihelbewegung des Merkur aus der allgemeinen Relativitätstheorie. *Sitzungsber. Kön. Preuß. Akad. Wiss. zu Berlin* **1915**, 831–839. (In German)
45. Dyson, F.W.; Eddington, A.S.; Davidson, C. A Determination of the Deflection of Light by the Sun's Gravitational Field, from Observations Made at the Total Eclipse of May 29, 1919. *Philos. Trans. R. Soc. London. Series A.* **1920**, *220*, 291–333. [CrossRef]
46. Will, C.M. The 1919 measurement of the deflection of light. *Class. Quantum Gravit.* **2015**, in press. [CrossRef]
47. Klein, M.J.; Kox, A.J.; Renn, J.; Schulman, R. Appendix A. Scratch Notebook, 1909–1914? In *The Collected Papers of Albert Einstein. Volume 3: The Swiss Years: Writings 1909–1911*; Princeton University Press: Princeton, NJ, USA, 1994; p. 585.
48. Renn, J.; Sauer, T.; Stachel, J. The Origin of Gravitational Lensing: A Postscript to Einstein's 1936 Science paper. *Science* **1997**, *275*, 184–186. [CrossRef] [PubMed]
49. Walsh, D.; Carswell, R.F.; Weymann, R.J. 0957 + 561 A, B: Twin quasistellar objects or gravitational lens? *Nature* **1979**, *279*, 381–384. [CrossRef] [PubMed]
50. Einstein, A. Gibt es eine Gravitationswirkung, die der elektrodynamischen Induktionswirkung analog ist? *Vierteljahrschrift für gerichtliche Medizin und öffentliches Sanitätswesen* **1912**, *44*, 37–40. (In German)
51. Einstein, A. Zum gegenwärtigen Stande des Gravitationsproblem. *Phys. Z.* **1913**, *14*, 1249–1266. (In German)
52. Klein, M.J.; Kox, A.J.; Schulmann, R. Einstein and Besso: Manuscript on the Motion of the Perihelion of Mercury. In *The Collected Papers of Albert Einstein. Volume 4. The Swiss Years: Writings, 1912–1914*; Princeton University Press: Princeton, NJ, USA, 1995; pp. 360–473.
53. Schulmann, R.; Kox, A.J.; Janssen, M.; Illy, J. To Hans Thirring. In *The Collected Papers of Albert Einstein. Volume 8. The Berlin Years: Correspondence, 1914–1918*; Princeton University Press: Princeton, NJ, USA, 1998; pp. 500–501 and pp. 564–566.
54. Lense, J.; Thirring, H. Über den Einfluß der Eigenrotation der Zentralkörper auf die Bewegung der Planeten und Monde nach der Einsteinschen Gravitationstheorie. *Phys. Z.* **1918**, *19*, 156–163. (In German)
55. Thirring, H. Über die formale Analogie zwischen den elektromagnetischen Grundgleichungen und den Einsteinschen Gravitationsgleichungen erster Näherung. *Phys Z.* **1918**, *19*, 204–205. (In German)
56. Thirring, H. Über die Wirkung rotierender ferner Massen in der Einsteinschen Gravitationstheorie. *Phys. Z* **1918**, *19*, 33–39. (In German)

57. Thirring, H. Berichtigung zu meiner Arbeit: "Über die Wirkung rotierender ferner Massen in der Einsteinschen Gravitationstheorie". *Phys. Z* **1921**, *22*, 29–30. (In German)
58. Mashhoon, B. On the gravitational analogue of Larmor's theorem. *Phys. Lett. A* **1993**, *173*, 347–354. [CrossRef]
59. Iorio, L. A gravitomagnetic effect on the orbit of a test body due to the earth's variable angular momentum. *Int. J. Mod. Phys. D* **2002**, *11*, 781–787. [CrossRef]
60. Bini, D.; Cherubini, C.; Chicone, C.; Mashhoon, B. Gravitational induction. *Class. Quantum Gravit.* **2008**, *25*, 225014. [CrossRef]
61. Mashhoon, B.; Hehl, F.W.; Theiss, D.S. On the Influence of the Proper Rotation of Central Bodies on the Motions of Planets and Moons According to Einstein's Theory of Gravitation. *Gen. Relativ. Gravit.* **1984**, *16*, 727–741.
62. Pfister, H. On the history of the so-called Lense-Thirring effect. *Gen. Relativ. Gravit.* **2007**, *39*, 1735–1748. [CrossRef]
63. Pfister, H. Editorial note to: Hans Thirring, On the formal analogy between the basic electromagnetic equations and Einstein' s gravity equations in first approximation. *Gen. Relativ. Gravit.* **2012**, *44*, 3217–3224. [CrossRef]
64. Everitt, C.W.F.; Debra, D.B.; Parkinson, B.W.; Turneaure, J.P.; Conklin, J.W.; Heifetz, M.I.; Keiser, G.M.; Silbergleit, A.S.; Holmes, T.; Kolodziejczak, J.; *et al.* Gravity Probe B: Final Results of a Space Experiment to Test General Relativity. *Phys. Rev. Lett.* **2011**, *106*, 221101. [CrossRef] [PubMed]
65. Ciufolini, I. Dragging of inertial frames. *Nature* **2007**, *449*, 41–47. [CrossRef] [PubMed]
66. Iorio, L.; Lichtenegger, H.I.M.; Ruggiero, M.L.; Corda, C. Phenomenology of the Lense-Thirring effect in the solar system. *Astrophys. Space Sci.* **2011**, *331*, 351–395. [CrossRef]
67. Renzetti, G. History of the attempts to measure orbital frame-dragging with artificial satellites. *Cent. Eur. J. Phys.* **2013**, *11*, 531–544. [CrossRef]
68. Rindler, W. The Lense-Thirring effect exposed as anti-Machian. *Phys. Lett. A* **1994**, *187*, 236–238. [CrossRef]
69. Bondi, H.; Samuel, J. The Lense-Thirring effect and Mach's principle. *Phys. Lett. A* **1997**, *228*, 121–126. [CrossRef]
70. Rindler, W. The case against space dragging. *Phys. Lett. A* **1997**, *233*, 25–29. [CrossRef]
71. Lichtenegger, H.I.M.; Mashhoon, B. Mach's Principle. In *The Measurement of Gravitomagnetism: A Challenging Enterprise*; Iorio, L., Ed.; Nova: Hauppauge, NY, USA, 2007; pp. 13–25.
72. Einstein, A.; Grossmann, M. Kovarianzeigenschaften der Feldgleichungen der auf die verallgemeinerte Relativitätstheorie gegründeten Gravitationstheorie. *Z. Math. Phys.* **1914**, *63*, 215–225. (In German)
73. Landau, L.D.; Lifšitis, E.M. *Teoriya Polia*; Nauka: Moscow, Russia, 1973. (In Russian)
74. Hilbert, D. Die Grundlagen der Physik (Erste Mitteilung). *Nachrichten von der Königlichen Gesellschaft der Wissenschaften zu Göttingen. Mathematisch-physikalische Klasse* **1915**, 395–407. (In German)
75. Corry, L.; Renn, J.; Stachel, J. Belated Decision in the Hilbert-Einstein Priority Dispute. *Science* **1997**, *278*, 1270–1273. [CrossRef]
76. Le Verrier, U.; Lettre de, M.; Le Verrier à, M. Faye sur la théorie de Mercure et sur le mouvement du périhélie de cette planète. *Cr. Hebd. Acad. Sci.* **1859**, *49*, 379–383. (In French)
77. Schwarzschild, K. Über das Gravitationsfeld eines Massenpunktes nach der Einsteinschen Theorie. *Sitzungsber. Kön. Preuß. Akad. Wiss. zu Berlin* **1916**, 189–196. (In German)
78. Droste, J. Het veld van een enkel centrum in Einstein's theorie der zwaartekracht, en de beweging van een stoffelijk punt in dat veld. *Koninklijke Akademie van Wetenschappen te Amsterdam. Wis-en Natuurkundige Afdeeling. Verslagen van de Gewone Vergaderingen* **1916**, *25*, 163–180. (In Dutch)
79. Weyl, H. Zur Gravitationstheorie. *Ann. Phys.* **1917**, *359*, 117–145. (In German) [CrossRef]
80. Schwarzschild, K. Über das Gravitationsfeld einer Kugel aus inkompressibler Flüssigkeit nach der Einsteinschen Theorie. *Sitzungsber. Kön. Preuß. Akad. Wiss. zu Berlin* **1916**, 424–434. (In German)
81. Birkhoff, G.D. *Relativity and Modern Physics*; Harvard University Press: Cambridge, MA, USA, 1923.
82. Brush, S.G. Prediction and Theory Evaluation: The Case of Light Bending. *Science* **1989**, *246*, 1124–1129. [CrossRef] [PubMed]
83. Weinberg, S. *Dreams of a Final Theory*; Pantheon Books: New York, NY, USA, 1992; pp. 96–97.
84. Einstein, A. Die Grundlage der allgemeinen Relativitätstheorie. *Ann. Phys.* **1916**, *354*, 769–822. (In German) [CrossRef]

85. De Sitter, W. On Einstein's Theory of Gravitation and its Astronomical Consequences. Second Paper. *Mon. Not. R. Astron. Soc.* **1916**, *77*, 155–184. [CrossRef]

86. Schouten, J.A. Over het ontstaan eener praecessiebeweging tengevolge van het niet euklidisch zijn der ruimte in de nabijheid van de zon. *Koninklijke Akademie van Wetenschappen te Amsterdam. Wis-en Natuurkundige Afdeeling. Verslagen van de Gewone Vergaderingen* **1918**, *27*, 214–220. (In Dutch)

87. Fokker, A.D. De geodetische precessie: Een uitvloeisel van Einstein's gravitatietheorie. *Koninklijke Akademie van Wetenschappen te Amsterdam. Wis-en Natuurkundige Afdeeling. Verslagen van de Gewone Vergaderingen* **1920**, *29*, 611–621. (In Dutch)

88. Bertotti, B.; Ciufolini, I.; Bender, P.L. New test of general relativity-Measurement of de Sitter geodetic precession rate for lunar perigee. *Phys. Rev. Lett.* **1987**, *58*, 1062–1065. [CrossRef] [PubMed]

89. Williams, J.G.; Newhall, X.X.; Dickey, J.O. Relativity parameters determined from lunar laser ranging. *Phys. Rev. D* **1996**, *53*, 6730–6739. [CrossRef]

90. Shapiro, I.I. Fourth Test of General Relativity. *Phys. Rev. Lett.* **1964**, *13*, 789–791. [CrossRef]

91. Shapiro, I.I.; Pettengill, G.H.; Ash, M.E.; Stone, M.L.; Smith, W.B.; Ingalls, R.P.; Brockelman, R.A. Fourth Test of General Relativity: Preliminary Results. *Phys. Rev. Lett.* **1968**, *20*, 1265–1269. [CrossRef]

92. Bertotti, B.; Iess, L.; Tortora, P. A test of general relativity using radio links with the Cassini spacecraft. *Nature* **2003**, *425*, 374–376. [CrossRef] [PubMed]

93. Einstein, A. Näherungsweise Integration der Feldgleichungen der Gravitation. *Sitzungsber. Kön. Preuß. Akad. Wiss. zu Berlin* **1916**, 688–696. (In German)

94. Will, C.M. Propagation Speed of Gravity and the Relativistic Time Delay. *Astrophys. J.* **2003**, *590*, 683–690. [CrossRef]

95. Carlip, S. Model-dependence of Shapiro time delay and the "speed of gravity/speed of light" controversy. *Class. Quantum Gravit.* **2005**, *21*, 3803–3812. [CrossRef]

96. Kopeikin, S.M. Comment on 'Model-dependence of Shapiro time delay and the "speed of gravity/speed of light" controversy'. *Class. Quantum Gravit.* **2005**, *22*, 5181–5186. [CrossRef]

97. Kopeikin, S.M. Testing the Relativistic Effect of the Propagation of Gravity by Very Long Baseline Interferometry. *Astrophys. J.* **2001**, *556*, L1–L5. [CrossRef]

98. Fomalont, E.B.; Kopeikin, S.M. The Measurement of the Light Deflection from Jupiter: Experimental Results. *Astrophys. J.* **2003**, *598*, 704–711. [CrossRef]

99. Einstein, A. Über Gravitationswellen. *Sitzungsber. Kön. Preuß. Akad. Wiss. zu Berlin* **1918**, 154–167. (In German)

100. Taylor, J.H.; Fowler, L.A.; McCulloch, P.M. Measurements of general relativistic effects in the binary pulsar PSR 1913+16. *Nature* **1979**, *277*, 437–440. [CrossRef]

101. Taylor, J.H.; Weisberg, J.M. A new test of general relativity-Gravitational radiation and the binary pulsar PSR 1913+16. *Astrophys. J.* **1982**, *253*, 908–920. [CrossRef]

102. Weisberg, J.M.; Taylor, J.H. Observations of Post-Newtonian Timing Effects in the Binary Pulsar PSR 1913+16. *Phys. Rev. Lett.* **1984**, *52*, 1348–1350. [CrossRef]

103. Hulse, R.A.; Taylor, J.H. Discovery of a pulsar in a binary system. *Astrophys. J.* **1975**, *195*, L51–L53. [CrossRef]

104. Damour, T. 1974: The discovery of the first binary pulsar. *Class. Quantum Gravit.* **2015**. In press. [CrossRef]

105. Pitkin, M.; Reid, S.; Rowan, S.; Hough, J. Gravitational Wave Detection by Interferometry (Ground and Space). *Living Rev. Relativ.* **2011**, *14*, 5. [CrossRef]

106. Freise, A.; Strain, K.A. Interferometer Techniques for Gravitational-Wave Detection. *Living Rev. Relativ.* **2010**, *13*, 1. [CrossRef]

107. Losurdo, G. Ground-based gravitational wave interferometric detectors of the first and second generation: An overview. *Class. Quantum Gravit.* **2012**, *29*, 124005. [CrossRef]

108. Acernese, F.; *et al.* Advanced Virgo: A second-generation interferometric gravitational wave detector. *Class. Quantum Gravit.* **2015**, *31*, 024001. [CrossRef]

109. Abbott, B.P.; Abbott, R.; Adhikari, R.; Ajith, P.; Allen, B.; Allen, G.; Amin, R.S.; Anderson, S.B.; Anderson, W.G.; Arain, M.A.; *et al.* LIGO: The Laser Interferometer Gravitational-Wave Observatory. *Rep. Progr. Phys.* **2009**, *72*, 076901. [CrossRef]

110. Hammond, G.; Hild, S.; Pitkin, M. Advanced technologies for future ground-based, laser-interferometric gravitational wave detectors. *J. Mod. Optics* **2014**, *61*, S10–S45. [CrossRef] [PubMed]

111. Gair, J.R.; Vallisneri, M.; Larson, S.L.; Baker, J.G. Testing General Relativity with Low-Frequency, Space-Based Gravitational-Wave Detectors. *Liv. Rev. Relativ.* **2013**, *16*, 7. [CrossRef]

112. Vitale, S. Space-borne gravitational wave observatories. *Gen. Relativ. Gravit.* **2014**, *46*, 1730. [CrossRef]

113. Armstrong, J.W. Low-Frequency Gravitational Wave Searches Using Spacecraft Doppler Tracking. *Living Rev. Relativ.* **2006**, *9*, 1. [CrossRef]

114. Porter, E.K. The Challenges in Gravitational Wave Astronomy for Space-Based Detectors. In *Gravitational Wave Astrophysics*; Sopuerta, C., Ed.; Springer: Berlin, Germany, 2015; pp. 267–279.

115. Amaro-Seoane, P.; Aoudia, S.; Babak, S.; Binétruy, P.; Berti, E.; Bohé, A.; Caprini, C.; Colpi, M.; Cornish, N.J.; Danzmann, K.; *et al.* eLISA: Astrophysics and cosmology in the millihertz regime. *GW Notes* **2013**, *6*, 4–110.

116. Ni, W.-T. Astrod-Gw Overview and Progress. *Int. J. Mod. Phys. D* **2013**, *22*, 1341004. [CrossRef]

117. Loeb, A.; Maoz, D. Using Atomic Clocks to Detect Gravitational Waves. 2015; arXiv:1501.00996.

118. Thorne, K.S. Sources of Gravitational Waves and Prospects for their Detection. In *Recent Advances in General Relativity*; Janis, A.I., Porter, J.R., Eds.; Birkhäuser: Boston, MA, USA, 1992; pp. 196–229.

119. Gertsenshtein, M.E.; Pustovoit, V.I. K voprosu ob obnaruzhenii gravitatsionnykh valn malykh chastot. *Zh. Eksp. Teor. Fiz.* **1962**, *43*, 605–607. (In Russian)

120. Weber, J. Evidence for Discovery of Gravitational Radiation. *Phys. Rev. Lett.* **1969**, *22*, 1320–1324. [CrossRef]

121. Weber, J. Gravitational-Wave-Detector Events. *Phys. Rev. Lett.* **1968**, *20*, 1307–1308. [CrossRef]

122. Damour, T. An Introduction to the Theory of Gravitational Radiation. In *Gravitation in Astrophysics*; Carter, B., Hartle, J.B., Eds.; Plenum Press: New York, NY, USA, 1987; pp. 3–62.

123. Corda, C. Interferometric Detection of Gravitational Waves: the Definitive Test for General Relativity. *Int. J. Mod. Phys. D* **2009**, *18*, 2275–2282. [CrossRef]

124. Sathyaprakash, B.S.; Schutz, B.F. Physics, Astrophysics and Cosmology with Gravitational Waves. *Living Rev. Relativ.* **2009**, *12*, 2. [CrossRef]

125. Press, W.H.; Thorne, K.S. Gravitational-Wave Astronomy. *Ann. Rev. Astron. Astrophys.* **1972**, *10*, 335–374. [CrossRef]

126. Gasperini, M. *Relatività Generale e Teoria della Gravitazione*, 2nd ed.; Springer: Milan, Italy, 2015. (In Italian)

127. Penzias, A.A.; Wilson, R.W. A Measurement of Excess Antenna Temperature at 4080 Mc/s. *Astrophys. J.* **1965**, *142*, 419–421. [CrossRef]

128. Dodelson, S. *Modern Cosmology*; Academic Press: San Diego, CA, USA, 2003.

129. Kamionkowski, M.; Kosowsky, A.; Stebbins, A. Statistics of cosmic microwave background polarization. *Phys. Rev. D* **1997**, *55*, 7368–7388. [CrossRef]

130. Seljak, U.; Zaldarriaga, M. Signature of Gravity Waves in the Polarization of the Microwave Background. *Phys. Rev. Lett.* **1997**, *78*, 2054–2057. [CrossRef]

131. BICEP2 Collaboration. Detection of B-Mode Polarization at Degree Angular Scales by BICEP. *Phys. Rev. Lett.* **2014**, *112*, 241101.

132. BICEP2/Keck, Planck Collaborations. A Joint Analysis of BICEP2/Keck Array and Planck Data. *Phys. Rev. Lett.* **2015**, in press.

133. Newton, I. *Scholium to the Definitions in Philosophiae Naturalis Principia Mathematica, Bk. 1 (1689); trans. Andrew Motte (1729), rev. Florian Cajori*; University of California Press: Berkeley, CA, USA, 1934; pp. 6–12. Available online: http://plato.stanford.edu/entries/newton-stm/scholium.html (accessed on 20 April 2015).

134. Friedländer, B.; Friedländer, I. *Absolute Oder Relative Bewegung?* Simion: Berlin, Germany, 1896. (In German)

135. Pugh, G.E. Proposal for a Satellite Test of the Coriolis Prediction of General Relativity. In *Weapons Systems Evaluation Group. Research Memorandum No. 11*; The Pentagon: Washington, DC, USA, 1959.

136. Schiff, L.I. Possible New Experimental Test of General Relativity Theory. *Phys. Rev. Lett.* **1960**, *4*, 215–217. [CrossRef]

137. Everitt, C.W.F. The Gyroscope Expriment. I: General Description and Analysis of Gyroscope Performance. In Proceedings of the International School of Physics "Enrico Fermi". Course LVI. Experimental Gravitation; Bertotti, B., Ed.; Academic Press: New York, NY, USA, 1974; pp. 331–402.

138. Reissner, H. Über die Eigengravitation des elektrischen Feldes nach der Einsteinschen Theorie. *Ann. Phys.* **1916**, *355*, 106–120. (In German) [CrossRef]

139. Nordström, G. Een an Ander Over de Energie van het Zwaartekrachtsveld Volgens de Theorie van Einstein. *Koninklijke Akademie van Wetenschappen te Amsterdam. Wis-en Natuurkundige Afdeeling. Verslagen van de Gewone Vergaderingen* **1918**, *26*, 1201–1208. (In Dutch)

140. Rindler, W. Visual Horizons in World Models. *Mon. Not. R. Astron. Soc.* **1956**, *116*, 662–677. [CrossRef]
141. Finkelstein, D. Past-Future Asymmetry of the Gravitational Field of a Point Particle. *Phys. Rev.* **1958**, *110*, 965–967. [CrossRef]
142. Brill, D. Black Hole Horizons and How They Begin. *Astron. Rev.* **2012**, *7*, 1–12.
143. Wheeler, J.A.; Gearhart, M. FORUM: From the Big Bang to the Big Crunch. *Cosmic Search* **1979**, *1*, 2–8.
144. Thorne, K.S. *From Black Holes to Time Warps: Einstein's Outrageous Legacy*; W.W. Norton & Company: New York, NY, USA, 1994; pp. 255–256.
145. Wheeler, J.A. Our universe: The known and the unknown. *Am. Sci.* **1968**, *56*, 1–20. [CrossRef]
146. Ruffini, R.J.; Wheeler, J.A. Introducing the black hole. *Phys. Today* **1971**, *24*, 30–41. [CrossRef]
147. Eddington, A.S. *The Internal Constitution of the Stars*; Cambridge University Press: Cambridge, UK, 1926; p. 6.
148. Einstein, A. On a Stationary System with Spherical Symmetry Consisting of Many Gravitating Masses. *Ann. Math.* **1939**, *40*, 922–936. [CrossRef]
149. Oppenheimer, J.R.; Snyder, H. On Continued Gravitational Contraction. *Phys. Rev.* **1939**, *56*, 455–459. [CrossRef]
150. Eddington, A.S. A Comparison of Whitehead's and Einstein's Formulæ. *Nature* **1924**, *113*, 192. [CrossRef]
151. 't Hooft, G. *Introduction to the Theory of Black Holes*; Spinoza Institute: Utrecht, The Netherlands, 2009; pp. 47–48.
152. Penrose, R. Gravitational collapse and space-time singularities. *Phys. Rev. Lett.* **1965**, *14*, 57–59. [CrossRef]
153. Kerr, R.P. Gravitational Field of a Spinning Mass as an Example of Algebraically Special Metrics. *Phys. Rev. Lett.* **1963**, *11*, 237–238. [CrossRef]
154. Boyer, R.H.; Lindquist, R.W. Maximal Analytic Extension of the Kerr Metric. *J. Math. Phys.* **1967**, *8*, 265–281. [CrossRef]
155. Krasiński, A. Ellipsoidal Space-Times, Sources for the Kerr Metric. *Ann. Phys.-New York* **1978**, *112*, 22–40. [CrossRef]
156. Thorne, K.S. Relativistic Stars, Black Holes and Gravitational Waves (Including an In-Depth Review of the Theory of Rotating, Relativistic Stars). In Proceedings of the International School of Physics "Enrico Fermi". Course XLVII. General Relativity and Cosmology; Sachs, B.K., Ed.; Academic Press: New York, NY, USA, 1971; pp. 237–283.
157. Boshkayev, K.; Quevedo, H.; Ruffini, R. Gravitational field of compact objects in general relativity. *Phys. Rev. D* **2012**, *86*, 064043. [CrossRef]
158. Teukolsky, S.A. The Kerr Metric. *Class. Quantum Gravit.* **2015**, in press. [CrossRef]
159. Misner, C.W.; Thorne, K.S.; Wheeler, J.A. *Gravitation*; W.H. Freeman: San Francisco, CA, USA, 1973; pp. 875–876.
160. Heusler, M. Uniqueness Theorems for Black Hole Space-Times. In *Black Holes: Theory and Observation*; Hehl, F.W., Kiefer, C., Metzler, R.J.K., Eds.; Springer: Berlin, Germany, 1998; pp. 157–186.
161. Geroch, R. Multipole Moments. II. Curved Space. *J. Math. Phys.* **1970**, *11*, 2580–2588. [CrossRef]
162. Hansen, R.O. Multipole moments of stationary space-times. *J. Math. Phys.* **1974**, *15*, 46–52. [CrossRef]
163. Hartle, J.B.; Thorne, K.S. Slowly Rotating Relativistic Stars. II. Models for Neutron Stars and Supermassive Stars. *Astrophys. J.* **1968**, *153*, 807–834. [CrossRef]
164. Matthews, T.A.; Sandage, A.R. Optical Identification of 3C 48, 3C 196, and 3C 286 with Stellar Objects. *Astrophys. J.* **1963**, *138*, 30–58. [CrossRef]
165. Chiu, H.-Y. Gravitational Collapse. *Phys. Today* **1964**, *17*, 21–34. [CrossRef]
166. Heinicke, C.; Hehl, F.W. Schwarzschild and Kerr solutions of Einstein's field equation: An Introduction. *Int. J. Mod. Phys. D* **2015**, *24*, 1530006. [CrossRef]
167. Newman, E.T.; Couch, R.; Chinnapared, K.; Exton, A.; Prakash, A.; Torrence, R. Metric of a Rotating, Charged Mass. *J. Math. Phys.* **1965**, *6*, 918–919. [CrossRef]
168. Newman, E.T.; Adamo, T. Kerr-Newman metric. *Scholarpedia* **2014**, *9*, 31791. [CrossRef]
169. Newman, E.T.; Janis, A.I. Note on the Kerr spinning particle metric. *J. Math. Phys.* **1965**, *6*, 915–917. [CrossRef]
170. Ryan, M.P.; Shepley, L.C. *Homogeneous Relativistic Cosmologies*; Princeton University Press: Priceton, NJ, USA, 1975.
171. Newton, Is. *Original letter from Isaac Newton to Richard Bentley, dated 10 December 1692*; Trinity College Cambridge 189.R.4.47, ff. 4A-5; Trinity College Library: Cambridge, UK, 1692; Available online: http://www.newtonproject.sussex.ac.uk/view/texts/normalized/THEM00254 (accessed on 20 April 2015).

172. Olbers, H.W. Ueber die Durchsichtigkeit des Weltraums, von Hrn. Dr. Olbers in Bremen, unterm 7. Mai 1823. eingesandt. In *Astronomisches Jahrbuch für das Jahr 1826*; Bode, J.E., Ed.; Ferd. Dümmler: Berlin, Germany, 1823; pp. 110–121. (In German)

173. IOANNIS KEPLERI Mathematici Cæsarei. *DISSERTATIO Cum NVNCIO SIDEREO nuper ad mortales misso à GALILÆO GALILÆO Mathematico Patauino*; Sedesanus: Prague, Czech Republic, 1610. (In Latin)

174. Halley, E. On the Infinity of the Sphere of the Fix'd Stars. *Philos. Trans.* **1720**, *31*, 22–24. [CrossRef]

175. Halley, E. Of the Number, Order and Light of the Fix'd Stars. *Philos. Trans.* **1720**, *31*, 24–26. [CrossRef]

176. De Chéseaux, J.P.L. *Traité de la comète: qui a paru en décembre 1743 & en janvier, février & mars 1744 / contenant outre les observations de l'auteur, celles qui ont été faites à Paris par Mr. Cassini, & à Genève par Mr. Calandrini; on y a joint diverses observations & dissertations astronomiques; le tout accompagné de figures en taille douce*; Marc-Michel Bousquet & compagnie: Lausanne, Switzerland, 1744; Appendix 2. (In French)

177. Von Seeliger, H. Ueber das Newton'sche Gravitationsgesetz. *Astron. Nachr.* **1895**, *137*, 129–136. (In German) [CrossRef]

178. Einstein, A. Kosmologische Betrachtungen zur allgemeinen Relativitätstheorie. *Sitzungsber. Kön. Preuß. Akad. Wiss. zu Berlin* **1917**, 142–152. (In German)

179. Von Seeliger, H. Über das Newton'sche Gravitationsgesetz. *Sitzungsberichte der Königlich Bayerische Akademie der Wissenschaften zu München. Mathematisch-physikalische Klasse* **1896**, *126*, 373–400. (In German)

180. Norton, J.D. The Cosmological Woes of Newtonian Gravitation Theory. In *The Expanding Worlds of General Relativity*; Goenner, H., Renn, J., Ritter, T., Sauer, T., Eds.; Birkhäuser: Basel, Switzerland, 1999; pp. 271–322.

181. Gasperini, M. *Lezioni di Cosmologia Teorica*; Springer: Milan, Italy, 2012. (In Italian)

182. Planck Collaboration. Planck 2013 results. XVI. Cosmological parameters. *Astron. Astrophys.* **2014**, *571*, A16.

183. Luminet, J.-P. *L'Unives chiffonné*; Gallimard: Paris, France, 2005. (In French)

184. Cartan, E. Sur les équations de la gravitation d'Einstein. *J. Math. Pures Appl.* **1922**, *9*, 141–203.

185. Eddington, A.S. On the Instability of Einstein's Spherical World. *Mon. Not. R. Astron. Soc.* **1930**, *90*, 668–678. [CrossRef]

186. de Sitter, W. Over de relativiteit der traagheid: Beschouwingen naar aanleiding van Einstein's laatste hypothese. *Koninklijke Akademie van Wetenschappen te Amsterdam. Wis-en Natuurkundige Afdeeling. Verslagen van de Gewone Vergaderingen* **1917**, *25*, 1268–1276. (In Dutch)

187. de Sitter, W. Over de kromming der ruimte. *Koninklijke Akademie van Wetenschappen te Amsterdam. Wis-en Natuurkundige Afdeeling. Verslagen van de Gewone Vergaderingen* **1917**, *26*, 222–236. (In Dutch)

188. de Sitter, W. Over de mogelijkheid van statistisch evenwicht van het heelal. *Koninklijke Akademie van Wetenschappen te Amsterdam. Wis-en Natuurkundige Afdeeling. Verslagen van de Gewone Vergaderingen* **1920**, *29*, 651–653. (In Dutch)

189. Eddington, A.S. *The Expanding Universe*; The University of Michigan Press: Ann Arbor, MI, USA, 1958; p. 46.

190. Starobinsky, A.A. Spektr reliktovogo gravitatsionnogo izlucheniya i nachal'noe sostoyanie Vselennoi. *Zh. Eksp. Teor. Fiz.* **1979**, *30*, 719–723. (In Russian)

191. Guth, A.H. Inflationary universe: A possible solution to the horizon and flatness problems. *Phys. Rev. D* **1981**, *23*, 347–356. [CrossRef]

192. Linde, A.D. A new inflationary universe scenario: A possible solution of the horizon, flatness, homogeneity, isotropy and primordial monopole problems. *Phys. Lett. B* **1982**, *108*, 389–393. [CrossRef]

193. Friedman, A. Über die Krümmung des Raumes. *Zeitschr. Phys.* **1922**, *10*, 377–386. (In German) [CrossRef]

194. Friedman, A. Über die Möglichkeit einer Welt mit konstanter negativer Krümmung des Raumes. *Zeitschr. Phys.* **1924**, *21*, 326–332. (In German) [CrossRef]

195. Lachieze-Rey, M.; Luminet, J.-P. Cosmic topology. *Phys. Rep.* **1995**, *254*, 135–214. [CrossRef]

196. Luminet, J.-P. Cosmic topology: Twenty years after. *Grav. Cosmol.* **2014**, *20*, 15–20. [CrossRef]

197. Slipher, V.M. On the spectrum of the nebula in the Pleiades. *Lowell Obs. Bull.* **1912**, *2*, 26–27.

198. Slipher, V.M. The radial velocity of the Andromeda Nebula. *Lowell Obs. Bull.* **1913**, *1*, 56–57.

199. Slipher, V.M. Nebulæ. *Proc. Am. Philos. Soc.* **1917**, *56*, 403–409.

200. Hubble, E. A Relation between Distance and Radial Velocity among Extra-Galactic Nebulae. *Proc. Natl. Acad. Sci. USA* **1929**, *15*, 168–173. [CrossRef] [PubMed]

201. Humason, M.L. Apparent Velocity-Shifts in the Spectra of Faint Nebulae. *Astrophys. J.* **1931**, *74*, 35–42. [CrossRef]

202. Einstein, A. Bemerkung zu der Arbeit von A. Friedmann "Über die Krümmung des Raumes". *Zeitschr. Phys.* **1922**, *11*, 326. (In German) [CrossRef]

203. Einstein, A. Notiz zu der Arbeit von A. Friedmann "Über die Krümmung des Raumes". *Zeitschr. Phys.* **1923**, *16*, 228. (In German) [CrossRef]

204. Lemaître, G. Un Univers homogène de masse constante et de rayon croissant rendant compte de la vitesse radiale des nébuleuses extra-galactiques. *Annales de la Société Scientifique de Bruxelles* **1927**, *47*, 49–59. (In French)

205. Hubble, E. NGC 6822, a remote stellar system. *Astrophys. J.* **1925**, *62*, 409–433. [CrossRef]

206. Einstein, A. Zum kosmologischen Problem der allgemeinen Relativitätstheorie. *Sitzungsber. Kön. Preuß. Akad. Wiss. zu Berlin* **1931**, 235–237. (In German)

207. Lemaître, G.; L' étrangeté de l'Univers. 1960; *La Revue Generale Belge*, 1–14. (In French)

208. Luminet, J.-P. *L'invention du Big Bang*; Editions du Seuil: Paris, France, 2004. (In French)

209. Robertson, H.P. On the Foundations of Relativistic Cosmology. *Proc. Natl. Acad. Sci. USA* **1929**, *15*, 822–829. [CrossRef] [PubMed]

210. Robertson, H.P. Relativistic Cosmology. *Rev. Mod. Phys.* **1933**, *5*, 62–90. [CrossRef]

211. Robertson, H.P. Kinematics and World-Structure. *Astrophys. J.* **1935**, *82*, 284–301. [CrossRef]

212. Walker, A.G. On the formal comparison of Milne's kinematical system with the systems of general relativity. *Mon. Not. R. Astron. Soc.* **1935**, *95*, 263–269. [CrossRef]

213. Einstein, A.; de Sitter, W. On the Relation between the Expansion and the Mean Density of the Universe. *Proc. Natl. Acad. Sci. USA* **1932**, *18*, 213–214. [CrossRef] [PubMed]

214. Eddington, A.S. Forty Years of Astronomy. In *Background to Modern Science*; Needham, J., Page, W., Eds.; Cambridge University Press: Cambridge, UK, 1938; p. 128.

215. Barrow, J.D. *The Book of Universes*; The Bodley Head: London, UK, 2011.

216. Wesson, P.S. The real reason the night sky is dark: Correcting a myth in astronomy teaching. *J. Br. Astron. Ass.* **1989**, *99*, 10–13.

217. Harrison, E. *Cosmology. The Science of the Universe*, 2nd ed.; Cambridge University Press: Cambridge, UK, 2000; pp. 491–514.

218. Weinberg, S. *Cosmology*; Oxford University Press: Oxford, UK, 2008.

219. Weyl, H. Zur allgemeinen Relativitätstheorie. *Zeitschr. Phys.* **1923**, *24*, 230–232. (In German)

220. Riess, A.G.; Filippenko, A.V.; Challis, P.; Clocchiatti, A.; Diercks, A.; Garnavich, P.M.; Gilliland, R.L.; Hogan, C.J.; Jha, S.; Kirshner, R.P.; *et al.* Observational Evidence from Supernovae for an Accelerating Universe and a Cosmological Constant. *Astron. J.* **1998**, *116*, 1009–1038. [CrossRef]

221. Schmidt, B.P.; Suntzeff, N.B.; Phillips, M.M.; Schommer, R.A.; Clocchiatti, A.; Kirshner, R.P.; Garnavich, P.; Challis, P.; Leibundgut, B.; Spyromilio, J.; *et al.* The High-Z Supernova Search: Measuring Cosmic Deceleration and Global Curvature of the Universe Using Type IA Supernovae. *Astrophys. J.* **1998**, *507*, 46–63. [CrossRef]

222. Perlmutter, S.; Aldering, G.; Goldhaber, G.; Knop, R.A.; Nugent, P.; Castro, P.G.; Deustua, S.; Fabbro, S.; Goobar, A.; Groom, D.E.; *et al.* Measurements of Ω and Λ from 42 High-Redshift Supernovae. *Astrophys. J.* **1999**, *517*, 565–586. [CrossRef]

223. Padmanabhan, N.; Xu, X.; Eisenstein, D.J.; Scalzo, R.; Cuesta, A.J.; Mehta, K.T.; Kazin, E. A 2 per cent distance to $z = 0.35$ by reconstructing baryon acoustic oscillations - I. Methods and application to the Sloan Digital Sky Survey. *Mon. Not. R. Astron.* **2012**, *427*, 2132–2145. [CrossRef]

224. Hinshaw, G.; Larson, D.; Komatsu, E.; Spergel, D.N.; Bennett, C.L.; Dunkley, J.; Nolta, M.R.; Halpern, M.; Hill, R.S.; Odegard, N.; *et al.* Nine-year Wilkinson Microwave Anisotropy Probe (WMAP) Observations: Cosmological Parameter Results. *Astrophys. J. Suppl. Ser.* **2013**, *208*, 19. [CrossRef]

225. Hawking, S.W.; Penrose, R. The Singularities of Gravitational Collapse and Cosmology. *Proc. R. Soc. Lond. A* **1970**, *314*, 529–548. [CrossRef]

226. Hawking, S.W.; Ellis, G.F.R. *The Large Scale Structure of Space-time*; Cambridge University Press: Cambridge, UK, 1973.

227. Stephani, A.; Kramer, D.; Mac Callum, M.A.H.; Hoenselaers, C.; Herlt, E. *Exact Solutions of Einstein's Field Equations*, 2nd ed.; Cambridge University Press: Cambridge, UK, 2003.

228. Mac Callum, M.A.H. Exact solutions of Einstein's equations. *Scholarpedia* **2013**, *8*, 8584. [CrossRef]

229. Kopeikin, S.; Efroimsky, M.; Kaplan, G. Relativistic Reference Frames. In *Relativistic Celestial Mechanics of the Solar System*; Kopeikin, S., Efroimsky, M., Kaplan, G., Eds.; Wiley-VCH: Berlin, Germany, 2011; pp. 371–428.

230. Kopeikin, S.; Efroimsky, M.; Kaplan, G. Post-Newtonian Coordinate Transformations. In *Relativistic Celestial Mechanics of the Solar System*; Kopeikin, S., Efroimsky, M., Kaplan, G., Eds.; Wiley-VCH: Berlin, Germany, 2011; pp. 249–462.

231. Ashby, N. Relativity in the Global Positioning System. *Living Rev. Relativ.* **2003**, *6*, 1. [CrossRef]

232. Guinot, B. Metrology and general relativity. *Comptes Rendus Phys.* **2004**, *5*, 821–828. [CrossRef]

233. Larson, K.M.; Ashby, N.; Hackman, C.; Bertiger, W. An assessment of relativistic effects for low Earth orbiters: The GRACE satellites. *Metrologia* **2007**, *44*, 484–490. [CrossRef]

234. Reynaud, S.; Salomon, C.; Wolf, P. Testing General Relativity with Atomic Clocks. *Space Sci. Rev.* **2009**, *148*, 233–247. [CrossRef]

235. Ahmadi, M.; Bruschi, D.E.; Sabín, C.; Adesso, G.; Fuentes, I. Relativistic Quantum Metrology: Exploiting relativity to improve quantum measurement technologies. *Nat. Sci. Rep.* **2014**, *4*, 4996. [CrossRef] [PubMed]

236. Müller, M.; Soffel, M.; Klioner, S.A. Geodesy and relativity. *J. Geodesy* **2008**, *82*, 133–145.

237. Kopeikin, S.; Efroimsky, M.; Kaplan, G. Relativistic Geodesy. In *Relativistic Celestial Mechanics of the Solar System*; Kopeikin, S., Efroimsky, M., Kaplan, G., Eds.; Wiley-VCH: Berlin, Germany, 2011; pp. 671–714.

238. Combrinck, L. General Relativity and Space Geodesy. In *Sciences of Geodesy - II Innovations and Future Developments*; Xu, G., Ed.; Springer: Berlin, Germany, 2013; pp. 53–95.

239. De Sabbata, V.; Fortini, P.; Gualdi, C.; Petralia, S. A Proposal for Combined Efforts Regarding Geophysical Research and Detection of Gravitational Waves. *Ann. Geophys.* **1970**, *23*, 21–25.

240. Grishchuk, L.; Kulagin, V.; Rudenko, V.; Serdobolski, A. Geophysical studies with laser-beam detectors of gravitational waves. *Class. Quantum Grav.* **2005**, *22*, 245–269. [CrossRef]

241. Gusev, A.V.; Rudenko, V.N.; Tsybankov, I.V.; Yushkin, V.D. Detection of gravitational geodynamic effects with gravitational-wave interferometers. *Grav. Cosmol.* **2011**, *17*, 76–79. [CrossRef]

242. Soffel, M.H. *Relativity in Astrometry, Celestial Mechanics and Geodesy*; Springer: Berlin, Germany, 1989.

243. Brumberg, V.A. *Essential Relativistic Celestial Mechanics*; Adam Hilger: Bristol, UK, 1991.

244. Kopeikin, S. Relativistic astrometry. *Scholarpedia* **2011**, *6*, 11382. [CrossRef]

245. Kopeikin, S.; Efroimsky, M.; Kaplan, G. Relativistic Celestial Mechanics. In *Relativistic Celestial Mechanics of the Solar System*; Kopeikin, S., Efroimsky, M., Kaplan, G., Eds.; Wiley-VCH: Berlin, Germany, 2011; pp. 463–518.

246. Kopeikin, S.; Efroimsky, M.; Kaplan, G. Relativistic Astrometry. In *Relativistic Celestial Mechanics of the Solar System*; Kopeikin, S., Efroimsky, M., Kaplan, G., Eds.; Wiley-VCH: Berlin, Germany, 2011; pp. 519–669.

247. Gai, M.; Vecchiato, A.; Ligori, S.; Sozzetti, A.; Lattanzi, M.G. Gravitation astrometric measurement experiment. *Exp. Astron.* **2012**, *34*, 165–180. [CrossRef]

248. Chandrasekhar, S. *The Mathematical Theory of Black Holes*; Oxford University Press: New York, NY, USA, 1983.

249. Harrison, B.K.; Thorne, K.S.; Wakano, M.; Wheeler, J.A. *Gravitation Theory and Gravitational Collapse*; University of Chicago Press: Chicago, IL, USA, 1965.

250. Zeldovich, Ya.B.; Novikov, I.D. *Stars and Relativity*; University of Chicago Press: Chicago, IL, USA, 1971.

251. Colpi, M.; Casella, P.; Gorini, V.; Moschella, U.; Possenti, A. *Physics of Relativistic Objects in Compact Binaries: From Birth to Coalescence*; Springer: Berlin, Germany, 2009.

252. Rezzolla, L.; Zanotti, O. *Relativistic Hydrodynamics*; Oxford University Press: Oxford, UK, 2013.

253. Mukhanov, V. *Physical Foundations of Cosmology*; Cambridge University Press: Cambridge, UK, 2005.

254. Plebański, J.; Krasiński, A. *An Introduction to General Relativity and Cosmology*; Cambridge University Press: Cambridge, UK, 2006.

255. Ellis, G.F.R.; Maartens, R.; Mac Callum, M.A.H. *Relativistic Cosmology*; Cambridge University Press: Cambridge, UK, 2012.

256. Fairbank, W.M. The Use of Low-Temperature Technology in Gravitational Experiments. In Proceedings of the International School of Physics "Enrico Fermi". Course LVI. Experimental Gravitation; Bertotti, B., Ed.; Academic Press: New York, NY, USA, 1974; pp. 280–293.

257. Lipa, A.; Fairbank, W.M.; Everitt, C.W.F. The Gyroscope Experiment - II: Development of the London-Moment Gyroscope and of Cryogenic Technology for Space. In Proceedings of the International School of Physics "Enrico Fermi". Course LVI. Experimental Gravitation; Bertotti, B., Ed.; Academic Press: New York, NY, USA, 1974; pp. 361–380.

258. Stedman, G.E. Ring-laser tests of fundamental physics and geophysics. *Rep. Prog. Phys.* **1997**, *60*, 615–688. [CrossRef]

259. Stedman, G.E.; Schreiber, K.U.; Bilger, H.R. On the detectability of the Lense-Thirring field from rotating laboratory masses using ring laser gyroscope interferometers. *Class. Quantum Grav.* **2003**, *20*, 2527–2540. [CrossRef]

260. Stedman, G.E.; Hurst, R.B.; Schreiber, K.U. On the potential of large ring lasers. *Opt. Commun.* **2007**, *279*, 124–129. [CrossRef]

261. van Zoest, T.; Gaaloul, N.; Singh, Y.; Ahlers, H.; Herr, W.; Seidel, S.T.; Ertmer, W.; Rasel, E.; Eckart, M.; Kajari, E.; *et al.* Bose-Einstein Condensation in Microgravity. *Science* **2010**, *328*, 1540–1543. [CrossRef] [PubMed]

262. Belfi, J.; Beverini, N.; Cuccato, D.; Di Virgilio, A.; Maccioni, E.; Ortolan, A.; Santagata, R. Interferometric length metrology for the dimensional control of ultra-stable ring laser gyroscopes. *Class. Quantum Gravit.* **2014**, *31*, 225003. [CrossRef]

263. Koenneker, C. Even Einstein Could Not Have Imagined Technology Used to Directly Detect Gravitational Waves. *Scientific American.* 21 March 2014. Available online: http://www.scientificamerican.com/article/even-einstein-could-not-have-imagined-technology-used-to-directly-detect-gravitational-waves/ (accessed on 16 February 2015).

264. Bassan, M. *Advanced Interferometers and the Search for Gravitational Waves*; Springer: Berlin, Germany, 2014.

265. Di Virgilio, A.; Allegrini, M.; Beghi, A.; Belfi, J.; Beverini, N.; Bosi, F.; Bouhadef, B.; Calamai, M.; Carelli, G.; Cuccato, D.; *et al.* A ring lasers array for fundamental physics. *Phys. Rev. D* **2014**, *15*, 866–874. [CrossRef]

266. Pegna, R.; Nobili, A.M.; Shao, M.; Turyshev, S.G.; Catastini, G.; Anselmi, A.; Spero, R.; Doravari, S.; Comandi, G.L.; de Michele, A. Abatement of Thermal Noise due to Internal Damping in 2D Oscillators with Rapidly Rotating Test Masses. *Phys. Rev. Lett.* **2011**, *107*, 200801. [CrossRef] [PubMed]

267. Islam, M.R.; Ali, M.M.; Lai, M.-H.; Lim, K.-S.; Ahmad, H. Chronology of Fabry-Perot Interferometer Fiber-Optic Sensors and Their Applications: A Review. *Sensors* **2014**, *14*, 7451–7488. [CrossRef] [PubMed]

268. Graham, P.W.; Hogan, J.M.; Kasevich, M.A.; Rajendran, S. New Method for Gravitational Wave Detection with Atomic Sensors. *Phys. Rev. Lett.* **2013**, *110*, 171102. [CrossRef] [PubMed]

269. Maselli, A.; Gualtieri, L.; Pani, P.; Stella, L.; Ferrari, V. Testing Gravity with Quasi Periodic Oscillations from accreting Black Holes: The Case of Einstein-Dilaton-Gauss-Bonnet Theory. *Astrophys. J.* **2015**. In press. [CrossRef]

270. The Extraordinary Technologies of GP-B. Available online: https://einstein.stanford.edu/TECH/technology1.html (accessed on 20 April 2015).

271. Everitt, C.W.F.; Huber, M.C.H.; Kallenbach, R.; Schaefer, G.; Schutz, B.F.; Treumann, R.A. *Probing the Nature of Gravity. Confronting Theory and Experiment in Space*; Springer: New York, NY, USA, 2010.

272. Gould, A.; Loeb, A. Discovering planetary systems through gravitational microlenses. *Astrophys. J.* **1992**, *396*, 104–114. [CrossRef]

273. Beaulieu, J.-P.; Bennett, D.P.; Fouqué, P.; Williams, A.; Dominik, M.; Jørgensen, U.G.; Kubas, D.; Cassan, A.; Coutures, C.; Greenhill, J.; *et al.* Discovery of a cool planet of 5.5 Earth masses through gravitational microlensing. *Nature* **2006**, *439*, 437–440. [CrossRef] [PubMed]

274. Massey, R.; Rhodes, J.; Ellis, R.; Scoville, N.; Leauthaud, A.; Finoguenov, A.; Capak, P.; Bacon, D.; Aussel, H.; Kneib, J.-P.; *et al.* Dark matter maps reveal cosmic scaffolding. *Nature* **2007**, *445*, 286–290. [CrossRef] [PubMed]

275. Munshi, D.; Valageas, P.; van Waerbeke, L.; Heavens, A. Cosmology with weak lensing surveys. *Phys. Rep.* **2008**, *462*, 67–121. [CrossRef]

276. Hoekstra, H.; Jain, B. Weak Gravitational Lensing and Its Cosmological Applications. *Ann. Rev. Nucl. Part. Syst.* **2008**, *58*, 99–123. [CrossRef]

277. Iorio, L. Juno, the angular momentum of Jupiter and the Lense-Thirring effect. *New Astron.* **2010**, *15*, 554–560. [CrossRef]

278. Helled, R.; Anderson, J.D.; Schubert, G.; Stevenson, D.J. Jupiter's moment of inertia: A possible determination by Juno. *Icarus* **2011**, *216*, 440–448. [CrossRef]

279. Helled, R. Constraining Saturn's Core Properties by a Measurement of Its Moment of Inertia-Implications to the Cassini Solstice Mission. *Astrophys. J.* **2011**, *735*, L16. [CrossRef]

280. Iorio, L. Constraining the Angular Momentum of the Sun with Planetary Orbital Motions and General Relativity. *Sol. Phys.* **2012**, *281*, 815–826. [CrossRef]
281. Alexander, S.; Finn, L.S.; Yunes, N. Gravitational-wave probe of effective quantum gravity. *Phys. Rev. D* **2008**, *78*, 066005. [CrossRef]
282. de Laurentis, M.; Capozziello, S.; Nojiri, S.; Odintsov, S.D. PPN limit and cosmological gravitational waves as tools to constrain f(R)-gravity. *Ann. Phys.* **2010**, *19*, 347–350. [CrossRef]
283. Damour, T.; Vilenkin, A. Gravitational radiation from cosmic (super)strings: Bursts, stochastic background, and observational windows. *Phys. Rev. D* **2005**, *71*, 063510. [CrossRef]
284. Stroeer, A.; Vecchio, A. The LISA verification binaries. *Class. Quantum Gravit.* **2006**, *23*, S809–S817. [CrossRef]
285. Pössel, M. Chirping neutron stars. *Einstein Online* **2005**, *1*, 1013. Available online: http://www.einstein-online.info/spotlights/chirping_neutron_stars (accessed on 20 April 2015).
286. Krolak, A. White Dwarf binaries as gravitational wave sources. *Einstein Online* **2010**, *4*, 1001. Available online: http://www.einstein-online.info/spotlights/galactic-binaries (accessed on 20 April 2015).
287. Nelemans, G. The Galactic gravitational wave foreground. *Class. Quantum Gravit.* **2009**, *26*, 094030. [CrossRef]
288. Sesana, A.; Vecchio, A.; Colacino, C.N. The stochastic gravitational-wave background from massive black hole binary systems: Implications for observations with Pulsar Timing Arrays. *Mon. Not. R. Astron. Soc.* **2008**, *390*, 192–209. [CrossRef]
289. Berry, C.P.L.; Gair, J.R. Observing the Galaxy's massive black hole with gravitational wave bursts. *Mon. Not. R. Astron. Soc.* **2012**, *429*, 589–612. [CrossRef]
290. Kotake, K.; Sato, K.; Takahashi, K. Explosion mechanism, neutrino burst and gravitational wave in core-collapse supernovae. *Phys. Rep.* **2006**, *69*, 971–1143. [CrossRef]
291. Choquet-Bruhat, Y. Results and Open Problems in Mathematical General Relativity. *Milan J. Math.* **2007**, *75*, 273–289. [CrossRef]
292. Geroch, R. What is a singularity in general relativity? *Annals Phys.* **1968**, *48*, 526–540. [CrossRef]
293. Natario, J. Relativity and Singularities - A Short Introduction for Mathematicians. *Resenhas* **2005**, *6*, 309–335.
294. Uggla, C. Spacetime singularities. *Einstein Online* **2006**, *2*, 1002. Available online: http://www.einstein-online.info/spotlights/singularities/?set_language=en (accessed on 20 April 2015).
295. Curiel, E.; Bokulich, P. Singularities and Black Holes. In *The Stanford Encyclopedia of Philosophy*; Zalta, E.N., Ed.; The Metaphysics Research Lab: Stanford, CA, USA, 2012; Available online: http://plato.stanford.edu/archives/fall2012/entries/spacetime-singularities/ (accessed on 16 February 2015).
296. Rovelli, C. *Quantum Gravity*; Cambridge University Press: Cambridge, UK, 2004.
297. Ashtekar, A. The winding road to quantum gravity. *Curr. Sci.* **2005**, *89*, 2064–2074.
298. Kiefer, C. Quantum Gravity: General Introduction and Recent Developments. *Ann. Phys.* **2005**, *15*, 129–148. [CrossRef]
299. Das, S.; Vagenas, E.C. Universality of Quantum Gravity Corrections. *Phys. Rev. Lett.* **2008**, *101*, 221301. [CrossRef] [PubMed]
300. Weinstein, S.; Rickles, D. Quantum Gravity. In *The Stanford Encyclopedia of Philosophy*; Zalta, E.N., Ed.; The Metaphysics Research Lab: Stanford, CA, USA, 2011; Available online: http://plato.stanford.edu/archives/spr2011/entries/quantum-gravity/ (accessed on 16 February 2015).
301. De Haro, S.; Dieks, D.; 't Hooft, G.; Verlinde, E. Special Issue: Forty Years of String Theory: Reflecting on the Foundations. *Found. Phys.* **2013**, *43*, 1–200. [CrossRef]
302. Gambini, R.; Pullin, J. Loop Quantization of the Schwarzschild Black Hole. *Phys. Rev. Lett.* **2013**, *110*, 211301. [CrossRef] [PubMed]
303. Bosma, A. 21-cm Line Studies of Spiral Galaxies. II - The Distribution and Kinematics of Neutral Hydrogen in Spiral Galaxies of Various Morphological Types. *Astron. J.* **1981**, *86*, 1825–1846. [CrossRef]
304. Rubin, V.C. The Rotation of Spiral Galaxies. *Science* **1983**, *220*, 1339–1344. [CrossRef] [PubMed]
305. Salucci, P.; De Laurentis, M. Dark Matter in galaxies: leads to its nature. In Proceedings of 8th International Workshop on the Dark Side of the Universe, Buzios, Brazil, 10–15 June 2012.
306. Clifton, T.; Barrow, J.D. The power of general relativity. *Phys. Rev. D* **2005**, *72*, 103005. [CrossRef]
307. Alimi, J.-M.; Füzfa, A. The abnormally weighting energy hypothesis: The missing link between dark matter and dark energy. *J. Cosmol. Astropart. Phys.* **2008**, *9*, 14. [CrossRef]
308. De Felice, A.; Tsujikawa, S. f(R) Theories. *Living Rev. Relativ.* **2010**, *13*, 3. [CrossRef]

309. Capozziello, S.; de Laurentis, M. Extended Theories of Gravity. *Phys. Rep.* **2011**, *509*, 167–321. [CrossRef]

310. Clifton, T.; Ferreira, P.G.; Padilla, A.; Skordis, C. Modified gravity and cosmology. *Phys. Rep.* **2012**, *513*, 1–189. [CrossRef]

311. Milgrom, M. The MOND paradigm of modified dynamics. *Scholarpedia* **2014**, *9*, 31410. [CrossRef]

312. Amendola, L.; Bertone, G.; Profumo, S.; Tait, T. Next Decade in Dark Matter and Dark Energy— Next Decade in Dark Matter and Dark Energy. *Phys. Dark Univ.* **2012**, *1*, 1–218.

313. Kopeikin, S.M. *Frontiers in Relativistic Celestial Mechanics. Volume 1 Theory*; De Gruyter: Berlin, Germany, 2014.

314. Kopeikin, S.M. *Frontiers in Relativistic Celestial Mechanics. Volume 2 Applications and Experiments*; De Gruyter: Berlin, Germany, 2014.

universe

MDPI

Article

General Relativity and Cosmology: Unsolved Questions and Future Directions

Ivan Debono [1,*,†] and George F. Smoot [1,2,3,†]

1   Paris Centre for Cosmological Physics, APC, AstroParticule et Cosmologie, Université Paris Diderot, CNRS/IN2P3, CEA/lrfu, Observatoire de Paris, Sorbonne Paris Cité, 10, rue Alice Domon et Léonie Duquet, 75205 Paris CEDEX 13, France
2   Physics Department and Lawrence Berkeley National Laboratory, University of California, Berkeley, 94720 CA, USA; gfsmoot@lbl.gov
3   Helmut and Anna Pao Sohmen Professor-at-Large, Hong Kong University of Science and Technology, Clear Water Bay, Kowloon, 999077 Hong Kong, China
*   Correspondence: ivan.debono@apc.univ-paris7.fr; Tel.: +33-1-57276991
†   These authors contributed equally to this work.

Academic Editors: Lorenzo Iorio and Elias C. Vagenas
Received: 21 August 2016; Accepted: 14 September 2016; Published: 28 September 2016

Abstract: For the last 100 years, General Relativity (GR) has taken over the gravitational theory mantle held by Newtonian Gravity for the previous 200 years. This article reviews the status of GR in terms of its self-consistency, completeness, and the evidence provided by observations, which have allowed GR to remain the champion of gravitational theories against several other classes of competing theories. We pay particular attention to the role of GR and gravity in cosmology, one of the areas in which one gravity dominates and new phenomena and effects challenge the orthodoxy. We also review other areas where there are likely conflicts pointing to the need to replace or revise GR to represent correctly observations and consistent theoretical framework. Observations have long been key both to the theoretical liveliness and viability of GR. We conclude with a discussion of the likely developments over the next 100 years.

Keywords: General Relativity; gravitation; cosmology; Concordance Model; dark energy; dark matter; inflation; large-scale structure

1. Perspective

Scientists have been fascinated by General Relativity ever since it was developed. It has been described as poetic, beautiful, elegant, and, at times, as impossible to understand.

General Relativity is often described as a simple theory. It is hard to define simplicity in science. One can always construct an entire theory encapsulated in one equation. Richard Feynman famously demonstrated this in a thought experiment where he rewrote all the laws of physics as $\vec{U} = 0$, where each element of $\vec{U}$ contained the hidden structure [1]. His point was that simplicity does not automatically bring truth.

An examination of the mathematical structure of General Relativity gives us a more sober definition of "simplicity". Under certain assumptions about the structure of physical theories, and of the properties of the gravitational field, General Relativity is the only theory that describes gravity. Alternative theories introduce additional interactions and fields.

General Relativity is also unique among theories of fundamental interactions in the Standard Model. Like electromagnetism, but unlike the strong and weak interactions, its domain of validity covers the entire range of length scales from zero to infinity. However, unlike the other forces, gravity as described by General Relativity acts on all particles. This implies that the theory does not fail below the Planck scale. All gravitational phenomena, from infinitesimal scales to distances beyond the observable universe, may be modelled by General Relativity. We may therefore formulate a mathematically rigorous description of General Relativity: it is the most complete theory of gravity ever developed.

All gravitational phenomena that have ever been observed can be modelled by General Relativity. It describes everything from falling apples, to the orbit of planets, the bending of light, the dynamics of galaxy clusters, and even black holes and gravitational waves. The domain of validity of the theory covers a wide range of energy levels and scales. That is why it has survived so long, and that is why it survives today, one hundred years after it was formulated, in an age in which the amount of data and knowledge increases by orders of magnitude every few years.

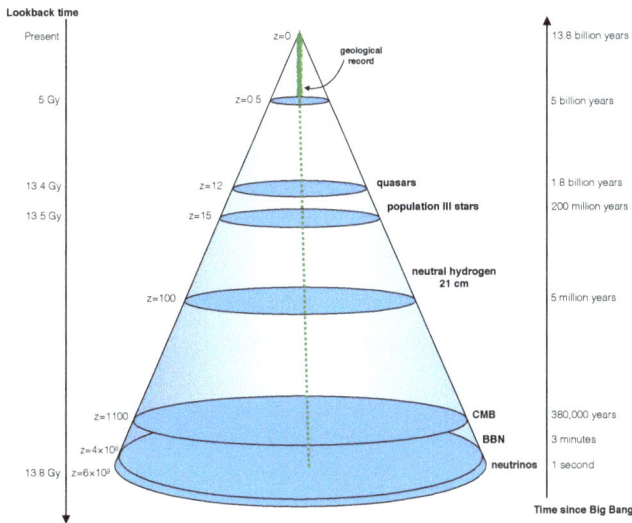

Figure 1. How we observe the universe. The lookback time is the difference between the age of the universe now, and the age of the universe when photons from an object were emitted. The more distant an object, the farther in its past we are observing its light. This distance in both space and time is expressed by the cosmological redshift z. We obtain most of our astrophysical information from the surface of our past light cone, because it is carried by photons. The only information from within the cone come from local experiments and observations, such as geological records. The green dotted line is the world-line of the atoms and nuclei providing the material for our geological data. Local experiments are carried out along this bundle of world-lines. They provide a useful test of physical constants. One example is the observation of the Oklo phenomenon [2]. The earliest information we have collected so far comes from the cosmic microwave background (CMB). Earlier than the CMB time-like slice is the cosmic neutrino background. We observe Big Bang nucleosynthesis (BBN) indirectly, through observations of the abundances of chemical elements.

Why, then, are we still testing General Relativity? Why do we still develop, discuss, test and fine-tune alternative theories? Because there are some very fundamental open questions in physics, particularly in cosmology. Moreover, the big questions in cosmology happen to be the ones that are not answered by General Relativity: the accelerated expansion of the universe, the presence of a mysterious form of matter which cannot be observed directly, and the initial conditions in the early universe.

The theoretical completeness described above is both a necessary and aesthetic feature of a fundamental theory. However, it creates experimental difficulties, for it compels us to test the theory at extreme scales, where experimental errors may be large enough to allow several alternative theories.

At extremely small scales below the Planck length, classical mechanics should break down. This compels us to question whether General Relativity is still accurate at these scales, whether it needs to be modified, and whether a quantum description of gravity can be formulated. At the other end of the scale, at cosmological distances, we may question whether General Relativity is valid, given that the universe cannot be modelled sufficiently accurately by General Relativity without invoking either a cosmological constant, or some additional, unknown component of the universe. Finally, we may question the accuracy of our solutions to the equations of General Relativity, which depend on some approximation scheme. These approximations provided analytical solutions which enabled most of the early progress in General Relativistic cosmology and astrophysics. However, one century after the formulation of the theory, we now have a flood of data from increasingly accurate observations (as shown in Figure 1), coupled with computing power which was hitherto unheard of. Tests of the higher-order effects predicted by General Relativity and some of its competitors are now within reach.

The purpose of this review is to examine the motivation for the development of alternative theories throughout the history of GR, to give an overview of the state of the art in General Relativistic cosmology, and to look ahead. In the next few decades, some of the open questions in cosmology may well be answered by a new generation of experiments, and GR may be challenged by alternative theories.

2. A Brief History

Let us start this review by breaking our own rule about unscientific adjectives. General Relativity is a beautiful theory of gravity. It has not only thrilled us, but has survived 100 years of challenges, both by experimental tests and by alternative theories. The beauty of the theory was clear at the beginning, but the initial focus was on whether it was right. When General Relativity provided an explanation for the 43 seconds of arc per century discrepancy in the advance of the perihelion of Mercury [3], it got the attention of the scientific community. However, it was the prediction and the observation of the bending of light by the Sun [4] that confirmed GR's place as the new reigning theory of gravity [5].

The setting at the Royal Society under the portrait of Newton for the report of the eclipse light bending observations led by Arthur Eddington, and reported by the great writer Aldous Huxley, was perfect to describe to the world the ascendancy of a new theory replacing Newton's gravity (see, e.g., [6]). From this point onwards, the scientific community started to take General Relativity seriously, and theorists worked hard to understand this new theory, beguilingly simple but hard to apply, and to advance its predictions.

Shortly after its publication, GR quickly became the framework for astrophysics, and for the Standard or Concordance Model of cosmology. However, it was still challenged by alternative theories. Initially, the alternatives were motivated by theoretical considerations. This early period led to a fuller understanding of GR and its predictions. Some of the predictions, such as black holes and gravitational waves, divided the scientific community. Did they exist as physical objects, or just as mathematical artifacts of the theory?

By the time GR turned 50, the model of cosmology had been established, GR had been tested, and things had started to stagnate. However, advances in observations led to new discoveries, which in turn led to renewed challenges.

First came the missing mass in the universe. Could GR be modified to account for it? Then came the theories about the very early universe, and the behaviour of the quantum-scale, tiny initial universe. Finally, twenty years ago, came the confirmation of cosmic acceleration. This had a twofold effect. On one hand, it spurred the development of a whole range of alternative theories of gravity. On the other hand, it confirmed GR like never before, for General Relativity, with a cosmological constant, can account for the observations perfectly.

In 2015, on the 100th birthday of General Relativity, gravitational waves were observed for the first time. This had been the last major untested prediction of General Relativity. It was a remarkable achievement, and in many ways it heralds a new age of astrophysical observations. The experimental capabilities and the computing power have finally caught up with the theory. Cosmology and astrophysics have now entered the era of Big Data, and much of the theoretical effort is now driven by data. However, the foundation for almost the entire scientific endeavour is still this theory of chronogeometrodynamics, developed 100 years ago when today's instruments and computers were still a distant dream.

From Aristotle to Einstein

General Relativity is the basis for the Standard Model of physical cosmology, and here we shall discuss the development of General Relativity (GR). The history of cosmology and GR are intertwined. We shall discuss why the theory has been so successful, and the criteria that must be satisfied by any alternative theory of physics, and by cosmological models.

Cosmology, in its broadest definition, is the study of the cosmos. It aims to provide an accurate description of the universe. Throughout much of the history of science, the development of cosmology was hampered by the lack of a universal physical theory. Observational tools were extremely limited, and there was no mathematical formulation for physical laws. The cosmos was described in metaphysical, rather than physical terms.

Discussions on the history of physics often refer to Karl Popper's concept of 'Falsifizierbarkeit' (falsifiability) [7]. In this formalism, scientific discovery proceeds by successive falsifications of theories. A falsifiable theory that covers observations, and that has not yet been proven false can be regarded as provisionally acceptable. Yet we know that in reality it is not quite as straightforward. A theory that is considered to be correct acquires this status by accumulation of evidence rather than by a single falsification of a previous theory [8]. This is especially true in cosmology, where the selection of theoretical models often depends on the outcome of statistical calculations.

The scientific revolution which brought about the development of a precise mathematical language for physical theories heralded the scientific age of cosmology. Physical laws, tested here on Earth and later in the Solar System, could be applied to the 'entire universe', and could thus provide a precise physical description of the cosmos. Modern cosmology is based upon this epistemological framework. Cosmology depends upon a fundamental premise. As a science, it must deal strictly with what can be observed, but the observable universe forms only a fraction of the whole cosmos. One is forced to make the fundamental but unverifiable assumption that the portion of the universe which can be observed is representative of the whole, and that the laws of physics are the same throughout the whole universe [9]. Once we make this assumption, we can construct a model of the universe based on a description of its observable part.

Any cosmological model which assumes the universality of physical laws must be based upon some physical theory. Since cosmology aims to describe the universe on the largest possible scales, it must be based upon an application long-range physical interactions. Since the theory of gravitation is the physical theory at the basis of standard cosmology, and is also at the centre of the big questions facing modern cosmology, we shall give an overview of the development of theories of gravitation.

The development of physical theories of gravity was far from smooth, nor did it always conform to Popper's scheme. Before the logical tools (mathematics) for the phenomenological description (physics) were invented, progress was rather haphazard.

According to Popper's scheme, this development should be driven by the search for ever more general principles. Yet Aristotelian theory, to take one example, considered itself to be general enough—its claimed region of validity was the entire universe, except that rising smoke, floating feathers, falling apples and orbiting celestial spheres each had their own rules.

The real revolution came when it was realised that the behaviour of all bodies could be described by a single rule—a universal theory of gravitation. This theory is a description of the long range forces that electrically neutral bodies exert on one another because of their matter content.

Whether they choose to or not, scientists will always stand on the shoulders of giants. No theory is invented in a scientific vacuum. This goes all the way back to the cosmology of the Euro-Mediterranean Ancient World, codified in the Aristotelian teachings of the 4th century B.C. This Hellenic "natural philosophy" provided qualitative rather than quantitative descriptions for what we would call today the free parameters of the theory [10]. It stands to reason—the instruments had not yet been invented that could test the theory of gravity to within numerical accuracy. Without accurate timekeeping instruments, processes could at best be described as "slower than" or "faster than". However, instruments to measure the movement of the celestial bodies, such as sundials, quadrants and astrolabes, were invented and improved upon, and measurements were carried out [11]. Astronomy flourished.

There is a certain logic to the development of physical theories from the Ancient World, to the Middle Ages, and right up to the Renaissance [12–14]. The basic tenet of the physics of Aristotle is that actions follow logically from causes. He distinguished between natural and violent motion. Natural motion implies falling at a speed proportional to the weight of the object and inversely proportional to the density of the medium. Violent motion happens whenever there is a force acting on an object, and the speed of the object is proportional to this force. Strato of Lampsacus replaced Aristotle's explanation of 'unnatural' motion with one that is very close to the modern notion of inertia. He identified natural motion as a form of acceleration, and demonstrated experimentally that falling bodies accelerate. In the 14th century, Jean Buridan came up with the notion of impetus, where the initial force imparts motion to the object, which gradually diminishes as gravity and air resistance act against this initial force. Concurrently, Nicole Oresme was using a crude early form of graph to describe motion, and unwittingly showing the complicated notions of differentiation and integration in pictorial form[15,16].

The cosmological observations, limited to the innermost five planets of the Solar System (Mercury, Venus, Mars, Jupiter, and Saturn) and the sphere of stars, seemed to confirm the Aristotelian-Ptolemaic theory. Celestial bodies moved in regular patterns made up of repeating circles. Small discrepancies were explained by circles within circles.

The fact that the theories were based on these regular patterns is no accident. Patterns are the keyword in all of physics. Human beings are wired to recognise patterns. We can only build theories because we recognise patterns in the universe. This characteristic of valid theories has been called sloppiness. The patterns fall within some hyper-ribbon of stability in the theory [17].

The revolution in physics came with the development of mathematical, quantitative, models to describe physical reality. Starting in the 1580 Galileo carried out a series of observations in which he subjected kinematics to rigorous experiment, and showed that naturally-falling objects really do accelerate. Crucially, he showed that the composition of the body has no effect whatsoever on this acceleration. He also realised that for violent motion, the speed is constant in the absence of friction. Galileo also took rigorous observations of astronomical objects. In 1610 he made the first observation of Jupiter's satellites, and the first observation of the phases of Venus, which is impossible according to the Ptolemaic geocentric model. His observations were important in putting to rest the Aristotelian theory of perfect and unchanging heavens.

By the time Newton came along, telescopes had been invented. Galileo had observed moons orbiting the Solar System planets, and hundreds of stars invisible to the naked eye. His 1610 treatise, aptly called *Siderus Nuncius* ("Starry Message", or "Astronomical Report" in modern language) [18], was the first scientific work based on observations through a telescope. Mechanical clocks had been invented. The sphere of observed data had expanded [19]. Calculus provided the tool to make sense of this new flood of data. Thus, physicists of Newton's generation found a very different scientific environment than the one in which Galileo had started off.

In 1687, Isaac Newton published in his "Mathematical Principles of Natural Philosophy", known by its abbreviated Latin title as *Principia* [20]. This was a significant milestone in physics. Newton's model of gravitation was, in his own words, a "universal" law. It applied to all bodies in the universe, whether it was cannonballs on Earth, or planets orbiting the Sun. For more than two centuries, Newton's theory, was the standard physical description of gravity. There was no other attempt to find a different theory for the gravitational force, although the intervening years between Newtonian gravity and Relativity produced some important physical concepts such as de Maupertuis's "Principle of Least Action" [21], further developed by Euler [22], Lagrange [23] and Hamilton [24,25]. The path of each particle is assigned a number called an action, which is the integral of the Lagrangian. In classical mechanics, the action principle is equivalent to Newton's Laws. Lagrangian field theory is an important cornerstone of modern physics. The Lagrangian of any physical interaction, when subjected to an action principle, give us field equations and conservation laws for the theory. It is an expression of the symmetries in physical laws.

Newtonian gravity was the great success story of nineteenth century physics, the golden age of mathematical astronomy. It allowed astronomers to calculate the position of planets and asteroids with ever greater precision, and to confirm their calculations by observation. Thus the size of the known universe grew. Evidence started to accumulate suggesting that there might be other galaxies in the universe besides our own. In 1845, the planet Neptune was discovered, after Urbain le Verrier suggested pointing telescopes in a region of the Solar System which he predicted by Newtonian calculations [26,27]. The search was motivated in the first place by an anomaly in the orbit of Uranus which could not be otherwise explained using Newtonian theory [28]. The discovery of Neptune showed that Newtonian theory was valid even in the very farthest limits of the Solar System.

There was another anomaly which could not be explained—the excessive perihelion precession of Mercury by 43 arcseconds per century, confirmed by le Verrier himself. Urbain le Verrier thus holds the distinction of being one of the few experimentalists to have confirmed Newton's theory and then disproved it. Astronomers attempted to explain this perihelion anomaly using Newtonian mechanics, which led them to speculate on the existence of Vulcan, a hypothetical planet whose orbit was even closer to the Sun [29].

The first doubts on Newtonian theory began to take shape just at the time when theorists were examining the full implications of the theory for complex, multi-body dynamical systems such as the Solar System. In 1890, Henri Poincaré published his magnum opus on the three-body problem [30], a masterpiece of celestial mechanics. At the time, Poincaré was working on another open question in physics: the aether. This led him to formulate a theory which was very close to Special Relativity [31], but which did not quite fit with Maxwell's electromagnetism [32], and was ultimately flawed.

By the end of the 19th century, the necessary mathematical tools were in place which would enable the development of Special and then General Relativity. There is an intimate connection between physics and the language of mathematics which is often overlooked. The former, especially in modern times, depends on the latter. Could Aristotle have developed General Relativity? No. Because he had not the mathematical language. Equations and mathematical formulations are relatively recent in the history of physics. Even Newton, for all his fame as a mathematical genius, never wrote the equation $F = -GMm/r^2$. He wrote a series of statements implying this law in (Latin) words: "*Gravitatem, quæ Planetam unumquemque respicit, ese reciprocæ ut quadratum distantiæ locorum ab ipsius centro*", and so on. It is hard to imagine how human beings could manipulate tensors and solve the field equations of Relativity in anything but numbers and symbols. Theories and physics do not happen in a cultural and scientific vacuum. They are human creations, and they depend intimately on tools for the transmission and communication of human knowledge.

The physical theory of gravity—the laws that govern gravitational interactions—remained unchanged until Einstein's time. In 1905, Einstein published his Theory of Special Relativity (SR) [33]. Soon after, he turned to the problem of including gravitation within four-dimensional spacetime [34–37].

Newton's formulation of the gravitational laws is expressed by the equations:

$$\frac{d^2 x^i}{dt^2} = -\frac{\partial \Phi}{\partial x^i}, \tag{1}$$

$$\triangle \Phi = 4\pi G \rho, \tag{2}$$

where Φ is the gravitational potential, G is the universal gravitational constant, ρ is the mass density, and $\triangle = \nabla^2$ is the Laplace operator. These equations cannot be incorporated into Special Relativity as they stand. The equation of motion (1) for a particle is in three-dimensional form, so it must be modified into a four-dimensional vector equation for $d^2 x^\mu / d\tau^2$. Similarly, the field Equation (2) is not Lorentz-invariant, since the three-dimensional Laplacian operator instead of the four-dimensional d'Alembertian $\square = \partial^\mu \partial_\mu$ means that the gravitational potential Φ responds instantaneously to changes in the density ρ at arbitrarily large distances. The conclusion is that Newtonian gravitational fields propagate with infinite velocity. In other words, instantaneous action in Newtonian theory implies action at a distance when reconsidered in the light of Special Relativity. This violates one of the postulates of SR. How do we reconcile gravity and Special Relativity?

3. The Development of General Relativity

3.1. From Special to General Relativity

The simplest relativistic generalisation of Newtonian gravity is obtained by representing the gravitational field by a scalar Φ. Since matter is described in Relativity by the stress-energy tensor $T_{\mu\nu}$, the only scalar with dimensions of mass density (which corresponds to ρ) is T^μ_μ. A consistent scalar relativistic theory of gravity would thus have the field equation

$$\square \Phi = 4\pi G T^\mu_\mu. \tag{3}$$

However, when the equation of motion from this theory are applied to a static, spherically symmetric field Φ, such as that of the sun, acting on an orbiting planet, they would result in a negative precession, or retardation of the perihelion. Experimental evidence since the time of Le Verrier and his observation of the orbit of Mercury [38] clearly shows that planets experience a prograde precession of the perihelion. Moreover, in the limit of a zero rest-mass particle, such as a photon, the equations of motion show that the particle experiences no geodesic deviation. The existence of an energy-momentum tensor due to an electromagnetic field would also be impossible, since $(T_{\text{electromagnetic}})^\mu_\mu = 0$. The theory therefore allows neither gravitational redshift, nor deviation of light by matter, both of which are clearly observable phenomena [39]. Another route to generalisation could be to represent the gravitational field by a vector field Φ_μ, analogous to electromagnetism. Following through with this strategy, the "Coulomb" law in this theory gives a repulsion between two massive particles, which clearly contradicts observations. The theory also predicts that gravitational waves should carry negative energy, and, like the scalar theory, predicts no deviation of light. Like the scalar theory, then, the vector theory must be discarded.

What about a flat-space tensor theory? The gravitational field in this theory is described by a symmetric tensor $h_{\mu\nu} = h_{\nu\mu}$. The choice of the Lagrangian in this theory is dictated by the requirement that $h_{\mu\nu}$ be a Lorentz-covariant, massless, spin-two field.

In the 1930s, Wolfgang Pauli and Markus Fierz [40] were the first to write down this Lagrangian and investigate the resulting theory. The predictions of the theory for deviation of light agree with those of General Relativity, and are consistent with observations. Since the field equations and gauge properties are identical to those of Einsein's linearised theory, the predictions for the properties of gravitational waves, including positive energy, agree with those obtained using the linearised theory in General Relativity. However, the theory differs from General Relativity in its predicted value for the perihelion precession, which is $\frac{4}{3}$ of that given by GR. This disagrees with the value obtained from observations of Mercury's orbit.

The theory has an even worse deficiency. If two gravitating bodies (that is, not test particles) are considered, and the field equations are applied to them, then the theory predicts that gravitating bodies cannot be affected by gravity, since they all move along straight lines in a global Lorentz reference frame. This holds for bodies made of arbitrary stress-energy, and since all bodies gravitate, then one must conclude that no body can be accelerated by gravity, which is a obvious self-inconsistency in the theory.

The only way in which a consistent theory of gravity can be constructed within Special Relativity is to consider the geometry of spacetime as the gravitational field itself. In other words, *all matter* moves in an effective Riemann space of metric $g^{\mu\nu} \equiv \eta^{\mu\nu} + h^{\mu\nu}$, where $\eta^{\mu\nu}$ is the Minkowski metric. The requirement of consistency leads us to universal coupling, which implies the Equivalence Principle.

The existence of curved spacetime can be deduced from purely physical arguments. In 1911, before he had fully developed General Relativity, Einstein [34] showed that a photon must be affected by a gravitational field, using conservation of energy applied to Newtonian gravitation theory. Schild [41–43] showed by a simple thought experiment, formulated within Special Relativity, that a consistent theory of gravity cannot be constructed within this framework. His argument is based upon a gravitational redshift experiment carried out in the field of the Earth, using a global Lorentz frame tied to the Earth's centre. Successive pulses of light rising to the same height should experience a redshift, and therefore the pulse rate at the top should be slower than that at the bottom. But light rays are drawn at 45 degrees in Minkowski spacetime diagrams, so that top and bottom time intervals are equal, which is impossible if redshift occurs. Hence the spacetime must be curved. One therefore concludes that in the presence of gravity, Special Relativity cannot be valid over any sufficiently extended region.

General Relativity may be understood as a generalisation of Special Relativity over extended regions. Since Special Relativity can comfortably be described using tensor calculus, it was only natural to extend the flat Minkowski spacetime of Special Relativity to the curved spacetime of General Relativity. This was a physical application of Riemannian geometry [44,45], which had been developed in the second half of the 19th century. The idea of tensor calculus on curved manifolds was already mathematically well-established. Einstein's innovation lay in identifying the Einstein tensor, itself related to the Riemann curvature tensor, as the "gravitational field" in the theory.

Einstein had been working on the problem for some years, starting in 1907. He arrived at the final, correct form in 1915 [46,47]. He was well-aware of the significance of his publication, and he gave it the succinct title of "The Field Equations of Gravitation" (*Feldgleichungen der Gravitation*). The correct field equations for the theory contained in this publication served as the starting point or subsequent derivations.

3.2. The Formalism of General Relativity

General Relativity is based on two independent but mutually supporting postulates.

The first postulate is sometimes referred to collectively as the *Einstein Equivalence Principle*:

- *The Strong Equivalence Principle:* The laws of physics take the same form in a freely-falling reference frame as in Special Relativity
- *The Weak Equivalence Principle:* An observer in freefall should experience no gravitational field. That is to say, an observer cannot determine from a local experiment whether the his laboratory is being accelerated by a rocket of static at the surface of a gravitating body. Gravity is erased up to tidal forces, which are determined by the size of the laboratory and its distance to the centre of the gravitational attraction.

The Equivalence Principle allows us to construct the metric and the equation of motion by transforming from a freely-falling to an accelerating frame. It can be mathematically expressed by the assuming that all matter fields are minimally coupled to a single metric tensor $g_{\mu\nu}$. The distance between two points in 4-dimensional spacetime, called events, is:

$$ds^2 = c^2\, d\tau^2 = g_{\mu\nu}\, dx^\mu\, dx^\nu\,. \tag{4}$$

Throughout the text, we follow the Einstein summation convention for repeated indices, so that $c_i x^i = \sum_{i=1}^{n} c_i x^i$ for $i = 1, \ldots, n$. Greek indices are used for space and time components, while Latin indices are spatial ones only. We use the following metric signature: $(-+++)$.

The metric defines lengths and times measured by laboratory rods and clocks. This metric implies that the action for any matter field ψ is of the form

$$S_{\text{matter}}[\psi, g_{\mu\nu}]\,, \tag{5}$$

which gives us three important results. First, it implies the universality of freefall. Second, it implies that all non-gravitational constants are spacetime independent. Third, it implies that the laws of physics are isotropic. This equation defines how matter behaves in a given curved geometry, how light rays propagate, how stars, planets and galaxies move, and gives us verifiable observational consequences.

The second postulate is related to the dynamics of the gravitational interaction. This is assumed to be governed by the Einstein-Hilbert action:

$$S_{\text{gravity}} = \frac{c^3}{16\pi G} \int d^4x \sqrt{-g_*} R_* \tag{6}$$

where $g_{\mu\nu}^*$ is a massless spin-2 field called the Einstein metric. General Relativity identifies the Einstein metric with the physical metric, that is: $g_{\mu\nu} = g_{\mu\nu}^*$. This implements the Strong Equivalence Principle.

The Einstein-Hilbert action defines the dynamics of gravity itself. Relativity is thus a geometrical approach to fundamental interactions. These are realised though continuous classical fields which are inseparably connected to the geometrical structures of spacetime, such as the metric, affine connection, and curvature.

The General Relativistic equation of motion is simply parallel transport on curved spacetime. It is given by

$$\frac{d^2 x^\mu}{d\tau^2} + \Gamma^\mu_{\alpha\beta} \frac{dx^\alpha}{d\tau} \frac{dx^\beta}{d\tau} = 0\,, \tag{7}$$

where x^μ is some set of coordinates for a point in spacetime. $\Gamma^\mu_{\alpha\beta}$ are the components of the affine connection (or metric connection). The fundamental theorem of Riemannian geometry states that the affine connection can be expressed entirely in terms of the metric:

$$\Gamma^\alpha_{\lambda\nu} = \frac{1}{2} g^{\alpha\nu} (g_{\mu\nu,\lambda} + g_{\lambda\nu,\mu} - g_{\mu\lambda,\nu})\,, \tag{8}$$

where the comma denotes a derivative, i.e., $g_{\mu\nu,\lambda} = \frac{\partial g_{\mu\nu}}{\partial x^\lambda}$.

We need to construct invariant quantities in GR (quantities that are the same for all observers). To achieve this, we need to contract the covariant A_μ and contravariant A^μ components of a vector or tensor A by using the metric to raise or lower indices: $A_\mu = g_{\mu\nu} A^\mu$. Thus the equation of motion (7) can be made covariant by recasting it as the covariant derivative of the 4-velocity $U^\mu = \gamma(c, \mathbf{v})$:

$$\frac{D_\mu U^\mu}{d\tau} = 0\,, \tag{9}$$

where the covariant derivative is defined as

$$D_\mu A^\mu = dA^\mu + \Gamma^\mu_{\alpha\beta} A^\alpha\, dx^\beta\,. \tag{10}$$

The quantity γ is the Lorentz factor:

$$\gamma = \frac{1}{\sqrt{1 - v^2/c^2}}.$$ (11)

The transformation from SR to GR is then carried out by mapping the Minkowski metric to a general metric: $\eta \to g$ and by mapping $\partial \to D$.

In GR, freely-falling bodies travel along a geodesic. Geometrically, this is the shortest distance between two points in spacetime. The path length along a geodesic is given by

$$S = \int (g_{\mu\nu}\, \mathrm{d}x^\mu\, \mathrm{d}x^\nu)^{1/2}.$$ (12)

In cosmology, it is essential for us to be able to describe spacetime which is not "empty". In the presence of a perfect fluid (an inviscid fluid with density ρ and isotropic pressure p), the energy and momentum of spacetime is described by the energy-momentum tensor (or stress-energy tensor)

$$T^{\mu\nu} = \left(\rho + \frac{p}{c^2}\right) U^\mu U^\nu - p g^{\mu\nu}.$$ (13)

Classical energy and momentum conservation are generalized in GR as the four conservation laws

$$D_\mu T^{\mu\nu} = 0.$$ (14)

In other words, the stress-energy tensor has a vanishing covariant divergence. In the absence of a component possessing pressure or density, or both, the energy-momentum tensor is zero.

The central notion in General Relativity is that gravitation can be described by a metric. The Einstein equations give us the relation between the metric and the matter and energy in the universe:

$$G^{\mu\nu} = -\frac{8\pi G}{c^4} T^{\mu\nu}.$$ (15)

The left-hand side of this equation is a function of the metric: $G^{\mu\nu}$ is the Einstein tensor, defined as:

$$G^{\mu\nu} = R^{\mu\nu} - \frac{1}{2} g^{\mu\nu} R,$$ (16)

where $R^{\mu\nu}$ is the Ricci tensor, which depends on the metric and its derivatives, and the Ricci scalar R is the contraction of the Ricci tensor ($R = g_{\mu\nu} R^{\mu\nu}$). The right-hand side of Equation (15) is a function of the energy: G is Newton's constant, and $T^{\mu\nu}$ is the energy-momentum tensor.

Einstein's Relativity has three main distinguishing characteristics:

- it agrees with experiment
- it describes gravity entirely in terms of geometry
- it is free of any "prior geometry"

These characteristics are lacking in most of the other theories [48,49]. Apart from the issue of agreement with experiment, Einstein's theory is unique in its physical simplicity.

Every other theory introduces auxiliary gravitational fields, or involves prior geometry. Prior geometry is any aspect of the geometry of spacetime which is fixed immutably, that is, it cannot be changed by changing the distribution of gravitating sources.

A rigorous mathematical definition of the unique simplicity of General Relativity is given by Lovelock's theorem [50–52]. This is a generalisation of an earlier theorem by Élie Cartan [53], and may be formulated as follows:

In 4 spacetime dimensions, the only divergence-free symmetric rank-2 tensor constructed solely from the metric g and its derivatives up to second differential order, and preserving diffeomorphism invariance, is the Einstein tensor plus a cosmological term.

In simple terms, the theorem states that GR emerges as the unique theory of gravity if the conditions of the theorem are followed. In fact, Lovelock's theorem provides a useful scheme for classifying alternatives to General Relativity.

Einstein described both the demand for "no prior geometry" and for a "geometric, coordinate-independent formulation of physics" by the single phrase "general covariance", but the two concepts are not quite the same.

While many physical theories can be formulated in a generally covariant way, General Relativity is actually based on the "no prior geometry" demand. This distinction was not always made, especially in the first decades after Einstein's publications [54,55]. Erich Kretschmann's famous objection in 1917 [56] concerned this point, since he regarded general covariance merely as formal feature that any theory could have, not as a special feature belonging to GR.

3.3. Newtonian Nostalgia: The First Wave of Alternative Theories

Newton's theories had predicted observations of Solar System objects, comets and asteroids, with astounding precision. Why should they be tampered with? The first wave of alternative theories were driven more by theoretical considerations than by observations. Equations (1) and (2) can be generalized so that they are consistent with the postulates of Special and General Relativity. Several generalisations of this kind were attempted in the first few decades following the development of GR, motivated by lingering resistance to any deviation from Newtonian gravity.

One early theory, involving prior geometry, was formulated by Nordstrøm in 1913 [57]. In this theory, the physical metric of spacetime g is generated by a background flat spacetime metric η, and by a scalar gravitational field ϕ. Stress-energy generates ϕ:

$$\eta^{\alpha\beta}\phi_{,\alpha\beta} = -4\pi\phi\eta^{\alpha\beta}T_{\alpha\beta} \tag{17}$$

and g is constructed from ϕ and η:

$$g_{\alpha\beta} = \phi^2\eta_{\alpha\beta}. \tag{18}$$

Prior geometry cannot be removed by rewriting Nordstrøm's equations in a form devoid of η and ϕ [58]. Mass only influences one degree of freedom in the spacetime geometry, while the other degrees of freedom are fixed *a priori*. This prior geometry, if it existed, could be detected by physical experiments.

In the 1920s, Alfred North Whitehead [59] formulated a two-tensor theory of gravity in which the prior geometry is quite different from later theories such as Ni's [48]. Whitehead's theory is remarkable in that it agrees with Einstein's in its predictions for the four standard tests (bending of light, gravitational redshift, perihelion shift, and time delay). It was accepted as a viable alternative for Einstein's theory until Clifford Martin Will [60] showed that it predicts velocity-independent anisotropies in the Cavendish constant (the gravitational constant G in Newtonian theory). This would produce time-dependent Earth tides which are clearly contradicted by everyday observations. Any valid theory of gravity must not only agree with relativistic experiments, but also with past experiments in the Newtonian regime.

One theory which disagrees violently with non-relativistic experiments is due to George David Birkhoff [61]. It was developed in the 1940s, and it predicts the same redshift, perihelion shift, deflection and time-delay as General Relativity, but it requires that the pressure inside gravitating bodies should be equal to the total density of mass-energy ($p = \rho$). This means that sound waves travel with the speed of light. This clearly contradicts everyday experiments.

Most of the early alternative theories were abandoned either because they were contradicted by observations, or because of internal inconsistencies in the theories themselves. One notable exception is Dicke-Brans-Jordan theory, sometimes called Brans-Dicke, or Jordan-Fierz-Brans-Dicke theory [62,63], developed in the 1960s by Robert H. Dicke and Carl H. Brans following earlier work by Pascual Jordan and Markus Fierz. The different names arise from the fact that the theory is a special case of Jordan's, with $\eta = -1$. An alternative mathematical representation of the theory is given by [64].

This theory introduced auxiliary gravitational fields. Brans and Dicke took the equivalence principle as the starting point of their theory, and thus they describe gravity in terms of spacetime curvature, but their gravitational field, unlike Einstein's, is a scalar-tensor combination. In this way it overcomes the difficulties associated with tensor or scalar-only theories mentioned earlier. The trace of the energy-momentum tensor $(T_M)_{\mu\nu}$ (representing matter) and a coupling constant λ generate the long-range scalar field ϕ via the equation

$$\Box^2 \phi = 4\pi\lambda (T_M)^\mu_\mu. \tag{19}$$

The scalar field ϕ fixes the value of G, which is therefore not a constant, but simply the coupling strength of matter to gravity. The gravitational field equations relate the curvature to the energy-momentum tensors of the scalar field and matter:

$$R_{\mu\nu} - \tfrac{1}{2}g_{\mu\nu}R = -\frac{8\pi}{c^4\phi}\left[(T_M)_{\mu\nu} + (T_\Phi)_{\mu\nu}\right], \tag{20}$$

where $(T_M)_{\mu\nu}$ is the energy-momentum tensor of matter and $(T_\Phi)_{\mu\nu}$ is the energy-momentum tensor of the scalar field ϕ. For historical reasons, it is usual to write the coupling constant as

$$\lambda = \frac{2}{3 + 2\omega}, \tag{21}$$

where ω is the dimensionless 'Dicke coupling constant'. In the limit $\omega \to \infty$, we have $\lambda \to 0$, so ϕ is not affected by the matter distribution, and can be set to a constant $\phi = 1/G$. Hence Dicke-Brans-Jordan theory reduces to Einstein's theory in the limit $\omega \to \infty$.

The equivalence principle is satisfied in this theory since the special-relativistic laws are valid in the local Lorentz frames of the metric g of spacetime. The scalar field does not exert any direct influence on matter. It only enters the field equations that determine the geometry of spacetime. On a conceptual level, Brans-Dicke theory can be seen as more fully Machian than Einstein's theory. Einstein himself attempted to incorporate Mach's Principle into his theory, but in Einstein's General Relativity, the inertial mass of an object will always be independent of the mass distribution in the universe. In Brans-Dicke theory, the long-range scalar field is an indirectly coupling field, so it does not directly influence matter, but the Einstein tensor is determined partly by the energy-momentum tensor, and partly by the long-range scalar field.

Dicke-Brans-Jordan theory is self-consistent and complete, but experimental evidence based on Solar System tests, shows that $\omega \geq 600$ [65], as a conservative estimate. Some calculations raise this limit even higher, with $\omega \gtrsim 10^4$ [66]. The Cassini mission set a comparable limit of $\omega > 40,000$ [67]. Recent cosmological data from the *Planck* probe show that $\omega \geq 890$ [68,69]. This is consistent with the Solar System bounds. Future cosmological experiments and data from large-scale structure could provide even better constraints [70].

Brans-Dicke theory is a special case of general scalar-tensor theories with $\omega(\phi) = $ constant, where ϕ is a value depending on the cosmological epoch. In these theories, the function $\omega(\phi)$ could be such that the theory is very different from GR in the early universe or in future epochs, but very close to GR in the present. In fact, it has been shown that GR is a natural attractor for such scalar-tensor theories, since cosmological evolution naturally drives the fields towards large values of ω [71,72].

3.4. Self-Consistency, Completeness, and Agreement with Experiment

Any viable theory must satisfy three fundamental criteria: self-consistency, completeness, and agreement with past experiment.

To be self-consistent, a theory must not contain any internal contradictions. The spin-two field theory of gravity [40] is equivalent to linearised General Relativity but it is internally inconsistent since it predicts that gravitating bodies should have their motion unaffected by gravity. When one tries to remedy this inconsistency, the resulting theory is nothing but General Relativity. Another self-inconsistent theory is due to Paul Kustaainheimo [73,74]. It predicts zero gravitational redshift when the wave version of light (Maxwell theory) is used, and nonzero redshift when the particle version (photon) is used.

To be complete, a theory must be able to analyse the outcome of any experiment. This means that it must be compatible with other physical theories which describe any other forces that are present in experiments. This can only be achieved if the theory is derived from first principles, since the theoretical postulates must be as general as possible if the theory is to cover the widest range of phenomena.

A viable theory must agree with past experiment, which includes experiments in the Newtonian regime, and the standard tests of General Relativity. Its results must agree with those obtained from Newtonian theory in the weak field limit, and with GR in relativistic situations. It also means that the theory must agree with cosmological observations.

The experimental criterion also works the other way. Any alternative to General Relativity that claims to have a smaller set of limiting cases must be experimentally distinguishable, perhaps by future experiment. At some point, the divergence between GR and other theories must manifest itself physically, in the form of predictions which can be verified by experiment. This is perhaps the greatest challenge of current alternatives to GR.

3.5. Metric Theories and Quantum Gravity

Most theories of gravity incorporate two principles: *spacetime possesses a metric; and that metric satisfies the equivalence principle.* Such theories are called metric theories. There are some exceptions.

Soon after the publication of the theory of General Relativity, it became apparent that its formulation is incompatible with a Quantum Mechanical description of the gravitational field. It was Einstein himself who pointed out that quantum effects must lead to a modification of General Relativity [75]. Back then, the first successful applications of Quantum Mechanics to electromagnetism were starting to give useful results. These developments led to the question of whether General Relativity can be quantized.

This is a difficult question. First, Einstein's field equations are much more complicated than Maxwell's equations, and in fact are nonlinear. The physical reason for this is that the gravitational field is coupled to itself—the stress-energy tensor acts as the source for spacetime curvature, which in general contributes to the stress-energy tensor. This means that the equations seem to violate the superposition principle, which requires the existence of a linear vector space (see, e.g., [76,77]). This is the mathematical expression of wave-particle duality—a central tenet of Quantum Theory.

Second, to quantize the gravitational field we would have to quantize spacetime itself. The physical meaning of this is not completely clear.

Finally, there are experimental problems. Maxwell's equations predict electromagnetic radiation, which was first observed by Hertz [78]. Quantization of the field results in being able to observe individual photons, and these were first seen in the photoelectric effect predicted by Einstein [79]. Similarly, Einstein's equations for the gravitational field predict gravitational radiation [75], so there should be, in principle, the possibility of observing individual gravitons, which are the quanta of the field. The direct observation of gravitational waves was finally achieved in September 2015 by the LIGO instrument [80]. The detection of individual gravitons is more difficult and is beyond the capability of current experiments.

To develop a quantum theory of General Relativity, the fundamental interactions in GR would have to follow quantum rules. In Quantum Theory, particle interactions are described by gauge theories, so GR would have to follow the gauge principle. Although the gauge principle was first recognized in electromagnetism, modern gauge theory, formulated initially by Chen Ning Yang and Robert Mills [81,82], emerged entirely within the framework of the quantum field programme. As more particles were discovered after the 1940s, various possible couplings between those elementary particles were being proposed. It was therefore necessary to have some principle to choose a unique form out of the many possibilities suggested. The principle suggested by Yang and Mills in 1954 is based on the concept of gauge invariance, and is hence called the gauge principle.

3.6. The Gauge Approach and Non-Metric Theories

The idea of gauge invariance, and the term itself, originated earlier, from the following consideration due to Hermann Weyl in 1918 [83,84]. In addition to the requirement of General Relativity that coordinate systems have to be defined only locally, so likewise the standard of length, or scale, should only be defined locally. It is therefore necessary to set up a separate unit of length at every spacetime point. Weyl called such a system of unit-standards a gauge system (analogous to the standard width, or "gauge", of a railway track).

The gauge principle therefore may be formulated as follows: If a physical system is invariant with respect to some global (spacetime independent) group of continuous transformations G, then it remains invariant when that group is considered locally (spacetime dependent), that is $G \mapsto G(x)$. Partial derivatives are replaced by covariant ones, which depend on some new vector field.

In Weyl's view, a gauge system is as necessary for describing physical events as a coordinate system. Since physical events are independent of our choice of descriptive framework, Weyl maintained that gauge invariance, just like general covariance, must be satisfied by any physical theory.

In Euclidean geometry, we know that translation of a vector preserves its length and direction. In Riemannian geometry, the Christoffel connection [85] (or affine connection) guarantees length preservation, but a vector's orientation is path dependent. However, the angle between two vectors, following the same path, is preserved under translation. Weyl wondered why the remnant of planar geometry, length preservation, persisted in Riemannian geometry. After all, our measuring standards (rigid rods and clocks), are known only at one point in spacetime. To measure lengths at another point, we must carry our measuring tools along with us. Weyl maintained that only the *relative* lengths of any two vectors at the same point, and the angle between them, are preserved under parallel transport. The length of any single vector is arbitrary. To encode this mathematically, Weyl made the following substitution:

$$g_{\mu\nu}(x) \mapsto \lambda(x)g_{\mu\nu}(x), \tag{22}$$

where the conformal factor $\lambda(x)$ is an arbitrary, positive, smooth function of position. Weyl required that in addition to GR's coordinate invariance, formulae must remain invariant under the substitution (Equation (22)). He called this a *gauge transformation*. The scale therefore becomes a local property of the metric.

Weyl's theory enabled him to unify gravity and electromagnetism, the only two forces known at the time. However, Weyl's original scale invariance was abandoned soon after it was proposed, since its physical implications seemed to contradict experiments. In particular, if two identical clocks C_1 and C_2 are transported on two different paths, which both end at the same point Q, the time-like vectors l_1 and l_2 given by C_1 and C_2 at Q would be different in the presence of an electromagnetic field. Therefore the two clock rates would differ. As Einstein (probably the only expert who could keep an eye on Weyl's theory at the time) pointed out, this concept meant that spectral lines with definite frequencies could not exist, since the frequency of the spectral lines of atomic clocks would depend on the atom's location, both past and present. However, we know the atomic spectral lines to be definite, and independent of spacetime position [86–89].

Despite its initial failures, Weyl's idea of a local gauge symmetry survived, and acquired new meaning with the development of Quantum Mechanics. According to Quantum Mechanics, interactions are realized through quantum (that is, non-continuous) fields which underlie the local coupling and propagation of field quanta, but which have nothing to do with the geometry of spacetime.

The question is whether General Relativity can be formulated as a gauge theory. This question has been discussed by ever since it was first posed [90–96].

If features of General Relativity could be recovered from a gauging argument, then that would show that the two formulations are not inconsistent. The first to succeed in this was Kibble [91], who elaborated on an earlier, unsuccessful attempt by Utiyama [90]. Kibble arrived at a set of gravitational field equations, although not the Einstein equations, constructing a slightly more general theory, known as "spin-torsion" theory. The inclusion of torsion in Einstein's General Relativity had long been theorized. In fact the necessary modifications to General Relativity were first suggested by Élie Cartan in the 1920s [97–100], who identified torsion as a possible physical field.

The connection between torsion and quantum spin was only made later [91,101,102], once it became clear that the stress-energy tensor for a massive fermion field must be asymmetric [103,104]. The Einstein-Cartan (1920s) and the Kibble-Sciama (late 1950s) developments occurred independently. For historical reasons, spin-torsion theories are sometimes referred to as Einstein-Cartan-Kibble-Sciama (ECKS) theories, but Einstein-Cartan Theory (ECT) is the term more commonly employed.

The Einstein-Cartan Theory of gravity is a modification of GR allowing spacetime to have torsion in addition to curvature, and, more importantly, relating torsion to the density of intrinsic angular momentum. This modification was put forward by Cartan before the discovery of quantum spin, so the physical motivation was anything but quantum theoretic. Cartan was influenced by the works of the Cosserat brothers [105] who considered a rotation stress tensor in a generalized continuous medium besides a force tensor.

Cartan assumed the linear connection to be metric and derived, from a variational principle, a set of gravitational field equations. However, Cartan required, without justification, that the covariant divergence of the energy-momentum tensor be zero, which led to algebraic constraint equations, thus severely restricting the geometry. This probably discouraged Cartan from pursuing his theory. It is now known that the conservation laws in relativistic theories of gravitation follow from the Bianchi identities and in the presence of torsion, the divergence of the energy-momentum tensor need not vanish.

In simple mathematical terms, a non-zero torsion tensor means that

$$T^{\mu}_{\nu\sigma} = \Gamma^{\mu}_{\nu\sigma} - \Gamma^{\mu}_{\sigma\nu} \neq 0 \,. \tag{23}$$

Geometrically, it means that an infinitesimal geodesic parallelogram forms a non-closed loop. Torsion is therefore a local property of the metric. The Lagrangian action of Einstein-Cartan theory takes the usual Einstein-Hilbert form:

$$S = \int \mathrm{d}^4 x \sqrt{-g} \left(-\frac{g^{\mu\nu} R_{\mu\nu}(\Gamma)}{16\pi G} + \mathcal{L}_m \right) \,, \tag{24}$$

where Γ is a general affine connection and $\mathcal{L}_m$ is the matter Lagrangian. The theory differs from GR in the structure of Γ, leading to a field theory with additional interactions.

Torsion vanishes in the absence of spin and the Einstein-Cartan field equation is then the classical Einstein field equation. In particular, there is no difference between the Einstein and Einstein-Cartan theories in empty space. Since practically all tests of relativistic theory are based on free space experiments, the two theories are, to all effects, indistinguishable via the standard tests of GR. The inclusion of torsion only results in a slight change in the energy-momentum tensor. Cartan's theory holds the distinction of being complete, self-consistent and in agreement with experiment, but of being a non-metric theory of gravitation. The link between torsion and quantum spin means that it could be possible to study the divergence between the GR and ECKS theories at the quantum level. Such experiments have recently been proposed [106].

Kibble's theory contains some features which were criticized [107]. It is now accepted that torsion is an inevitable feature of a gauge theory based upon the Poincaré group. Classical GR must be modified by the introduction of a spin-torsion interaction if it is to be viewed as a gauge theory. The gauge principle alone fails to provide a conceptual framework for GR as a theory of gravity.

In the 1990s, Anthony Lasenby, Chris Doran and Stephen Gull proposed an alternative formulation of General Relativity which is derived from gauge principles alone [108–113]. Their treatment is very different from earlier ones where only infinitesimal translations are considered [91,107]. There are a few other theories similar in their approach to that of Lasenby, Doran and Gull (e.g., [114,115]).

4. Why Consider Alternative Theories?

The motivation for considering alternatives to GR comes mainly from theoretical arguments, like scale invariance of the gravitational theory, additional scalar fields that emerge from string theories, Dark Matter, dark energy or inflation, or additional degrees of freedom that arise in the framework of brane-world theories.

In Table 1, we draw up a list of some of the more well-known alternatives to General Relativity. This list is far from exhaustive, but it serves to highlight the major elements which differentiate these theories. There are several works containing a more detailed listing and discussion of the various alternative theories (e.g., [39,116,117]).

Table 1. A "comparative morphology" of some of the major alternatives to General Relativity, in approximate chronological order. We have only listed the theories of particular historical significance. The current landscape, in which cosmologists seek to explain Dark Matter, dark energy, and inflation, offers far more theories. It is generally easier to incorporate the non-gravitational laws of physics within metric theories, since other theories would result in greater complexity, rendering calculations difficult. The only way in which metric theories significantly differ from each other is in their laws for the generation of the metric. Abbreviations: Tensor (T), V (Vector), S (Scalar), P (Potential), Dy (Dynamic), Einstein Equivalence Principle (EEP), i.e., uniqueness of freefall, Local Lorentz Invariance (LLI), Local Position Invariance (LPI), param (Parameter), ftn (Function).

Theory	Metric	Other Fields	Free Elements	Status
Newton 1687 [20]	Nonmetric	P	None	Nonrelativistic, implicit action at a distance
Poincaré 1890s–1900s [31,118]				Fails; does not mesh with electromagnetism
Nordstrøm 1913 [57]	Minkowski	S	None	Fails to predict observed light detection
General Relativity 1915 [46]	Dy	None	None	Viable
Whitehead 1922 [59]				Violates LLI; contradiction by everyday observation of tides
Cartan 1922–1925 [98]	ST			Still viable; introduces matter spin
Kaluza-Klein 1920s [119,120]	T	S	Extra dimensions	Violates Equivalence Principle
Birkhoff 1943 [61]	T			Fails Newtonian test; demands speed of sound equal to speed of light
Milne 1948 [121]	Machian background			Incomplete; no gravitational redshift prediction; contradicts cosmological observations.
Thiry 1948 [122]	ST			Unlikely; extremely constrained by results on γ^{PPN}
Belifante-Swihart 1957 [123]	Nonmetric	T	K param	Violates EEP; contradicted by Dicke–Braginsky experiments
Brans–Dicke 1961 [63]	Generic S	Dy	S	Viable for $\omega > 500$
Ni 1972 [48]	Minkowski	T, V, S	1 param, 3 ftns	Violates LPI; predicts preferred-frame effects
Will-Nordtvedt 1972 [124]	Dy T	V		Viable but can only be significant at high energy regimes
Barker 1978 [125]	ST			Unlikely; severely constrained.
Rosen 1973 [126,127]	Fixed	T	None	Contradicted by binary pulsar data
Rastall 1976 [128]	Minkowski	S, V	None	Contradicted by gravitational wave data
$f(R)$ models 1970s [129,130]	$n + 1$ST	S	Free ftn	Consistent with Solar System tests; viable but severely constrained
MOND 1983 [131–133]	Nonmetric	P	Free ftn	Nonrelativistic theory
DGP 2000 [134]	ST/Quantum			Appears to be contradicted by BAOs, CMB and Supernovae Ia unless DE added
TeVeS 2004 [135]	T,V,S	Dy S	Free ftn	Highly unstable [136]; ruled out by SDSS data [137]

5. From General Relativity to Standard Cosmology

When Einstein published his seminal GR papers it became almost immediately apparent that the theory could be applied to the whole universe, under certain assumptions, to obtain a relativistic cosmological description. If the content of the universe is known, then the energy-momentum tensor can be constructed, and the metric derived using Einstein's equations. Einstein himself was the first to apply GR to cosmology in 1917 [138]. The first expanding-universe solutions to the relativistic field equations, describing a universe with positive, zero and negative curvature, were discovered by Alexander Friedmann [139,140]. This occurred before Edwin Hubble's observations and the empirical confirmation, in 1929, that the redshift of a galaxy is proportional to its distance. Hubble formulated the law which bears his name: $v = H_0 r$, where H_0 is the constant of proportionality [141]. The problem of an expanding universe was independently followed up during the 1930s by Georges Lemaître [142], and by Howard P. Robertson [143–145] and Arthur Geoffrey Walker [146].

These exact solutions define what came to be known as the Friedmann-Lemaître-Robertson-Walker (FLRW) metric, also referred to as the FRW, RW, or FL metric. This metric starts with the assumption of spatial homogeneity and isotropy, allowing for time-dependence of the spatial component of the metric. Indeed, it is the only metric which can exist on homogeneous and isotropic spacetime. The assumption of homogeneity and isotropy, known as the Cosmological Principle, follows from the Copernican Principle, which states that we are not privileged observers in the universe. This is no longer true below a certain observational scale of around 100 Mpc (sometimes called the "End of Greatness"), but it does simplify the description of the distribution of mass in the universe.

The FLRW metric describes a homogeneous, isotropic universe, with matter and energy uniformly distributed as a perfect fluid. Using the definition of the metric in Equation (4), it is written as:

$$- ds^2 = c \, d\tau^2 - R^2(t)[dr^2 + S_k^0(r)(d\theta^2 + \sin^2\theta \, d\phi^2)], \tag{25}$$

where r is a time independent comoving distance, θ and ϕ are the transverse polar coordinates, and t is the cosmic or physical time. $R(t)$ is the scale factor of the universe. The function $S_k^0(r)$ is defined as:

$$S_k^0(r) = \begin{cases} \sin(r) & (k = +1) \\ r & (k = 0) \\ \sinh(r) & (k = -1) \end{cases} \tag{26}$$

where k is the geometric curvature of spacetime, the values 0, +1, and −1 indicating flat, positively curved, and negatively curved spacetime, respectively.

Another common form of the metric defines the comoving distance as $S_k^0(r) \rightarrow r$, so that

$$- ds^2 = c \, dt^2 - R^2(t) \left[\frac{dr^2}{1 - kr^2} + r^2(d\theta^2 + \sin^2\theta \, d\phi^2) \right], \tag{27}$$

where t is again the physical time, and r, θ and ϕ are the spatial comoving coordinates, which label the points of the 3-dimensional constant-time hypersurface.

The dimensionless scale factor $a(t)$ is defined as

$$a(t) \equiv \frac{R(t)}{R_0}, \tag{28}$$

where R_0 is the present scale factor (i.e., $a = 1$ at present). The scale factor is therefore a function of time, so it can be abbreviated to a. The metric can then be written in a dimensionless form:

$$- ds^2 = c^2 \, d\tau^2 = c^2 \, dt^2 - a^2 \left[dr^2 + S_k^2(r)(d\theta^2 + \sin^2\theta \, d\phi^2) \right], \tag{29}$$

where $S_k(r)$ can be redefined as

$$S_k(r) = \begin{cases} R_0 \sin(r/R_0) & (k = +1) \\ r & (k = 0) \\ R_0 \sinh(r/R_0) & (k = -1). \end{cases} \tag{30}$$

Equivalently, using the definition in Equation (27),

$$- ds^2 = c^2 \, dt^2 - a^2 \left[\frac{dr^2}{1 - k(r/R_0)^2} + r^2 (d\theta^2 + \sin^2 \theta \, d\phi^2) \right]. \tag{31}$$

The comoving distance is distance between two points measured along a path defined at the present cosmological time. It means that for objects moving with the Hubble flow, the comoving distance remains constant in time. The proper distance, on the other hand, is dynamic and changes in time. At the current age of the universe, therefore, the proper and comoving distances are numerically equal, but they differ in the past and in the future. The comoving distance from an observer to a distant object such as a galaxy can be computed by the following formula:

$$\chi = \int_{t_e}^{t} c \frac{dt'}{a(t')} \tag{32}$$

where $a(t')$ is the scale factor, t_e is the time of emission of photons from the distant object, and t is the present time.

The comoving distance defines the comoving horizon, or particle horizon. This is the maximum distance from which particles could have travelled to the observer since the beginning of the universe. It represents the boundary between the observable and the unobservable regions of the universe.

If we take the time at the Big Bang as $t = 0$, we can define a quantity called the conformal time η at a time t as:

$$\eta = \int_{0}^{t} \frac{dt'}{a(t')}. \tag{33}$$

This is useful, because the particle horizon for photons is then simply the conformal time multiplied by the speed of light c. The conformal time is not the same as the age of the universe. In fact it is much larger. It is rather the amount of time it would take a photon to travel from the furthest observable regions of the universe to us. Because the universe is expanding, the conformal time is continuously increasing.

The concept of particle horizons is important. It defines causal contact. The only objects not in causal contact are those for which there is no event in the history of the universe that could have sent a beam of light to both. This is at the origin of some of the big questions about the universe associated with the Big Bang model, which gave rise to the Inflationary paradigm (see [147]). We shall discuss this later.

5.1. Cosmological Expansion and Evolution Histories

The FLRW metric relates the spacetime interval ds to the cosmic time t and the comoving coordinates through the scale factor $R(t)$. The scale factor is the key quantity of any cosmological model, since it describes the evolution of the universe. The notion of distance is fairly straightforward in Euclidean geometry. In General Relativity, however, where we work with generally curved spacetime, the meaning of distance is no longer unique. The separation between events in spacetime depends on the definition of the distance being used.

By combining the GR field equation (Equation (15)) and the definition of the metric (Equation (31)), we obtain two independent Einstein equations, known as the Friedmann equations:

$$\left(\frac{\dot{a}}{a}\right)^2 + \frac{kc^2}{a^2} = \frac{8\pi G}{3}\rho \tag{34}$$

and

$$2\left(\frac{\ddot{a}}{a}\right) + \left(\frac{\dot{a}}{a}\right) + \frac{kc^2}{a^2} = -\frac{8\pi G}{c^2}p. \tag{35}$$

The Friedmann equations relate the total density ρ of the universe, including all contributions, to its global geometry. There exists a critical density ρ_c for which $k = 0$. By rearranging the Friedmann equation and using the definition of the Hubble parameter we then obtain

$$\rho_c(t) = \frac{3H^2(t)}{8\pi G}. \tag{36}$$

A universe whose density is above this value will have a positive curvature, that is, it will be spatially closed (k = +1); one whose density is less than or equal to this value will be spatially open ($k = 0$ or $k = -1$).

A dimensionless density parameter for any fluid component of the universe (i.e., a component for whose gravitational field is produced entirely by the mass, momentum, and stress density) can be defined by

$$\Omega(t) = \frac{\rho(t)}{\rho_c(t)} = \frac{8\pi G\rho(t)}{3H^2(t)}. \tag{37}$$

The current value of the density parameter is denoted Ω_0.

Subtracting Equation (34) from Equation (35) yields the acceleration equation:

$$\frac{\ddot{a}}{a} = -4\pi G\left(\frac{\rho}{3} + \frac{p}{c^2}\right). \tag{38}$$

The geodesic Equation (12) allows us to compute the evolution in time of the energy and momentum of the various components particles which make up the universe. From this evolution, we can construct the fluid equation, or continuity equation, which describes the relation between the density and pressure:

$$\dot{\rho} + 3\frac{\dot{a}}{a}\left(\rho + \frac{p}{c^2}\right) = 0. \tag{39}$$

This is valid for any fluid component of the universe, such as baryonic and nonbaryonic matter, or radiation.

The foundations of the Concordance Model of cosmology depend on General Relativity. Any modification to the theory that changes the Einstein equations will have solutions that differ from the Friedmann equations.

The FLRW universe contains different mass-energy components which are assumed to evolve independently. This is physically valid at late cosmological times, when the components are decoupled, so the density evolutions are distinct. In Table 2, we give the equation of state and the evolution of the density and scale factor for different components of the universe. The quantities in this table are explained in detail in the following sections.

Table 2. The evolution of the various cosmological components. The quantities are the equation of state $w \equiv p/\rho c^2$, the density ρ, the pressure p, and the scale factor $a(t)$.

Component	w	$\rho = a^{3(1+w)}$	$a(t) = t^{2/3(1+w)}$
Radiation (photons and relativistic neutrinos)	$1/3$	$\sim a^{-4}$	$\sim t^{1/2}$
Dust (includes CDM, baryons and non-relativistic neutrinos)	0	$\sim a^{-3}$	$\sim t^{2/3}$
Curvature	$-1/3$	$\sim a^{-2} \to a^{-4}$	t
Cosmic strings	$-1/3$	$\sim a^{-2} \to a^{-4}$	t
Domain walls	$-2/3$	a^{-1}	$\sim t^2$
Inflation	$\to -1$	$\frac{1}{2}\dot{\phi}^2 + V(\phi)$	$\sim e^{Ht}$
Vacuum energy	-1	constant	$\sim e^{Ht}$

5.2. Matter (Dust)

Matter which is pressureless is referred to as "dust". This is a useful approximation for cosmological structures which do not interact, such as individual galaxies. Substituting $p_m = 0$ in the equation of state for dust shows that the density of this component scales as:

$$\rho_m(a) = \frac{\rho_{m,0}}{a^3}, \tag{40}$$

where $\rho_{m,0}$ is the current density. Assuming spatial flatness, the time evolution of the scale factor is then

$$a(t) = \left(\frac{t}{t_0}\right)^{2/3}, \tag{41}$$

which gives us

$$H(t) = \frac{2}{3t}. \tag{42}$$

This is known as the Einstein-de Sitter (EdS) solution, and it describes the evolution of H in a constant-curvature homogeneneous universe with a pressureless fluid as the only component. It was first described by Einstein and Willem de Sitter in 1932 [148].

5.3. Radiation

In the early universe, the energy content was dominated by photons and relativistic particles (especially neutrinos). The expansion of the universe dilutes the radiation fluid, and the wavelength is increased by the expansion so that the energy decreases. From thermodynamics,

$$E_{rad} = \rho_{rad}c^2 = \alpha T^4, \tag{43}$$

where T is the radiation temperature and α is the Stefan-Boltzmann constant. The equation of state for radiation can then be derived from the fluid Equation (39):

$$\rho_{rad}(a) = \frac{\rho_{rad,0}}{a^4} \quad ; \quad p_{rad} = \frac{\rho_{rad}c^2}{3}. \tag{44}$$

Combining this with the Friedmann equations, and assuming flatness ($k = 0$), we obtain the time dependence of the scale factor and the Hubble parameter:

$$a(t) = \left(\frac{t}{t_0}\right)^{1/2} \quad ; \quad H(t) = \frac{1}{2t}. \tag{45}$$

6. The Components and Geometry of the Universe and Cosmic Expansion

How do we relate the expansion of the universe to its contents? The total density of the universe in terms of its constituent components can be written as the sum of the densities of these components at any given time or scale factor:

$$\rho = \rho_m + \rho_{\text{rad}} + \rho_{\text{DE}},\tag{46}$$

where the subscript "DE" denotes another component of the universe, called Dark Energy.

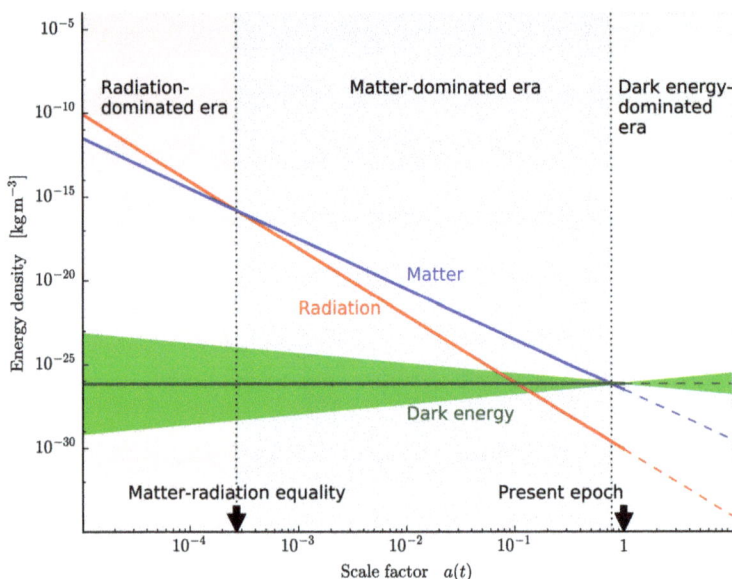

Figure 2. The density evolution of the main components of the universe. The early universe was radiation-dominated, until the temperature dropped enough for matter density to being to dominate. The energy density of dark energy is constant if its equation of state parameter $w = 1$. Because the matter energy density drops as the scale factor increased, dark energy began to dominate in the recent past. At the present time ($a(t) = 1$), we live in a universe dominated by dark energy. For dark energy, the green band represents an equation of state parameter $w = -1 \pm 0.2$, showing how a small change in the value of this parameter can give very different evolution histories for dark energy. If the Concordance Model is correct, the universe will be completely dominated by dark energy in future epochs (shown by the dashed lines). The matter density will keep decreasing as the universe expands. Our Milky Way will merge with the Andromeda Galaxy, and eventually, the entire Local Group will coalesce into one galaxy. The luminosities of galaxies will begin to decrease as the stars run out of fuel and the supply of gas for star formation is exhausted. In the very far future, this galaxy will be in the only one in our Hubble patch, as all the other galaxies will pass behind the cosmological horizon. The night sky, save for the stars in the Local Group, will be very dark indeed. Stellar remnants will either escape galaxies or fall into the central supermassive black hole. Eventually, baryonic matter may disappear altogether as all nucleons including protons decay, or all matter may decay into iron. In either scenario, the universe will end up being dominated by black holes, which will evaporate by Hawking radiation. The end result is a Dark Era with an almost empty universe, and the entire universe in an extremely low energy state, with a possible heat death as entropy production ceases (see, e.g., [149,150]) What happens after that is speculative.

The total dimensionless density can then be written:

$$\Omega = \Omega_m + \Omega_{rad} + \Omega_{DE} , \qquad (47)$$

where we have dropped the subscript for clarity, i.e., $\Omega_{m,0} = \Omega_m$, etc. The Friedmann Equation (34) can now be rewritten using the equations of state for the different components:

$$H^2(a) = \frac{8\pi G}{3} \left(\rho_m a^{-3} + \rho_{rad} a^{-4} + \rho_{DE} e^{-3 \int_a^1 [1+w(a')] \, d\ln a'} \right) - \frac{kc^2}{a^2} . \qquad (48)$$

This can be rearranged to give:

$$H^2(a) = H_0^2 \left[\Omega_m a^{-3} + \Omega_{rad} a^{-4} + \Omega_{DE} e^{-3 \int_a^1 [1+w(a')] \, d\ln a'} + (1 - \Omega)a^{-2} \right], \qquad (49)$$

or, in terms of redshift:

$$H^2(z) = H_0^2 \left[\Omega_m (1+z)^3 + \Omega_{rad}(1+z)a^4 + \Omega_{DE} e^{-3 \int_0^z [1+w(z')]/(1+z') \, d\ln z'} + (1 - \Omega)(1+z)^2 \right]. \qquad (50)$$

The term $1 - \Omega$ is sometimes replaced by Ω_k, the density due to the intrinsic geometry of spacetime. Equation (50) is of central importance since it relates the redshift of an object to the global density components and geometry of the universe.

The density evolution of the various components of the universe is shown in Figure 2.

7. The Hot Big Bang

In the Standard Model, it is generally accepted that the universe arose from an initial singularity, often termed the "Big Bang", which occurred some 13.8 billion years ago (as measured by *Planck* [151]). This is not discussed here, but it should be noted that there are several proposals for the mechanism of this singularity. During this epoch, we are dealing with Planck scale physics, so most of these mechanisms involve quantum gravity. Other proposals (such as some superstring and braneworld theories) do away with the need for an initial singularity altogether.

7.1. The Cosmic Microwave Background

The radiation density $\rho_{rad} \propto a^{-4}$, so the temperature evolution of the universe from an initial T_0 is:

$$T = \frac{T_0}{a} . \qquad (51)$$

In other words, the universe cools down as it expands. Conversely, this means at early times, when the scale factor was close to zero, the temperature was very high (hence the term "Hot Big Bang"). The radiation left from the early hot universe, cooled by expansion, is known as the Cosmic Microwave Background, or CMB.

The properties of atomic and nuclear processes in an expanding universe provided the first clue for the existence of a hot Big Bang. This was a remarkable achievement of the Big Bang model, because it provided an explanation for the observed abundances of chemical elements in terms of nucleosynthesis. The processes that created nuclei and atoms could only have been possible in an early universe in thermal equilibrium, with black-body spectrum which cooled down as the universe expanded. This allowed Ralph Alpher, Robert Herman, Hans Bethe and George Gamow to predict the existence and temperature of the CMB in 1948 [152–155]. The universe therefore has a thermal as well as an expansion history. Hence the 'Hot Big Bang'.

The first direct evidence for the Hot Big Bang came two decades later, with the observation of the CMB by Arno Penzias and Robert Woodrow Wilson in 1964 [156].

The confirmation of the thermal history of the universe, together with the discovery of charge parity violation in 1964 [157], provided clues about baryogenesis and the observed matter-antimatter imbalance in the universe. This inspired the first proposals for a mechanism for baryogenesis by Andrei Sakharov in1967 [158], followed by electroweak symmetry breaking by Vadim Kuzmin in 1970 [159].

This is a remarkable demonstration of the success of the Concordance Model. The cosmological model fits very well with the predictions of particle physics, which in turn can be tested by cosmological observations. The Concordance Model of the structure and evolution of the universe requires a mechanism for baryogenesis as well as an explanation for Dark Matter and dark energy. The challenge for physical theories beyond (or within) the Standard Model is to explain the preference of matter over antimatter, and to explain the magnitude of this asymmetry. Cosmological observations can be used to address these challenges [160].

The CMB is an extremely isotropic source of microwave radiation, with a spectrum corresponding to a perfect blackbody at a temperature $T_0 = 2.7260 \pm 0.0013 \, \text{K}$ [161]. Using the current temperature and $E_{rad} = \rho_{rad} c^2 = a T^4$, the radiation density today is given by:

$$\Omega_{rad} = 2.47 \times 10^{-5} h^{-2}. \tag{52}$$

At some time in the early universe, the ambient radiation temperature corresponded to the ionisation potential of hydrogen, which is 13.6 eV. During this epoch, the universe was filled with a sea of highly energetic particles and photons—a hot ionised plasma. The particles were mainly electrons and protons. Other fundamental particles (quarks) existed earlier when the ambient energy corresponded to their rest mass. At some point, as the universe expanded and cooled, the energy of the photons was no longer sufficient to ionise the hydrogen, and within a relatively short time, all of the electrons and protons combined to form neutral hydrogen. The photons were then free to move through the universe. This process is known as decoupling and it occurred at a temperature of ~2500 K, when the universe was approximately 380, 000 years old [162]. It is these decoupled photons which make up the CMB. The surface on the sky from which these photons originate is known as the surface of last scattering.

7.2. Matter-Radiation Equality

At the present epoch, neglecting dark energy, the universe is dominated by matter. This component is characterised by the fact that the matter particles can be treated in a non-relativistic regime, whereas photons and relativistic neutrinos both behave like radiation. The total contribution to the energy density from non-relativistic components (matter) and relativistic components (radiation and relativistic neutrinos) can be written as Ω_{NR} and $\Omega_R = \Omega_{rad} + \Omega_\nu$, respectively. Using the fact that $\Omega_m = \Omega_{m,0} a^{-3}$, the ratio of the contributions of the components is a function of the scale factor a:

$$\frac{\Omega_R}{\Omega_{NR}} = \frac{\Omega_{rad} + \Omega_\nu}{\Omega_m} = \frac{4.15 \times 10^{-5}}{\Omega_{m,0} h^2 a^{-3}}, \tag{53}$$

where we explicitly use the subscript 0 for the present-day values.

Then there must exist a scale factor for which the ratio is unity. This is given by:

$$a_{eq} = \frac{4.15 \times 10^{-5}}{\Omega_{m,0} h^2}, \tag{54}$$

or, in terms of redshift,

$$1 + z_{eq} = 2.4 \times 10^4 \Omega_{m,0} h^2. \tag{55}$$

The epoch at which the matter energy density equals the radiation energy density is called matter-radiation equality, and it has a special role in large-scale structure formation.

7.3. Neutrinos

Neutrinos have particular properties which give rise to a distinct evolution history. They are known to exist from the Standard Model of particle physics, and the Hot Big Bang model predicts the amount of neutrinos in the universe. Neutrinos can be thought of as "dark" matter because of their very small reaction cross-section, which implies negligible self-interaction. However, they are not *cold* Dark Matter. They are simply extremely light particle that can stream out of high-density regions. They therefore cause the suppression of perturbations on scales smaller than the free-streaming scale. Unlike photons and baryons, cosmic neutrinos have not been observed. However, particle physics allows us to chart the history of this particle during nucleosynthesis, and to relate the neutrino temperature to the photon temperature today [163–165].

The scale on which perturbations are damped by neutrinos is determined by the comoving distance that a neutrino can travel in one Hubble time at equality. For a neutrino mass ~1 eV, the average velocity, T_ν / m_ν is of order unity at equality. This leads to a suppression of power on all scales smaller than k_{eq}. Note that this phenomenon depends on the individual neutrino mass, rather than the total neutrino mass. A lighter neutrino can free-stream out of larger scales, so the suppression begins at lower k for the lighter neutrino species. Heavier neutrinos constitute more of the total neutrino density, and so suppress small-scale power more than lighter neutrino species, which means that we need at least two parameters to model massive neutrino phenomenology to sufficient accuracy: the neutrino mass fraction Ω_ν, or some expression of this quantity in terms of the total neutrino mass $\sum m_\nu$, and the number of massive neutrino species N_ν.

Neutrinos introduce a redshift and scale dependence in the transfer function. We know that the perturbation modes of a certain wavelength λ can grow if they are greater than the Jeans wavelength. Above the Jeans scale, perturbations grow at the same rate independently of the scale. For the baryonic and cold Dark Matter components, the time and scale dependence of the power spectrum can therefore be separated at low redshifts. This is not the case with massive neutrinos, which introduce a new length scale given by the size of the comoving Jeans length when the neutrinos become non-relativistic. In terms of the comoving wavenumber k_{nr}, this scale is given by:

$$k_{nr} = 0.026 \left(\frac{m_\nu}{1\,\text{eV}} \right)^{1/2} \Omega_m^{1/2}\, h\,\text{Mpc}^{-1} \tag{56}$$

for three neutrinos of equal mass, each with mass m_ν. The growth of Fourier modes with $k > k_{nr}$ is suppressed because of neutrino free-streaming. From the equation above, it is evident that the free-streaming scale varies with the cosmological epoch (since there is a dependence on Ω_m), and therefore the scale and time dependence of the power spectrum cannot be separated.

Neutrinos are fermions, with a Fermi-Dirac distribution with assumed zero chemical potential. When they decoupled from the plasma, their distribution remained Fermi-Dirac, with their temperature falling as a^{-1}. This decoupling occurred slightly before the annihilation of electrons and positrons, which occurred when the cosmic temperature was of the order of the electron mass ($T \approx m_e$). Neutrinos decoupled when the cosmic plasma had a temperature of around 1 MeV. The energy associated with this annihilation was therefore not inherited by the neutrinos, and the entropy was completely transferred to the entropy of the photon background. Thus:

$$(S_e + S_\gamma)_{\text{before}} = (S_\gamma)_{\text{after}}, \tag{57}$$

where S_e and S_γ are respectively the entropy of the electron-positron pairs and the photon background, and 'before' and 'after' refer to the annihilation time.

The entropy per particle species, ignoring constant factors, is $S \propto g T^3$, where g is the statistical weight of the species. For bosons, $g = 1$ and for fermions, $g = 7/8$ per spin state. According to the Standard Model, the neutrino has one spin degree of freedom, each neutrino has an antiparticle, and there are three generations of neutrinos, also called "families" or "species" (μ, τ and electron neutrinos). This means that the degeneracy factor of neutrinos is equal to 6. Before annihilation, the fermions are electrons (2 spin states), positrons (2), neutrinos and antineutrinos (6 spin states). The bosons are photons (2 spin states). We therefore have $g_{before} = 4(7/8) + 2 = 11/2$, while after annihilation $g = 2$ because only photons remain. Applying entropy conservation and counting relativistic degrees of freedom, the ratio of neutrino and photon temperatures below m_e is therefore:

$$\frac{T_\nu}{T_\gamma} = \left(\frac{4}{11}\right)^{1/3},$$ (58)

so that the present neutrino temperature is

$$T_{\nu,0} = \left(\frac{4}{11}\right)^{1/3} T_{CMB} = 1.945 \, \text{K}.$$ (59)

The number density of neutrinos is then

$$n_\nu = \frac{6\zeta(3)}{11\pi^2} T_{CMB}^3,$$ (60)

where $\zeta(3) \approx 1.202$, which gives $n_\nu \approx 112 \, \text{cm}^{-3}$ at the present epoch [166]. In the early universe, neutrinos are relativistic and behave like radiation. So they contribute to the total radiation energy density ρ_{rad}, which includes the photon energy density ρ_γ:

$$\rho_{rad} = \left[1 + \left(\frac{7}{8}\right)\left(\frac{4}{11}\right)^{4/3} N_{eff}\right] \rho_\gamma,$$ (61)

where N_{eff} is the effective number of neutrino species. At late times, when massive neutrinos become non-relativistic, their contribution to the mass density is $m_\nu n_\nu$, giving

$$\Omega_\nu = \frac{\rho_\nu}{\rho_c} \approx \frac{\sum_i^{N_\nu} m_{\nu,i}}{93.14 \, \text{eV} \, h^2},$$ (62)

where $m_{\nu,i}$ is the mass of individual neutrino species and N_ν is the number of massive neutrino species. This expression relates the total neutrino mass $\sum m_\nu$ to the neutrino fraction Ω_ν.

It can be seen from the above that this equation can be modified through a change in the effective number of neutrino species by many factors: a non-zero initial chemical potential, or a sizeable neutrino-antineutrino asymmetry, or even a fourth, 'sterile' neutrino [166–168]. The Standard Model predicts a value of $N_{eff} = 3.046$ for the effective number of neutrino species. This accounts for the three neutrino families together with relativistic degrees of freedom, since neutrinos are not completely decoupled at electron-positron annihilation and are subsequently slightly heated [169]. Any significant deviation from this value could be a signature of hidden physical effects, possibly requiring a modification of General Relativity [170].

Neutrino oscillation experiments do not, at present, determine absolute neutrino mass scales, since they only measure the difference in the squares of the masses between neutrino mass eigenstates. Cosmological observations, on the other hand, can constrain the neutrino mass fraction, and can distinguish between different mass hierarchies [166].

Observations of neutrino flavour oscillations in atmospheric and solar neutrinos, provide evidence of a difference between the masses of the different species or flavours, as well as for a non-zero mass. For three neutrino mass eigenstates m_1, m_2 and m_3, the squared mass differences are [171]:

$$|\Delta m_{21}^2| = m_2^2 - m_1^2| \cong 7.5 \times 10^{-5} \, \text{eV}^2$$
$$|\Delta m_{31}^2| = |m_3^2 - m_1^2| \cong 2.5 \times 10^{-3} \, \text{eV}^2$$
$$\frac{|\Delta m_{21}^2|}{|\Delta m_{31}^2|} \cong 0.03 \,. \tag{63}$$

The ambiguity in the sign of the mass differences Δm allows for two possible mass hierarchies: the normal hierarchy given by the scheme $m_3 \gg m_2 > m_1$, or the inverted hierarchy $m_2 > m_1 \gg m_3$. Given Equation (63), constraining the total neutrino mass to a small enough maximum value could exclude an inverted hierarchy. Conversely, a total neutrino mass $m_\nu \sim 2$ eV is only possible with a degenerate neutrino mass scheme. Hence the interest in finding cosmological neutrino mass bounds.

The fact that cosmological constraints could be stronger than constraints from particle accelerators was noticed quite early (see [172]). The 'closure limit' gives us $m_\nu < 90$ eV. This was first derived in the late 1960s and 1970s [173–176]. Since then, cosmological neutrino bounds have improved significantly, with different methods being used e.g., luminous red galaxies [177], CMB anisotropies [151,178], or weak lensing [179–181].

Joint *Planck* CMB and BAO observations give us $m_\nu < 0.23$ eV, but various data combinations can change this figure, and strong priors on the value of the Hubble constant can provide tighter constraints [151].

8. Inflation: The Second Wave of Alternative Theories

In the late 1970s, General Relativity had been largely accepted by the scientific community. But a series of cosmological considerations led to renewed interest in alternative theories. These were not so much attempt to solve problems in the theory itself, but to find explanations for observations that were not explained by the theory.

General Relativity applied to the universe gave us the Hot Big Bang model: a universe expanding out of an initial highly energetic, dense state. The Hot Big Bang model was successful in explaining many interlinked phenomena which were subsequently confirmed by observation: the Hubble Law and the expansion of the universe, the thermal history of the universe, primordial nucleosynthesis, the existence of the cosmic microwave background, the relation between the temperature and scale factor, and finally the blackbody nature of the CMB. The remarkable fact is that these phenomena occur on extremely different scales, and are observed via different physical processes, and yet they all fit neatly within one model.

However, there are some observations which the Hot Big Bang model fails to explain. These cosmological problems are linked to the primordial universe, the most obvious being the following (for details see [182–186], and references therein):

- The Horizon Problem
- The Flatness Problem
- The Monopole Problem

The horizon problem arises from the structure of spacetime. In the standard cosmological model described by the FLRW equations, different regions of the universe observed today could have not been in causal contact with because of the great distances between them which are greater than the distance that could have been traversed by light since the Big Bang. The transfer of information (i.e., any physical interaction) or energy can occur, at most, at the speed of light, but these regions have the same temperature and other physical characteristics. In particular, we observe causally-disconnected regions of the CMB to be in thermal equilibrium. How could this have happened?

The horizon problem was first identified in the late 1960s. This led to to early attempts to model chaotic solutions to Einstein's field equations near the initial singularity [187,188]. In the late 1970s, Alexei Starobinsky noted that quantum corrections to General Relativity should be important in the very early universe. These corrections would lead to a modification of gravity, which induces an inflationary phase [189]. Starobinsky's was the first model of inflation.

The flatness problem is one the so-called "coincidence problems" of modern cosmology. By the 1960s, observations had determined that the density of matter in the universe is comparable to the critical density necessary for a flat universe. So the contribution of curvature had to be of the same order of magnitude as the contribution of matter throughout the history of the universe. This represents a fine-tuning problem. Observations of the CMB have confirmed that the universe is spatially flat to within a few percent. Why is the global geometry of the universe so flat?

The magnetic monopole problem arises from the Hot Big Bang model. Grand Unified Theories predict the production of a large number of magnetic monopoles [190,191] in the early, extremely hot universe. Why have none ever been observed? If they exist at all, they are much more rare than the Big Bang theory predicts. This was noted by Zel'dovich and others in the late 1970s [192,193].

Hot Big Bang Plus Inflation

This gave rise to the idea of a model in which the early universe undergoes a period of exponential accelerated expansion. This theory, called "inflation", was first formulated by Alan Guth in the 1980s [194] while he was trying to investigate why no magnetic monopoles are observed. It was realised that inflation solves the horizon and flatness problems, as well as explaining the absence of relic monopoles. Better still, it explains the origin of structure in the universe.

In the Standard ΛCDM Model, the initial perturbations from which structure evolved are assumed to have been seeded by the inflationary potential. Reconstructing the primordial power spectrum is no easy task, and poses two main problems. Observationally, we want to extract the amplitude and scale variation from the data. Theoretically, we seek to explain the origin of the perturbations. At present, the leading theoretical paradigm for the primordial fluctuations is inflation, which provides initial conditions for both large-scale structure and the cosmic microwave background radiation. The theory of inflation offers a plethora of models, each of which predicts a certain power spectrum of primordial fluctuations $\mathcal{P}(k)$. Since the inflationary paradigm is linked to the theoretical description of the primordial power spectrum, it is necessary to briefly explain some of the main concepts here (for the full details, see [183,185]).

The precise definition of inflation is any period during which the scale factor of the universe is accelerating, that is, $\ddot{a} > 0$. This expression is equivalent to other definitions of inflation:

$$\frac{\mathrm{d}}{\mathrm{d}t}\frac{H^{-1}}{a} < 0 \implies \epsilon \equiv -\frac{\dot{H}}{H^2} < 1 \iff \frac{\mathrm{d}^2 a}{\mathrm{d}t^2} > 0 \iff \rho + 3p < 0. \tag{64}$$

The first expression above has a remarkable physical interpretation. It means that the observable universe becomes smaller during inflation.

The basic theory of inflation states that from the initial Big Bang singularity to approximately 10^{-37} s, there existed a set of highly energetic scalar fields. By definition, Ω is driven towards 1 during inflation. Inflationary theories assume that gravity is described by GR, which means that the component driving inflation must satisfy $\rho + 3P < 0$. If for example, the universe was dominated during the inflationary phase by a scalar field (or set of fields) ϕ with a self-interaction potential $V(\phi)$. It is the form of this potential which differentiates the various inflationary theories. Most theories assume a "Mexican hat" potential, with a single field, while chaotic inflation assumes a simple power law potential with a slowly varying field [195]. The action for this potential is then [196]

$$S = -\int \mathrm{d}^4 x \sqrt{-g} \left[\frac{m_{\mathrm{Pl}}^2 R}{16\pi} - \frac{1}{2}(\nabla\phi)^2 + V(\phi) \right], \tag{65}$$

where m_{Pl} is the Planck mass. As the universe cooled, the scalar field became trapped in a false vacuum, so its energy density became constant. The potential energy, however, is nonzero, so the pressure is negative. The scale factor during inflation has the de Sitter form:

$$a(t) = e^{(\Lambda_{\mathrm{I}}/3)^{1/2}t}, \tag{66}$$

where Λ_{I} represents the energy density of the inflationary field (sometimes called the inflaton).

Since the energy density of the inflaton field was very high, the associated magnitude of the negative pressure would have been very large as well. The scale factor is thought to have increased during inflation by $\sim e^{65}$, and any point in the universe which found itself in a false vacuum state would have undergone inflation. The accelerated expansion lasted until the field rolled down to a minimum, when it decays into the familiar particles of the Standard Model, and the universe can then be described by an FLRW model.

The inflationary paradigm provides an explanation for the origin of structure and for the observed geometry of the universe, in addition to solving the aforementioned cosmological problems.

First, inflation solves the flatness problem. Using Equation (66), the evolution of Ω during inflation can be written as:

$$|\Omega(t) - 1| \propto e^{-(4\Lambda_{\mathrm{I}}/3)^{1/2}t}, \tag{67}$$

so that $|\Omega - 1|$ is driven very close to 0 as t increases. This explains why the universe is flat. It also means that this value has not deviated significantly from its initial value right after expansion. We can therefore safely assume spatial flatness throughout the history of the universe. Given the observational difficulties, this provides a theoretical motivation for taking the idea of a large Ω_{DE} seriously.

Second, inflation solves the horizon problem (Figure 3). Regions of the universe which are causally disconnected today evolved out of the same causally-connected region in the early universe. The observed uniformity of the CMB is no longer a problem.

Third, inflation explains why we have never observed magnetic monopoles. Due to rapid expansion of the universe during inflation, they become so rare in any given volume of space that we would be very unlikely to ever encounter one. Nor would they have sufficient density to alter the gravity and thereby the normal expansion of the universe following inflation. The problem of magnetic monopoles motivated the Guth's development of his theory in 1981 [194]. The solution of the monopole problem, and problems related to other relics, was an early success of the inflationary paradigm, and inspired similar theories [197,198].

Fourth, the inflationary scenario provides a natural explanation for the origin of structure, providing a link between quantum mechanics and relativistic cosmological paradigm. This was realised soon after the development of the theory of inflation, and the details were worked out in the early 1980s [199–204].

An initially smooth background needs seed fluctuations around which gravitational collapse can occur. The inflationary scenario attributes their origin to quantum fluctuations in the inflaton field potential, so that the universe is not perfectly symmetric. Different points in the universe inflate from slightly different points on the potential, separated by $\delta\phi$. Inflation for these two points ends at different times, separated by $\delta t = \delta\phi/\dot{\phi}$. This induces a density fluctuation $\delta = H\delta t$ (see [184]). Since all the points undergoing inflation are part of the same potential field, the initial fluctuations are nearly scale invariant. This means that the density amplitude on the horizon scale will also be constant:

$$\delta_{\mathrm{H}} = H\delta t = \frac{H^2}{2\pi\dot{\phi}} = \text{constant}. \tag{68}$$

In summary, inflation solves the three cosmological problems listed above:

- The Horizon Problem. Solution: the entire universe evolved out of the same causally-connected region.

- The Flatness Problem. Solution: any initial curvature is diluted by the inflationary epoch and driven to zero.
- The Monopole Problem. Solution: the rapid expansion of the universe drastically reduces the predicted density of magnetic monopoles, if they exist.

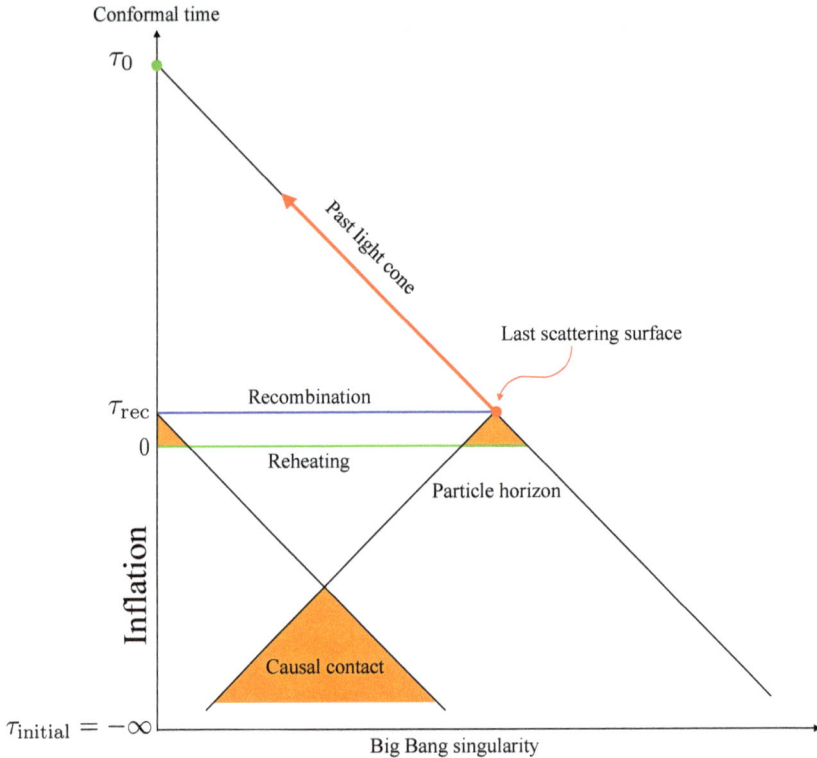

Figure 3. How inflation solves the horizon problem. The light cones on the causal diagram of an inflationary FLRW model are at ±45°. The worldlines of comoving matter are vertical on this kind of diagram. The particle horizons are horizontal lines. Here we have shown the particle horizon for the CMB. Without inflation, conformal time would only go back to τ_0, and different regions of the CMB which we observe today along our past light cone would never have been in causal contact. Because of inflation, conformal time is extended to the Big Bang singularity, so these regions would have been in causal contact at some point in our past light cone.

How long did inflation last? The answer is given by looking closely at Equation (67) above. A convenient measure of expansion is the so-called *e*-fold number, defined as:

$$N \equiv \ln\left(\frac{a_f}{a_i}\right) = \int_{t_i}^{t_f} H \, dt. \tag{69}$$

Here, a_i and a_f are the values of the scale factor at the beginning and end of inflation, while t_i and t_f are the corresponding proper times. The scale factor a is only physically meaningful up to a normalisation constant, so the *e*-fold number is defined with respect to some chosen origin. The reason is that in cosmology, what is fixed is not the initial condition, but the current expansion—we cannot measure any H, but we measure H_0 then extrapolate backwards.

We can search for the minimum duration of inflation required to solve the horizon problem. At the very least, we require that the observable universe today fits in the comoving Hubble radius at the beginning of inflation:

$$(a_0 H_0)^{-1} < (a_i H_i)^{-1}. \tag{70}$$

The condition is the same for the horizon and flatness problems.

If we assume that the universe was radiation-dominated since the end of inflation (giving us $H \propto a^{-2}$), and ignore the relatively recent matter- and dark energy-dominated epochs, we obtain

$$\frac{a_f}{a_i} = \frac{a_0}{a_f}. \tag{71}$$

In the general case, the condition becomes:

$$\frac{a_f}{a_i} \geq \frac{a_0}{a_f}, \tag{72}$$

or in terms of the number of *e*-folds,

$$N_f - N_i \geq N_0 - N_f. \tag{73}$$

In other words, there should be as much expansion during inflation as after inflation.

The solution to the horizon problem is the same as the solution to the flatness problem. Taking into account the present energy density of the universe, we need a minimum of about 50 to 60 *e*-folds. This already gives us a useful criterion for realistic inflation models. The most recent *Planck* results show a preference for a higher number of *e*-folds: $78 < N < 157$ [205].

Most models of inflation are slow-roll models, in which the Hubble rate varies slowly [185,206,207]. This model was first developed by Andrei Linde in 1982 [208]. It solved a major problem in Guth's early theory. Instead of tunnelling out of a false vacuum state, inflation occurrs by a scalar field rolling down a potential energy gradient. When the field rolls very slowly compared to the expansion of the universe, inflation occurs. Hence the name "slow-roll inflation". However, when the gradient becomes steeper, inflation ends and reheating can occur. It is beyond the scope of this review to go into the detail of the theory, but it is necessary for us to briefly refer to the link between this theory and the spectral index of primordial fluctuations, which is an important observational parameter in the Concordance Model of cosmology.

To quantify slow roll, cosmologists typically use two parameters ϵ and η which vanish in the limit that ϕ becomes constant. The first parameter is defined as:

$$\epsilon \equiv \frac{d}{dt}\left(\frac{1}{H}\right) = \frac{-\dot{H}}{aH^2}, \tag{74}$$

which is always positive, since H is always decreasing. The second complementary variable which defines how slowly the field is rolling is:

$$\eta \equiv \frac{1}{aH\dot{\phi}^{(0)}}\left[3aH\dot{\phi}^{(0)} + a^2 V'\right], \tag{75}$$

where $\phi^{(0)}$ is the zero-order field, and V is the potential.

The scalar spectral index can be defined in terms of some function, usually a polynomial, involving the two slow-roll parameters ϵ and η. As an example we shall give two such parameterisations: $n = 1 - 4\epsilon - 2\eta$ [209] and $n = 1 - 6\epsilon + 2\eta$ [206]. The rate of change of n can also be expressed in terms of inflationary parameters: $dn/d\ln k = 16\epsilon\eta + 24\epsilon^2 + 2\xi^2$ [210], where

$$\xi^2 \equiv m_{Pl}^4 \frac{V'(d^3/d\phi^3)}{V^2}, \tag{76}$$

m_{Pl} being the reduced Planck mass (4.342×10^{-6} g).

Therefore, by extracting the values of ϵ and η from the data, using methods such as weak lensing, we can directly probe the potential of of the inflaton field. Likewise for the tilt or spectral index of the primordial power spectrum. Slow-roll inflation predicts that the spectral index of primordial fluctuations should be slightly less than 1. The reason for this is simple. For inflation to end, the Hubble parameter H has to change in time. This time-dependence changes the conditions at the time when each fluctuation mode exits the Hubble horizon and therefore gets translated into a scale-dependence.

Inflation accounts for the observed spatial flatness of the universe, and the absence of magnetic monopoles. These predictions have been confirmed by various probes, most notably by precision measurements of CMB anisotropies, starting with the Cosmic Background Explorer (COBE) in 1992 [211–213], then with the Wilkinson Microwave Anisotropy Probe (WMAP) which ran for nine years from 2001 to 2010. WMAP data placed tight constraints on the predicted burst of growth in the very early universe, providing compelling evidence that the large-scale fluctuations are slightly more intense than the small-scale ones, which is a subtle prediction of many inflation models [162,214–219]. Significantly, WMAP found evidence that the scalar spectral index is less than 1 (a 2σ deviation), implying a deviation from scale invariance for the primordial power spectrum. As explained above, this is a major prediction of inflation, and this observation reinforced the evidence in favour of the theory. Conclusive proof of a scale-dependent primordial power spectrum (a 5σ deviation from $n_s = 1$) was provided by the *Planck* CMB anisotropy probe in 2013 [220,221] and confirmed in 2015 [151,205].

One current experimental challenge is to observe the B-modes of polarisation of the CMB caused by primordial gravitational waves produced by inflation. Their detection by the BICEP2 experiment was announced in early 2014. However, more accurate modelling of the signal over the next few months, which allowed the observation to be explained by polarised dust emission in our Galaxy, decreased the statistical confidence of the initial result [222]. This was confirmed by *Planck* data in 2016 [223]. Upcoming large-scale structure surveys, such as the *Euclid* satellite mission, or 21-cm radiation surveys such as the Square Kilometre Array, may measure the power spectrum with greater precision than current CMB probes, and could provide further evidence in favour of the inflationary paradigm [224–226].

9. The First Unknown Component: Dark Matter

The first evidence for Dark Matter came from astronomy rather than cosmology. Newtonian physics and General Relativity both provide very precise rules for the dynamics of galaxies: the mass determines the rotation velocity. Starting in the 1920s, stronomers noticed that amount of visible matter in galaxies did not match the observed rotation curves. These curves relate the tangential velocity of the constituent stars (or gas) about the centre of the galaxy to their radial distance. Observations of the velocities of globular clusters about galaxies showed that at large radii the velocities are approximately constant, implying that the amount of mass in the galaxies is much higher than the visible mass.

The first suggestion of the existence of hidden matter, motivated by stellar velocities, was made by Jacobus Kapteyn in 1922 [227]. Radio astronomy pioneer Jan Oort also hypothesized the existence of Dark Matter in 1932 [228]. Oort was studying stellar motions in the local galactic neighbourhood and found that the mass in the galactic plane must be greater than what was observed. This measurement was later determined to be erroneous.

In 1933, Fritz Zwicky, who studied galactic clusters while working at the California Institute of Technology, made a similar inference [229]. Zwicky applied the virial theorem to the Coma galaxy cluster and obtained evidence of unseen mass that he called *dunkle Materie* in German, or "Dark Matter". Zwicky estimated its mass based on the motions of galaxies near its edge and compared that to an estimate based on its brightness and number of galaxies. He estimated that the cluster had about 400 times more mass than was visually observable. The gravity effect of the visible galaxies was far too small for such fast orbits, thus mass must be hidden from view. Based on these conclusions, Zwicky inferred that some unseen matter provided the mass and associated gravitation attraction to hold the cluster together.

In 1937, Zwicky made the bold assertion that galaxies would be unbound without some form of invisible matter [230]. Zwicky's estimates were off by more than an order of magnitude, mainly due to an obsolete value of the Hubble constant. The same calculation today shows a smaller fraction, using greater values for luminous mass. However, Zwicky did correctly infer that the bulk of the matter was dark.

More evidence started to accumulate for the existence of some non-emitting component which was now being called Dark Matter [231]. In 1959, Kahn and Woltjer [232] pointed out that the motion of Andromeda towards us implied that there must be Dark Matter in our Local Group of galaxies. Dynamical evidence for massive Dark Matter halos around individual galaxies came later, starting in the 1970s, when rotation curve data from multiple galaxies confirmed the Dark Matter halo hypothesis [233–236]. Like baryonic matter, Dark Matter is a fluid with vanishingly small pressure. Unlike baryonic matter, it has no interaction with photons, making it both dark and transparent. It also has a vanishingly small self-interaction beside gravity. One result of this is that the Dark Matter halos surrounding galaxies are rounder than the galaxies themselves [237] .

In the last few decades, cosmology has contributed one important piece of information: the amount of Dark Matter. The observed value of the matter density in the universe is $\Omega_m = 0.3089 \pm 0.0062$. However, the density of baryonic matter is $\Omega_b = 0.0486 \pm 0.0010$ [151]. The missing mass is made up of Dark Matter.

The name "Dark Matter" is an indication of its nonbaryonic nature: it cannot be observed by emission of photons, so observers need to find a way around this problem. Current evidence for the existence of Dark Matter comes from a variety of sources besides galactic dynamics [231]. The two most important ones are CMB anisotropies and gravitational lensing. In addition, Big Bang nucleosynthesis provides evidence that some of the Dark Matter may be baryonic. The inventory of observed baryons in the local universe falls short of the total anticipated abundance from Big Bang nucleosynthesis, implying that most of the baryons in the universe are unseen [238].

Anisotropies in the CMB are related to anisotropies in the baryonic density field by the Sachs-Wolfe effect [239]. This means that the baryon density field variation at the time of decoupling can be linked to CMB anisotropies. If all matter were made of baryons, the amplitude of the density fluctuations should have reached $\delta \sim 10^{-2}$ at the present epoch. However, we observe structures with $\delta \gg 1$ at the present epoch (e.g., galaxies and galaxy clusters). The discrepancy can only be explained by the presence of additional matter, which created potential wells for the baryons to fall into after decoupling. These potential wells would have had to be formed by a weakly interacting fluid that decoupled well before baryons and began to cluster much earlier. Such a fluid would only interact via the gravitational and possibly the weak nuclear force. As the baryons accumulated in the potential wells, their pressure would have built up, leading to oscillations in the baryon fluid, termed "baryon acoustic oscillations" (BAO) [240,241]. These oscillations leave an imprint on the CMB power spectrum, which has been confirmed observationally, and which constrains the mass density, leading to a further confirmation of the existence of this missing mass.

The phenomenon of gravitational lensing includes cosmic shear, weak lensing, cosmic magnification. Although the theory of cosmic shear had been worked out from the 1960s to the early 1990s [242], the first detection had to await the development of instruments sensitive enough to make the required observations, and image analysis software to accurately correct for unwanted effects when measuring the shapes of galaxies. In March 2000, four groups independently announced the first discovery of cosmic shear [243–246]. Since then, cosmic shear has established itself as an important technique in observational cosmology.

Gravitational lensing shows that the amount of lensing of galaxies around galaxy clusters is too high to be caused by the visible matter. Apart from the stars themselves, a galaxy cluster also has a gas component, but X-ray observations show that this is still not enough to account for the extra mass. The cluster must therefore have a non-emitting halo of Dark Matter around it.

Various Dark Matter candidates have been proposed [247–251]. However, all these candidates have one common characteristic: a very small reaction cross-section, making them extremely difficult to detect directly [252–255]. Experiments have, however, placed limits on the mass of Weakly Interacting Massive Particles (WIMPs), which are the current best candidate for Dark Matter (together with axions). WIMPs are an entire new class of fundamental particle outside of the Standard Model that result from supersymmetry [256,257]. These results show that even the lightest Dark Matter particle should have a mass which is not below $\sim$10 MeV. We also know that $\Omega_{CDM} = 0.2589 \pm 0.0057$ [151]. The conclusion is that $\Omega_\nu \ll \Omega_m$, implying that hot Dark Matter (i.e., neutrinos) cannot account for the Dark Matter density Ω_{CDM}.

An alternative to Dark Matter is to explain the missing mass by means of a modification of gravity at large distances or more specifically at small accelerations. In 1983, Morderhai Milgrom proposed a phenomenological modification of Newton's law which fits galaxy rotation curves [131–133]. The theory, known as Modified Newtonian Dynamics (MOND) automatically recovers the Tully-Fischer law. The theory modifies the acceleration of a particle below a small acceleration $a_0 \sim 10^{-10}$ ms^{-2}, which therefore enters the theory as a universal constant. The gravitational acceleration at large distances then reads $a = \sqrt{GMa_0}/r$ at large distances, instead of the Newtonian law $a = \sqrt{GM/r^2}$.

There are two main difficulties with MOND. First, it does not explain how galaxy clusters can be bound without the presence of some hidden mass [258,259]. Second, attempts to derive MOND from a consistent relativistic field theory have failed. One such attempt is the Tensor-Vector-Scalar Theory (TeVeS) [135] is more successful and actually relativistic but not apparently necessary since it still requires dark energy and Dark Matter. Many models are unstable [260], or require actions which depend on the mass M of the galaxy, thereby giving a different theory for each galaxy. Moreover, modified gravity theories have serious difficulties reproducing the CMB power spectrum and the evolution of large-scale structure [261–263].

The greatest challenge to modified gravity theories, and also the clearest direct evidence of Dark Matter, comes from observations of a pair of colliding galaxy clusters known as the Bullet Cluster [264] in which the stars and Dark Matter separate from the substantial mass of ionised gas. The Dark Matter follows the less substantial stars and not the more massive gas.

The modifications of gravity proposed as alternative to the Dark Matter paradigm illustrate the need for tests of GR at large distances and low accelerations. They also illustrate the problems faced by models which favour goodness of fit over parsimony. Modified gravity theories can give an excellent phenomenological fit through an adjustment of the values of the extra parameters, but there is no universal principle to determine these values. This requirement for simplicity and predictivity is met by General Relativity.

10. The Second Unknown: Dark Energy and the New Wave Alternative Theories

The current motivation for alternative theories seems to be the search for an explanation of the observed accelerating expansion of the universe. Let us consider the justification for the dark energy paradigm within the inflationary ΛCDM model, and the process which led to its acceptance by the scientific community (see [265]).

Round about the time that GR was developed, the universe was thought to be static. There was no compelling reason to think otherwise. Einstein realised that his equations implied a non-static universe, so in 1917 he revised his field equations of GR to read [138]:

$$G_{\mu\nu} - \Lambda g_{\mu\nu} = G_{\mu\nu} - 8\pi G \rho_\Lambda g_{\mu\nu} = 8\pi G T_{\mu\nu} \tag{77}$$

where $\rho_\Lambda = \Lambda/8\pi G$ is proportional to the cosmological constant Λ.

It can be seen from this equation that Einstein did not consider the cosmological constant to be part of the stress-energy term. One could, of course, put $\rho_\Lambda g_{\mu\nu}$ on the right-hand side of the equation and count it as part of the source term of the stress-energy tensor and simply consider ρ_Λ to be the vacuum energy. This is not just a semantic distinction. When ρ_Λ takes part in the dynamics of the universe, then the field equation is properly written with ρ_Λ, or its generalisation, as part of the stress-energy tensor:

$$G_{\mu\nu} = 8\pi G(T_{\mu\nu} + \rho_\Lambda g_{\mu\nu}). \qquad (78)$$

The equation describing gravity is then unchanged from its original form—there is no new physical theory. Instead, there is a new component in the content of the universe.

This component must satisfy Special Relativity (that is, an observer can choose coordinates so that the metric tensor has Minkowskian form). An observer moving in spacetime in such a way that the universe is observed to be homogeneous and isotropic would measure the stress-energy tensor to be

$$T_{\mu\nu} = \begin{pmatrix} \rho & 0 & 0 & 0 \\ 0 & p & 0 & 0 \\ 0 & 0 & p & 0 \\ 0 & 0 & 0 & p \end{pmatrix}. \qquad (79)$$

This means that the new component in the stress-energy tensor looks like an ideal fluid with negative pressure:

$$p_\Lambda = -\rho_\Lambda. \qquad (80)$$

In modern concordance cosmology, this component is usually termed "dark energy". If the equation of state parameter of dark energy is constant, i.e., $w(z) = -1$, then its energy density will be constant regardless of the expansion of the universe.

Einstein inserted the cosmological constant because he felt that the non-static universe predicted by the formalism of GR was incorrect, given the data available in 1917 [138]. At the time, observations of the universe were limited primarily to stars in our own galaxy, with observed low velocities, so there was solid observational evidence justifying the assumption that the universe was static. Einstein's goal was to obtain a universe that satisfied Mach's principle of the relativity of inertia. However, observational evidence started to accumulate for another paradigm. In 1917, Vesto Slipher [266] published his measurements of the spectra of spiral nebulae, which showed that most were shifted towards the red. The breakthrough came when the linear redshift-distance relation was formulated by Hubble [141], who showed that the universe was expanding. Einstein then dropped his support for the cosmological constant.

In the FLRW cosmological model, the expansion history of the universe is determined by the mass density of the different components, whose sum is normalised to unity:

$$\Omega_{m,0} + \Omega_{rad,0} + \Omega_{X,0} + \Omega_{k,0} = 1, \qquad (81)$$

where the 0 subscript indicates the present epoch. We use the term Ω_X to show that this equation does not assume anything about the nature of the additional energy component (dark energy). In fact we could have used Ω_Λ or Ω_{DE} in the current Concordance Model.

Big bang nucleosynthesis and observations of large scale structure provide a good determination of the mass content of the universe, allowing Ω_m and Ω_{rad} to be fixed. However, observations in the 1980s and 1990s started to show inconsistencies with the cosmological model at the time, which was a matter-dominated, expanding universe with a present-epoch Hubble constant of $H_0 \simeq 0.7 \, \text{kms}^{-1} \, \text{Mpc}$ and $\Omega_\Lambda = 0$ [267,268]. This was the so-called "age problem", where the predicted age of the universe seemed to be younger than the age of the oldest stars. Angular-diameter distances to the last scattering surface at $z = 1100$ measured from the CMB are in fact 1.7 times smaller than those predicted by an isotropic and homogeneous universe containing only pressureless matter (see [269]). Since the inflationary scenario, which by then was well established, predicts a flat $\Omega_{total} = 1$ universe, there was a problem with the cosmological model.

It was realised that one of the three assumptions of the cosmological model had to be wrong. Either the universe contains exotic matter with a negative pressure, or standard General Relativity is wrong, or the universe is not homogeneous and isotropic. The solution could also lie in some combination of the three. Most of the research since the late 1990s has followed the first approach, and the term "Concordance Model" refers to an FLRW universe, following General Relativistic cosmology, containing dark energy.

Within the FLRW framework, two main proposals were put forward: one was ΛCDM, in which there is a contribution to the energy density from a term similar to the cosmological constant (or the cosmological constant itself), and the other was $\nu + $CDM, where the missing mass came from massive neutrinos ($m_\nu \simeq 7$eV) (e.g., [270,271]).

The first strong evidence of dark energy came in 1998 and 1999, when observations of the luminosities of type Ia supernovae indicated that the expansion of the universe is accelerating [272,273]. Concurrently, other observations constrained the neutrino mass to $m_\nu \ll 7$eV, thus discounting the $\nu + $CDM model and confirming ΛCDM as the Concordance Model (e.g., [265,274]). It is not clear when the term "dark energy" was first used, but it seems to have been around 1998. The term is analogous to "Dark Matter", which had been in use for some time [275].

Since then, numerous observations have confirmed cosmic acceleration, including supernovae, the cosmic microwave background, large-scale structure and baryon acoustic oscillations (see, e.g., [276–278]). The values of the present epoch matter and radiation components are well established:

$$\Omega_{m,0} \equiv \frac{8\pi G \rho_{m,0}}{3H_0^2} \sim 0.3, \quad \Omega_{\mathrm{rad},0} \equiv \frac{8\pi G \rho_{\mathrm{rad},0}}{3H_0^2} \sim 1 \times 10^{-4}, \tag{82}$$

where H_0 is the present value of the Hubble parameter $H(a = 1)$.

The data also indicate that the universe is currently nearly spatially flat:

$$|\Omega_K| \ll 1. \tag{83}$$

This is normally taken to imply that the spatial curvature $K = 0$, since

$$\Omega_{k,0} = 0 \equiv \frac{-K}{a_0^2 H_0^2} \sim 0. \tag{84}$$

Thus it also justifies the inflationary paradigm. However, inflation only tells us that $\Omega_K \to 0$, so that the curvature may have had a nonzero value in the past. In the present universe, however, the distinction is negligible. In any case, Equation (81) implies that there has to be a nonzero Λ (a constant term added to the Einstein equation) such that

$$\Omega_{\Lambda,0} \equiv \frac{\Lambda}{3H_0^2} \sim 0.7. \tag{85}$$

Inserting these values into the Friedmann equation leads to the dramatic conclusion that the expansion of the universe is accelerating:

$$\ddot{a}_0 = H_0^2 \left(\Omega_\Lambda - \frac{1}{2}\Omega_m - \Omega_{\mathrm{rad}} \right) > 0, \tag{86}$$

where a_0 is the present value of the scale factor $a(t)$.

Note that this conclusion only holds if the universe is homogeneous and isotropic (i.e., a Friedmann-Lemaître model). In such a universe, the distance to a given redshift z and the time elapsed since that redshift are tightly related via the only free function, $a(t)$. If the universe is isotropic around us, but not homogeneous, that is, a non-Copernican Tolman-Bondi-Lemaître model [279], then this relation would be lost and present data might not imply acceleration. A Copernican model where this relation again breaks down is the inhomogeneous universe, where the acceleration can be produced via nonlinear averaging—the backreaction of inhomogeneities.

Dark energy is a fluid component whose equation of state is:

$$p_{DE} = wc^2 \rho_{DE} . \tag{87}$$

This is the equation of state in its most general form, since w can be any function of redshift, scale factor or cosmic time, with the constraint that $w \leq 0$ (i.e., the fluid has a negative pressure). Assuming that $w = w(a)$, we have the following density-scale relation:

$$\rho_{DE}(a) = \rho_{DE,0} e^{-3 \int_a^1 [1+w(a')] \, d(\ln a')} . \tag{88}$$

It can be seen that in the special case of a constant $w = -1$, the fluid equation implies that the density is constant.

If dark energy is a cosmological constant, it still leaves the question of its physical nature. Its observed value of $\Lambda \approx 3 \times 10^{-122} c^3 / \hbar G$ is so small that it is hard to interpret as the vacuum energy.

One possibility is that the final value of the cosmological constant is zero, and that cosmic acceleration is due to the potential energy of a scalar field, with some sort of mechanism to dynamically relax it to a small value. This notion leads to models of dark energy which invoke a slowly-rolling cosmological scalar field to source accelerated expansion, similar to cosmological inflation.

How can a scalar field drive cosmic acceleration? The action of a scalar field minimally coupled to Einstein gravity is

$$S = \int d^4 x \sqrt{-g} \left(\frac{m_{Pl}^2 R}{2} - \frac{1}{2} (\partial \phi)^2 - V(\phi) \right) . \tag{89}$$

The stress-energy tensor for the scalar field is given by

$$T_{\mu\nu}^{\phi} = \partial_\mu \phi \partial_\nu \phi - g_{\mu\nu} \left(\frac{1}{2} (\partial \phi)^2 + V(\phi) \right) . \tag{90}$$

For a homogeneous field such that $\phi = \phi(t)$, a cosmological scalar acts like a perfect fluid with equation of state $w = P/\rho$ given by

$$w_\phi = \frac{\frac{1}{2} \dot{\phi}^2 - V(\phi)}{\frac{1}{2} \dot{\phi}^2 + V(\phi)} . \tag{91}$$

The observed expansion leads us to a value of $w_\phi \simeq -1$, which requires a very slowly-rolling field: $\dot{\phi}^2 \ll V(\phi)$. There has also been a lot of interest in constructing quintessence models which can produce an equation of state of the "phantom" type ($w_\phi < -1$).

The equation of state of dark energy that has the potential to distinguish between dark energy candidates. The most important distinction that can be made between different dark energy models is whether the energy density of this component is constant, filling space homogeneously, or whether it is some form of quintessence field whose energy density can vary in time and space. There is a multitude of alternative models, such as $f(R)$ [280–282], or Chameleon Models [283,284]. It is therefore useful to consider the redshift evolution of w, so that w is an arbitrary function of redshift z. There are a number of different parameterisations of $w(z)$ ([285], and references therein).

The most common parameterisation is sometimes termed the Chevallier-Linder-Polarski (or CPL) parameterisation [285,286]:

$$w(z) = w_0 - \frac{dw}{da} (1 - a) , \tag{92}$$

where the scale factor $a = 1/(1 + z)$. If we define

$$w_a = -\frac{dw}{a \, d \ln a} , \tag{93}$$

then the equation becomes

$$w(a) = w_0 + w_a(1 - a) \,, \tag{94}$$

which is the most commonly used form.

It has been shown that this parameterisation is stable and robust over large redshift ranges [287]. A wide range of functional forms of $w(a)$ can be parameterised by the $w_0 - w_a$ combination. However, there are some dark energy models which it cannot reproduce (see [288–291]).

The problem with the dark energy paradigm, stated simply, is that the parameters are not constrained well enough to rule out certain models. We have fairly good bounds on the dark energy density, but the dark energy equation of state is still poorly constrained. Even for a constant w model, corresponding to ΛCDM, the bounds are such that a time-varying $w(a)$ could mimic a constant w, thereby disguising underlying physics (see [292–294]).

11. The Evolution of Large-Scale Structure

After the epoch of matter-radiation equality, and before the onset of dark energy domination, the mass-energy content of the universe became dominated by matter. From an initially smooth background (as evidenced by CMB observations), structures have evolved to a scale of more than 100 Mpc, with the term "large scale structure" being used to refer to objects modelled on this scale. At this scale, the mass-energy inhomogeneities can be modelled as perturbations on a homogeneous and isotropic unperturbed background spacetime. Below this scale we observe galaxy clusters, individual galaxies, and stars. The model of structure formation must be accurate enough to provide an good description of the universe on a wide range of scales.

The standard model for the formation of structure assumes that at some early time there existed small fluctuations, which grew by gravitational instability. The origins of these fluctuations are unclear, but they are thought to arise from quantum fluctuations of the primordial universe, uncorrelated and with Gaussian amplitudes, which were then amplified during a later inflationary phase [185]. The assumption that the amplitudes of the relative density contrasts is much smaller than unity means that we can think of the primordial fluctuations as small perturbations on a homogeneous and isotropic background density. This ensures that we can describe them using linear theory.

Heuristically, the mechanism of structure formation can be understood in terms of gravitational self-collapse. Matter collapses gravitationally around initial mass overdensities. This increases the relative density of that region, causing further collapse of more matter, and amplifying the effect. The linear theory of structure formation needs to be relativistic, because the perturbations on any length scale are comparable or larger than the horizon size at sufficiently early times. The horizon size is defined as the distance ct which light can travel in time t since the Big Bang. Dissipative effects and pressure also affect structure formation, as explained below (for details of the theory, see [295,296]).

The relative density is the density ρ at a particular point in space x relative to the mean $\bar{\rho}$ at some time parameterised by the scale factor a, and can be expressed as a dimensionless density contrast:

$$\delta(\mathbf{x}, a) = \frac{\rho(\mathbf{x}, a) - \bar{\rho}(a)}{\bar{\rho}(a)} \,. \tag{95}$$

This quantity can be understood as the dimensionless density perturbation of some background matter distribution.

There are two types of density perturbations that can occur within a matter-radiation fluid. If the fluid could be compressed adiabatically in space, the perturbations have a constant matter-to-radiation ratio everywhere. Since the energy density of radiation is proportional to T^4, and the number density is proportional to T^3, the energy densities of radiation and matter are related by:

$$\delta_{\text{rad}} = \frac{4}{3}\delta_{\text{m}} \,. \tag{96}$$

Isocurvature perturbations occur when the entropy density is perturbed, but not the energy density. Since the total energy density remains constant, there is no change in the spatial curvature and

$$\rho_{rad}\delta_{rad} = \rho_m \delta_m \,. \tag{97}$$

Perturbations can occur at different scales, or 'modes'. The latter term is used when the amount of perturbation on a particular scale is expressed using Fourier analysis. The Fourier transform pair of $\delta(\mathbf{x})$ is:

$$\hat{\delta}(\mathbf{k}) = \int d^3x \delta(\mathbf{x}) e^{i\mathbf{k}.\mathbf{x}} \quad ;$$

$$\delta(\mathbf{x}) = \int \frac{d^3k}{(2\pi)^3} \hat{\delta}(\mathbf{k}) e^{-i\mathbf{k}.\mathbf{x}}, \tag{98}$$

with each mode assumed to evolve independently. In the Einstein-de Sitter regime, linear adiabatic perturbations scale with time as follows:

$$\delta \propto \begin{cases} a(t)^2 & \text{(radiation domination)} \\ a(t) & \text{(matter domination)} \end{cases} \tag{99}$$

while isocurvature perturbations are initially constant and then decline:

$$\delta \propto \begin{cases} \text{constant} & \text{(radiation domination)} \\ a(t)^{-1} & \text{(matter domination)}\,. \end{cases} \tag{100}$$

In both cases, the overall shape of the spectrum of the perturbations over all modes is preserved, while the amplitude changes with time. The evolution described above is affected on small scales by a number of processes.

11.1. Evolution on Small Scales

During the radiation-dominated epoch the growth of certain modes is suppressed. This behaviour can be modelled in terms of the horizon scale $\lambda_H(a)$, which is the distance ct that light could have travelled since the initial singularity (a comoving horizon size). A mode k is said to enter the horizon when $\lambda = \lambda_H(a)$, where $\lambda = (2\pi)/k$. If $\lambda < \lambda_H(a_{eq})$ then a mode enters the horizon during the radiation-dominated epoch. The time scale for collapse of matter during this epoch is larger than the typical expansion time scale ($t \sim 1/H(a)$) due to the relatively rapid expansion $\rho_{rad} \propto a^{-4}$. The growth of these modes is therefore suppressed. After the epoch of matter-radiation equality ($a = a_{eq}$), these perturbations can then start to collapse gravitationally. We can define the suppression factor for a particular mode as the factor by which the amplitude is reduced had it not entered the horizon:

$$f_{sup} = \left(\frac{a_{enter}}{a_{eq}}\right)^2 = \left(\frac{k_0}{k}\right)^2 \tag{101}$$

where the mode evolves as $\propto a^2$ until it enters the horizon at a_{enter} and is suppressed until a_{eq}, when its evolution resumes as $\propto a$. The second equality in the above equation comes from applying an Einstein-de Sitter approximation where $k_0 = 1/\lambda_H(a_{eq})$ (see [297]).

Pressure opposes gravitational collapse for modes with a wavelength less than the Jeans length [298], sometimes called the free-streaming scale, defined as

$$\lambda_J = c_s \sqrt{\frac{\pi}{G\rho}}\,. \tag{102}$$

During the radiation-dominated epoch, the sound speed $c_s = c/\sqrt{3}$ and the Jeans length is always close to the horizon size. The Jeans length then reaches a maximum at $a = a_{eq}$ and then begins to decrease as the sound speed declines. This means that on scales larger than the comoving horizon size, perturbations are only affected by gravity, and the spectrum starts to turn over at this point (where the effects of pressure begin to dominate). The comoving horizon size at z_{eq} is given by:

$$R_0 r_H(z_{eq}) \approx \frac{16.0}{\Omega_m h^2} \text{Mpc}. \tag{103}$$

Another important scale occurs where photon diffusion erases perturbations in the matter-radiation fluid. This process is termed Silk damping [299]. The scale at which it occurs is characterised by the distance travelled by the photon in a random walk by the time of last scattering:

$$\lambda_S \approx 16.3(1+z)^{-5/4}(\Omega_b^2 \Omega_m h^6)^{-1/4} \text{Gpc}. \tag{104}$$

All of the effects mentioned above are particularly important where the behaviour of massive neutrinos is concerned. Heuristically we can understand the complexity of their behaviour by considering them as a component whose equation of state changes as the universe evolves. From a component which behaves like photons (since the particles have a very small mass and relativistic speeds), massive neutrinos lose energy and start behaving like baryonic matter [300,301].

11.2. Growth oF Perturbations in the Presence of Dark Energy

All of the above effects were described in an Einstein-de Sitter universe. In a universe with a smooth non-clustering dark energy component below the horizon scale, the matter perturbation fields evolves according to:

$$\ddot{\delta} + 2H\dot{\delta} - (3/2)H^2\Omega_m\delta = 0$$
$$\delta'' + (2-q)a^{-1}\delta' - (3/2)\Omega_m a^{-2}\delta = 0, \tag{105}$$

where a dot denotes a time derivative and a dash denotes a derivative with respect to a. The term q is the deceleration parameter. This can be interpreted in the following way: the perturbations grow according to a source term which involves the amount of matter (Ω_m) but the growth is suppressed by the friction term due to the expansion of the universe. The latter is also known as the Hubble drag.

If we define the growth as the ratio of the amplitude of a perturbation at a time a to some initial amplitude, i.e.,

$$D(a) = \frac{\delta(a)}{\delta(a_{initial})}, \tag{106}$$

the equation becomes, for a general dark energy scenario where $w = w(a)$ (see [302])

$$D'' + \frac{3}{2}\left(1 - \frac{w(a)}{1+X(a)}\right)\frac{D'(a)}{a} - \frac{3}{2}\left(\frac{X(a)}{1+X(a)}\right)\frac{D}{a^2} = 0, \tag{107}$$

where

$$X(a) = \frac{\Omega_m}{\Omega_{DE}} e^{-3\int_a^1 d\ln a' w(a')} \tag{108}$$

is the ratio of the matter density to the dark energy density. For large X (i.e., $\Omega_m \sim 1$ where $\Omega_{DE} \sim 1 - \Omega_m$) we recover the matter-dominated behaviour ($D \sim a$). To parameterise deviations from this behaviour we define the "normalised growth" as $G = D/a$. The evolution equation is then:

$$G'' + \left[\frac{7}{2} - \frac{3}{2}\left(\frac{w(a)}{1+X(a)}\right)\right]\frac{G''}{a} + \frac{3}{2}\left(\frac{1-w(a)}{1+X(a)}\right)\frac{G}{a^2} = 0. \tag{109}$$

This equation allows us to physically interpret the effects of dark energy. In the presence of dark energy, the Hubble drag term is increased, so that growth is suppressed in a universe with an accelerating expansion. This is similar to the suppression due to radiation dominance.

11.3. The Power Spectrum of Matter

In an FLRW universe, the homogeneity and isotropy assumption means that any statistical properties must also be homogeneous and isotropic. The implication for the matter perturbation field is that its Fourier modes must be uncorrelated (due to homogeneity). Usually, we assume that the mode amplitudes are Gaussian. This assumption is well motivated since the theory for the seed fluctuations assumes that they have a quantum origin. Due to the central limit theorem, the sum of a sufficiently large number of mode amplitudes will tend towards a Gaussian distribution (see, e.g., [303,304]).

Such a field, with uncorrelated modes, and a Gaussian distribution of mode amplitudes is called a Gaussian random field, and can be entirely described by its two-point correlation function:

$$\langle \delta(\mathbf{x}) \delta^*(\mathbf{y}) \rangle = C_{\delta\delta}(|\mathbf{x} - \mathbf{y}|). \tag{110}$$

The angled brackets denote an ensemble average (an average over a multitude of realisations). The value of δ at a given point in the universe will have a different value in each realisation, with a variance $\langle \delta^2 \rangle$. Since we can only observe one realisation of our universe (in other words, at most only a finite region in this one universe), we apply the ergodic principle: The average over a sufficiently large volume is equal to the ensemble average.

In Fourier space, the correlation function can be written as:

$$\langle \hat{\delta}(\mathbf{k}) \hat{\delta}^*(\mathbf{k}') \rangle = \int d^3 x e^{i\mathbf{k}.\mathbf{x}} \int d^3 x' e^{-i\mathbf{k}'.\mathbf{x}'} \langle \delta(\mathbf{x}) \delta^*(\mathbf{x}') \rangle. \tag{111}$$

Replacing $\mathbf{x}' = \mathbf{x} + \mathbf{y}$, and substituting Equation (110), this can be written as:

$$\langle \hat{\delta}(\mathbf{k}) \hat{\delta}^*(\mathbf{k}') \rangle = \int d^3 x e^{i\mathbf{k}.\mathbf{x}} \int d^3 y e^{-i\mathbf{k}'.(\mathbf{x}+\mathbf{y})} C_{\delta\delta}(|\mathbf{y}|) \tag{112}$$

$$= (2\pi)^3 \delta_D(\mathbf{k} - \mathbf{k}') \int d^3 y e^{-i\mathbf{k}.(\mathbf{y})} C_{\delta\delta}(|\mathbf{y}|) \tag{113}$$

$$= (2\pi)^3 \delta_D(\mathbf{k} - \mathbf{k}') P_\delta(|\mathbf{k}|). \tag{114}$$

The power spectrum has been defined as the Fourier transform of the correlation function:

$$P_\delta(|\mathbf{k}|) = \int d^3 y e^{i\mathbf{k}.(\mathbf{y})} C_{\delta\delta}(|\mathbf{y}|). \tag{115}$$

The standard convention in cosmology is to abbreviate $P_\delta(|\mathbf{k}|)$ to $P(k)$, where $k = |\mathbf{k}|$. The power spectrum can be expressed in dimensionless form as the variance per $\ln k$, so that:

$$\Delta^2(k) = \frac{k^3 P(k)}{2\pi^2}. \tag{116}$$

11.3.1. Nonlinear Evolution

The power spectrum gives us the evolution of the initial matter density fluctuations. However, the linear evolution breaks down at small scales, when complex structures begin to form, and overdensities can no longer treated as perturbations on a smooth background. This is the nonlinear regime of gravitational evolution. The scale above which nonlinearities cannot be ignored is approximately set by $\Delta(k_{\mathrm{NL}}) \simeq 1$, which corresponds to $k_{\mathrm{NL}} \simeq 0.2\, h\,\mathrm{Mpc}^{-1}$ in most cosmological models. The standard way to model nonlinear evolution is by using phenomenological fits based on *N*-body simulations.

One strategy is to build a models based on a stable clustering hypothesis [305–307], which assumes that the nonlinear collapsed objects form isolated, virialised systems that are decoupled from the expansion of the universe.

A different approach is used in the halo model [308]. Here, the density field is decomposed into individual clumps of matter with some density profile and varying mass. By using this model to calculate the number of clumps within a given volume, the galaxy halo profile can be calculated. This is the equivalent of the power spectrum for these matter halos. A functional relation between the linear power spectrum and this halo profile is then derived and calibrated using large *N*-body simulations. This relation is then used to calculate the nonlinear power spectrum, using a fitting formula, for instance (see, e.g., [309,310]).

Whichever approach is used, it must account for a range of small-scale physical processes, such as baryonic physics, stellar formation, galactic magnetic fields, and Dark Matter and neutrino properties, which have become more significant with the ability of future astrophysical experiments to probe the nonlinear regime with ever increasing precision (see, e.g., [311–317]).

The need for realistic models of the universe has come to the fore in recent years, due the massive improvement in the quality and volume of cosmological measurements. Most of the *N*-body simulations rely on a perturbative approach. They use an FLRW cosmological background (perfectly homogeneous and isotropic)), and assume that any sub-horizon inhomogeneous structure of the universe will contribute to an average expansion on horizon-sized volumes driven by the horizon-averaged density. With the next generation of cosmological probes, these may not be accurate enough to model the universe realistically. There are several ongoing efforts to build fully relativistic, nonlinear, inhomogeneous and asymmetric models using numerical methods [318–320]. This is a significant contribution to the study of the backreaction effect and the question of the expansion rate of the universe.

11.3.2. The Primordial Perturbations

In the very early universe, the tiny initial perturbations are thought to have formed a Gaussian random field whose covariance function is diagonal and nearly scale-invariant. This form, known as the Harrison-Peebles-Zel'dovich spectrum, was assumed in most cases within the Standard Model, as it corresponds very closely to the observed power spectrum in the universe.

This type of spectrum was first proposed in the 1970s by Edward Robert Harrison [321], Yakov Zel'dovich [322], and Phillip James Edwin Peebles [323], who were working independently, as the spectrum for initial density fluctuations. This hypothesis was subsequently closely borne out by observations. The defining characteristic of a Harrison-Peebles-Zel'dovich spectrum spectrum is that it describes a fractal metric, where the degree of perturbation is the same on all scales (hence the term "scale-invariant"), so that $P(k) \propto k$. If we assume scale invariance for the power spectrum on large scales, and combine this with Equation (101), this implies the following general shape for the matter power spectrum in the Einstein-de Sitter scenario:

$$P(k) \propto \begin{cases} k & \text{for} \quad k \ll k_0 \\ k^{-3} & \text{for} \quad k \gg k_0. \end{cases} \tag{117}$$

The actual form of the spectrum depends in non-trivial ways on the parameters in the cosmological model, including the 'slope' of the initial power spectrum n_s, where $P(k) \propto k^{n_s}$. In a scale-invariant spectrum in the linear regime, the fiducial value of n_s is taken to be 1.

Cosmological observations were consistent with scale invariance up until the early data releases from the *Wilkinson Microwave Anisotropy Probe* (WMAP) [214,324], but showed some tension. Subsequent observations cast further doubts on scale invariance [216]. In 2013, the *Planck* probe, mapping the anisotropies in the CMB, led to an important and conclusive result: it ruled out scale invariance at over 5σ. The primordial power spectrum was found to be scale-dependent, with $n_s = 0.9603 \pm 0.0073$ [221]. This was confirmed by the 2015 data release, which found that $n_s = 0.968 \pm 0.006$ [205]. This significant result is a powerful demonstration of the importance of multiple sources of data, joint observations from different probes, and the increasing reliance on complex statistical techniques for cosmological model selection [325–328].

Thus it can be seen that the current Concordance Model, consisting of ΛCDM with an inflationary epoch in the early universe, was built in stages over the last 100 years. It is the result of a process of accumulation of evidence and testing of competing models and theories. At each step, theory provides the basis for adjustments to the model, and observations from different probes provided the evidence (Figure 4). The current model should in no way be seen as "true". It is merely the best model that fits all the data available so far. Future data may very well require an adjustment to the Concordance Model.

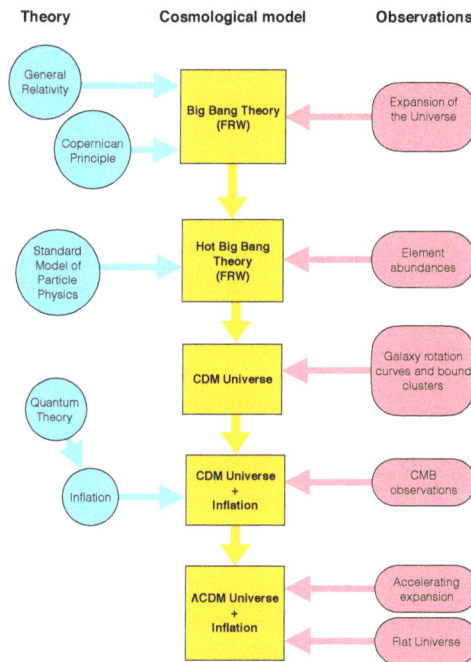

Figure 4. How the Concordance Model of Cosmology was developed. Theories and observations motivated the development of cosmological models, which were adjusted as new observations challenged the older models.

12. How Do We Test General Relativity?

The assumption of metric coupling given by Equation (5) has been tested accurately many times over the last 100 years, at scales from 10^{-4} m in laboratories [329,330], up to 10^{14} m in the Solar System [331]. These experiment test the implications of the metric coupling: the spacetime independence of non-gravitational constants, the isotropy of the coupling of all matter field to a unique metric tensor, the universality of free-fall, and gravitational redshift.

The standard way to test constraints on S_{gravity} and therefore the dynamics of General Relativity is by using the parameterised post-Newtonian (PPN) formalism [65]. This assumes that gravity is described by a metric over all scales. The idea is to write the most general form that $g_{\mu\nu}$ can take in the presence of matter, when considering correction of order $1/c^2$ with respect to the Newtonian limit. This method was first used in 1923 by Arthur Eddington [332]. In its simplest form, it provides us with two phenomenological parameters β^{PPN} and γ^{PPN} entering the Schwarzschild metric in isotropic coordinates:

$$g_{00} = -1 + \frac{2Gm}{rc^2} + 2\beta^{\text{PPN}} \left(\frac{2Gm}{rc^2}\right)^2 \quad \text{and} \quad g_{ij} = \left(1 + 2\gamma^{\text{PPN}} \frac{2Gm}{rc^2}\right)\delta_{ij}. \tag{118}$$

According to General Relativity, $\beta^{\text{PPN}} = \gamma^{\text{PPN}} = 1$. The experimental constraints on these parameters are summarised in Table 3.

Table 3. Some constraints on parameterised post-Newtonian (PPN) parameters from recent tests.

Method	Constraint	Experiment
Shift of perihelion of Mercury	$\|2\gamma^{\text{PPN}} - \beta^{\text{PPN}} - 1\| < 3 \times 10^{-3}$	Data to 1990 [333]
Lunar laser ranging	$\|4\beta^{\text{PPN}} - \gamma^{\text{PPN}} - 3\| = (4.4 \pm 4.5) \times 10^{-4}$	Data to 2004 [334,335]
Very long baseline interferometry	$\|\gamma^{\text{PPN}} - 1\| = 4 \times 10^{-4}$	Data from 1979 to 1999 [336]
Time-delay variation	$\gamma^{\text{PPN}} - 1 = (1.2 \pm 2.3) \times 10^{-5}$	Cassini spacecraft [67]
Planetary perihelion precessions	$\beta^{\text{PPN}} - 1 = (-2 \pm 3) \times 10^{-5}$ $\gamma^{\text{PPN}} - 1 = (4 \pm 6) \times 10^{-5}$	Solar System ephemerides to 2013 [337]

It should be noted that the PPN formalism assumes that there is no characteristic length scale for the gravitational interaction, and therefore it does not allow testing of finite-range effects. These too have been constrained to be very close their General Relativistic value of zero. The deviation of amplitude α from a Newton potential on a characteristic scale λ is typically $\alpha < 10^{-2}$ on scales ranging from a few millimetres to Solar System size [329]. This implies no deviation from GR over more than 15 orders of magnitude in length scale.

The Solar System tests constrain Eddington's two PPN parameters to a tiny region very close to 1. The formalism has been generalized to include eight additional phenomenological parameters [65] to describe any possible deviation from GR at the first post-Newtonian order. They have all been constrained to be very close to their General Relativistic values. The latest data use Solar System ephemerides, which include perihelion measurements for Mercury and the other planets, as well as lunar ephemerides. Independent teams of astronomers have estimated corrections to the standard first-order post-Newtonian General Relativistic formalism to be all statistically compatible with zero [337,338]. This leads us to conclude that General Relativity is the only theory consistent with Solar System experiments at the post-Newtonian order.

It should be noted that, so far, it is only the first-order post-Newtonian, static, Schwarzschild-like part of the spacetime metric that has been modelled and tested using Solar System dynamics. The first-order post-Newtonian gravitomagnetic or Lense-Thirring part of the spacetime metric has neither been modelled nor tested yet in the Solar System. However, it has recently been pointed out that this could be possible in the next few years by focussing on particular models (Sun and planets, planets and spacecraft, or planets and planets) [339]. This would open the field to the possibility of constraining the PPN parameters using a more complete model of General Relativistic effects.

What about larger or smaller scales? The size of the universe was around 10^{-35} m in the beginning, and the present size is around 10^{26} m. GR remains untested at these extreme scales. But one fundamental assumption of GR is that it describes gravity at all scales (i.e., the theory assumes the analytical continuity of solutions).

The evolution of the universe itself therefore provides a useful test of General Relativity. The challenge lies in testing a wide range of potentials (weak and strong field regimes) over cosmological scales. The strong field regime is tested using compact objects. However, over cosmological scales, the kind of objects that would produce gravitational potentials approaching the strong field regime simply do not exist. One solution is to observe the evolution of the universe and check whether the evolution of large-scale structure corresponds to the predictions of General Relativity.

13. Cosmological Tests

General Relativity has been submitted to 100 years of Solar System tests, which it has passed with flying colours. The discoveries of the last two decades: cosmic acceleration, the scale dependence of the primordial power spectrum, and also the open question of Dark Matter, have made it clear that GR must also be tested at astrophysical and cosmological scales.

This presents a serious challenge. We derive our knowledge of the universe from measurements of distances and times, and statistical properties like the distribution of matter. The main difficulty in extending the Solar System tests to cosmological scales is that these measurements depends strongly on the construction of cosmological models.

These models depend on four hypotheses:

(1) A theory of gravitational interactions.
(2) A description of the matter in the universe and the non-gravitational interactions such as electromagnetic emissions.
(3) A hypothesis on the symmetry.
(4) A hypothesis on the topology, or the global structure of the universe.

Some of the hypotheses are hard to verify, and some have unverifiable implications. Assuming the symmetry of our solutions means that we also assume the laws of physics are the same throughout the universe, including its unobservable part outside the cosmological horizon (delimited by a radius of around 15.7 Gpc). This is a very strong assumption, but one that is unverifiable.

Any cosmological model needs all four hypotheses. The first two are the physical theories, but their equations cannot be solved without some kind of assumption on the symmetry of the solutions, given by the third hypothesis. The fourth hypothesis is then an assumption on the global properties of these solutions.

The simplest ΛCDM Concordance Model assumes that gravity is described by General Relativity (Hypothesis 1), and that the universe contains the particles and fields of the Standard Model of particle physics [171], together with Dark Matter and a cosmological constant (Hypothesis 2). The Einstein equations require an effective stress-energy tensor averaged out on large scales, so the model requires an extra assumption on the averaging procedure. This averaging and the validity or otherwise of this assumption is the subject of research on the backreaction effect. The model assumes the Copernican Principle (Hypothesis 3), and the continuity of the solutions of Einstein equations across all spatial sections (Hypothesis 4).

The number of alternative cosmological models proposed, especially since the 1990s, is too numerous to list here. Many of them are impossible to rule out simply by fitting them to the data, because their predictions are so close to those of ΛCDM. It is easier to test the four hypotheses rather than trying to test the observables predicted by the models. As with any null test, a significant violation would indicate the need to modify one or more hypotheses.

As far as tests of GR (the theory of gravity in ΛCDM) are concerned, many experiments test more than one assumption. We list the main experiment types, and the assumptions they test, in Table 4.

Table 4. What we are actually testing. Experimental tests of General Relativity (GR) often probe more than one assumption, and isolating the effects is a challenge in itself.

Experiment	Assumption Tested
Solar System tests	Metric coupling
Quadrupolar shift of nuclear energy levels	Isotropy [340–342]
Lunar laser ranging and orbiting gyroscopes	Universality of freefall [343] and structure of metric
Space-borne clocks	Gravitational redshift [344]
Shift in perihelion of planets	Structure of metric [345]
Time invariance of physical constants	Metric coupling [346]
Detection of gravitational waves	Lorenz gauge condition and inhomogeneous wave equation

13.1. Testing the Description of Matter and Non-Gravitational Interactions

General Relativity can be tested by measuring the fundamental constants of the theory [346–349]. Any variation would require a modification of GR [350,351]. A local measurement of a fundamental constant, such as a determination of the fine-structure constant from the Oklo phenomenon [352–354], is actually a cosmological-scale measurement along the time dimension. Astrophysical probes such as 21-cm radiation [355] or the cosmic microwave background [356], can be used to test the constancy of the fine-structure constant α, or to constrain simultaneous variations of α and Newton's gravitational constant G [357]. A non-constant G would have serious implications both for Newtonian physics and for General Relativity. This motivates ongoing efforts to devise new methods to measure this quantity, and to push the limits of experimental accuracy in order to test the constancy of G [358–362].

13.2. Testing the Assumption of Symmetry

The assumption of symmetry can be tested by checking for any deviations from isotropy. This requires statistical ensembles of data, so the best observables are the CMB, and, on a smaller scale, large-scale structure [363–365].

13.3. Testing the Gravitational Interactions

The gravitational interactions have been tested many times at laboratory scales and all the way up to Solar System scales. The challenge today is to test them at cosmological scales.

Tests of General Relativity may be placed in three broad categories: laboratory, astrophysical, and cosmological. Experiments cannot span all length scales, and so they cannot test all theories. We are forced to design experiments which can test alternative theories, or the effects of GR, at particular length scales (Table 5).

Laboratory tests probe effects from sub-millimetre scales to a few hundred metres. Galileo's experiments on weights dropped from the Leaning Tower of Pisa [366] are one example of a laboratory-scale test. The torsion balance experiments of Cavendish and Eötvös [367], and their modern versions [368–371], are another.

Astrophysical tests probe gravity from Solar System scales all the way up to galactic cluster scales. They include experiments such as laser ranging off the Moon (lunar laser ranging) and several of the Earth's artificial satellites [372–375], and proposals to extend laser ranging to other planets in the Solar System (planetary laser ranging) [376–380], radar astrometry on near-Earth objects [381], observations of the precession of the perihelion as well as higher-order effects for Mercury and other planets [338,382,383], observations of spacecraft in the Solar System [337], actual or proposed measurements of the acceleration of the Pioneer [384–386] and New Horizons probes [387–389], observations of the orbits of compact objects such as neutron stars [390–392], and measurements of galaxy rotation curves [393,394]. In the last decade, observations of extrasolar planetary systems have been proposed and used in order to test GR [395–404]. All of these systems are characterised by their spherical symmetry. Many can be approximated by a test object orbiting at a distance r around a central mass M. The gravitational field of the more massive central object is then probed by observing the orbit of the test body.

Table 5. Bridging the length scales to test the cosmological model. Experimental tests of gravity and dark sector couplings, at their typical length scales. Massive gravity (MG) screening mechanisms would show up at short ranges, while smooth dark energy manifests itself at cosmological scales. We give the experimental accuracy from current and future experiments planned over the next decade. The comparison between growth and expansion history comes from combined BAO, supernova, weak lensing, redshift distortion, cosmic microwave background (CMB) lensing, and cluster data. Lensing effects and dynamical mass comparisons can be carried out over a range of scales: inside galaxies, using strong lensing and stellar velocities, and on cosmological scales, using cross-correlations. This is a route to testing screening effects in alternative theories. Laboratory and Solar System tests can also probe dark sector couplings besides short-range effects, but many of the constraints obtained depend on the cosmological model.

Test	Length Scale	Theories Probed	Current Status (and Future)
Growth vs. expansion history	100 Mpc – 1 Gpc	GR with smooth dark energy	10% accuracy (2%–4%)
Lensing vs. Dynamical mass	0.01 – 100 Mpc	Test of GR	20% accuracy (5%)
Astrophysical tests	0.01 AU – 1 Mpc	MG screening mechanisms	~10% (Up to 10 times improvement)
Laboratory and Solar System tests	1 mm – 1 AU	PPN parameters in MG	Constraints are model dependent (Up to a tenfold improvement)

For such systems, the deviation of the metric from the Minkowski form is characterised by the magnitude of the Newtonian gravitational potential

$$\epsilon = \frac{GM}{rc^2}. \tag{119}$$

The strongest gravitational fields accessible to an observer occur in the limit $\epsilon \to \mathcal{O}(1)$. In this limit, the central object is a black hole and the test object is close to the event horizon.

We know that the Riemann curvature tensor is an essential quantity in General Relativity. The approximate magnitude of this tensor is expressed by the Kretschmann scalar (the fully contracted Ricci scalar) for the Schwarzchild metric:

$$\zeta \propto \frac{GM}{r^3 c^2}. \tag{120}$$

Note that this expression is more complicated for rotating objects [405]. For the purposes of our description, however, we may use this simple expression, which we call the curvature.

The third broad category is the cosmological tests. First, the measurements are made on a statistical ensemble. The masses must be treated as power spectra. Second, the gravitational field assigned to each wavenumber k is very weak. The description of the system often depends on the cosmological model. At this scale, the description must take into account the background perturbations and the expansion of the universe.

The position of the various systems on a potential-curvature parameter space is shown in Figure 5. The potential accessible to observers is bounded by the line $\epsilon \simeq 1$. There is no limit to the maximum curvature that can be observed except for the Planck limit, where $r \simeq 1.6 \times 10^{-33}$ cm. This lies many orders of magnitude above the boundaries of the figure. General Relativity is a complete theory, and does not fail below the Planck scale. To test or even probe GR below the Planck scale is a different matter, and a goal that has not yet been achieved. Many alternative theories of gravity that attempt to explain cosmic acceleration do so through modifications to GR at the Planck scale, so it is important to explore ways in which GR can be tested at these scales.

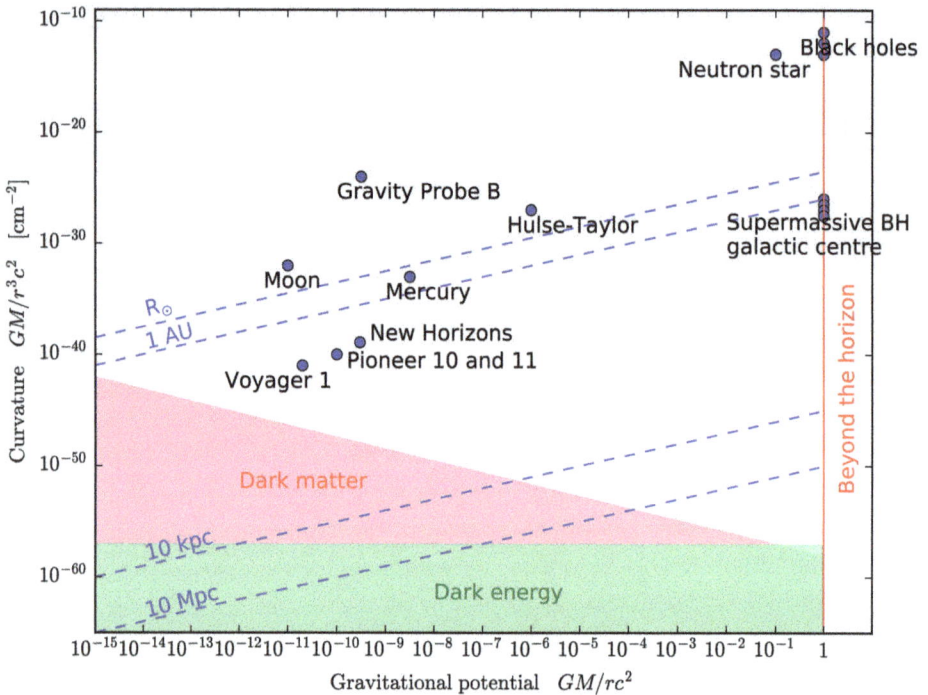

Figure 5. The parameter space for quantifying the strength of a gravitational field. The horizontal axis measures the potential. The vertical axis measures the spacetime curvature of the gravitational field at a radius r away from a central object of mass M. Regions of this parameter space with potential greater than 1 represent distances from a gravitating object that are smaller than the event horizon radius and are therefore inaccessible to observers. The red vertical line on the right-hand side of the plot marks the horizon limit. This is a schematic plot, and in no way do we show an exhaustive list of objects and systems that have been used or could be used to test GR. The region of Solar System-scale tests is broadly bounded by the Moon, Gravity probe B [406,407], Mercury, and the Pioneer and New Horizons spacecraft. We have included the Voyager spacecraft in the diagram, even though, unlike the Pioneer probe, it was never suitable for tests of GR. The famous Hulse-Taylor binary pulsar [408], although a neutron star binary, is also roughly in the region of Solar System tests. Black holes and neutron stars are in the strong field regime, and the former are at at the limit of the event horizon boundary. Adapted from [409].

What is the minimum curvature? The unperturbed FLRW metric is isotropic, and the unperturbed Kretschmann scalar is a function of time only. The curvature for the homogeneous universe drops as the universe expands. The present universe has a curvature which is just above the boundary of the region marked 'Dark energy', since dark energy is not yet completely dominant over pressureless matter. However, the curvature will approach this limit asymptotically. In this paradigm, this represents a fundamental minimum curvature scale. It is shown in the figure by the region labelled "Dark energy".

Galaxies are astrophysical probes of GR. Their innermost regions can be modelled as test particles orbiting a central supermassive black hole. Galactic rotation curves, which can only be explained by introducing Dark Matter, describe galaxy velocities up to the outermost regions. Systems below a constant acceleration scale of approximately 1.2×10^{-10} ms^{-2} cannot be modelled without adding a contribution to the gravitational filed in the form of Dark Matter. This constant acceleration is the diagonal boundary of the region labelled "Dark matter" in the figure.

Note that the regions of parameter space occupied by Dark Matter and dark energy overlap—there is a degeneracy between these two unknown components of the cosmological model. However, it is not impossible to distinguish between their effects, since their properties are different. In particular, Dark Matter forms clumps just like baryonic matter, while dark energy does not.

GR has not been tested in the region between curvatures of $\sim 10^{-40}$ and $\sim 10^{-50}$. This corresponds to the region between Solar System scales and cosmological scales probed by galaxy surveys and the CMB. The challenge lies in finding systems which span these scales. In theory they do exist, in the form of galaxies, clusters and superclusters. Their rotation curves transition from Schwarzchild orbits in their innermost regions, to outer regions dominated by Dark Matter. However, our observations are hampered by the fact that we are limited by the resolution of our telescopes, so we can only observe the outer regions. In addition, the untested region is situated between a region where GR is extremely well-constrained (by Solar System tests), and a region where Dark Matter and dark energy have to be invoked, and where we must take the cosmological model into account, because the effect of large-scale structure on background dynamics becomes non-negligible. It has been suggested that this backreaction effect may be at the origin of the observed cosmic acceleration.

Cosmological tests of General Relativity may be broadly classified as follows:

- Tests of the consistency between the expansion history and the growth of structure. A discrepancy in the equation of state parameter of dark energy w, inferred from the two approaches can indicate a breakdown of the GR-based smooth dark energy cosmological paradigm.
- Detailed measurements of the linear growth factor across different scales and redshifts.
- Comparison of the cosmological mass distribution inferred from different probes, especially redshift space distortions and lensing.

14. Possible Modifications of GR and Cosmological Implications

It is useful to identify the regimes in which modifications to GR may appear. This enables us to get a clearer picture of the capabilities and limitations of current and future experiments to tests these alternative theories.

14.1. Weak and Strong-Field Regimes

In order to test gravity in the strong-field regime, we need to observe compact objects with a very high density [410–412]. Black holes are good candidates. Their compactness and mass takes GM/rc^2 close to the maximum limit of unity. However, they have a serious drawback. Because of the "no hair theorem", they are not characterised by any coupling to a scalar field and therefore cannot be used to test for this effect, and to discriminate between scalar-tensor theories and GR.

Neutron stars, on the other hand, are still very compact bodies, but they can be strongly coupled to a hypothetical scalar field. This property has been used to test relativistic parameters by observing the Hulse-Taylor binary pulsar PSR B1913+16 [413], and the neutron star–white dwarf binary PSR J1141-6545 [414]. Gravitational time delay has been tested using other binary pulsars [415–419], and pulsars have also been suggested as ideal probes of the Lense-Thirring effect [420,421]. General Relativity passes these tests with flying colours.

14.2. Small and Large Distances

Distance-dependent modifications can be induced by a massive degree of freedom, which will cause a Yukawa-like coupling. General Relativity is very well constrained on the size of the Solar System, and there are several tests constraining Yukawa interactions at Solar System scales [422–424], but there are no constraints on scales larger than $10\,h\,\mathrm{Mpc}^{-1}$, at least without assuming some cosmological model. Some theories put forward to explain cosmic acceleration, such as Chameleon mechanisms [425], are essentially modifications of GR at cosmological distance scales.

14.3. Low and High Accelerations

Galaxy rotation curves and galaxy dynamics motivated the Dark Matter paradigm. The Tully-Fischer law [426] tells us that Dark Matter cannot be explained by a modification of General Relativity at a fixed distance. MOND instead explains it by modifying gravity at low accelerations below the typical acceleration $a_0 \sim 10^{-8}$ cm $\cdot$ s^{-2}.

14.4. Low and High Curvature

Curvature R is important in distinguishing possible extensions of the Einstein-Hilbert action. A curvature-dependent may become important even if the potential Φ remains small. In the Solar System, the curvature $R_\odot \sim 4 \times 10^{-28}$ cm^{-2}. The curvature of the homogeneous universe according to the Friedmann equation is

$$R_{\text{FLRW}}(z) = 3H_0^2 \left(\Omega_m (1+z)^3 + 4\Omega_\Lambda \right) , \tag{121}$$

from which we can see that the curvature of the universe evolved with time, from $\sim 10^{-33}$ cm^{-2} at the time of nucleosynthesis, to $\sim 10^{-56}$ cm^{-2} at $z = 1$. The curvature scale associated with the cosmological constant is $R_\Lambda = (1/6)\Lambda$, so the phenomenology of the cosmological constant occurs in low curvature regime

$$R < R_\Lambda \sim 1.2 \times 10^{-30} R_\odot . \tag{122}$$

This is of particular interest to paradigms which seek to explain cosmic acceleration through the backreaction effect. For cosmological-scale perturbations, we are always in the weak field regime. However, the curvature perturbation associated with large-scale structure becomes of the order of the background curvature at redshift $z \sim 0$, even if we are still in the weak field limit. This means that the effect of large-scale structure on the background dynamics may be non-negligible [427].

In summary, in order to explain the dark energy or Dark Matter problem by modifying General Relativity, the modifications have to be either at large scales (typically Hubble scales), low accelerations (typically below a_0, or small curvatures (typically R_Λ). The regions corresponding to dark energy and Dark Matter curvature and potentials are shown in Figure 5.

14.5. Cosmological Probes

The idea of testing General Relativity using large-scale structure was first proposed in 2001 [428]. This relies on the ingenious idea that if gravity is well-described by General Relativity, and the universe well-described by ΛCDM, then on sub-Hubble scales, and considering only scalar perturbations, the spacetime metric can be written as

$$ds^2 = -(1 + 2\Phi) \, dt^2 + (1 - 2\Psi)a^2(t)\gamma_{ij} \, dx^i \, dx^j \tag{123}$$

where Φ and Ψ are the two potentials, and γ_{ij} is the metric of the spatial section [429–431]. The Einstein equations reduce to the Poisson equation

$$\triangle \Psi = 4\pi G \rho_{\text{matter}} a^2 \delta_{\text{matter}} \tag{124}$$

and

$$\Psi - \Phi = 0 , \tag{125}$$

since the matter anisotropic stress is negligible. The spectrum of the two gravitational potentials has to be proportional to the matter power spectrum. The scale dependence of the gravitational potential P_Φ and of the matter distribution $P(k)$ are related by:

$$k^4 P_\Phi(k, a) = \frac{9}{4} \Omega_m H^2 a^{-2} P(k, a) . \tag{126}$$

If the Poisson equation is modified by some modification of gravity, the matter power spectrum changes shape according to the cosmological model assumed, so the above relation is not model-independent. However, the fact that the two spectra differ is independent of the cosmological model. It therefore provides a test of the underlying gravitational theory. Such a test can be carried out by comparing weak lensing data to galaxy surveys.

Various similar approaches have been proposed [432]. This approach allows us to test various classes of alternative theories by means of large-scale structure. Some of these theories include Dvali-Gabadadze-Porrati models [134,433,434], quintessence [435–439], and scalar-tensor theories [440–443]. The difficulty lies in finding a parameterisation of the perturbation equations which is consistent with the one used for the background evolution, since both assume the same theory for gravitational interactions.

The cosmic microwave background is another potential testing ground for General Relativity. The amplitude and position of the peaks of the cosmic microwave background allows us to probe the potential wells present during the recombination era by extracting information on the primordial power spectrum created during inflation. *Planck* data, alone or in combination with weak lensing has been used to test GR [444,445] and modified gravity [446–448].

Despite the fact that GR is extremely well tested on laboratory and Solar System scales, cosmology provides plenty of scope for alternatives to General Relativity. Let us consider cosmic acceleration, which is now a confirmed observation. It requires an explanation. The cosmological constant paradigm rests on three assumptions: that the observations are correct, that GR is the correct theory of gravity, and that our FLRW model of the universe is correct.

More than two decades of precision measurements have removed any doubt that the observed acceleration may be due to incorrect modelling of experimental errors. So in order to do away with the cosmological constant, we must seek possible answers in the other two assumptions.

Backreaction is an "alternative" to the alternative theories, or to the Dark Matter paradigm. It is firmly within the General Relativity paradigm, and seeks to explain cosmic acceleration by modelling the universe and the structures within it, and therefore its expansion history, in more detail than is the case with the FLRW model. It keeps the first and last assumption, but does away with the assumption of an FLRW universe. However, is it enough to explain the observed acceleration?

If GR is assumed to be the correct theory together with the other two assumptions, cosmological measurements are usually interpreted as providing evidence for Dark Matter and a nonzero cosmological constant or dark energy. This poses conceptual problems. Why is the observed value of the cosmological constant so small in Planck units? It also poses a coincidence problem. Why is the energy density of the cosmological constant so close to the present matter density? No dynamical solution of the cosmological constant problem is possible within GR—the cosmological constant is not the attractor of some dynamical function.

This opens the field to possible modifications of GR. Should GR be modified at low and high energies? This is a serious challenge for theorists. Einstein's theory is the unique interacting theory of a Lorentz-invariant massless spin-2 particle. New physics in the gravitational sector must introduce additional degrees of freedom. These additional degrees of freedom must modify the theory at low or high energies, or both, while being consistent with GR in the intermediate-energy regime, that is, at length scales $1\mu \leqslant \ell \leqslant 10^{11}$ m, where the theory is extremely well tested.

Figure 6 illustrates some of the difficulties in testing the completeness of General Relativity. There is an obvious "scale desert" between Solar System scales, and cosmological scales. At the other end of the scale, there is a gap which is often overlooked: between sub-millimetres scales, which is the current limit where GR has been tested, and the Planck scale. A modification of gravity at very small scales would be apparent in this regime. Even ignoring quantum effects, it is a serious challenge to test gravity in a regime in which vastly stronger forces come into play.

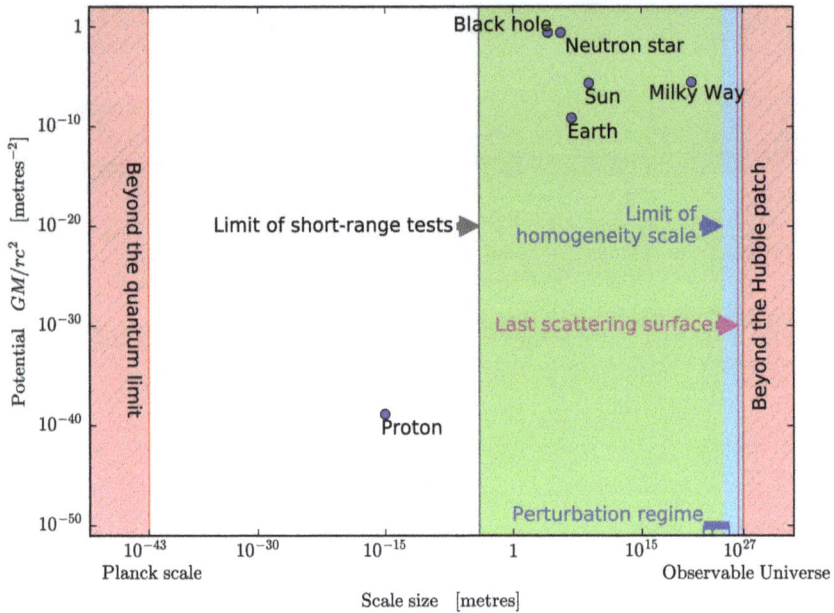

Figure 6. The parameter space for experiments. The horizontal axis is the typical length scale of the object in question. The vertical axis measures the gravitational potential. The red vertical line on the left-hand side marks the Planck scale. The vertical line on right-hand side of the plot marks radius of the observable universe, or Hubble radius. Experimental verification of GR is impossible beyond these limits. The radius of the surface of last scattering is only slightly smaller than our Hubble radius. Assuming GR implies assumptions far beyond the range that has been experimentally tests. For instance, if we define the Planck mass as $m_{\text{Planck}} = \sqrt{\hbar c / G}$, we are assuming that the gravitational constant remains constant down to the Planck length. This extrapolates the inverse square law over a scale of more than 10^{30} from what ha s been tested. The green region is where Solar System tests have been carried out. Beyond $\sim$100 Mpc, assumption of a homogeneous and isotropic metric becomes accurate enough to use in physical models. The blue region shows length scales at which the FLRW metric is valid. By way of comparison to this parameter space, the nonlinear regime for perturbation theory, which gives use the matter power spectrum, covers $\sim$$10^{22}$ to $\sim$$10^{23}$ m, the linear, quasi-static regime covers $\sim$$10^{23}$ to $\sim$$10^{25}$ m, and beyond that is the superhorizon regime. These length scales ranging from 1 Mpc to above 1 Gpc fit on a tiny part of the horizontal axis above, shown by the thick blue horizontal line.

15. The Nature of Dark Energy and the Implications for General Relativity

The Concordance Model of cosmology assumes that General Relativity is correct, an assumption which is justified by the tests which GR has undergone. Within this model, $\sim$95% of the content of the universe is unaccounted for. Dark matter, which makes up around 25% of the mass-energy of the universe is a matter-like component which is cold (sub-relativistic) and weakly interacting. The discrepancy between the observed acceleration of the expansion of the universe and the predictions of GR leads to the conclusion that there must be a cosmological component with a negative equation of state parameter making up around 70% of the mass-energy content of the universe: dark energy.

But the dark energy paradigm does not fix the nature of this component. There exist many theories which attempt to explain its nature. In particular, one can ask whether the basic assumptions of the Concordance Model—homogeneity and isotropy—are correct.

If the universe is not homogeneous and isotropic, the FLRW equations are no longer valid. Over the last decade, one line of research has attempted to explain the accelerated expansion by exploring the implications of an inhomogeneous universe on a General Relativistic cosmology. This effect is the backreaction. This approach does away with dark energy.

There are various ongoing investigations on the effect of the backreaction due to an inhomogeneous universe [269,427,449–455], with different lines of research offering different interpretations of the Buchert equations, where the Friedmann equations are supplemented by an additional backreaction term [456]. Whether one can explain all of the observed expansion history of the universe as a consequence of the growth of inhomogeneities without invoking some additional fluid component is the subject of ongoing debate [452,457]. The backreaction, even if it turns out to be incapable of replacing the dark energy paradigm, is still a subject worth investigating, if anything as a correction to the homogeneity assumption, which obviously breaks down at small scales.

If we assume that the Copernican Principle holds, then the universe is well described by a Friedmann-Robertson-Walker spacetime. The dynamics of the background expansion are determined by the content of the universe: the list of fluids (perfect fluids, due to the assumption of the Copernican Principle), with their equations of state. Within the dark energy paradigm, we can distinguish two main strategies for formulating hypotheses:

(1) there is some new kind of component in the universe, or
(2) there is some new property of gravity.

Let us first recall that General Relativity rests on two assumptions: the gravitational interaction is described by a massless spin-2 field, and matter is minimally coupled to the metric, which implies the weak equivalence principle. The Einstein-Hilbert action described by Equation (6) implies that

$$S_{\text{gravity}} = \frac{c^3}{16\pi G} \int R\sqrt{-g}\, \text{d}^4x + S_{\text{matter}}[\text{matter}; g_{\mu\nu}],$$ (127)

where R is the Ricci scalar of the metric tensor $g_{\mu\nu}$, and $S_{\text{matter}}[\text{matter}; g_{\mu\nu}]$ is the action of the matter fields. If we only consider field theories, this gives us a useful classification scheme for the different theories that seek to explain the nature of dark energy.

The first strategy listed above assumes that gravitation is described by General Relativity, and introduces new forms of gravitating components beyond the Standard Model of particle physics. This adds a new term $S_{\text{DE}}[\psi; g_{\mu\nu}]$ to the action in Equation (127) while keeping the Einstein-Hilbert action of all the standard and Dark Matter unchanged.

The second strategy modifies gravity, and therefore extends the action, either by modifying the Einstein-Hilbert action of the coupling of matter. These theories also involve new forms of matter.

A cosmological constant Λ is the simplest modification which can be made to gravity, and it is equivalent to dark energy with a constant equation of state. To explain the observed acceleration, the new form of matter must have an equation of state $w < -1/3$. Dark energy can also be attributed to the energy of the vacuum, although the energy predicted by the Standard Model of particle physics is either 0 (using super-symmetry), or 10^{120} orders of magnitude larger than the observed cosmological value [458]. There are ongoing attempts to solve this 'fine-tuning problem' using string theory [459–461], causal sets [462,463], or by using anthropic arguments [464,465].

The other approach is to attribute dark energy to a scalar field whose potential has evolved in some way that it currently exerts a negative pressure. Such fields, in theories within the framework of GR, are termed Quintessence. Their distinguishing feature is that they allow the equation of state of dark energy to evolve. Alternatives to Quintessence within the same approach include K-essence, Phantom Fields, or the Chaplygin Gas [466–469].

Another strategy is to depart from General Relativity and modify the laws of gravity and posit dark energy as the manifestation of an effect arising from extra dimensions, or higher-order corrections. Within this category, the more successful theories have been of two types. Dvali-Gabadadze-Porrati

(DGP) dark energy considers the universe as a 4D brane within a 5D Minkowskian bulk [134]. The weakness of gravity relative to the other forces is explained by gravity "leaking" into the higher dimensions as it acts through the bulk [470], whereas the other forces act within the brane. The other class of theories is $f(R)$ gravity, where the Ricci scalar R in the Lagrangian is replaced by some function $f(R)$. Such theories correspond to scalar-tensor gravity with vanishing Brans-Dicke parameter [471].

We can therefore identify four main classes of theories, as shown in Figure 7. Classes 1 and 2 assume GR and introduce new forms of gravitating matter, the difference being that in class 2, the distance-duality relationship may be violated due to mechanisms such as photon decay. Classes 3 and 4 modify gravity in some way. In class 3, a new field introduces a long-range force so that gravity is no longer described by a massless spin-2 graviton. In this class, there may be a variation of the fundamental constants. Class 4 includes theories in which there may exist an infinite number of new degrees of freedom, such as brane models or multidimensional models. These models predict a violation of the Poisson equation on large scales. The constraints which may be placed on these theories, and the experiments to test them, correspond to the regimes in which gravity is modified, as summarised in Section 14.

An exhaustive list of dark energy or cosmic acceleration theories is beyond the scope of this review. We simply note that most of them give different predictions for the equation of state of dark energy, its evolution, or the expansion history of the universe . To distinguish between these proposals we need to track the evolution of these parameters throughout the history of the universe.

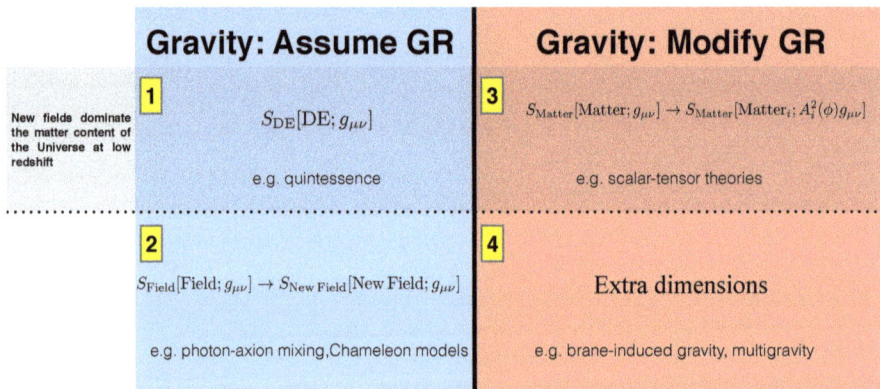

Gravity: Assume GR	Gravity: Modify GR
1 $S_{\rm DE}[{\rm DE}; g_{\mu\nu}]$ e.g. quintessence	**3** $S_{\rm Matter}[{\rm Matter}; g_{\mu\nu}] \rightarrow S_{\rm Matter}[{\rm Matter}_i; A_i^2(\phi)g_{\mu\nu}]$ e.g. scalar-tensor theories
2 $S_{\rm Field}[{\rm Field}; g_{\mu\nu}] \rightarrow S_{\rm New\ Field}[{\rm New\ Field}; g_{\mu\nu}]$ e.g. photon-axion mixing, Chameleon models	**4** Extra dimensions e.g. brane-induced gravity, multigravity

(New fields dominate the matter content of the Universe at low redshift)

Figure 7. The four main classes of dark energy theories, within the two broad strategies, classified as modifications of the General Relativistic action. Classes 1 and 2 assume the gravitational metric coupling of GR, whereas classes 3 and 4 modify this metric coupling, and are therefore modifications of gravity. In the upper line of classes (1 and 3), new fields dominate the matter content of the recent universe. Adapted from [431].

16. The Current Status of General Relativity

General Relativity has been subjected to a multitude of tests in its 100 years of existence. As of 2016, the main predictions of GR have been tested and confirmed. Whereas it is sufficient for most purposes at ordinary accelerations and energy scales to use Newtonian calculations, General Relativity has found its way into daily life, in the Global Positioning System and geodesy. The postulates of General Relativity have been confirmed with ever-increasing accuracy. Deviations from the Einstein Equivalence Principle are now constrained to below $\sim 10^{-14}$, deviations from Local Lorentz Invariance down to $\sim 10^{-20}$, and deviations from Local Position Invariance down to $\sim 10^{-6}$.

General Relativity has been probed down to scales of $\sim 10^{-6}$ metres in laboratories, and up to 1000 AU in space-based experiments and observations. Astrophysical and cosmological observations have probed GR at scales from 1 Mpc up to gigaparsec scales. In the Solar System, the dynamics of GR have been tested with radar and laser-ranging. We can track the ephemerides of the planets in the Solar System, right up to the minor outlying planets (such as Sedna, with an aphelion just short of 1000 AU). The farthest objects whose trajectory has been followed from the moment they were "thrown" are now outside the Solar System. They are the Pioneer 10 and 11 spacecraft, and are now around 70 AU distant, and Voyager 1, which is at a distance of 135 AU, making it the only object to have reached interstellar space. Communication with the Pioneer spacecraft ceased in 2003. The Pioneer spacecraft exhibited an anomalous constant acceleration towards the Sun which could not be explained using GR. This prompted a reexamination of all the recorded data. It is now generally accepted that the anomaly is caused by thermal radiation, and that once this is accounted for, there is no remaining anomalous acceleration [472].

Gravitational waves, among the last untested predictions of GR, were first detected in late 2015, ushering a whole new era of observational astrophysics, in which the strong field regime can be probed and where the predictions of GR can be tested.

The evidence for General Relativity is extremely strong. The theory has passed all tests in the weak-field limit at Solar System scales, including tests of the assumptions (the Equivalence Principle) and the predictions specific to GR (frame-dragging, gravitational time dilation), and in the strong field with the observation of a binary black hole merger and the resulting gravitational waves. We give a summary of the experimental milestones in Table 6. However, there are still issues that allow room for speculation.

The first is the question of the completeness of GR. Is it valid at all scales? There is a scale gap in our tests of GR between laboratory, Solar System and galactic scales, and cosmological ones. This gives rise to a multitude of domains of validity for different alternative theories, whereas ideally we should seek a universal theory that can explain phenomena at all scales.

The second issue is the accuracy of our approximations. General Relativity may be conceptually simple in that it is based on a minimum number of postulates. But the resulting field equations, when applied to real physical systems, can be very hard to solve. We get around this by making approximations such as spherical symmetry, or homogeneous and isotropic perfect fluids, which allows us to obtain analytical solutions. However, the accuracy of these approximations may not always be good enough. This is evident in the case of cosmological perturbations and large-scale structure. At which scales is it valid to use a Friedmann-Lemaître-Robertson-Walker metric? Can the backreaction explain some or all of the observed cosmic acceleration? For smaller systems such as aspherical collapsing bodies, we still need accurate models in order to match theory and observation. Such questions have spurred the development of numerical methods in General Relativity. Physicists now have the necessary computing power to go far beyond simple first-order approximations.

The remaining open questions are of a cosmological nature: Dark Matter and dark energy, and inflation. Dark matter and dark energy account for around 25% and 70% of the mass-energy content of the universe, respectively. They are not a problem for the theory of gravity itself, but it does mean that we do not know the nature of 95% of the content of the universe. Inflation solves a number of cosmological problems, but whatever theory we choose to explain inflation, we still need to introduce new physics beyond GR.

Table 6. The major predictions of GR, and their experimental status.

Effect	Milestones	Current, Future and Proposed Experiments	Status
Mass equivalence	Galileo [366,473] Eötvös [367]	Eöt-Wash Group [371] Lunar Laser Ranging [343] MICROSCOPE[474] STEP [475] Galileo Galilei satellite [476]	Confirmed
Gravitational time dilation [478]	Eddington solar eclipse [4] Pound-Rebka experiment [479] Space-borne hydrogen masers [481]	Quantum interference of atoms [477] ACES [480] Galileo 5 and 6 satellites[344] Einstein Gravity Explorer [482]	Confirmed
Precession of orbits	Orbit of Mercury (Einstein [3])	Binary pulsar observations [483–485] Solar System and extrasolar planets [337,402]	Confirmed
De Sitter precession	Gravity Probe B [486] Lunar laser ranging [488–490] Binary pulsars[493,494]	Binary pulsars [487] Improved lunar laser ranging [491,492]	Confirmed
Lense-Thirring precession	Gravity Probe B [486] LARES [420,421,496]	LARASE [495] Laboratory tests [497] Solar System bodies [339] Binary pulsars [420,421] Black holes [478,496,498,499]	Confirmed
Gravitational waves	LIGO [80]	Advanced LIGO [500] eLISA https://www.elisascience.org	Recently confirmed in two events [80,501]
Strong field effects	PSR J0348 + 0432 [483]	Black holes [411,502] Binary pulsars [117]	Recently confirmed
Orbital precession due to oblateness of central body		Low-orbit satellites [503–506] Stars orbiting black holes [506] Juno spacecraft around Jupiter [507]	Not yet observed

17. Future Developments

The current Concordance Model of cosmology was built in successive (and sometimes concurrent) steps. General Relativity applied to a spacetime under the Copernican Principle, filled with pressureless matter, produced the Einstein-de Sitter model. Motivated by the observed Hubble expansion, it resulted in the Big Bang model. Clues from element abundances, baryon assymmetry, and knowledge of nucleosynthesis from the Standard Model of particle physics meant that the universe had to have a thermal history, which resulted in the Hot Big Bang Model. When evidence for missing mass became incontrovertible, cold Dark Matter had to be added to the inventory of cosmic components. This model worked well, but not well enough. It could not explain the observed homogeneity of the universe across regions which were causally disconnected, nor its flatness. So Inflation was introduced. Observations of an accelerated cosmic expansion motivated a search for explanations within the then current paradigm, which resulted in various hypotheses: a curved geometry, supermassive neutrinos, or perhaps a particular cosmological topology. In the end, the paradigm had to be shifted yet again with the introduction of dark energy.

The Concordance Model can explain the observations with just six parameters: the physical baryon density parameter $\Omega_b h^2$, where h is the Hubble parameter, the physical Dark Matter density parameter $\Omega_c h^2$, the age of the universe t_0, the scalar spectral index n_s, the curvature fluctuation amplitude Δ_R^2, and the reionisation optical depth τ. That such a degree of fit is offered by such a simple model is remarkable.

The success of the Concordance Model has been its ability to include physical effects at extremely different scales, from primordial nucleosynthesis to large-scale structure evolution, in one coherent theory. However, this does not allow us to state that the ΛCDM model is correct. It merely implies that deviations from ΛCDM are too small compared to the current observational uncertainties to be inferred from cosmological data alone. This leaves room for some very fundamental open questions, which we have described in this review.

The science of cosmology finds itself at a critical point where it has to make sense of the vast quantity of data that has become available. Different probes have allowed us to piece together interlocking information which, so far, confirms the Concordance Model. The cosmic microwave background has provided conclusive evidence of a flat geometry, super-horizon features, the correct harmonic peaks, adiabatic fluctuations, Gaussian random fields, and most recently, a departure from scale invariance. We have not yet observed primordial inflationary gravitational waves. Large-scale structure observations, which probe the recent universe, provide firm evidence in favour of the Concordance Model's explanation of the evolution of density perturbations and the growth of structure, and provide a bridge between the effects of long-range gravitational interactions and shorter-range forces.

The recent detection of gravitational waves in two events (possibly three), one hundred years after they were predicted by Einstein, directly validates General Relativity in several ways. It shows that GR is correct in the strong-field regime, that black holes really exist, and black hole binaries too, and it proves that gravitational waves are a real physical phenomenon, and not just a mathematical artefact of GR.

We are fast approaching the point where cosmological observations will be limited only by cosmic variance, i.e., no more data will be available from our Hubble patch [508]. What does that imply for the development of new theories? How do we test the predictivity of these theories without new data?

We will have to look for new effects in old data, and for new correlations in future data. Large-scale structure is extremely useful in testing General Relativity and cosmological models, but the future may bring other observational windows. In particular, the next fifty to one hundred years may see the development of a gravitational wave astronomy. This is likely to be an even more significant development than even CMB astronomy. The observation of primordial gravitational waves would provide vital information on the inflationary epoch [509,510]. B-mode polarisation of the CMB offers an indirect pathway to the observation of this gravitational wave background [511,512]. The second major development could be the observation of a cosmic neutrino background [513], which is the result of neutrino decoupling in the lepton era. This would push the observations along our past light cone even further back in redshift, providing information on the universe before recombination and the CMB.

17.1. Plausible Conclusions from Incomplete Information

The statistical questions facing cosmologists pose some particular problems. We observe a finite region of our universe, which is itself a single realisation of the cosmological theory. We can only observe whatever is on or inside our past light cone, as shown in Figure 1. Not only are we limited by cosmic variance, we also have just a single data point for the cosmological model.

There exist several alternatives to GR and to ΛCDM that have not been ruled out by experiment. Constructing viable physical models is not just a question of fitting the model to the data. It is a question of model selection, which requires robust statistical techniques that allow us to make sensible decisions using our incomplete information. Bayesian inference provides a quantitative framework for plausible conclusions [514,515]. We can identify three levels of Bayesian inference.

(1) Parameter inference (estimation). We assume that a model M is true, and we select a prior for the parameters θ, or the $\text{Prob}(\theta|M)$.
(2) Model comparison. There are several possible models M_i. We find the relative plausibility of each in the light of the data D, that is we calculate the ratio $\text{Prob}(D|M_i)/\text{Prob}(D|M_0)$.
(3) Model averaging. There is no clear evidence for a best model. We find the inference on the parameters which accounts for the model uncertainty.

At the first level of Bayesian inference, we can estimate the allowed parameter values of the theory. If we assume General Relativity as our theory, we still need to fix the values of the various constants. This is the rationale behind ongoing efforts to measure quantities like Newton's constant G ever more accurately, despite the fact that the parameter has been around for three centuries. What are the energy densities of the various components of a ΛCDM universe? If we include a dark energy equation of state parameter w, what value does it take?

Next, we can ask which parameters we should include in the theory. Should we include a cosmological constant in General Relativity? Or should we include dynamical dark energy parameter? Although current data are consistent with the six-parameter ΛCDM model based on GR, there are more than twenty candidate parameters which might be required by future data (see [516]). We cannot simply include all possible parameters to fit the data, since each one will give rise to degeneracies that weaken constraints on other parameters, including the ΛCDM parameter set (e.g., [517–519]). The landscape of alternative cosmological models is even larger if we relax our assumptions on the theory of gravity [520].

The goal in data analysis is usually to decide which parameters need to be included in order to explain the data. For physicists, those extra parameters must be physically motivated. That is, we need to know the physical effects to which our data are sensitive, so that we can relate these effects to physics. At the current state of knowledge, we have to acknowledge the possibility of more than one model. We therefore require a consistent method to discard or include parameters. This is the second level of Bayesian inference—model selection.

Bayesian model selection penalises models which introduce wasted parameter space. We can always construct a theory that fits the data perfectly, even better than GR, but we would need to introduce extra free parameters (e.g., extra fields or couplings between matter and the metric). This is the mathematical equivalent of Occam's razor. We seek a balance between goodness of fit (the degree of complexity) and predictive power (consistency with prior knowledge). General Relativity fits all the data with the minimum number of parameters. In cosmology, ΛCDM is the best model because it only involves one new parameter and no new fields.

The problem of model selection in relation to GR is as old as the theory itself. But only recently have cosmologists have started to use Bayesian methods for cosmological model selection (e.g.,

[521–524]), when the astrophysical data began to have the necessary statistical power to enable model testing. Bayesian techniques are starting to be applied to General Relativity itself [525]. With the next generation of astrophysical probes in the pipeline, model selection is likely to grow in importance [514].

Bayesian model selection cannot be completely free of assumptions. In cosmology, there is some model structure which depends on a number of unverifiable hypotheses about the nature of the universe. The Copernican Principle is one such hypothesis [9,526,527].

The third level of Bayesian inference is model averaging. In the current scenario, there is firm evidence for General Relativity as the best model for gravitational interactions. However, it is still useful to quantify our degree of certainty (or doubt), for one simple reason: we do not have the final list of alternative theories. In other words, when we choose GR against any number of alternative theories, we have no knowledge about other alternatives outside that list (such as theories yet to be developed, for example). At best, we know which alternatives are ruled out by the data. This level of Bayesian inference is the application of a principle that has been called 'Cromwell's Rule' [528]: even if all the data show our theory to be correct, we should allow a non-zero probability, even if tiny, that the theory is false.

The utility of alternative theories becomes evident when we apply Bayesian model selection. For science to advance by falsification, it is not enough to claim that the present theory is false. We need to know which alternative theory is favoured instead. Newtonian gravity would have likely have survived the 1919 eclipse if General Relativity had not been formulated.

17.2. Experimental Progress

There has been rapid progress in constraining cosmological parameters and models over the last two decades, with a multitude of experiments observing the CMB, large-scale structure, galaxies and supernovae. We will just provide a summary of the most recent data sets.

The first are anisotropies in the CMB, where the main statistic is the angular power spectrum of fluctuations C_ℓ, and polarisation of the CMB. The most recent and current are: WMAP (http://map.gsfc.nasa.gov), *Planck* (http://www.cosmos.esa.int/web/planck), Atacama Cosmology Telescope (ACT) (http://act.princeton.edu), South Pole Telescope (SPT) (https://pole.uchicago.edu), Atacama Cosmology Telescope polarisation-sensitive receiver (ACTPol) [529], SPTPol, Spider (http://spider.princeton.edu), Polarbear (http://bolo.berkeley.edu/polarbear), Background Imaging of Cosmic Extragalactic Polarization (BICEP2) (http://bicepkeck.org,https://www.cfa.harvard.edu/CMB/bicep2), Keck Array (http://bicepkeck.org,https://www.cfa.harvard.edu/CMB/keckarray).

The second source of data are surveys cataloguing the angular positions and redshifts of individual galaxies, leading to the power spectrum of fluctuations $P(k, z)$, or the two-point correlation function $\xi(r)$. The recent and current experiments are: Baryon Oscillation Spectroscopic Survey (BOSS) using Sloan Digital Sky Survey (SDSS) data (http://www.sdss3.org/surveys/boss.php), Dark Energy Survey (DES) (https://www.darkenergysurvey.org), Weave (http://www.ing.iac.es/weave/science.html), Hobby-Eberly Telescope Dark Energy Experiment (HETDEX) (http://www.hetdex.org), Extended Baryon Oscillation Spectroscopic Survey (eBOSS) (https://www.sdss3.org/future/eboss.php), Mid-Scale Dark Energy Spectroscopic Instrument (MS-DESI) (https://www.skatelescope.org), Canadian Hydrogen Intensity Mapping Experiment (CHIME) (http://chime.phas.ubc.ca), Baobab, MeerKAT (http://www.ska.ac.za/science-engineering/meerkat), and ASKAP (http://www.atnf.csiro.au/projects/askap/index.html).

The third source of data are weak lensing, which use the fact that images of distant galaxies are distorted and correlated by intervening gravitational potential wells to produce statistics such as the convergence power spectrum C_ℓ^κ [530,531]. Some current experiments are: Dark Energy Survey (DES) (https://www.darkenergysurvey.org), Red Cluster Sequence Lensing Survey (RCSLens) (http://www.rcslens.org), Canada-France Hawaii Telescope Lensing Survey (CFHTLenS) (http://www.cfhtlens.org), New Instrument of Kids Arrays (NIKA2) (http://ipag.osug.fr/nika2), and Hyper Suprime-Cam (HSC) (http://hsc.mtk.nao.ac.jp/ssp).

The final source of data are catalogues of peculiar velocities. By measuring redshifts and radial distances of galaxies and clusters it is possible to reconstruct a radially projected map of large-scale motions. Progress in this field will come from all three data sets above.

The main science goal of the next generation of cosmological probes is to test the Concordance Model of cosmology. Some major experiments will be operational in the next decade.

Planck, decommissioned in 2013, marked a major milestone in CMB experiments. Its proposed successor is the ground-based programme CMB-S4 [532], which should reach sensitivities below $10^{-3}\mu$ K whose main aim is to achieve higher resolutions, probe larger scales, and measure new observables such as polarisation.

There are various future experiments to map the mass-energy content of the universe (including baryons and Dark Matter), either in the planning phase, or close to completion. The Large Synoptic Space Telescope (LSST) (https://www.lsst.org), which should achieve first light in 2019, is a ground-based telescope which will map the entire sky.

The *Euclid* space telescope (http://www.euclid-ec.org), due for launch in 2020, will map galaxies and large-scale structure over the whole sky at visible and near-infrared wavelengths, providing a catalogue of 12 billion sources at 50 million . *Euclid* will probe the recent universe, when galaxies have formed and dark energy starts to dominate. Its main scientific objective is to understand the origin of the accelerated expansion of the universe by probing the nature of dark energy using weak-lensing observables (which include cosmic shear, higher-order distortions, and cosmic magnification), and galaxy-clustering observations [225].

The Square Kilometre Array (SKA) (https://www.skatelescope.org), which will begin operations in 2020, is a multi-wavelength radio telescope, built across multiple sites to achieve the largest collecting area ever [70,533]. It should provide the highest resolution images of the radio sky, thus providing maps of large-scale structure, and observing pulsars which should provide direct tests of General Relativity. The SKA will observe the epoch between the emission of the CMB and the formation of the first galaxies. Neutral hydrogen surveys (or 21 cm intensity mapping) [534], offer yet another promising probe, as do galaxy redshift surveys [535].

Joint observables will be key in the next generation of experiments. These include the Sachs-Wolfe effect, and the Sunyaev-Zel'dovich effect [536]. We will also need to extract more observables from the CMB. We need information on in order to constrain inflationary models.

As experiments probe larger scales at better resolutions (low ℓ and high ℓ), the data analysis and the cosmological tests (dark energy, Dark Matter, the properties of cosmic neutrinos) will require accurate calculations on the growth of large-scale structure, which can only be achieved using N-body simulations [537–544].

At the other end of the length scale, there is particle physics. The Very Large Hadron Collider, the successor to the Large Hadron Collider (https://home.cern/topics/large-hadron-collider) (which can achieve energies around 14 TeV), is still in its conceptual phase. It could, if built, probe energies around 100 TeV, allowing it to test physics beyond the Standard Model, possibly including Supersymmetry and Grand Unified Theories, and thus provide clues on the nature of Dark Matter and dark energy (see, e.g., [545]).

Tremendous progress is also being made in Milky Way astrophysics. With the launch on 2013 of the Gaia mission (http://www.cosmos.esa.int/web/gaia), we will soon have improved data on Solar System ephemerides, and on the orbits and tidal streams in the Milky Way. This will allow precise tests of GR at Galactic scales. In particular, tidal streams provide an opportunity to probe GR at late cosmological times and to close the gap between astronomical and cosmological scales.

Now that gravitational waves have been detected, gravitational wave physics is set to become an established branch of astrophysics. The Laser Interferometer Gravitational-Wave Observatory (LIGO) (https://www.ligo.caltech.edu), Advanced LIGO (aLigo) (https://www.advancedligo.mit.edu) and Virgo (http://www.virgo-gw.eu) will subject GR to a battery of test in the strong regime. Also targeting the regime of strong curvatures and potentials, the planned Event Horizon Telescope (http://www.eventhorizontelescope.org) is a network of millimetre and sub-millimetres telescopes being used for very-long baseline interferometry to directly image supermassive black holes in galactic centres.

17.3. Theoretical and Computational Progress

On the theoretical front, there are four main lines of development in theories of gravity. They are all motivated by current open questions in physics. The ongoing attempt to find a Grand Unified Theory continues to motivate the development of string theory [546] as a framework for gravity. Aside from the theoretical difficulties of a mathematically complex theory, the challenge for string theory is to produce physical predictions which can be experimentally tested. The second approach is brane theories or supergravity, in which spacetime has more than four dimensions. The resulting field theory combines Supersymmetry (from particle physics) and General Relativity [547]. The third approach is quantum gravity [548], in which spacetime, as a dynamical field, is a quantum object. This implies a violation of Lorentz Invariance near the Planck scale, which in turn means that some particle decays forbidden by Special Relativity are allowed, and possibly charge-parity-time violations too. This motivates the search for signatures of quantum gravity in particle physics experiments.

The final approach, which is closer to the theoretical framework of the Concordance Model of cosmology, is a phenomenological use of General Relativity. This includes the various alternative theories that seek to explain cosmic acceleration and the missing mass, and also the various theories which provide dark energy and Dark Matter candidates.

The early alternatives to General Relativity were motivated by theoretical considerations. The current alternatives are mainly motivated by the open questions in cosmology. Cosmology is now in the age of Big Data. In the last decade, the data finally caught up with the theory, and we are now in a position to test many of the current alternative theories using statistical techniques.

The principles that underpin statistical techniques are simple enough. Under minimal assumptions about signal and noise, it is simply a question of maximising the Gaussian likelihoods. However, in practice they are extremely complex. The foreground contamination, the signal, the possibly non-Gaussian noise and the systematics all have to be modelled. This requires data simulation, so the number of maps that is generated is far larger than the number that is actually observed by the instrument (see, e.g., [549–554]).

A low signal-to-noise ratio, and higher-resolution observations of fainter signals over a larger frequency range, result in massively big data sets. For statistical probes such as the CMB and large-scale structure, these data sets have to be analysed as a whole, in order to correlate data. In addition to the volume of data collected, this generates a huge amount of computational data, and requires the appropriate computing power to carry out multiple complex calculations. We are witnessing a Moore's Law in cosmology, where the data volume of experiments increases by a factor of around 1000 every ten years, as shown in Figure 8.

In addition to the data volume challenge, we have an algorithmic challenge. The number of computations for cosmological data analysis depends on four main factors: the number of observations N_t, the number of pixels N_p, the number of multipoles (a function of the resolution and frequency range) N_ℓ, and the number of iterations N_i. The first three quantities determine the data volume. The data simulation scales at least as $\mathcal{O}(N_t)$. The map-making scales as $\mathcal{O}(N_i N_t \log N_t)$. The maximum-likelihood power spectrum estimation scales as $\mathcal{O}(N_i N_l N_p^3)$.

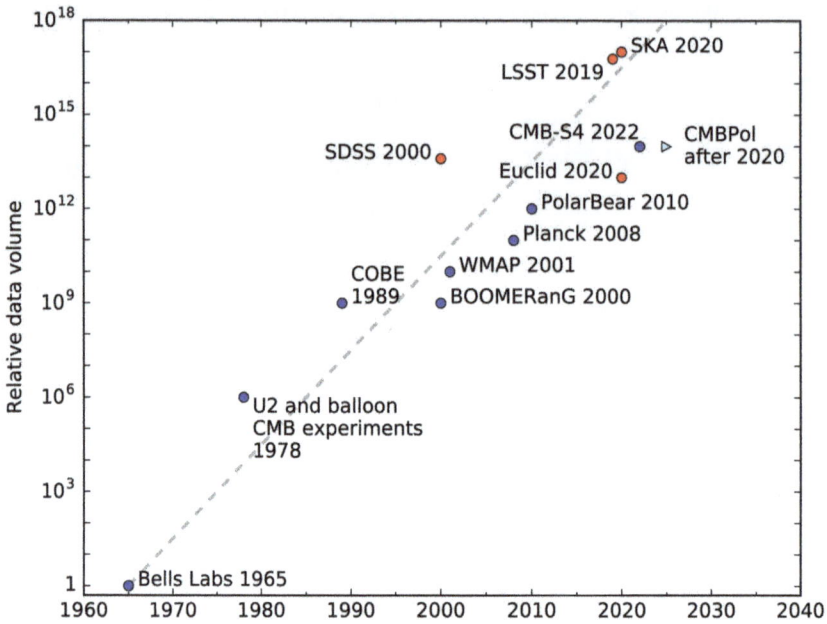

Figure 8. The growing data volume of experiments. The data volume of each experiment is shown as an order-of-magnitude multiple of the data volume of the 1965 Bells Labs experiment which detected the CMB. The labels show the year of 'first light' for each experiment. CMB surveys are marked by blue dots, while red dots show large-scale structure surveys. A first light date for CMBPol (light blue triangle) has not yet been fixed. Note that the vertical axis is logarithmic: the date volume increases about a thousandfold every ten years (grey dashed line). Note too that we plot CMB, large-scale structure, space and ground-based probes on the same graph. Ground-based probes will always tend to have a larger data volume than space-borne probes, due to the bandwidth limit on data transmission from spacecraft. Longer-running experiments will also have a larger data volume.

Planck marked a milestone in data science. It was the first CMB experiment in which the whole data treatment process was parallelised, and where Monte Carlo methods were used in order to cut down on the number of data realisation iterations that were carried out.

There are two main considerations in data analysis: the amount of data, and the complexity of the theory. Future experiments will require sophisticated techniques, and considerable computing power to process the vast amounts of data. Gaia is already collecting data [555], while Euclid, the SKA, Enhanced LIGO, and CMB-S4 will soon be operational. Farther ahead, the Evolved Laser Interferometer Space Antenna (eLISA) (https://www.elisascience.org), due for launch in 2034, will also generate huge volumes of data, and will require the necessary computing power to test GR directly, by comparing the data to simulated black hole collapse, inspiralling binaries.

What direction will experimental cosmology take over the coming decade? Extracting fainter signals, such as CMB polarisation, or going to very high resolutions requires larger data volumes to provide a higher signal-to-noise ratio, and it requires more complex models to control fainter systematic effects. Even the Solar System tests of General Relativity will depend upon vastly greater data volumes and computational complexity to get the full relativistic ephemerides, given the ever-increasing number of objects being tracked in the Solar System and the significantly greater precision of the data for each object.

This enables us to make a sensible prediction on future developments. The science we are able to extract from present and future data sets will be determined by the limits on our computational capability, and our ability and willingness to exploit it.

17.4. Conclusions

GR may well survive for another 100 years. After all, Newtonian gravity was around for 200 years. GR has just reached its peak, when data and computing power have caught up with the theory. We are at a pivotal moment in the history of GR. We are on the point of confirming beyond reasonable doubt all its predictions throughout its entire domain of validity.

We have seen how modern cosmology is faced with big questions which touch the very foundations of physics. What is this form of matter which interacts only with gravity and apparently with nothing else? Why is the expansion of the universe accelerating? What caused the universe to undergo a period of rapid expansion soon after the Big Bang? These questions, motivated by cosmological observations, lead to questions about fundamental physics. Are there forces and interactions besides the four we know of, that is, gravity, electromagnetism, and the strong and weak nuclear forces? Are there particles beyond the Standard Model? What determines the value of the fundamental constants of nature? What is the real structure of spacetime? Are there extra dimensions?

Science needs data, so each of these questions must be addressed through careful experiment. The challenge of modern experimental physics is to probe nature at extreme distances and energies, well outside the capabilities of the instruments that were available to Einstein. It has certainly come a long way, as shown by the detection of gravitational waves in 2015, a feat which was thought to be impossible by many of Einstein's contemporaries.

General Relativity is not the final theory of gravity, for there is no such thing. As General Relativity turns 100, we would do well to celebrate it with a healthy does of scientific scepticism. Long live General Relativity, and a big welcome to its eventual replacement, whether in our lifetime or not.

Acknowledgments: George F. Smoot acknowledges support through his Chaire d'Excellence Université Sorbonne Paris Cité and the financial support of the UnivEarthS Labex programme at Université Sorbonne Paris Cité (ANR-10-LABX-0023 and ANR-11-IDEX-0005-02). I.D. acknowledges that the research work disclosed in this publication is partially funded by the REACH HIGH Scholars Programme—Post-Doctoral Grants. The grant is part-financed by the European Union, Operational Programme II—Cohesion Policy 2014–2020.

Author Contributions: All the authors conceived the idea and contributed equally.

Conflicts of Interest: The authors declare no conflict of interest.

Abbreviations

The following abbreviations are used in this work:

AU	Astronomical Unit
CDM	Cold Dark Matter
CMB	Cosmic Microwave Background
EEP	Einstein Equivalence Principle
ECKS	Einstein-Cartan-Kibble-Sciama
eV	electronvolt
FLRW	Friedmann-Lemaître-Robertson-Walker
GR	General Relativity
Gy	Gigayear (10^9 years)
ΛCDM	Λ Cold Dark Matter
LLI	Local Lorentz invariance
LLR	Lunar laser ranging
LPI	Local position invariance
Mpc	Megaparsec
PPN	Parameterised post-Newtonian
SR	Special Relativity
TeV	teraelectronvolt (10^{12} electronvolts)

References

1. Feynman, R.; Leighton, R.B.; Sands, M.L. *The Feynman Lectures on Physics*; Addison-Wesley: Redwood City, CA, USA, 1989; Volume 2.
2. Gauthier-Lafaye, F.; Holliger, P.; Blanc, P.L. Natural fission reactors in the Franceville basin, Gabon: A review of the conditions and results of a "critical event" in a geologic system. *Geochim. Cosmochim. Acta* **1996**, *60*, 4831–4852.
3. Einstein, A. Erklärung der Perihelbewegung des Merkur aus der allgemeinen Relativitätstheorie. *Sitzungsber. Preuss. Akad. Wiss.* **1915**, *1915*, 831–839. (In German)
4. Dyson, F.W.; Eddington, A.S.; Davidson, C. A Determination of the deflection of light by the sun's gravitational field, from observations made at the total eclipse of 29 May 1919. *Philos. Trans. R. Soc. Lond. Ser. A* **1920**, *220*, 291–333.
5. Will, C.M. The 1919 measurement of the deflection of light. *Class. Quantum Gravity* **2015**, *32*, 124001.
6. Coles, P. Einstein, Eddington and the 1919 Eclipse. In *Historical Development of Modern Cosmology*; Martínez, V.J., Trimble, V., Pons-Bordería, M.J., Eds.; Astronomical Society of the Pacific: San Francisco, CA, USA, 2001; Volume 252, p. 21.
7. Popper, K.R. *Logik der Forschung*; Mohr Siebeck: Tübingen, Germany, 1934.
8. Lahav, O.; Massimi, M. Dark energy, paradigm shifts, and the role of evidence. *Astron. Geophys.* **2014**, *55*, 3.12–3.15.
9. Ellis, G.F.R. Cosmology and verifiability. *Q. J. R. Astron. Soc.* **1975**, *16*, 245–264.
10. Crombie, A.C. Quantification in medieval physics. *Isis* **1961**, *52*, 143–160.
11. Bennett, J. Early modern mathematical instruments. *Isis* **2011**, *102*, 697–705.
12. Franklin, A. Principle of inertia in the middle ages. *Am. J. Phys.* **1976**, *44*, 529–543.
13. Sorabji, R. *Matter, Space and Motion: Theories in Antiquity and Their Sequel*; Duckworth: London, UK, 1988.
14. Cushing, J.T. *Philosophical Concepts in Physics*; Cambridge University Press: Cambridge, UK, 1998.
15. Grant, E. Scientific Thought in Fourteenth-Century Paris: Jean Buridan and Nicole Oresme. *Ann. N. Y. Acad. Sci.* **1978**, *314*, 105–126.
16. Babb, J. Mathematical concepts and proofs from nicole oresme: Using the History of calculus to teach mathematics. *Sci. Educ.* **2005**, *14*, 443–456.
17. Transtrum, M.K.; Machta, B.B.; Brown, K.S.; Daniels, B.C.; Myers, C.R.; Sethna, J.P. Perspective: Sloppiness and emergent theories in physics, biology, and beyond. *J. Chem. Phys.* **2015**, *143*, 010901.
18. Galilei, G. *Sidereus Nuncius Magna, Longeque Admirabilia Spectacula Pandens*; Tommaso Baglioni: Venice, Italy, 1610.
19. Perryman, M. The history of astrometry. *Eur. Phys. J. H* **2012**, *37*, 745–792.
20. Newton, I. *Philosophiae Naturalis Principia Mathematica*, 1st ed.; Joseph Streater: London, UK, 1687.
21. De Maupertuis, P.L.M. Les Loix du mouvement et du repos déduites d'un principe metaphysique. In *Histoire de l'Académie Royale des Sciences et des Belles Lettres*; Académie Royale des Sciences et des Belles Lettres de Berlin: Berlin, Germany, 1746.
22. Euler, L. Réfléxions sur quelques loix générales de la nature qui s'observent dans les effets des forces quelconques. *Acad. R. Sci. Berl.* **1750**, *4*, 189–218.
23. Lagrange, L. *Méchanique Analytique*; Chez La Veuve Desaint: Paris, France, 1788.
24. Hamilton, W.R. On a general method in dynamics. *Philos. Trans. R. Soc. Lond.* **1834**, *124*, 247–308.
25. Hamilton, W.R. Second essay on a general method in dynamics. *Philos. Trans. R. Soc. Lond.* **1835**, *125*, 95–144.
26. Galle, J.G. Account of the discovery of Le Verrier's planet Neptune, at Berlin, Sept. 23, 1846. *Mon. Not. R. Astron. Soc.* **1846**, *7*, 153.
27. Danjon, A. Le centenaire de la découverte de Neptune. *Ciel Terre* **1946**, *62*, 369.
28. Lexell, A.J. Recherches sur la nouvelle planete, decouverte par M. Herschel & nominee Georgium Sidus. *Acta Acad. Sci. Imp. Petropolitanae* **1783**, *1*, 303–329.
29. Sheehan, W. News from Front (of the Solar System): The problem with Mercury, the Vulcan hypothesis, and General Relativity's first astronomical triumph. In Proceedings of the 227th AAS Meeting, Kissimmee, FL, USA, 4–8 January 2016; Volume 227.
30. Poincaré, H. Sur le problème des trois corps et les équations de la dynamique. *Acta Math.* **1890**, *13*, 1–270.

31. Poincaré, H. *Thorie Mathématique de la Lumiére*; Carré & C. Naud: Paris, France, 1889.

32. Maxwell, J.C. A dynamical theory of the electromagnetic field. *Philos. Trans. R. Soc. Lond.* **1865**, *155*, 489–512.

33. Einstein, A. Zur Elektrodynamik bewegter Körper. *Ann. Phys. (Germany)* **1905**, *17*, 891–921.

34. Einstein, A. Über den Einfluss der Schwerkraft auf die Ausbreitung des Lichtes. *Ann. Phys. (Germany)* **1911**, *35*, 898–908.

35. Einstein, A. Lichtgeschwindigkeit und Statik des Gravitationsfeldes. *Ann. Phys. (Germany)* **1912**, *38*, 355–369.

36. Einstein, A. Theorie des statischen Gravitationsfeldes. *Ann. Phys. (Germany)* **1912**, *38*, 443–458.

37. Einstein, A.; Grossmann, M. Entwurf einer verallgemeinerten Relativitätstheorie und eine Theorie der Gravitation. I. Physikalischer Teil von A. Einstein II. Mathematischer Teil von M. Grossmann. *Z. Math. Phys.* **1913**, *62*, 225–244 and 245–261.

38. Le Verrier, U. Lettre de M. Le Verrier à M. Faye sur la théorie de Mercure et sur le mouvement du périhélie de cette planète. *C. R. Hebd. Séances L'Acad. Sci.* **1859**, *49*, 379–383.

39. Will, C.M. The confrontation between general relativity and experiment. *Living Rev. Relativ.* **2014**, *17*, 4.

40. Fierz, M.; Pauli, W. Relativistic wave equations for particles of arbitrary spin in an electromagnetic field. *Proc. R. Soc. Lond.* **1939**, *173*, 211–232.

41. Schild, A. Time. *Tex. Q.* **1960**, *3*, 42–62.

42. Schild, A. Gravitational Theories of the Whitehead type and the principle of equivalence. In *Evidence for Gravitational Theories*; Møller, C., Ed.; Academic Press: New York, NY, USA, 1962.

43. Schild, A. Lectures on General Relativity Theory. In *Relativity Theory and Astrophysics: I, Relativity and Cosmology: II, Galactic Structure: III, Stellar Structure*; Ehlers, J., Ed.; American Mathematical Society: Providence, RI, USA, 1967.

44. Riemann, B. *Bernhard Riemann's Gesammelte mathematische Werke und wissenschaftlicher Nachlass*, 1st ed; Weber, H., Ed.; Teubner: Leipzig, Germany, 1876.

45. Riemann, B. Ueber die Hypothesen, welche der Geometrie zu Grunde liegen. *Abh. Kön. Ges. Wiss. Gött.* **1868**, *13*, 133–150.

46. Einstein, A. Die Feldgleichungen der Gravitation. *Sitzungsber. Preuss. Akad. Wiss.* **1915**, *1915*, 844–847. (In German)

47. Einstein, A. Zur Allgemeinen Relativitätstheorie. *Sitzungsber. Preuss. Akad. Wiss.* **1915**, *1915*, 799–801. (In German)

48. Ni, W.T. Theoretical Frameworks for Testing Relativistic Gravity. IV. A Compendium of Metric Theories of Gravity and Their Post-Newtonian Limits. *Astrophys. J.* **1972**, *176*, 769–796.

49. Thorne, K.S.; Ni, W.T.; Will, C.M. Theoretical frameworks for testing relativistic gravity: A review. In Proceedings of the Conference on Experimental Tests of Gravitational Theories, Pasadena, CA, USA, 11–13 November 1970; pp. 10–31.

50. Lovelock, D. The uniqueness of the Einstein field equations in a four-dimensional space. *Arch. Ration. Mech. Anal.* **1969**, *33*, 54–70.

51. Lovelock, D. The Einstein Tensor and Its Generalizations. *J. Math. Phys.* **1971**, *12*, 498–501.

52. Lovelock, D. The Four-Dimensionality of Space and the Einstein Tensor. *J. Math. Phys.* **1972**, *13*, 874–876.

53. Cartan, E. Sur les équations de la gravitation d'Einstein. *J. Math. Pures Appl.* **1922**, *1*, 141–204.

54. Anderson, J.L. *Principles of Relativity Physics*; Academic Press: New York, NY, USA, 1967.

55. Norton, J.D. General covariance, gauge theories and the Kretschmann objection. In *Symmetries in Physics: Philosophical Reflections*; Brading, K.A., Castellani, E., Eds.; Cambridge University Press: Cambridge, UK, 2003.

56. Kretschmann, E. Über den physikalischen Sinn der Relativitätspostulate, A. Einsteins neue und seine ursprüngliche Relativitätstheorie. *Ann. Phys. (Germany)* **1917**, *53*, 575–614.

57. Nordstrøm, G. Zur Theorie der Gravitation vom Standpunkt des Relativitäsprinzip. *Ann. Phys. (Germany)* **1913**, *42*, 533–554.

58. Einstein, A.; Fokker, A.D. The North current gravitation theory from the viewpoint of absolute differential calculus. *Ann. Phys. (Germany)* **1914**, *44*, 321–328.

59. Whitehead, A.N. *The Principle of Relativity*; Cambridge University Press: Cambridge, UK, 1922.

60. Will, C.M. Relativistic gravity in the solar system, II: Anisotropy in the Newtonian gravitational constant. *Astrophys. J.* **1971**, *169*, 141–156.

61. Birkhoff, G.D. Matter, electricity and gravitation in flat spacetime. *Proc. Natl. Acad. Sci. USA* **1943**, *29*, 231–239.
62. Jordan, P. Zum gegenwärtigen Stand der Diracschen kosmologischen Hypothesen. *Z. Phys.* **1959**, *157*, 112–121.
63. Brans, C.; Dicke, R.H. Mach's principle and a relativistic theory of gravitation. *Phys. Rev.* **1961**, *124*, 925–935.
64. Dicke, R.H. Mach's principle and invariance under transformation of units. *Phys. Rev.* **1962**, *125*, 2163–2167.
65. Will, C.M. *Theory and Experiment in Gravitational Physics*, 2nd ed.; Cambridge University Press: Cambridge, UK, 1993.
66. Psaltis, D. Constraining Brans-Dicke Gravity with Accreting Millisecond Pulsars in Ultracompact Binaries. *Astrophys. J.* **2008**, *688*, 1282–1287.
67. Bertotti, B.; Iess, L.; Tortora, P. A test of general relativity using radio links with the Cassini spacecraft. *Nature* **2003**, *425*, 374–376.
68. Avilez, A.; Skordis, C. Cosmological Constraints on Brans-Dicke Theory. *Phys. Rev. Lett.* **2014**, *113*, 011101.
69. Ooba, J.; Ichiki, K.; Chiba, T.; Sugiyama, N. Planck constraints on scalar-tensor cosmology and the variation of the gravitational constant. *Phys. Rev. D* **2016**, *93*, 122002.
70. Bull, P. Extending Cosmological Tests of General Relativity with the Square Kilometre Array. *Astrophys. J.* **2016**, *817*, 26.
71. Damour, T.; Nordtvedt, K. General relativity as a cosmological attractor of tensor-scalar theories. *Phys. Rev. Lett.* **1993**, *70*, 2217–2219.
72. Damour, T.; Nordtvedt, K. Tensor-scalar cosmological models and their relaxation toward general relativity. *Phys. Rev. D* **1993**, *48*, 3436–3450.
73. Kustaanheimo, P. Route dependence of the gravitational redshift. *Phys. Lett.* **1966**, *23*, 75–77.
74. Kustaanheimo, P.E.; Nuotio, V.S. *Relativistic Theories of Gravitation I: The One-Body Problem*; University of Helsinki: Helsinki, Finland, 1967; unpublished.
75. Einstein, A. Näherungsweise Integration der Feldgleichungen der Gravitation. *Sitzungsber. Preuss. Akad. Wiss.* **1916**, *1916 Pt 1*, 688–696. (In German)
76. Kumar, D.; Soni, V. Single particle Schrödinger equation with gravitational self-interaction. *Phys. Lett. A* **2000**, *271*, 157–166.
77. Deser, S. Gravity from self-interaction redux. *Gen. Relativ. Gravit.* **2010**, *42*, 641–646.
78. Hertz, H. *Untersuchungen über die Ausbreitung der Elektrischen Kraft*; J.A. Barth: Leipzig, Germany, 1894.
79. Einstein, A. Über einen die Erzeugung und Verwandlung des Lichtes betreffenden heuristischen Gesichtspunkt. *Ann. Phys. (Germany)* **1905**, *17*, 132–148.
80. LIGO Collaboration. Observation of Gravitational Waves from a Binary Black Hole Merger. *Phys. Rev. Lett.* **2016**, *116*, 061102.
81. Yang, C.L.; Mills, R.L. Conservation of isotopic spin and isotopic gauge invariance. *Phys. Rev.* **1954**, *96*, 191–195.
82. Yang, C.L.; Mills, R.L. Isotopic Spin Conservation and a Generalized Gauge Invariance. *Phys. Rev.* **1954**, *95*, 631.
83. Weyl, H. Gravitation und Elektrizität. In *Das Relativitätsprinzip*; J. A. Barth: Leipzig, Germany, 1918.
84. Weyl, H. *Raum-Zeit-Materie*; Springer: Berlin, Germany, 1918.
85. Christoffel, E.B. Über die Transformation der homogenen Differentialausdrücke zweiten Grades. *J. Reine Angew. Math.* **1869**, *70*, 46–70.
86. Goenner, H.F.M. On the History of Unified Field Theories. *Living Rev. Relativ.* **2004**, *7*, 2.
87. Smiciklas, M.; Brown, J.M.; Cheuk, L.W.; Smullin, S.J.; Romalis, M.V. New test of local lorentz invariance using a [21]Ne-Rb-**K** comagnetometer. *Phys. Rev. Lett.* **2011**, *107*, 171604.
88. Peck, S.K.; Kim, D.K.; Stein, D.; Orbaker, D.; Foss, A.; Hummon, M.T.; Hunter, L.R. Limits on local Lorentz invariance in mercury and cesium. *Phys. Rev. A* **2012**, *86*, 012109.
89. Hohensee, M.A.; Leefer, N.; Budker, D.; Harabati, C.; Dzuba, V.A.; Flambaum, V.V. Limits on violations of lorentz symmetry and the einstein equivalence principle using radio-frequency spectroscopy of atomic dysprosium. *Phys. Rev. Lett.* **2013**, *111*, 050401.
90. Utiyama, R. Invariant theoretical interpretation of interaction. *Phys. Rev.* **1956**, *101*, 1597–1607.
91. Kibble, T.W.B. Lorentz invariance and the gravitational field. *J. Math. Phys.* **1961**, *2*, 212–221.

92. Carmeli, M.; Malin, S. Reformulation of general relativity as a gauge theory. *Ann. Phys. (Germany)* **1977**, *103*, 208–232.

93. Ivanenko, D.; Sardanashvily, G. The gauge treatment of gravity. *Phys. Rep.* **1983**, *94*, 1–45.

94. Sardanashvily, G.; Zakharov, O. *Gauge Gravitation Theory*; World Scientific: Singapore, 1992.

95. Jackson, J.D.; Okun, L.B. Historical roots of gauge invariance. *Rev. Mod. Phys.* **2001**, *73*, 663–680.

96. Sardanashvily, G. Classical gauge theory of gravity. *Theor. Math. Phys.* **2002**, *132*, 1163–1171.

97. Cartan, É. Sur une généralisation de la notion de courboure de Riemann et les espaces à torsion. *C. R. Acad. Sci. (Paris)* **1922**, *174*, 593–595.

98. Cartan, É. Sur les variétés à connexion affine et la théorie de la relativité généralisée (première partie). *Ann. Sci. École Norm. Super.* **1923**, *40*, 325–412.

99. Cartan, É. Sur les variétés à connexion affine et la théorie de la relativité généralisée (suite). *Ann. Sci. École Norm. Super.* **1924**, *41*, 1–25.

100. Cartan, É. Sur les variétés à connexion affine et la théorie de la relativité généralisée (deuxième partie). *Ann. Sci. École Norm. Super.* **1925**, *42*, 17–88.

101. Weyl, H. A remark on the coupling of gravitation and electron. *Phys. Rev.* **1950**, *77*, 699–701.

102. Sciama, D. The physical structure of general relativity. *Rev. Mod. Phys.* **1964**, *36*, 1103.

103. Weyssenhoff, J.; Raabe, A. Relativistic Dynamics of spin-fluids and spin-particles. *Acta Phys. Pol.* **1947**, *9*, 7–18.

104. Costa de Beauregard, O. Translational inertial spin effect. *Phys. Rev.* **1963**, *129*, 466–471.

105. Cosserat, E.; Cosserat, F. *Théorie des Corps déformables*; Hermann: Paris, France, 1909.

106. Puetzfeld, D.; Obukhov, Y.N. Prospects of detecting spacetime torsion. *Int. J. Mod. Phys. D* **2014**, *23*, 1442004.

107. Hehl, F.W.; von der Heyde, P.; Kerlick, G.D.; Nester, J.M. General relativity with spin and torsion: Foundations and prospects. *Rev. Mod. Phys.* **1976**, *48*, 393–416.

108. Lasenby, A.N.; Doran, C.J.; Gull, S.F. Cosmological consequences of a flat-space theory of gravity. In *Clifford Algebras and Their Applications in Mathematical Physics*; Brackx, F., Delanghe, R., Serras, H., Eds.; Kluwer: Dordrecht, The Netherlands, 1993; p. 387.

109. Lasenby, A.N.; Doran, C.J.; Gull, S.F. Grassmann calculus, pseudoclassical mechanics and geometric algebra. *J. Math. Phys.* **1993**, *34*, 3683–3712.

110. Lasenby, A.N.; Doran, C.J.L.; Gull, S.F. A multivector derivative approach to Lagrangian field theory. *Found. Phys.* **1993**, *23*, 1295–1327.

111. Lasenby, A.N.; Doran, C.J.L.; Gull, S.F. 2-spinors, twistors and supersymmetry in the spacetime algebra. In *Spinors, Twistors, Clifford Algebras and Quantum Deformations*; Oziewicz, Z., Jancewicz, B., Borowiec, A., Eds.; Kluwer: Dordrecht, The Netherlands, 1993; p. 233.

112. Lasenby, A.N.; Doran, C.J.L.; Gull, S.F. Astrophysical and cosmological consequences of a gauge theory of gravity. In *Current Topics in Astrofundamental Physics*; Sánchez, N., Zichichi, A., Eds.; World Scientific: Singapore, 1995; p. 359.

113. Lasenby, A.N.; Doran, C.J.; Gull, S.F. Gravity, gauge theories and geometric algebra. *Philos. Trans. R. Soc. Lond.* **1998**, *A356*, 487–582.

114. Carmeli, M.; Leibowitz, E.; Nissani, N. *Gravitation: $SL(2, \mathbb{C})$ Gauge Theory and Conservation Laws*; World Scientific: Singapore, 1990.

115. Mukunda, N. An elementary introduction to the gauge theory approach to gravity. In *Gravitation, Gauge Theories and the Early Universe*; Iyer, B.R., Mukunda, N., Vishvershwara, C.V., Eds.; Kluwer: Dordrecht, The Netherlands, 1989.

116. Misner, C.W.; Thorne, K.S.; Wheeler, J.A. *Gravitation*; Freeman: San Francisco, CA, USA, 1973.

117. Berti, E.; Barausse, E.; Cardoso, V.; Gualtieri, L.; Pani, P.; Sperhake, U.; Stein, L.C.; Wex, N.; Yagi, K.; Baker, T.; et al. Testing general relativity with present and future astrophysical observations. *Class. Quantum Gravity* **2015**, *32*, 243001.

118. Poincaré, H. L'état actuel et l'avenir de la physique mathématique. *Bull. Sci. Math.* **1904**, *2*, 302–324.

119. Kaluza, T. *Zum Unitätsproblem in der Physik*; Sitzungsberichte der Königlich Preußischen Akademie der Wissenschaften (Berlin): Berlin, Germany, 1921; pp. 966–972.

120. Klein, O. Quantentheorie und fünfdimensionale Relativitätstheorie. *Z. Phys.* **1926**, *37*, 895–906.

121. Milne, E.A. *Kinematic Relativity; A Sequel to Relativity, Gravitation and World Structure*; Clarendon Press: Oxford, UK, 1948.

122. Thiry, Y. Sur la régularité des champs gravitationels et électromagnétiques dans les théories unitaires. *C. R. Hebd. Séances Acad. Sci.* **1948**, *226*, 1881–1882.

123. Belinfante, F.; Swihart, J. Phenomenological linear theory of gravitation: Part I. Classical mechanics. *Ann. Phys. (Germany)* **1957**, *1*, 168–195

124. Will, C.M.; Nordtvedt, K., Jr. Conservation laws and preferred frames in relativistic gravity. I. Preferred-frame theories and an extended PPN formalism. *Astrophys. J.* **1972**, *177*, 757–774.

125. Barker, B.M. General scalar-tensor theory of gravity with constant G. *Astrophys. J.* **1978**, *219*, 5–11.

126. Rosen, N. A bi-metric theory of gravitation. *Gen. Relativ. Gravitat.* **1973**, *4*, 435–447.

127. Rosen, N. A bi-metric theory of gravitation. II. *Gen. Relativ. Gravitat.* **1975**, *6*, 259–268.

128. Rastall, P. A theory of gravity. *Can. J. Phys.* **1976**, *54*, 66–75.

129. Buchdahl, H.A. Non-linear lagrangians and cosmological theory. *Mon. Not. R. Astron. Soc.* **1970**, *150*, 1–8.

130. Starobinsky, A. A new type of isotropic cosmological models without singularity. *Phys. Lett. B* **1980**, *91*, 99–102.

131. Milgrom, M. A modification of the Newtonian dynamics as a possible alternative to the hidden mass hypothesis. *Astrophys. J.* **1983**, *270*, 365–370.

132. Milgrom, M. A modification of the newtonian dynamics—Implications for galaxies. *Astrophys. J.* **1983**, *270*, 371–389.

133. Milgrom, M. A modification of the newtonian dynamics—Implications for galaxy systems. *Astrophys. J.* **1983**, *270*, 384–415.

134. Dvali, G.; Gabadadze, G.; Porrati, M. 4D gravity on a brane in 5D Minkowski space. *Phys. Lett. B* **2000**, *485*, 208–214.

135. Bekenstein, J.D. Relativistic gravitation theory for the modified Newtonian dynamics paradigm. *Phys. Rev. D* **2004**, *70*, 083509.

136. Seifert, M.D. Stability of spherically symmetric solutions in modified theories of gravity. *Phys. Rev. D* **2007**, *76*, 064002.

137. Reyes, R.; Mandelbaum, R.; Seljak, U.; Baldauf, T.; Gunn, J.E.; Lombriser, L.; Smith, R.E. Confirmation of general relativity on large scales from weak lensing and galaxy velocities. *Nature* **2010**, *464*, 256–258.

138. Einstein, A. Kosmologische Betrachtungen zur allgemeinen Relativitätstheorie. *Sitzungsber. Preuss. Akad. Wiss.* **1917**, *1917*, 142–152.

139. Friedmann, A. Über die Krümmung des Raumes. *Z. Phys.* **1922**, *10*, 377–386.

140. Friedmann, A. Über die Möglichkeit einer Welt mit konstanter negativer Krümmung des Raumes. *Z. Phys.* **1924**, *21*, 326–332.

141. Hubble, E. A relation between distance and radial velocity among extra-galactic nebulae. *Proc. Natl. Acad. Sci. USA* **1929**, *15*, 168–173.

142. Lemaître, G. Un univers homogène de masse constante et de rayon croissant rendant compte de la vitesse radiale des nébuleuses extra-galactiques. *Ann. Soc. Sci. Brux.* **1927**, *A47*, 49–59.

143. Robertson, H.P. Kinematics and world structure. *Astrophys. J.* **1935**, *82*, 248–301.

144. Robertson, H.P. Kinematics and world structure II. *Astrophys. J.* **1936**, *83*, 187–201.

145. Robertson, H.P. Kinematics and world structure III. *Astrophys. J.* **1936**, *83*, 257–271.

146. Walker, A.G. On Milne's theory of world-structure. *Proc. Lond. Math. Soc.* **1937**, *42*, 90–127.

147. Ellis, G.F.R.; Stoeger, W. Horizons in inflationary universes. *Class. Quantum Gravity* **1988**, *5*, 207–220.

148. Einstein, A.; de Sitter, W. On the relation between the expansion and the mean density of the universe. *Proc. Natl. Acad. Sci. USA* **1932**, *18*, 213–214.

149. Adams, F.C.; Laughlin, G. A dying universe: The long-term fate and evolution of astrophysical objects. *Rev. Mod. Phys.* **1997**, *69*, 337–372.

150. Caldwell, R.R.; Kamionkowski, M.; Weinberg, N.N. Phantom energy: Dark energy with $w < -1$ causes a cosmic doomsday. *Phys. Rev. Lett.* **2003**, *91*, 071301.

151. Planck Collaboration. Planck 2015 Results. XIII. Cosmological Parameters. *Astron. Astrophys.* **2016**, *594*, A13.

152. Alpher, R.A.; Bethe, H.; Gamow, G. The origin of chemical elements. *Phys. Rev.* **1948**, *73*, 803–804.

153. Alpher, R.A.; Herman, R. Evolution of the universe. *Nature* **1948**, *162*, 774–775.

154. Gamow, G. The evolution of the universe. *Nature* **1948**, *162*, 680–682.
155. Gamow, G. The origin of elements and the separation of galaxies. *Phys. Rev.* **1948**, *74*, 505–506.
156. Penzias, A.A.; Wilson, R.W. A measurement of excess antenna temperature at 4080 Mc/s. *Astrophys. J.* **1965**, *142*, 419–421.
157. Christenson, J.H.; Cronin, J.W.; Fitch, V.L.; Turlay, R. Evidence for the 2π decay of the K_2^0 meson. *Phys. Rev. Lett.* **1964**, *13*, 138–140.
158. Sakharov, A.D. Violation of CP invariance, c asymmetry, and baryon asymmetry of the universe. *JETP Lett.* **1967**, *5*, 24–27.
159. Kuzmin, V.A. CP-noninvariance and baryon asymmetry of the universe. *J. Exp. Theor. Phys. Lett.* **1970**, *12*, 228–230.
160. Khlopov, M. Cosmological reflection of particle symmetry. *Symmetry* **2016**, *8*, 81.
161. Fixsen, D.J. The temperature of the cosmic microwave background. *Astrophys. J.* **2009**, *707*, 916–920.
162. Dunkley, J.; Komatsu, E.; Nolta, M.R.; Spergel, D.N.; Larson, D.; Hinshaw, G.; Page, L.; Bennett, C.L.; Gold, B.; Jarosik, N.; et al. Five-year Wilkinson Microwave Anisotropy Probe observations: Likelihoods and parameters from the WMAP data. *Astrophys. J. Suppl.* **2009**, *180*, 306–329.
163. Hannestad, S. Primordial neutrinos. *Annu. Rev. Nucl. Part. Sci.* **2006**, *56*, 137–161.
164. Lesgourgues, J.; Pastor, S. Massive neutrinos and cosmology. *Phys. Rep.* **2006**, *429*, 307–379.
165. Lesgourgues, J.; Pastor, S. Neutrino cosmology and Planck. *New J. Phys.* **2014**, *16*, 065002.
166. Elgarøy, Ø.; Lahav, O. Neutrino masses from cosmological probes. *New J. Phys.* **2005**, *7*, 61.
167. Abazajian, K.; Fuller, G.M.; Patel, M. Sterile neutrino hot, warm, and cold Dark Matter. *Phys. Rev. D* **2001**, *64*, 023501.
168. Boyarsky, A.; Ruchayskiy, O.; Shaposhnikov, M. The role of sterile neutrinos in cosmology and astrophysics. *Annu. Rev. Nucl. Part. Sci.* **2009**, *59*, 191–214.
169. Mangano, G.; Miele, G.; Pastor, S.; Peloso, M. A precision calculation of the effective number of cosmological neutrinos. *Phys. Lett. B* **2002**, *534*, 8–16.
170. Feeney, S.M.; Peiris, H.V.; Verde, L. Is there evidence for additional neutrino species from cosmology? *J. Cosmol. Astropart. Phys.* **2013**, *4*, 036.
171. Olive, K.; Group, P.D. Review of particle physics. *Chin. Phys. C* **2014**, *38*, 090001.
172. Primack, J.R.; Gross, M.A.K. Hot Dark Matter in cosmology. In *Current Aspects of Neutrino Physics*; Caldwell, D.O., Ed.; Springer: Berlin, Germany, 2001; pp. 287–308.
173. Gerstein, G.; Zel'dovich, Y.B. Rest mass of the muonic neutrino and cosmology (English translation). *Lett. J. Exp. Theor. Phys.* **1966**, *4*, 120–122.
174. Marx, G.; Szalay, A.S. Cosmological limit on the neutretto rest mass. In Proceedings of the Neutrino 72, Balatonfured, Hungary, 11–17 June 1972; Volume 1, p. 123.
175. Cowsik, R.; McClelland, J. An upper limit on the neutrino rest mass. *Phys. Rev. Lett.* **1972**, *29*, 669–670.
176. Szalay, A.S.; Marx, G. Neutrino rest mass from cosmology. *Astron. Astrophys.* **1976**, *49*, 437–441.
177. Tegmark, M.; Eisenstein, D.J.; Strauss, M.A.; Weinberg, D.H.; Blanton, M.R.; Frieman, J.A.; Fukugita, M.; Gunn, J.E.; Hamilton, A.J.S.; Knapp, G.R.; et al. Cosmological constraints from the SDSS luminous red galaxies. *Phys. Rev. D* **2006**, *74*, 123507.
178. Shiraishi, M.; Ichikawa, K.; Ichiki, K.; Sugiyama, N.; Yamaguchi, M. Constraints on neutrino masses from WMAP5 and BBN in the lepton asymmetric universe. *J. Cosmol. Astropart. Phys.* **2009**, *7*, 5.
179. Tereno, I.; Schimd, C.; Uzan, J.P.; Kilbinger, M.; Vincent, F.H.; Fu, L. CFHTLS weak-lensing constraints on the neutrino masses. *Astron. Astrophys.* **2009**, *500*, 657–665.
180. Ichiki, K.; Takada, M.; Takahashi, T. Constraints on neutrino masses from weak lensing. *Phys. Rev. D* **2009**, *79*, 023520.
181. Xia, J.Q.; Granett, B.R.; Viel, M.; Bird, S.; Guzzo, L.; Haehnelt, M.G.; Coupon, J.; McCracken, H.J.; Mellier, Y. Constraints on massive neutrinos from the CFHTLS angular power spectrum. *J. Cosmol. Astropart. Phys.* **2012**, *6*, 010.
182. Olive, K.A. Inflation. *Phys. Rep.* **1990**, *190*, 307–403.
183. Guth, A.H. (Ed.) *The Inflationary Universe. The Quest for a New Theory of Cosmic Origins*; Addison-Wesley: Reading, MA, USA, 1997.
184. Peacock, J.A. *Cosmological Physics*; Cambridge University Press: Cambridge, UK, 1999.

185.   Liddle, A.R.; Lyth, D.H. *Cosmological Inflation and Large-Scale Structure*; Cambridge University Press: Cambridge, UK, 2000.
186.   Uzan, J.P. Inflation in the standard cosmological model. *C. R. Phys.* **2015**, *16*, 875–890.
187.   Misner, C.W. Mixmaster universe. *Phys. Rev. Lett.* **1969**, *22*, 1071–1074.
188.   Belinskij, V.A.; Khalatnikov, I.M.; Lifshits, E.M. Oscillatory approach to a singular point in the relativistic cosmology. *Adv. Phys.* **1970**, *19*, 525–573.
189.   Starobinsky, A.A. Spectrum of relict gravitational radiation and the early state of the universe. *Sov. J. Exp. Theor. Phys. Lett.* **1979**, *30*, 682–685.
190.   't Hooft, G. Magnetic monopoles in unified gauge theories. *Nucl. Phys. B* **1974**, *79*, 276–284.
191.   Preskill, J. Magnetic monopoles. *Annu. Rev. Nucl. Part. Sci.* **1984**, *34*, 461–530.
192.   Zel'dovich, Y.B.; Khlopov, M.Y. On the concentration of relic monopoles in the universe. *Phys. Lett. B* **1979**, *79*, 239–241.
193.   Preskill, J.P. Cosmological production of superheavy magnetic monopoles. *Phys. Rev. Lett.* **1979**, *43*, 1365–1368.
194.   Guth, A.H. Inflationary universe: A possible solution to the horizon and flatness problems. *Phys. Rev. D* **1981**, *23*, 347–356.
195.   Linde, A. *Inflation and Quantum Cosmology*; Academic Press: New York, NY, USA, 1989.
196.   Lidsey, J.E.; Liddle, A.R.; Kolb, E.W.; Copeland, E.J.; Barreiro, T.; Abney, M. Reconstructing the inflaton potential—An overview. *Rev. Mod. Phys.* **1997**, *69*, 373–410.
197.   Sato, K. Cosmological baryon-number domain structure and the first order phase transition of a vacuum. *Phys. Lett. B* **1981**, *99*, 66–70.
198.   Einhorn, M.B.; Sato, K. Monopole production in the very early universe in a first-order phase transition. *Nucl. Phys. B* **1981**, *180*, 385–404.
199.   Mukhanov, V.F.; Chibisov, G.V. Energy of vacuum and the large-scale structure of the universe. *Zhurnal Eksper. Teor. Fiz.* **1982**, *83*, 475–487.
200.   Mukhanov, V.F.; Chibisov, G.V. Quantum fluctuations and a nonsingular universe. *Sov. J. Exp. Theor. Phys. Lett.* **1981**, *33*, 532–535.
201.   Guth, A.H.; Pi, S.Y. Fluctuations in the new inflationary universe. *Phys. Rev. Lett.* **1982**, *49*, 1110–1113.
202.   Hawking, S.W. The development of irregularities in a single bubble inflationary universe. *Phys. Lett. B* **1982**, *115*, 295–297.
203.   Starobinsky, A.A. Dynamics of phase transition in the new inflationary universe scenario and generation of perturbations. *Phys. Lett. B* **1982**, *117*, 175–178.
204.   Bardeen, J.M.; Steinhardt, P.J.; Turner, M.S. Spontaneous creation of almost scale-free density perturbations in an inflationary universe. *Phys. Rev. D* **1983**, *28*, 679–693.
205.   Planck Collaboration. Planck 2015 Results. XX. Constraints on Inflation *Astron. Astrophys.* **2016**, *594*, A20.
206.   Liddle, A.R.; Lyth, D.H. COBE, gravitational waves, inflation and extended inflation. *Phys. Lett. B* **1992**, *291*, 391–398.
207.   Liddle, A.R.; Lyth, D.H. The cold Dark Matter density perturbation. *Phys. Rep.* **1993**, *231*, 1–105.
208.   Linde, A.D. A new inflationary universe scenario: A possible solution of the horizon, flatness, homogeneity, isotropy and primordial monopole problems. *Phys. Lett. B* **1982**, *108*, 389–393.
209.   Dodelson, S. *Modern Cosmology*; Academic Press: New York, NY, USA, 2003.
210.   Kosowsky, A.; Turner, M.S. CBR anisotropy and the running of the scalar spectral index. *Phys. Rev. D* **1995**, *52*, 1739–1743.
211.   Smoot, G.F.; Bennett, C.L.; Kogut, A.; Wright, E.L.; Aymon, J.; Boggess, N.W.; Cheng, E.S.; de Amici, G.; Gulkis, S.; Hauser, M.G.; et al.; Structure in the COBE differential microwave radiometer first-year maps. *Astrophys. J.* **1992**, *396*, L1–L5.
212.   Fixsen, D.J.; Cheng, E.S.; Cottingham, D.A.; Eplee, R.E., Jr.; Isaacman, R.B.; Mather, J.C.; Meyer, S.S.; Noerdlinger, P.D.; Shafer, R.A.; Weiss, R.; et al. Cosmic microwave background dipole spectrum measured by the COBE FIRAS instrument. *Astrophys. J.* **1994**, *420*, 445–449.
213.   Dwek, E.; Arendt, R.G.; Hauser, M.G.; Fixsen, D.; Kelsall, T.; Leisawitz, D.; Pei, Y.C.; Wright, E.L.; Mather, J.C.; Moseley, S.H.; et al. The COBE diffuse infrared background experiment search for the cosmic infrared background. IV. Cosmological implications. *Astrophys. J.* **1998**, *508*, 106–122.

214. Spergel, D.N.; Verde, L.; Peiris, H.V.; Komatsu, E.; Nolta, M.R.; Bennett, C.L.; Halpern, M.; Hinshaw, G.; Jarosik, N.; Kogut, A.; et al. First-year Wilkinson Microwave Anisotropy Probe (WMAP) observations: Determination of cosmological parameters. *Astrophys. J. Suppl.* **2003**, *148*, 175–194.

215. Peiris, H.V.; Komatsu, E.; Verde, L.; Spergel, D.N.; Bennett, C.L.; Halpern, M.; Hinshaw, G.; Jarosik, N.; Kogut, A.; Limon, M.; et al. First-year Wilkinson Microwave Anisotropy Probe (WMAP) observations: Implications for inflation. *Astrophys. J. Suppl.* **2003**, *148*, 213–231.

216. Spergel, D.N.; Bean, R.; Doré, O.; Nolta, M.R.; Bennett, C.L.; Dunkley, J.; Hinshaw, G.; Jarosik, N.; Komatsu, E.; Page, L.; et al. Three-year Wilkinson Microwave Anisotropy Probe (WMAP) observations: Implications for cosmology. *Astrophys. J. Suppl.* **2007**, *170*, 377–408.

217. Komatsu, E.; Dunkley, J.; Nolta, M.R.; Bennett, C.L.; Gold, B.; Hinshaw, G.; Jarosik, N.; Larson, D.; Limon, M.; Page, L.; et al. Five-year Wilkinson Microwave Anisotropy Probe observations: Cosmological interpretation. *Astrophys. J. Suppl.* **2009**, *180*, 330–376.

218. Komatsu, E.; Smith, K.M.; Dunkley, J.; Bennett, C.L.; Gold, B.; Hinshaw, G.; Jarosik, N.; Larson, D.; Nolta, M.R.; Page, L.; et al. Seven-year Wilkinson Microwave Anisotropy Probe (WMAP) observations: Cosmological interpretation. *Astrophys. J. Suppl.* **2011**, *192*, 18.

219. Hinshaw, G.; Larson, D.; Komatsu, E.; Spergel, D.N.; Bennett, C.L.; Dunkley, J.; Nolta, M.R.; Halpern, M.; Hill, R.S.; Odegard, N.; et al. Nine-year Wilkinson Microwave Anisotropy Probe (WMAP) observations: Cosmological parameter results. *Astrophys. J. Suppl.* **2013**, *208*, 19.

220. Planck Collaboration. Planck 2013 results. XVI. Cosmological parameters. *Astron. Astrophys.* **2014**, *571*, A16.

221. Planck Collaboration. Planck 2013 results. XXII. Constraints on inflation. *Astron. Astrophys.* **2014**, *571*, A22.

222. BICEP2 Collaboration. Detection of B-mode polarization at degree angular scales by BICEP2. *Phys. Rev. Lett.* **2014**, *112*, 241101.

223. Planck Collaboration. Planck intermediate results XXX. The angular power spectrum of polarized dust emission at intermediate and high Galactic latitudes. *Astron. Astrophys.* **2016**, *586*, A133.

224. Huang, Z.; Verde, L.; Vernizzi, F. Constraining inflation with future galaxy redshift surveys. *J. Cosmol. Astropart. Phys.* **2012**, *4*, 005.

225. Amendola, L.; Appleby, S.; Bacon, D.; Baker, T.; Baldi, M.; Bartolo, N.; Blanchard, A.; Bonvin, C.; Borgani, S.; Branchini, E.; et al. Cosmology and fundamental physics with the Euclid satellite. *Living Rev. Relativ.* **2013**, *16*, 6.

226. Namikawa, T.; Yamauchi, D.; Sherwin, B.; Nagata, R. Delensing cosmic microwave background B modes with the Square Kilometre Array Radio Continuum Survey. *Phys. Rev. D* **2016**, *93*, 043527.

227. Kapteyn, J.C. First attempt at a theory of the arrangement and motion of the sidereal system. *Astrophys. J.* **1922**, *55*, 302–328.

228. Oort, J.H. The force exerted by the stellar system in the direction perpendicular to the galactic plane and some related problems. *Bull. Astron. Inst. Neth.* **1932**, *6*, 249–287.

229. Zwicky, F. Die Rotverschiebung von extragalaktischen Nebeln. *Helv. Phys. Acta* **1933**, *6*, 110–127.

230. Zwicky, F. On the Masses of Nebulae and of Clusters of Nebulae. *Astrophys. J.* **1937**, *86*, 217–246.

231. Freese, K. Review of observational evidence for Dark Matter in the universe and in upcoming searches for dark stars. In Proceedings of the Dark Energy and Dark Matter: Observations, Experiments and Theories, Lyons, France, 7–11 July 2008; Volume 36, pp. 113–126.

232. Kahn, F.D.; Woltjer, L. Intergalactic matter and the galaxy. *Astrophys. J.* **1959**, *130*, 705–717.

233. Roberts, M.S.; Rots, A.H. Comparison of rotation curves of different galaxy types. *Astron. Astrophys.* **1973**, *26*, 483–485.

234. Einasto, J.; Kaasik, A.; Saar, E. Dynamic evidence on massive coronas of galaxies. *Nature* **1974**, *250*, 309–310.

235. Ostriker, J.P.; Peebles, P.J.E.; Yahil, A. The size and mass of galaxies, and the mass of the universe. *Astrophys. J.* **1974**, *193*, L1–L4.

236. Rubin, V.C.; Thonnard, N.; Ford, W.K., Jr. Extended rotation curves of high-luminosity spiral galaxies. IV—Systematic dynamical properties, SA through SC. *Astrophys. J.* **1978**, *225*, L107–L111.

237. Navarro, J.F.; Frenk, C.S.; White, S.D.M. The structure of cold Dark Matter halos. *Astrophys. J.* **1996**, *462*, 563.

238. Fukugita, M.; Hogan, C.J.; Peebles, P.J.E. The cosmic baryon budget. *Astrophys. J.* **1998**, *503*, 518–530.

239. Sachs, R.K.; Wolfe, A.M. Perturbations of a cosmological model and angular variations of the microwave background. *Astrophys. J.* **1967**, *147*, 73–90.

240. Eisenstein, D.J. Dark energy and cosmic sound. *New Astron. Rev.* **2005**, *49*, 360–365.

241. Bassett, B.; Hlozek, R. Baryon acoustic oscillations. In *Dark Energy: Observational and Theoretical Approaches*; Ruiz-Lapuente, P., Ed.; Cambridge University Press: Cambridge, UK, 2010; p. 246.

242. Schneider, P. Weak gravitational lensing. In *Gravitational Lensing: Strong, Weak and Micro, Saas-Fee Advanced Courses 33*; Schneider, P., Kochanek, C.S., Wambsganss, J., Eds.; Springer: Berlin/Heidelberg, Germany, 2006; p. 269.

243. Bacon, D.J.; Réfrégier, A.R.; Ellis, R.S. Detection of weak gravitational lensing by large-scale structure. *Mon. Not. R. Astron. Soc.* **2000**, *318*, 625–640.

244. Kaiser, N. A new shear estimator for weak-lensing observations. *Astrophys. J.* **2000**, *537*, 555–577.

245. Van Waerbeke, L.; Mellier, Y.; Erben, T.; Cuillandre, J.C.; Bernardeau, F.; Maoli, R.; Bertin, E.; Mc Cracken, H.J.; Le Fèvre, O.; Fort, B.; et al. Detection of correlated galaxy ellipticities from CFHT data: First evidence for gravitational lensing by large-scale structures. *Astron. Astrophys.* **2000**, *358*, 30–44.

246. Wittman, D.M.; Tyson, J.A.; Kirkman, D.; Dell'Antonio, I.; Bernstein, G. Detection of weak gravitational lensing distortions of distant galaxies by cosmic Dark Matter at large scales. *Nature* **2000**, *405*, 143–148.

247. Bertone, G.; Hooper, D.; Silk, J. Particle Dark Matter: Evidence, candidates and constraints. *Phys. Rep.* **2005**, *405*, 279–390.

248. Salati, P. The bestiary of Dark Matter species. In Proceedings of the Dark Energy and Dark Matter: Observations, Experiments and Theories, Lyons, France, 7–11 July 2008; Volume 36, pp. 175–186.

249. Garrett, K.; Dūda, G. Dark matter: A primer. *Adv. Astron.* **2011**, *2011*, 968283.

250. Kappl, R.; Winkler, M.W. New limits on Dark Matter from Super-Kamiokande. *Nucl. Phys. B* **2011**, *850*, 505–521.

251. Arina, C.; Hamann, J.; Wong, Y.Y.Y. A Bayesian view of the current status of Dark Matter direct searches. *J. Cosmol. Astropart. Phys.* **2011**, *9*, 022.

252. Porter, T.A.; Johnson, R.P.; Graham, P.W. Dark matter searches with astroparticle data. *Annu. Rev. Astron. Astrophys.* **2011**, *49*, 155–194.

253. Calore, F.; de Romeri, V.; Donato, F. Conservative upper limits on WIMP annihilation cross section from Fermi-LAT γ rays. *Phys. Rev. D* **2012**, *85*, 023004.

254. Klasen, M.; Pohl, M.; Sigl, G. Indirect and direct search for Dark Matter. *Prog. Part. Nucl. Phys.* **2015**, *85*, 1–32.

255. Mayet, F.; Green, A.M.; Battat, J.B.R.; Billard, J.; Bozorgnia, N.; Gelmini, G.B.; Gondolo, P.; Kavanagh, B.J.; Lee, S.K.; Loomba, D.; et al. A review of the discovery reach of directional Dark Matter detection. *Phys. Rep.* **2016**, *627*, 1–49.

256. Jungman, G.; Kamionkowski, M.; Griest, K. Supersymmetric Dark Matter. *Phys. Rep.* **1996**, *267*, 195–373.

257. IceCube Collaboration. Search for Dark Matter annihilations in the Sun with the 79-string IceCube detector. *Phys. Rev. Lett.* **2013**, *110*, 131302.

258. Aguirre, A.; Schaye, J.; Quataert, E. Problems for modified Newtonian dynamics in clusters and the Lyα forest? *Astrophys. J.* **2001**, *561*, 550–558.

259. Pointecouteau, E.; Silk, J. New constraints on modified Newtonian dynamics from galaxy clusters. *Mon. Not. R. Astron. Soc.* **2005**, *364*, 654–658.

260. Contaldi, C.R.; Wiseman, T.; Withers, B. TeVeS gets caught on caustics. *Phys. Rev. D* **2008**, *78*, 044034.

261. Lue, A.; Starkman, G.D. Squeezing MOND into a cosmological scenario. *Phys. Rev. Lett.* **2004**, *92*, 131102.

262. Dodelson, S. The real problem with MOND. *Int. J. Mod. Phys. D* **2011**, *20*, 2749–2753.

263. Ferreras, I.; Mavromatos, N.E.; Sakellariadou, M.; Yusaf, M.F. Confronting MOND and TeVeS with strong gravitational lensing over galactic scales: An extended survey. *Phys. Rev. D* **2012**, *86*, 083507.

264. Clowe, D.; Bradač, M.; Gonzalez, A.H.; Markevitch, M.; Randall, S.W.; Jones, C.; Zaritsky, D. A direct empirical proof of the existence of Dark Matter. *Astrophys. J.* **2006**, *648*, L109–L113.

265. Peebles, P.J.; Ratra, B. The cosmological constant and dark energy. *Rev. Mod. Phys.* **2003**, *75*, 559–606.

266. Slipher, V.M. Nebulae. *Proc. Am. Philos. Soc.* **1917**, *56*, 403–409.

267. Paal, G.; Horvath, I.; Lukacs, B. Inflation and compactification from galaxy redshifts? *Astrophys. Space Sci.* **1992**, *191*, 107–124.

268. Krauss, L.M. The end of the age problem, and the case for a cosmological constant revisited. *Astrophys. J.* **1998**, *501*, 461–466.

269. Räsänen, S. Structure formation as an alternative to dark energy and modified gravity. In Proceedings of the Dark Energy and Dark Matter: Observations, Experiments and Theories, Lyons, France, 7–11 July 2008; Volume 36, pp. 63–74.

270. Blasone, M.; Capolupo, A.; Capozziello, S.; Carloni, S.; Vitiello, G. Neutrino mixing contribution to the cosmological constant. *Phys. Lett. A* **2004**, *323*, 182–189.

271. Capolupo, A.; Capozziello, S.; Vitiello, G. Neutrino mixing as a source of dark energy. *Phys. Lett. A* **2007**, *363*, 53–56.

272. Riess, A.G.; Filippenko, A.V.; Challis, P.; Clocchiatti, A.; Diercks, A.; Garnavich, P.M.; Gilliland, R.L.; Hogan, C.J.; Jha, S.; Kirshner, R.P.; et al. Observational evidence from supernovae for an accelerating universe and a cosmological constant. *Astron. J.* **1998**, *116*, 1009–1038.

273. Perlmutter, S.; Aldering, G.; Goldhaber, G.; Knop, R.A.; Nugent, P.; Castro, P.G.; Deustua, S.; Fabbro, S.; Goobar, A.; Groom, D.E.; et al. The supernova cosmology project. Measurements of Omega and Lambda from 42 high-redshift supernovae. *Astrophys. J.* **1999**, *517*, 565–586.

274. Sahni, V.; Starobinsky, A. The case for a positive cosmological Λ-term. *Int. J. Mod. Phys. D* **2000**, *9*, 373–443.

275. Turner, M.S. Dark Matter and Dark Energy in the universe. In *The Third Stromlo Symposium: The Galactic Halo*; Gibson, B.K., Axelrod, R.S., Putman, M.E., Eds.; Astronomical Society of the Pacific Conference Series; The Astronomical Society of the Pacific: San Francisco, CA, USA 1999; Volume 165, p. 431.

276. Eisenstein, D.J.; Zehavi, I.; Hogg, D.W.; Scoccimarro, R.; Blanton, M.R.; Nichol, R.C.; Scranton, R.; Seo, H.J.; Tegmark, M.; Zheng, Z.; et al. Detection of the baryon acoustic peak in the large-scale correlation function of SDSS luminous red galaxies. *Astrophys. J.* **2005**, *633*, 560–574.

277. Frieman, J.A.; Turner, M.S.; Huterer, D. Dark Energy and the Accelerating Universe. *Annu. Rev. Astron. Astrophys.* **2008**, *46*, 385–432.

278. Astier, P.; Pain, R. Observational evidence of the accelerated expansion of the universe. *C. R. Phys.* **2012**, *13*, 521–538.

279. Enqvist, K. Lemaitre Tolman Bondi model and accelerating expansion. *Gen. Relativ. Gravit.* **2008**, *40*, 451–466.

280. Sotiriou, T.P.; Faraoni, V. $f(R)$ theories of gravity. *Rev. Mod. Phys.* **2010**, *82*, 451–497.

281. Felice, A.D.; Tsujikawa, S. $f(R)$ theories. *Living Rev. Relativ.* **2010**, *13*, 3.

282. Capozziello, S.; de Laurentis, M. Extended theories of gravity. *Phys. Rep.* **2011**, *509*, 167–321.

283. Khoury, J.; Weltman, A. Chameleon cosmology. *Phys. Rev. D* **2004**, *69*, 044026.

284. Durrer, R.; Maartens, R. Dark energy and dark gravity: Theory overview. *Gen. Relativ. Gravit.* **2008**, *40*, 301–328..

285. Linder, E.V. Exploring the expansion history of the universe. *Phys. Rev. Lett.* **2003**, *90*, 091301.

286. Chevallier, M.; Polarski, D. Accelerating Universes with scaling Dark Matter. *Int. J. Mod. Phys. D* **2001**, *10*, 213–223.

287. Linden, S.; Virey, J.M. Test of the Chevallier-Polarski-Linder parametrization for rapid dark energy equation of state transitions. *Phys. Rev. D* **2008**, *78*, 023526.

288. Wang, Y.; Freese, K. Probing dark energy using its density instead of its equation of state. *Phys. Lett. B* **2006**, *632*, 449–452.

289. Johri, V.B.; Rath, P.K. Parametrization of the dark energy equation of state. *Int. J. Mod. Phys. D* **2007**, *16*, 1581–1591.

290. Avelino, P.P.; Martins, C.J.A.P.; Nunes, N.J.; Olive, K.A. Reconstructing the dark energy equation of state with varying couplings. *Phys. Rev. D* **2006**, *74*, 083508.

291. Scherrer, R.J. Mapping the Chevallier-Polarski-Linder parametrization onto physical dark energy models. *Phys. Rev. D* **2015**, *92*, 043001.

292. Chongchitnan, S.; King, L. Imprints of dynamical dark energy on weak-lensing measurements. *Mon. Not. R. Astron. Soc.* **2010**, *407*, 1989–1997.

293. Astashenok, A.V.; Odintsov, S.D. Confronting dark energy models mimicking ΛCDM epoch with observational constraints: Future cosmological perturbations decay or future Rip? *Phys. Lett. B* **2013**, *718*, 1194–1202.

294. Debono, I. Bayesian model selection for dark energy using weak lensing forecasts. *Mon. Not. R. Astron. Soc.* **2014**, *437*, 887–897.

295. Bardeen, J.M. Gauge-invariant cosmological perturbations. *Phys. Rev. D* **1980**, *22*, 1882–1905.

296. Lifshitz, E.M. On the gravitational stability of the expanding universe. *J. Phys. (USSR)* **1946**, *46*, 587–602.

297. Bartelmann, M.; Schneider, P. Weak gravitational lensing. *Phys. Rep.* **2001**, *340*, 291–472.

298. Jeans, J.H. The stability of a spherical nebula. *Philos. Trans. R. Soc. Lond. A Math. Phys. Eng. Sci.* **1902**, *199*, 1–53.

299. Silk, J. Cosmic black-body radiation and galaxy formation. *Astrophys. J.* **1968**, *151*, 459–471.

300. Bond, J.R.; Efstathiou, G.; Silk, J. Massive neutrinos and the large-scale structure of the universe. *Phys. Rev. Lett.* **1980**, *45*, 1980–1984.

301. Silk, J.; Mamon, G.A. The current status of galaxy formation. *Res. Astron. Astrophys.* **2012**, *12*, 917–946.

302. Linder, E.V.; Jenkins, A. Cosmic structure growth and dark energy. *Mon. Not. R. Astron. Soc.* **2003**, *346*, 573–583.

303. Bernardeau, F.; Colombi, S.; Gaztañaga, E.; Scoccimarro, R. Large-scale structure of the Universe and cosmological perturbation theory. *Phys. Rep.* **2002**, *367*, 1–248.

304. Percival, W.J. Cosmological structure formation in a homogeneous dark energy background. *Astron. Astrophys.* **2005**, *443*, 819–830.

305. Hamilton, A.J.S.; Kumar, P.; Lu, E.; Matthews, A. Reconstructing the primordial spectrum of fluctuations of the universe from the observed nonlinear clustering of galaxies. *Astrophys. J.* **1991**, *374*, L1–L4.

306. Jain, B.; Mo, H.J.; White, S.D.M. The evolution of correlation functions and power spectra in gravitational clustering. *Mon. Not. R. Astron. Soc.* **1995**, *276*, L25–L29.

307. Peacock, J.A.; Dodds, S.J. Non-linear evolution of cosmological power spectra. *Mon. Not. R. Astron. Soc.* **1996**, *280*, L19–L26.

308. Smith, R.E.; Peacock, J.A.; Jenkins, A.; White, S.D.M.; Frenk, C.S.; Pearce, F.R.; Thomas, P.A.; Efstathiou, G.; Couchman, H.M.P. Stable clustering, the halo model and non-linear cosmological power spectra. *Mon. Not. R. Astron. Soc.* **2003**, *341*, 1311–1332.

309. Seljak, U. Analytic model for galaxy and Dark Matter clustering. *Mon. Not. R. Astron. Soc.* **2000**, *318*, 203–213.

310. Peacock, J.A.; Smith, R.E. Halo occupation numbers and galaxy bias. *Mon. Not. R. Astron. Soc.* **2000**, *318*, 1144–1156.

311. Huterer, D.; Takada, M. Calibrating the nonlinear matter power spectrum: Requirements for future weak lensing surveys. *Astropart. Phys.* **2005**, *23*, 369–376.

312. Van Daalen, M.P.; Schaye, J.; Booth, C.M.; Dalla Vecchia, C. The effects of galaxy formation on the matter power spectrum: A challenge for precision cosmology. *Mon. Not. R. Astron. Soc.* **2011**, *415*, 3649–3665.

313. Semboloni, E.; Hoekstra, H.; Schaye, J.; van Daalen, M.P.; McCarthy, I.G. Quantifying the effect of baryon physics on weak lensing tomography. *Mon. Not. R. Astron. Soc.* **2011**, *417*, 2020–2035.

314. Bird, S.; Viel, M.; Haehnelt, M.G. Massive neutrinos and the non-linear matter power spectrum. *Mon. Not. R. Astron. Soc.* **2012**, *420*, 2551–2561.

315. Hearin, A.P.; Zentner, A.R.; Ma, Z. General requirements on matter power spectrum predictions for cosmology with weak lensing tomography. *J. Cosmol. Astropart. Phys.* **2012**, *4*, 034.

316. Takahashi, R.; Sato, M.; Nishimichi, T.; Taruya, A.; Oguri, M. Revising the halofit model for the nonlinear matter power spectrum. *Astrophys. J.* **2012**, *761*, 152.

317. Villaescusa-Navarro, F.; Viel, M.; Datta, K.K.; Choudhury, T.R. Modeling the neutral hydrogen distribution in the post-reionization Universe: Intensity mapping. *J. Cosmol. Astropart. Phys.* **2014**, *9*, 050.

318. Bentivegna, E.; Bruni, M. Effects of nonlinear inhomogeneity on the cosmic expansion with numerical relativity. *Phys. Rev. Lett.* **2016**, *116*, 251302.

319. Mertens, J.B.; Giblin, J.T.; Starkman, G.D. Integration of inhomogeneous cosmological spacetimes in the BSSN formalism. *Phys. Rev. D* **2016**, *93*, 124059.

320. Giblin, J.T.; Mertens, J.B.; Starkman, G.D. Departures from the Friedmann-Lemaitre-Robertston-Walker cosmological model in an inhomogeneous universe: A numerical examination. *Phys. Rev. Lett.* **2016**, *116*, 251301.

321. Harrison, E.R. Fluctuations at the threshold of classical cosmology. *Phys. Rev. D* **1970**, *1*, 2726–2730.

322. Zel'dovich, Y.B. A hypothesis, unifying the structure and the entropy of the Universe. *Mon. Not. R. Astron. Soc.* **1972**, *160*, 1–3.

323. Peebles, P.J.E.; Yu, J.T. Primeval adiabatic perturbation in an expanding universe. *Astrophys. J.* **1970**, *162*, 815–836.

324. Mukherjee, P.; Wang, Y. Model-independent reconstruction of the primordial power spectrum from Wilkinson Microwave Anistropy Probe data. *Astrophys. J.* **2003**, *599*, 1–6.

325. Trotta, R. Forecasting the Bayes factor of a future observation. *Mon. Not. R. Astron. Soc.* **2007**, *378*, 819–824.

326. Trotta, R. Applications of Bayesian model selection to cosmological parameters. *Mon. Not. R. Astron. Soc.* **2007**, *378*, 72–82.

327. Bridges, M.; Feroz, F.; Hobson, M.P.; Lasenby, A.N. Bayesian optimal reconstruction of the primordial power spectrum. *Mon. Not. R. Astron. Soc.* **2009**, *400*, 1075–1084.

328. Vázquez, J.A.; Bridges, M.; Hobson, M.P.; Lasenby, A.N. Model selection applied to reconstruction of the Primordial Power Spectrum. *J. Cosmol. Astropart. Phys.* **2012**, *6*, 006.

329. Hoyle, C.D.; Kapner, D.J.; Heckel, B.R.; Adelberger, E.G.; Gundlach, J.H.; Schmidt, U.; Swanson, H.E. Submillimeter tests of the gravitational inverse-square law. *Phys. Rev. D* **2004**, *70*, 042004.

330. Kapner, D.J.; Cook, T.S.; Adelberger, E.G.; Gundlach, J.H.; Heckel, B.R.; Hoyle, C.D.; Swanson, H.E. Tests of the gravitational inverse-square law below the dark-energy length scale. *Phys. Rev. Lett.* **2007**, *98*, 021101.

331. Baker, T.; Psaltis, D.; Skordis, C. Linking tests of gravity on all scales: From the strong-field regime to cosmology. *Astrophys. J.* **2015**, *802*, 63.

332. Eddington, A.S. *The Mathematical Theory of Relativity*; Cambridge University Press: Cambridge, UK, 1923.

333. Shapiro, I.I. Solar system tests of general relativity: Recent results and present plans. In *General Relativity and Gravitation, 1989*; Ashby, N., Bartlett, D.F., Wyss, W., Eds.; Cambridge University Press: Cambridge, UK, 2005; p. 313.

334. Williams, J.G.; Turyshev, S.G.; Boggs, D.H. Progress in lunar laser ranging tests of relativistic gravity. *Phys. Rev. Lett.* **2004**, *93*, 261101.

335. Williams, J.G.; Turyshev, S.G.; Boggs, D.H. Lunar laser ranging tests of the equivalence principle with the Earth and Moon. *Int. J. Mod. Phys. D* **2009**, *18*, 1129–1175.

336. Shapiro, S.S.; Davis, J.L.; Lebach, D.E.; Gregory, J.S. Measurement of the solar gravitational deflection of radio waves using geodetic very-long-baseline interferometry data, 1979–1999. *Phys. Rev. Lett.* **2004**, *92*, 121101.

337. Pitjeva, E.V.; Pitjev, N.P. Relativistic effects and Dark Matter in the Solar system from observations of planets and spacecraft. *Mon. Not. R. Astron. Soc.* **2013**, *432*, 3431–3437.

338. Fienga, A.; Laskar, J.; Kuchynka, P.; Manche, H.; Desvignes, G.; Gastineau, M.; Cognard, I.; Theureau, G. The INPOP10a planetary ephemeris and its applications in fundamental physics. *Celest. Mech. Dyn. Astron.* **2011**, *111*, 363–385.

339. Iorio, L.; Lichtenegger, H.I.M.; Ruggiero, M.L.; Corda, C. Phenomenology of the Lense-Thirring effect in the solar system. *Astrophys. Space Sci.* **2011**, *331*, 351–395.

340. Hughes, V.W.; Robinson, H.G.; Beltran-Lopez, V. Upper Limit for the anisotropy of inertial mass from nuclear resonance experiments. *Phys. Rev. Lett.* **1960**, *4*, 342–344.

341. Drever, R.W.P. A search for anisotropy of inertial mass using a free precession technique. *Philos. Mag.* **1961**, *6*, 683–687.

342. Allmendinger, F.; Heil, W.; Karpuk, S.; Kilian, W.; Scharth, A.; Schmidt, U.; Schnabel, A.; Sobolev, Y.; Tullney, K. New limit on Lorentz-invariance- and CPT-violating neutron spin interactions using a free-spin-precession He3-Xe129 comagnetometer. *Phys. Rev. Lett.* **2014**, *112*, 110801.

343. Williams, J.G.; Turyshev, S.G.; Boggs, D.H. Lunar laser ranging tests of the equivalence principle. *Class. Quantum Gravity* **2012**, *29*, 184004.

344. Delva, P.; Hees, A.; Bertone, S.; Richard, E.; Wolf, P. Test of the gravitational redshift with stable clocks in eccentric orbits: Application to Galileo satellites 5 and 6. *Class. Quantum Gravity* **2015**, *32*, 232003.

345. Iorio, L. Gravitational anomalies in the solar system? *Int. J. Mod. Phys. D* **2015**, *24*, 1530015.

346. Uzan, J.P. Varying constants, gravitation and cosmology. *Living Rev. Relativ.* **2011**, *14*, 2.

347. Flambaum, V.V. Variation of the fundamental constants: Theory and observations. *Int. J. Mod. Phys. A* **2007**, *22*, 4937–4950.

348. Lea, S.N. Limits to time variation of fundamental constants from comparisons of atomic frequency standards. *Rep. Prog. Phys.* **2007**, *70*, 1473–1523.

349. Rich, J. Which fundamental constants for cosmic microwave background and baryon-acoustic oscillation? *Astron. Astrophys.* **2015**, *584*, A69.

350. Barrow, J.D.; Sandvik, H.B.; Magueijo, J.A. Behavior of varying-alpha cosmologies. *Phys. Rev. D* **2002**, *65*, 063504.
351. Barrow, J.D.; Graham, A.A.H. General dynamics of varying-alpha universes. *Phys. Rev. D* **2013**, *88*, 103513.
352. Fujii, Y.; Iwamoto, A.; Fukahori, T.; Ohnuki, T.; Nakagawa, M.; Hidaka, H.; Oura, Y.; Möller, P. The nuclear interaction at Oklo 2 billion years ago. *Nucl. Phys. B* **2000**, *573*, 377–401.
353. Uzan, J.P. The fundamental constants and their variation: Observational and theoretical status. *Rev. Mod. Phys.* **2003**, *75*, 403–455.
354. Lamoreaux, S.K.; Torgerson, J.R. Neutron moderation in the Oklo natural reactor and the time variation of α. *Phys. Rev. D* **2004**, *69*, 121701.
355. Khatri, R.; Wandelt, B.D. 21-cm radiation: A new probe of variation in the fine-structure constant. *Phys. Rev. Lett.* **2007**, *98*, 111301.
356. Nakashima, M.; Ichikawa, K.; Nagata, R.; Yokoyama, J. Constraining the time variation of the coupling constants from cosmic microwave background: Effect of ΛQCD. *J. Cosmol. Astropart. Phys.* **2010**, *2010*, 030.
357. Martins, C.J.A.P.; Menegoni, E.; Galli, S.; Mangano, G.; Melchiorri, A. Varying couplings in the early universe: Correlated variations of α and G. *Phys. Rev. D* **2010**, *82*, 023532.
358. Anderson, J.D.; Schubert, G.; Trimble, V.; Feldman, M.R. Measurements of Newton's gravitational constant and the length of day. *Europhys. Lett.* **2015**, *110*, 10002.
359. Pitkin, M. Comment on "Measurements of Newton's gravitational constant and the length of day". *Europhys. Lett.* **2015**, *111*, 30002.
360. Anderson, J.D.; Schubert, G.; Trimble, V.; Feldman, M.R. Reply to the comment by M. Pitkin. *Europhys. Lett.* **2015**, *111*, 30003.
361. Iorio, L. Does Newton's gravitational constant vary sinusoidally with time? Orbital motions say no. *Class. Quantum Gravity* **2016**, *33*, 045004.
362. Feldman, M.R.; Anderson, J.D.; Schubert, G.; Trimble, V.; Kopeikin, S.M.; Lämmerzahl, C. Deep space experiment to measure G. *Class. Quantum Gravity* **2016**, *33*, 125013.
363. Lahav, O. Observational tests of FRW world models. *Class. Quantum Gravity* **2002**, *19*, 3517–3526.
364. Hansen, F.K.; Banday, A.J.; Górski, K.M. Testing the cosmological principle of isotropy: Local power-spectrum estimates of the WMAP data. *Mon. Not. R. Astron. Soc.* **2004**, *354*, 641–665.
365. Schwarz, D.J.; Bacon, D.; Chen, S.; Clarkson, C.; Huterer, D.; Kunz, M.; Maartens, R.; Raccanelli, A.; Rubart, M.; Starck, J.L. Testing foundations of modern cosmology with SKA all-sky surveys. In Proceedings of the Advancing Astrophysics with the Square Kilometre Array (AASKA14), Giardini Naxos, Italy, 9–13 June 2014; p. 32.
366. Segre, M. Galileo, Viviani and the tower of Pisa. *Stud. Hist. Philos. Sci. Part A* **1989**, *20*, 435–451.
367. Eötvos, R. Über die Anziehung der Erde auf Verschiedene Substanzen. *Math. Naturwissenschaft. Ber. Ung.* **1890**, *8*, 65–68.
368. Gundlach, J.H.; Smith, G.L.; Adelberger, E.G.; Heckel, B.R.; Swanson, H.E. Short-range test of the equivalence principle. *Phys. Rev. Lett.* **1997**, *78*, 2523–2526.
369. Schlamminger, S.; Choi, K.Y.; Wagner, T.A.; Gundlach, J.H.; Adelberger, E.G. Test of the equivalence principle using a rotating torsion balance. *Phys. Rev. Lett.* **2008**, *100*, 041101.
370. Adelberger, E.; Gundlach, J.; Heckel, B.; Hoedl, S.; Schlamminger, S. Torsion balance experiments: A low-energy frontier of particle physics. *Prog. Part. Nucl. Phys.* **2009**, *62*, 102–134.
371. Wagner, T.A.; Schlamminger, S.; Gundlach, J.H.; Adelberger, E.G. Torsion-balance tests of the weak equivalence principle. *Class. Quantum Gravity* **2012**, *29*, 184002.
372. Dickey, J.O.; Bender, P.L.; Faller, J.E.; Newhall, X.X.; Ricklefs, R.L.; Ries, J.G.; Shelus, P.J.; Veillet, C.; Whipple, A.L.; Wiant, J.R.; et al. Lunar Laser ranging: A continuing legacy of the Apollo program. *Science* **1994**, *265*, 482–490.
373. Murphy, T.W. Lunar laser ranging: The millimeter challenge. *Rep. Prog. Phys.* **2013**, *76*, 076901.
374. Pearlman, M.R.; Degnan, J.J.; Bosworth, J.M. The international laser ranging service. *Adv. Space Res.* **2002**, *30*, 135–143.
375. Appleby, G.; Rodríguez, J.; Altamimi, Z. Assessment of the accuracy of global geodetic satellite laser ranging observations and estimated impact on ITRF scale: Estimation of systematic errors in LAGEOS observations 1993–2014. *J. Geod.* **2016**, doi:10.1007/s00190-016-0929-2.

376. Degnan, J.J. Laser transponders for high-accuracy interplanetary laser ranging and time transfer. In *Lasers, Clocks and Drag-Free Control: Exploration of Relativistic Gravity in Space*; Dittus, H., Lammerzahl, C., Turyshev, S.G., Eds.; Astrophysics and Space Science Library; Springer: Berlin/Heidelberg, Germany, 2008; Volume 349, p. 231.

377. Iorio, L. Effects of standard and modified gravity on interplanetary ranges. *Int. J. Mod. Phys. D* **2011**, *20*, 181–232.

378. Dirkx, D.; Noomen, R.; Visser, P.N.A.M.; Bauer, S.; Vermeersen, L.L.A. Comparative analysis of one- and two-way planetary laser ranging concepts. *Planet. Space Sci.* **2015**, *117*, 159–176.

379. Smith, D.; Zuber, M.; Sun, X.; Neumann, G.; Cavanaugh, J.; McGarry, J.; Zagwodzki, T. Two-way laser link over interplanetary distance. *Science* **2006**, *311*, 53.

380. Chen, Y.; Birnbaum, K.M.; Hemmati, H. Active laser ranging over planetary distances with millimeter accuracy. *Appl. Phys. Lett.* **2013**, *102*, 241107.

381. Margot, J.L.; Giorgini, J.D. Probing general relativity with radar astrometry in the inner solar system. *Proc. Int. Astron. Union* **2010**, *261*, 183–188.

382. Fienga, A.; Laskar, J.; Manche, H.; Gastineau, M. Tests of GR with INPOP15a Planetary Ephemerides: Estimations of Possible Supplementary Advances of Perihelia for Mercury and Saturn. 2016, arXiv:1601.00947.

383. Iorio, L. The Solar Lense-Thirring Effect: Perspectives for a Future Measurement. 2016, arXiv:1601.01382.

384. Anderson, J.D.; Laing, P.A.; Lau, E.L.; Liu, A.S.; Nieto, M.M.; Turyshev, S.G. Study of the anomalous acceleration of Pioneer 10 and 11. *Phys. Rev. D* **2002**, *65*, 082004.

385. Nieto, M.M.; Anderson, J.D. Using early data to illuminate the Pioneer anomaly. *Class. Quantum Gravity* **2005**, *22*, 5343–5354.

386. Turyshev, S.G.; Toth, V.T. The pioneer anomaly. *Living Rev. Relativ.* **2010**, *13*, 4.

387. Nieto, M.M. New Horizons and the onset of the Pioneer anomaly. *Phys. Lett. B* **2008**, *659*, 483–485.

388. Iorio, L. Perspectives on effectively constraining the location of a massive trans-Plutonian object with the New Horizons spacecraft: A sensitivity analysis. *Celest. Mech. Dyn. Astron.* **2013**, *116*, 357–366.

389. Iorio, L.; Ruggiero, M.L.; Radicella, N.; Saridakis, E.N. Constraining the Schwarzschild-de Sitter solution in models of modified gravity. *Phys. Dark Universe* **2016**, *13*, 111–120.

390. Damour, T.; Taylor, J.H. Strong-field tests of relativistic gravity and binary pulsars. *Phys. Rev. D* **1992**, *45*, 1840–1868.

391. Lyne, A.G.; Burgay, M.; Kramer, M.; Possenti, A.; Manchester, R.N.; Camilo, F.; McLaughlin, M.A.; Lorimer, D.R.; D'Amico, N.; Joshi, B.C.; et al. A double-pulsar system: A rare laboratory for relativistic gravity and plasma physics. *Science* **2004**, *303*, 1153–1157.

392. Kramer, M.; Wex, N. The double pulsar system: A unique laboratory for gravity. *Class. Quantum Gravity* **2009**, *26*, 073001.

393. McGaugh, S.S.; Schombert, J.M.; Bothun, G.D.; de Blok, W.J.G. The baryonic tully-fisher relation. *Astrophys. J.* **2000**, *533*, L99–L102.

394. Famaey, B.; McGaugh, S.S. Modified Newtonian Dynamics (MOND): Observational phenomenology and relativistic extensions. *Living Rev. Relativ.* **2012**, *15*, 10.

395. Adams, F.C.; Laughlin, G. Relativistic effects in extrasolar planetary systems. *Int. J. Mod. Phys. D* **2006**, *15*, 2133–2140.

396. Adams, F.C.; Laughlin, G. Effects of secular interactions in extrasolar planetary systems. *Astrophys. J.* **2006**, *649*, 992–1003.

397. Adams, F.C.; Laughlin, G. Long-term evolution of close planets including the effects of secular interactions. *Astrophys. J.* **2006**, *649*, 1004–1009.

398. Iorio, L. Are we far from testing general relativity with the transitting extrasolar planet HD 209458b 'Osiris'? *New Astron.* **2006**, *11*, 490–494.

399. Jordán, A.; Bakos, G.Á. Observability of the general relativistic precession of periastra in exoplanets. *Astrophys. J.* **2008**, *685*, 543–552.

400. Pál, A.; Kocsis, B. Periastron precession measurements in transiting extrasolar planetary systems at the level of general relativity. *Mon. Not. R. Astron. Soc.* **2008**, *389*, 191–198.

401. Iorio, L. Classical and relativistic node precessional effects in WASP-33b and perspectives for detecting them. *Astrophys. Space Sci.* **2011**, *331*, 485–496.

402. Iorio, L. Classical and relativistic long-term time variations of some observables for transiting exoplanets. *Mon. Not. R. Astron. Soc.* **2011**, *411*, 167–183.

403. Iorio, L. Accurate characterization of the stellar and orbital parameters of the exoplanetary system WASP-33 b from orbital dynamics. *Mon. Not. R. Astron. Soc.* **2016**, *455*, 207–213.

404. Iorio, L. Post-Keplerian corrections to the orbital periods of a two-body system and their measurability. *Mon. Not. R. Astron. Soc.* **2016**, *460*, 2445–2452.

405. Henry, R.C. Kretschmann scalar for a kerr-newman black hole. *Astrophys. J.* **2000**, *535*, 350–353.

406. Will, C.M. Focus Issue: Gravity Probe B. *Class. Quantum Gravity* **2015**, *32*, 220301.

407. Everitt, C.W.F.; Muhlfelder, B.; DeBra, D.B.; Parkinson, B.W.; Turneaure, J.P.; Silbergleit, A.S.; Acworth, E.B.; Adams, M.; Adler, R.; Bencze, W.J.; et al. The Gravity Probe B test of general relativity. *Class. Quantum Gravity* **2015**, *32*, 224001.

408. Hulse, R.A.; Taylor, J.H. Discovery of a pulsar in a binary system. *Astrophys. J.* **1975**, *195*, L51–L53.

409. Psaltis, D. Probes and tests of strong-field gravity with observations in the electromagnetic spectrum. *Living Rev. Relativ.* **2008**, *11*, 9.

410. Psaltis, D. Two approaches to testing general relativity in the strong-field regime. *J. Phys. Conf. Ser.* **2009**, *189*, 012033.

411. Johannsen, T. Testing general relativity in the strong-field regime with observations of black holes in the electromagnetic spectrum. *Publ. Astron. Soc. Pac.* **2012**, *124*, 1133–1134.

412. Kramer, M. Precision tests of theories of gravity using pulsars. In *Thirteenth Marcel Grossmann Meeting: On Recent Developments in Theoretical and Experimental General Relativity, Astrophysics and Relativistic Field Theories*; Rosquist, K., Ed.; World Scientific: Singapore, 2015; pp. 315–332.

413. Weisberg, J.M.; Taylor, J.H. The Relativistic Binary Pulsar B1913+16: Thirty Years of Observations and Analysis. *Astron. Soc. Pac. Conf. Ser.* **2005**, *328*, 25–32.

414. Bailes, M.; Ord, S.M.; Knight, H.S.; Hotan, A.W. Self-consistency of relativistic observables with general relativity in the white dwarf-neutron star binary PSR J1141-6545. *Astrophys. J.* **2003**, *595*, L49–L52.

415. Stairs, I.H.; Thorsett, S.E.; Taylor, J.H.; Wolszczan, A. Studies of the relativistic binary pulsar PSR B1534+12. I. Timing analysis. *Astrophys. J.* **2002**, *581*, 501–508.

416. Kramer, M.; Lorimer, D.R.; Lyne, A.G.; McLaughlin, M.; Burgay, M.; D'Amico, N.; Possenti, A.; Camilo, F.; Freire, P.C.C.; Joshi, B.C.; et al. Testing GR with the double pulsar: Recent results. In Proceedings of the 22nd Texas Symposium on Relativistic Astrophysics, Stanford, CA, USA , 13–17 December 2004; pp. 142–148.

417. Kramer, M.; Stairs, I.H.; Manchester, R.N.; McLaughlin, M.A.; Lyne, A.G.; Ferdman, R.D.; Burgay, M.; Lorimer, D.R.; Possenti, A.; D'Amico, N.; et al. Tests of general relativity from timing the double pulsar. *Science* **2006**, *314*, 97–102.

418. Desai, S.; Kahya, E.O. Galactic one-way Shapiro delay to PSR B1937+21. *Mod. Phys. Lett. A* **2016**, *31*, 1650083.

419. Huang, Y.; Weisberg, J.M. Timing the relativistic binary pulsar PSR B1913+16. In Proceedings of the 228th American Astronomical Society Meeting, San Diego, CA, USA, 12–16 June 2016.

420. Iorio, L. Prospects for measuring the moment of inertia of pulsar J0737-3039A. *New Astron.* **2009**, *14*, 40–43.

421. Kehl, M.S.; Wex, N.; Kramer, M.; Liu, K. Future Measurements of the Lense-Thirring Effect in the Double Pulsar. 2016, arXiv:1605.00408.

422. Iorio, L. Constraints from orbital motions around the Earth of the environmental fifth-force hypothesis for the OPERA superluminal neutrino phenomenology. *J. High Energy Phys.* **2012**, *5*, 73.

423. Deng, X.M.; Xie, Y. Yukawa effects on the clock onboard a drag-free satellite. *Mon. Not. R. Astron. Soc.* **2013**, *431*, 3236–3239.

424. Li, Z.W.; Yuan, S.F.; Lu, C.; Xie, Y. New upper limits on deviation from the inverse-square law of gravity in the solar system: A Yukawa parameterization. *Res. Astron. Astrophys.* **2014**, *14*, 139–143.

425. Khoury, J.; Weltman, A. Chameleon fields: Awaiting surprises for tests of gravity in space. *Phys. Rev. Lett.* **2004**, *93*, 171104.

426. Tully, R.B.; Fisher, J.R. A new method of determining distances to galaxies. *Astron. Astrophys.* **1977**, *54*, 661–673.

427. Dunsby, P.; Goheer, N.; Osano, B.; Uzan, J.P. How close can an inhomogeneous universe mimic the concordance model? *J. Cosmol. Astropart. Phys.* **2010**, *6*, 017.

428. Uzan, J.P.; Bernardeau, F. Lensing at cosmological scales: A test of higher dimensional gravity. *Phys. Rev. D* **2001**, *64*, 083004.

429. Peebles, P.J.E. *The Large-Scale Structure of the Universe*; Princeton Series in Physics: Princeton, NJ, USA, 1980.

430. Rich, J. *Fundamentals of Cosmology*; Springer: Berlin, Germany, 2001.

431. Uzan, J.P. The acceleration of the universe and the physics behind it. *Gen. Relativ. Gravit.* **2007**, *39*, 307–342.

432. Uzan, J.P. Tests of general relativity on astrophysical scales. *Gen. Relativ. Gravit.* **2010**, *42*, 2219–2246.

433. Lue, A.; Scoccimarro, R.; Starkman, G.D. Probing Newton's constant on vast scales: Dvali-Gabadadze-Porrati gravity, cosmic acceleration, and large scale structure. *Phys. Rev. D* **2004**, *69*, 124015.

434. Song, Y.S. Large scale structure formation of the normal branch in the DGP brane world model. *Phys. Rev. D* **2008**, *77*, 124031.

435. Benabed, K.; Bernardeau, F. Testing quintessence models with large-scale structure growth. *Phys. Rev. D* **2001**, *64*, 083501.

436. Koivisto, T.; Mota, D.F. Gauss-Bonnet quintessence: Background evolution, large scale structure, and cosmological constraints. *Phys. Rev. D* **2007**, *75*, 023518.

437. Tsujikawa, S. Quintessence: A review. *Class. Quantum Gravity* **2013**, *30*, 214003.

438. Baldi, M.; Pettorino, V.; Amendola, L.; Wetterich, C. Oscillating non-linear large-scale structures in growing neutrino quintessence. *Mon. Not. R. Astron. Soc.* **2011**, *418*, 214–229.

439. Koivisto, T.S.; Saridakis, E.N.; Tamanini, N. Scalar-fluid theories: Cosmological perturbations and large-scale structure. *J. Cosmol. Astropart. Phys.* **2015**, *2015*, 047.

440. Schimd, C.; Uzan, J.P.; Riazuelo, A. Weak lensing in scalar-tensor theories of gravity. *Phys. Rev. D* **2005**, *71*, 083512.

441. Rodríguez-Meza, M.A. Power spectrum of large-scale structure cosmological models in the framework of scalar-tensor theories. *J. Phys. Conf. Ser.* **2010**, *229*, 012063.

442. Goenner, H. Some remarks on the genesis of scalar-tensor theories. *Gen. Relativ. Gravit.* **2012**, *44*, 2077–2097.

443. Takushima, Y.; Terukina, A.; Yamamoto, K. Bispectrum of cosmological density perturbations in the most general second-order scalar-tensor theory. *Phys. Rev. D* **2014**, *89*, 104007.

444. Dossett, J.N.; Ishak, M.; Parkinson, D.; Davis, T.M. Constraints and tensions in testing general relativity from Planck and CFHTLenS data including intrinsic alignment systematics. *Phys. Rev. D* **2015**, *92*, 023003.

445. Di Valentino, E.; Melchiorri, A.; Silk, J. Cosmological hints of modified gravity? *Phys. Rev. D* **2016**, *93*, 023513.

446. Hu, B.; Liguori, M.; Bartolo, N.; Matarrese, S. Parametrized modified gravity constraints after Planck. *Phys. Rev. D* **2013**, *88*, 123514.

447. Pettorino, V. Testing modified gravity with Planck: The case of coupled dark energy. *Phys. Rev. D* **2013**, *88*, 063519.

448. Planck Collaboration. Planck 2015 Results. XIV. Dark Energy and Modified Gravity. *Astron. Astrophys.* **2016**, *594*, A14.

449. Räsänen, S. Dark energy from back-reaction. *J. Cosmol. Astropart. Phys.* **2004**, *2*, 003.

450. Wiltshire, D.L. Gravitational energy as dark energy: Towards concordance cosmology without Lambda. In Proceedings of the Dark Energy and Dark Matter: Observations, Experiments and Theories, Lyons, France, 7–11 July 2008; Volume 36, pp. 91–98.

451. Räsänen, S. Backreaction: Directions of progress. *Class. Quantum Gravity* **2011**, *28*, 164008.

452. Buchert, T.; Carfora, M.; Ellis, G.F.R.; Kolb, E.W.; MacCallum, M.A.H.; Ostrowski, J.J.; Räsänen, S.; Roukema, B.F.; Andersson, L.; Coley, A.A.; et al. Is there proof that backreaction of inhomogeneities is irrelevant in cosmology? *Class. Quantum Gravity* **2015**, *32*, 215021.

453. Räsänen, S.; Bolejko, K.; Finoguenov, A. New test of the Friedmann-Lemaître-Robertson-Walker metric using the distance sum rule. *Phys. Rev. Lett.* **2015**, *115*, 101301.

454. Räsänen, S.; Väliviita, J.; Kosonen, V. Testing distance duality with CMB anisotropies. *J. Cosmol. Astropart. Phys.* **2016**, *4*, 050.

455. Boehm, C.; Räsänen, S. Violation of the FRW consistency condition as a signature of backreaction. *J. Cosmol. Astropart. Phys.* **2013**, *9*, 003.

456. Buchert, T. On average properties of inhomogeneous fluids in general relativity: Dust cosmologies. *Gen. Relativ. Gravit.* **2000**, *32*, 105–126.

457. Buchert, T. Dark Energy from structure: A status report. *Gen. Relativ. Gravit.* **2008**, *40*, 467–527.
458. Banks, T. Supersymmetry breaking and the cosmological constant. *Int. J. Mod. Phy. A* **2014**, *29*, 1430010.
459. Arkani-Hamed, N.; Dimopoulos, S. Supersymmetric unification without low energy supersymmetry and signatures for fine-tuning at the LHC. *J. High Energy Phys.* **2005**, *6*, 073.
460. Susskind, L. The anthropic landscape of string theory. In *Universe or Multiverse?* Carr, B., Ed.; Cambridge University Press: Cambridge, UK, 2007.
461. Bauer, F.; Solà, J.; Štefancić, H. Dynamically avoiding fine-tuning the cosmological constant: The "Relaxed Universe". *J. Cosmol. Astropart. Phys.* **2010**, *12*, 029.
462. Bombelli, L.; Lee, J.; Meyer, D.; Sorkin, R.D. Space-time as a causal set. *Phys. Rev. Lett.* **1987**, *59*, 521–524.
463. Ahmed, M.; Dodelson, S.; Greene, P.B.; Sorkin, R. Everpresent Λ. *Phys. Rev. D* **2004**, *69*, 103523.
464. Weinberg, S. Anthropic bound on the cosmological constant. *Phys. Rev. Lett.* **1987**, *59*, 2607–2610.
465. Garriga, J.; Linde, A.; Vilenkin, A. Dark energy equation of state and anthropic selection. *Phys. Rev. D* **2004**, *69*, 063521.
466. Armendariz-Picon, C.; Mukhanov, V.; Steinhardt, P.J. Essentials of k-essence. *Phys. Rev. D* **2001**, *63*, 103510.
467. Kamenshchik, A.; Moschella, U.; Pasquier, V. An alternative to quintessence. *Phys. Lett. B* **2001**, *511*, 265–268.
468. de Putter, R.; Linder, E.V. Kinetic k-essence and quintessence. *Astropart. Phys.* **2007**, *28*, 263–272.
469. Durrer, R.; Maartens, R. Dark Energy and Modified Gravity. In *Dark Energy: Observational and Theoretical Approaches*; Ruiz-Lapuente, P., Ed.; Cambridge University Press: Cambridge, UK, 2008; pp. 48–91.
470. Deffayet, C.; Dvali, G.; Gabadadze, G. Accelerated universe from gravity leaking to extra dimensions. *Phys. Rev. D* **2002**, *65*, 044023.
471. Amendola, L.; Polarski, D.; Tsujikawa, S. Are f(R) Dark Energy Models Cosmologically Viable? *Phys. Rev. Lett.* **2007**, *98*, 131302.
472. Turyshev, S.G.; Toth, V.T.; Kinsella, G.; Lee, S.C.; Lok, S.M.; Ellis, J. Support for the thermal origin of the pioneer anomaly. *Phys. Rev. Lett.* **2012**, *108*, 241101.
473. Adler, C.G.; Coulter, B.L. Galileo and the Tower of Pisa experiment. *Am. J. Phys.* **1978**, *46*, 199–201.
474. Bergé, J.; Touboul, P.; Rodrigues, M.; The MICROSCOPE Team. Status of MICROSCOPE, a mission to test the Equivalence Principle in space. *J. Phys. Conf. Ser.* **2015**, *610*, 012009.
475. Overduin, J.; Everitt, F.; Worden, P.; Mester, J. STEP and fundamental physics. *Class. Quantum Gravity* **2012**, *29*, 184012.
476. Nobili, A.M.; Comandi, G.L.; Doravari, S.; Bramanti, D.; Kumar, R.; Maccarrone, F.; Polacco, E.; Turyshev, S.G.; Shao, M.; Lipa, J.; et al. "Galileo Galilei" (GG) a small satellite to test the equivalence principle of Galileo, Newton and Einstein. *Exp. Astron.* **2009**, *23*, 689–710.
477. Muller, H.; Peters, A.; Chu, S. A precision measurement of the gravitational redshift by the interference of matter waves. *Nature* **2010**, *463*, 926–929.
478. Bardeen, J.M.; Petterson, J.A. The lense-thirring effect and accretion disks around kerr black holes. *Astrophys. J.* **1975**, *195*, L65.
479. Pound, R.V.; Rebka, G.A. Apparent weight of photons. *Phys. Rev. Lett.* **1960**, *4*, 337–341.
480. Turyshev, S.G.; Yu, N.; Toth, V.T. General relativistic observables for the ACES experiment. *Phys. Rev. D* **2016**, *93*, 045027.
481. Vessot, R.F.C.; Levine, M.W.; Mattison, E.M.; Blomberg, E.L.; Hoffman, T.E.; Nystrom, G.U.; Farrel, B.F.; Decher, R.; Eby, P.B.; Baugher, C.R. Test of relativistic gravitation with a space-borne hydrogen maser. *Phys. Rev. Lett.* **1980**, *45*, 2081–2084.
482. Schiller, S.; Tino, G.M.; Gill, P.; Salomon, C.; Sterr, U.; Peik, E.; Nevsky, A.; Görlitz, A.; Svehla, D.; Ferrari, G.; et al. Einstein Gravity Explorer—A medium-class fundamental physics mission. *Exp. Astron.* **2009**, *23*, 573–610.
483. Antoniadis, J.; Freire, P.C.C.; Wex, N.; Tauris, T.M.; Lynch, R.S.; van Kerkwijk, M.H.; Kramer, M.; Bassa, C.; Dhillon, V.S.; Driebe, T.; et al. A massive pulsar in a compact relativistic binary. *Science* **2013**, *340*, 448.
484. Iorio, L. Constraining the preferred-frame α_1, α_2 parameters from solar system planetary precessions. *Int. J. Mod. Phys. D* **2014**, *23*, 1450006.
485. Fienga, A.; Laskar, J.; Exertier, P.; Manche, H.; Gastineau, M. Numerical estimation of the sensitivity of INPOP planetary ephemerides to general relativity parameters. *Celest. Mech. Dyn. Astron.* **2015**, *123*, 325–349.

486. Everitt, C.W.F.; Debra, D.B.; Parkinson, B.W.; Turneaure, J.P.; Conklin, J.W.; Heifetz, M.I.; Keiser, G.M.; Silbergleit, A.S.; Holmes, T.; Kolodziejczak, J.; et al. Gravity Probe B: Final results of a space experiment to test general relativity. *Phys. Rev. Lett.* **2011**, *106*, 221101.

487. Kramer, M. Pulsars as probes of gravity and fundamental physics. *Int. J. Mod. Phys. D* **2016**, *25*, 1630029.

488. Bertotti, B.; Ciufolini, I.; Bender, P.L. New test of general relativity—Measurement of de Sitter geodetic precession rate for lunar perigee. *Phys. Rev. Lett.* **1987**, *58*, 1062–1065.

489. Shapiro, I.I.; Reasenberg, R.D.; Chandler, J.F.; Babcock, R.W. Measurement of the de Sitter precession of the moon—A relativistic three-body effect. *Phys. Rev. Lett.* **1988**, *61*, 2643–2646.

490. Williams, J.G.; Newhall, X.X.; Dickey, J.O. Relativity parameters determined from lunar laser ranging. *Phys. Rev. D* **1996**, *53*, 6730–6739.

491. Merkowitz, S.M. Tests of gravity using lunar laser ranging. *Living Rev. Relativ.* **2010**, *13*, 7.

492. Martini, M.; Dell'Agnello, S.; Currie, D.; Delle Monache, G.O.; Vittori, R.; Berardi, S.; Boni, A.; Cantone, C.; Ciocci, E.; Lops, C.; et al. Moonlight: A new lunar laser ranging retroreflector and the lunar geodetic precession. *Acta Polytech.* **2013**, *53*, 745–750.

493. Kramer, M. Determination of the geometry of the PSR B1913 + 16 System by geodetic precession. *Astrophys. J.* **1998**, *509*, 856–860.

494. Breton, R.P.; Kaspi, V.M.; Kramer, M.; McLaughlin, M.A.; Lyutikov, M.; Ransom, S.M.; Stairs, I.H.; Ferdman, R.D.; Camilo, F.; Possenti, A. Relativistic spin precession in the double pulsar. *Science* **2008**, *321*, 104–107.

495. Lucchesi, D.M.; Anselmo, L.; Bassan, M.; Pardini, C.; Peron, R.; Pucacco, G.; Visco, M. Testing the gravitational interaction in the field of the Earth via satellite laser ranging and the Laser Ranged Satellites Experiment (LARASE). *Class. Quantum Gravity* **2015**, *32*, 155012.

496. Ingram, A.; van der Klis, M.; Middleton, M.; Done, C.; Altamirano, D.; Heil, L.; Uttley, P.; Axelsson, M. A quasi-periodic modulation of the iron line centroid energy in the black hole binary H 1743-322. *Mon. Not. R. Astron. Soc.* **2016**, *461*, 1967–1980.

497. Bosi, F.; Cella, G.; di Virgilio, A.; Ortolan, A.; Porzio, A.; Solimeno, S.; Cerdonio, M.; Zendri, J.P.; Allegrini, M.; Belfi, J.; et al. Measuring gravitomagnetic effects by a multi-ring-laser gyroscope. *Phys. Rev. D* **2011**, *84*, 122002.

498. Stone, N.; Loeb, A. Observing lense-thirring precession in tidal disruption flares. *Phys. Rev. Lett.* **2012**, *108*, 061302.

499. Franchini, A.; Lodato, G.; Facchini, S. Lense-Thirring precession around supermassive black holes during tidal disruption events. *Mon. Not. R. Astron. Soc.* **2016**, *455*, 1946–1956.

500. Martynov, D.V.; Hall, E.D.; Abbott, B.P.; Abbott, R.; Abbott, T.D.; Adams, C.; Adhikari, R.X.; Anderson, R.A.; Anderson, S.B.; Arai, K.; et al. Sensitivity of the advanced LIGO detectors at the beginning of gravitational wave astronomy. *Phys. Rev. D* **2016**, *93*, 112004.

501. LIGO Scientific Collaboration and Virgo Collaboration. GW151226: Observation of gravitational waves from a 22-solar-mass binary black hole coalescence. *Phys. Rev. Lett.* **2016**, *116*, 241103.

502. Johannsen, T. Sgr A* and general relativity. *Class. Quantum Gravity* **2016**, *33*, 113001.

503. Soffel, M.; Wirrer, R.; Schastok, J.; Ruder, H.; Schneider, M. Relativistic effects in the motion of artificial satellites: The oblateness of the central body I. *Celest. Mech.* **1987**, *42*, 81–89.

504. Heimberger, J.; Soffel, M.; Ruder, H. Relativistic effects in the motion of artificial satellites: The oblateness of the central body II. *Celest. Mech. Dyn. Astron.* **1989**, *47*, 205–217.

505. Soffel, M. *Relativity in Astrometry, Celestial Mechanics and Geodesy*; Springer-Verlag: Berlin, Germany, 1989.

506. Iorio, L. Post-Newtonian direct and mixed orbital effects due to the oblateness of the central body. *Int. J. Mod. Phys. D* **2015**, *24*, 1550067.

507. Iorio, L. A possible new test of general relativity with Juno. *Class. Quantum Gravity* **2013**, *30*, 195011.

508. Starkman, G.D.; Trotta, R.; Vaudrevange, P.M. The virtues of frugality—Why cosmological observers should release their data slowly. *Mon. Not. R. Astron. Soc.* **2010**, *401*, L15–L18.

509. Domcke, V.; Pieroni, M.; Binétruy, P. Primordial gravitational waves for universality classes of pseudoscalar inflation. *J. Cosmol. Astropart. Phys.* **2016**, *6*, 031.

510. Ito, A.; Soda, J. MHz gravitational waves from short-term anisotropic inflation. *J. Cosmol. Astropart. Phys.* **2016**, *4*, 035.

511. Fidler, C.; Pettinari, G.W.; Beneke, M.; Crittenden, R.; Koyama, K.; Wands, D. The intrinsic B-mode polarisation of the Cosmic Microwave Background. *J. Cosmol. Astropart. Phys.* **2014**, *7*, 011.

512. Namikawa, T.; Nagata, R. Non-Gaussian structure of B-mode polarization after delensing. *J. Cosmol. Astropart. Phys.* **2015**, *10*, 004.

513. Abazajian, K.N.; Arnold, K.; Austermann, J.; Benson, B.A.; Bischoff, C.; Bock, J.; Bond, J.R.; Borrill, J.; Calabrese, E.; Carlstrom, J.E.; et al. Neutrino physics from the cosmic microwave background and large scale structure. *Astropart. Phys.* **2015**, *63*, 66–80.

514. Trotta, R. Bayes in the sky: Bayesian inference and model selection in cosmology. *Contemp. Phys.* **2008**, *49*, 71–104.

515. Hobson, M.P.; Jaffe, A.H.; Liddle, A.R.; Mukeherjee, P.; Parkinson, D. *Bayesian Methods in Cosmology*; Cambridge University Press: Cambridge, UK, 2010.

516. Liddle, A.R. How many cosmological parameters? *Mon. Not. R. Astron. Soc.* **2004**, *351*, L49–L53.

517. Metcalf, R.B.; Silk, J. On breaking cosmic degeneracy. *Astrophys. J.* **1998**, *492*, L1–L4.

518. Debono, I.; Rassat, A.; Réfrégier, A.; Amara, A.; Kitching, T.D. Weak lensing forecasts for dark energy, neutrinos and initial conditions. *Mon. Not. R. Astron. Soc.* **2010**, *404*, 110–119.

519. Howlett, C.; Lewis, A.; Hall, A.; Challinor, A. CMB power spectrum parameter degeneracies in the era of precision cosmology. *J. Cosmol. Astropart. Phys.* **2012**, *4*, 027.

520. Joyce, A.; Jain, B.; Khoury, J.; Trodden, M. Beyond the cosmological standard model. *Phys. Rep.* **2015**, *568*, 1–98.

521. Jaffe, A. H_0 and odds on cosmology. *Astrophys. J.* **1996**, *471*, 24.

522. Mukherjee, P.; Parkinson, D.; Liddle, A.R. A nested sampling algorithm for cosmological model selection. *Astrophys. J.* **2006**, *638*, L51–L54.

523. Kilbinger, M.; Wraith, D.; Robert, C.P.; Benabed, K.; Cappé, O.; Cardoso, J.F.; Fort, G.; Prunet, S.; Bouchet, F.R. Bayesian model comparison in cosmology with Population Monte Carlo. *Mon. Not. R. Astron. Soc.* **2010**, *405*, 2381–2390.

524. Wraith, D.; Kilbinger, M.; Benabed, K.; Cappé, O.; Cardoso, J.; Fort, G.; Prunet, S.; Robert, C.P. Estimation of cosmological parameters using adaptive importance sampling. *Phys. Rev. D* **2009**, *80*, 023507.

525. Del Pozzo, W.; Veitch, J.; Vecchio, A. Testing general relativity using Bayesian model selection: Applications to observations of gravitational waves from compact binary systems. *Phys. Rev. D* **2011**, *83*, 082002.

526. Ellis, G.F.R. *Dark Matter and Dark Energy Proposals: Maintaining Cosmology as a True Science*; EAS Publications Series; European Astronomical Society: Versoix, Switzerland, 2009; Volume 36, pp. 325–336.

527. Ellis, G.F.R. Issues in the Philosophy of Cosmology. In *Philosophy of Physics*; Butterfield, J., Earman, J., Eds.; Handbook of the Philosophy of Science; Elsevier: Amsterdam, The Netherlands, 2006; pp. 1183–1285.

528. Jackman, S. *Bayesian Analysis for the Social Sciences*; Wiley: Hoboken, NJ, USA, 2009.

529. Niemack, M.D.; Ade, P.A.R.; Aguirre, J.; Barrientos, F.; Beall, J.A.; Bond, J.R.; Britton, J.; Cho, H.M.; Das, S.; Devlin, M.J.; et al. ACTPol: A polarization-sensitive receiver for the Atacama Cosmology Telescope. *Proc. SPIE* **2010**, *7741*, 77411S.

530. Beynon, E.; Bacon, D.J.; Koyama, K. Weak lensing predictions for modified gravities at non-linear scales. *Mon. Not. R. Astron. Soc.* **2010**, *403*, 353–362.

531. Harnois-Déraps, J.; Munshi, D.; Valageas, P.; van Waerbeke, L.; Brax, P.; Coles, P.; Rizzo, L. Testing modified gravity with cosmic shear. *Mon. Not. R. Astron. Soc.* **2015**, *454*, 2722–2735.

532. Carlstrom, J.E.; CMB-S4 Collaboration. The next generation ground-based CMB experiment, CMB-S4. In Proceedings of the 228th AAS Meeting, San Diego, CA, USA, 12–16 June 2016.

533. Santos, M.; Bull, P.; Alonso, D.; Camera, S.; Ferreira, P.; Bernardi, G.; Maartens, R.; Viel, M.; Villaescusa-Navarro, F.; Abdalla, F.B.; et al. Cosmology from a SKA HI intensity mapping survey. In Proceedings of the Advancing Astrophysics with the Square Kilometre Array (AASKA14), Giardini Naxos, Italy, 9–13 June 2014.

534. Hall, A.; Bonvin, C.; Challinor, A. Testing general relativity with 21-cm intensity mapping. *Phys. Rev. D* **2013**, *87*, 064026.

535. Lahav, O.; Suto, Y. Measuring our Universe from Galaxy Redshift Surveys. *Living Rev. Relativ.* **2004**, *7*, 8.

536. Dupé, F.X.; Rassat, A.; Starck, J.L.; Fadili, M.J. Measuring the integrated Sachs-Wolfe effect. *Astron. Astrophys.* **2011**, *534*, A51.

537. Stabenau, H.F.; Jain, B. N-body simulations of alternative gravity models. *Phys. Rev. D* **2006**, *74*, 084007.

538. Adamek, J.; Daverio, D.; Durrer, R.; Kunz, M. General relativistic N-body simulations in the weak field limit. *Phys. Rev. D* **2013**, *88*, 103527.

539. He, J.H.; Li, B.; Hawken, A.J. Effective Dark Matter power spectra in $f(R)$ gravity. *Phys. Rev. D* **2015**, *92*, 103508.

540. Winther, H.A.; Schmidt, F.; Barreira, A.; Arnold, C.; Bose, S.; Llinares, C.; Baldi, M.; Falck, B.; Hellwing, W.A.; Koyama, K.; et al. Modified gravity N-body code comparison project. *Mon. Not. R. Astron. Soc.* **2015**, *454*, 4208–4234.

541. Eingorn, M. First-order cosmological perturbations engendered by point-like masses. *Astrophys. J.* **2016**, *825*, 84.

542. Hahn, O.; Paranjape, A. General Relativistic "Screening" in Cosmological Simulations. 2016, arXiv:1602.07699.

543. Kamdar, H.M.; Turk, M.J.; Brunner, R.J. Machine learning and cosmological simulations—I. Semi-analytical models. *Mon. Not. R. Astron. Soc.* **2016**, *455*, 642–658.

544. Kamdar, H.M.; Turk, M.J.; Brunner, R.J. Machine learning and cosmological simulations—II. Hydrodynamical simulations. *Mon. Not. R. Astron. Soc.* **2016**, *457*, 1162–1179.

545. Smoot, G.F. See Saw Inflation/Dark Matter/Dark Energy/Baryogenesis. 2014, arXiv:1405.2776.

546. Rickles, D. A "Second Superstring Revolution" and the Future of String Theory. In *A Brief History of String Theory: From Dual Models to M-Theory*; Springer: Berlin, Germany, 2014; pp. 207–242.

547. Maartens, R.; Koyama, K. Brane-world gravity. *Living Rev. Relativ.* **2010**, *13*, 5.

548. Woodard, R.P. How far are we from the quantum theory of gravity? *Rep. Prog. Phys.* **2009**, *72*, 126002.

549. Norris, R.P. Data challenges for next-generation radio telescopes. In Proceedings of the Sixth IEEE International Conference on eScience, Brisbane, Australia, 7–10 December 2010; pp. 21–24.

550. Planck Collaboration. Planck 2015 Results. II. Low Frequency Instrument Data Processing. *Astron. Astrophys.* **2016**, *594*, A2.

551. Planck Collaboration. Planck 2015 Results. VIII. High Frequency Instrument Data Processing: Calibration and Maps. *Astron. Astrophys.* **2016**, *594*, A8.

552. Jurić, M.; Kantor, J.; Lim, K.; Lupton, R.H.; Dubois-Felsmann, G.; Jenness, T.; Axelrod, T.S.; Aleksić, J.; Allsman, R.A.; AlSayyad, Y.; et al. The LSST Data Management System. 2015, arXiv:1512.07914.

553. Desai, S.; Mohr, J.J.; Henderson, R.; Kümmel, M.; Paech, K.; Wetzstein, M. CosmoDM and its application to Pan-STARRS data. *J. Instrum.* **2015**, *10*, C06014.

554. Dodson, R.; Vinsen, K.; Wu, C.; Popping, A.; Meyer, M.; Wicenec, A.; Quinn, P.; van Gorkom, J.; Momjian, E. Imaging SKA-scale data in three different computing environments. *Astron. Comput.* **2016**, *14*, 8–22.

555. Arviset, C.; González, J.; Gutiérrez, R.; Hernández, J.; Salgado, J.; Segovia, J.C. The Gaia archive: VO in action in the big data era. In Proceedings of the International Symposium Dedicated to the 50th Anniversary of the Markarian Survey and the 10th Anniversary of the Armenian Virtual Observatory, Byurakan, Armenia, 5–8 October 2015; Volume 505, p. 218.

![universe logo] *universe*

MDPI

Review

A Brief History of Gravitational Waves

Jorge L. Cervantes-Cota [1], Salvador Galindo-Uribarri [1] and George F. Smoot [2,3,4,*]

[1] Department of Physics, National Institute for Nuclear Research, Km 36.5 Carretera Mexico-Toluca, Ocoyoacac, C.P. 52750 Mexico, Mexico; jorge.cervantes@inin.gob.mx (J.L.C.-C.); salvador.galindo@inin.gob.mx (S.G.-U.)
[2] Helmut and Ana Pao Sohmen Professor at Large, Institute for Advanced Study, Hong Kong University of Science and Technology, Clear Water Bay, Kowloon, 999077 Hong Kong, China
[3] Université Sorbonne Paris Cité, Laboratoire APC-PCCP, Université Paris Diderot, 10 rue Alice Domon et Leonie Duquet, 75205 Paris Cedex 13, France
[4] Department of Physics and LBNL, University of California; MS Bldg 50-5505 LBNL, 1 Cyclotron Road Berkeley, 94720 CA, USA
* Correspondence: gfsmoot@lbl.gov; Tel.:+1-510-486-5505

Academic Editors: Lorenzo Iorio and Elias C. Vagenas
Received: 21 July 2016; Accepted: 2 September 2016; Published: 13 September 2016

Abstract: This review describes the discovery of gravitational waves. We recount the journey of predicting and finding those waves, since its beginning in the early twentieth century, their prediction by Einstein in 1916, theoretical and experimental blunders, efforts towards their detection, and finally the subsequent successful discovery.

Keywords: gravitational waves; General Relativity; LIGO; Einstein; strong-field gravity; binary black holes

1. Introduction

Einstein's General Theory of Relativity, published in November 1915, led to the prediction of the existence of gravitational waves that would be so faint and their interaction with matter so weak that Einstein himself wondered if they could ever be discovered. Even if they were detectable, Einstein also wondered if they would ever be useful enough for use in science. However, exactly 100 years after his theory was born, on 14 September 2015, these waves were finally detected and are going to provide scientific results.

In fact at 11:50:45 a.m. CET on 14 September 2015 Marco Drago—a postdoc—was seated in front of a computer monitor at the Max Planck Institute for Gravitational Physics in Hanover, Germany, when he received an e-mail, automatically generated three minutes before from the monitors of LIGO (for its acronym Laser Interferometer Gravitational wave Observatory). Marco opened the e-mail, which contained two links. He opened both links and each contained a graph of a signal similar to that recorded by ornithologists to register the songs of birds. One graph came from a LIGO station located at Hanford, in Washington State, and the other from Livingston Station in the state of Louisiana [1].

Marco is a member of a team of 30 physicists working in Hanover, analyzing data from Hanford and Livingston. Marco's duty is to be aware of and analyze the occurrence of an "event" that records the passage of a gravitational wave, in one of the four lines that automatically track the signals from the detectors in the two LIGO observatories on the other side of the Atlantic.

Marco noticed that the two graphs were almost identical, despite having been registered independently in sites separated by 1900 km (see Figure 1a); for comparison we include sonograms from animals (see Figure 1b,c). The time that elapsed between the two signals differed by about 7 milliseconds. These almost simultaneous records of signals coming from sites far away from each other, the similarity of their shapes and their large size, could not be anything but, either: a possible record of a gravitational wave traveling at the speed of light or, a "signal" artificially "injected" to the detectors by one of the four members of the LIGO program who are allowed to "inject" dummy signals. The reason why artificial signals are injected to the system is to the test whether the operation of the detectors is correct and if the duty observers are able to identify a real signal.

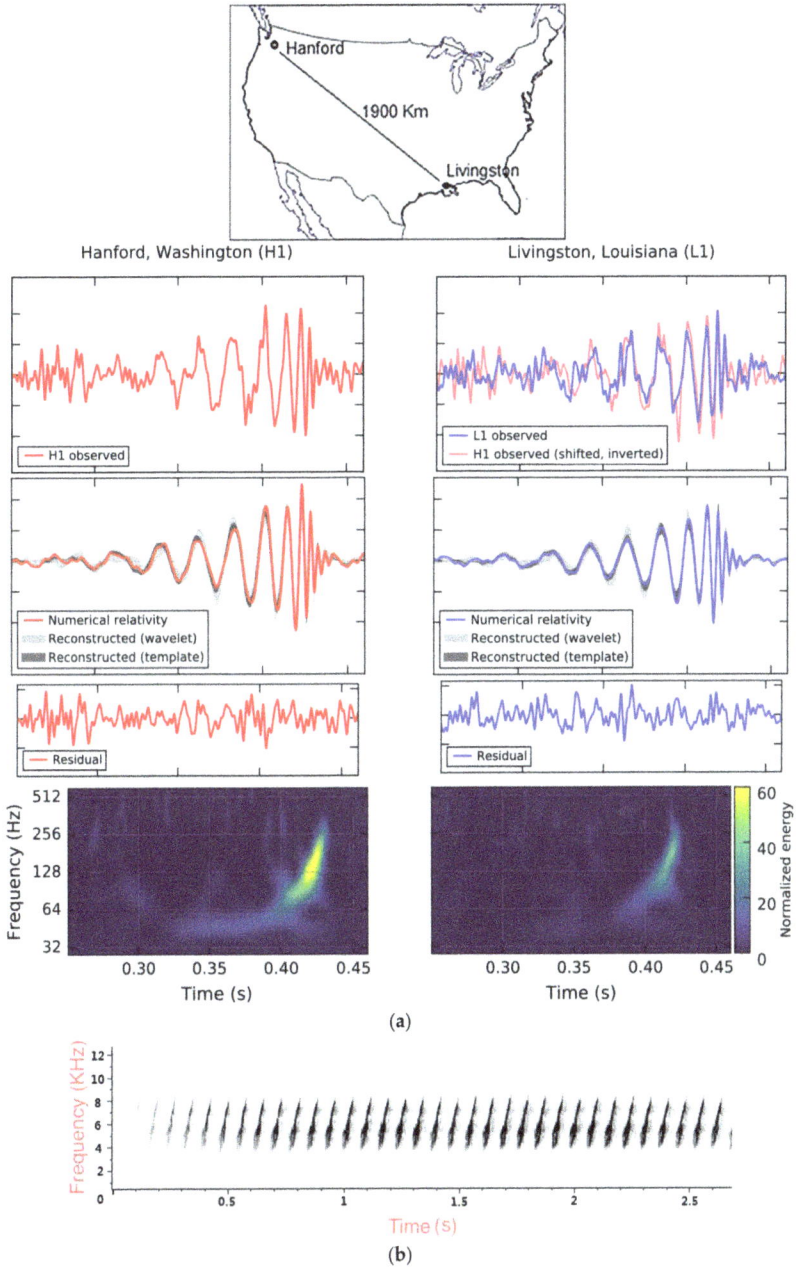

Figure 1. *Cont.*

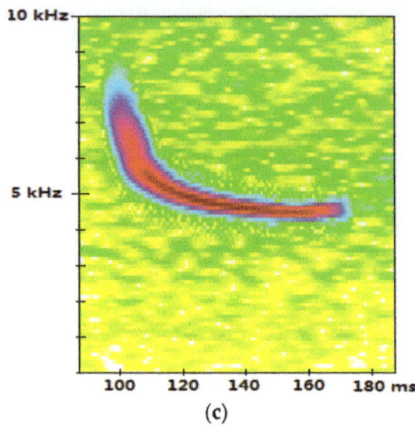

(c)

Figure 1. (**a**) Waveforms from LIGO sites [2] and their location and sonograms. Figure from https://losc.ligo.org/events/GW150914/. The gravitational-wave event GW150914 observed by the LIGO Hanford (H1, left column panels) and Livingston (L1, right column panels) detectors. Times are shown relative to 14 September 2015 at 09:50:45 UTC. For visualization, all time series are filtered with a 35–350 Hz band-pass filter to suppress large fluctuations outside the detectors' most sensitive frequency band, and band-reject filters to remove the strong instrumental spectral lines; (**b**) Chirping Sparrow (Spizella passerina) song: frequency versus time (in seconds), showing a song made up of a series of chirps; (**c**) A Pipistrelle bat call for echolocation. Bats use ultrasound for "seeing" and for social calls. This spectrogram is a graphic representation of frequencies against time. The color represents the loudness of each frequency. This spectrogram shows a falling call, which becomes a steady note. The yellow and green blotches are noise. This is nearly a time reverse of black hole merger sonograms.

Following the pre-established protocols, Marco tried to verify whether the signals were real or the "event" was just a dummy injected signal. Since aLIGO was still in engineering mode, there was no way to inject fake signals, i.e., hardware injections. Therefore, everyone was nearly 100% certain that this was a detection. However, it was necessary to go through the protocols of making sure that this was the case.

Marco asked Andrew Lundgren, another postdoc at Hanover, to find out if the latter was the case. Andrew found no evidence of a "dummy injection." On the other hand, the two signals detected were so clear, they did not need to be filtered to remove background noise. They were obvious. Marco and Andrew immediately phoned the control rooms at Livingston and Hanford. It was early morning in the United States and only someone from Livingston responded. There was nothing unusual to report. Finally, one hour after receiving the signal, Marco sent an e-mail to all collaborators of LIGO asking if anyone was aware of something that might cause a spurious signal. No one answered the e-mail.

Days later, LIGO leaders sent a report stating that there had not been any "artificial injection". By then the news had already been leaked to some other members of the world community of astrophysicists. Finally after several months, the official news of the detection of gravitational waves was given at a press conference on 11 February 2016, after the team had ascertained that the signals were not the result of some experimental failure, or any signal locally produced, earthquake, or electromagnetic fluctuation. This announcement is the most important scientific news so far this century. Gravitational waves were detected after 60 years of searching and 100 years since the prediction of their existence. The scientific paper was published in *Physical Review Letters* [2]. This discovery not only confirmed one of the most basic predictions of General Relativity but also opened a new window of observation of the universe, and we affirm without exaggeration that a new era in astronomy has been born.

In what follows we shall narrate the journey experienced in search of gravitational waves, including their conception in the early twentieth century, their prediction by Einstein in 1916, the theoretical controversy, efforts towards detection, and the recent discovery.

2. Lost and Found Gravitational Waves

On 5 July 1905 the *Comptes Rendus of the French Academy of Sciences* published an article written by Henri Poincare entitled "Sur la dynamique d' l'électron". This work summarized his theory of relativity [3]. The work proposed that gravity was transmitted through a wave that Poincaré called a gravitational wave (*onde gravifique*).

It would take some years for Albert Einstein to postulate in 1915 in final form the Theory of General Relativity [4]. His theory can be seen as an extension of the Special Theory of Relativity postulated by him 10 years earlier in 1905 [5]. The General Theory explains the phenomenon of gravity. In this theory, gravity is not a force—a difference from Newton's Law—but a manifestation of the curvature of space–time, this curvature being caused by the presence of mass (and also energy and momentum of an object). In other words, Einstein's equations match, on the one side, the curvature of local space–time with, on the other side of the equation, local energy and momentum within that space–time.

Einstein's equations are too complicated to be solved in full generality and only a few very specific solutions that describe space–times with very restrictive conditions of symmetry are known. Only with such restrictions it is possible to simplify Einstein's equations and so find exact analytical solutions. For other cases one must make some simplifications or approximations that allow a solution, or there are cases where equations can be solved numerically using computers, with advanced techniques in the field called Numerical Relativity.

Shortly after having finished his theory Einstein conjectured, just as Poincaré had done, that there could be gravitational waves similar to electromagnetic waves. The latter are produced by accelerations of electric charges. In the electromagnetic case, what is commonly found is dipolar radiation produced by swinging an electric dipole. An electric dipole is formed by two (positive and negative) charges that are separated by some distance. Oscillations of the dipole separation generate electromagnetic waves. However, in the gravitational case, the analogy breaks down because there is no equivalent to a negative electric charge. There are no negative masses. In principle, the expectation of theoretically emulate gravitational waves similar to electromagnetic ones faded in Einstein's view. This we know from a letter he wrote to his colleague, Karl Schwarzschild, on 19 February 1916. In this letter, Einstein mentioned in passing:

> "Since then [November 14] I have handled Newton's case differently, of course, according to the final theory [the theory of General Relativity]. Thus there are no gravitational waves analogous to light waves. This probably is also related to the one-sidedness of the sign of the scalar T, incidentally [this implies the nonexistence of a "gravitational dipole"] [6]

However, Einstein was not entirely convinced of the non-existence of gravitational waves; for a few months after having completed the General Theory, he refocused efforts to manipulate his equations to obtain an equation that looked like the wave equation of electrodynamics (Maxwell's wave equation), which predicts the existence of electromagnetic waves. However, as mentioned, these equations are complex and Einstein had to make several approximations and assumptions to transform them into something similar to Maxwell's equation. For some months his efforts were futile. The reason was that he used a coordinate system that hindered his calculations. When, at the suggestion of a colleague, he changed coordinate systems, he found a solution that predicted three different kinds of gravitational waves. These three kinds of waves were baptized by Hermann Weyl as longitudinal-longitudinal, transverse-longitudinal, and transverse-transverse [7].

These approaches made by Einstein were long open to criticism from several researchers and even Einstein had doubts. In this case Einstein had manipulated his field equations into a first approximation for wave-emitting bodies whose own gravitational field is negligible and with waves that propagate in empty and flat space.

Yet the question of the existence of these gravitational waves dogged Einstein and other notable figures in the field of relativity for decades to come. By 1922 Arthur Eddington wrote an article entitled "The propagation of gravitational waves" [8]. In this paper, Eddington showed that two of the three types of waves found by Einstein could travel at any speed and this speed depends on the coordinate system; therefore, they actually were spurious waves. The problem Eddington found in Einstein's original calculations is that the coordinate system he used was in itself a "wavy" system and therefore two of the three wave types were simply flat space seen from a wavy coordinates system; i.e., mathematical artifacts were produced by the coordinate system and were not really waves at all. So the existence of the third wave (the transverse-transverse), allegedly traveling at the speed of light, was also questioned. Importantly, Eddington did prove that this last wave type propagates at the speed of light in all coordinate systems, so he did not rule out its existence.

In 1933 Einstein emigrated to the United States, where he had a professorship at the Institute for Advanced Study in Princeton. Among other projects, he continued to work on gravitational waves with the young American student Nathan Rosen.

In 1936 Einstein wrote to his friend, renowned physicist Max Born, "Together with a young collaborator [Rosen], I arrive at the interesting result that *gravitational waves do not exist*, though they have been assumed a certainty to the first approximation" (emphasis added) [9] (p. 121, Letter 71).

That same year, Einstein and Rosen sent on 1 June an article entitled "Are there any gravitational waves?" to the prestigious journal *Physical Review*, whose editor was John T. Tate [10]. Although the original version of the manuscript does not exist today, it follows from the abovementioned letter to Max Born that the answer to the title of the article was "they do not exist".

The editor of the *Physical Review* sent the manuscript to Howard Percy Robertson, who carefully examined it and made several negative comments. John Tate in turn wrote to Einstein on 23 July, asking him to respond to the reviewer's comments. Einstein's reaction was anger and indignation; he sent the following note to Tate [10]:

July 27, 1936

Dear Sir.

"We (Mr. Rosen and I) had sent you our manuscript for publication and had not authorized you to show it to specialists before it is printed. I see no reason to address the—in any case erroneous—comments of your anonymous expert. On the basis of this incident I prefer to publish the paper elsewhere."

Respectfully

Einstein

P.S. Mr. Rosen, who has left for the Soviet Union, has authorized me to represent him in this matter.

On July 30th, John Tate replied to Einstein that he very much regretted the withdrawal of the article, saying "I could not accept for publication in *The Physical Review* a paper which the author was unwilling I should show to our Editorial Board before publication" [10].

During the summer of 1936 a young physicist named Leopold Infeld replaced Nathan Rosen as the new assistant to Einstein. Rosen had departed a few days before for the Soviet Union. Once he arrived at Princeton, Infeld befriended Robertson (the referee of the Einstein–Rosen article). During one of their encounters the topic of gravitational waves arose. Robertson confessed to Infeld his skepticism about the results obtained by Einstein. Infeld and Robertson discussed the point and reviewed together the Einstein and Rosen manuscript, confirming the error. Infeld in turn informed Einstein about the conversation with Robertson.

An anecdote illustrating the confused situation prevailing at that time is given in Infeld's autobiography. Infeld refers to the day before a scheduled talk that Einstein was to give at Princeton on the "Nonexistence of gravitational waves". Einstein was already aware of the error in his manuscript, which was previously pointed out by Infeld. There was no time to cancel the talk. The next day Einstein gave his talk and concluded, *"If you ask me whether there are gravitational waves or not, I must answer that I don't know. But it is a highly interesting problem"* [10].

After having withdrawn the Einstein–Rosen paper from the *Physical Review*, Einstein had summited the very same manuscript to the *Journal of the Franklin Society* (Philadelphia). This journal accepted the paper for publication without modifications. However, after Einstein learned that the paper he had written with Rosen was wrong, he had to modify the galley proofs of the paper. Einstein sent a letter to the editor on 13 November 1936 explaining the reasons why he had to make fundamental changes to the galley proofs. Einstein also renamed the paper, entitling it "On gravitational waves", and modified it to include different conclusions [10]. It should be noted that this would not have happened if Einstein had accepted in the first instance Robertson's valid criticisms. Tellingly, the new conclusions of his rewritten article read [11]:

"Rigorous solution for Gravitational cylindrical waves is provided. For convenience of the reader the theory of gravitational waves and their production, known in principle, is presented in the first part of this article. After finding relationships that cast doubt on the existence of gravitational fields rigorous wavelike solutions, we have thoroughly investigated the case of cylindrical gravitational waves. As a result, there are strict solutions and the problem is reduced to conventional cylindrical waves in Euclidean space".

Furthermore, Einstein included this explanatory note at the end of his paper [11],

> *"Note—The second part of this article was considerably altered by me after the departure to Russia of Mr. Rosen as we had misinterpreted the results of our formula. I want to thank my colleague Professor Robertson for their friendly help in clarifying the original error. I also thank Mr. Hoffmann your kind assistance in translation."*

In the end, Einstein became convinced of the existence of gravitational waves, whereas Nathan Rosen always thought that they were just a formal mathematical construct with no real physical meaning.

3. Pirani's Trip to Poland; the Effect of a Gravitational Wave

To prove the existence of a gravitational wave it becomes necessary to detect its effects. One of the difficulties presented by the General Theory of Relativity resides in how to choose the appropriate coordinate system in which one observer may calculate an experimentally measurable quantity, which could, in turn, be compared to a real observation. Coordinate systems commonly used in past calculations were chosen for reasons of mathematical simplification and not for reasons of physical convenience. In practice, a real observer in each measurement uses a local Cartesian coordinate system relative to its state of motion and local time. To remedy this situation, in 1956 Felix A. E. Pirani published a work that became a classic article in the further development of the Theory of Relativity. The article title was "On the physical significance of the Riemann tensor" [12]. The intention of this work was to demonstrate a mathematical formalism for the deduction of physical observable quantities applicable to gravitational waves. Curiously, the work was published in a Polish magazine. The reason for this was that Pirani, who at that time worked in Ireland, went to Poland to visit his colleague Leopold Infeld, of whom we have already spoken; the latter had returned to his native Poland in 1950 to help boost devastated Polish postwar physics. Because Infeld went back to Poland, and because of the anti-communist climate of that era, Infeld was stripped of his Canadian citizenship. In solidarity Pirani visited Poland and sent his aforementioned manuscript to *Acta Physica Polonica*. The importance of Pirani's Polish paper is that he used a very practical approach that got around this whole problem of the coordinate system, and he showed that the waves would move particles back and forth as they pass by.

4. Back and Forth as Waves Pass by

One of the most famous of Einstein's collaborators, Peter Bergmann, wrote a well-known popular book *The Riddle of Gravitation*, which describes the effect a gravitational wave passing over a set of particles would have [13]. Following Bergmann, we shall explain this effect.

When a gravitational wave passes through a set of particles positioned in an imaginary circle and initially at rest, the passing wave will move these particles. This motion is perpendicular (transverse) to the direction in which the gravitational wave travels. For example, suppose a gravitational pulse passes in a direction perpendicular to this page, Figure 2 shows how a set of particles, initially arranged in a circle, would sequentially move (a, b, c, d).

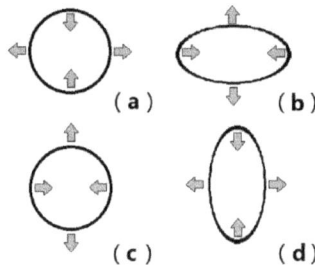

Figure 2. (a,b,c,d) Sequential effect of a gravitational wave on a ring of particles. In the image of Figure 2a is observed as the particles near horizontal move away from each other while those are near vertical move together to reach finally the next moment as shown in (Figure 2b). At that moment all the motions are reversed and so on. This is shown in Figure 2c,d. All these motions occur successively in the plane perpendicular (transverse) to the direction of wave propagation.

At first sight the detection of gravitational waves now seems very simple. One has to compare distances between perpendicularly placed pairs of particles and wait until a gravitational wave transits. However, one has to understand in detail how things happen. For instance, a ruler will not stretch, in response to a gravitational wave, in the same way as a free pair of particles, due to the elastic properties of the ruler, c.f. note 11, p. 19 in [14]. Later it was realized that changes produced in such disposition of particle pairs (or bodies) can be measured if, instead of the distances, we measure the time taken by light to traverse them, as the speed of light is constant and unaltered by a gravitational wave.

Anyway, Pirani's 1956 work remained unknown among most physicists because scientists were focusing their attention on whether or not gravitational waves carry energy. This misperception stems from the rather subtle matter of defining energy in General Relativity. Whereas in Special Relativity energy is conserved, in General Relativity energy conservation is not simple to visualize. In physics, a conservation law of any quantity is the result of an underlying symmetry. For example, linear momentum is conserved if there is spatial translational symmetry, that is, if the system under consideration is moved by a certain amount and nothing changes. In the same way, energy is conserved if the system is invariant under time. In General Relativity, time is part of the coordinate system, and normally it depends on the position. Therefore, globally, energy is not conserved. However, any curved space–time can be considered to be locally flat and, locally, energy is conserved.

During the mid-1950s the question of whether or not gravitational waves would transmit energy was still a hot issue. In addition, the controversy could not be solved since there were no experimental observations that would settle this matter. However, this situation was finally clarified thanks to the already mentioned work by Pirani [12], and the comments suggested by Richard Feynman together with a hypothetical experiment he proposed. The experiment was suggested and comments were delivered by Feynman during a milestone Congress held in 1957 in Chapel Hill, North Carolina.

We will come back to this experiment later, but first we shall speak about the genesis of the Chapel Hill meeting.

5. What Goes Up Must Come Down

The interest in the search for gravitational waves began at a meeting occurred in Chapel Hill, North Carolina in 1957. The meeting brought together many scientists interested in the study of gravity. What is unusual is that this meeting would not have been possible without the funding of an eccentric American millionaire named Roger W. Babson.

On 19 January 1949 Roger W. Babson founded the Gravity Research Foundation (GRF), which still exists today. Babson's motivation for establishing the foundation was a "debt" that he thought he owed to Newton's laws—which, according to his understanding, led him to become a millionaire [15]. Babson earned the greatest part of his fortune in the New York Stock Exchange by applying his own version of Newton's Gravity law, "What goes up must come down". Thus he bought cheap shares on their upward route and sold them before their price collapsed. His ability to apply the laws of Newton was surprising because he anticipated the 1929 Wall Street crash. "To every action there is a reaction", he used to preach.

Babson's interest in gravity arose when he was a child, following a family tragedy. Babson's older sister drowned when he was still an infant. In his version of the unfortunate accident, he recalls, "... she could not fight gravity." The story of this eccentric millionaire is detailed on the website of the GRF foundation [15]. Babson became obsessed with finding a way to control the force of gravity and therefore he established the aforementioned foundation, which had as its main activity arranging a yearly essay contest that dealt with "the chances of discovering a partial insulation, reflector, or absorber of gravity". An annual award of $1000 (a considerable amount at the time) was offered to the best essay. The essays submitted for the competition were limited to 2000 words. This annual award attracted several bizarre competitors and was awarded several times to risible submissions. However, in 1953 Bryce DeWitt, a young researcher at Lawrence Livermore Laboratory in California, decided to write an essay and enter the contest because he needed the money to pay his home's mortgage.

The essay presented by DeWitt in the 1953 competition was a devastating critique of the belief that it is possible to control gravity. In DeWitt's own words, his writing "essentially nagged [the organizers] for that stupid idea" [16]. To his surprise, his essay was the winner despite having been written in one night. DeWitt notes those were "the faster 1000 dollars earned in my whole life!" [16].

But DeWitt never imagined he would earn many thousands more dollars than he won with his essay. The reason for it might be found in the final paragraph of his essay,

"In the near future, external stimuli to induce young people to engage in gravitational physics research, despite its difficulties, are urgently needed" [16].

This final paragraph of DeWitt's essay echoed in Babson's mind. Perhaps, he thought, why not focus my philanthropy to support serious studies of gravitation? Perhaps he thought his GRF could refocus its activities onto the scientific study of gravitation.

Babson shared this new enthusiasm with a friend, Agnew Bahnson, also a millionaire and also interested in gravity. Bahnson was a little more practical than Babson and convinced him to found an independent institute separate from GRF. Thus arose the idea of founding a new Institute of Field Physics (IOFP), whose purpose would be pure research in the gravitational fields. The idea of founding the IOFP was clever because the old GRF was severely discredited among the scientific community. For example, one of the promotional brochures GRF mentioned as an example of the real possibility of gravity control the biblical episode where Jesus walks on water. Such was the ridicule and vilification of the GRF in scientific circles that a famous popularizer of mathematics, Martin Gardner, devoted an entire chapter of one of his books to ridiculing the GRF. In this work Gardner claims that the GRF "is perhaps the most useless project of the twentieth century" [17]. So Bahnson knew that to research on gravity in the discredited GRF had very little chance of attracting serious scientists. We must mention that today GRF enjoys good prestige and many well-known scientists have submitted their essays to its annual competition. That proves that it is worth trying for a thousand dollars.

In order to start the new IOFP institute off on the right foot, Bahnson contacted a famous Princeton physicist, John Archibald Wheeler, who supported the idea of hiring Bryce DeWitt to lead the new institute, whose headquarters would be established in Chapel Hill, NC, Bahnson's hometown and headquarters of the University of North Carolina. Wheeler, knowing the vast fortune of the couple of millionaires, hastened to send a telegram to DeWitt. In one of his lines the telegram said "Please do not give him a 'no' for answer from the start" [16]. That's how DeWitt won more than one thousand dollars, actually much more. In January 1957 the IOFP was formally inaugurated, holding a scientific conference on the theme "The role of gravitation in Physics". As we shall review below, the Chapel Hill conference rekindled the crestfallen and stagnant study of gravitation prevailing in those days.

6. The Chapel Hill Conference 1957

The 1957 Chapel Hill conference was an important event for the study of gravity. Attendance was substantial: around 40 speakers from institutions from 11 countries met for six days, from 18 to 23 January 1957 on the premises of the University of North Carolina at Chapel Hill. Participants who attended the meeting were predominantly young physicists of the new guard: Feynman, Schwinger, Wheeler, and others. During the six-day conference, discussions focused on various topics: classical gravitational fields, the possibility of unification of gravity with quantum theory, cosmology, measurements of radio astronomy, the dynamics of the universe, and gravitational waves [18].

The conference played a central role in the future development of classical and quantum gravity. It should be noted that the Chapel Hill 1957 conference today is known as the GR1 conference. That is, the first of a series of GR meetings that have been held regularly in order to discuss the state of the art in matters of Gravitation and General Relativity (GR = General Relativity). The conference has been held in many countries and possesses international prestige. The last was held in New York City in 2016.

In addition to the issues and debates on the cosmological models and the reality of gravitational waves, during the conference many questions were formulated, including ideas that are topical even today. To mention a few, we can say that one of the assistants, named Hugh Everett, briefly alluded to his parallel universes interpretation of quantum physics. On the other hand, DeWitt himself pointed out the possibility of solving gravitation equations through the use of electronic computers and warned of the difficulties that would be encountered in scheduling them for calculations, thus foreseeing the future development of the field of Numerical Relativity. However, what concerns us here is that gravitational waves were also discussed at the conference; chiefly, the question was whether gravitational waves carrying energy or not.

Hermann Bondi, a distinguished physicist at King's College London, presided over session III of Congress entitled "General Relativity not quantized". In his welcome address to the participants he warned "...still do not know if a transmitter transmits energy radiation ..." [18] (p. 95). With these words Bondi marked the theme that several of the speakers dealt with in their presentations and subsequent discussion. Some parts of the debate focused on a technical discussion to answer the question about the effect a gravitational pulse would have on a particle when passing by, i.e., whether or not the wave transmits energy to the particle.

During the discussions, Feynman came up with an argument that convinced most of the audience.

His reasoning is today known as the "sticky bead argument". Feynman's reasoning is based on a thought experiment that can be described briefly as follows: Imagine two rings of beads on a bar (see Figure 3, upper part). The bead rings can slide freely along the bar. If the bar is placed transversely to the propagation of a gravitational wave, the wave will generate tidal forces with respect to the midpoint of the bar. These forces in turn will produce longitudinal compressive stress on the bar. Meanwhile, and because the bead rings can slide on the bar and also in response to the tidal forces, they will slide toward the extreme ends first and then to the center of the bar (Figure 3, bottom). If contact between the beads and the bar is "sticky", then both pieces (beads and bar) will be heated by friction. This heating implies that energy was transmitted to the bar by the gravitational wave, showing that gravitational waves carry energy [18].

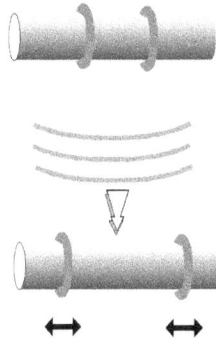

Figure 3. Sketch of the "sticky bead argument".

In a letter to Victor Weisskopf, Feynman recalls the 1957 conference in Chapel Hill and says, "I was surprised to find that a whole day of the conference was spent on this issue and that 'experts' were confused. That's what happens when one is considering energy conservation tensors, etc. instead of questioning, can waves do work?" [19].

Discussions on the effects of gravitational waves introduced at Chapel Hill and the "sticky bead argument" convinced many—including Hermann Bondi, who had, ironically, been among the skeptics on the existence of gravitational waves. Shortly after the Chapel Hill meeting Bondi issued a variant of the "sticky bead argument" [20].

Among the Chapel Hill audience, Joseph Weber was present. Weber was an engineer at the University of Maryland. He became fascinated by discussions about gravitational waves and decided to design a device that could detect them. Thus, while discussions among theoretical physicists continued in subsequent years, Weber went even further because, as discussed below, he soon began designing an instrument to make the discovery.

7. The First Gravitational Wave Detector

The year following the meeting at Chapel Hill, Joseph Weber began to speculate how he could detect gravitational waves. In 1960 he published a paper describing his ideas on this matter [21]. Basically he proposed the detection of gravitational waves by measuring vibrations induced in a mechanical system. For this purpose, Weber designed and built a large metal cylinder as a sort of "antenna" to observe resonant vibrations induced in this antenna that will eventually be produced by a transit of a gravitational wave pulse. This is something like waiting for someone to hit a bell with a hammer to hear its ring.

It took his team several years to build the "antenna", a task that ended by the mid-sixties. In 1966, Weber, in a paper published in *Physical Review*, released details of his detector and provided evidence of its performance [22]. His "antenna" was a big aluminum cylinder about 66 cm in diameter and 153 cm in length, weighing 3 tons. The cylinder was hanging by a steel wire from a support built to isolate vibrations of its environment (see Figure 4). In addition, the whole arrangement was placed inside a vacuum chamber. To complete his instrument, Weber placed a belt of detectors around the cylinder. The detectors were piezoelectric crystals to sense cylinder vibrations induced by gravitational waves. Piezoelectric sensors convert mechanical vibrations into electrical impulses.

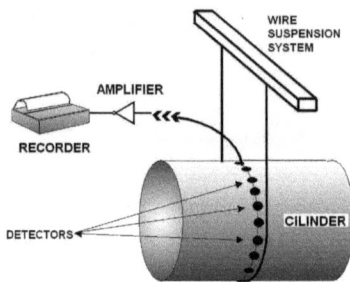

Figure 4. Sketch of Weber's cylinder detector and photo of Joseph Weber at the antenna.

Weber built two detectors. The first one was at the University of Maryland and the other was situated 950 km away, in Argonne National Laboratory near Chicago. Both detectors were connected to a registration center by a high-speed phone line. The idea of having two antennas separated by a large distance allowed Weber to eliminate spurious local signals, that is, signals produced by local disturbances such as thunderstorms, cosmic rays showers, power supply fluctuations, etc. In other words, if a detected signal was not recorded simultaneously in both laboratories, the signal should be discarded because it was a local signal and therefore spurious.

For several years, Weber made great efforts to isolate his cylinders from spurious vibrations, local earthquakes, and electromagnetic interference, and argued that the only significant source of background noise was random thermal motions of the atoms of the aluminum cylinder. This thermal agitation caused the cylinder length to vary erratically by about 10^{-16} meters, less than the diameter of a proton; however, the gravitational signal he anticipated was not likely to get much greater than the threshold stochastic noise caused by thermal agitation.

It took several years for Weber and his team to begin detecting what they claimed were gravitational wave signals. In 1969 he published results announcing the detection of waves [23]. A year later, Weber claimed that he had discovered many signals that seemed to emanate from the center of our galaxy [24]. This meant that in the center of the Milky Way a lot of stellar mass became energy ($E = mc^2$) in the form of gravitational waves, thus reducing the mass of our galaxy. This "fact" presented the problem that a mass conversion into energy as large as Weber's results implied involved a rapid decrease of the mass that gravitationally keeps our galaxy together. If that were the case, our galaxy would have already been dispersed long ago. Theoretical physicists Sciama, Field, and Rees calculated that the maximum conversion of mass into energy for the galaxy, so as not to expand more than what measurements allowed, corresponded to an upper limit of 200 solar masses per year [25]. However, Weber's measurements implied that a conversion of 1000 solar masses per year was taking place. Something did not fit. Discussions took place to determine what mechanisms could make Weber's measurements possible. Among others, Charles Misner, also from the University of Maryland, put forward the idea that signals, if stemming from the center of the Milky Way, could have originated by gravitational synchotron radiation in narrow angles, so as to avoid the above constraints considered for isotropic emission. Some others, like Peter Kafka of the Max Planck Institute in Munich, claimed in an essay for the Gravity Research Foundation's contest in 1972 (in which he won the second prize) that Weber's measurements, if they were isotropically emitted, and taking into account the inefficiency of bars, would imply a conversion of three million solar masses per year in the center of the Milky Way [26]. It soon became clear that Weber's alleged discoveries were not credible. Weber's frequent observations of gravitational waves related to very sporadic events and raised many suspicions among some scientists. It seemed that Weber was like those who have a hammer in hand and to them everything looks like a nail to hit.

Despite Weber's doubtful measurements, he began to acquire notoriety. In 1971, the famous magazine *Scientific American* invited him to write an article for their readers entitled "The detection of gravitational waves" [27].

Whether it was the amazing—for some—findings of Weber, or doubtful findings of others, or the remarks made by Sciama, Field, Rees, and Kafka, the fact is that many groups of scientists thought it was a good idea to build their own gravitational wave detectors to repeat and improve on Weber's measurements. These first-generation antennas were aluminum cylinders weighing about 1.5 tons and operating at room temperature [28]. Joseph Weber is considered a pioneer in experimental gravitation and therefore he is honored by the American Astronomical Society, which awards every year the Joseph Weber Award for Astronomical Instrumentation.

By the mid-seventies, several detectors were already operative and offered many improvements over Weber's original design; some cylinders were even cooled to reduce thermal noise. These experiments were operating in several places: at Bell Labs Rochester-Holmdel; at the University of Glasgow, Scotland; in an Italo-German joint program in Munich and Frascati; in Moscow; in Tokyo; and at the IBM labs in Yorktown Heights [28]. As soon as these new instruments were put into operation, a common pattern emerged: there were no signals. In the late seventies, everyone except Weber himself agreed that his proclaimed detections were spurious. However, the invalidation of Weber's results urged other researchers to redouble the search for gravitational waves or devise indirect methods of detection.

At that time great pessimism and disappointment reigned among the "seekers" of gravitational waves. However, in 1974 an event occurred that raised hopes. In that year Joseph Hooten Taylor and Alan Russell Hulse found an object in the sky (a binary pulsar) that revealed that an accelerated mass radiated gravitational energy. While this observation did not directly detect gravitational waves, it pointed to their existence. The announcement of the detection of gravitational radiation effects was made in 1979 [29].

This announcement sparked renewed interest in the future discovery of gravitational waves, and urged other researchers to redouble the search for the lost waves and devise other methods of detection. Some were already trying the interferometric method (see Figure 5).

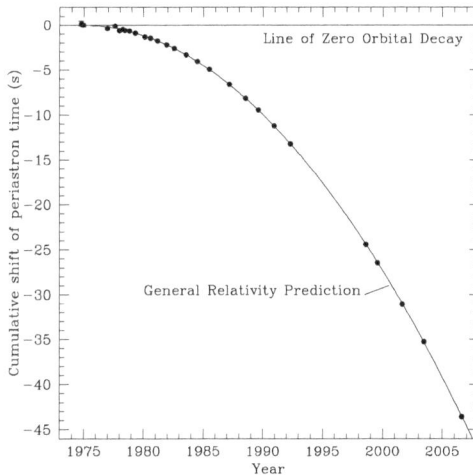

Figure 5. Binary Pulsar Advance of the Periastron (point of closest approach of the stars) versus Time. General Relativity predicts this change because of the energy radiated away by gravitational waves. Hulse and Taylor were awarded the Nobel Prize for this observation in 1993. Figure taken from (Living Rev.Rel.11:8, 2008).

To proceed with the description of the method that uses interferometers, it is necessary to know the magnitude of the expected effects of a gravitational wave on matter. This magnitude is properly quantified by the "h" parameter.

8. The Dimensionless Amplitude, h

The problem with gravitational waves, as recognized by Einstein ever since he deduced for the first time their existence, is that their effect on matter is almost negligible. Among other reasons, the value of the gravitational constant is very, very small, which makes a possible experimental observation extremely difficult.

Furthermore, not all waves are equal, as this depends on the phenomenon that generates them; nor is the effect of a wave on matter has the same intensity. To evaluate the intensity of the effect that a particular wave produces on matter, a dimensionless factor, denoted by the letter "h," has been defined. The dimensionless amplitude "h" describes the maximum displacement per unit length that would produce waves on an object. To illustrate this definition we refer to Figure 6. This figure shows two particles represented by gray circles. The pair is shown originally spaced by a distance "l" and locally at rest.

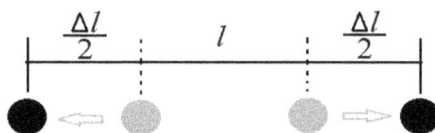

Figure 6. Definition of dimensionless amplitude h = $\Delta l / L$.

By impinging a gravitational wave perpendicularly on the sheet of paper, both particles are shifted respective to the positions marked by black circles. This shift is denoted by $\Delta l / 2$, which means a relative shift between the pair of particles is now equal to $\Delta l / L \approx h$, where Δl is the change in the spacing between particles due to gravitational wave, l is the initial distance between particles, and h is the dimensionless amplitude. In reality the factor h is more complex and depends upon the geometry of the measurement device, the arrival direction, and the frequency and polarization of the gravitational wave [14]. Nature sets a natural amplitude of h~10^{-21}.

This factor h is important when considering the design of a realistic gravitational wave detector. We must mention that the value of h depends on the kind of wave to be detected and this in turn depends on how the wave was produced and how far its source is from an observer. Later we shall return to the subject and the reader shall see the practicality of the factor h.

To identify the sources that produce gravitational waves it is important to consider their temporal behavior. Gravitational waves are classified into three types: stochastic, periodic, and impulsive (bursts) [28]. Stochastic waves contribute to the gravitational background noise and possibly have their origin in the Big Bang. There are also expected stochastic backgrounds due to Black Hole-Black Hole coalescences. These types of waves fluctuate randomly and would be difficult to identify and separate due to the background noise caused by the instruments themselves. However, their identification could be achieved by correlating data from different detectors; this technique applies to other wave types too [14]. The second type of wave, periodic, corresponds to those whose frequency is more or less constant for long periods of time. Their frequency can vary up to a limit (quasiperiodic). For example, these waves may have their origin in binary neutron stars rotating around their center of mass, or from a neutron star that is close to absorb material from another star (accreting neutron star). The intensity of the generated waves depends on the distance from the binary source to the observer. The third type of wave comes from impulsive sources such as bursts that emit pulses of intense gravitational radiation. These may be produced during the creation of Black Holes in a supernova explosion or through the merging of two black holes. The greater the mass, the more intense the signal. They radiate at a frequency inversely proportional to their mass. Such sources are more intense and are expected to have higher amplitudes. Figure 7 shows various examples of possible sources of gravitational waves in which the three different wave types appear in different parts of the spectrum.

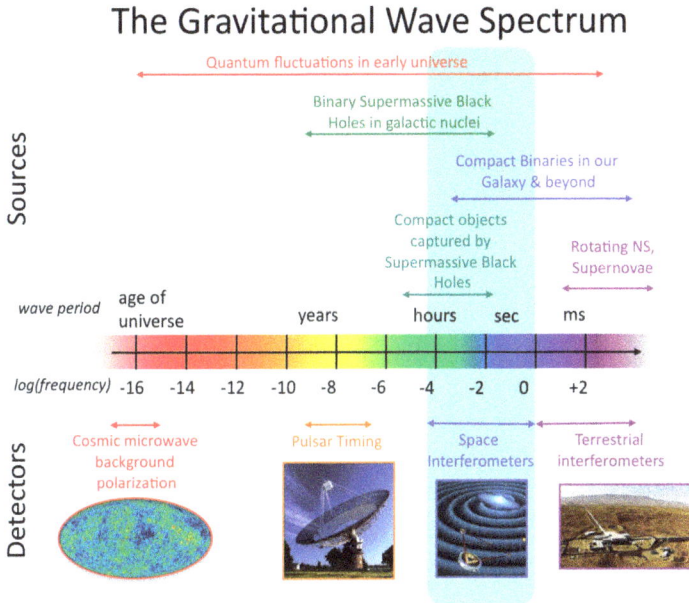

Figure 7. Gravitational wave spectrum showing wavelength and frequency along with some anticipated sources and the kind of detectors one might use. Figure credit: NASA Goddard Space Flight Center.

Different gravitational phenomena give rise to different gravitational wave emissions. We expect primordial gravitational waves stemming from the inflationary era of the very early universe. Primordial quantum fields fluctuate and yield space–time ripples at a wide range of frequencies. These could in principle be detected as B-mode polarization patterns in the Cosmic Microwave Background radiation, at large angles in the sky. Unsuccessful efforts have been reported in recent years, due to the difficulty of disentangling the *noisy* dust emission contribution of our own galaxy, the BICEP2 and PLANCK projects. On the other hand, waves of higher frequencies but still very long wavelengths arising from the slow inspiral of massive black holes in the centers of merged galaxies will cause a modified pulse arrival timing, if very stable pulsars are monitored. Pulsar timing also places the best limits on potential gravitational radiation from cosmic string residuals from early universe phase transitions. Other facilities are planned as space interferometers, such as the Laser Interferometer Space Antenna (LISA), which is planned to measure frequencies between 0.03 mHz and 0.1 Hz. LISA plans to detect gravitational waves by measuring separation changes between fiducial masses in three spacecrafts that are supposed to be 5 million kilometers apart! The expected sources are merging of very massive Black Holes at high redshifts, which corresponds to waves emitted when the universe was 20 times smaller than it is today. It should also detect waves from tens of stellar-mass compact objects spiraling into central massive Black Holes that were emitted when the universe was one half of its present size. Last but not least, Figure 7 shows terrestrial interferometers that are planned to detect waves in the frequency from Hertz to 10,000 Hertz. The most prominent facilities are those of LIGO in the USA, VIRGO in Italy, GEO600 in Germany, and KAGRA in Japan, which are all running or expected to run soon. They are just beginning to detect Black Hole mergings, as was the case of the 14 September event [2]. We will go into detail about the interferometric technique later.

9. The Origin of the Interferometric Method

It is not known for sure who invented the interferometer method to detect gravitational waves, possibly because the method had several precursors. After all, an idea can arise at the same time among various individuals and this indeed seems to be the case here. However, before going into detail about the historical origins of the method, we shall briefly discuss the basics of this technique.

Figure 8 shows a very simplified interferometer. It consists of a light source (a laser), a pair of reflective mirrors attached to a pair of test masses (not shown in the figure), a beam splitter (which can be a semi-reflecting mirror or half-silvered mirror), and a light detector or photodetector.

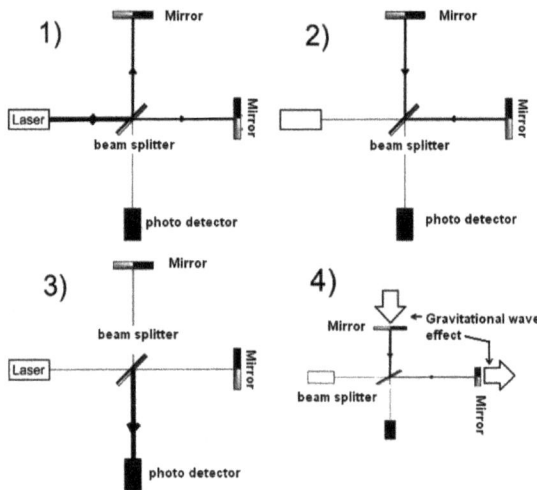

Figure 8. Schematic of an interferometer for detecting gravitational waves.

The laser source emits a beam of monochromatic light (i.e., at a single frequency) that hits the beam splitter surface. This surface is partially reflective, so part of the light is transmitted through to the mirror at the right side of the diagram while some is reflected to the mirror at the upper side of the sketch (Figure 8(1)). Then, as seen in Figure 8(2), both beams recombine when they meet at the splitter and the resulting beam is reflected toward the detector (Figure 8(3)). Finally, the photodetector measures the light intensity of the recombined beam. This intensity is proportional to the square of the height of the recombined wave.

Initially, both reflecting mirrors are positioned at nearly the same distance from the beam splitter. In reality what is needed is that the interferometer is locked on a dark fringe. Deviations from a dark fringe are then measured with the passage of a gravitational wave.

If the distance between one of the mirrors to the light splitter varies by an amount Δl with respect to distance to the same splitter of the second mirror, then the recombined beam will change its intensity. From measuring the intensity change of the recombined light beam, it is possible to obtain Δl.

When a gravitational wave passes through the interferometer at a certain direction, for example perpendicular to the plane where the pair of mirrors lies, both mirrors shift positions. One of the mirrors slightly reduces its distance to the beam splitter, while the second mirror slightly increases its distance to the splitter (see Figure 8(4)). The sum of the two displacements is equal to Δl. The photodetector records a variation in the intensity of the recombined light, thereby detecting the effect of gravitational waves.

A very important specific feature of the interferometer effectiveness is given by the length of its arms. This is the distance "l" between the wave splitter and its mirrors. On the other hand, the wavelength of the gravitational wave sets the size of the detector L needed. The optimal size of the arms turns out to be one-fourth of the wavelength. For a typical gravitational wave frequency of 100 Hz, this implies L = 750 km, which is actually too long to make except by folding the beams back and forth via the Fabry–Pérot technique, which helps to achieve the desired optimal size. In practice it is of the utmost importance to have an interferometer with very long arms matching the frequencies one plans to observe.

The importance of a long arm "l" is easy to explain if we remember the definition of the dimensionless amplitude h = Δl/L, which we already discussed above (see Figure 6). If a gravitational wave produces a displacement Δl for the distance between the mirrors, according to the definition of h, the resulting change Δl will be greater the longer the interferometer arm l is, since they are directly proportional (Δl = L × h). Therefore, the reader can notice that interferometric arm lengths must be tailored depending on what type of gravitational source is intended for observation. The explanation just given corresponds to a very basic interferometer, but in actuality these instruments are more complex. We will come back to this.

10. Genesis of the Interferometer Method (Or, Who Deserves the Credit?)

Let us now turn our attention to the origin of the interferometer method. The first explicit suggestion of a laser interferometer detector was outlined in the former USSR by Gertsenshtein and Pustovoid in 1962 [30]. The idea was not carried out and eventually was resurrected in 1966 behind the "Iron Curtain" by Vladimir B. Braginskiĭ, but then again fell into oblivion [31].

Some year before, Joseph Weber returned to his laboratory at the University of Maryland after having attended the 1957 meeting at Chapel Hill. Weber came back bringing loads of ideas. Back at his university, Weber outlined several schemes on how to detect gravitational waves. As mentioned, one of those was the use of a resonant "antenna" (or cylinder), which, in the end, he finally built. However, among other various projects, he conceived the use of interferometer detectors. He did not pursue this conception, though, and the notion was only documented in the pages of his laboratory notebook [28] (p. 414). One can only say that history produces ironies.

By the end of 1959 Weber began the assembly of his first "antenna" with the help of his students Robert L. Forward and David M. Zipoy [32]. Forward would later (in 1978) turn out to be the first scientist to build an interferometric detector [33].

In the early seventies Robert L. Forward, a former student of Joseph Weber at that time working for Hughes Research Laboratory in Malibu, California, decided, with the encouragement of Rainer Weiss, to build a laboratory interferometer with Hughes' funds. Forward's interest in interferometer detectors had evolved some years before when he worked for Joseph Weber at his laboratory in the University of Maryland on the development and construction of Weber's antennas.

By 1971 Forward reported the design of the first interferometer prototype (which he called a "Transducer Laser"). In his publication Forward explained, "The idea of detecting gravitational radiation by using a laser to measure the differential motion of two isolated masses has often been suggested in past[5]" The footnote reads, "To our knowledge, the first suggestion [of the interferometer device] was made by J. Weber in a telephone conversation with one of us (RLF) [Forward] on 14 September 1964" [34].

After 150 hours of observation with his 8.5-m arms interferometer, Robert Forward reported "an absence of significant correlation between the interferometer and several Weber bars detectors, operating at Maryland, Argonne, Glasgow and Frascati". In short, Robert Forward did not observe gravitational waves. Interestingly, in the acknowledgments of his article, Forward recognizes the advice of Philip Chapman and Weiss [33].

Also in the 1970s, Weiss independently conceived the idea of building a Laser Interferometer, inspired by an article written by Felix Pirani, the theoretical physicist who, as we have already mentioned, developed in 1956 the necessary theory to grow the conceptual framework of the method [12]. In this case it was Weiss who developed the method. However, it was not only Pirani's paper that influenced Weiss; he also held talks with Phillip Chapman, who had glimpsed, independently, the same scheme [35]. Chapman had been a member of staff at MIT, where he worked on electro-optical systems and gravitational theory. He left MIT to join NASA, where he served from 1967 to 1972 as a scientist–astronaut (he never went to space). After leaving NASA, Chapman was employed as a researcher in laser propulsion systems at Avco Everett Research Laboratory in Malibu, California. It was at this time that he exchanged views with Weiss. Chapman subsequently lost interest in topics related to gravitation and devoted himself to other activities.

In addition, Weiss also held discussions with a group of his students during a seminar on General Relativity he was running at MIT. Weiss gives credit to all these sources in one of his first publications on the topic: "The notion is not new; It has appeared as a gedanken experiment in F.A.E. Pirani's studies of the measurable properties of the Riemann tensor. However, the realization that with the advent of lasers it is feasible to detect gravitational waves by using this technique [interferometry], grew out of an undergraduate seminar that I ran at MIT several years ago, and has been independently discovered by Dr. Phillip Chapman of the the National Aeronautics and Space Administration, Houston" [35].

Weiss recalled, in a recent interview, that the idea was incubated in 1967 when he was asked by the head of the teaching program in physics at MIT to give a course of General Relativity. At that time Weiss's students were very interested in knowing about the "discoveries" made by Weber in the late sixties. However, Weiss recalls that "I couldn't for the life of me understand the thing he was doing" and "I couldn't explain it to the students". He confesses "that was my quandary at the time" [36].

A year later (in 1968) Weiss began to suspect the validity of Weber's observations because other groups could not verify them. He thought something was wrong. In view of this he decided to spend a summer in a small cubicle and worked the whole season on one idea that had occurred to him during discussions with his students at the seminar he ran at MIT [36].

After a while, Weiss started building a 1.5-m long interferometer prototype, in the RLE (Research Laboratory of Electronics) at MIT using military funds. Some time later, a law was enacted in the United States (the "Mansfield amendment"), which prohibited Armed Forces financing projects that were not of strictly military utility. Funding was suddenly suspended. This forced Weiss to seek financing from other U.S. government and private agencies [36].

11. Wave Hunters on a Merry-Go-Round (GEO)

In 1974, NSF asked Peter Kafka of the Max Planck Institute in Munich to review a project. The project was submitted by Weiss, who requested $53,000 in funds for enlarging the construction of a prototype interferometer with arms nine meters in length [37]. Kafka agreed to review the proposal. Being a theoretician himself, Kafka showed the proposal documents to some experimental physicists at his institute for advice. To Kafka's embarrassment, the local group currently working on Weber bars became very enthusiastic about Weiss's project and decided to build their own prototype [38], headed by Heinz Billing.

This German group had already worked in collaboration with an Italian group in the construction of "Weber antennas". The Italian–German collaboration found that Weber was wrong [36]. The Weiss proposal fell handily to the Germans as they were in the process of designing a novel Weber antenna that was to be cooled to temperatures near absolute zero to reduce thermal noise in the new system. However, learning of Weiss's proposal caused a shift in the research plans of the Garching group. They made the decision to try the interferometer idea. Germans contacted Weiss for advice and they also offered a job to one of his students on the condition that he be trained on the Weiss 1.5-m prototype. Eventually Weiss sent David Shoemaker, who had worked on the MIT prototype, to join the Garching group. Shoemaker later helped to build a German 3-m prototype and later a 30-m interferometer [39]. This interferometer in Garching served for the development of noise suppression methods that would later be used by the LIGO project.

It is interesting to mention that Weiss's proposal may seem modest nowadays (9-m arms), but he already had in mind large-scale interferometers. His prototype was meant to lead to a next stage featuring a one-kilometer arm length device, which his document claimed it would be capable of detecting waves from the Crab pulsar (PSR B0531 + 21) if its periodic signal were integrated over a period of months. Furthermore, the project envisaged a future development stage where long baseline interferometers in outer space could eventually integrate Crab pulsar gravitational waves in a matter of few hours.

NSF then supported Weiss's project and funds were granted in May 1975 [37,38].

At that time Ronald Drever, then at the University of Glasgow, attended the International School of Cosmology and Gravitation in Erice, *La Città della Scienza*, Sicily in March 1975. There, a lecture entitled "Optimal detection of signals through linear devices with thermal source noises, an application to the Munich–Frascati Weber-type gravitational wave detectors" was delivered by the same Peter Kafka of Munich [40]. His lecture was again very critical of Weber's results and went on to showing that the current state of the art of Weber bars including the Munich–Frascati experiment, was far from the optimal sensitivity required for detection. In fact, the conclusion of his notes reads: "It seems obvious that only a combination of extremely high quality and extremely low temperature will bring resonance detectors [Weber bars] near the range where astronomical work is possible. Another way which seems worth exploring is *Laser interferometry with long free mass antennas*" (emphasis ours) [40].

Ronald Drever was part of Kafka's lecture audience. The lecture probably impressed him as he started developing interferometric techniques on his return to Glasgow. He began with simple tasks, a result of not having enough money. One of them was measuring the separation between two massive bars with an interferometer monitoring the vibrations of aluminum bar detectors. The bars were given to him by the group at the University of Reading, United Kingdom. The bars were two halves of the Reading group's split bar antenna experiment [41]. By the end of the 1970s he was leading a team at Glasgow that had completed a 10-meter interferometer. Then in 1979 Drever was invited to head up the team at Caltech, where he accepted a part-time post. James Hough took his place in Glasgow.

Likewise, in 1975 the German group at Munich (Winkler, Rüdiger, Schilling, Schnupp, and Maischberger), under the leadership of Heinz Billing, built a prototype with an arm length of 3 m [42]. This first prototype displayed unwanted effects such as laser frequency instabilities, lack of power, a shaky suspension system, etc. The group worked hard to reduce all these unwanted effects by developing innovative technologies that modern-day gravitational interferometers embraced. In 1983 the same group, now at the Max Planck Institute of Quantum Optics (MPQ) in Garching, improved their first prototype by building a 30 m arm length instrument [43]. To "virtually" increase the optical arm length of their apparatus, they "folded" the laser beam path by reflecting the beam backwards and forwards between the mirrors many times, a procedure that is known as "delay line." In Weiss's words, "the Max Planck group actually did most of the very early interesting development. They came up with a lot of what I would call the practical ideas to make this thing [gravitational interferometers] better and better" [44].

After a couple of years of operating the 30-m model, the Garching group was prepared to go for Big-Science. In effect, in June 1985, they presented a document "Plans for a large Gravitational wave antenna in Germany" at the Marcel Grossmann Meeting in Rome [45]. This document contains the first detailed proposal for a full-sized interferometer (3 km). The project was submitted for funding to the German authorities but there was not sufficient interest in Germany at that time, so it was not approved.

In the meantime, similar research was undertaken by the group at Glasgow, now under Jim Hough after Drever's exodus to Caltech. Following the construction of their 10-m interferometer, the Scots decided in 1986 to take a further step by designing a Long Baseline Gravitational Wave Observatory [46]. Funds were asked for, but their call fell on deaf ears.

Nevertheless, similar fates bring people together, but it is still up to them to make it happen. So, three years later, the Glasgow and the Garching groups decided to unite efforts to collaborate in a plan to build a large detector. It did not take long for both groups to jointly submit a plan for an underground 3-km installation to be constructed in the Harz Mountains in Germany, but again their proposal was not funded [47]. Although reviewed positively, a shortage of funds on both ends (the British Science and Engineering Research Council (SERC) and the Federal Ministry of Research and Technology (Bundesministerium für Forschung und Technologie BMFT)) prevented the approval. The reason for the lack of funds for science in Germany was a consequence of the German re-unification (1989–1990), as there was a need to boost the Eastern German economy; since private, Western funds lagged, public funds were funneled to the former German Democratic Republic. Incidentally the Harz Mountains are the land of German fairy tales.

In spite of this disheartening ruling, the new partners decided to try for a shorter detector and compensate by employing more advanced and clever techniques [48]. A step forward was finally taken in 1994 when the University of Hanover and the State of Lower Saxony donated ground to build a 600-m instrument in Ruthe, 20 km south of Hanover. Funding was provided by several German and British agencies. The construction of GEO 600 started on 4 September 1995.

The following years of continuous hard work by the British and Germans brought results. Since 2002, the detector has been operated by the Center for Gravitational Physics, of which the Max Planck Institute is a member, together with Leibniz Universität in Hanover and Glasgow and Cardiff Universities.

The first stable operation of the Power Recycled interferometer was achieved in December 2001, immediately followed by a short coincidence test run with the LIGO detectors, testing the stability of the system and getting acquainted with data storage and exchange procedures. The first scientific data run, again together with the LIGO detectors, was performed in August and September of 2002. In November 2005, it was announced that the LIGO and GEO instruments began an extended joint science run [49]. In addition to being an excellent observatory, the GEO 600 facility has served as a development and test laboratory for technologies that have been incorporated in other detectors all over the world.

12. The (Nearly) ... Very Improbable Radio Gravitational Observatory—VIRGO

In the late 70s, when Allain Brillet was attracted to the detection of gravitational waves, the field was ignored by a good number of his colleagues after the incorrect claims of Joe Weber. However, Weiss's pioneering work on laser interferometers in the early 1970s seemed to offer more chances of detection beyond those of Weber bars.

Brillet's interest in the field started during a postdoc stay at the University of Colorado, Boulder under Peter L. Bender of the Laboratory of Astrophysics, who, together with Jim Faller, first proposed the basic concept behind LISA (the Laser Interferometer Space Antenna). Brillet also visited Weiss at MIT in 1980 and 1981 where he established good links with him that produced, as we shall see, a fruitful collaboration in the years to come.

Upon Brillet's return to France in 1982 he joined a group at Orsay UPMC that shared the same interests. This small group (Allain Brillet, Jean Yves Vinet, Nary Man, and two engineers) experienced difficulties and had to find refuge in the nuclear physics department of the Laboratoire de Physique de l'Institute Henri Poincaré, led by Philippe Tourrenc [50].

On 14 November 1983 a meeting on Relativity and Gravitation was organized by the Direction des Etudes Recherches et Techniques de la Délégation Générale pour l'Armement. One of the objectives of the meeting was the development of a French project of gravitational wave detectors [51]. There, Brillet gave a lecture that advocated for the use of interferometers as the best possible detection method [52]. His lecture raised some interest, but French agencies and academic departments were not willing to invest money or personnel in this area. The technology was not yet available, mainly in terms of power laser stability, high-quality optical components, and seismic and thermal noise isolation. In addition, at that point in time there was no significant experimental research on gravitation in France.

However, interest in gravitational wave detection began to change when Hulse and Taylor demonstrated the existence of gravitational waves. It was at the Marcel Grossman meeting of 1985 in Rome that Brillet met Adalberto Giazotto, an Italian scientist working at the Universita di Pisa on the development of suspension systems. At that meeting Giazotto put forward his ideas and the first results of his super-attenuators, devices that serve as seismic isolators to which interferometer mirrors could be attached. During the same meeting Jean-Yves Vinet (Brillet's colleague) gave a talk about his theory of recycling, a technique invented by Ronald Drever to reduce by a large factor the laser power required by gravitational interferometers. Conditions for a partnership were given.

Both scientists then approached the research leaders of a German project (the Max Planck Institute of Quantum Optics in Garching), hoping to collaborate on a big European detector, but they were told that their project was "close to being financed" and the team at Garching did not accept the idea of establishing this international collaboration, because it "would delay project approval" [50].

So they decided to start their own parallel project, the VIRGO Interferometer, named for the cluster of about 1500 galaxies in the Virgo constellation about 50 million light-years from Earth. As no terrestrial source of gravitational wave is powerful enough to produce a detectable signal, VIRGO must observe far enough out into the universe to see many of the potential source sites; the Virgo Cluster is the nearest large cluster.

At that time Brillet was told that CNRS (Le Centre National de la Recherche Scientifique) would not be able to finance VIRGO's construction on the grounds that priority was given to the Very Large Telescope in Chile. Even so, both groups (Orsay and Pisa) did not give in to dismay and continued their collaboration; in 1989 they were joined by the groups of Frascati and Naples. This time they decided to submit the VIRGO project to the CNRS (France) and the INFN (Istituto Nazionale di Fisica Nucleare, Italy) [50].

The VIRGO project was approved in 1993 by the French CNRS and in 1994 by the Italian INFN. The place chosen for VIRGO was the alluvial plain of Cascina near Pisa. The first problem INFN encountered was persuading the nearly 50 land title holders to cooperate and sell their parcels to the government. Gathering the titles took a long time. The construction of the premises started in 1996. To complicate matters further, VIRGO's main building was constructed on a very flat alluvial plain, so it was vulnerable to flooding. That took additional time to remedy [53].

From the beginning it was decided to use the VIRGO interferometer as its own prototype, in contrast to LIGO, which used MIT and Caltech and the German–British (Geo 600) installations to test previous designs before integrating them into the main instrument. This strategy was decided on the grounds that it would be faster to solve problems in actual size directly, rather than spend years on a smaller prototype and only then face the real difficulties.

Between 1996 and 1999, VIRGO had management problems as the construction was handled by an association of separate laboratories without a unified leadership, so it was difficult to ensure proper coordination [42]. As a result, in December 2000 the French CNRS and the Italian INFN created the European Gravitational Observatory (EGO consortium), responsible for the VIRGO site, the construction, maintenance, and operation of the detector, and its upgrades.

The construction of the initial VIRGO detector was completed in June 2003 [54]. It was not until 2007 that VIRGO and LIGO agreed to join in a collaborative search for gravitational waves. This formal agreement between VIRGO and LIGO comprises the exchange of data, and joint analysis and co-authorship of all publications concerned. Several joint data-taking periods followed between 2007 and 2011.

Even though a formal cooperation has been established, continued informal cooperation has been running for years ever since Alain Brillet visited the MIT laboratories back in 1980–1981. As a matter of fact, VIRGO and LIGO have exchanged a good number of students and postdocs. Just to name one, David Shoemaker, the current MIT LIGO Laboratory Director, received his PhD on the Nd-YAG lasers and recycling at Orsay before joining LIGO. Also, in 1990 Jean Yves Vinet provided

LIGO with a computer simulation program necessary to specify its optical system. LIGO adapted this computer code.

The year 2016 will be an important milestone for the construction of the advanced VIRGO detector. After a months-long commissioning period, the advanced VIRGO detector will join the two advanced LIGO detectors ("aLIGO") for a first common data-taking period that should include on the order of one gravitational wave event per month. With all three detectors operating, data can be further correlated and the direction of the gravity waves' source should be much more localized.

13. The Origin of the LIGO Project

In the summer of 1975 Weiss went to Dulles Airport in Washington, D.C. to pick up Kip Thorne, a renowned theoretical physicist from Caltech. The reason for visiting Washington was attending a NASA meeting on uses of space research in the field of cosmology and relativity. (One of us, GFS, attended this meeting as a young postdoc and remembers a presentation by a tired Rai Weiss on the concept of a "Laser Interferometer Space Array" for detecting gravitational waves. It was a very naïve and ambitious space project presentation and only in about 2035 (60 years later) does it appear likely to be a working realization. The meeting did open up to me the idea of doing science in space.) Weiss recalls, "I picked Kip up at the airport on a hot summer night when Washington, D.C., was filled with tourists. He did not have a hotel reservation so we shared a room for the night" [37].

They did not sleep that night because both spent the night discussing many topics, among them how to search for gravitational waves.

Weiss remembers that night "We made a huge map on a piece of paper of all the different areas in gravity. Where was there a future? Or what was the future, or the thing to do?" [55]. Thorne decided that night that the thing they ought to do at Caltech was interferometric gravitational wave detection. However, he would need help from an experimental physicist.

Thorne first thought of bringing to the United States his friend Vladimir B. Braginskiĭ, a Russian scientist who had closely worked with him and, moreover, had already acquired experience in the search for gravitational waves [56]. However, the Cold War prevented his transfer. Meanwhile, Weiss suggested another name, Ronald Drever. Weiss had only known Drever from his papers, not in person. Drever was famous for the Hughes–Drever Experiments, spectroscopic tests of the isotropy of mass and space confirming the Lorentz invariance aspects of the theory of Relativity, and had also been the leader of the group that built a "Weber cylinder" at the University of Glasgow [57]. At that time Drever was planning the construction of an interferometer.

In 1978 Thorne offered a job to Drever at Caltech for the construction of an interferometer. Drever accepted the offer in 1979, dividing his time between the Scottish university and Caltech. Hiring Drever half-time soon paid dividends because in 1983 he had already built his first instrument at Caltech, an interferometer whose "arms" measured 40 m. The instrument was noisier than expected and new ingenious solutions ranging from improving seismic isolation and laser power increase to stabilization were attempted. In 1983 Drever began full-time work at Caltech with the idea of gradually improving and increasing the size of the prototype as it was built and run. In contrast to the Caltech apparatus, as already mentioned, at MIT Weiss had built a modest 1.5-m prototype with a much smaller budget than the Californian instrument. In late 1979 the NSF granted modest funds to the Caltech interferometer group and gave a much smaller amount of money to the MIT team. Soon Drever and Weiss began to compete to build more sensitive and sophisticated interferometers.

The sensitivity of the interferometers can be enhanced by boosting the power of the lasers and increasing the optical path of the light beam as it travels through the interferometer arms.

To increase the sensitivity of the interferometer, Weiss put forward the use of an optical delay line. In the optical delay method, the laser light passes through a small hole in an adjacent wave divider mirror and the beam is reflected several times before emerging through the inlet port. Figure 9 shows a simplified diagram of the method. In this figure only a couple of light "bounces" are shown between the mirrors to maintain clarity of the scheme, but in reality the beam is reflected multiple times. This method effectively increases the length of the interferometer.

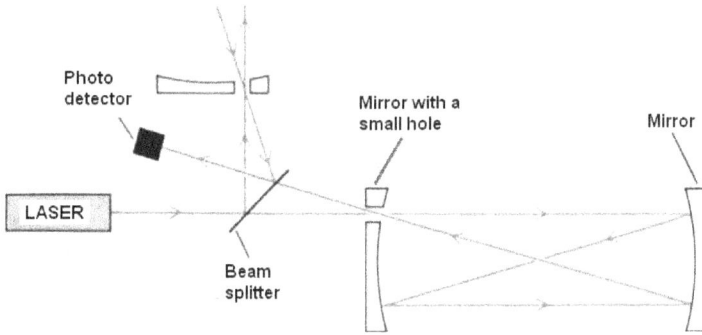

Figure 9. Optical delay method.

In the meantime Drever developed an arrangement that utilized "Fabry–Pérot cavities." In this method the light passes through a partially transmitting mirror to enter a resonant cavity flanked at the opposite end by a fully reflecting mirror. Subsequently, the light escapes through the first mirror, as shown in Figure 10.

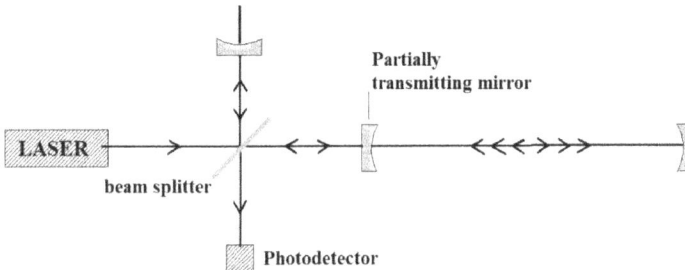

Figure 10. Fabry–Pérot method.

As already mentioned, Weiss experimented with an interferometer whose two L-shaped "arms" were 1.5 m long. Drever, meanwhile, had already built and operated a 40-m interferometer. With the Caltech group appearing to be taking the lead, Weiss decided in 1979 to "do something dramatic". That year Weiss held talks with Richard Isaacson, who at the time served as program director of the NSF Gravitational Physics division. Isaacson had a very strong professional interest in the search for gravitational waves as he had developed a mathematical formalism to approximate gravitational waves solutions from Einstein's equations of General Relativity in situations where the gravitational fields are very strong [58]. Weiss offered to conduct a study in collaboration with industry partners to determine the feasibility and cost of an interferometer whose arms should measure in kilometers. In turn Isaacson receive a document to substantiate a device on a scale of kilometers, with possible increased funding from the NSF. The study would be funded by NSF. The study by Weiss and colleagues took three years to complete. The produced document was entitled "A study of a long Baseline Gravitational Wave Antenna System," co-authored by Peter Saulson and Paul Linsay [59]. This fundamental document is nowadays known as the "The Blue Book" and covers very many important issues in the construction and operation of such a large interferometer. The Blue Book was submitted to the NSF in October 1983. The proposed budget was just under $100 million to build two instruments located in the United States.

Before "The Blue Book" was submitted for consideration to the NSF, Weiss met with Thorne and Drever at a Relativity congress in Italy. There they discussed how they could work together—this was mandatory, because the NSF would not fund two megaprojects on the same subject and with the same objective. However, from the very beginning it was clear that Drever did not want to collaborate with Weiss, and Thorne had to act as mediator. In fact, the NSF settled matters by integrating the MIT and Caltech groups together in a "shotgun wedding" so the "Caltech–MIT" project could be jointly submitted to the NSF [55].

14. The LIGO Project

The Caltech–MIT project was funded by NSF and named the "Laser Interferometer Gravitational-Wave Observatory", known by its acronym LIGO. The project would be led by a triumvirate of Thorne, Weiss, and Drever. Soon interactions between Drever and Weiss became difficult because, besides the strenuous nature of their interaction, both had differing opinions on technical issues.

During the years 1984 and 1985 the LIGO project suffered many delays due to multiple discussions between Drever and Weiss, mediated when possible by Thorne. In 1986 the NSF called for the dissolution of the triumvirate of Thorne, Drever, and Weiss. Instead Rochus E. Vogt was appointed as a single project manager [60].

In 1988 the project was finally funded by the NSF. From that date until the early 1990s, project progress was slow and underwent a restructuring in 1992. As a result, Drever stopped belonging to the project and in 1994 Vogt was replaced by a new director, Barry Clark Barish, an experimental physicist who was an expert in high-energy physics. Barish had experience in managing big projects in physics. His first activity was to review and substantially amend the original five-year old NSF proposal. With its new administrative leadership, the project received good financial support. Barish's plan was to build the LIGO as an evolutionary laboratory where the first stage, "initial LIGO" (or iLIGO), would aim to test the concept and offer the possibility of detecting gravitational waves. In the second stage ("aLIGO" or advanced LIGO), wave detection would be very likely (see Figure 11).

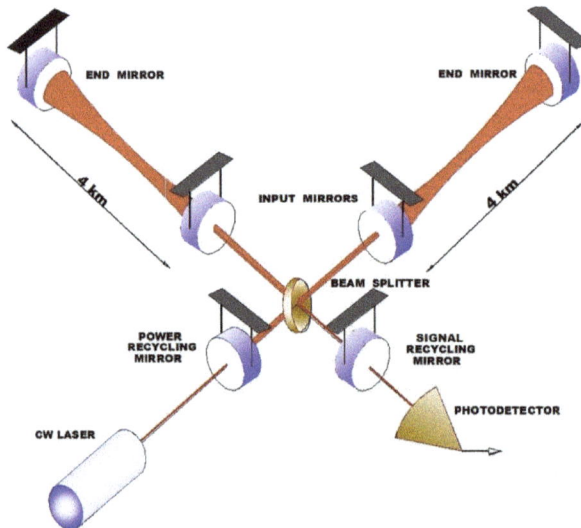

Figure 11. Advanced LIGO interferometer design concept. Figure made after T.F. Carruthers and D.H. Reitze, "LIGO", *Optics & Photonics News*, March 2015.

Two observatories, one in Hanford in Washington State and one in Livingston, Louisiana, would be built. Construction began in late 1994 and early 1995, respectively, and ended in 1997. Once the construction of the two observatories was complete, Barish suggested two organizations to be funded: the laboratory LIGO and scientific collaboration LIGO (LIGO Scientific Collaboration (LSC)). The first of these organizations would be responsible for the administration of laboratories. The second organization would be a scientific forum headed by Weiss, responsible for scientific and technological research. LSC would be in charge of establishing alliances and scientific collaborations with Virgo and GEO600.

Barish's idea to make LIGO an evolutionary apparatus proved in the end to pay dividends. The idea was to produce an installation whose parts (vacuum system, optics, suspension systems, etc.) could always be readily improved and buildings that could house those ever-improving interferometer components.

In effect, LIGO was incrementally improved by advances made in its own laboratories and those due to associations with other laboratories (VIRGO and GEO600). To name some: Signal-recycling mirrors were first used in the GEO600 detector, as well as the monolithic fiber-optic suspension system that was introduced into advanced LIGO. In brief, LIGO detection was the result of a worldwide collaboration that helped LIGO evolve into its present remarkably sensitive state.

15. Looking Back Over the Trek

The initial LIGO operated between 2002 and 2010 and did not detect gravitational waves. The upgrade of LIGO (advanced LIGO) began in 2010 to replace the detection and noise suppression and improve stability operations at both facilities. This upgrade took five years and had contribution from many sources. For example, the seismic suspension used in aLIGO is essentially the design that has been used in VIRGO since the beginning. While advanced VIRGO is not up and running yet, There have been many technological contributions from both LSC scientists on the European side and VIRGO.

aLIGO began in February 2015 [61]. The team operated in "engineering mode"—that is, in test mode—and in late September began scientific observation [62]. It did not take many days for LIGO to detect gravitational waves [2]. Indeed, LIGO detected the collision of two black holes of about 30 solar masses collapsed to 1300 million light years from Earth.

Even the latest search for gravitational waves was long and storied. The upgrade to aLIGO cost $200 million, and preparing it took longer than expected, so the new and improved instrument's start date was pushed back to 18 September 2015.

16. The Event: 14 September 2015

On Sunday September 13th the LIGO team performed a battery of last-minute tests. "We yelled, we vibrated things with shakers, we tapped on things, we introduced magnetic radiation, we did all kinds of things", one of the LIGO members said. "And, of course, everything was taking longer than it was supposed to". At four in the morning, with one test still left to do—a simulation of a truck driver hitting his brakes nearby—we stopped for the night. We went home, leaving the instrument to gather data in peace and quiet. The signal arrived not long after, at 4:50 a.m. local time, passing through the two detectors within seven milliseconds of each other. It was four days before the start of Advanced LIGO's first official run. It was still during the time meant for engineering tests, but nature did not wait.

The signal had been traveling for over a billion years, coming from a pair of 30-solar mass black holes orbiting around their common center of mass and slowly drawing into a tighter and tighter orbit from the energy being lost by gravitational radiation. The event was the end of this process —formed from final inspiral, the merger of the black holes to form a larger one, and the ring down of that new massive black hole. All of this lasted mere thousandths of a second, making the beautiful signals seen by both aLIGO detectors which immediately answered several questions: Are there gravitational waves and can we detect them? Is General Relativity likely right for strong fields? Are large black holes common?

The waveform detected by both LIGO observatories matched the predictions of General Relativity for a gravitational wave emanating from the inward spiral and merger of a pair of black holes of around 36 and 29 solar masses and the subsequent "ringdown" of the single resulting black hole. See Figures 12 and 13 for the signals and GR theoretical predictions.

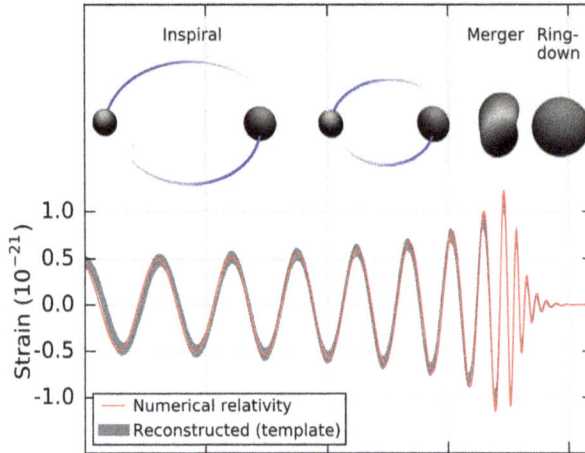

Figure 12. Showing the phases: binary orbits inspiraling, black holes merging, and final black hole ringing down to spherical or ellipsoidal shape [2].

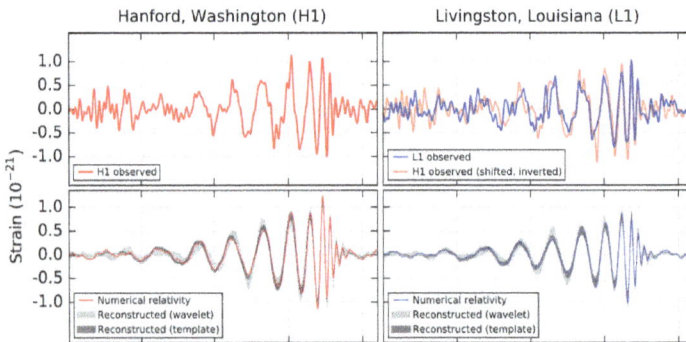

Figure 13. LIGO measurement of the gravitational waves at the Livingston (**right**) and Hanford (**left**) detectors, compared with the theoretical predicted values from General Relativity [2].

The signal was named GW150914 (from "Gravitational Wave" and the date of observation). It appeared 14 September 2015 and lasted about 0.2 s. The estimated distance to the merged black holes is 410 $^{+160}_{-180}$ Mpc or 1.3 billion light years, which corresponds to a redshift of about z = 0.09. The interesting thing is that the estimated energy output of the event in gravitational waves is about $(3.0 +/- 0.5)\ M_\odot \times c^2$. That is three times the rest mass of our sun converted into energy.

More recently, the aLIGO team announced the detection and analysis of another binary black hole merger event, GW151226. This event is apparently the merger of a 14 solar mass black hole with a 7.5 solar mass black hole, again at a distance of about 440 Mpc (about 1.4 billion light years). An interesting feature is that one of the black holes is measured to have significant spin, s = 0.2. There is also a report of a candidate event that occurred between the two confirmed events and had black holes and energy intermediate to the two confirmed events. These additional reports also show the beginning of measuring rates and the distribution of black hole binary systems. It has long been expected that as aLIGO improves its design sensitivity (about three times better than this first run), that a stable set of events would include the merger of binary neutron stars, for whose inventory we have better estimates. These early events predict that there will be a whole distribution of gravitational wave events observed in the future, from neutron star mergers, black hole mergers, and even neutron star–black hole mergers, see Figure 14.

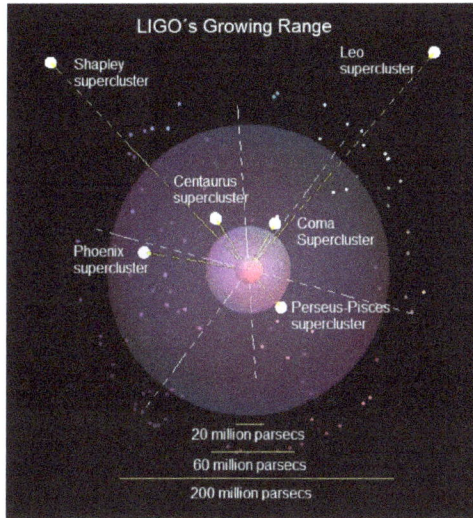

Figure 14. Showing the initial range of LIGO and the anticipated range of aLIGO. The volume is much greater and the anticipated rate of events and detections are expected to scale up with the volume. Note the large number of galaxies included in the observational volume. The anticipated factor of 3 in sensitivity should correspondingly increase the event rate by up to nine times.

Albert Einstein originally predicted the existence of gravitational waves in 1916, based upon General Relativity, but wrote that it was unlikely that anyone would ever find a system whose behavior would be measurably influenced by gravitational waves. He was pointing out that the waves from a typical binary star system would carry away so little energy that we would never even notice that the system had changed—and that is true. The reason we can see it from the two black holes is that they are closer together than two stars could ever be. The black holes are so tiny and yet so massive that they can be close enough together to move around each other very, very rapidly. Still, to get such a clear signal required a very large amount of energy and the development of extraordinarily sensitive instruments. This clearly settles the argument about whether gravitational waves really exist; one major early argument was about whether they carried any energy. They do! That was proved strongly and clearly.

Some analyses have been carried out to establish whether or not GW150914 matches with a binary black hole configuration in General Relativity [63]. An initial consistency test encompasses the mass and spin of the end product of the coalescence. In General Relativity, the final black hole product of a binary coalescence is a Kerr black hole, which is completely described by its mass and spin. It has been verified that the remnant mass and spin from the late-stage coalescence deduced by numerical relativity simulations, inferred independently from the early stage, are consistent with each other, with no evidence for disagreement from General Relativity. There is even some data on the ring down phase, but we can hope for a better event to provide quality observations to test this phase of General Relativity.

GW150914 demonstrates the existence of black holes more massive than $\simeq 25 M_\odot$, and establishes that such binary black holes can form in nature and merge within a Hubble time. This is of some surprise to stellar theorists, who predicted smaller mass black holes would be much more common.

17. Conclusions

This observation confirms the last remaining unproven prediction of General Relativity (GR)—gravitational waves—and validates its predictions of space–time distortion in the context of large-scale cosmic events (known as strong field tests). It also inaugurates the new era of gravitational-wave astronomy, which promises many more observations of interesting and energetic objects as well as more precise tests of General Relativity and astrophysics. While it is true that we can never rule out deviations from GR at the 100% level, all three detections so far agree with GR to an extremely high level (>96%). This will put constraints on some non-GR theories and their predictions.

With such a spectacular early result, others seem sure to follow. In the four-month run, 47 days' worth of coincident data was useful for scientific analysis, i.e., this is data taken when both LIGOs were in scientific observation mode. The official statement is that these 47 days' worth of data have been fully analyzed and no further signals lie within them. We can expect many more events once the detectors are running again.

For gravitational astronomy, this is just the beginning. Soon, aLIGO will not be alone. By the end of the year VIRGO, a gravitational-wave observatory in Italy, should be operating to join observations and advanced modes. Another detector is under construction in Japan and talks are underway to create a fourth in India. Most ambitiously, a fifth, orbiting, observatory, the Evolved Laser Interferometer Space Antenna, or e-LISA, is on the cards. The first pieces of apparatus designed to test the idea of e-LISA are already in space and the first LISA pathfinder results are very encouraging.

Together, by jointly forming a telescope that will permit astronomers to pinpoint whence the waves come, these devices will open a new vista onto the universe. (On the science side, the data is analyzed jointly by members of both LIGO and VIRGO, even though these data only come from LIGO. This is due to the analysis teams now being fully integrated. As this is not widely known, people do not realize that there is a large contribution from VIRGO scientists to the observations and to the future.) As technology improves, waves of lower frequency—corresponding to events involving larger masses—will become detectable. Eventually, astronomers should be able to peer at the first 380,000 years after the Big Bang, an epoch of history that remains inaccessible to every other kind of telescope yet designed.

The real prize, though, lies in proving Einstein wrong. For all its prescience, the theory of relativity is known to be incomplete because it is inconsistent with the other great 20th-century theory of physics, quantum mechanics. Many physicists suspect that it is in places where conditions are most extreme—the very places that launch gravitational waves—that the first chinks in relativity's armor will be found, and with them we will get a glimpse of a more all-embracing theory.

Gravitational waves, of which Einstein remained so uncertain, have provided direct evidence for black holes, about which he was long uncomfortable, and may yet yield a peek at the Big Bang, an event he knew his theory was inadequate to describe. They may now lead to his theory's unseating. If so, its epitaph will be that in predicting gravitational waves, it predicted the means of its own demise.

Acknowledgments: GFS acknowledges Laboratoire APC-PCCP, Université Paris Diderot, Sorbonne Paris Cité (DXCACHEXGS), and also the financial support of the UnivEarthS Labex program at Sorbonne Paris Cité (ANR-10-LABX-0023 and ANR-11-IDEX-0005-02). JLCC acknowledges financial support from conacyt Project 269652 and Fronteras Project 281.

Author Contributions: J.L.C.-C., S.G.-U., and G.F.S. conceived the idea and equally contributed.

Conflicts of Interest: The authors declare no conflict of interest.

References

1. Davide, C. LIGO's path to victory. *Nature* **2016**, *530*, 261–262.
2. Abbot, B.P.; Abbott, R.; Abbott, T.D.; Abernathy, M.R.; Acernese, F.; Ackley, K.; Adams, C.; Adams, T.; Addesso, P.; Adhikari, P.X.; et al. Observation of Gravitational Waves from a Binary Black Hole Merger. *Phys. Rev. Lett.* **2016**, *116*, 061102. [CrossRef] [PubMed]
3. Henri, P. "Sur la Dynamique de l'électron". *Proc. Acad. Sci.* **1905**, *140*, 1504–1508. (In French)
4. Albert, E. Integration Neherunsweise Feldgleichungen der der Gravitation. *Sitzungsber. Preuss. Akad. Wiss. Berlin (Math.Phys.)* **1916**, *22*, 688–696. (In Germany)
5. Albert, E. Zur Elektrodynamik bewegter Körper. *Annalen Phys.* **1905**, *17*, 891–921. (In Germany)
6. Albert, E. The Collected Papers of Albert Einstein, Volume 8: The Berlin Years: Correspondence, 1914–1918 (English Translation Supplement, Translated by Ann M. Hentschel) Page 196, Doc 194. Available online: http://einsteinpapers.press.princetonedu/VOL8-trans/224 (accessed on 22 June 2016).
7. Hermann, W. Space, Time, Matter. Methuen & Co. Ltd.: London, 1922; Chapter 4; p. 376, Translated from German by Henry l. Brose. Available online: http://www.gutenberg.org/ebooks/43006 (accessed on 22 June 2016).
8. Stanley, E.A. The propagation of Gravitational waves. *Proc. R. Soc. Lond.* **1922**, *102*, 268–282.
9. Albert, E.; Max, B. *The Born-Einstein Letters. Friendship, Politics and Physics in Uncertain Times*; Macmillan: London, UK, 2005; p. 121.
10. Daniel, K. Einstein versus the Physical Review. *Phys. Today* **2005**, *58*, 43–48.
11. Albert, E.; Rosen, N. On gravitational waves. *J. Frank. Inst.* **1937**, *223*, 43–54.
12. Pirani Felix, A.E. On the Physical Significance of the Riemann Tensor. *Gen. Relativ. Gravit.* **2009**, *41*, 1215–1232. [CrossRef]
13. Bergmann Peter, G. *The Riddle of Gravitation*; Charles Scribner's Sons: New York, NY, USA, 1968.
14. Maggiore, M. *Gravitational Waves*; Oxford University Press: Oxford, UK, 2008.
15. Gravity Research Foundation. Available online: http://www.gravityresearchfoundation.org/origins.html (accessed on 21 June 2016).
16. Interview of Bryce DeWitt and Cecile DeWitt-Morette by Kenneth W. In *Ford on 1995 February 28, Niels Bohr Library & Archives*; American Institute of Physics: College Park, MD, USA. Available online: http://www.aip.org/history-programs/niels-bohr-library/oral-histories/23199 (accessed on 21 June 2016).
17. Martin, G. Chapter 8 "Sir Isaac Babson". In *Fads and Fallacies in the Name of Science*; Dover Publications: New York, NY, USA, 1957; p. 93.
18. DeWitt, M.C.; Rickles, D. The role of Gravitation in Physics. Report from the 1957 Chapel Hill Conference. Max Planck Research Library for the History and Development of Knowledge. Open Access Edition (2011). In particular see Chapter 27 "An Expanded Version of the Remarks by R.P. Feynman on the Reality of Gravitational Waves" also see comments Feymann on page 252. Available online: http://www.edition-open-sources.org/sources/5/index.html (accessed on 21 June 2016).
19. Feynman, R.P. Unpublished letter to Victor F. Weisskopf, January 4–February 11, 1961; in Box 3 File 8 of The Papers of Richard P. Feynman, the Archives, California Institute of Technology. Available online: http://www.oac.cdlib.org/findaid/ark:/13030/kt5n39p6k0/dsc/?query=lettertoWeisskopf1961#c02--1.7.7.3.2 (accessed on 21 June 2016).
20. Bondi, H. Plane gravitational waves in the general relativity. *Nature* **1957**, *179*, 1072–1073. [CrossRef]
21. Weber, J. Detection and Generation of Gravitational Waves. *Phys. Rev.* **1960**, *117*, 306–313. [CrossRef]
22. Weber, J. Observation of the Thermal Fluctuations of a Gravitational-Wave Detector. *Phys. Rev. Lett.* **1966**, *17*, 1228–1230. [CrossRef]
23. Weber, J. Evidence for Discovery of Gravitational Radiation. *Phys. Rev. Lett.* **1969**, *22*, 1320–1324. [CrossRef]
24. Weber, J. Anisotropy and Polarization in the Gravitational-Radiation Experiments. *Phys. Rev. Lett.* **1970**, *25*, 180–184. [CrossRef]
25. Sciama, D.; Field, G.; Rees, M. Upper Limit to Radiation of Mass Energy Derived from Expansion of Galaxy. *Phys. Rev. Lett.* **1969**, *23*, 1514–1515. [CrossRef]
26. Kafka, P. *Are Weber's Pulses Illegal?*; Gravity Research Foundation: Babson Park, MA, USA, 1972.
27. Weber, J. The Detection of Gravitational Waves. *Sci. Am.* **1971**, *224*, 22–29. [CrossRef]

28. Thorne Kip, S. Gravitational Radiation. In *Three Hundred Years of Gravitation*; Hawking, S., Israel, W.W., Eds.; Cambridge University Press: Cambridge, UK, 1987; pp. 330–458.

29. Taylor, J.H.; Fowler, L.A.; McCulloch, P.M. Overall measurements of relativistic effects in the binary pulsar PSR 1913 + 16. *Nature* **1979**, *277*, 437–440. [CrossRef]

30. Gertsenshtein, M.E.; Pustovoit, V.I. On the detection of low frequency gravitational waves. *Sov. J. Exp. Theor. Phys.* **1962**, *43*, 605–607.

31. Braginskiĭ Vladimir, B. Gravitational radiation and the prospect of its experimental discovery. *Sov. Phys. Usp.* **1966**, *8*, 513–521. [CrossRef]

32. Weber, J. How I discovered Gravitational Waves. *Pop. Sci.* **1962**, *5*, 106–107.

33. Forward, R.L. Wide band Laser-Interferometer Gravitational-radiation experiment. *Phys. Rev. D* **1978**, *17*, 379–390. [CrossRef]

34. Moss, G.E.; Miller, R.L.; Forward, R.L. Photon-noise-Limited Laser Transducer for Gravitational Antenna. *Appl. Opt.* **1971**, *10*, 2495–2498. [CrossRef] [PubMed]

35. Weiss, R. Quarterly Progress Report 1972, No 105, 54-76. Research Laboratory of Electronics, MIT. Available online: http://dspace.mit.edu/bitstream/handle/1721.1/56271/RLE_QPR_105_V.pdf?sequence= 1 (accessed on 21 June 2016).

36. Chu Jennifer "Rainer Weiss on LIGO's origins" MIT Q & A NEWS February 11, 2016. Available online: http://news.mit.edu/2016/rainer-weiss-ligo-origins-0211 (accessed on 23 June 2016).

37. Weiss, R. Interferometric Broad Band Gravitational Antenna. Grant identification number MPS75–04033. Available online: https://dspace.mit.edu/bitstream/handle/1721.1/56655/RLE_PR_119_ XV.pdf?sequence=1 (accessed on 23 June 2016).

38. Collins Harry, interview with Peter Kafka. See WEBQUOTE under "Kafka Referees Weiss's Proposal". Available online: http://sites.cardiff.ac.uk/harrycollins/webquote/ (accessed on 21 June 2016).

39. Shoemaker, D.; Schilling, R.; Schnupp, L.; Winkler, W.; Maischberger, K.; Rüdiger, A. Noise behavior of the Garching 30-meter gravitational wave detector prototype. *Phys. Rev. D* **1988**, *38*, 423–432. [CrossRef]

40. De Sabbata, V.; Weber, J. Topics in Theoretical and Experimental Gravitation Physics. In Proceedings of the International School of Cosmology and Gravitation, Sicily, Italy, 13–25 March 1975.

41. Allen, W.D.; Christodoulides, C. Gravitational radiation experiments at the University of Reading and the Rutherford Laboratory. *J. Phys. A Math. Gen.* **1975**, *8*, 1726–1733. [CrossRef]

42. Collins, H. *Gravity's Shadow*; University of Chicago Press: Chicago, IL, USA, 2004.

43. Shoemaker, D.; Winkler, W.; Maischberger, K.; Ruediger, A.; Schilling, R.; Schnupp, L. Progress with the Garching (West Germany) 30 Meter Prototype for a Gravitational Wave Detector. In Proceedings of the 4th Marcel Grossmann Meeting, Rome, Italy, 17–21 June 1985.

44. Cohen, S.K. Rainer Weiss Interviewed. 10 May 2000. Available online: http://resolver.caltech.edu/ CaltechOH:OH_Weiss_R (accessed on 21 June 2016).

45. Winkler, W.; Maischberger, K.; Rudiger, A.; Schilling, R.; Schnupp, L.; Shoemaker, D. Plans for a large gravitational wave antenna in Germany. In Proceedings of the 4th Marcel Grossmann Meeting, Rome, Italy, 17–21 June 1985.

46. Hough, J.; Meers, B.J.; Newton, G.P.; Robertson, N.A.; Ward, H.; Schutz, B.F.; Drever, R.W.P. *A British Long Baseline Gravitational Wave Observatory: Design Study Report*; Rutherford Appleton Laboratory Gravitational Wave Observatory: Oxon, UK, 1986.

47. Leuchs, G.; Maischberger, K.; Rudiger, A.; Roland, S.; Lise, S.; Walter, W. Vorschlag zum Bau eines groBen Laser-Interferometers zur Messung von Gravitationswellen –erweiterte Fasssung-. Available online: http://www.mpq.mpg.de/4464435/mpq_reports#1987 (accessed on 23 June 2016).

48. Hough, J.; Meers, B.J.; Newton, G.P.; Robertson, N.A.; Ward, H.; Leuchs, G.; Niebauer, T.M.; Rudiger, A.; Schilling, R.; Schnupp, L.; et al. Proposal for a Joint German-British Interferometric Gravitational Wave Detector. Available online: http://eprints.gla.ac.uk/114852/7/114852.pdf (accessed on 23 June 2016).

49. Biennial Reports 2004/05 Max Planck Institute for Gravitational Physics (PDF). Available online: https://www.aei.mpg.de/148353/biennial2004_05.pdf (accessed on 23 June 2016).

50. Brillet, A. Avant VIRGO, ARTEMIS UMR 7250, jeudi 26 mai 2016. Available online: https://artemis.oca.eu/ spip.php?article463 (accessed on 23 June 2016).

51. Bordé, Ch.J. Méthodes optiques de détection des ondes gravitationnelles—Préface. *Ann. Phys. Fr.* **1985**, *10*, R1–R2. (In French) [CrossRef]

52. Brillet, A. Interferometric gravitational wave antennae. *Ann. Phys. Fr.* **1985**, *10*, 219–226. [CrossRef]
53. Accadia, T.; Acernese, F.; Alshourbagy, M.; Amico, P.; Antonucci, F.; Aoudia, S.; Arnaud, N.; Arnault, C.; Arun, K.G.; Astone, P.; et al. VIRGO: A laser interferometer to detect gravitational waves. *J. Instrum.* **2012**, *7*, P03012. [CrossRef]
54. Ondes Gravitationnelles Inauguration du Détecteur Franco-Italien VIRGO. Available online: http://www2. cnrs.fr/presse/communique/206.htm?&debut=40 (accessed on 21 June 2016). (In French).
55. Janna, L. *Black Hole Blues and Other Songs from Outer Space*; Knopf: New York, NY, USA, 2016.
56. Braginski, V.B.; Manukin, A.B.; Popov, E.I.; Rudenko, V.N.; Korev, A.A. Search for Gravitational Radiation of Extraterrestrial Origin. *Sov. Phys. Usp.* **1937**, *15*, 831–832. [CrossRef]
57. Drever, R.W.P. A search for anisotropy of inertial mass using a free precession technique. *Philos. Mag.* **1961**, *6*, 683–687. [CrossRef]
58. Isaacson Richard, A. Gravitational Radiation in the Limit of High Frequency. I. The Linear Approximation and Geometrical Optics. *Phys. Rev.* **1968**, *166*, 1263–1271. [CrossRef]
59. Linsay, P.; Saulson, P.; Weiss, R. A Study of a Long Baseline Gravitational Wave Antenna System. Available online: https://dcc.ligo.org/public/0028/T830001/000/NSF_bluebook_1983.pdf (accessed on 23 June 2016).
60. Russell, R. Catching the wave. *Sci. Am.* **1992**, *266*, 90–99.
61. Davide, C. Hunt for gravitational waves to resume after massive upgrade: LIGO experiment now has better chance of detecting ripples in space-time. *Nature* **2015**, *525*, 301–302.
62. Jonathan, A. Advanced Ligo: Labs Their Ears Open to the Cosmos. Available online: http://www.bbc.com/news/science-environment-34298363 (accessed on 23 June 2016).
63. Abbott, B.P. Tests of General Relativity with GW150914. *Phys. Rev. Lett.* **2016**, *116*, 221101. [CrossRef] [PubMed]

universe

[MDPI]

Review

The Status of Cosmic Topology after Planck Data

Jean-Pierre Luminet [1,2]

1   Laboratoire d'Astrophysique de Marseille (LAM), CNRS UMR 7326, F-13388 Marseille, France
2   LUTH (Observatoire de Paris), CNRS UMR 8102, F-92195 Meudon, France; jean-pierre.luminet@lam.fr

Academic Editors: Stephon Alexander, Jean-Michel Alimi, Elias C. Vagenas and Lorenzo Iorio
Received: 19 November 2015; Accepted: 7 January 2016; Published: 15 January 2016

Abstract: In the last decade, the study of the overall shape of the universe, called Cosmic Topology, has become testable by astronomical observations, especially the data from the Cosmic Microwave Background (hereafter CMB) obtained by WMAP and Planck telescopes. Cosmic Topology involves both global topological features and more local geometrical properties such as curvature. It deals with questions such as whether space is finite or infinite, simply-connected or multi-connected, and smaller or greater than its observable counterpart. A striking feature of some relativistic, multi-connected small universe models is to create multiples images of faraway cosmic sources. While the last CMB (Planck) data fit well the simplest model of a zero-curvature, infinite space model, they remain consistent with more complex shapes such as the spherical Poincaré Dodecahedral Space, the flat hypertorus or the hyperbolic Picard horn. We review the theoretical and observational status of the field.

Keywords: cosmology; general relativity; topology; cosmic microwave background; Planck telescope

1. Introduction

The idea that the universe might have a non-trivial topology and, if sufficiently small in extent, display multiple images of faraway sources, was first discussed in 1900 by Karl Schwarzschild (see [1] for reference and English translation). With the advent of Einstein's general relativity theory (see, e.g., the recent historical overview in [2]) and the discoveries of non-static universe models by Friedmann and Lemaître in the decade 1922–1931, the face of cosmology definitively changed. While Einstein's cosmological model of 1917 described space as the simply-connected, positively curved hypersphere S^3, de Sitter in 1917 and Lemaître in 1927 used the multi-connected projective sphere P^3 (obtained by identifying opposite points of S^3) for describing the spatial part of their universe models. In 1924, Friedmann [3] pointed out that Einstein's equations are not sufficient for deciding if space is finite or infinite: Euclidean and hyperbolic spaces, which in their trivial (*i.e.*, simply-connected) topology are infinite in extent, can become finite (although without an edge) if one identifies different points—an operation which renders the space multi-connected. This opens the way to the existence of phantom sources, in the sense that at a single point of space an object coexists with its multiple images. The whole problem of cosmic topology was thus posed, but as the cosmologists of the first half of 20th century had no experimental means at their disposal to measure the topology of the universe, the vast majority of them lost all interest in the question. A revival of interest in multi-connected cosmologies arose in the 1970s, under the lead of theorists who investigated several kinds of topologies (see [4] for an exhaustive review and references, in which the term "Cosmic Topology" was coined). However, most cosmologists either remained completely ignorant of the possibility, or regarded it as unfounded, until the 1990s when data on the CMB provided by space telescopes gave access to the largest possible volume of the observable universe. Since then, hundreds of articles have considerably enriched the field of theoretical and observational cosmology.

2. A Theoretical Reminder

At very large scale our Universe seems to be correctly described by one of the Friedmann-Lemaître (hereafter FL) models, namely homogeneous and isotropic solutions of Einstein's equations, of which the spatial sections have constant curvature. They fall into three general classes, according to the sign of curvature. In most cosmological studies, the spatial topology is assumed to be that of the corresponding simply-connected space: the hypersphere, Euclidean space or 3D-hyperboloid, the first being finite and the other two infinite. However, there is no particular reason for space to have a trivial topology: the Einstein field equations are local partial differential equations which relate the metric and its derivatives at a point to the matter-energy contents of space at that point. Therefore, to a metric element solution of Einstein field equations, there are several, if not an infinite number, of compatible topologies, which are also possible models for the physical universe. For example, the hypertorus T^3 and the usual Euclidean space E^3 are locally identical, and relativistic cosmological models describe them with the same FL equations, even though the former is finite in extent while the latter is infinite. Only the boundary conditions on the spatial coordinates are changed. The multi-connected FL cosmological models share exactly the same kinematics and dynamics as the corresponding simply-connected ones; in particular, the time evolutions of the scale factor are identical.

In FL models, the curvature of physical space (averaged on a sufficiently large scale) depends on the way the total energy density of the universe may counterbalance the kinetic energy of the expanding space. The normalized density parameter Ω_0, defined as the ratio of the actual energy density to the critical value that strictly Euclidean space would require, characterizes the present-day contents (matter, radiation and all forms of energy) of the universe. If Ω_0 is greater than 1, then the space curvature is positive and the geometry is spherical; if Ω_0 is smaller than 1, the curvature is negative and geometry is hyperbolic; eventually Ω_0 is strictly equal to 1 and space is locally Euclidean (currently said flat, although the term can be misleading).

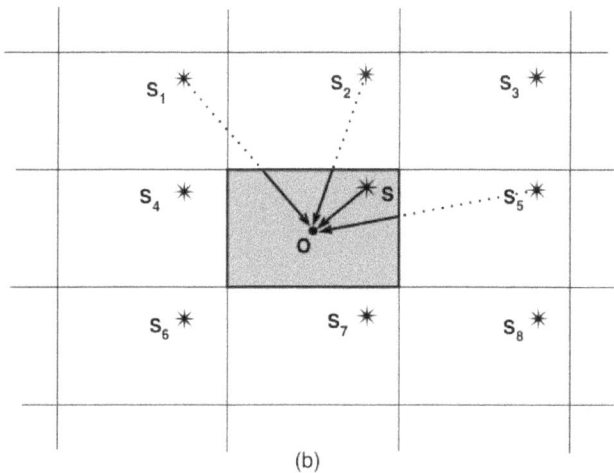

(b)

Figure 1. The Illusion of the Universal Covering Space. In the case of a 2D torus space, the fundamental domain, which represents real space, is the interior of a rectangle, whose opposite edges are identified. The observer O sees rays of light from the source S coming from several directions. He has the illusion of seeing distinct sources S_1, S_2, S_3, *etc.*, distributed along a regular canvas which covers the UC space—an infinite plane.

Independently of curvature, a much discussed question in the history of cosmology (and also philosophy) is to know whether space is finite or infinite in extent. Of course no physical measure can ever prove that space is infinite, but a sufficiently small, finite space model could be testable. Although the search for space topology does not necessarily solve the question of finiteness, it provides many multi-connected universe models of finite volume.

The effect of a non-trivial topology on a cosmological model is equivalent to considering the observed space as a simply-connected 3D-slice of space-time, known as the "universal covering space" (hereafter UC) being filled with repetitions of a given shape, the "fundamental domain", which is finite in some or all directions, for instance a convex polyhedron; by analogy with the two-dimensional case, we say that the fundamental domain tiles the UC space.

There is a subgroup of isometries acting on the UC which produces its tiling by these copies (for such group action, see the basic works [5,6]). Physical fields repeat their configuration in every copy and thus can be viewed as defined on the UC space, but subject to periodic boundary conditions, which are the matching rules between neighbouring tiles. The copies around a fixed one carry the multiple images of objects from the cosmos. By analogy with crystallography, the UC plays the role of the macroscopic crystal, the cosmos plays the role of the fundamental unit (see Figure 1). But in contrast to crystallography, the UC in topology can be Euclidean, spherical or hyperbolic. For the flat and hyperbolic geometries, there are infinitely many copies of the fundamental domain; for the spherical geometry with a finite volume, there are a finite number of tiles.

Figure 2. The Poincaré Dodecahedral Space can be described as the interior of a spherical ball whose surface is tiled by 12 curved regular pentagons. When one leaves through a pentagonal face, one returns to the ball through the opposite face after having turned by 36°. As a consequence, the space is finite but without boundaries, therefore one can travel through it indefinitely. One has the impression of living in a UC space 120 times larger, paved with dodecahedra that multiply the images like a hall of mirrors. The return of light rays that cross the walls produces optical mirages: a single object has several images. This numerical simulation calculates the closest phantom images of the Earth, which would be seen in the UC space (Image courtesy of J. Weeks).

There are seventeen multi-connected Euclidean spaces (for an exhaustive study, see [7]), the simplest of which being the hypertorus T^3, whose fundamental domain is a parallelepiped of which opposite faces are identified by translations. Seven of these spaces have an infinite volume, ten are of finite volume, six of them being orientable hypertori. All of them could correctly describe the spatial part of the flat universe models, as they are consistent with recent observational data which constrain the space curvature to be very close to zero. Note that in current inflationary scenarios for the big bang, one can always have a nearly flat universe at present without fine-tuning the initial value of the spatial curvature, while considering exactly flat models corresponds to fine-tuning the initial curvature to be strictly zero.

In spaces with non-zero curvature, the presence of a length scale (the curvature radius) precludes topological compactification at an arbitrary scale. The size of the space must now reflect its curvature, linking topological properties to the total energy density Ω_0. All spaces of constant positive curvature are finite whatever be their topology. The reason is that the universal covering space—the simply-connected hypersphere S^3—is itself finite. There is a countable infinity of spherical spaceforms (for a complete classification, see [8]), but there is only a finite set of "well-proportioned" topologies, *i.e.*, those with roughly comparable sizes in all directions, which are of a particular interest for cosmology. As a now celebrated example, let us mention the Poincaré Dodecahedral Space (hereafter PDS), obtained by identifying the opposite pentagonal faces of a regular spherical dodecahedron after rotating by $36°$ in the clockwise or counterclockwise direction around the axis orthogonal to the face, depending on which handedness the physical nature favors [9], see Figure 2. Its volume is 120 times smaller than that of the hypersphere with the same curvature radius.

Eventually there is also an infinite (but non countable) number of hyperbolic manifolds, with finite or infinite volumes. Their classification is not well understood, but the volumes of the compact hyperbolic space forms are bounded below by $V = 0.94271$ (in units of the curvature radius), which correspond to the so-called Weeks manifold.

The computer program *CurvedSpaces* [10] is especially useful to depict the rich structure of multi-connected manifolds with any curvature.

3. Probing Cosmic Topology

The observable universe is the interior of a sphere centered on the observer and whose radius is that of the cosmological horizon—roughly the radius of the last scattering surface (hereafter LSS), presently estimated at 14.4 Gpc. Cosmic Topology aims to describe the shape of the *whole* universe. One could think that the whole universe is necessarily greater than the observable one, as it would obviously be the case if space was infinite, for instance the simply-connected flat or hyperbolic space. Then the observable universe would be an infinitesimal patch of the whole universe and, although it has long been the standard "mantra" of many theoretical cosmologists, this is not and will never be a testable hypothesis.

The whole universe can also be finite (without an edge), e.g., a hypersphere or a closed multi-connected space, but greater than the observable universe. In that case, one easily figures out that if whole space widely encompasses the observable one, no signature of its finiteness will show in the experimental data. But if space is not too large, or if space is not globally homogeneous (as is permitted in many space models with multi-connected topology), and if the observer occupies a special position, some imprints of the space finiteness could be observable.

Surprisingly enough, the whole space could be smaller than the observable universe, due to the fact that space can be both multi-connected, have a small volume and produce topological lensing. This is the only case where there are a lot of testable possibilities, whatever the curvature of space.

The present observational constraints on the Ω_0 parameter favor a spatial geometry that is nearly flat with a 0.4% margin of error [11]. Note that the constraints on the curvature parameter can be looser if we consider a general form of dark energy (not the cosmological constant), which leaves rooms to consider positively or negatively curved cosmological models that are usually regarded as being excluded. However, even with the curvature so severely constrained by cosmological data, there are still possible multi-connected topologies that support positively curved, negatively curved, or flat metrics. Sufficiently "small" universe models would generate multiple images of some light sources, in such a way that the hypothesis of multi-connected topology can be tested by astronomical observations. The smaller the space, the easier it is to observe the multiple images of luminous sources in the sky (generally not seen at the same age, except for the CMB spots). Note, however, the coincidence problem that occurs in order to get an observable non-trivial topology: for flat space, we need to have the topology scale length near the horizon scale, while for curved spaces, the curvature radius needs to be near the horizon scale. However, there are so many other, non-explained coincidence problems in standard cosmology that it should not deviate our attention from the possibility of a detectable topology.

How do the present observational data constrain the possible multi-connectedness of the universe and, more generally, what kinds of tests are conceivable (see [12] for a non-technical book about all aspects of topology and its applications to cosmology)?

Different approaches have been proposed for extracting information about the topology of the universe from experimental data. One approach is to use the 3D distribution of astronomical objects such as galaxies, quasars and galaxy clusters: if the whole universe is finite and small enough, we should be able to see "all around" it because the photons might have crossed it once or more times. In such a case, any observer might recognize multiple images of the same light source, although distributed in different directions of the sky and at various redshifts, or to detect specific statistical properties in the distribution of faraway sources. Various methods of "cosmic crystallography", initially proposed in [13], have been widely developed by other groups ([14] and references therein). However, for plausible small universe models, the first signs of topological lensing would appear only at pretty high redshift, say $z \approx 2$. The main limitation of cosmic crystallography is that the presently available catalogs of observed sources at high redshift are not complete enough to perform convincing tests for topology. For instance, looking for nontoroidal topological lensing, [15] applied the crystallographic method to the SDSS quasar sample; though they found no robust signature, cosmological interpretation of the result was prohibited by the data incompleteness and by the uncertainty in quasar physics. On the other hand, [16] proposed to use deep surveys of distant (at redshifts $z \sim 6$) starburst galaxies for an independent test of the cubic hypertorus model. Their calculation showed that even photometric redshifts would suffice in this purpose, which makes their strategy a realistic and interesting one.

The other approach uses the 2D CMB maps (for a review, [17]). The last scattering surface (LSS) from which the CMB is released represents the most distant source of photons in the universe, hence the largest scales with which the topology of the universe can be probed.

The idea that a small universe model could lead to a suppression of power on large angular scales in the fluctuation spectrum of the CMB had been proposed in the 1980s [18]: in some way, space would be not big enough to sustain long wavelengths. After the release of COBE data in 1992 and the higher resolution and sensitivity of WMAP (2003), there were indeed indications of low power on large scales which could have had a topological origin, and many authors used it to constrain the models. The best fits between theoretical power spectra computed for various topologies and the observed one were obtained with the positively curved Poincaré Dodecahedral Space [9,19,20] and the flat hypertorus [21]. In addition, it was shown [22] that the low-order multipoles tended to be relatively weak in "well-proportioned" spaces, *i.e.*, whose dimensions are approximately equal in all directions. Some globally inhomogeneous topologies can also reproduce the large-angle CMB power suppression if the location of the observer is so adjusted that his fundamental domain becomes well-proportioned [23]. However, this possibility was not borne out by detailed real- and harmonic-space analyses in two dimensions, so that the arguments based on the angular power spectrum and favoring small universe models failed [24]. In any case, to gain all the possible information from the correlations of CMB anisotropies, one has to consider the full covariance matrix rather than just the power spectrum.

Indeed the main imprint of a non trivial topology on the CMB is well-known in the case when the characteristic topological length scale (called the injectivity radius) is smaller than the radius of the LSS: the crossings of the LSS with its topological images give rise to pairs of matched circles of equal radii, centered at different points on the CMB sky, and exhibiting correlated patterns of temperature variations [25], see Figure 3. For instance, the PDS model predicts six pairs of antipodal circles with an angular radius comprised between $5°$ and $55°$ (sensitively depending on the cosmological parameters) and a relative phase of $36°$.

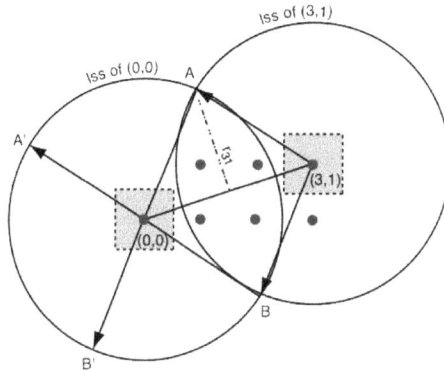

Figure 3. The circles-in-the-sky method is illustrated here in a 2D torus space. The fundamental domain is a square (with a dotted outline), all of the red points are copies of the same observer. The two large circles (which are normally spheres in a three-dimensional space) represent the last scattering surfaces (LSS) centered on two copies of the same observer. One is in position (0,0), its copy is in position (3,1) in the universal covering space. The intersection of the circles is made up of the two points A and B (in three dimensions, this intersection is a circle). The observers (0,0) and (3,1), who see the two points (A,B) from two opposite directions, are equivalent to a unique observer at (0,0) who sees two identical pairs (A,B) and (A′,B′) in different directions. In three dimensions, the pairs of points (A,B) and (A′,B′) become a pair of identical circles, whose radius r_{31} depends on the size of the fundamental domain and the topology.

4. Results and Discussion

Such "circles-in-the-sky" searches have been looked for in WMAP maps by several groups, using various statistical indicators and massive computer calculations, and interpreting their results differently. Some authors [26] claimed that most of non-trivial topologies, including PDS and T^3, were ruled out: they searched for antipodal or nearly antipodal pairs of circles in the WMAP data and found no such circles. However, their analysis could not be applied to more complex topologies, for which the matched circles deviate strongly from being antipodal. On the other hand, other groups claimed to have found hints of multi-connected topology, using different statistical indicators [27–29].

Most studies have emphasized searches for fundamental domains with antipodal correlations. The search for matched circle pairs that are not back-to-back has nevertheless been carried out recently, with no obvious topological signal appearing in the seven-year WMAP data [30]. The statistical significance of such results still has to be clarified. In any case, a lack of nearly matched circles does not exclude a multi-connected topology on scale less than the horizon radius: detectable topologies may produce circles of small radii which are statistically hard to detect and current analysis of CMB sky maps could have missed even antipodal matching circles, because various effects may damage or even destroy the temperature matching.

Other methods for experimental detection of non-trivial topologies have thus been proposed and used to analyze the experimental data, such as the multipole vectors and the likelihood (Bayesian) method. The latter ameliorates some of the spoiling effects of the temperature correlations such as the integrated Sachs-Wolfe and Doppler contributions [31].

The most up-to-date study based on CMB temperature correlations used the Planck 2013 intensity data [32]. In that work, they applied two techniques: first, a direct likelihood calculation of (a very few) specific topological models; second, a search for the expected repeated "circles in the sky", calibrated by simply-connected simulations. Both of these showed that the scale of any possible topology must exceed roughly the comoving distance to the LSS, χ_{rec}. For the cubic torus, they found that the radius of the largest sphere inscribed in the topological fundamental domain must be $R_i > 0.92$

χ_{rec}. The matched-circle limit on topologies predicting back-to-back circles larger than $15°$ in angular radius and assuming that the relative orientation of the fundamental domain and mask allows its detection was $R_i > 0.94 \chi_{rec}$ at the 99% confidence level.

Finally, it is now widely understood that the polarization of the CMB can provide a lot of additional informations for reconstructing the cosmological model. Riazuelo *et al.* [33] were the first to show how the polarization could also be used to put additional constraints on space topology and a little bit tighter than those coming from temperature intensity. Maps of CMB polarization from the 2015 release of Planck data [34] provided the highest-quality full-sky view of the LSS available to date. However their study specialized only to flat spaces with cubic toroidal (T^3) and slab (T^1) topologies. These searches yield no detection of a compact topology with a scale below the diameter of the LSS. More precisely, $R_i > 0.97 \chi_{rec}$ for the T^3 cubic torus and $R_i > 0.56 \chi_{rec}$ for the T^1 slab.

5. Conclusions

The overall topology of the universe has become an important concern in astronomy and cosmology. Even if particularly simple and elegant models such as the PDS and the hypertorus are now claimed to be ruled out at a subhorizon scale, many more complex models of multi-connected space cannot be eliminated as such. In addition, even if the size of a multi-connected space is larger (but not too much) than that of the observable universe, we could still discover an imprint in the CMB, even while no pair of circles, much less ghost galaxy images, would remain. The topology of the universe could therefore provide information on what happens outside of the cosmological horizon [35].

Whatever the observational constraints, a lot of unsolved theoretical questions remain. The most fundamental one is the expected link between the present-day topology of space and its quantum origin, since classical general relativity does not allow for topological changes during the course of cosmic evolution. Theories of quantum gravity should allow us to address the problem of a quantum origin of space topology. For instance, in quantum cosmology, the question of the topology of the universe is completely natural. Quantum cosmologists seek to understand the quantum mechanism whereby our universe (as well as other ones in the framework of multiverse theories) came into being, endowed with a given geometrical and topological structure. We do not yet have a correct quantum theory of gravity, but there is no sign that such a theory would *a priori* demand that the universe have a trivial topology. Wheeler first suggested that the topology of space-time might fluctuate at a quantum level, leading to the notion of a space-time foam [36]. Additionally, some simplified solutions of the Wheeler-de Witt equations for quantum cosmology show that the sum over all topologies involved in the calculation of the wavefunction of the universe is dominated by spaces with small volumes and multi-connected topologies [37]. In the approach of brane worlds in string/M-theories, the extra-dimensions are often assumed to form a compact Calabi-Yau manifold; in such a case, it would be strange that only the ordinary, large dimensions of our 3-brane would not be compact like the extra ones. However, still at an early stage of development, string quantum cosmology can only provide heuristic indications on the way multi-connected spaces would be favored.

Acknowledgments: The author thanks the anonymous referees for suggesting additional references and improvements of the original manuscript.

Conflicts of Interest: The author declares no conflict of interest.

References

1. Starkman, G.D. Topology and Cosmology. *Class. Quantum Gravity* **2008**, *15*, 2529–2538. [CrossRef]
2. Iorio, L. Editorial for the Special Issue 100 Years of Chronogeometrodynamics: The Status of the Einstein's Theory of Gravitation in Its Centennial Year. *Universe* **2015**, *1*, 38–81. [CrossRef]

3. Friedmann, A. Über die Möglichkeit einer Welt mit konstanter negativer Krümmung des Raumes. *Zeitsfricht Phys.* **1924**, *21*, 326–332. (English Translation in Bernstein, J.; Feinberg, G. *Cosmological Constants. Papers in Modern Cosmology*; Columbia University Press: New York, NY, USA, 1986; pp. 59–65.)

4. Lachièze-Rey, M.; Luminet, J.-P. Cosmic topology. *Phys. Rep.* **1995**, *254*, 135–214. [CrossRef]

5. Wolf, J.A. *Spaces of Constant Curvature*; Publish or Perish Inc.: Wilmington, DC, USA, 1984.

6. Kramer, P. Topology of Platonic Spherical Manifolds: From Homotopy to Harmonic Analysis. *Symmetry* **2015**, *7*, 305–326. [CrossRef]

7. Riazuelo, A.; Weeks, J.; Uzan, J.-P.; Lehoucq, R.; Luminet, J.-P. Cosmic microwave background anisotropies in multiconnected flat spaces. *Phys. Rev. D* **2004**, *69*, 103518. [CrossRef]

8. Gausmann, E.; Lehoucq, R.; Luminet, J.-P.; Uzan, J.-P.; Weeks, J. Topological lensing in spherical spaces. *Class. Quantum Gravity* **2001**, *18*, 5155–5186. [CrossRef]

9. Luminet, J.-P.; Weeks, J.; Riazuelo, A.; Lehoucq, R.; Uzan, J.-P. Dodecahedral space topology as an explanation for weak wide-angle temperature correlations in the cosmic microwave background. *Nature* **2003**, *425*, 593–595. [CrossRef] [PubMed]

10. Geometrygames. Available online: http://www.geometrygames.org (accessed on 8 January 2016).

11. Hinshaw, G.; Larson, D.; Komatsu, E.; Spergel, D.N.; Bennett, C.L.; Dunkley, J.; Nolta, M.R.; Halpern, M.; Hill, R.S.; Odegard, N.; *et al*. Nine-year WMAP Observations: Cosmological Parameter Results. *Astrophys. J. Suppl.* **2013**, *208*, 19.

12. Luminet, J.-P. *The Wraparound Universe*; AK Peters: New York, NY, USA, 2008.

13. Lehoucq, R.; Lachièze-Rey, M.; Luminet, J.-P. Cosmic crystallography. *Astron. Astrophys.* **1996**, *313*, 339–346.

14. Fujii, H.; Yoshii, Y. An improved cosmic crystallography method to detect holonomies in flat spaces. *Astron. Astrophys.* **2011**, *529*, A121. [CrossRef]

15. Fujii, H.; Yoshii, Y. A Search for Nontoroidal Topological Lensing in the Sloan Digital Sky Survey Quasar Catalog. *Astrophys. J.* **2013**, *773*, 152–160. [CrossRef]

16. Roukema, B.; France, M.; Kazimierczak, T.; Buchert, T. Deep redshift topological lensing: Strategies for the T^3 candidate. *Mon. Not. R. Astron. Soc.* **2014**, *437*, 1096–1108. [CrossRef]

17. Levin, J. Topology and the cosmic microwave background. *Phys. Rep.* **2002**, *365*, 251–333. [CrossRef]

18. Fagundes, H.V. The Quadrupole Component of the Relic Radiation in a Quasi-Hyperbolic Cosmological Model. *Astrophys. Lett.* **1983**, *23*, 161.

19. Caillerie, S.; Lachièze-Rey, M.; Luminet, J.-P. A new analysis of the Poincaré dodecahedral space model. *Astron. Astrophys.* **2007**, *476*, 691–696. [CrossRef]

20. Roukema, B.F.; Bulinski, Z.; Gaudin, N.E. Poincaré dodecahedral space parameter estimates. *Astron. Astrophys.* **2008**, *492*, 657–673. [CrossRef]

21. Aurich, R. A spatial correlation analysis for a toroidal universe. *Class. Quantum Gravity* **2008**, *25*, 225017. [CrossRef]

22. Weeks, J.; Luminet, J.-P.; Riazuelo, A.; Lehoucq, R. Well-proportioned universes suppress the cosmic microwave background quadrupole. *Mon. Not. R. Astron. Soc.* **2004**, *352*, 258–262. [CrossRef]

23. Aurich, R.; Lustig, S. How well proportioned are lens and prism spaces? *Class. Quantum Gravity* **2012**, *29*, 175003. [CrossRef]

24. Bielewicz, P.; Riazuelo, A. The study of topology of the Universe using multipole vectors. *Mon. Not. R. Astron. Soc.* **2009**, *396*, 609–623. [CrossRef]

25. Cornish, N.J.; Spergel, D.N.; Starkman, G.D. Circles in the sky: Finding topology with the microwave background radiation. *Class. Quantum Gravity* **1998**, *15*, 2657–2670. [CrossRef]

26. Cornish, N.J.; Spergel, D.N.; Starkman, G.D.; Komatsu, E. Constraining the Topology of the Universe. *Phys. Rev. Lett.* **2004**, *92*, 201302. [CrossRef] [PubMed]

27. Roukema, B.; Lew, B.; Cechowska, M.; Marecki, A.; Bajtlik, S. A Hint of Poincaré Dodecahedral Topology in the WMAP First Year Sky Map. *Astron. Astrophys.* **2004**, *423*, 821–831. [CrossRef]

28. Lew, B.; Roukema, B. A test of the Poincaré dodecahedral space topology hypothesis with the WMAP CMB data. *Astron. Astrophys.* **2008**, *482*, 747–753. [CrossRef]

29. Aurich, R.; Lustig, S.; Steiner, F. The circles-in-the-sky signature for three spherical universes. *Mon. Not. R. Astron. Soc.* **2006**, *369*, 240–248. [CrossRef]

30. Vaudrevange, P.; Starkman, G.; Cornish, N.; Spergel, D. Constraints on the topology of the Universe: Extension to general geometries. *Phys. Rev. D* **2012**, *86*, 083526. [CrossRef]

31. Kunz, M.; Aghanim, N.; Cayon, L.; Forni, O.; Riazuelo, A.; Uzan, J.-P. Constraining topology in harmonic space. *Phys. Rev. D* **2006**, *73*, 023511. [CrossRef]
32. Ade, P.A.R.; Aghanim, N.; Armitage-Caplan, C.; Arnaud, M.; Ashdown, M.; Atrio-Barandela, F.; Aumont, J.; Baccigalupi, C.; Banday, A.J.; Barreiro, R.B.; *et al.* Planck 2013 results XXVI. Background geometry and topology of the Universe. *Astron. Astrophys.* **2014**, *571*, A26.
33. Riazuelo, A.; Caillerie, S.; Lachièze-Rey, M.; Lehoucq, R.; Luminet, J.-P. Constraining Cosmic Topology with CMB Polarization. **2006**, arXiv:astro-ph/0601433.
34. Ade, P.A.R.; Aghanim, N.; Arnaud, M.; Ashdown, M.; Aumont, J.; Baccigalupi, C.; Banday, A.J.; Barreiro, R.B.; Bartolo, N.; Battaner, E.; *et al.* Planck 2015 results XVIII. Background geometry and topology. **2015**, arXiv:1502.01593.
35. Fabre, O.; Prunet, S.; Uzan, J.-P. Topology beyond the horizon: How far can it be probed? *Phys. Rev. D* **2015**, *92*, 04003. [CrossRef]
36. Wheeler, J. On the nature of quantum geometrodynamics. *Ann. Phys.* **1957**, *2*, 604–614. [CrossRef]
37. e Costa, S.S.; Fagundes, H.V. On the Birth of a Closed Hyperbolic Universe. *Gen. Relativ. Gravit.* **2001**, *33*, 1489–1494. [CrossRef]

![universe]

universe

MDPI

Review

The Singularity Problem in Brane Cosmology

Ignatios Antoniadis [1,2,*] and Spiros Cotsakis [3,†]

1   Laboratoire de physique théorique et hautes energies (LPTHE), Les Unités Mixtes de Recherche (UMR)
     CNRS 7589, Sorbonne Universités, UPMC Paris 6, 4 place Jussieu, T13-14, 75005 Paris, France
2   Albert Einstein Center for Fundamental Physics, ITP, University of Bern, Sidlerstrasse 5,
     CH-3012 Bern, Switzerland
3   Department of Mathematics, American University of the Middle East, P. O. Box 220, 15453, Dasman, Kuwait;
     skot@aegean.gr
*   Correspondence: antoniad@lpthe.jussieu.fr
†   On leave from the University of the Aegean, Greece.

Academic Editors: Lorenzo Iorio, Elias C. Vagenas, Stephon Alexander and Jean-Michel Alimi
Received: 31 December 2016; Accepted: 7 February 2017; Published: 16 February 2017

Abstract: We review results about the development and asymptotic nature of singularities in
"brane–bulk" systems. These arise for warped metrics obeying the five-dimensional Einstein equations
with fluid-like sources, and including a brane four-metric that is either Minkowski, de Sitter, or Anti-de
Sitter. We characterize all singular Minkowski brane solutions, and look for regular solutions with
nonzero curvature. We briefly comment on matching solutions, energy conditions, and finite Planck
mass criteria for admissibility, and we briefly discuss the connection of these results to ambient theory.

Keywords: brane cosmological models; singularities; asymptotic analysis

1. Introduction

The singularity problem in the setup of brane cosmological models is concerned with the existence
and nature of the dynamical singularities that may arise when one considers the evolution of metrics
and fields propagating in spaces with "large extra dimensions" containing certain lower-dimensional
slices. Such systems obey higher-dimensional Einstein (or possibly similar string gravity) equations,
with the standard interactions usually confined in a four-dimensional slice (the brane) sitting in
a five-(or higher-)dimensional spacetime (the bulk). Such "brane–bulk" systems are used in an essential
way as a means to overcome the hierarchy problem [1–3], and in a crucial way in approaches to solve
the cosmological constant problem [4,5].

In this paper we provide a concise overview of the various ramifications and results that have
been obtained in recent years about the singularity problem in such contexts, basically using the
methods developed in References [6–9]. Previous work on this subject can be found in [10]. We also
briefly discuss the connection of these results with the ambient approach to the singularity problem
towards the end of this work.

We write the *bulk metric* in the form,

$$g_5 = a^2(Y)g_4 + dY^2, \tag{1}$$

where g_4 represents the *brane metric*, taken to be either Minkowski, de Sitter (dS) or anti-de Sitter (AdS),

$$g_4 = -dt^2 + f_k^2 g_3, \tag{2}$$

with

$$g_3 = dr^2 + h_k^2 g_2, \quad g_2 = d\theta^2 + \sin^2\theta d\varphi^2, \tag{3}$$

where $f_k = \cosh(Ht)/H$ or $\cos(Ht)/H$ (H^{-1} is the de Sitter (or AdS) curvature radius) and $h_k = \sin r$ or $\sinh r$, for dS or AdS respectively.

There are two interesting interpretations of the metric (1) that are relevant in the present context. The first is of course the standard one; namely, to view the brane as a domain wall solution, a hypersurface in the five-space, the bulk. This is the most common interpretation of the geometric setup, across the entire braneworld literature (cf. [11] and references therein). There is, however, a different one that is useful in certain contexts (cf. especially the discussion towards the end of the present paper), namely, to view the metric (1) as a cone metric, or a warped product metric [12,13].

Whatever the geometric interpretation, we impose the five-dimensional Einstein equations on the metric (1),

$$G_{AB} = \kappa_5^2 T_{AB}, \tag{4}$$

where we shall usually take the energy–momentum tensor to be that of an analog of a five-dimensional (5d) fluid (with the Y coordinate playing the role of time), or a combination of fluids, possibly exchanging energy. In fact, it is an interesting result that our 5d-fluid must by necessity be an anisotropic pressure fluid (such fluids have recently emerged as important instability factors in other contexts in string cosmology, for example in the possible disruption of the isotropic fluid stability of simple ekpyrotic cyclic models [14]). To see this, we start with the standard energy–momentum tensor for the 5d-fluid in the form,

$$T_{AB} = (\rho + P)u_A u_B - P g_{AB}, \tag{5}$$

where $A, B = 1, 2, 3, 4, 5$ and $u_A = (0,0,0,0,1)$, with the fifth coordinate corresponding to Y, and seek an anisotropic pressure form,

$$T_{AB} = (\rho^0 + p^0)u_A^0 u_B^0 + p^0 g_{\alpha\beta}\delta_A^\alpha \delta_B^\beta + p_Y g_{55}\delta_A^5 \delta_B^5, \tag{6}$$

where $u_A^0 = (a(Y),0,0,0,0)$ and $\alpha, \beta = 1,2,3,4$. When we combine (5) with (6), we find that the 5d-fluid has an anisotropic energy–momentum tensor of the form [7,9],

$$T_{AB} = -P g_{\alpha\beta}\delta_A^\alpha \delta_B^\beta + \frac{P}{\gamma} g_{55}\delta_A^5 \delta_B^5, \tag{7}$$

when $P = \gamma\rho$. We see that isotropic fluids in this context correspond to the limiting case of a cosmological constant-like equation of state, $\gamma \to -1$. We can then satisfy the various energy conditions by restricting γ to take values in certain intervals [7,9].

It is important to further point out that in this work—except for a fixation of the braneworld four-geometry (either Minkowski, dS, or AdS, respecting 4d maximal symmetry)—we do not fix the bulk five-geometry other than take it to be of the above warped type near the (presumed) singularity. Hence, only the *asymptotic* geometry of the bulk is found and dictated by the five-dimensional Einstein equations with the fluid source discussed above. Away from such an open neighborhood around the singularity, the bulk space geometry remains compatible with that requirement. This is in sharp contrast with other approaches, such as in [15], where the bulk is fixed rigidly to be of some preassigned form (e.g., AdS_5).

2. Flat Branes

The prototype case for the evolution of the brane–bulk system near its finite-distance singularities is when the brane is described by Minkowski space and there is a single free scalar field ϕ in the bulk. In this case, the five-dimensional Einstein Equation (4) in the bulk with source ϕ can be symbolically written as an autonomous dynamical system in the form,

$$\dot{X} = f(X). \tag{8}$$

All solutions of this system then become the integral curves of the three-dimensional, non-polynomial vector field [6],

$$f(X) = \left(y, -\lambda A z^2 x, -4yz/x\right),\tag{9}$$

subject to the constraint,

$$\frac{y^2}{x^2} = \frac{A\lambda}{3}z^2.\tag{10}$$

Here, $X = (x, y, z)$, A, λ are constraints, while we have introduced new variables by setting

$$x = a, \quad y = a', \quad z = \phi',\tag{11}$$

with a prime denoting differentiation with respect to the extra dimension Y. Then, we have the following result ([6], Section 2.1).

Theorem 1 (Minkowski brane-massless dilaton). *With the setup of a flat three-brane in a five-dimensional bulk spacetime filled with a free scalar field as described above, let Y_s denote the position of the finite-distance singularity from the brane position. Then, there is only one possible asymptotic behaviour of the solutions of the field equations towards singularity, given by,*

$$a \to 0, \quad a' \to \infty, \quad \phi' \to \infty,\tag{12}$$

as $Y \to Y_s$.

This result means that all solutions asymptote towards a state wherein the flat brane collapses after "traveling" a finite distance in the bulk, starting from its initial position, with the energy of the scalar field blowing up there. This implies that any initial configuration involving a Minkowski three-brane coupled to a bulk massless dilaton satisfying the five-dimensional Einstein equations will gradually evolve to the collapse state described in the Theorem above. This result completely fixes the nature of the singularity in this simple case and the behaviour of all solutions near the singularity. (An exact particular solution with these properties was first found in [4,5]. One may view the result contained in the Theorem above as implying that the exact solution found in those references is a stable one in the sense that all other solutions of the system approach this form asymptotically towards the singular point.)

However, one naturally wonders whether the above result has some degree of genericity; in other words, whether and how the existence and nature of the singularity in this simplest Minkowski brane model persists when one passes on to more general ones, while keeping the flatness assumption (the extension to branes with curvature is separately discussed in the next section of this paper). There are at least three ways to treat the flat brane problem in a more general setting:

- Add self-interaction to the dilaton
- Add a perfect fluid in the bulk
- Add a mixture of a fluid and a (possibly interacting) dilaton field.

When we turn to a Minkowski brane-fluid bulk system instead of a massless dilaton bulk, is that although the existence of the finite-distance singularity remains (except perhaps moved on to the envelope—see below), its nature depends on the range of the equation of state fluid parameter γ defined by the equation $P = \gamma\rho$. For a Minkowski brane, the Einstein Equation (8) gives

$$\mathbf{f} = \left(y, -2A\frac{(1+2\gamma)}{3}wx, -4(1+\gamma)\frac{y}{x}w\right),\tag{13}$$

subject to the constraint,

$$\frac{y^2}{x^2} = \frac{w}{\delta}, \quad \delta = 3/2A, \tag{14}$$

with $A = \kappa_5^2/4$, and where for the new variables of this problem we introduce the definitions

$$x = a, \quad y = a', \quad w = \rho, \tag{15}$$

with ρ being the fluid energy density. Then we have the following result ([6], Section 3)

Theorem 2 (Minkowski brane-Single bulk fluid). *With the setup of a flat three-brane in a five-dimensional bulk spacetime filled with a fluid as described above, let Y_s denote the position of the finite-distance singularity. Then, the possible asymptotic behaviours of the solutions of the field equations are all singular, have the required number of arbitrary constants to qualify as corresponding to a general solution, and are given by,*

- *Collapse-type I:* $\gamma > -1/2$
$$a \to 0, \quad a' \to \infty, \quad \rho \to \infty, \tag{16}$$

- *Collapse-type II:* $\gamma = -1/2$
$$a \to 0, \quad a' \to const., \quad \rho \to \infty, \tag{17}$$

- *Big rip:* $\gamma < -1$
$$a \to \infty, \quad a' \to -\infty, \quad \rho \to \infty, \tag{18}$$

- *At envelope:* $\gamma \in (-1, -1/2)$
$$a \to 0, \quad a' \to 0, \quad \rho \to \infty, \tag{19}$$

as $Y \to Y_s$.

Generally speaking, this result implies that the situation described by Theorem 1 is still valid when we pass to the more general fluid content of Theorem 2: Minkowski brane-fluid systems are generically singular and behave basically like the massless dilaton case. For example, item 1 in Theorem 1 means that the runaway situation of the Theorem 2 remains valid for any fluid having $\gamma > -1/2$. A slightly milder singularity is approached by the flat brane-fluid systems when $\gamma = -1/2$, the singular point is attained with bounded speed. The approach to the singularity at a finite distance from the brane changes its nature to that of a big rip when the bulk fluid is phantom-like.

The last item in Theorem 2 requires a separate more involved analysis based on the observation that when solutions of a differential equation have an envelope, the dominant balance picks the envelope and not the general solution, and therefore instead of looking at enveloping solutions from the general solution, we may proceed to construct such solutions directly from the field equations [7]. Equation (19) then implies that all solutions of the field equations having $\gamma \in (-1, -1/2)$ asymptote to the singular first component Σ_1 of an "enveloping brane" defined as a disjoint (we use the term "disjoint" because their common element—namely, $(0, 0, 0)$, is not a realizable state asymptotically) union (cf. [7], Section 2, where this result is proved),

$$\Sigma = \Sigma_1 \bigsqcup \Sigma_2, \tag{20}$$

where

$$\Sigma_1: \quad x = 0, \quad y = \pm H\sqrt{k}, \tag{21}$$

$$\Sigma_2: \quad y = \pm H\sqrt{k}, \quad w = 0. \tag{22}$$

The impossibility of regular solutions away from a Minkowski brane with a massless dilaton or single fluid as sources in the bulk prompts us to search for such solutions further, by considering *mixtures* of the two in the bulk. This is an on-going project with many open problems, the non-interacting, co-existing fluid case is treated in detail in Reference [8]. For the massless scalar, we then take an energy–momentum tensor of the form $T_{AB}^1 = (\rho_1 + P_1)u_A u_B - P_1 g_{AB}$, where $A, B = 1, 2, 3, 4, 5$, $u_A = (0, 0, 0, 0, 1)$ and ρ_1, P_1 are its density and pressure, which we take as $P_1 = \rho_1 = \lambda \phi'^2 / 2$, with λ a parameter. For the second fluid, we assume that $T_{AB}^2 = (\rho_2 + P_2)u_A u_B - P_2 g_{AB}$, and an equation of state of the form $P_2 = \gamma \rho_2$. Here ρ_1, ρ_2, and P_1, P_2 are functions of the fifth dimension Y only. The five-dimensional Einstein field Equation (4) in the case of a flat (Minkowski) brane assume a more complicated form, basically a neat problem in bifurcation theory. Namely,

$$x' = y \tag{23}$$

$$y' = -A\lambda z^2 x - \frac{2}{3} A(1 + 2\gamma)wx \tag{24}$$

$$z' = -\left(4 + \frac{\nu}{2}\right)\frac{yz}{x} + \frac{\sigma}{\lambda}\frac{yw}{xz} \tag{25}$$

$$w' = -(4(\gamma + 1) + \sigma)\frac{yw}{x} + \frac{\lambda\nu}{2}\frac{yz^2}{x}, \tag{26}$$

with the constraint,

$$\frac{y^2}{x^2} = \frac{A\lambda}{3}z^2 + \frac{2A}{3}w. \tag{27}$$

We write $(x, y, z, w) = (a, a', \phi', \rho_2)$, and the new system has four parameters $\lambda, \gamma, \sigma, \nu$, the last two describing the possible exchange of energy between the two components, no exchange of energy corresponding to the case $\nu = \sigma = 0$. This is the main case analyzed in Reference [8]. The main result in this case is this.

Theorem 3 (Minkowski brane: Non-interacting pair of massless dilaton–fluid). *With the setup as described above, let Y_s denote the position of the finite-distance singularity. Then, the possible asymptotic behaviours of the solutions of the field equations are all singular, have the required number of arbitrary constants to qualify as corresponding to a general solution, and are given by,*

- *Collapse-type I: any γ*

$$a \to 0, \quad a' \to \infty, \quad \phi' \to \infty, \quad \rho_2 \to 0, \rho_s, \infty, \tag{28}$$

- *Big rip: $\gamma < -1$*

$$a \to \infty, \quad a' \to -\infty, \quad \phi' \to 0, \quad \rho_2 \to \infty. \tag{29}$$

as $Y \to Y_s$.

We observe that in these asymptotic solutions the final states are characterized by the asymptotic dominance of the dilaton over the fluid component. This is the reason why we obtain singularities for all possible asymptotic balances but of a similar character as the massless dilaton case. In particular, there cannot be any stable asymptotic situation wherein the fluid attains some finite asymptotic value with vanishing dilaton. This is reminiscent of the generic early behaviour of scalar–tensor cosmologies where there is a complete dominance of the scalar field over matter.

The inclusion of an interacting pair of dilaton–fluid could in principle lead to regular solutions away from the Minkowski brane. In Reference [8], it was noticed that suitably choosing the exchange parameters ν, σ, and analyzing the resulting dynamical system has the effect of moving these singular points to infinity. There is an intricate structure of the eigenvalues of the asymptotic matrix that controls the behaviour of the solutions in this case, and this structure leads to the interesting result that for the same interval of the fluid parameter as in the massless dilaton case—namely, $\gamma \in (-1, -1/2)$—the singularities are seen to move to infinity. However, we expect that they are just moved to the singular envelope as before, therefore not being true regular solutions. The generic problem, however, is entirely open at present.

3. Curved Branes

Making the brane positively or negatively curved has the apparent effect of moving the singularities to infinite distance away from the original brane position. However, this may just mean that we are looking at the enveloping brane. The problem then is to determine the precise extent of the singular and regular parts of the enveloping set. This may be an intricate problem. Below, we call a solution *regular* if the scale factor is non-collapsing or divergent in a finite distance away from the brane. This does not exclude the density from having a singularity at the envelope.

With just a massless dilaton support in the bulk, one may indeed get regular curved brane solutions with a decaying dilaton. However, to get this, one has to sacrifice an arbitrary constant, ending up with a family that does not correspond to a general solution of the field equations—at least for de Sitter branes, (for AdS branes there may be no such restriction) cf. [6], Section 2.2.

However, there is one case where we generically reach the regular part of the envelope, as described by the following result.

Theorem 4 (dS or AdS brane: Single bulk fluid with $\gamma \geq -1/2$). *In the above setup, there are two possible nonsingular asymptotic behaviours corresponding to general (three arbitrary constants) solutions of the field equations, having the following properties:*

- $\gamma > -1/2$

$$
\begin{aligned}
x &= \alpha Y + c_{-11} - A\alpha/3c_{-23}Y^{-1} + \cdots, & (30)\\
y &= \alpha + A\alpha/3c_{-23}Y^{-2} + \cdots, & (31)\\
w &= c_{-23}Y^{-4} + \cdots, & (32)
\end{aligned}
$$

where c_{-11} and c_{-23} are arbitrary constants. For $Y \to \infty$, we see that this is on Σ_2 given by Equation (22).

- $\gamma = -1/2$

$$
\begin{aligned}
x &= \alpha Y + c_{-11} \cdots, & (33)\\
y &= \alpha \cdots, & (34)\\
w &= c_{-23}Y^{-2} + \cdots, & (35)
\end{aligned}
$$

where c_{-11} and c_{-23} are arbitrary constants. Taking $Y \to \infty$ demonstrates that this is on Σ_2, given by Equation (22).

This result has two parts. The asymptotic behaviour was found in Reference [6], Section 3.5. The envelope was derived in Reference [7], App. A. We see that these universes look emptier at long distances into the bulk.

Further, there are curved brane solutions with regular support in the bulk coming not from the enveloping set, but from the general solution of the field equations. The following result is shown in complete detail in Reference [9].

Theorem 5. *No collapse singularity can arise in any brane model that comprises either*

- *a de Sitter brane in a single bulk fluid with negative energy density and $\gamma > -1/2$, or*
- *an Anti de Sitter brane in a single bulk fluid with positive energy density and $\gamma \in (-1, -1/2)$,*

as the bulk scale factor is bounded from below and never vanishes.

We note that this result does not exclude the possibility of a big rip singularity in a finite distance from the brane position where the scale factor diverges.

However, whether or not such solutions are singular one may arrange for those solutions that allow for a jump discontinuity on the first derivative of the scale factor across the brane and satisfy the null energy condition, to match for certain ranges of γ producing non-singular universes [9]. Unfortunately, the γ-ranges for the existence of such universes do not quite match other conditions on the fluid to localize gravity on the brane by requiring finiteness of the Planck mass there. This problem lies in the frontier of the singularity problem for such models.

There are a host of other regular asymptotic solutions describing a curved brane sitting in a bulk with a coexisting—even slightly interacting—dilaton–fluid system, cf. [8]. It is an open problem to identify the structure of the enveloping brane in all these solutions, and so we do not discuss them any further here.

Another problem that is beyond our present results is what types of global bulk geometry are compatible with the asymptotic forms discussed here. If we suppose that away from the singularity the bulk space metric g_5' is a kind of perturbation of the metric form assumed here; for instance,

$$g_5' = g_5 + \delta g_5, \tag{36}$$

with g_5 given by Equation (1), then what are the types of geometries which tend to the present ones discussed in previous sections? This is somewhat reminiscent of the isotropization problem in inflationary cosmology.

Finally, we briefly comment about a different approach to the singularities in general brane–bulk systems. There is a basic issue of principle involved in any discussion of singularities in the geometric context of a braneworld embedded in extra dimensions, because the singularities present in the bulk away from the position of the brane—whose existence and nature we discussed in this paper—appear to be totally disconnected from the standard spacetime singularities predicted by the standard singularity theorems for the general relativistic metrics g_4. The same unconnectedness also holds for the cosmic censorship hypothesis (presumably valid on the brane) which, in a truly higher-dimensional theory, ought to be perhaps an emerging property from structures which do not have a four-dimensional counterpart. A way to connect the two is to extend the brane–bulk geometry in such a way as to allow our universe to be the conformal infinity of a certain five-dimensional geometry—the ambient cosmological metric. One then finds that the existence of the four-dimensional spacetime singularities is constrained by the long-term, asymptotic properties of the ambient cosmological metric, while cosmic censorship holds true provided that ambient space remains non-degenerate. For more details of this theory, we invite the reader to consult the recent review [16] (which also includes the original papers).

Author Contributions: Both authors are equally responsible for the main text of this paper.

Conflicts of Interest: The authors declare no conflict of interest.

References

1. Arcani-Hamed, N.; Dimopoulos, S.; Dvali, G.R. The hierarchy problem and New Dimensions at a millimeter. *Phys. Lett. B* **1998**, *429*, 263–272.
2. Arcani-Hamed, N.; Antoniadis, I.; Dimopoulos, S.; Dvali, G.R. New dimensions at a millimeter to a fermi and superstrings at a TeV. *Phys. Lett. B* **1998**, *436*, 257–263.
3. Antoniadis, I. A new approach to supersymmetry breaking in superstring models. *Phys. Lett. B* **1990**, *246*, 377–384.
4. Arkani-Hamed, N.; Dimopoulos, S.; Kaloper, N.; Sundrum, R. A small cosmological constant from a large extra dimension. *Phys. Lett. B* **2000**, *480*, 193–199.
5. Kachru, S.; Schulz, M.; Silverstein, E. *Bounds on curved domain walls in 5d gravity*, *Phys. Rev. D* **2000**, *62*, 085003.
6. Antoniadis, I.; Cotsakis, S.; Klaoudatou, I. Brane singularities and their avoidance. *Class. Quantum Gravity* **2010**, *27*, 235018.

7.   Antoniadis, I.; Cotsakis, S.; Klaoudatou, I. Enveloping branes and braneworld singularities. *Eur. Phys. J. C* **2014**, *74*, 3192.

8.   Antoniadis, I.; Cotsakis, S.; Klaoudatou, I. Brane singularities with mixtures in the bulk. *Fortschr. Phys.* **2013**, *61*, 20–49.

9.   Antoniadis, I.; Cotsakis, S.; Klaoudatou, I. Curved branes with regular support. *Eur. Phys. J. C* **2016**, *76*, 511.

10. Gubser, S.S. Curvature singularities: The good, the bad, and the naked. *Adv. Theor. Math. Phys.* **2000**, *4*, 679–745.

11. Gasperini, M. *Elements of String Cosmology*; Cambridge University Press: Cambridge, UK, 2007; Chapter 10.

12. Peterson, P. *Riemannian Geometry*; Springer: New York, NY, USA, 2006.

13. O'Neill, B. *Semi-Riemannian Geometry with Applications to Relativity*; Academic Press: New York, NY, USA, 1983.

14. Barrow, J.D.; Yamamoto, K. Anisotropic Pressures at Ultra-stiff Singularities and the Stability of Cyclic Universes. *Phys. Rev. D* **2010**, *82*, 063516.

15. Randall, L.; Sundrum, S. An alternative to compactification. *Phys. Rev. Lett.* **1999**, *83*, 4960.

16. Antoniadis, I.S.; Cotsakis, S. The large-scale structure of the ambient boundary. In Proceedings of the Fourteenth Marcel Grossman Meeting on General Relativity, Rome, Italy, 12–18 July 2015; Bianchi, M., Jantzen, R.T., Ruffini, R., Eds.; World Scientific: Singapore, 2017.

Review

Tests of Lorentz Symmetry in the Gravitational Sector

Aurélien Hees [1,*], **Quentin G. Bailey** [2], **Adrien Bourgoin** [3], **Hélène Pihan-Le Bars** [3],
Christine Guerlin [3,4] **and Christophe Le Poncin-Lafitte** [3]

1   Department of Physics and Astronomy, University of California, Los Angeles, CA 90095, USA
2   Physics Department, Embry-Riddle Aeronautical University, Prescott, AZ 86301, USA; baileyq@erau.edu
3   SYRTE, Observatoire de Paris, PSL Research University, CNRS, Sorbonne Universités, UPMC Univ. Paris 06,
     LNE, 61 avenue de l'Observatoire, 75014 Paris, France; adrien.bourgoin@obspm.fr (A.B.);
     helene.pihan-lebars@obspm.fr (H.P.-L.B.); christine.guerlin@obspm.fr (C.G.);
     christophe.leponcin@obspm.fr (C.L.P.-L.)
4   Laboratoire Kastler Brossel, ENS-PSL Research University, CNRS, UPMC-Sorbonne Universités,
     Collège de France, 75005 Paris, France
*   Correspondence: ahees@astro.ucla.edu; Tel.: +1-310-825-8345

Academic Editors: Lorenzo Iorio and Elias C. Vagenas
Received: 15 October 2016; Accepted: 22 November 2016; Published: 1 December 2016

Abstract: Lorentz symmetry is one of the pillars of both General Relativity and the Standard Model of particle physics. Motivated by ideas about quantum gravity, unification theories and violations of CPT symmetry, a significant effort has been put the last decades into testing Lorentz symmetry. This review focuses on Lorentz symmetry tests performed in the gravitational sector. We briefly review the basics of the pure gravitational sector of the Standard-Model Extension (SME) framework, a formalism developed in order to systematically parametrize hypothetical violations of the Lorentz invariance. Furthermore, we discuss the latest constraints obtained within this formalism including analyses of the following measurements: atomic gravimetry, Lunar Laser Ranging, Very Long Baseline Interferometry, planetary ephemerides, Gravity Probe B, binary pulsars, high energy cosmic rays, ... In addition, we propose a combined analysis of all these results. We also discuss possible improvements on current analyses and present some sensitivity analyses for future observations.

Keywords: experimental tests of gravitational theories; Lorentz and Poincaré invariance; modified theories of gravity; celestial mechanics; atom interferometry; binary pulsars

---

1. Introduction

The year 2015 was the centenary of the theory of General Relativity (GR), the current paradigm for describing the gravitational interaction (see e.g., the Editorial of this special issue [1]). Since its creation, this theory has passed all experimental tests with flying colors [2,3]; the last recent success was the discovery of gravitational waves [4], summarized in [5]. On the other hand, the three other fundamental interactions of Nature are described within the Standard Model of particle physics, a framework based on relativistic quantum field theory. Although very successful so far, it is commonly admitted that these two theories are not the ultimate description of Nature but rather some effective theories. This assumption is motivated by the construction of a quantum theory of gravitation that has not been successfully developed so far and by the development of a theory that would unify all the fundamental interactions. Moreover, observations requiring the introduction of Dark Matter and Dark Energy also challenge GR and the Standard Model of particle physics since they cannot be explained by these two paradigms altogether [6]. It is therefore extremely important to test our current description of the four fundamental interactions [7].

Lorentz invariance is one of the fundamental symmetry of relativity, one of the corner stones of both GR and the Standard Model of particle physics. It states that the outcome of any local experiment is independent of the velocity and of the orientation of the laboratory in which the experiment is performed. If one considers non-gravitational experiments, Lorentz symmetry is part of the Einstein Equivalence Principle (EEP). A breaking of Lorentz symmetry implies that the equations of motion, the particle thresholds, etc... may be different when the experiment is boosted or rotated with respect to a background field [8]. More precisely, it is related to a violation of the invariance under "particle Lorentz transformations" [8] which are the boosts and rotations that relate the properties of two systems within a specific oriented inertial frame (or in other words they are boosts and rotations on localized fields but not on background fields). On the other hand, the invariance under coordinates transformations known as "observer Lorentz transformations" [8] which relate observations made in two inertial frames with different orientations and velocities is always preserved. Considering the broad field of applicability of this symmetry, searches for Lorentz symmetry breaking provide a powerful test of fundamental physics. Moreover, it has been suggested that Lorentz symmetry may not be a fundamental symmetry of Nature and may be broken at some level. While some early motivations came from string theories [9–11], breaking of Lorentz symmetry also appears in loop quantum gravity [12–15], non commutative geometry [16,17], multiverses [18], brane-world scenarios [19–21] and others (see for example [22,23]).

Tests of Lorentz symmetry have been performed since the time of Einstein but the last decades have seen the number of tests increased significantly [24] in all fields of physics. In particular, a dedicated effective field theory has been developed in order to systematically consider all hypothetical violations of the Lorentz invariance. This framework is known as the Standard-Model Extension (SME) [8,25] and covers all fields of physics. It contains the Standard Model of particle physics, GR and all possible Lorentz-violating terms that can be constructed at the level of the Lagrangian, introducing a large numbers of new coefficients that can be constrained experimentally.

In this review, we focus on the gravitational sector of the SME which parametrizes deviations from GR. GR is built upon two principles [2,26,27]: (i) the EEP; and (ii) the Einstein field equations that derive from the Einstein-Hilbert action. The EEP gives a geometric nature to gravitation allowing this interaction to be described by spacetime curvature. From a theoretical point of view, the EEP implies the existence of a spacetime metric to which all matter minimally couples [28]. A modification of the matter part of the action will lead to a breaking of the EEP. In SME, such a breaking of the EEP is parametrized (amongst others) by the matter-gravity coupling coefficients $\bar{a}_\mu$ and $\bar{c}_{\mu\nu}$ [29,30]. From a phenomenological point of view, the EEP states that [2,27]: (i) the universality of free fall (also known as the weak equivalence principle) is valid; (ii) the outcome of any local non-gravitational experiment is independent of the velocity of the free-falling reference frame in which it is performed; and (iii) the outcome of any local non-gravitational experiment is independent of where and when in the universe it is performed. The second part of Einstein theory concerns the purely gravitational part of the action (the Einstein-Hilbert action) which is modified in SME to introduce hypothetical Lorentz violations in the gravitational sector. This review focuses exclusively on this kind of Lorentz violations and not on breaking of the EEP.

A lot of tests of GR have been performed in the last decades (see [2] for a review). These tests rely mainly on two formalisms: the parametrized post-Newtonian (PPN) framework and the fifth force formalism. In the former one, the weak gravitational field spacetime metric is parametrized by 10 dimensionless coefficients [27] that encode deviations from GR. This formalism therefore provides a nice interface between theory and experiments. The PPN parameters have been constrained by a lot of different observations [2] confirming the validity of GR. In particular, three PPN parameters encode violations of the Lorentz symmetry: the $\alpha_{1,2,3}$ PPN coefficients. In the fifth force formalism, one is looking for a deviation from Newtonian gravity where the gravitational potential takes the form of a Yukawa potential characterized by a length λ and a strength α of interaction [31–34]. These two parameters are very well constrained as well except at very small and large distances (see [35]).

The gravitational sector of SME offers a new framework to test GR by parametrizing deviations from GR at the level of the action, introducing new terms that are breaking Lorentz symmetry. The idea is to extend the standard Einstein-Hilbert action by including Lorentz-violating terms constructed by contracting new fields with some operators built from curvature tensors and covariant derivatives with increasing mass dimension [36]. The lower mass dimension (dimension 4) term is known as the minimal SME and its related new fields can be split into a scalar part u, a symmetric trace free part $s^{\mu\nu}$ and a traceless piece $t^{\kappa\lambda\mu\nu}$. In order to avoid conflicts with the underlying Riemann geometry, the Lorentz violating coefficients can be assumed to be dynamical fields and the Lorentz violation to arise from a spontaneous symmetry breaking [37–42]. The Lorentz violating fields therefore acquire a non-vanishing vacuum expectation value (denoted by a bar). It has been shown that in the linearized gravity limit the fluctuations around the vacuum values can be integrated out so that only the vacuum expectation values of the SME coefficients influence observations [39]. In the minimal SME, the coefficient $\bar{u}$ corresponds to a rescaling of the gravitational constant and is therefore unobservable and the coefficients $\bar{t}^{\kappa\lambda\mu\nu}$ do not play any role at the post-Newtonian level, a surprising phenomenon known as the t-puzzle [43,44]. The $\bar{s}^{\mu\nu}$ coefficients lead to modifications from GR that have thoroughly been investigated in [39]. In particular, the SME framework extends standard frameworks such as the PPN or fifth force formalisms meaning that "standard" tests of GR cannot directly be translated into this formalism.

In the last decade, several measurements have been analyzed within the gravitational sector of the minimal SME framework: Lunar Laser Ranging (LLR) analysis [45,46], atom interferometry [47,48], planetary ephemerides analysis [49,50], short-range gravity [51], Gravity Probe B (GPB) analysis [52], binary pulsars timing [53,54], Very Long Baseline Interferometry (VLBI) analysis [55] and Čerenkov radiation [56]. In addition to the minimal SME, there exist some higher order Lorentz-violating curvature couplings in the gravity sector [43] that are constrained by short-range experiments [57–59], Čerenkov radiation [30,56] and gravitational waves analysis [60,61]. Finally, some SME experiments have been used to derive bounds on spacetime torsion [62,63]. A review for these measurements can be found in [30]. The classic idea to search for or to constrain Lorentz violations in the gravitational sector is to search for orientation or boost dependence of an observation. Typically, one will take advantage of modulations that will occur through an orientation dependence of the observations due to the Earth's rotation, the motion of satellites around Earth (the Moon or artificial satellites), the motion of the Earth (or other planets) around the Sun, the motion of binary pulsars, ... The main goal of this communication is to review all the current analyses performed in order to constrain Lorentz violation in the pure gravitational sector.

Two distinct procedures have been used to analyze data within the SME framework. The first procedure consists in deriving analytically the signatures produced by the SME coefficients on some observations. Then, the idea is to fit these signatures within residuals obtained by a data analysis performed in pure GR. This approach has the advantage to be relatively easy and fast to perform. Nevertheless, when using this postfit approach, correlations with other parameters fitted in the data reduction are completely neglected and may lead to overoptimistic results. A second way to analyze data consists in introducing the Lorentz violating terms directly in the modeling of observables and in the global data reduction. In this review, we highlight the differences between the two approaches.

In this communication, a brief theoretical review of the SME framework in the gravitational sector is presented in Section 2. The two different approaches to analyze data within the SME framework (postfit analysis versus full modeling of observables within the SME framework) are discussed and compared in Section 3. Section 4 is devoted to a discussion of the current measurements analyzed within the SME framework. This discussion includes a general presentation of the measurements, a brief review of the effects of Lorentz violation on each of them, the current analyses performed with real data and a critical discussion. A "grand fit" combining all existing analyses is also presented. In Section 5, some future measurements that are expected to improve the current analyses are developed. Finally, our conclusion is presented in Section 6.

2. The Standard-Model Extension in the Gravitational Sector

Many of the tests of Lorentz and CPT symmetry have been analyzed within an effective field theory framework which generically describes possible deviations from exact Lorentz and CPT invariance [8,25] and contains some traditional test frameworks as limiting cases [64,65]. This framework is called, for historical reasons, the Standard-Model Extension (SME). One part of the activity has been a resurgence of interest in tests of relativity in the Minkowski spacetime context, where global Lorentz symmetry is the key ingredient. Numerous experimental and observational constraints have been obtained on many different types of hypothetical Lorentz and CPT symmetry violations involving matter [24]. Another part, which has been developed more recently, has seen the SME framework extended to include the curved spacetime regime [37]. Recent work shows that there are many ways in which the spacetime symmetry foundations of GR can be tested [29,39].

In the context of effective field theory in curved spacetime, violations of these types can be described by an action that contains the usual Einstein-Hilbert term of GR, a matter action, plus a series of terms describing Lorentz violation for gravity and matter in a generic way. While the fully general coordinate invariant version of this action has been studied in the literature, we focus on a limiting case that is valid for weak-field gravity and can be compactly displayed. Using an expansion of the spacetime metric around flat spacetime, $g_{\mu\nu} = \eta_{\mu\nu} + h_{\mu\nu}$, the effective Lagrange density to quadratic order in $h_{\mu\nu}$ can be written in a compact form as

$$\mathcal{L} = \mathcal{L}_{\text{EH}} + \frac{c^3}{32\pi G} h^{\mu\nu} \bar{s}^{\alpha\beta} \mathcal{G}_{\alpha\mu\nu\beta} + ..., \tag{1}$$

where $\mathcal{L}_{\text{EH}}$ is the standard Einstein-Hilbert term, $\mathcal{G}_{\alpha\mu\nu\beta}$ is the double dual of the Einstein tensor linearized in $h_{\mu\nu}$, G the bare Newton constant and c the speed of light in a vacuum. The Lorentz-violating effects in this expression are controlled by the 9 independent coefficients in the traceless and dimensionless $\bar{s}^{\mu\nu}$ [39]. These coefficients are treated as constants in asymptotically flat cartesian coordinates. The ellipses represent additional terms in a series including terms that break CPT symmetry for gravity; such terms are detailed elsewhere [43,56,60] and are part of the so-called nonminimal SME expansion. Note that the process by which one arrives at the effective quadratic Lagrangian (1) is consistent with the assumption of the spontaneous breaking of local Lorentz symmetry, which is discussed below.

Also of interest are the matter-gravity couplings. This form of Lorentz violation can be realized in the classical point-mass limit of the matter sector. In the minimal SME the point-particle action can be written as

$$S_{\text{Matter}} = \int d\lambda \, c \left(-m \sqrt{-(g_{\mu\nu} + 2c_{\mu\nu})u^\mu u^\nu} - a_\mu u^\mu \right), \tag{2}$$

where the particle's worldline tangent is $u^\mu = dx^\mu/d\lambda$ [29]. The coefficients controlling local Lorentz violation for matter are $c_{\mu\nu}$ and a_μ. In contrast to $\bar{s}^{\mu\nu}$, these coefficients depend on the type of point mass (particle species) and so they can also violate the EEP. When the coefficients $\bar{s}_{\mu\nu}$, $c_{\mu\nu}$, and a_μ vanish perfect local Lorentz symmetry for gravity and matter is restored. It is also interesting to mention that this action with fixed (but not necessarily constant) a_μ and $c_{\mu\nu}$ represents motion in a Finsler geometry [66,67].

It has been shown that explicit local Lorentz violation is generically incompatible with Riemann geometry [37]. One natural way around this is assumption of spontaneous Lorentz-symmetry breaking. In this scenario, the tensor fields in the underlying theory acquire vacuum expectation values through a dynamical process. Much of the literature has been devoted to studying this possibility in the last decades [9,38,68–78], including some original work on spontaneous Lorentz-symmetry breaking in string field theory [10,11]. For the matter-gravity couplings in Equation (2), the coefficient fields $c_{\mu\nu}$, and a_μ are then expanded around their background (or vacuum) values $\bar{c}_{\mu\nu}$, and $\bar{a}_\mu$. Both a modified spacetime metric $g_{\mu\nu}$ and modified point-particle equations of motion result from the spontaneous breaking of Lorentz symmetry. In the linearized gravity limit these results rely only on the vacuum values $\bar{c}_{\mu\nu}$, and $\bar{a}_\mu$. The dominant signals for Lorentz violation controlled by these coefficients are revealed in the calculation of observables in the post-Newtonian limit.

Several novel features of the post-Newtonian limit arise in the SME framework. It was shown in Ref. [39] that a subset of the $\bar{s}^{\mu\nu}$ coefficients can be matched to the PPN formalism [2,27], but others lie outside it. For example, a dynamical model of spontaneous Lorentz symmetry breaking can be constructed from an antisymmetric tensor field $B_{\mu\nu}$ that produces $\bar{s}^{\mu\nu}$ coefficients that cannot be reduced to an isotropic diagonal form in any coordinate system, thus lying outside the PPN assumptions [78]. We can therefore see that the SME framework has a partial overlap with the PPN framework, revealing new directions to explore in analysis via the $\bar{s}^{\mu\nu}$, $\bar{c}_{\mu\nu}$, and $\bar{a}_\mu$ coefficients. The equations of motion for matter are modified by the matter-gravity coefficients for Lorentz violation $\bar{c}_{\mu\nu}$ and $\bar{a}_\mu$, which can depend on particle species, thus implying that these coefficients also control EEP violations. One potentially important class of experiments from the action (2) concerns the Universality of Free Fall of antimatter whose predictions are discussed in [29,79]. In addition, the post-Newtonian metric itself receives contributions from the matter coefficients $\bar{c}_{\mu\nu}$ and $\bar{a}_\mu$. So for example, two (chargeless) sources with the same total mass but differing composition will yield gravitational fields of different strength.

For solar-system gravity tests, the primary effects due to the nine coefficients $\bar{s}^{\mu\nu}$ can be obtained from the post-Newtonian metric and the geodesic equation for test bodies. A variety of ground-based and space-based tests can measure these coefficients [80–82]. Such tests include Earth-laboratory tests with gravimeters, lunar and satellite laser ranging, studies of the secular precession of orbital elements in the solar system, and orbiting gyroscope experiments, and also classic effects such as the time delay and bending of light around the Sun and Jupiter. Furthermore, some effects described by the Lagrangian (1) can be probed by analyzing data from binary pulsars and measurements of cosmic rays [56].

For the matter-gravity coefficients $\bar{c}_{\mu\nu}$ and $\bar{a}_\mu$, which break Lorentz symmetry and EEP, several experiments can be used for analysis in addition to the ones already mentioned above including ground-based gravimeter and WEP experiments. Dedicated satellite EEP tests are among the most sensitive where the relative acceleration of two test bodies of different composition is the observable of interest. Upon relating the satellite frame coefficients to the standard Sun-centered frame used for the SME, oscillations in the acceleration of the two masses occur at a number of different harmonics of the satellite orbital and rotational frequencies, as well as the Earth's orbital frequency. Future tests of particular interest include the currently flying MicroSCOPE experiment [83,84].

While the focus of the discussion to follow are the results for the minimal SME coefficients $\bar{s}^{\mu\nu}$, recent work has also involved the nonminimal SME coefficients in the pure-gravity sector associated with mass dimension 5 and 6 operators. One promising testing ground for these coefficients is sensitive short-range gravity experiments. The Newtonian force between two test masses becomes modified in the presence of local Lorentz violation by an anisotropic quartic force that is controlled by a subset of coefficients from the Lagrangian organized as the totally symmetric $(\bar{k}_{\text{eff}})_{jklm}$, which has dimensions of length squared [43]. This contains 14 measurable quantities and any one short-range experiment is sensitive to 8 of them. Two key experiments, from Indiana University and Huazhong University of Science and Technology, have both reported analysis in the literature [57,58] . A recent work combines the two analyses to place new limits on all 14, a priori independent, $(\bar{k}_{\text{eff}})_{jklm}$ coefficients [59]. Other higher mass dimension coefficients play a role in gravitational wave propagation [60] and gravitational Čerenkov radiation [56].

To conclude this section, we ask: what can be said about the possible sizes of the coefficients for Lorentz violation? A broad class of hypothetical effects is described by the SME effective field theory framework, but it is a test framework and as such does not make specific predictions concerning the sizes of these coefficients. One intriguing suggestion is that there is room in nature for violations of spacetime symmetry that are large compared to other sectors due to the intrinsic weakness of gravity. Considering the current status of the coefficients $\bar{s}^{\mu\nu}$, the best laboratory limits are at the 10^{-10}–10^{-11} level, with improvements of four orders of magnitude in astrophysical tests on these coefficients [56]. However, the limits are at the $10^{-8}\,\text{m}^2$ level for the mass dimension 6 coefficients

$(\bar{k}_{\text{eff}})_{jklm}$ mentioned above. Comparing this to the Planck length 10^{-35} m, we see that symmetry breaking effects could still have escaped detection that are not Planck suppressed. This kind of "countershading" was first pointed out for the $\bar{a}_{\mu}$ coefficients [85], which, having dimensions of mass, can still be as large as a fraction of the electron mass and still lie within current limits.

In addition, any action-based model that breaks local Lorentz symmetry either explicitly or spontaneously can be matched to a subset of the SME coefficients. Therefore, constraints on SME coefficients can directly constrain these models. Matches between various toy models and coefficients in the SME have been achieved for models that produce effective $\bar{s}^{\mu\nu}$, $\bar{c}_{\mu\nu}$, $\bar{a}_{\mu}$, and other coefficients. This includes vector and tensor field models of spontaneous Lorentz-symmetry breaking [29,39,75–78], models of quantum gravity [12,65] and noncommutative quantum field theory [17]. Furthermore, Lorentz violations may also arise in the context of string field theory models [86].

3. Postfit Analysis Versus Full Modeling

Since the last decade, several studies aimed to find upper limits on SME coefficients in the gravitational sector. A lot of these studies are based on the search of possible signals in post-fit residuals of experiments. This was done with LLR [45], GPB [52], binary pulsars [53,54] or Solar System planetary motions [49,50]. However, two new works focused on a direct fit to data with LLR [46] and VLBI [55], which are more satisfactory.

Indeed, in the case of a post-fit analysis, a simple modeling of extra terms containing SME coefficients are least square fitted in the residuals, attempting to constrain the SME coefficients of a testing function in residual noise obtained from a pure GR analysis, where of course Lorentz symmetry is assumed. It comes out correlations between SME coefficients and other global parameters previously fitted (masses, position and velocity...) cannot be assessed in a proper way. In others words, searching hypothetical SME signals in residuals, i.e., in noise, can lead to an overestimated formal error on SME coefficients, as illustrated in the case of VLBI [55], and without any chance to learn something about correlations with other parameters, as for example demonstrated in the case of LLR [46]. Let us consider the VLBI example to illustrate this fact. The VLBI analysis is described in Section 4.2. Including the SME contribution within the full VLBI modeling and estimating the SME coefficient $\bar{s}^{TT}$ altogether with the other parameters fitted in standard VLBI data reduction leads to the estimate $\bar{s}^{TT} = (-5 \pm 8) \times 10^{-5}$. A postfit analysis performed by fitting the SME contribution within the VLBI residuals obtained after a pure GR analysis leads to $\bar{s}^{TT} = (-0.6 \pm 2.1) \times 10^{-8}$ [55]. This example shows that a postfit analysis can lead to results with overoptimistic uncertainties and one needs to be extremely careful when using such results.

4. Data Analysis

In this section, we will review the different measurements that have already been used in order to constrain the SME coefficients. The different analyses are based on quite different types of observations. In order to compare all the corresponding results, we need to report them in a canonical inertial frame. The standard canonical frame used in the SME framework is a Sun-centered celestial equatorial frame [64], which is approximately inertial over the time scales of most observations. This frame is asymptotically flat and comoving with the rest frame of the Solar System. The cartesian coordinates related to this frame are denoted by capital letters

$$X^{\Xi} = (cT, X^J) = (cT, X, Y, Z). \tag{3}$$

The Z axis is aligned with the rotation axis of the Earth, while the X axis points along the direction from the Earth to the Sun at vernal equinox. The origin of the coordinate time T is given by the time when the Earth crosses the Sun-centered X axis at the vernal equinox. These conventions are depicted in Figure 2 from [39].

In the following subsections, we will present the different measurements used to constrain the SME coefficients. Each subsection contains a brief description of the principle of the experiment, how it can be used to search for Lorentz symmetry violations, what are the current best constraints obtained with such measurements and eventually how it can be improved in the future.

4.1. Atomic Gravimetry

The most sensitive experiments on Earth searching for Lorentz Invariance Violation (LIV) in the minimal SME gravity sector are gravimeter tests. As Earth rotates, the signal recorded in a gravimeter, i.e., the apparent local gravitational acceleration g of a laboratory test body, would be modulated in the presence of LIV in gravity. This was first noted by Nordtvedt and Will in 1972 [87] and used soon after with gravimeter data to constrain preferred-frame effects in the PPN formalism [88,89] at the level of 10^{-3}.

This test used a superconducting gravimeter, based on a force comparison (the gravitational force is counter-balanced by an electromagnetic force maintaining the test mass at rest). While superconducting gravimeters nowadays reach the best sensitivity on Earth, force comparison gravimeters intrinsically suffer from drifts of their calibration factor (with e.g., aging of the system). Development of other types of gravimeters has evaded this drawback: free fall gravimeters. Monitoring the motion of a freely falling test mass, they provide an absolute measurement of g. State-of-the art free fall gravimeters use light to monitor the mass free fall. Beyond classical gravimeters that drop a corner cube, the development of atom cooling and trapping techniques and atom interferometry has led to a new generation of free fall gravimeters, based on a quantum measurement: atomic gravimeters.

Atomic gravimeters use atoms in gaseous phase as a test mass. The atoms are initially trapped with magneto-optical fields in vacuum, and laser cooled (down to 100 nK) in order to control their initial velocity (down to a few mm/s). The resulting cold atom gas, containing typically a million atoms, is then launched or dropped for a free fall measurement. Manipulating the electronic and motional state of the atoms with two counterpropagating lasers, it is possible to measure, using atom interferometry, their free fall acceleration with respect to the local frame defined by the two lasers [90]. This sensitive direction is aligned to be along the local gravitational acceleration noted $\hat{z}$; the atom interferometer then measures the phase $\varphi = k a^{\hat{z}} T^2$, where T is half the interrogation time, $k \simeq 2(2\pi/\lambda)$ with λ the laser wavelength, and $a^{\hat{z}}$ is the free fall acceleration along the laser direction. The free fall time is typically on the order of 500 ms, corresponding to a free fall distance of about a meter. A new "atom preparation—free fall—detection" cycle is repeated every few seconds. Each measurement is affected by white noise, but averaging leads to a typical sensitivity on the order of or below 10^{-9} g [91–93].

Such an interferometer has been used by H. Müller et al. in [47] and K. Y. Chung et al. in [48] for testing Lorentz invariance in the gravitational sector with Caesium atoms, leading to the best terrestrial constraints on the $\bar{s}^{\mu\nu}$ coefficients. The analysis uses three data sets of respectively 2.5 days for the first two and 10 days for the third, stretched over 4 years, which allows one to observe sidereal and annual LIV signatures. The gravitational SME model used for this analysis can be found in [39,47,48]; its derivation will be summarized hereunder. Since the atoms in free fall are sensitive to the local phase of the lasers, LIV in the interferometer observable could also come from the pure electromagnetic sector. This contribution has been included in the experimental analysis in [48]. Focusing here on the gravitational part of SME, we ignore it in the following.

The gravitational LIV model adjusted in this test restricts to modifications of the Earth-atom two-body gravitational interaction. The Lagrangian describing the dynamics of a test particle at a point on the Earth's surface can be approximated by a post-Newtonian series as developed in [39]. At the Newtonian approximation, the two bodies Lagrangian is given by

$$\mathcal{L} = \frac{1}{2}mV^2 + G_N \frac{Mm}{R} \left(1 + \frac{1}{2}\bar{s}_t^{JK}\hat{R}^J\hat{R}^K - \frac{3}{2}\bar{s}^{TJ}\frac{V^J}{c} - \bar{s}^{TJ}\hat{R}^J\frac{V^K}{c}\hat{R}^k \right), \qquad (4)$$

where R and V are the position and velocity expressed in the standard SME Sun-centered frame and $\hat{R} = R/R$ with $R = |R|$. In addition, we have introduced G_N the observed Newton constant measured by considering the orbital motion of bodies and defined by (see also [39,50] or Section IV of [52])

$$G_N = G \left(1 + \frac{5}{3} \bar{s}^{TT} \right) , \tag{5}$$

and the 3-dimensional traceless tensor

$$\bar{s}^{JK}_t = \bar{s}^{JK} - \frac{1}{3} \bar{s}^{TT} \delta^{JK} . \tag{6}$$

From this Lagrangian one can derive the equations of motion of the free fall mass in a laboratory frame (see the procedure in Section V.C.1. from [39]). It leads to the modified local acceleration in the presence of LIV [39] given by

$$a^{\hat{z}} = g \left(1 - \frac{1}{6} i_4 \bar{s}^{TT} + \frac{1}{2} i_4 \bar{s}^{\hat{z}\hat{z}} \right) - \omega_\oplus^2 R_\oplus \sin^2 \chi - g i_4 \bar{s}^{T\hat{z}} \beta_\oplus^{\hat{z}} - 3 g i_1 \bar{s}^{TJ} \beta_\oplus^J , \tag{7}$$

where $g = G_N M_\oplus / R_\oplus^2$, $\omega_\oplus$ is the Earth's angular velocity, $\beta_\oplus = \frac{V_\oplus}{c} \sim 10^{-4}$ is the Earth's boost, $R_\oplus$ is the Earth radius, $M_\oplus$ is the Earth mass and χ the colatitude of the lab whose reference frame's $\hat{z}$ direction is the sensitive axis of the instrument as previously defined here. This model includes the shape of the Earth through its spherical moment of inertia $I_\oplus$ which appears in $i_\oplus = \frac{I_\oplus}{M_\oplus R_\oplus^2}$, $i_1 = 1 + \frac{1}{3} i_\oplus$ and $i_4 = 1 - 3 i_\oplus$. In [48], Earth has been approximated as spherical and homogeneous leading to $i_\oplus = \frac{1}{5}$, $i_1 = \frac{7}{6}$ and $i_4 = -\frac{1}{2}$.

The sensing direction of the experiment precesses around the Earth rotation axis with sidereal period, and the lab velocity varies with sidereal period and annual period. At first order in $V_\oplus$ and $\omega_\oplus$ and as a function of the SME coefficients, the LIV signal takes the form of a harmonic series with sidereal and annual base frequencies (denoted resp. $\omega_\oplus$ and Ω) together with first harmonics. The time dependence of the measured acceleration $a^{\hat{z}}$ from Equation (7) arises from the terms involving the $\hat{z}$ indices. It can be decomposed in frequency according to [39]

$$\frac{\delta a^{\hat{z}}}{a^{\hat{z}}} = \sum_l C_l \cos(\omega_l t + \phi_l) + D_l \sin(\omega_l t + \phi_l) . \tag{8}$$

The model contains seven frequencies $l \in \{\Omega, \omega_\oplus, 2\omega_\oplus, \omega_\oplus \pm \Omega, 2\omega_\oplus \pm \Omega\}$. The 14 amplitudes C_l and D_l are linear combinations of 7 $\bar{s}^{\mu\nu}$ components: $\bar{s}^{JK}$, $\bar{s}^{TJ}$ and $\bar{s}^{XX} - \bar{s}^{YY}$ which can be found in Table 1 of [48] or Table IV from [39].

In order to look for tiny departures from the constant Earth-atom gravitational interaction, a tidal model for $a^{\hat{z}}$ variations due to celestial bodies is removed from the data before fitting to Equation (8). This tidal model consists of two parts. One part is based on a numerical calculation of the Newtonian tide-generating potential from the Moon and the Sun at Earth's surface based on ephemerides. It uses here the Tamura tidal catalog [94] which gives the frequency, amplitude and phase of 1200 harmonics of the tidal potential. These arguments are used by a software (ETGTAB) that calculates the time variation of the local acceleration in the lab and includes the elastic response of Earth's shape to the tides, called "solid Earth tides", also described analytically e.g., by the DDW model [95]. A previous SME analysis of the atom gravimeter data using only this analytical tidal correction had been done, but it led to a degraded sensitivity of the SME test [47]. Indeed, a non-negligible contribution to $a^{\hat{z}}$ is not covered by this non-empirical tidal model: oceanic tide effects such as ocean loading, for which good global analytical models do not exist. They consequently need to be adjusted from measurements. For the second analysis, reported here, additional local tidal corrections fitted on altimetric data have been removed [96] allowing to improve the statistical uncertainty of the SME test by one order of magnitude.

After tidal subtraction, signal components are extracted from the data using a numerical Fourier transform (NFT). Due to the finite data length, Fourier components overlap, but the linear combinations of spectral lines that the NFT estimates can be expressed analytically. Since the annual component $\omega_l = \Omega$ has not been included in this analysis, the fit provides 12 measurements. From there, individual constraints on the 7 SME coefficients and their associated correlation coefficients can be estimated by a least square adjustment. The results obtained are presented in Table 1.

Table 1. Atom-interferometry limits on Lorentz violation in gravity from [48]. The correlation coefficients can be derived from Table III of [48].

Coefficient	
$\bar{s}^{TX}$	$(-3.1 \pm 5.1) \times 10^{-5}$
$\bar{s}^{TY}$	$(0.1 \pm 5.4) \times 10^{-5}$
$\bar{s}^{TZ}$	$(1.4 \pm 6.6) \times 10^{-5}$
$\bar{s}^{XX} - \bar{s}^{YY}$	$(4.4 \pm 11) \times 10^{-9}$
$\bar{s}^{XY}$	$(0.2 \pm 3.9) \times 10^{-9}$
$\bar{s}^{XZ}$	$(-2.6 \pm 4.4) \times 10^{-9}$
$\bar{s}^{YZ}$	$(-0.3 \pm 4.5) \times 10^{-9}$

Correlation Coefficients						
1						
0.05	1					
0.11	−0.16	1				
−0.82	0.34	−0.16	1			
−0.38	−0.86	0.10	−0.01	1		
−0.41	0.13	−0.89	0.38	0.02	1	
−0.12	−0.19	−0.89	0.04	0.20	0.80	1

All results obtained are compatible with null Lorentz violation. As expected from boost suppressions in Equation (7) and from the measurement uncertainty, on the order of a few 10^{-9} g [97], typical limits obtained are in the 10^{-9} range for purely spatial $\bar{s}^{\mu\nu}$ components and 4 orders of magnitude weaker for the spatio-temporal components $\bar{s}^{TJ}$. It can be seen e.g., with the purely spatial components that these constraints do not reach the intrinsic limit of acceleration resolution of the instrument (which has a short term stability of 11×10^{-9} $g/\sqrt{\text{Hz}}$) because the coefficients are still correlated. Their marginalized uncertainty is broadened by their correlation.

Consequently, improving the uncertainty could be reached through a better decorrelation, by analyzing longer data series. In parallel, the resolution of these instruments keeps increasing and has nowadays improved by about a factor 10 since this experiment. However, increasing the instrument's resolution brings back to the question of possible accidental cancelling in treating "postfit" data. Indeed, it should be recalled here that local tidal corrections subtracted prior to analysis are based on adjusting a model of ocean surface from altimetry data. In principle, this observable would as well be affected by gravity LIV; fitting to these observations thus might remove part of SME signatures from the atom gravimeter data. This was mentioned in the first atom gravimeter SME analysis [47]. The adjustment process used to assess local corrections in gravimeters is not made directly on the instrument itself, but it always involves a form of tidal measurement (here altimetry data, or gravimetry data from another instrument in [98]). All LIV frequencies match to the main tidal frequencies. Further progress on SME analysis with atom gravimeters would thus benefit from addressing in more details the question of possible signal cancelling.

4.2. Very Long Baseline Interferometry

VLBI is a geometric technique measuring the time difference in the arrival of a radio wavefront, emitted by a distant quasar, between at least two Earth-based radio-telescopes. VLBI observations are done daily since 1979 and the database contains nowadays almost 6000 24 h sessions, corresponding to 10 millions group-delay observations, with a present precision of a few picoseconds. One of the principal goals of VLBI observations is the kinematical monitoring of Earth rotation with respect to a global inertial frame realized by a set of defining quasars, the International Celestial Reference Frame [99], as defined by the International Astronomical Union [100]. The International VLBI Service for Geodesy and Astrometry (IVS) organizes sessions of observation, storage of data and distribution of products, in particular the Earth Orientation parameters. Because of this precision, VLBI is also a very interesting tool to test gravitation in the Solar System. Indeed, the gravitational fields of the Sun and the planets are responsible of relativistic effects on the quasar light beam through the propagation of the signal to the observing station and VLBI is able to detect these effects very accurately. By using the complete VLBI observations database, it was possible to obtain a constraint on the γ PPN parameter at the level of 1.2×10^{-4} [101,102]. In its minimal gravitational sector, SME can also be investigated with VLBI and obtaining a constrain on the $\bar{s}^{TT}$ coefficient is possible.

Indeed, the propagation time of a photon emitted at the event $(cT_e, \mathbf{X}_e)$ and received at the position $\mathbf{X}_r$ can be computed in the SME formalism using the time transfer function formalism [103–107] and is given by [39,80]

$$\mathcal{T}(\mathbf{X}_e, T_e, \mathbf{X}_r) = T_r - T_e = \frac{R_{er}}{c} + 2\frac{G_N M}{c^3}\left[1 - \frac{2}{3}\bar{s}^{TT} - \bar{s}^{TJ}N_{er}^J\right]\ln\frac{R_e - N_{er}.\mathbf{X}_e}{R_r - N_{er}.\mathbf{X}_r}$$

$$+ \frac{G_N M}{c^3}\left(\bar{s}^{TJ}P_{er}^J - \bar{s}^{JK}N_{er}^J P_{er}^K\right)\frac{R_e - R_r}{R_e R_r} + \frac{G_N M}{c^3}\left[\bar{s}^{TJ}N_{er}^J + \bar{s}^{JK}\hat{P}_{er}^J \hat{P}_{er}^K - \bar{s}^{TT}\right](N_r.N_{er} - N_e.N_{er})$$

(9)

where the terms a_1 and a_2 from [80] are taken as unity (which corresponds to using the harmonic gauge, which is the one used for VLBI data reduction), $R_e = |\mathbf{X}_e|$, $R_r = |\mathbf{X}_r|$, $R_{er} = |\mathbf{X}_r - \mathbf{X}_e|$ with the central body located at the origin and where we introduce the following vectors

$$\mathbf{K} = \frac{\mathbf{X}_e}{R_e}, \quad \mathbf{N}_{ij} \equiv \frac{\mathbf{X}_{ij}}{R_{ij}} = \frac{\mathbf{X}_j - \mathbf{X}_i}{|\mathbf{X}_{ij}|}, \quad \mathbf{N}_i = \frac{\mathbf{X}_i}{|\mathbf{X}_i|}, \quad \mathbf{P}_{er} = \mathbf{N}_{er} \times (\mathbf{X}_r \times \mathbf{N}_{er}), \quad \text{and} \quad \hat{\mathbf{P}}_{er} = \frac{\mathbf{P}_{er}}{|\mathbf{P}_{er}|}, \quad (10)$$

and where G_N is the observed Newton constant measured by considering the orbital motion of bodies and is defined in Equation (5). This equation is the generalization of the well-known Shapiro time delay including Lorentz violation. The VLBI is actually measuring the difference of the time of arrival of a signal received by two different stations. This observable is therefore sensitive to a differential time delay (see [108] for a calculation in GR). Assuming a radio-signal emitted by a quasar at event $(T_e, \mathbf{X}_e)$ and received by two different VLBI stations at events $(T_1, \mathbf{X}_1)$ and $(T_2, \mathbf{X}_2)$ (all quantities being expressed in a barycentric reference frame), respectively, the VLBI group-delay $\Delta\tau_{(\text{SME})}$ in SME formalism can be written [55]

$$\Delta\tau_{(\text{SME})} = 2\frac{G_N M}{c^3}\left(1 - \frac{2}{3}\bar{s}^{TT}\right)\ln\frac{R_1 + \mathbf{K}.\mathbf{X}_1}{R_2 + \mathbf{K}.\mathbf{X}_2} + \frac{2}{3}\frac{G_N M}{c^3}\bar{s}^{TT}(\mathbf{N}_2.\mathbf{K} - \mathbf{N}_1.\mathbf{K}), \quad (11)$$

where we only kept the $\bar{s}^{TT}$ contribution (see Equation (7) from [55] for the full expression) and we use the same notations as in [108] by introducing three unit vectors

$$\mathbf{K} = \frac{\mathbf{X}_e}{|\mathbf{X}_e|}, \quad \mathbf{N}_1 = \frac{\mathbf{X}_1}{|\mathbf{X}_1|}, \quad \text{and} \quad \mathbf{N}_2 = \frac{\mathbf{X}_2}{|\mathbf{X}_2|}. \quad (12)$$

Ten million VLBI delay observations between August 1979 and mid-2015 have been used to estimate the $\bar{s}^{TT}$ coefficient. First, VLBI observations are corrected from delay due to the radio wave crossing of dispersive media by using 2 GHz and 8 GHz recordings. Then, we used only the 8 GHz delays and the Calc/Solve geodetic VLBI analysis software, developed at NASA Goddard Space Flight Center and coherent with the latest standards of the International Earth Rotation and Reference Systems Service [109]. We added the partial derivative of the VLBI delay with respect to $\bar{s}^{TT}$ from Equation (11) to the software package using the USERPART module of Calc/Solve. We turned to a global solution in which we estimated $\bar{s}^{TT}$ as a global parameter together with radio source coordinates. We obtained

$$\bar{s}^{TT} = (-5 \pm 8) \times 10^{-5}, \tag{13}$$

with a postfit root mean square of 28 picoseconds and a χ^2 per degree of freedom of 1.15. Correlations between radio source coordinates and $\bar{s}^{TT}$ are lower than 0.02, the global estimate being consistent with the mean value obtained with the session-wise solution with a slightly lower error.

In conclusion, VLBI is an incredible tool to test Lorentz symmetry, especially the $\bar{s}^{TT}$ coefficient. This coefficient has an isotropic impact on the propagation speed of gravitational waves as can be noticed from Equation (27) below (or see Equation (9) from [56] or Equation (11) from [60]). The analysis performed in [55] includes the SME contribution in the modeling of VLBI observations and includes the $\bar{s}^{TT}$ parameter in the global fit with other parameters. It is therefore a robust analysis that produces the current best estimate on the $\bar{s}^{TT}$ parameter. In the future, the accumulation of VLBI data in the framework of the permanent geodetic monitoring program leads us expect improvement of this constraint.

4.3. Lunar Laser Ranging

On 20 August 1969, after ranging to the lunar retro-reflector placed during the Apollo 11 mission, the first LLR echo was detected at the McDonald Observatory in Texas. Currently, there are five stations spread over the world which have realized laser shots on five lunar retro-reflectors. Among these stations four are still operating: Mc Donald Observatory in Texas, Observatoire de la Côte d'Azur in France, Apache point Observatory in New Mexico and Matera in Italy while one on Maui, Hawaii has stopped lunar ranging since 1990. Concerning the lunar retro-reflectors three are located at sites of the Apollo missions 11, 14 and 15 and two are French-built array operating on the Soviet roving vehicle Lunakhod 1 and 2.

LLR is used to conduct high precision measurements of the light travel time of short laser pulses emitted at time t_1 by a LLR station, reflected at time t_2 by a lunar retro-reflector and finally received at time t_3 at a station receiver. The data are presented as normal points which combine time series of measured light travel time of photons, averaged over several minutes to achieve a higher signal-to-noise ratio measurement of the lunar range at some characteristic epoch. Each normal-point is characterized by one emission time (t_1 in universal time coordinate—UTC), one time delay (Δt_c in international atomic time—TAI) and some additional observational parameters as laser wavelength, atmospheric temperature and pressure *etc*. According to [110], the theoretical pendent of the observed time delay ($\Delta t_c = t_3 - t_1$ in TAI) is defined as

$$\Delta t_c = \left[T_3 - \Delta \tau_t(T_3) \right] - \left[T_1 - \Delta \tau_t(T_1) \right], \tag{14}$$

where T_1 is the emission time expressed in barycentric dynamical time (TDB) and $\Delta \tau_t$ is a relativistic correction between the TDB and the terrestrial time (TT) at the level of the station. The reception time T_3 expressed in TDB is defined by the following two relations

$$T_3 = T_2 + \frac{1}{c} \|\boldsymbol{X}_{o'}(T_3) - \boldsymbol{X}_r(T_2)\| + \Delta \mathcal{T}_{\text{(grav)}} + \Delta \tau_a, \tag{15a}$$

$$T_2 = T_1 + \frac{1}{c} \|\boldsymbol{X}_r(T_2) - \boldsymbol{X}_o(T_1)\| + \Delta \mathcal{T}_{\text{(grav)}} + \Delta \tau_a, \tag{15b}$$

with T_2 the time in TDB at the reflection point $\boldsymbol{X}_o$ and $\boldsymbol{X}_{o'}$ are respectively the barycentric position vector at the emitter and the reception point, $\boldsymbol{X}_r$ is the barycentric position vector at the reflection point, $\Delta \mathcal{T}_{\text{(grav)}}$ is the one way gravitational time delay correction and $\Delta \tau_a$ is the one way tropospheric correction.

LLR measurements are used to produce the Lunar ephemeris but also provide a unique opportunity to study the Moon's rotation, the Moon's tidal acceleration, the lunar rotational dissipation, etc. [111]. In addition, LLR measurements have turn the Earth-Moon system into a laboratory to study fundamental physics and to conduct tests of the gravitation theory. Nordtvedt was the first to suggest that LLR can be used to test GR by testing one of its pillar: the Strong Equivalence Principle [112–114]. He showed that precise laser ranging to the Moon would be capable of measuring precisely the ratio of gravitational mass to inertial mass of the Earth to an accuracy sufficient to constrain a hypothetical dependence of this ratio on the gravitational self-energy. He concluded that such a measurement could be used to test Einstein's theory of gravity and others alternative theories as scalar tensor theories. The best current test of the Strong Equivalence Principle is provided by a combination of torsion balance measurements with LLR analysis and is given by [115–117]

$$\eta = (4.4 \pm 4.5) \times 10^{-4}, \tag{16}$$

where η is the Nordtvedt parameter that is defined as $m_G/m_I = 1 + \eta U/mc^2$ with m_G the gravitational mass, m_I the inertial mass and U the gravitational self-energy of the body. Using the Cassini constraint on the γ PPN parameter [118] and the relation $\eta = 4\beta - \gamma - 3$ leads to a constraint on β PPN parameter at the level $\beta - 1 = (1.2 \pm 1.1) \times 10^{-4}$ [116].

In addition to tests of the Strong Equivalence Principle, many other tests of fundamental physics were performed with LLR analysis. For instance, LLR data can be used to search for a temporal evolution of the gravitational constant $\dot{G}/G$ [115] and to constrain the fifth force parameters [119]. In addition, LLR has been used to constrain violation of the Lorentz symmetry in the PPN framework. Müller et al. [119] deduced from LLR data analysis constraints on the preferred frame parameters α_1 and α_2 at the level $\alpha_1 = (-7 \pm 9) \times 10^{-5}$ and $\alpha_2 = (1.8 \pm 2.5) \times 10^{-5}$.

Considering all the successful GR tests performed with LLR observations, it is quite natural to use them to search for Lorentz violations in the gravitation sector. In the SME framework, Battat et al. [45] used the lunar orbit to provide estimates on the SME coefficients. Using a perturbative approach, the main signatures produced by SME on the lunar orbit have analytically been computed in [39]. These computations give a first idea of the amplitude of the signatures produced by a breaking of Lorentz symmetry. Nevertheless, these analytical signatures have been computed assuming the lunar orbit to be circular and fixed (i.e., neglecting the precession of the nodes for example). These analytical signatures have been fitted to LLR residuals obtained from a data reduction performed in pure GR [45]. They determined a *"realistic"* error on their estimates from a similar postfit analysis performed in the PPN framework. The results obtained by this analysis are presented in Table 2. It is important to note that this analysis uses projections of the SME coefficients into the lunar orbital plane $\bar{s}^{11}, \bar{s}^{22}, \bar{s}^{0i}$ (see Section V.B.2 of [39]) while the standard SME analyses uses coefficients defined in a Sun-centered equatorial frame (and denoted by capital letter $\bar{s}^{IJ}$).

Table 2. Estimation of Standard-Model Extension (SME) coefficients from Lunar Laser Ranging (LLR) postfit data analysis from [45]. No correlations coefficients have been derived in this analysis. The coefficients $\bar{s}^{ij}$ are projections of the $\bar{s}^{IJ}$ into the lunar orbital plane (see Equation (107) from [39]) while the linear combinations $\bar{s}_{\Omega_\oplus c}$ and $\bar{s}_{\Omega_\oplus s}$ are given by Equation (108) from [39].

	Coefficient
$\bar{s}^{11} - \bar{s}^{22}$	$(1.3 \pm 0.9) \times 10^{-10}$
$\bar{s}^{12}$	$(6.9 \pm 4.5) \times 10^{-11}$
$\bar{s}^{01}$	$(-0.8 \pm 1.1) \times 10^{-6}$
$\bar{s}^{02}$	$(-5.2 \pm 4.8) \times 10^{-7}$
$\bar{s}_{\Omega_\oplus c}$	$(0.2 \pm 3.9) \times 10^{-7}$
$\bar{s}_{\Omega_\oplus s}$	$(-1.3 \pm 4.1) \times 10^{-7}$

However, as discussed in Section 3 and in [46,55], a postfit search for SME signatures into residuals of a data reduction previously performed in pure GR is not fully satisfactory. First of all, the uncertainties obtained by a postfit analysis based on a GR data reduction can be underestimated by up to two orders of magnitude. This is mainly due to correlations between SME coefficients and others global parameters (masses, positions and velocities, ...) that are neglected in this kind of approach. In addition, in the case of LLR data analysis, the oscillating signatures derived in [39] and used in [45] to determine pseudo-constraints are computed only accounting for short periodic oscillations, typically at the order of magnitude of the mean motion of the Moon around the Earth. Therefore, this analytic solution remains only valid for few years while LLR data spans over 45 years (see also the discussions in footnote 2 from [50] and page 22 from [39]).

Regarding LLR data analysis, a more robust strategy consists in including the SME modeling in the complete data analysis and to estimate the SME coefficients in a global fit along with others parameters by taking into account short and long period terms and also correlations (see [46]). In order to perform such an analysis, a new numerical lunar ephemeris named "Éphéméride Lunaire Parisienne Numérique" (ELPN) has been developed within the SME framework. The dynamical model of ELPN is similar to the DE430 one [120] but includes the Lorentz symmetry breaking effects arising on the orbital motion of the Moon. The SME contribution to the lunar equation of motion has been derived in [39] and is given by

$$
\begin{aligned}
a_{\mathrm{SME}}^{J} &= \frac{G_N M}{r^3}\left[\bar{s}_t^{JK} r^K - \frac{3}{2}\bar{s}_t^{KL}\hat{r}^K\hat{r}^L r^J + 2\frac{\delta m}{M}\left(\bar{s}^{TK}\hat{v}^K r^J - \bar{s}^{TJ}\hat{v}^K r^K\right)\right. \\
&\left. + 3\bar{s}^{TK}\hat{V}^K r^J - \bar{s}^{TJ}\hat{V}^K r^K - \bar{s}^{TK}\hat{V}^J r^K + 3\bar{s}^{TL}\hat{V}^K\hat{r}^K\hat{r}^L r^J\right],
\end{aligned}
\tag{17}
$$

where G_N is the observed Newtonian constant defined by Equation (5), M is the mass of the Earth-Moon barycenter, δm is the difference between the Earth and the lunar masses; $\hat{r}^J$ being the unit position vector of the Moon with respect to the Earth; $\hat{v}^J = v^J/c$ with v^J being the relative velocity vector of the Moon with respect to the Earth; $\hat{V}^J = V^J/c$ with V^J being the Heliocentric velocity vector of the Earth-Moon barycenter and the 3-dimensional traceless tensor defined by Equation (6). These equations of motion as well as their partial derivatives are integrated numerically in ELPN.

In addition to the orbital motion, effects of a violation of Lorentz symmetry on the light travel time of photons is also considered. More precisely, the gravitational time delay $\Delta \mathcal{T}_{(\mathrm{grav})}$ appearing in Equation (14) is given by the gravitational part of Equation (9) [80].

Estimates on the SME coefficients are obtained by a standard chi-squared minimization: the LLR residuals are minimized by an iterative weighted least squares fit using partial derivatives previously computed from variational equations in ELPN. After an adjustment of 82 parameters including the SME coefficients a careful analysis of the covariance matrix shows that LLR data does not allow to estimate independently all the SME coefficients but that they are sensitive to the following three linear combinations:

$$
\bar{s}^{XX} - \bar{s}^{YY}, \qquad \bar{s}^{TY} + 0.43\bar{s}^{TZ}, \qquad \bar{s}^{XX} + \bar{s}^{YY} - 2\bar{s}^{ZZ} - 4.5\bar{s}^{YZ}.
\tag{18}
$$

The estimations on the 6 SME coefficients derived in [46] is summarized in Table 3. In particular, it is worth emphasizing that the quoted uncertainties are the sum of the statistical uncertainties obtained from the least-square fit with estimations of systematics uncertainties obtained with a Jackknife resampling method [121,122].

Table 3. Estimation of SME coefficients from a full LLR data analysis from [46] and associated correlation coefficients.

Coefficient	Estimates
$\bar{s}^{TX}$	$(-0.9 \pm 1.0) \times 10^{-8}$
$\bar{s}^{XY}$	$(-5.7 \pm 7.7) \times 10^{-12}$
$\bar{s}^{XZ}$	$(-2.2 \pm 5.9) \times 10^{-12}$
$\bar{s}^{XX} - \bar{s}^{YY}$	$(0.6 \pm 4.2) \times 10^{-11}$
$\bar{s}^{TY} + 0.43\,\bar{s}^{TZ}$	$(6.2 \pm 7.9) \times 10^{-9}$
$\bar{s}^{XX} + \bar{s}^{YY} - 2\bar{s}^{ZZ} - 4.5\,\bar{s}^{YZ}$	$(2.3 \pm 4.5) \times 10^{-11}$

Correlation Coefficients					
1					
−0.06	1				
−0.04	0.29	1			
0.58	−0.12	−0.16	1		
0.16	−0.01	−0.09	0.25	1	
0.07	−0.10	−0.13	−0.10	0.03	1

In summary, LLR is a powerful experiment to constrain gravitation theory and in particular hypothetical violation of the Lorentz symmetry. A first analysis based on a postfit estimations of the SME coefficients have been performed [45] which is not satisfactory regarding the neglected correlations with other global parameters as explained in Section 3. A full analysis including the integration of the SME equations of motion and the SME contribution to the gravitational time delay has been done in [46]. The resulting estimates on some SME coefficients are presented in Table 3. In addition, some SME coefficients are still correlated with parameters appearing in the rotational motion of the Moon as the principal moment of inertia, the quadrupole moment, the potential Stokes coefficient C_{22} and the polar component of the velocity vector of the fluid core [46]. A very interesting improvement regarding this analysis would be to produce a joint GRAIL (Gravity Recovery And Interior Laboratory) [123–125] and LLR data analysis that would help in decorrelating the SME parameters from the lunar potential Stokes coefficients of degree 2 and therefore improve marginalized estimations of the SME coefficients. Finally, in [45,46], the effects of SME on the translational lunar equations of motion are considered and used to derive constraints on the SME coefficients. It would be also interesting to extend these analyses by considering the modifications due to SME on the rotation of the Moon. A first attempt has been proposed in Section V. A. 2. of [39] but needs to be extended.

4.4. Planetary Ephemerides

The analysis of the motion of the planet Mercury around the Sun was historically the first evidence in favor of GR with the explanation of the famous advance of the perihelion in 1915. From there, planetary ephemerides have always been a very powerful tool to constrain GR and alternative theories of gravitation. Currently, three groups in the world are producing planetary ephemerides: the NASA Jet Propulsion Laboratory with the DE ephemerides [120,126–131], the French INPOP (Intégrateur Numérique Planétaire de l'Observatoire de Paris) ephemerides [132–137] and the Russian EPM ephemerides [138–142]. These analyses use an impressive number of different observations to produce high accurate planetary and asteroid trajectories. The observations used to produce ephemerides comprise radioscience observations of spacecraft that orbited around Mercury, Venus, Mars and Saturn, flyby tracking of spacecraft close to Mercury, Jupiter, Uranus and Neptune and optical observations of all planets. This huge set of observations have been used to constrain the γ and β post-Newtonian parameter at the level of 10^{-5} [136,137,141–143], the fifth force interaction (see [32] and Figure 31 from [143]), the quantity of Dark Matter in our Solar System [144], the Modified Newtonian Dynamics [131,145–147], . . .

A violation of Lorentz symmetry within the gravity sector of SME induces different types of effects that can have implications on planetary ephemerides analysis: effects on the orbital dynamics and effects on the light propagation. Simulations using the Time Transfer Formalism [104,106,107] based on the software presented in [148] have shown that only the $\bar{s}^{TT}$ coefficients produce a non-negligible effect on the light propagation (while it has impact only at the next post-Newtonian level on the orbital dynamics [29,39]). On the other hand, the other coefficients produce non-negligible effects on the orbital dynamics [39] and can therefore be constrained using planetary ephemerides data. In the linearized gravity limit, the contribution from SME to the 2-body equations of motion within the gravitational sector of SME are given by the first line of Equation (17) (i.e., for a vanishing V^k). The coefficient $\bar{s}^{TT}$ is completely unobservable in this context since absorbed in a rescaling of the gravitational constant (see the discussion in [39,52]).

Ideally, in order to perform a solid estimation of the SME coefficients using planetary ephemerides, one should include the full SME equations in the integration of the planets motion and fit them simultaneously with the other estimated parameters (positions and velocities of planets, J_2 of the Sun, ...). This solid analysis within the SME formalism has not been performed so far.

As a first step, a postfit analysis has been performed [49,50]. The idea of this analysis is to derive the analytical expression for the secular evolution of the orbital elements produced by the SME contribution to the equations of motion. Using the Gauss equations, secular perturbations induced by SME on the orbital elements have been computed in [39] (see also [49] for a similar calculations done for the $\bar{s}^{TJ}$ coefficients only). In particular, the secular evolution of the longitude of the ascending node Ω and the argument of the perihelion ω is given by

$$\left\langle \frac{d\Omega}{dt} \right\rangle = \frac{n}{\sin i (1 - e^2)^{1/2}} \left[\frac{\varepsilon}{e^2} \bar{s}_{kP} \sin \omega + \frac{(e^2 - \varepsilon)}{e^2} \bar{s}_{kQ} \cos \omega - \frac{\delta m}{M} \frac{2na\varepsilon}{ec} \bar{s}^k \cos \omega \right], \qquad (19a)$$

$$\left\langle \frac{d\omega}{dt} \right\rangle = -\cos i \left\langle \frac{d\Omega}{dt} \right\rangle - n \left[-\frac{\varepsilon^2}{2e^4} (\bar{s}_{PP} - \bar{s}_{QQ}) + \frac{\delta m}{M} \frac{2na(e^2 - \varepsilon)}{ce^3 (1 - e^2)^{1/2}} \bar{s}^Q \right], \qquad (19b)$$

where a is the semimajor axis, e the eccentricity, i the orbit inclination (with respect to the ecliptic), $n = (G_N m_\odot / a^3)^{1/2}$ is the mean motion, $\varepsilon = 1 - (1 - e^2)^{1/2}$, δm the difference between the two masses and M their sum (in the cases of planets orbiting the Sun, one has $M \approx \delta m$). In all these expressions, the coefficients for Lorentz violation with subscripts P, Q, and k are understood to be appropriate projections of $\bar{s}^{\mu\nu}$ along the unit vectors P, Q, and k, respectively. For example, $\bar{s}^k = k^i \bar{s}^{Ti}$, $\bar{s}_{PP} = P^i P^j \bar{s}^{ij}$. The unit vectors P, Q and k define the orbital plane (see [39] or Equation (8) from [50]).

Instead of including the SME equations of motion in planetary ephemerides, the postfit analysis uses estimations of supplementary advances of perihelia and nodes derived from ephemerides analysis [135,140,144] to fit the SME coefficients through Equation (19). In [50], estimations of supplementary advances of perihelia and longitude of nodes from INPOP (see Table 5 from [135]) are used to fit a posteriori the SME coefficients. This analysis suffers from large correlations due to the fact that the planetary orbits are very similar to each other: nearly eccentric orbit and very low inclination orbital planes. In order to deal properly with these correlations a Bayesian Monte Carlo inference has been used [50]. The posterior probability distribution function can be found on Figure 1 from [50]. The intervals corresponding to the 68% Bayesian confidence levels are given in Table 4 as well as the correlation matrix. It is interesting to mention that a decomposition of the normal matrix in eigenvectors allows one to find linear combinations of SME coefficients that are uncorrelated with the planetary ephemerides analysis (see Equation (15) and Table IV from [50]).

Table 4. Estimations of the SME coefficients from a postfit data analysis based on planetary ephemerides from [50]. The uncertainties correspond to the 68% Bayesian confidence levels of the marginal posterior probability distribution function. The associated correlation coefficients can be found in Table III from [50].

Coefficient	
$\bar{s}^{XX} - \bar{s}^{YY}$	$(-0.8 \pm 2.0) \times 10^{-10}$
$\bar{s}^{XX} + \bar{s}^{YY} - 2\,\bar{s}^{ZZ}$	$(-0.8 \pm 2.7) \times 10^{-10}$
$\bar{s}^{XY}$	$(-0.3 \pm 1.1) \times 10^{-10}$
$\bar{s}^{XZ}$	$(-1.0 \pm 3.5) \times 10^{-11}$
$\bar{s}^{YZ}$	$(5.5 \pm 5.2) \times 10^{-12}$
$\bar{s}^{TX}$	$(-2.9 \pm 8.3) \times 10^{-9}$
$\bar{s}^{TY}$	$(0.3 \pm 1.4) \times 10^{-8}$
$\bar{s}^{TZ}$	$(-0.2 \pm 5.0) \times 10^{-8}$

Correlation coefficients							
1							
0.99	1						
0.99	0.99	1					
0.98	0.98	0.99	1				
−0.32	−0.24	−0.26	−0.26	1			
0.99	0.98	0.98	0.98	−0.32	1		
0.62	0.67	0.62	0.59	0.36	0.60	1	
−0.83	−0.86	−0.83	−0.81	−0.14	−0.82	−0.95	1

In summary, planetary ephemerides offer a great opportunity to constrain hypothetical violations of Lorentz symmetry. So far, only postfit estimations of the SME coefficients have been performed [49,50]. In this analysis, estimations of secular advances of perihelia and longitude of nodes obtained with the INPOP planetary ephemerides [135] are used to fit a posteriori the SME coefficients using the Equations (19). The 68% marginalized confidence intervals are given in Table 4. This analysis suffers highly from correlations due to the fact that the planetary orbits are very similar. A very interesting improvement regarding this analysis would be to perform a full analysis by integrating the planetary equations of motion directly within the SME framework and by fitting the SME coefficients simultaneously with the other parameters fitted during the ephemerides data reduction.

4.5. Gravity Probe B

In GR, a gyroscope in orbit around a central body undergoes two relativistic precessions with respect to a distant inertial frame: (i) a geodetic drift in the orbital plane due to the motion of the gyroscope in the curved spacetime [149]; and (ii) a frame-dragging due to the spin of the central body [150]. In GR, the spin of a gyroscope is parallel transported, which at the post-Newtonian approximation gives the relativistic drift

$$R = \frac{d\hat{S}}{dt} = \Omega_{GR} \times S, \tag{20a}$$

$$\Omega_{GR} = \frac{3GM}{2c^2 r^3} r \times v + \frac{3\hat{r}(\hat{r}.J) - J}{c^2 r^3}, \tag{20b}$$

where $\hat{S}$ is the unit vector pointing in the direction of the spin S of the gyroscope, r and v are the position and velocity of the gyroscope, $\hat{r} = r/r$ and J is the angular momentum of the central body. In 1960, it has been suggested to use these two effects to perform a new test of GR [151,152]. In April 2004, GPB, a satellite carrying 4 cryogenic gyroscopes was launched in order to measure these two precessions. GPB was orbiting Earth on a polar orbit such that the two relativistic drifts are orthogonal to each other [153]: the geodetic effect is directed along the NS direction (North-South, i.e., parallel to the satellite motion) while the frame-dragging effect is directed on the WE direction (West-East, see [52,153] for further details about the axes conventions in the GPB data reduction).

A year of data gives the following measurements of the relativistic drift: (i) the geodetic drift $R_{NS} = -6601.8 \pm 18.3$ mas/yr (milliarcsecond per year) to be compared to the GR prediction of -6606.1 mas/yr; and (ii) the frame-dragging drift $R_{WE} = -37.2 \pm 7.2$ mas/yr to be compared with the GR prediction of -39.2 mas/yr. In other word, the GPB results can be written as a measurement of a deviation from GR given by

$$\Delta R_{NS} = 4.3 \pm 18.3 \text{ mas/yr} \qquad \text{and} \qquad \Delta R_{WE} = 2 \pm 7.2 \text{ mas/yr}. \tag{21}$$

Within the SME framework, if one considers only the $\bar{s}^{\mu\nu}$ coefficients, the equation of parallel transport in term of the spacetime metric is not modified (see Equation (143) from [39]). Nevertheless, the expression of the spacetime metric is modified leading to a modification of the relativistic drift given by Equation (150) from [39]. In order to focus only on the dominant secular part of the evolution of the spin orientation, the relativistic drift equation has been averaged over a period. The SME contribution to the precession can be written as [39]

$$\Delta \Omega^J = \frac{G_N M}{r^2} v \left[\left(-\frac{4}{3} \bar{s}^{TT} - \frac{9}{8} \tilde{i}_{(-5/3)} \bar{s}_t^{JK} \hat{\sigma}^J \hat{\sigma}^K \right) \hat{\sigma}^J + \frac{5}{4} \tilde{i}_{(-3/5)} \bar{s}_t^{JK} \hat{\sigma}^K \right], \tag{22}$$

where G_N is the effective gravitational constant defined by Equation (5), the coefficients $\tilde{i}$ are defined by $\tilde{i}_{(\beta)} = 1 + \beta I_\oplus/(M_\oplus r^2)$, $\hat{\sigma}^J$ is a unit vector normal to the gyroscope orbital plane, r and v are the norm of the position and velocity of the gyroscope and $\bar{s}_t^{JK}$ is the traceless part of $\bar{s}^{JK}$ as defined by Equation (6). Using the geometry of GPB into the last equation and using Equation (20a), one finds that the gyroscope anomalous drift is given by

$$\Delta R_{NS} = 5872 \bar{s}^{TT} + 794 \left(\bar{s}^{XX} - \bar{s}^{YY} \right) - 317 \left(\bar{s}^{XX} + \bar{s}^{YY} - 2 \bar{s}^{ZZ} \right) - 1050 \bar{s}^{XY}, \tag{23a}$$

$$\Delta R_{WE} = -368 (\bar{s}^{XX} - \bar{s}^{YY}) - 1112 \bar{s}^{XY} + 1269 \bar{s}^{XZ} + 4219 \bar{s}^{YZ}, \tag{23b}$$

where the units are mas/yr. These are the SME modifications to the relativistic drift arising from the modification of the equations of evolution of the gyroscope axis (i.e., modification of the parallel transport equation due to the modification of the underlying spacetime metric).

In addition to modifying the evolution of the spin axis, a breaking of Lorentz symmetry will impact the orbital motion of the gyroscope. As a result, the position and velocity of the gyroscope will depend on the SME coefficients and therefore also impact the evolution of the spin axis through the GR contribution given by Equation (20b). The best way to deal with this effect is to use the GPB tracking measurements (GPS) in order to constrain the gyroscope orbital motion and eventually constrain the SME coefficients through the equations of motion. In [52], these tracking observations are not used and only the gyroscope drift is used in order to constrain the SME contributions coming from both the modification of the parallel transport and from the modification of GPB orbital motion. In order to do this, the contribution of SME on the evolution of the orbital elements given by Equations (19) and (26) are used, averaged over a period and in the low eccentricity approximation. This secular evolution for the osculating elements is introduced in the relativistic drift equation for the gyroscope from Equation (20b) and averaged over the measurement time using Equation (20a). Using the GPB geometry, this contribution to the relativistic drift is given by

$$\Delta R'_{NS} = 5.7 \times 10^6 (\bar{s}^{XX} - \bar{s}^{YY}) + 1.7 \times 10^7 \bar{s}^{XY} - 1.9 \times 10^7 \bar{s}^{XZ} - 6.6 \times 10^7 \bar{s}^{YZ}, \tag{24a}$$

$$\Delta R'_{WE} = -1.89 \times 10^7 (\bar{s}^{XX} - \bar{s}^{YY}) - 5.71 \times 10^7 \bar{s}^{XY} - 5.96 \times 10^6 \bar{s}^{XZ} - 1.98 \times 10^7 \bar{s}^{YZ}, \tag{24b}$$

with units of mas/yr.

The sum of the two SME contributions to the gyroscope relativistic drift given by Equations (23) and (24) can be compared to the GPB estimations given by Equation (21). The result is given in Table 5. The main advantage of GPB comes from the fact that it is sensitive to the $\bar{s}^{TT}$ coefficient. The constraint on this coefficient is at the level of 10^{-3}, a little bit less good than the one obtained with VLBI or with binary pulsars but relying on a totally different type of observations. The constraints on the spatial part of the SME coefficients ($\bar{s}^{IJ}$) are at the level of 10^{-7} and are superseded by the other measurements. The constraints on these coefficients come mainly from the contribution arising from the orbital dynamics of GPB and not from a direct modification of the spin evolution. Constraining the orbital motion from GPB by using the gyroscope observations only is not optimal and tracking observations may help to improve the corresponding constraints (in this case, a dedicated satellite may be more appropriate as discussed in Section 5.3).

Table 5. Estimations of the SME coefficients from a postfit data analysis based on Gravity Probe B (GPB) [52].

Coefficient		
$\bar{s}^{(1)}_{\text{GPB}}$	$= \bar{s}^{TT} + 970\left(\bar{s}^{XX} - \bar{s}^{YY}\right) - 0.05\left(\bar{s}^{XX} + \bar{s}^{YY} - 2\bar{s}^{ZZ}\right)$ $+2895\,\bar{s}^{XY} - 3235\,\bar{s}^{XZ} - 11\,240\,\bar{s}^{YZ}$	$(0.7 \pm 3.1) \times 10^{-3}$
$\bar{s}^{(2)}_{\text{GPB}}$	$= \bar{s}^{XX} - \bar{s}^{YY} + 3.02\,\bar{s}^{XY} + 0.32\,\bar{s}^{XZ} + 1.05\,\bar{s}^{YZ}$	$(-1.1 \pm 3.8) \times 10^{-7}$

In summary, the GPB measurement of a gyroscope relativistic drifts due to geodetic precession or frame-dragging can be used to search for a breaking of Lorentz symmetry. The main advantage of this technique comes from its sensitivity to $\bar{s}^{TT}$. As already mentioned, this coefficient has an isotropic impact on the propagation velocity of gravitational waves as can be noticed from Equation (27) below (see also Equation (9) from [56] or Equation (11) from [60]). A preliminary result based on a post-fit analysis performed after a GR data reduction of GPB measurements gives a constraint on $\bar{s}^{TT}$ at the level of 10^{-3} [52]. This should be investigated further since the Earth's quadrupole moment has been neglected and Lorentz-violating effects on the aberration terms can also change slightly the results. In addition, impacts from Lorentz violations on frame-dragging arising in other contexts such as satellite laser ranging (see Section 5.3) or signals from accretion disks around collapsed stars [154] would also be interesting to consider.

4.6. Binary Pulsars

The discovery of the first binary pulsars PSR 1913+16 by Hulse and Taylor in 1975 [155] has opened a new window to test the theory of gravitation. Observations of this pulsar have allowed one to measure the relativistic advance of the periastron [156] and more importantly to measure the rate of orbital decay due to gravitational radiation [157]. Pulsars are rotating neutron stars that are emitting very strong radiation. The periods of pulsars are very stable which allows us to consider them as "clocks" that are moving in an external gravitational field (typically in the gravitational field generated by a companion). The measurements of the pulse time of arrivals can be used to infer several parameters by fitting an appropriate timing model (see for example Section 6.1 from [2]): (i) non-orbital parameter such as the pulsar period and its rate of change; (ii) five Keplerian parameters; and (iii) some post-Keplerian parameters [158]. In GR, the expressions of these post-Keplerian parameters are related to the masses of the two bodies and to the Keplerian parameters. If more than 2 of these post-Keplerian parameters can be determined, they can be used to test GR [159]. Nowadays, more than 70 binary pulsars have been observed [160]. A description of the most interesting binary pulsars in order to test the gravitation theory can be found in Section 6.2 from [2] or in the supplemental material from [53].

The model fitted to the observations is based on a post-Newtonian analytical solution to the 2 body equations of motion [161] (see also [162]) and includes contribution from the Einstein time delay (i.e., the transformation between proper and coordinate time), the Shapiro time delay, the Roemer time delay [158]. The model also corrects for several systematics like atmospheric delay, Solar system dispersion, interstellar dispersion, motion of the Earth and the Solar System, . . . (see for example [163]).

Pulsars observations provide some of the best current constraints on alternative theories of gravitation (for a review, see [164,165]). In addition to the Hulse and Taylor pulsar, the double pulsar [166] now provides the best measurement of the pulsar orbital rate of change [165]. In addition, the post-Keplerian modeling has been fully derived in tensor-scalar theories [167–169] such that pulsars observations have provided some of the best constraints on this class of theory [165,170,171]. It is important to mention that non perturbative strong field effects may arise in binary pulsars system and needs to be taken into account [169,172].

In addition, binary pulsars have also been successfully used to test Lorentz symmetry. For example, analyses of the pulses time of arrivals provide a constraint on the $\alpha_{1,2,3}$ PPN parameters. Since non perturbative strong field effects may arise in binary pulsars system (see for example [173] for strong field effects in Einstein-Aether theory), the obtained constraints are interpreted as strong field version of the PPN parameters denoted by $\hat{\alpha}_i$. Estimates of these parameters should be compared carefully to the standard weak field constraints since they may depend on the gravitational binding energy of the neutron star. The best current constraint on $\hat{\alpha}_1 = -0.4^{+3.7}_{-3.1} \times 10^{-5}$ is obtained by considering the orbital dynamics of the binary pulsars PSR J1738+0333 [174,175]. The best current constraint on $\hat{\alpha}_2$ takes advantage from the fact that this parameter produces a precession of the spin axis of massive bodies [176]. The combination of observations of two solitary pulsars lead to the best current constraints on $|\hat{\alpha}_2| < 1.6 \times 10^{-9}$ [177]. Finally, the parameter $\hat{\alpha}_3$ produces a violation of the momentum conservation in addition to a violation of the Lorentz symmetry. This parameter will induce a self-acceleration for rotating body that can be constrained using binary pulsars [178]. The best current constraint uses a set of 5 pulsars (4 binary pulsars and one solitary pulsar) and is given by $\hat{\alpha}_3 < 5.5 \times 10^{-20}$ [179].

Furthermore, specific Lorentz violating theories have also been constrained with binary pulsars. In [72,73], binary pulsars observations are used to constrain Einstein-Aether and khronometric theory. In these theories, the low-energy limit Lorentz violations can be parametrized by four parameters: the α_1 and α_2 PPN parameters and two other parameters. It has been shown [72,73,173] that the orbital period decay depends on these four parameters. Assuming the solar system constraints on α_1 and α_2 [2], measurements of the rate of change of the orbital period of binary pulsars have been used to constrain the two other parameters (see for example Figure 2 from [72]). In this work, strong field effects have been taken into account by solving numerically the field equations in order to determine the neutron stars sensitivity [73].

Finally, binary pulsars have been used in order to derive constraints on the SME coefficients. As in the PPN formalism, constraints obtained from binary pulsars need to be considered as constraints on strong-field version of the SME coefficients that may include non perturbative effects. Two different types of effects have been used to determine estimates on the SME coefficients: (i) tests using the spin precession of solitary pulsars and (ii) tests using effects on the orbital dynamics of binary pulsars [53]. The SME contribution to the precession rate of an isolated spinning body has been derived in [39] and is given by

$$\Omega^k_{\text{SME}} = \frac{\pi}{P} \bar{s}^{kj} \hat{S}^j , \qquad (25)$$

where P is the spin period and $\hat{S}^j$ is the unit vector pointing along the spin direction. The effects from the pulsar spin precession on the pulse width can be found in [177,180]. Two solitary pulsars have been used to constrain the SME coefficients with this effect. The second type of tests come from the orbital dynamics of binary pulsars. As mentioned in Sections 4.3 and 4.4, the SME will modify the two-body equations of motion by including the term from Equation (17). At first order in the SME coefficients, this will produce several secular effects that have been computed in [39]. In particular, an additional advance in the argument of periastron and of the longitude of the nodes has been mentioned in Equation (19) and used to constrain the SME with planetary ephemerides. For binary pulsars, it is possible to constrain a secular evolution of two other orbital elements: the eccentricity and the projected semi-major axis x. The secular SME contributions to these quantities have been computed in [39,53,54] and are given by

$$\left\langle \frac{de}{dt} \right\rangle = -n\sqrt{1-e^2} \left[\frac{\varepsilon^2}{e^3} \bar{s}_{PQ} - 2\frac{\delta m}{M} \frac{na\varepsilon}{e^2} \bar{s}^P \right], \tag{26a}$$

$$\left\langle \frac{dx}{dt} \right\rangle = n\frac{m_C}{m_P + m_C} a \cos i \frac{\varepsilon}{e^2\sqrt{1-e^2}} \left[\bar{s}_{kP} \cos \omega - \sqrt{1-e^2}\bar{s}_{kQ} \sin \omega + 2\frac{\delta m}{M} nae\bar{s}^k \cos \omega \right], \tag{26b}$$

where m_P is the mass of the pulsar, m_C is the mass of the companion and all others quantities have been introduced after Eqs. (19). For each binary pulsar, in principle 3 tests can be constructed by using $\dot{\omega}, \dot{e}, \dot{x}$. In [53], 13 pulsars have been used to derive estimates on the SME coefficients. The combination of the observations from the solitary pulsars and from the 13 binary pulsars are reported in Table 6. Both orbital dynamics and spin precession are completely independent of $\bar{s}^{TT}$ whose constraint will be discussed later.

Table 6. Estimation of SME coefficients from binary pulsars data analysis from [53,54]. No correlations coefficients have been derived in this analysis. These estimates should be considered as estimates on the strong field version of the SME coefficients that may include non perturbative strong field effects due to the gravitational binding energy.

Coefficient	
$\left\| \bar{s}^{TT} \right\|$	$< 2.8 \times 10^{-4}$
$\bar{s}^{XX} - \bar{s}^{YY}$	$(0.2 \pm 9.9) \times 10^{-11}$
$\bar{s}^{XX} + \bar{s}^{YY} - 2\bar{s}^{ZZ}$	$(-0.05 \pm 12.25) \times 10^{-11}$
$\bar{s}^{XY}$	$(0.05 \pm 3.55) \times 10^{-11}$
$\bar{s}^{XZ}$	$(0.0 \pm 2.0) \times 10^{-11}$
$\bar{s}^{YZ}$	$(0.0 \pm 3.3) \times 10^{-11}$
$\bar{s}^{TX}$	$(0.05 \pm 5.25) \times 10^{-9}$
$\bar{s}^{TY}$	$(0.5 \pm 8.0) \times 10^{-9}$
$\bar{s}^{TZ}$	$(-0.05 \pm 5.85) \times 10^{-9}$

Several comments can be made about this analysis. First of all, it can be considered as a postfit analysis done after an initial fit performed in GR (or within the post-Keplerian formalism). In particular, correlations between the SME coefficients and other parameters (e.g., orbital parameters) are neglected. Secondly, for most of the pulsars, $\dot{x}$ $\dot{\omega}$ and $\dot{e}$ are not directly measured from the pulse time of arrivals but rather estimated from the uncertainties on x, ω and e divided by the time span of the observations. Further, it is important to mention that effects of Lorentz violations have been considered only for the orbital dynamics but never on the Einstein delay or on the Shapiro time delay in this analysis. The full timing model within SME can be found in Section V.E.3 from [39] (see also [181] for a similar derivation with the matter-gravity couplings). In addition, some parameters are not measured like for example the longitude of the ascending node Ω or the azimuthal angle of the spin. These parameters have been marginalized by using Monte Carlo simulations. It is unclear what type of prior probability distribution function has been used in this analysis and what is the impact of this choice. Nevertheless, the results obtained by this analysis (which does not include the $\bar{s}^{TT}$ parameter) are amongst the best ones currently available demonstrating the power of pulsars observations. The main advantages of using binary pulsars come from the fact that their orbital orientation vary which allows one to disentangle the different SME coefficients and to end up with low correlations. Furthermore, they are so far the only constraints on the strong field version of the SME coefficients.

In addition, a different analysis has been performed to constrain the parameter $\bar{s}^{TT}$ alone [54]. While the orbital dynamics and the spin precession is completely independent of $\bar{s}^{00}$ (i.e., the time component of $\bar{s}^{\mu\nu}$ in a local frame), the boost between the Solar System and the binary pulsar frame makes appear explicitly the $\bar{s}^{TT}$ coefficient. In [54], the assumption that there exists a preferred frame where the $\bar{s}^{\mu\nu}$ tensor is isotropic is made, which makes the results specific to that case (although the analysis can be done without this assumption). The analysis requires the knowledge of the pulsar velocity with respect to the preferred frame as well as the velocity of the Solar System with respect to the same frame. Three pulsars have their radial velocity measured, which combined with proper motion in the sky can be used to determine their velocity. The velocity of the Solar System is taken as its velocity with respect to the Cosmic Microwave Background (CMB) frame $w_\odot$ (with $|w_\odot| = 369$ km/s). The analysis is completely similar to the ones performed for the other SME coefficients (see the discussion in the previous paragraph). It is known that $\bar{s}^{TT}$ has a strong effect on the propagation of the light neglected in [54], which may impact the result. In addition, all correlations between $\bar{s}^{TT}$ and the other SME coefficients are neglected. Finally, two different scenarios have been considered regarding the preferred frame: (i) a scenario where the preferred frame is assumed to be the CMB frame and (ii) a scenario where the orientation of the preferred frame is left free and is marginalized over but the magnitude of the velocity of the Solar System with respect to that frame is still assumed to be the 369 km/s. The general case corresponding to a completely free preferred frame has not been considered. If the CMB frame is assumed to be the preferred frame, the constraint on $\bar{s}^{TT}$ is given by $|\bar{s}^{TT}| < 1.6 \times 10^{-5}$ which is a bit better than the one obtained with VLBI (see Equation (13)) although the VLBI analysis does not assume any preferred frame. The scenario where the orientation of the preferred frame is left as a free parameter leads to an upper bound on $|\bar{s}^{TT}| < 2.8 \times 10^{-4}$.

In summary, observations of binary pulsars are an incredible tool to test the gravitation theory. These tests are of the same order of magnitude (and sometimes better) than the ones performed in the Solar System. Moreover, observations of binary pulsars are sensitive to strong field effects. Observations of the pulse arrival times have been used to search for a breaking of Lorentz violation within the PPN framework by constraining the strong field version of the α_i parameters. The parameter $\hat{\alpha}_1$ is constrained at the level of 10^{-5}, $\hat{\alpha}_2$ at the level of 10^{-9} and $\hat{\alpha}_3$ at the level of 10^{-20} [164]. In addition, constraints on Einstein-Aether and khronometric theory have also been done by combining Solar System constraints with binary pulsars observations [72,73]. Finally, within the SME framework, a postfit analysis has been done by considering the spin precession of solitary pulsars and the orbital dynamics of binary pulsars. The obtained results are given in Table 6 and constrain the strong field version of the SME coefficients. The main advantage of using binary pulsars comes from the fact that they proved an estimate of all the SME coefficients with reasonable correlations. It has to be noted that the modification of the orbital period due to gravitational waves emission has not been computed so far in the SME formalism. In addition, the constraint on $\bar{s}^{TT}$ suffers from the assumption of the existence of a preferred frame. Moreover, the corresponding analysis has neglected all effects on the timing delay that may also impact the results and has neglected the other SME coefficients that may also impact this constraint.

4.7. Čerenkov Radiation

Gravitational Čerenkov radiation is an effect that occurs when the velocity of a particle exceeds the phase velocity of gravity. In this case, the particle will emit gravitational radiation until the particle loses enough energy to drop below the gravity speed [56]. In modified theory of gravity, the speed of gravity in a vacuum may be different from the speed of light and Čerenkov radiation may occur and produces energy losses for particles traveling over long distances. Observations of high energy cosmic rays that have not lost all their energy through Čerenkov radiation can be used to put constraints on models of gravitation that predicts gravitational waves that are propagating slower than light. This effect has been used to constrain some alternative gravitation theories [182,183]: a class of tensor-vector theories [184], a class of tensor-scalar theories [185], extended theories of gravitation [186] and some ghost-free bigravity [187].

The propagation of gravitational waves within the SME framework has been derived in [56,60] (including nonminimal SME contributions). In particular, in the minimal SME, the dispersion relation for the gravitational waves is given by [56]

$$l_0^2 = |I|^2 + \bar{s}^{\mu\nu} l_\mu l_\nu \,, \tag{27}$$

where l^α is the 4-momentum of the gravitational wave. A similar expression including nonminimal higher order SME terms can be found in [56,60]. If the minimal SME produces dispersion-free propagation, the higher order terms lead to dispersion and birefringence [60]. As can be directly inferred from the last equation, gravitational Čerenkov radiation can arise when the effective refractive index n is

$$n^2 = 1 - \bar{s}^{\mu\nu} \hat{l}_\mu \hat{l}_\nu > 1 \,, \tag{28}$$

where $\hat{l}_\mu = l_\mu / |l|$. The expression for the energy loss rate due to Lorentz-violating gravitational Čerenkov emission has been calculated from tree-level graviton emission for photons, fermions and scalar particles and is given by [56]

$$\frac{dE}{dt} = -F^w(d) G \left(\bar{s}^{(d)}(\hat{p}) \right)^2 |p|^{2d-4} \,, \tag{29}$$

where d is the dimension of the Lorentz violating operator ($d = 4$ for the minimal SME), $F^w(d)$ is a dimensionless factor depending on the flavor w of the particle emitting the radiation, p is the particle incoming momentum (with $\hat{p} = p/|p|$) and $\bar{s}^{(d)}$ is a direction-dependent combination of SME coefficients. In the minimal SME, $\bar{s}^{(4)}(\hat{p})$ is decomposed on spherical harmonics as

$$\bar{s}^{(4)}(\hat{p}) = \sum_{jm} Y_{jm}(\hat{p}) \bar{s}_{jm}^{(SH)} \,, \tag{30}$$

where we explicitly indicated the (SH) to specify that these coefficients are spherical harmonic decomposition of the SME coefficients. The calculation of the dimensionless factor $F^w(d)$ for scalar particles, fermions and photons has been done in [56]. The integration of Equation (29) shows that if a cosmic ray of specie w is observed on Earth with an energy E_f after traveling a distance L along the direction $\hat{p}$, this implies the following constraint on the SME coefficients

$$\bar{s}^{(d)}(\hat{p}) < \sqrt{\frac{\mathcal{F}^w(d)}{G E_f^{2d-5} L}} \,, \tag{31}$$

where $\mathcal{F}^w(d) = (2d-5)/F^w(d)$ is another dimensionless factor dependent on the matrix element of the tree-level process for graviton emission.

Using data for the energies and angular positions of 299 observed cosmic rays from different collaborations [188–195], Kostelecký and Tasson [56] derived lower and upper constraints on 80 SME coefficients, including the nine coefficients from the minimal SME whose constraints are given by the Table 7. In their analysis, they consider the coefficients from the different dimensions separately and did not fit all of them simultaneously. In addition, in the minimal SME, they did a fit for the $\bar{s}^{TT}$ parameter alone and another fit for the other 8 coefficients. The number of sources and their directional dependence across the sky allow one to disentangle the SME coefficients and to derive two-sided bounds from the Equation (31). The only coefficient that is one sided is $\bar{s}^{TT}$ because it produces isotropic effects. The bounds are severe for these coefficients, on the order of 10^{-13}. However, this analysis assumes that the matter sector coefficients vanish. Furthermore, several assumptions have been made in order to derive the bounds from Table 7. It is assumed that the cosmic ray primaries are nuclei of atomic weight $N = 56$ (iron), that the Čerenkov radiation is emitted by one of the fermionic partons in the nucleus that carries 10 % of the cosmic ray energy and that the travel distance of the cosmic ray is 10 Megaparsec (Mpc) [56]. Although only conservative assumptions are used for the astrophysical processes involved in the production of high-energy cosmic rays, the observations rely on the sources on the order of 10 Mpc distant, and thus the analysis is of a different nature than a controlled laboratory or even Solar-System test.

Table 7. Lower and upper limits on the SME coefficients decomposed in spherical harmonics derived from Čerenkov radiation [56].

Coefficient	Lower Bound	Upper Bound
$\bar{s}_{00}^{(SH)}$	-3×10^{-14}	
$\bar{s}_{10}^{(SH)}$	-1×10^{-13}	7×10^{-14}
Re $\bar{s}_{11}^{(SH)}$	-8×10^{-14}	8×10^{-14}
Im $\bar{s}_{11}^{(SH)}$	-7×10^{-14}	9×10^{-14}
$\bar{s}_{20}^{(SH)}$	-7×10^{-14}	1×10^{-13}
Re $\bar{s}_{21}^{(SH)}$	-7×10^{-14}	7×10^{-14}
Im $\bar{s}_{21}^{(SH)}$	-5×10^{-14}	8×10^{-14}
Re $\bar{s}_{22}^{(SH)}$	-6×10^{-14}	8×10^{-14}
Im $\bar{s}_{22}^{(SH)}$	-7×10^{-14}	7×10^{-14}

For the sake of completeness and to allow an easy comparison with the estimations of the other standard cartesian $\bar{s}^{\mu\nu}$ coefficients, the following relations give the links between the spherical harmonic decomposition and the standard cartesian decomposition of the SME coefficients:

$$\bar{s}_{00}^{(SH)} = \frac{4}{3}\sqrt{4\pi}\,\bar{s}^{TT}\,, \tag{32a}$$

$$\bar{s}_{10}^{(SH)} = -\sqrt{\frac{16\pi}{3}}\bar{s}^{TZ}\,, \qquad \text{Re } \bar{s}_{11}^{(SH)} = \sqrt{\frac{8\pi}{3}}\bar{s}^{TX}\,, \qquad \text{Im } \bar{s}_{11}^{(SH)} = -\sqrt{\frac{8\pi}{3}}\bar{s}^{TY}\,, \tag{32b}$$

$$\bar{s}_{20}^{(SH)} = -\sqrt{\frac{4\pi}{5}}\frac{1}{3}\left(\bar{s}^{XX} + \bar{s}^{YY} - 2\bar{s}^{ZZ}\right)\,, \qquad \text{Re } \bar{s}_{21}^{(SH)} = -\sqrt{\frac{8\pi}{15}}\bar{s}^{XZ}\,, \qquad \text{Im } \bar{s}_{21}^{(SH)} = \sqrt{\frac{8\pi}{15}}\bar{s}^{YZ}\,, \tag{32c}$$

$$\text{Re } \bar{s}_{22}^{(SH)} = \sqrt{\frac{2\pi}{15}}\left(\bar{s}^{XX} - \bar{s}^{YY}\right)\,, \qquad \text{Im } \bar{s}_{22}^{(SH)} = -2\sqrt{\frac{2\pi}{15}}\bar{s}^{XY}\,. \tag{32d}$$

In summary, observations of cosmic rays allow one to derive some stringent boundaries on the SME coefficients. The idea is that if Lorentz symmetry is broken, these high energy cosmic rays would have lost energy by emitting Čerenkov radiation that has not been observed. The boundaries on the spherical harmonic decomposition of the SME coefficients are given in the Table 7 (in order to compare these boundaries to other constraints, they have been transformed into boundaries on standard cartesian SME coefficients in Table 8). For the minimal SME, one can limit the isotropic $\bar{s}^{TT}$ (one sided bound) or the other eight other coefficients in $\bar{s}^{\mu\nu}$, but not all the nine simultaneously. These boundaries are currently the best available in the literature at the exception of $\bar{s}^{TT}$ whose constraint is only one sided. Nevertheless, several assumptions have been made in this analysis and the observations rely on sources located at very high distances. This analysis is therefore of a different nature than the other ones where more control on the measurements is possible.

Table 8. Summary of all estimations of the $\bar{s}^{\mu\nu}$ coefficients.

	Atomic Grav. [48]	LLR [46]	Planetary Eph. [50]	Pulsars [53,54]	Čerenkov rad. [56] Lower Bound	Čerenkov rad. [56] Upper Bound
$\bar{s}^{TT}$				$< 2.8 \times 10^{-4}$	$-6 \times 10^{-15} <$	
$\bar{s}^{XX} - \bar{s}^{YY}$	$(4.4 \pm 11) \times 10^{-9}$	$(0.6 \pm 4.2) \times 10^{-11}$	$(-0.8 \pm 2.0) \times 10^{-10}$	$(0.2 \pm 9.9) \times 10^{-11}$	$-9 \times 10^{-14} <$	$< 1.2 \times 10^{-13}$
$\bar{s}^{XX} + \bar{s}^{YY} - 2\bar{s}^{ZZ}$			$(-0.8 \pm 2.7) \times 10^{-10}$	$(-0.05 \pm 12.25) \times 10^{-11}$	$-1.9 \times 10^{-13} <$	$< 1.3 \times 10^{-13}$
$\bar{s}^{XY}$	$(0.2 \pm 3.9) \times 10^{-9}$	$(-5.7 \pm 7.7) \times 10^{-12}$	$(-0.3 \pm 1.1) \times 10^{-10}$	$(0.05 \pm 3.55) \times 10^{-11}$	$-3.9 \times 10^{-14} <$	$< 6.2 \times 10^{-14}$
$\bar{s}^{XZ}$	$(-2.6 \pm 4.4) \times 10^{-9}$	$(-2.2 \pm 5.9) \times 10^{-12}$	$(-1.0 \pm 3.5) \times 10^{-11}$	$(0.0 \pm 2.0) \times 10^{-11}$	$-5.4 \times 10^{-14} <$	$< 5.4 \times 10^{-14}$
$\bar{s}^{YZ}$	$(-0.3 \pm 4.5) \times 10^{-9}$		$(5.5 \pm 5.2) \times 10^{-12}$	$(0.0 \pm 3.3) \times 10^{-11}$	$-3.9 \times 10^{-14} <$	$< 6.2 \times 10^{-14}$
$\bar{s}^{TX}$	$(-3.1 \pm 5.1) \times 10^{-5}$	$(-0.9 \pm 1.0) \times 10^{-8}$	$(-2.9 \pm 8.3) \times 10^{-9}$	$(0.05 \pm 5.25) \times 10^{-9}$	$2.8 \times 10^{-14} <$	$< 2.8 \times 10^{-14}$
$\bar{s}^{TY}$	$(0.1 \pm 5.4) \times 10^{-5}$		$(0.3 \pm 1.4) \times 10^{-8}$	$(0.5 \pm 8.0) \times 10^{-9}$	$3.1 \times 10^{-14} <$	$< 2.4 \times 10^{-14}$
$\bar{s}^{TZ}$	$(1.4 \pm 6.6) \times 10^{-5}$		$(-0.2 \pm 5.0) \times 10^{-8}$	$(-0.05 \pm 5.85) \times 10^{-9}$	$1.7 \times 10^{-14} <$	$< 2.4 \times 10^{-14}$
$\bar{s}^{TY} + 0.43\,\bar{s}^{TZ}$		$(6.2 \pm 7.9) \times 10^{-9}$				
$\bar{s}^{XX} + \bar{s}^{YY} - 2\bar{s}^{ZZ} - 4.5\,\bar{s}^{YZ}$		$(2.3 \pm 4.5) \times 10^{-11}$				

VLBI [55]

$$\bar{s}^{TT} = (-5 \pm 8) \times 10^{-5}$$

GPB [52]

$$\bar{s}^{TT} + 970\left(\bar{s}^{XX} - \bar{s}^{YY}\right) - 0.05\left(\bar{s}^{XX} + \bar{s}^{YY} - 2\bar{s}^{ZZ}\right) + 2895\,\bar{s}^{XY} - 3235\,\bar{s}^{XZ} - 11\,240\,\bar{s}^{YZ} = (0.7 \pm 3.1) \times 10^{-3}$$

$$\bar{s}^{XX} - \bar{s}^{YY} + 3.02\,\bar{s}^{XY} + 0.32\,\bar{s}^{XZ} + 1.05\,\bar{s}^{YZ} = (-1.1 \pm 3.8) \times 10^{-7}$$

4.8. Summary and Combined Analysis

To summarize, several measurements have already successfully been used to constrain the minimal SME in the gravitational sector (i.e., the $\bar{s}^{\mu\nu}$ coefficients):

- Atom interferometry [47,48].
- Lunar Laser Ranging [45,46].
- Planetary ephemerides [49,50].
- Very Long Baseline Interferometry [55].
- Gravity Probe B [52].
- Pulsars timing [53,54].
- Čerenkov radiation [30,56].

A detailed description of all these analyses is provided in the previous subsections and the Table 8 summarizes the current estimates. It is also interesting to combine all these estimations together to provide the best estimates on the SME coefficients. In order to do this, we perform a large least-square fit including all the results from the Table 8 including the covariance matrices quoted in the previous subsections. The results from the Čerenkov radiation are not included since they rely on a very different type of observations. Two combined fits are presented: one without including the pulsars results and one including the pulsars results. This is due to the fact that pulsars are sensitive to a strong version of the SME coefficients that may include non perturbative strong field effects as described in Section 4.6. If this is the case, then the pulsars results cannot be directly combined with the weak gravitational field estimates on the SME coefficients. If no non perturbative strong field effect arises, then the right column from Table 9 presents a combined fit that includes these observations as well. The results from Table 9 include all the information currently available in the literature on the $\bar{s}^{\mu\nu}$ (estimations and correlation matrices). It can also be noted that the pulsars results improve significantly the marginalized estimations on $\bar{s}^{TY}$ and $\bar{s}^{TZ}$ by reducing strongly the correlation between these two coefficients.

In addition, several measurements have been used to constrain the non-minimal SME sectors:

- Short gravity experiment [57–59].
- Čerenkov radiation [56].
- Gravitational waves analysis [60].

A review of these measurements can be found in [30].

Table 9. Estimation of SME coefficients resulting from a fit combining results from: atomic gravimetry (see Table 1), VLBI (see Equation (13)), LLR (see Table 3), planetary ephemerides (see Table 4), Gravity Probe B (see Table 5). The correlation matrices from all these analyses have been used in the combined fit. The right column includes the pulsars results from Table 6 as well. The three estimates on $\bar{s}^{JJ}$ are obtained by using the traceless condition $\bar{s}^{TT} = \bar{s}^{XX} + \bar{s}^{YY} + \bar{s}^{ZZ}$.

Coefficient	Without Pulsars	With Pulsars
$\bar{s}^{TT}$	$(-5. \pm 8.) \times 10^{-5}$	$(-4.6 \pm 7.7) \times 10^{-5}$
$\bar{s}^{XX} - \bar{s}^{YY}$	$(-0.5 \pm 1.9) \times 10^{-11}$	$(-0.5 \pm 1.9) \times 10^{-11}$
$\bar{s}^{XX} + \bar{s}^{YY} - 2\bar{s}^{ZZ}$	$(1.6 \pm 3.1) \times 10^{-11}$	$(0.8 \pm 2.5) \times 10^{-11}$
$\bar{s}^{XY}$	$(-1.5 \pm 6.8) \times 10^{-12}$	$(-1.6 \pm 6.6) \times 10^{-12}$
$\bar{s}^{XZ}$	$(-1.0 \pm 4.1) \times 10^{-12}$	$(-0.8 \pm 3.9) \times 10^{-12}$
$\bar{s}^{YZ}$	$(2.6 \pm 4.7) \times 10^{-12}$	$(1.1 \pm 3.2) \times 10^{-12}$
$\bar{s}^{TX}$	$(-0.1 \pm 1.3) \times 10^{-9}$	$(-0.1 \pm 1.3) \times 10^{-9}$
$\bar{s}^{TY}$	$(0.5 \pm 1.1) \times 10^{-8}$	$(0.4 \pm 2.3) \times 10^{-9}$
$\bar{s}^{TZ}$	$(-1.2 \pm 2.7) \times 10^{-8}$	$(-0.6 \pm 5.5) \times 10^{-9}$
$\bar{s}^{XX}$	$(-1.7 \pm 2.7) \times 10^{-5}$	$(-1.5 \pm 2.6) \times 10^{-5}$
$\bar{s}^{YY}$	$(-1.7 \pm 2.7) \times 10^{-5}$	$(-1.5 \pm 2.6) \times 10^{-5}$
$\bar{s}^{ZZ}$	$(-1.7 \pm 2.7) \times 10^{-5}$	$(-1.5 \pm 2.6) \times 10^{-5}$

5. The future

In addition to all the improvements related to existing analysis suggested in the previous sections, there are a couple of sensitivity analyses that have been done within the SME framework. First of all, a thorough and detailed analysis of a lot of observables related to gravitation can be found in [39]. In addition, we will present in the next subsections a couple of analyses and ideas that may improve the SME coefficients estimates in the future.

5.1. The Gaia Mission

Launched in December 2013, the ESA Gaia mission [196] is scanning regularly the whole celestial sphere once every 6 months providing high precision astrometric data for a huge number ($\approx$1 billion) of celestial bodies. In addition to stars, it is also observing Solar System Objects (SSO), in particular asteroids. The high precision astrometry (at sub-mas level) will allow us to perform competitive tests of gravitation and to provide new constraints on alternative theories of gravitation.

First of all, the Gaia mission is expected to provide an estimate of the γ PPN parameter at the level of 10^{-6} [197] by measuring the deflection of the light on a 5 years timescale. Furthermore, in addition to this global determination of a global PPN parameter from observations of light deflection, it has been proposed to use Gaia observations to map the deflection angle in the sky and to look for a dependence of the γ PPN parameter with respect to the Sun impact parameter [198–202]. Such a dependence of the gravitational deflection with respect to the observation geometry is also a feature predicted by SME as shown in [82]. Therefore, the global mapping of the light deflection with Gaia can also be efficiently used to constrain some SME coefficients. A first sensitivity analysis can be found in [82] and is reported on Table 10. Note that proposals observations and missions like AGP [203] or LATOR [204] can in the long term improve these estimates further by improving the light deflection measurement.

Table 10. Sensitivity of the SME coefficients to the measurement of the light deflection by several space missions or proposals (these estimates are based on Table I from [82]).

Mission	$\bar{s}^{TT}$	$\bar{s}^{TJ}$	$\bar{s}^{IJ}$
Gaia [196]	10^{-6}	10^{-6}	10^{-5}
AGP [203]	10^{-7}	10^{-7}	10^{-6}
LATOR [204]	10^{-8}	10^{-8}	10^{-7}

In addition to gravitation tests performed by measuring the light deflection, Gaia also provides a unique opportunity to test gravitation by considering the orbital dynamics of SSO. One can estimate that about 360,000 asteroids will be regularly observed by Gaia at the sub-mas level, which will allow us to perform various valuable tests of gravitation [205,206]. In particular, realistic simulations of more than 250,000 asteroids have shown that Gaia will be able to constrain the β PPN parameter at the level of 10^{-3} [205]. The main advantage from Gaia is related to the huge number of bodies that will be observed with very different orbital parameters as illustrated on Figure 1. As a consequence, the huge correlations appearing in the planetary ephemerides analysis (see Section 4.4) will not appear when considering asteroids observations and the marginalized confidence intervals will be highly improved compared to planetary ephemerides analysis.

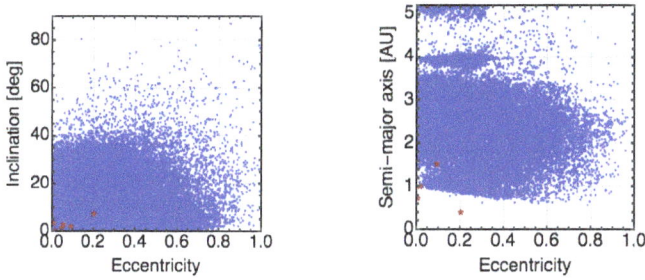

Figure 1. This figure represents the distribution of the orbital parameters for the Solar System Objects (SSOs) expected to be observed by the Gaia satellite. The red stars represent the innermost planets of the Solar System.

A realistic sensitivity analysis of Gaia SSOs observations within the SME framework has been performed (see also [206] for preliminary results). In this analysis, 360,000 asteroids have been considered over the nominal mission duration (i.e., five years) and a match between the SSO trajectories with the Gaia scanning law is performed to find the observation times for each SSO. Simultaneously with the equations of motion, we integrate the variational equations, the simulated SSO trajectories being transformed into astrometric observables as well as their partial derivatives with respect to the parameters considered in the covariance analysis. The covariance analysis leads to the estimated uncertainties presented in Table 11. These uncertainties are incredibly good, which is due to the variety of the asteroids orbital parameters as discussed above. Using our set of asteroids, the correlation matrix for the SME parameters is very reasonable: the most important correlation coefficients are 0.71, -0.68 and 0.46. All the other correlations are below 0.3. Therefore, Gaia offers a unique opportunity to constrain Lorentz violation through the SME formalism. Finally, the Gaia mission is likely to be extended to 10 years, therefore doubling the measurements baseline which will also impact significantly the expected uncertainties. Finally, it is worth mentioning that the Gaia dataset can be combined with radar observations [207] that are complementary in the time frame and orthogonal to astrometric telescopic observations.

Table 11. Sensitivity of the SME coefficients to the observations of 360,000 asteroids by the Gaia satellite during a period of 5 years.

SME Coefficients	Sensitivity $(1 - \sigma)$
$\bar{s}^{XX} - \bar{s}^{YY}$	3.7×10^{-12}
$\bar{s}^{XX} + \bar{s}^{YY} - 2\bar{s}^{ZZ}$	6.4×10^{-12}
$\bar{s}^{XY}$	1.6×10^{-12}
$\bar{s}^{XZ}$	9.2×10^{-13}
$\bar{s}^{YZ}$	1.7×10^{-12}
$\bar{s}^{TX}$	5.6×10^{-9}
$\bar{s}^{TY}$	8.8×10^{-9}
$\bar{s}^{TZ}$	1.6×10^{-8}

In summary, the Gaia space mission offers two opportunities to test Lorentz symmetry in the Solar System by looking at the deflection of light and by considering the orbital dynamics of SSO. The second type of observations is extremely interesting in the sense that the high number and the variety of orbital parameters of the observed SSO leads to decorrelate the SME coefficients.

5.2. Analysis of Cassini Conjunction Data

The space mission Cassini is exploring the Saturnian system since July 2004. During its cruising phase while the spacecraft was on its interplanetary journey between Jupiter and Saturn, a measurement of the gravitational time delay was performed [118]. This measurement occurred during a Solar conjunction in June 2002 and was made possible thanks to a multi-frequency radioscience link (at X and Ka-band) which allows a cancellation of the solar plasma noise [118]. The related data spans over 30 days and has been analyzed in the PPN framework leading to the best estimation of the γ PPN parameter so far given by $(2.1 \pm 2.3) \times 10^{-5}$ [118].

The exact same set of data can be reduced within the SME framework and is expected to improve our current $\bar{s}^{TT}$ estimation. The time delay within the SME framework has been derived in [80] and is given by Equation (9).

A simulation of the Cassini link during the 2002 conjunction within the full SME framework has been realized using the software presented in [148] (see also [208,209]). The signature produced by the $\bar{s}^{TT}$ coefficients on the 2-way Doppler link during the Solar conjunction is illustrated on Figure 2. In [80], a crude estimate of attainable sensitivities in estimate of the SME coefficients using the Cassini conjunction data is given (see Table I from [80]). It is shown that some combinations of the $\bar{s}^{IJ}$ coefficients can only be constrained at the level of 10^{-4}, which is 7 to 8 orders of magnitude worse than the current best constraints on these coefficients. It is therefore safe to neglect these and to concentrate only on the $\bar{s}^{TT}$ coefficient. A realistic covariance analysis performed over the 30 days of the Solar conjunction and assuming an uncertainty of the Cassini Doppler of 3 μm/s [210,211] shows that the $\bar{s}^{TT}$ parameter can be constrained at the level of 2×10^{-5} using the Cassini data allowing an improvement of a factor 4 with respect to the current best estimate coming from VLBI analysis (see Equation (13)). Therefore, a reanalysis of the 2002 Cassini data within the SME framework would be highly valuable.

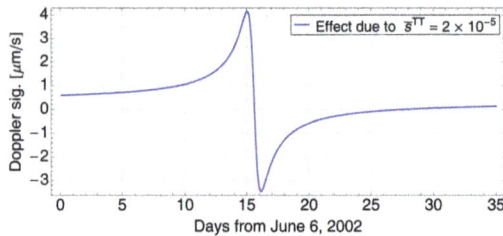

Figure 2. Doppler signature produced by $\bar{s}^{TT} = 2 \times 10^{-5}$ on the 2-way Doppler link Earth-Cassini-Earth during the 2002 Solar conjunction.

5.3. Satellite Laser Ranging (LAGEOS/LARES)

Searching for violations of Lorentz symmetry by using the orbital motion of planets (see Section 4.4), binary pulsars (see Section 4.6), the Moon (see Section 4.3) and asteroids (see Section 5.1) has turned out to be highly powerful. It is therefore logical to consider the motion of artificial satellite orbiting around Earth to search for Lorentz violations. In particular, laser ranging to the two LAGEOS and to the LARES satellites has successfully been used to test GR by measuring the impact of the Schwarzschild precession on the motion of the satellites [212–214]. It has also been claimed that the impact of the frame-dragging (or Lense-Thirring effect) due to the Earth's spin on the orbital motion of the satellites has been measured [215–220] although this claim remains controversial [221–226]. Similarly, the LAGEOS/LARES satellites can also be used to search for Lorentz violations. A sensitivity analysis has been done in [49] and it has been shown that the LAGEOS satellites are sensitive at the level of 10^{-4} to the $\bar{s}^{TJ}$ coefficients. Using LARES should improve significantly this value. Further numerical simulations are required in order to determine exactly the SME linear combinations to which the ranging to these satellites is sensitive to. A data analysis within the full

SME framework (i.e., including the integration of the SME equation of motion and including the SME coefficients with the other global parameters in the fit) would also be highly interesting. In addition, similar tests of Lorentz symmetry can also be included within the scientific goals of the LAser RAnged Satellites Experiment (LARASE) project [227] or within the OPTIS project [228].

5.4. Gravity-Matter Coefficients and Breaking of the Einstein Equivalence Principle

All the measurements mentioned in Section 4 can be analyzed by considering the gravity-matter coupling coefficients $\bar{a}_\mu$ and $\bar{c}_{\mu\nu}$ [29] that are breaking the EEP. Some atomic clocks measurements have already provided some constraints on the $\bar{a}_\mu$ coefficients [229–231]. In addition, in [50] the planetary ephemerides analysis is interpreted by considering the $\bar{a}_\mu$ coefficients and the atomic interferometry results from [48] and the LLR results from [45] are also reinterpreted by considering the gravity-coupling coefficients. Clearly this is a preliminary analysis that needs to be refined by more solid data reductions. Considering the increasing number of fitted parameters, it is of prime importance to increase the number of measurements used in the analysis and to produce combined analysis with as many types of observations as possible. The measurements developed in Section 4 are a first step in order to reach this goal. The gravity-coupling coefficients can also be constrained by more specific tests related to the EEP like for example tests of the Universality of Free Fall with MicroSCOPE [83,84], tests of the gravitational redshift with GNSS satellites [232], with the Atomic Clocks Ensemble in Space (ACES) project [233], or with the OPTIS project [228], ...

6. Conclusions

Lorentz symmetry is at the heart of both GR and the Standard Model of particle physics. This symmetry is broken in various scenarios of unification, of quantum gravity and even in some models of Dark Matter and Dark Energy. Searching for violations of Lorentz symmetry is therefore a powerful tool to test fundamental physics. The last decades have seen the number of tests of Lorentz invariance arise dramatically in all sectors of physics [24]. In this review, we focused on searches for Lorentz symmetry breaking in the pure gravitational sector. Mainly two frameworks exist to parametrize violations of Lorentz invariance in the gravitation sector. First of all, the three $\alpha_{1,2,3}$ PPN parameters phenomenologically encode a violation of Lorentz symmetry at the level of the spacetime metric [2]. These parameters are constrained by LLR (see Section 4.3) and by pulsars timing measurements (see Section 4.6). In addition, it is interesting to notice that the corresponding PPN metric parametrizes also Einstein-Aether and Khronometric theories in the weak gravitational field limit [72] while these theories have a more complex strong field limit (and can show non perturbative effects) that have been constrained by pulsars observations (see Section 4.6 and [72,73]).

In addition to the PPN formalism, the SME formalism has been developed by including systematically all possible Lorentz violations terms that can be constructed at the level of the action. In the pure gravitational sector, the gravitational action within the SME formalism contains the usual Einstein-Hilbert action but also new Lorentz violating terms constructed by contracting new fields with some operators built from curvature tensors and covariant derivatives with increasing mass dimension [36]. The lower mass dimension term is known as the minimal SME. In the limit of linearized gravity, the observations within the minimal SME formalism depend on 9 coefficients, the $\bar{s}^{\mu\nu}$ symmetric traceless tensor. This formalism offers a new opportunity to search for deviations from GR in a framework different from the standard PPN formalism. We reviewed the different observations that have been used so far to constrain the SME coefficients. The main idea is to search for a signature (usually periodic) that arises from a dependence on the orientation of the system measured (the dependence on the orientation is typically due to the Earth's rotation, the orbital motion of the planets around the Sun, etc...) or from a dependence on the boost of the system observed (so far, only the binary pulsars $\bar{s}^{TT}$ constraint comes from this type of dependence [54]). Most of SME analyses are postfit analyses in the sense that analytical signatures due to SME are fitted in residuals noise obtained in a previous data reduction performed in pure GR. In Section 3, we showed that this approach can sometimes lead to overoptimistic constraint on the SME coefficients and that one should be careful in interpreting results obtained using such an approach.

In Section 4, we discussed in details the different measurements used so far to constrain the $\bar{s}^{\mu\nu}$ coefficients: atomic gravimetry (Section 4.1), VLBI (Section 4.2), LLR (Section 4.3), planetary ephemerides (Section 4.4), Gravity Probe B (Section 4.5), pulsars timing (Section 4.6) and Čerenkov radiation (Section 4.7). In each of these subsections, we describe the current analyses performed in order to constrain the SME coefficients and provide a critical discussion from each of them. We also provide a summary of these constraints on Table 8. In addition, we used all these results to produce a combined analysis of the SME coefficients. This fit is done by taking into account the correlation matrices for each individual analysis. The results of this combined fit are presented in Table 9 and are the current best estimates of the SME coefficients that are possible to derive with all available analyses. In addition to the minimal SME, there exists higher order Lorentz violating terms that have been considered and constrained by short-range gravity experiments [57–59], gravitational waves analysis [60] and Čerenkov radiation [30,56].

In Section 5, we discussed some opportunities to improve the current constraints on the SME coefficients. In particular, the European space mission Gaia offers an excellent opportunity to probe Lorentz symmetry through the measurement of light deflection and through the orbital motion of asteroids. The Cassini conjunction data also offers a way to constrain the $\bar{s}^{TT}$ coefficient that impacts severely the propagation of light. Finally, existing satellite laser ranging data can also be analyzed within the SME framework.

In addition, as mentioned in Section 5.4, all the analyses presented in this review can include gravity-matter coefficients [29]. While considering these, the number of coefficients fitted increase significantly and it becomes crucial to produce a fit combining several kinds of experiments. A preliminary analysis considering these coefficients for planetary ephemerides, LLR and atomic gravimetry has been performed in [50] but needs to be refined. In addition, some atomic clocks experiments have already been used to constrain matter-gravity coefficients [229–231].

In conclusion, though no violation of Lorentz symmetry has been observed so far, an incredible number of opportunities still exists for additional investigations. There remains a large area of unexplored coefficients space that can be explored by improved measurements or by new projects aiming at searching for breaking of Lorentz symmetry. In addition, the increasing number of parameters fitted (by including the gravity-matter coupling coefficients simultaneously with the pure gravity coefficients in the analyses) will deter the marginalized estimates of each coefficient. This verdict emphasizes the need to increase the types of measurements that can be combined together to explore the vast parameters space as efficiently as possible. The current theoretical questions related to the quest for a unifying theory or for a quantum theory of gravitation suggests that Lorentz symmetry will play an important role in the search for new physics. Hopefully, future searches for Lorentz symmetry breaking will help theoreticians to unveil some of the mysteries about Planck-scale physics [22].

Acknowledgments: A.H. is thankful to P. Wolf, S. Lambert, B. Lamine, A. Rivoldini, F. Meynadier, S. Bouquillon, G. Francou, M.-C. Angonin, D. Hestroffer, P. David and A. Fienga for interesting discussions about some part of this work. Q.G.B. was supported in part by the National Science Foundation under Grant No. PHY-1402890. A.B. and C.L.P.L. are grateful for the CNRS/GRAM and "Axe Gphys" of Paris Observatory Scientific Council. C.G. and Q.G.B. acknowledge support from Sorbonne Universités Emergence grant.

Author Contributions: All the authors wrote the paper.

Conflicts of Interest: The authors declare no conflict of interest.

Abbreviations

The following abbreviations are used in this manuscript:

CMB	Cosmic Microwave Background
ELPN	Éphéméride Lunaire Parisienne Numérique
GPB	Gravity Probe B
GR	General Relativity
GRAIL	Gravity Recovery And Interior Laboratory
INPOP	Intégrateur Numérique Planétaire de l'Observatoire de Paris
IVS	International VLBI Service for Geodesy and Astrometry
LARASE	LAser RAnged Satellites Experiment
LLR	Lunar Laser Ranging
LIV	Lorentz Invariance Violation
mas	milliarcsecond
Mpc	Megaparsec
NFT	Numerical Fourier Transform
PPN	Parametrized Post-Newtonian
SME	Standard-Model Extension
SSO	Solar System Object
TAI	International Atomic Time
TDB	Barycentric Dynamical Time
TT	Terrestrial Time
UTC	Universal Time Coordinate
VLBI	Very Long Baseline Interferometry
yr	year

References

1. Iorio, L. Editorial for the Special Issue 100 Years of Chronogeometrodynamics: The Status of the Einstein's Theory of Gravitation in Its Centennial Year. *Universe* **2015**, *1*, 38–81.
2. Will, C.M. The Confrontation between General Relativity and Experiment. *Living Rev. Relativ.* **2014**, *17*, 4.
3. Turyshev, S.G. REVIEWS OF TOPICAL PROBLEMS: Experimental tests of general relativity: Recent progress and future directions. *Phys. Uspekhi* **2009**, *52*, 1–27.
4. Abbott, B.P.; Abbott, R.; Abbott, T.D.; Abernathy, M.R.; Acernese, F.; Ackley, K.; Adams, C.; Adams, T.; Addesso, P.; Adhikari, R.X.; et al. Observation of Gravitational Waves from a Binary Black Hole Merger. *Phys. Rev. Lett.* **2016**, *116*, 061102.
5. Cervantes-Cota, J.; Galindo-Uribarri, S.; Smoot, G. A Brief History of Gravitational Waves. *Universe* **2016**, *2*, 22.
6. Debono, I.; Smoot, G.F. General Relativity and Cosmology: Unsolved Questions and Future Directions. *Universe* **2016**, *2*, 23.
7. Berti, E.; Barausse, E.; Cardoso, V.; Gualtieri, L.; Pani, P.; Sperhake, U.; Stein, L.C.; Wex, N.; Yagi, K.; Baker, T. et al. Testing general relativity with present and future astrophysical observations. *Class. Quantum Gravity* **2015**, *32*, 243001.
8. Colladay, D.; Kostelecký, V.A. CPT violation and the standard model. *Phys. Rev. D* **1997**, *55*, 6760–6774.
9. Kostelecký, V.A.; Samuel, S. Gravitational phenomenology in higher-dimensional theories and strings. *Phys. Rev. D* **1989**, *40*, 1886–1903.
10. Kostelecký, V.A.; Samuel, S. Spontaneous breaking of Lorentz symmetry in string theory. *Phys. Rev. D* **1989**, *39*, 683–685.
11. Kostelecký, V.A.; Potting, R. CPT and strings. *Nucl. Phys. B* **1991**, *359*, 545.
12. Gambini, R.; Pullin, J. Nonstandard optics from quantum space-time. *Phys. Rev. D* **1999**, *59*, 124021.
13. Amelino-Camelia, G. Quantum-Spacetime Phenomenology. *Living Rev. Relativ.* **2013**, *16*, 5.
14. Mavromatos, N.E. *CPT Violation and Decoherence in Quantum Gravity*; Lecture Notes in Physics; Kowalski-Glikman, J., Amelino-Camelia, G., Eds.; Springer: Berlin, Germany, 2005.

15. Myers, R.C.; Pospelov, M. Ultraviolet Modifications of Dispersion Relations in Effective Field Theory. *Phys. Rev. Lett.* **2003**, *90*, 211601.

16. Hayakawa, M. Perturbative analysis on infrared aspects of noncommutative QED on R^4. *Phys. Lett. B* **2000**, *478*, 394–400.

17. Carroll, S.M.; Harvey, J.A.; Kostelecký, V.A.; Lane, C.D.; Okamoto, T. Noncommutative Field Theory and Lorentz Violation. *Phys. Rev. Lett.* **2001**, *87*, 141601.

18. Bjorken, J.D. Cosmology and the standard model. *Phys. Rev. D* **2003**, *67*, 043508.

19. Burgess, C.P.; Martineau, P.; Quevedo, F.; Rajesh, G.; Zhang, R.J. Brane-antibrane inflation in orbifold and orientifold models. *J. High Energy Phys.* **2002**, *3*, 052.

20. Frey, A.R. String theoretic bounds on Lorentz-violating warped compactification. *J. High Energy Phys.* **2003**, *4*, 12.

21. Cline, J.M.; Valcárcel, L. Asymmetrically warped compactifications and gravitational Lorentz violation. *J. High Energy Phys.* **2004**, *3*, 032.

22. Tasson, J.D. What do we know about Lorentz invariance? *Rep. Prog. Phys.* **2014**, *77*, 062901.

23. Mattingly, D. Modern Tests of Lorentz Invariance. *Living Rev. Relativ.* **2005**, *8*, 5.

24. Kostelecký, V.A.; Russell, N. Data tables for Lorentz and CPT violation. *Rev. Mod. Phys.* **2011**, *83*, 11–32.

25. Colladay, D.; Kostelecký, V.A. Lorentz-violating extension of the standard model. *Phys. Rev. D* **1998**, *58*, 116002.

26. Thorne, K.S.; Will, C.M. Theoretical Frameworks for Testing Relativistic Gravity. I. Foundations. *Astrophys. J.* **1971**, *163*, 595.

27. Will, C.M. *Theory and Experiment in Gravitational Physics*; Cambridge University Press: Cambridge, UK, 1993.

28. Thorne, K.S.; Lee, D.L.; Lightman, A.P. Foundations for a Theory of Gravitation Theories. *Phys. Rev. D* **1973**, *7*, 3563–3578.

29. Kostelecký, V.A.; Tasson, J.D. Matter-gravity couplings and Lorentz violation. *Phys. Rev. D* **2011**, *83*, 016013.

30. Tasson, J.D. The Standard-Model Extension and Gravitational Tests. *Symmetry* **2016**, *8*, 111.

31. Fischbach, E.; Sudarsky, D.; Szafer, A.; Talmadge, C.; Aronson, S.H. Reanalysis of the Eotvos experiment. *Phys. Rev. Lett.* **1986**, *56*, 3–6.

32. Talmadge, C.; Berthias, J.P.; Hellings, R.W.; Standish, E.M. Model-independent constraints on possible modifications of Newtonian gravity. *Phys. Rev. Lett.* **1988**, *61*, 1159–1162.

33. Fischbach, E.; Talmadge, C.L. *The Search for Non-Newtonian Gravity*; Aip-Press Series; Springer: New York, NY, USA, 1999.

34. Adelberger, E.G.; Heckel, B.R.; Nelson, A.E. Tests of the Gravitational Inverse-Square Law. *Annu. Rev. Nucl. Part. Sci.* **2003**, *53*, 77–121.

35. Reynaud, S.; Jaekel, M.T. Testing the Newton Law at Long Distances. *Int. J. Mod. Phys. A* **2005**, *20*, 2294–2303.

36. Bailey, Q.G. Gravity Sector of the SME. In Proceedings of the Seventh Meeting on CPT and Lorentz Symmetry, Bloomington, IN, USA, 20–24 June 2016.

37. Kostelecký, V.A. Gravity, Lorentz violation, and the standard model. *Phys. Rev. D* **2004**, *69*, 105009.

38. Bluhm, R.; Kostelecký, V.A. Spontaneous Lorentz violation, Nambu-Goldstone modes, and gravity. *Phys. Rev. D* **2005**, *71*, 065008.

39. Bailey, Q.G.; Kostelecký, V.A. Signals for Lorentz violation in post-Newtonian gravity. *Phys. Rev. D* **2006**, *74*, 045001.

40. Bluhm, R. Nambu-Goldstone Modes in Gravitational Theories with Spontaneous Lorentz Breaking. *Int. J. Mod. Phys. D* **2007**, *16*, 2357–2363.

41. Bluhm, R.; Fung, S.H.; Kostelecký, V.A. Spontaneous Lorentz and diffeomorphism violation, massive modes, and gravity. *Phys. Rev. D* **2008**, *77*, 065020.

42. Bluhm, R. Explicit versus spontaneous diffeomorphism breaking in gravity. *Phys. Rev. D* **2015**, *91*, 065034.

43. Bailey, Q.G.; Kostelecký, V.A.; Xu, R. Short-range gravity and Lorentz violation. *Phys. Rev. D* **2015**, *91*, 022006.

44. Bonder, Y. Lorentz violation in the gravity sector: The t puzzle. *Phys. Rev. D* **2015**, *91*, 125002.

45. Battat, J.B.R.; Chandler, J.F.; Stubbs, C.W. Testing for Lorentz Violation: Constraints on Standard-Model-Extension Parameters via Lunar Laser Ranging. *Phys. Rev. Lett.* **2007**, *99*, 241103.

46. Bourgoin, A.; Hees, A.; Bouquillon, S.; Le Poncin-Lafitte, C.; Francou, G.; Angonin, M.C. Testing Lorentz symmetry with Lunar Laser Ranging. *arXiv* **2016**, arXiv:gr-qc/1607.00294

47. Müller, H.; Chiow, S.W.; Herrmann, S.; Chu, S.; Chung, K.Y. Atom-Interferometry Tests of the Isotropy of Post-Newtonian Gravity. *Phys. Rev. Lett.* **2008**, *100*, 031101.

48. Chung, K.Y.; Chiow, S.W.; Herrmann, S.; Chu, S.; Müller, H. Atom interferometry tests of local Lorentz invariance in gravity and electrodynamics. *Phys. Rev. D* **2009**, *80*, 016002.

49. Iorio, L. Orbital effects of Lorentz-violating standard model extension gravitomagnetism around a static body: A sensitivity analysis. *Class. Quantum Gravity* **2012**, *29*, 175007.

50. Hees, A.; Bailey, Q.G.; Le Poncin-Lafitte, C.; Bourgoin, A.; Rivoldini, A.; Lamine, B.; Meynadier, F.; Guerlin, C.; Wolf, P. Testing Lorentz symmetry with planetary orbital dynamics. *Phys. Rev. D* **2015**, *92*, 064049.

51. Bennett, D.; Skavysh, V.; Long, J. Search for Lorentz Violation in a Short-Range Gravity Experiment. In Proceedings of the Fifth Meeting on CPT and Lorentz Symmetry, Bloomington, IN, USA, 28 June–2 July 2010.

52. Bailey, Q.G.; Everett, R.D.; Overduin, J.M. Limits on violations of Lorentz symmetry from Gravity Probe B. *Phys. Rev. D* **2013**, *88*, 102001.

53. Shao, L. Tests of Local Lorentz Invariance Violation of Gravity in the Standard Model Extension with Pulsars. *Phys. Rev. Lett.* **2014**, *112*, 111103.

54. Shao, L. New pulsar limit on local Lorentz invariance violation of gravity in the standard-model extension. *Phys. Rev. D* **2014**, *90*, 122009.

55. Le Poncin-Lafitte, C.; Hees, A.; lambert, S. Lorentz symmetry and Very Long Baseline Interferometry. *arXiv* **2016**, arXiv:gr-qc/1604.01663.

56. Kostelecký, V.A.; Tasson, J.D. Constraints on Lorentz violation from gravitational Čerenkov radiation. *Phys. Lett. B* **2015**, *749*, 551–559.

57. Shao, C.G.; Tan, Y.J.; Tan, W.H.; Yang, S.Q.; Luo, J.; Tobar, M.E. Search for Lorentz invariance violation through tests of the gravitational inverse square law at short ranges. *Phys. Rev. D* **2015**, *91*, 102007.

58. Long, J.C.; Kostelecký, V.A. Search for Lorentz violation in short-range gravity. *Phys. Rev. D* **2015**, *91*, 092003.

59. Shao, C.G.; Tan, Y.J.; Tan, W.H.; Yang, S.Q.; Luo, J.; Tobar, M.E.; Bailey, Q.G.; Long, J.C.; Weisman, E.; Xu, R.; et al. Combined Search for Lorentz Violation in Short-Range Gravity. *Phys. Rev. Lett.* **2016**, *117*, 071102.

60. Kostelecký, V.A.; Mewes, M. Testing local Lorentz invariance with gravitational waves. *Phys. Lett. B* **2016**, *757*, 510–514.

61. Yunes, N.; Yagi, K.; Pretorius, F. Theoretical physics implications of the binary black-hole mergers GW150914 and GW151226. *Phys. Rev. D* **2016**, *94*, 084002.

62. Kostelecký, V.A.; Russell, N.; Tasson, J.D. Constraints on Torsion from Bounds on Lorentz Violation. *Phys. Rev. Lett.* **2008**, *100*, 111102.

63. Heckel, B.R.; Adelberger, E.G.; Cramer, C.E.; Cook, T.S.; Schlamminger, S.; Schmidt, U. Preferred-frame and CP-violation tests with polarized electrons. *Phys. Rev. D* **2008**, *78*, 092006.

64. Kostelecký, V.A.; Mewes, M. Signals for Lorentz violation in electrodynamics. *Phys. Rev. D* **2002**, *66*, 056005.

65. Kostelecký, V.A.; Mewes, M. Electrodynamics with Lorentz-violating operators of arbitrary dimension. *Phys. Rev. D* **2009**, *80*, 015020.

66. Kostelecký, V.A. Riemann-Finsler geometry and Lorentz-violating kinematics. *Phys. Lett. B* **2011**, *701*, 137–143.

67. Kostelecký, V.A.; Russell, N. Classical kinematics for Lorentz violation. *Phys. Lett. B* **2010**, *693*, 443–447.

68. Jacobson, T.; Mattingly, D. Gravity with a dynamical preferred frame. *Phys. Rev. D* **2001**, *64*, 024028.

69. Jackiw, R.; Pi, S.Y. Chern-Simons modification of general relativity. *Phys. Rev. D* **2003**, *68*, 104012.

70. Hernaski, C.A.; Belich, H. Lorentz violation and higher derivative gravity. *Phys. Rev. D* **2014**, *89*, 104027.

71. Balakin, A.B.; Lemos, J.P.S. Einstein-aether theory with a Maxwell field: General formalism. *Ann. Phys.* **2014**, *350*, 454–484.

72. Yagi, K.; Blas, D.; Yunes, N.; Barausse, E. Strong Binary Pulsar Constraints on Lorentz Violation in Gravity. *Phys. Rev. Lett.* **2014**, *112*, 161101.

73. Yagi, K.; Blas, D.; Barausse, E.; Yunes, N. Constraints on Einstein-AEther theory and Hořava gravity from binary pulsar observations. *Phys. Rev. D* **2014**, *89*, 084067.

74. Hernaski, C.A. Quantization and stability of bumblebee electrodynamics. *Phys. Rev. D* **2014**, *90*, 124036.

75. Seifert, M.D. Vector models of gravitational Lorentz symmetry breaking. *Phys. Rev. D* **2009**, *79*, 124012.

76. Kostelecký, V.A.; Potting, R. Gravity from local Lorentz violation. *Gen. Relativ. Gravit.* **2005**, *37*, 1675–1679.

77. Kostelecký, V.A.; Potting, R. Gravity from spontaneous Lorentz violation. *Phys. Rev. D* **2009**, *79*, 065018.

78. Altschul, B.; Bailey, Q.G.; Kostelecký, V.A. Lorentz violation with an antisymmetric tensor. *Phys. Rev. D* **2010**, *81*, 065028.
79. Kostelecký, V.A.; Vargas, A.J. Lorentz and C P T tests with hydrogen, antihydrogen, and related systems. *Phys. Rev. D* **2015**, *92*, 056002.
80. Bailey, Q.G. Time delay and Doppler tests of the Lorentz symmetry of gravity. *Phys. Rev. D* **2009**, *80*, 044004.
81. Bailey, Q.G. Lorentz-violating gravitoelectromagnetism. *Phys. Rev. D* **2010**, *82*, 065012.
82. Tso, R.; Bailey, Q.G. Light-bending tests of Lorentz invariance. *Phys. Rev. D* **2011**, *84*, 085025.
83. Touboul, P.; Rodrigues, M. The MICROSCOPE space mission. *Class. Quantum Gravity* **2001**, *18*, 2487–2498.
84. Touboul, P.; Métris, G.; Lebat, V.; Robert, A. The MICROSCOPE experiment, ready for the in-orbit test of the equivalence principle. *Class. Quantum Gravity* **2012**, *29*, 184010.
85. Kostelecký, V.A.; Tasson, J.D. Prospects for Large Relativity Violations in Matter-Gravity Couplings. *Phys. Rev. Lett.* **2009**, *102*, 010402.
86. Kostelecký, V.A.; Lehnert, R. Stability, causality, and Lorentz and CPT violation. *Phys. Rev. D* **2001**, *63*, 065008.
87. Nordtvedt, K., Jr.; Will, C.M. Conservation Laws and Preferred Frames in Relativistic Gravity. II. Experimental Evidence to Rule Out Preferred-Frame Theories of Gravity. *Astrophys. J.* **1972**, *177*, 775.
88. Warburton, R.J.; Goodkind, J.M. Search for evidence of a preferred reference frame. *Astrophys. J.* **1976**, *208*, 881–886.
89. Nordtvedt, K., Jr. Anisotropic parametrized post-Newtonian gravitational metric field. *Phys. Rev. D* **1976**, *14*, 1511–1517.
90. Bordeé, C.J. Atomic interferometry with internal state labelling. *Phys. Lett. A* **1989**, *140*, 10–12.
91. Farah, T.; Guerlin, C.; Landragin, A.; Bouyer, P.; Gaffet, S.; Pereira Dos Santos, F.; Merlet, S. Underground operation at best sensitivity of the mobile LNE-SYRTE cold atom gravimeter. *Gyroscopy Navig.* **2014**, *5*, 266–274.
92. Hauth, M.; Freier, C.; Schkolnik, V.; Senger, A.; Schmidt, M.; Peters, A. First gravity measurements using the mobile atom interferometer GAIN. *Appl. Phys. B Lasers Opt.* **2013**, *113*, 49–55.
93. Hu, Z.K.; Sun, B.L.; Duan, X.C.; Zhou, M.K.; Chen, L.L.; Zhan, S.; Zhang, Q.Z.; Luo, J. Demonstration of an ultrahigh-sensitivity atom-interferometry absolute gravimeter. *Phys. Rev. A* **2013**, *88*, 043610.
94. Tamura, Y. Bulletin d'Information Marées Terrestres; Royal Observatory of Belgium: Uccle, Belgium, 1987; Volume 99, p. 6813.
95. Dehant, V.; Defraigne, P.; Wahr, J.M. Tides for a convective Earth. *J. Geophys. Res.* **1999**, *104*, 1035–1058.
96. Egbert, G.D.; Bennett, A.F.; Foreman, M.G.G. TOPEX/POSEIDON tides estimated using a global inverse model. *J. Geophys. Res.* **1994**, *99*, 24821.
97. Peters, A.; Chung, K.Y.; Chu, S. High-precision gravity measurements using atom interferometry. *Metrologia* **2001**, *38*, 25–61.
98. Merlet, S.; Kopaev, A.; Diament, M.; Geneves, G.; Landragin, A.; Pereira Dos Santos, F. Micro-gravity investigations for the LNE watt balance project. *Metrologia* **2008**, *45*, 265–274.
99. Fey, A.L.; Gordon, D.; Jacobs, C.S.; Ma, C.; Gaume, R.A.; Arias, E.F.; Bianco, G.; Boboltz, D.A.; Böckmann, S.; Bolotin, S.; et al. The Second Realization of the International Celestial Reference Frame by Very Long Baseline Interferometry. *Astron. J.* **2015**, *150*, 58.
100. Soffel, M.; Klioner, S.A.; Petit, G.; Wolf, P.; Kopeikin, S.M.; Bretagnon, P.; Brumberg, V.A.; Capitaine, N.; Damour, T.; Fukushima, T.; et al. The IAU 2000 Resolutions for Astrometry, Celestial Mechanics, and Metrology in the Relativistic Framework: Explanatory Supplement. *Astron. J.* **2003**, *126*, 2687–2706.
101. Lambert, S.B.; Le Poncin-Lafitte, C. Determining the relativistic parameter γ using very long baseline interferometry. *Astron. Astrophys.* **2009**, *499*, 331–335.
102. Lambert, S.B.; Le Poncin-Lafitte, C. Improved determination of γ by VLBI. *Astron. Astrophys.* **2011**, *529*, A70.
103. Le Poncin-Lafitte, C.; Linet, B.; Teyssandier, P. World function and time transfer: General post-Minkowskian expansions. *Class. Quantum Gravity* **2004**, *21*, 4463–4483.
104. Teyssandier, P.; Le Poncin-Lafitte, C. General post-Minkowskian expansion of time transfer functions. *Class. Quantum Gravity* **2008**, *25*, 145020.
105. Le Poncin-Lafitte, C.; Teyssandier, P. Influence of mass multipole moments on the deflection of a light ray by an isolated axisymmetric body. *Phys. Rev. D* **2008**, *77*, 044029.

106. Hees, A.; Bertone, S.; Le Poncin-Lafitte, C. Relativistic formulation of coordinate light time, Doppler, and astrometric observables up to the second post-Minkowskian order. *Phys. Rev. D* **2014**, *89*, 064045.
107. Hees, A.; Bertone, S.; Le Poncin-Lafitte, C. Light propagation in the field of a moving axisymmetric body: Theory and applications to the Juno mission. *Phys. Rev. D* **2014**, *90*, 084020.
108. Finkelstein, A.M.; Kreinovich, V.I.; Pandey, S.N. Relativistic reductions for radiointerferometric observables. *Astrophys. Space Sci.* **1983**, *94*, 233–247.
109. Petit, G.; Luzum, B. *IERS Conventions (2010)*; Bundesamt für Kartographie und Geodäsie: Frankfurt am Main, Germany, 2010.
110. Chapront, J.; Chapront-Touzé, M.; Francou, G. Determination of the lunar orbital and rotational parameters and of the ecliptic reference system orientation from LLR measurements and IERS data. *Astron. Astrophys.* **1999**, *343*, 624–633.
111. Dickey, J.O.; Bender, P.L.; Faller, J.E.; Newhall, X.X.; Ricklefs, R.L.; Ries, J.G.; Shelus, P.J.; Veillet, C.; Whipple, A.L.; Wiant, J.R.; et al. Lunar Laser Ranging: A Continuing Legacy of the Apollo Program. *Science* **1994**, *265*, 482–490.
112. Nordtvedt, K. Equivalence Principle for Massive Bodies. I. Phenomenology. *Phys. Rev.* **1968**, *169*, 1014–1016.
113. Nordtvedt, K. Equivalence Principle for Massive Bodies. II. Theory. *Phys. Rev.* **1968**, *169*, 1017–1025.
114. Nordtvedt, K. Testing Relativity with Laser Ranging to the Moon. *Phys. Rev.* **1968**, *170*, 1186–1187.
115. Williams, J.G.; Turyshev, S.G.; Boggs, D.H. Progress in Lunar Laser Ranging Tests of Relativistic Gravity. *Phys. Rev. Lett.* **2004**, *93*, 261101.
116. Williams, J.G.; Turyshev, S.G.; Boggs, D.H. Lunar Laser Ranging Tests of the Equivalence Principle with the Earth and Moon. *Int. J. Mod. Phys. D* **2009**, *18*, 1129–1175.
117. Merkowitz, S.M. Tests of Gravity Using Lunar Laser Ranging. *Living Rev. Relativ.* **2010**, *13*, 7.
118. Bertotti, B.; Iess, L.; Tortora, P. A test of general relativity using radio links with the Cassini spacecraft. *Nature* **2003**, *425*, 374–376.
119. Müller, J.; Williams, J.G.; Turyshev, S.G. Lunar Laser Ranging Contributions to Relativity and Geodesy. In *Lasers, Clocks and Drag-Free Control: Exploration of Relativistic Gravity in Space*; Astrophysics and Space Science Library; Dittus, H., Lammerzahl, C., Turyshev, S.G., Eds.; Springer: Berlin/Heidelberg, Germany, 2008; Volume 349, pp. 457–472.
120. Folkner, W.M.; Williams, J.G.; Boggs, D.H.; Park, R.; Kuchynka, P. The Planetary and Lunar Ephemeris DE 430 and DE431. *IPN Prog. Report.* **2014**, *42*, 196.
121. Lupton, R. Statistics in theory and practice. *Econ. J.* **1993**, *43*, 688–690.
122. Gottlieb, A.D. Asymptotic equivalence of the jackknife and infinitesimal jackknife variance estimators for some smooth statistics. *Ann. Inst. Stat. Math.* **2003**, *55*, 555–561.
123. Konopliv, A.S.; Park, R.S.; Yuan, D.N.; Asmar, S.W.; Watkins, M.M.; Williams, J.G.; Fahnestock, E.; Kruizinga, G.; Paik, M.; Strekalov, D.; et al. High-resolution lunar gravity fields from the GRAIL Primary and Extended Missions. *Gepphys. Res. Lett.* **2014**, *41*, 1452–1458.
124. Lemoine, F.G.; Goossens, S.; Sabaka, T.J.; Nicholas, J.B.; Mazarico, E.; Rowlands, D.D.; Loomis, B.D.; Chinn, D.S.; Neumann, G.A.; Smith, D.E.; et al. GRGM900C: A degree 900 lunar gravity model from GRAIL primary and extended mission data. *Gepphys. Res. Lett.* **2014**, *41*, 3382–3389.
125. Arnold, D.; Bertone, S.; Jäggi, A.; Beutler, G.; Mervart, L. GRAIL gravity field determination using the Celestial Mechanics Approach. *Icarus* **2015**, *261*, 182–192.
126. Standish, E.M. The JPL planetary ephemerides. *Celest. Mech.* **1982**, *26*, 181–186.
127. Newhall, X.X.; Standish, E.M.; Williams, J.G. DE 102—A numerically integrated ephemeris of the moon and planets spanning forty-four centuries. *Astron. Astrophys.* **1983**, *125*, 150–167.
128. Standish, E.M., Jr. The observational basis for JPL's DE 200, the planetary ephemerides of the Astronomical Almanac. *Astron. Astrophys.* **1990**, *233*, 252–271.
129. Standish, E.M. Testing alternate gravitational theories. *IAU Symp.* **2010**, *261*, 179–182.
130. Standish, E.M.; Williams, J.G. Orbital Ephemerides of the Sun, Moon, and Planets. In *Explanatory Supplement to the Astronomical Almanac*, 3rd ed.; Urban, S.E., Seidelmann, P.K., Eds.; Univeristy Science Books: Herndon, VA, USA, 2012; Chapter 8, pp. 305–346.
131. Hees, A.; Folkner, W.M.; Jacobson, R.A.; Park, R.S. Constraints on modified Newtonian dynamics theories from radio tracking data of the Cassini spacecraft. *Phys. Rev. D* **2014**, *89*, 102002.

132. Fienga, A.; Manche, H.; Laskar, J.; Gastineau, M. INPOP06: A new numerical planetary ephemeris. *Astron. Astrophys.* **2008**, *477*, 315–327.
133. Fienga, A.; Laskar, J.; Morley, T.; Manche, H.; Kuchynka, P.; Le Poncin-Lafitte, C.; Budnik, F.; Gastineau, M.; Somenzi, L. INPOP08, a 4-D planetary ephemeris: From asteroid and time-scale computations to ESA Mars Express and Venus Express contributions. *Astron. Astrophys.* **2009**, *507*, 1675–1686.
134. Fienga, A.; Laskar, J.; Kuchynka, P.; Le Poncin-Lafitte, C.; Manche, H.; Gastineau, M. Gravity tests with INPOP planetary ephemerides. *IAU Symp.* **2010**, *261*, 159–169.
135. Fienga, A.; Laskar, J.; Kuchynka, P.; Manche, H.; Desvignes, G.; Gastineau, M.; Cognard, I.; Theureau, G. The INPOP10a planetary ephemeris and its applications in fundamental physics. *Celest. Mech. Dyn. Astron.* **2011**, *111*, 363–385.
136. Verma, A.K.; Fienga, A.; Laskar, J.; Manche, H.; Gastineau, M. Use of MESSENGER radioscience data to improve planetary ephemeris and to test general relativity. *Astron. Astrophys.* **2014**, *561*, A115.
137. Fienga, A.; Laskar, J.; Exertier, P.; Manche, H.; Gastineau, M. Numerical estimation of the sensitivity of INPOP planetary ephemerides to general relativity parameters. *Celest. Mech. Dyn. Astron.* **2015**, *123*, 325–349.
138. Pitjeva, E.V. High-Precision Ephemerides of Planets EPM and Determination of Some Astronomical Constants. *Sol. Syst. Res.* **2005**, *39*, 176–186.
139. Pitjeva, E.V. EPM ephemerides and relativity. *IAU Symp.* **2010**, *261*, 170–178.
140. Pitjeva, E.V.; Pitjev, N.P. Relativistic effects and dark matter in the Solar system from observations of planets and spacecraft. *Mon. Not. R. Astrono. Soc.* **2013**, *432*, 3431–3437.
141. Pitjeva, E.V. Updated IAA RAS planetary ephemerides-EPM2011 and their use in scientific research. *Sol. Syst. Res.* **2013**, *47*, 386–402.
142. Pitjeva, E.V.; Pitjev, N.P. Development of planetary ephemerides EPM and their applications. *Celest. Mech. Dyn. Astron.* **2014**, *119*, 237–256.
143. Konopliv, A.S.; Asmar, S.W.; Folkner, W.M.; Karatekin, O.; Nunes, D.C.; Smrekar, S.E.; Yoder, C.F.; Zuber, M.T. Mars high resolution gravity fields from MRO, Mars seasonal gravity, and other dynamical parameters. *Icarus* **2011**, *211*, 401–428.
144. Pitjev, N.P.; Pitjeva, E.V. Constraints on dark matter in the solar system. *Astron. Lett.* **2013**, *39*, 141–149.
145. Milgrom, M. MOND effects in the inner Solar system. *Mon. Not. R. Astrono. Soc.* **2009**, *399*, 474–486.
146. Blanchet, L.; Novak, J. External field effect of modified Newtonian dynamics in the Solar system. *Mon. Not. R. Astrono. Soc.* **2011**, *412*, 2530–2542.
147. Hees, A.; Famaey, B.; Angus, G.W.; Gentile, G. Combined Solar system and rotation curve constraints on MOND. *Mon. Not. R. Astrono. Soc.* **2016**, *455*, 449–461.
148. Hees, A.; Lamine, B.; Reynaud, S.; Jaekel, M.T.; Le Poncin-Lafitte, C.; Lainey, V.; Füzfa, A.; Courty, J.M.; Dehant, V.; Wolf, P. Radioscience simulations in General Relativity and in alternative theories of gravity. *Class. Quantum Gravity* **2012**, *29*, 235027.
149. de Sitter, W. Einstein's theory of gravitation and its astronomical consequences. *Mon. Not. R. Astrono. Soc.* **1916**, *76*, 699–728.
150. Lense, J.; Thirring, H. Über den Einfluß der Eigenrotation der Zentralkörper auf die Bewegung der Planeten und Monde nach der Einsteinschen Gravitationstheorie. *Phys. Z.* **1918**, *19*, 156.
151. Schiff, L.I. Possible New Experimental Test of General Relativity Theory. *Phys. Rev. Lett.* **1960**, *4*, 215–217.
152. Pugh, G.E. Proposal for a Satellite Test of the Coriolis Prediction of General Relativity. In *Nonlinear Gravitodynamics: The Lense-Thirring Effect*; Word Scientific Publishing: Singapore, 1959.
153. Everitt, C.W.F.; Debra, D.B.; Parkinson, B.W.; Turneaure, J.P.; Conklin, J.W.; Heifetz, M.I.; Keiser, G.M.; Silbergleit, A.S.; Holmes, T.; Kolodziejczak, J.; et al. Gravity Probe B: Final Results of a Space Experiment to Test General Relativity. *Phys. Rev. Lett.* **2011**, *106*, 221101.
154. Stella, L.; Vietri, M. kHz Quasiperiodic Oscillations in Low-Mass X-Ray Binaries as Probes of General Relativity in the Strong-Field Regime. *Phys. Rev. Lett.* **1999**, *82*, 17–20.
155. Hulse, R.A.; Taylor, J.H. Discovery of a pulsar in a binary system. *Astrophys. J.* **1975**, *195*, L51–L53.
156. Taylor, J.H.; Hulse, R.A.; Fowler, L.A.; Gullahorn, G.E.; Rankin, J.M. Further observations of the binary pulsar PSR 1913+16. *Astrophys. J.* **1976**, *206*, L53–L58.
157. Taylor, J.H.; Fowler, L.A.; McCulloch, P.M. Measurements of general relativistic effects in the binary pulsar PSR 1913+16. *Nature* **1979**, *277*, 437–440.

158. Damour, T.; Deruelle, N. General relativistic celestial mechanics of binary systems. II. The post-Newtonian timing formula. *Ann. Inst. Henri Poincaré Phys. Théor.* **1986**, *44*, 263–292.
159. Stairs, I.H. Testing General Relativity with Pulsar Timing. *Living Rev. Relativ.* **2003**, *6*, 5.
160. Lorimer, D.R. Binary and Millisecond Pulsars. *Living Rev. Relativ.* **2008**, *11*, 8.
161. Damour, T.; Deruelle, N. General relativistic celestial mechanics of binary systems. I. The post-Newtonian motion. *Ann. Inst. Henri Poincaré Phys. Théor.* **1985**, *43*, 107–132.
162. Wex, N. The second post-Newtonian motion of compact binary-star systems with spin. *Class. Quantum Gravity* **1995**, *12*, 983–1005.
163. Edwards, R.T.; Hobbs, G.B.; Manchester, R.N. TEMPO2, a new pulsar timing package - II. The timing model and precision estimates. *Mon. Not. R. Astrono. Soc.* **2006**, *372*, 1549–1574.
164. Wex, N. Testing Relativistic Gravity with Radio Pulsars. In *Frontiers in Relativistic Celestial Mechanics; Applications and Experiments*; Kopeikin, S., Ed.; De Gruyter: Berlin, Germany, 2014; Volume 2.
165. Kramer, M. Pulsars as probes of gravity and fundamental physics. *Int. J. Mod. Phys. D* **2016**, *25*, 14.
166. Kramer, M.; Stairs, I.H.; Manchester, R.N.; McLaughlin, M.A.; Lyne, A.G.; Ferdman, R.D.; Burgay, M.; Lorimer, D.R.; Possenti, A.; D'Amico, N.; et al. Tests of General Relativity from Timing the Double Pulsar. *Science* **2006**, *314*, 97–102.
167. Damour, T.; Esposito-Farese, G. Tensor-multi-scalar theories of gravitation. *Class. Quantum Gravity* **1992**, *9*, 2093–2176.
168. Damour, T.; Taylor, J.H. Strong-field tests of relativistic gravity and binary pulsars. *Phys. Rev. D* **1992**, *45*, 1840–1868.
169. Damour, T.; Esposito-Farèse, G. Tensor-scalar gravity and binary-pulsar experiments. *Phys. Rev. D* **1996**, *54*, 1474–1491.
170. Freire, P.C.C.; Wex, N.; Esposito-Farèse, G.; Verbiest, J.P.W.; Bailes, M.; Jacoby, B.A.; Kramer, M.; Stairs, I.H.; Antoniadis, J.; Janssen, G.H. The relativistic pulsar-white dwarf binary PSR J1738+0333 - II. The most stringent test of scalar-tensor gravity. *Mon. Not. R. Astrono. Soc.* **2012**, *423*, 3328–3343.
171. Ransom, S.M.; Stairs, I.H.; Archibald, A.M.; Hessels, J.W.T.; Kaplan, D.L.; van Kerkwijk, M.H.; Boyles, J.; Deller, A.T.; Chatterjee, S.; Schechtman-Rook, A.; et al. A millisecond pulsar in a stellar triple system. *Nature* **2014**, *505*, 520–524.
172. Damour, T.; Esposito-Farese, G. Nonperturbative strong-field effects in tensor-scalar theories of gravitation. *Phys. Rev. Lett.* **1993**, *70*, 2220–2223.
173. Foster, B.Z. Strong field effects on binary systems in Einstein-aether theory. *Phys. Rev. D* **2007**, *76*, 084033.
174. Wex, N.; Kramer, M. A characteristic observable signature of preferred-frame effects in relativistic binary pulsars. *Mon. Not. R. Astrono. Soc.* **2007**, *380*, 455–465.
175. Shao, L.; Wex, N. New tests of local Lorentz invariance of gravity with small-eccentricity binary pulsars. *Class. Quantum Gravity* **2012**, *29*, 215018.
176. Nordtvedt, K. Probing gravity to the second post-Newtonian order and to one part in 10 to the 7th using the spin axis of the sun. *Astrophys. J.* **1987**, *320*, 871–874.
177. Shao, L.; Caballero, R.N.; Kramer, M.; Wex, N.; Champion, D.J.; Jessner, A. A new limit on local Lorentz invariance violation of gravity from solitary pulsars. *Class. Quantum Gravity* **2013**, *30*, 165019.
178. Bell, J.F.; Damour, T. A new test of conservation laws and Lorentz invariance in relativistic gravity. *Class. Quantum Gravity* **1996**, *13*, 3121–3127.
179. Gonzalez, M.E.; Stairs, I.H.; Ferdman, R.D.; Freire, P.C.C.; Nice, D.J.; Demorest, P.B.; Ransom, S.M.; Kramer, M.; Camilo, F.; Hobbs, G.; et al. High-precision Timing of Five Millisecond Pulsars: Space Velocities, Binary Evolution, and Equivalence Principles. *Astrophys. J.* **2011**, *743*, 102.
180. Lorimer, D.R.; Kramer, M. *Handbook of Pulsar Astronomy*; Cambridge University Press: Cambridge, UK, 2004.
181. Jennings, R.J.; Tasson, J.D.; Yang, S. Matter-sector Lorentz violation in binary pulsars. *Phys. Rev. D* **2015**, *92*, 125028.
182. Moore, G.D.; Nelson, A.E. Lower bound on the propagation speed of gravity from gravitational Cherenkov radiation. *J. High Energy Phys.* **2001**, *9*, 023.
183. Kiyota, S.; Yamamoto, K. Constraint on modified dispersion relations for gravitational waves from gravitational Cherenkov radiation. *Phys. Rev. D* **2015**, *92*, 104036.
184. Elliott, J.W.; Moore, G.D.; Stoica, H. Constraining the New Aether: Gravitational Cherenkov radiation. *J. High Energy Phys.* **2005**, *8*, 066.

185. Kimura, R.; Yamamoto, K. Constraints on general second-order scalar-tensor models from gravitational Cherenkov radiation. *J. Cosmol. Astropart. Phys.* **2012**, *7*, 050

186. De Laurentis, M.; Capozziello, S.; Basini, G. Gravitational Cherenkov Radiation from Extended Theories of Gravity. *Mod. Phys. Lett. A* **2012**, *27*, 1250136.

187. Kimura, R.; Tanaka, T.; Yamamoto, K.; Yamashita, Y. Constraint on ghost-free bigravity from gravitational Cherenkov radiation. *Phys. Rev. D* **2016**, *94*, 064059.

188. Takeda, M.; Hayashida, N.; Honda, K.; Inoue, N.; Kadota, K.; Kakimoto, F.; Kamata, K.; Kawaguchi, S.; Kawasaki, Y.; Kawasumi, N.; et al. Small-Scale Anisotropy of Cosmic Rays above 10^{19} eV Observed with the Akeno Giant Air Shower Array. *Astrophys. J.* **1999**, *522*, 225–237.

189. Bird, D.J.; Corbato, S.C.; Dai, H.Y.; Elbert, J.W.; Green, K.D.; Huang, M.A.; Kieda, D.B.; Ko, S.; Larsen, C.G.; Loh, E.C.; et al. Detection of a cosmic ray with measured energy well beyond the expected spectral cutoff due to cosmic microwave radiation. *Astrophys. J.* **1995**, *441*, 144–150.

190. Wada, M. *Catalogue of Highest Energy Cosmic Rays. Giant Extensive Air Showers. No._1. Volcano Ranch, Haverah Park*; Institute of Physical and Chemical Research: Tokyo, Japan, 1980.

191. High Resolution Fly'S Eye Collaboration.; Abbasi, R.U.; Abu-Zayyad, T.; Allen, M.; Amman, J.F.; Archbold, G.; Belov, K.; Belz, J.W.; BenZvi, S.Y.; Bergman, D.R.; et al. Search for correlations between HiRes stereo events and active galactic nuclei. *Astropart. Phys.* **2008**, *30*, 175–179.

192. Aab, A.; Abreu, P.; Aglietta, M.; Ahn, E.J.; Al Samarai, I.; Albuquerque, I.F.M.; Allekotte, I.; Allen, J.; Allison, P.; Almela, A.; et al. Searches for Anisotropies in the Arrival Directions of the Highest Energy Cosmic Rays Detected by the Pierre Auger Observatory. *Astrophys. J.* **2015**, *804*, 15.

193. Winn, M.M.; Ulrichs, J.; Peak, L.S.; McCusker, C.B.A.; Horton, L. The cosmic-ray energy spectrum above 10^{17} eV. *J. Phys. G Nucl. Phys.* **1986**, *12*, 653–674.

194. Abbasi, R.U.; Abe, M.; Abu-Zayyad, T.; Allen, M.; Anderson, R.; Azuma, R.; Barcikowski, E.; Belz, J.W.; Bergman, D.R.; Blake, S.A.; et al. Indications of Intermediate-scale Anisotropy of Cosmic Rays with Energy Greater Than 57 EeV in the Northern Sky Measured with the Surface Detector of the Telescope Array Experiment. *Astrophys. J.* **2014**, *790*, L21.

195. Pravdin, M.I.; Glushkov, A.V.; Ivanov, A.A.; Knurenko, S.P.; Kolosov, V.A.; Makarov, I.T.; Sabourov, A.V.; Sleptsov, I.Y.; Struchkov, G.G. Estimation of the giant shower energy at the Yakutsk EAS Array. *Int. Cosm. Ray Conf.* **2005**, *7*, 243.

196. De Bruijne, J.H.J. Science performance of Gaia, ESA's space-astrometry mission. *Astrophys. Space Sci.* **2012**, *341*, 31–41.

197. Mignard, F.; Klioner, S.A. Gaia: Relativistic modelling and testing. *IAU Symp.* **2010**, *261*, 306–314.

198. Jaekel, M.T.; Reynaud, S. Gravity Tests in the Solar System and the Pioneer Anomaly. *Mod. Phys. Lett. A* **2005**, *20*, 1047–1055.

199. Jaekel, M.T.; Reynaud, S. Post-Einsteinian tests of linearized gravitation. *Class. Quantum Gravity* **2005**, *22*, 2135–2157.

200. Jaekel, M.T.; Reynaud, S. Post-Einsteinian tests of gravitation. *Class. Quantum Gravity* **2006**, *23*, 777–798.

201. Reynaud, S.; Jaekel, M.T. Long Range Gravity Tests and the Pioneer Anomaly. *Int. J. Mod. Phys. D* **2007**, *16*, 2091–2105.

202. Reynaud, S.; Jaekel, M.T. Tests of general relativity in the Solar System. *Atom Opt. Space Phys.* **2009**, *168*, 203–207.

203. Gai, M.; Vecchiato, A.; Ligori, S.; Sozzetti, A.; Lattanzi, M.G. Gravitation astrometric measurement experiment. *Exp. Astron.* **2012**, *34*, 165–180.

204. Turyshev, S.G.; Shao, M. Laser Astrometric Test of Relativity: Science, Technology and Mission Design. *Int. J. Mod. Phys. D* **2007**, *16*, 2191–2203.

205. Mouret, S. Tests of fundamental physics with the Gaia mission through the dynamics of minor planets. *Phys. Rev. D* **2011**, *84*, 122001.

206. Hees, A.; Hestroffer, D.; Le Poncin-Lafitte, C.; David, P. Tests of gravitation with GAIA observations of Solar System Objects. In Proceedings of the Annual meeting of the French Society of Astronomy and Astrophysics (SF2A-2015), Toulouse, France, 2–5 June 2015; Martins, F., Boissier, S., Buat, V., Cambrésy, L., Petit, P., Eds.; pp. 125–131.

207. Margot, J.L.; Giorgini, J.D. Probing general relativity with radar astrometry in the inner solar system. *IAU Symp.* **2010**, *261*, 183–188.

208. Hees, A.; Lamine, B.; Poncin-Lafitte, C.L.; Wolf, P. How to Test the SME with Space Missions? In Proceedings of the Sixth Meeting CPT and Lorentz Symmetry, Bloomington, IN, USA, 17–21 June 2013; Kostelecky, A., Ed.; pp. 107–110.

209. Hees, A.; Lamine, B.; Reynaud, S.; Jaekel, M.T.; Le Poncin-Lafitte, C.; Lainey, V.; Füzfa, A.; Courty, J.M.; Dehant, V.; Wolf, P. Simulations of Solar System Observations in Alternative Theories of Gravity. In Proceedings of the Thirteenth Marcel Grossmann Meeting: On Recent Developments in Theoretical and Experimental General Relativity, Astrophysics and Relativistic Field Theories, Stockholm, Sweden, 1–7 July 2012; Rosquist, K., Ed.; pp. 2357–2359.

210. Iess, L.; Asmar, S. Probing Space-Time in the Solar System: From Cassini to Bepicolombo. *Int. J. Mod. Phys. D* **2007**, *16*, 2117–2126.

211. Kliore, A.J.; Anderson, J.D.; Armstrong, J.W.; Asmar, S.W.; Hamilton, C.L.; Rappaport, N.J.; Wahlquist, H.D.; Ambrosini, R.; Flasar, F.M.; French, R.G.; et al. Cassini Radio Science. *Space Sci. Rev.* **2004**, *115*, 1–70.

212. Iorio, L.; Ciufolini, I.; Pavlis, E.C. Measuring the relativistic perigee advance with satellite laser ranging. *Class. Quantum Gravity* **2002**, *19*, 4301–4309.

213. Lucchesi, D.M.; Peron, R. Accurate Measurement in the Field of the Earth of the General-Relativistic Precession of the LAGEOS II Pericenter and New Constraints on Non-Newtonian Gravity. *Phys. Rev. Lett.* **2010**, *105*, 231103.

214. Lucchesi, D.M.; Peron, R. LAGEOS II pericenter general relativistic precession (1993–2005): Error budget and constraints in gravitational physics. *Phys. Rev. D* **2014**, *89*, 082002.

215. Ciufolini, I.; Pavlis, E.C. A confirmation of the general relativistic prediction of the Lense-Thirring effect. *Nature* **2004**, *431*, 958–960.

216. Ciufolini, I.; Paolozzi, A.; Pavlis, E.C.; Koenig, R.; Ries, J.; Gurzadyan, V.; Matzner, R.; Penrose, R.; Sindoni, G.; Paris, C.; et al. A test of general relativity using the LARES and LAGEOS satellites and a GRACE Earth gravity model. Measurement of Earth's dragging of inertial frames. *Eur. Phys. J. C* **2016**, *76*, 120.

217. Ciufolini, I.; Paolozzi, A.; Pavlis, E.C.; Ries, J.C.; Koenig, R.; Matzner, R.A.; Sindoni, G.; Neumayer, H. Towards a One Percent Measurement of Frame Dragging by Spin with Satellite Laser Ranging to LAGEOS, LAGEOS 2 and LARES and GRACE Gravity Models. *Space Sci. Rev.* **2009**, *148*, 71–104.

218. Ciufolini, I.; Paolozzi, A.; Pavlis, E.; Ries, J.; Gurzadyan, V.; Koenig, R.; Matzner, R.; Penrose, R.; Sindoni, G. Testing General Relativity and gravitational physics using the LARES satellite. *Eur. Phys. J. Plus* **2012**, *127*, 133.

219. Ciufolini, I.; Pavlis, E.C.; Paolozzi, A.; Ries, J.; Koenig, R.; Matzner, R.; Sindoni, G.; Neumayer, K.H. Phenomenology of the Lense-Thirring effect in the Solar System: Measurement of frame-dragging with laser ranged satellites. *New Astron.* **2012**, *17*, 341–346.

220. Paolozzi, A.; Ciufolini, I.; Vendittozzi, C. Engineering and scientific aspects of LARES satellite. *Acta Astronaut.* **2011**, *69*, 127–134.

221. Iorio, L. Towards a 1% measurement of the Lense-Thirring effect with LARES? *Adv. Space Res.* **2009**, *43*, 1148–1157.

222. Iorio, L. Will the recently approved LARES mission be able to measure the Lense-Thirring effect at 1%? *Gen. Relativ. Gravit.* **2009**, *41*, 1717–1724.

223. Iorio, L. An Assessment of the Systematic Uncertainty in Present and Future Tests of the Lense-Thirring Effect with Satellite Laser Ranging. *Space Sci. Rev.* **2009**, *148*, 363–381.

224. Iorio, L.; Lichtenegger, H.I.M.; Ruggiero, M.L.; Corda, C. Phenomenology of the Lense-Thirring effect in the solar system. *Astrophys. Space Sci.* **2011**, *331*, 351–395.

225. Renzetti, G. Are higher degree even zonals really harmful for the LARES/LAGEOS frame-dragging experiment? *Can. J. Phys.* **2012**, *90*, 883–888.

226. Renzetti, G. First results from LARES: An analysis. *New Astron.* **2013**, *23*, 63–66.

227. Lucchesi, D.M.; Anselmo, L.; Bassan, M.; Pardini, C.; Peron, R.; Pucacco, G.; Visco, M. Testing the gravitational interaction in the field of the Earth via satellite laser ranging and the Laser Ranged Satellites Experiment (LARASE). *Class. Quantum Gravity* **2015**, *32*, 155012.

228. Lämmerzahl, C.; Ciufolini, I.; Dittus, H.; Iorio, L.; Müller, H.; Peters, A.; Samain, E.; Scheithauer, S.; Schiller, S. OPTIS–An Einstein Mission for Improved Tests of Special and General Relativity. *Gen. Relativ. Gravit.* **2004**, *36*, 2373–2416.

229. Hohensee, M.A.; Chu, S.; Peters, A.; Müller, H. Equivalence Principle and Gravitational Redshift. *Phys. Rev. Lett.* **2011**, *106*, 151102.
230. Hohensee, M.A.; Leefer, N.; Budker, D.; Harabati, C.; Dzuba, V.A.; Flambaum, V.V. Limits on Violations of Lorentz Symmetry and the Einstein Equivalence Principle using Radio-Frequency Spectroscopy of Atomic Dysprosium. *Phys. Rev. Lett.* **2013**, *111*, 050401.
231. Hohensee, M.A.; Müller, H.; Wiringa, R.B. Equivalence Principle and Bound Kinetic Energy. *Phys. Rev. Lett.* **2013**, *111*, 151102.
232. Delva, P.; Hees, A.; Bertone, S.; Richard, E.; Wolf, P. Test of the gravitational redshift with stable clocks in eccentric orbits: Application to Galileo satellites 5 and 6. *Class. Quantum Gravity* **2015**, *32*, 232003.
233. Cacciapuoti, L.; Salomon, C. Atomic clock ensemble in space. *J. Phys. Conf. Ser.* **2011**, *327*, 012049.

universe   MDPI

Article

Theoretical Tools for Relativistic Gravimetry, Gradiometry and Chronometric Geodesy and Application to a Parameterized Post-Newtonian Metric

Pacôme Delva [1],* and Jan Geršl [2]

[1] SYRTE, Observatoire de Paris, PSL Research University, CNRS, Sorbonne Universités, UPMC University, Paris 06, LNE, 61 avenue de l'Observatoire, 75014 Paris, France
[2] Czech Metrology Institute, Okružní 31, 63800 Brno, Czech Republic; jgersl@cmi.cz
* Correspondence: pacome.delva@obspm.fr

Academic Editors: Stephon Alexander, Jean-Michel Alimi, Elias C. Vagenas and Lorenzo Iorio
Received: 2 February 2017; Accepted: 8 March 2017; Published: 13 March 2017

Abstract: An extensive review of past work on relativistic gravimetry, gradiometry and chronometric geodesy is given. Then, general theoretical tools are presented and applied for the case of a stationary parameterized post-Newtonian metric. The special case of a stationary clock on the surface of the Earth is studied.

Keywords: general relativity; relativistic geodesy; relativistic gradiometry; relativistic gravimetry; relativistic time and frequency transfer; chronometric geodesy; parameterized post-Newtonian theory

1. Introduction

Physical geodesy is the study of the gravity field and of the figure of the Earth. One classical way to describe the "figure of the Earth" is to study the geoid, which is defined as one of the equipotentials of the Earth's gravity (Newtonian) potential, which best coincides with the (mean) surface of the oceans. Therefore, heights are both physically and geometrically defined, and the objects of study of physical geodesy are both the physical Earth (underground masses, Earth and ocean topography, Earth rotation, etc.) and the gravitational field it generates. It implies a high intricacy between the three main pillars of geodesy—the determination of variations of Earth's rotation, the geometric shape of the Earth and the spatial and temporal variations of its gravity field—but also with geodynamics and geophysics. Therefore, the International Association of Geodesy (IAG) established a "flagship" project named GGOS (Global Geodetic Observing System); it aims at connecting different communities in order to have a global understanding of the Earth system and to develop a common theoretical framework of high accuracy, which has to be based consistently (whenever this is necessary with respect to the measurement accuracy goal) on Einstein special and general relativity.

Many different techniques are used to monitor the Earth's system: space geodetic techniques (VLBI, SLR/LLR, GNSS, DORIS, altimetry, InSAR and gravity missions), as well as terrestrial techniques (leveling, absolute and relative gravimetry, gradiometry and tide gauges). Moreover, the advent of space and ground transportable atomic clocks [1,2] will bring a completely new observable in geodesy: the direct measurement of gravity potential differences [3–5]. The high accuracy of most of these techniques necessitates their description in a relativistic framework. A review of several space geodetic techniques in a relativistic framework can be found in [6], as well as a detailed relativistic model for VLBI observations in [7].

In this article, we develop a general framework for relativistic geodesy with a focus on relativistic gravimetry, gradiometry and chronometric geodesy. An extensive review of these scientific fields is given in Section 2. In Section 3, we introduce some theoretical tools necessary to do the calculation of relativistic geodesy observables: the local frame, the geodesic equation and a theoretical description of the observables of relativistic gravimetry, gradiometry and chronometric geodesy. In Section 4, we calculate these observables for the special case of a parameterized post-Newtonian (PPN) metric of a stationary spacetime. Finally, in Section 5, we apply further our calculations to the case of a static clock relative to the Earth surface and give orders of magnitude.

2. Review of Past Work in Relativistic Gravimetry, Gradiometry and Chronometric Geodesy

Probably the first author who began the theoretical investigation of relativistic effects in gravimetry was Will [8]. He determined the Newtonian gravitational constant G as measured locally by means of Cavendish experiments in a parametrized post-Newtonian (PPN) framework [9–11], showing that in such a theoretical framework, an anisotropy appears in the locally measured G. This anisotropy implies a variation in gravimeter readings, such that $\Delta g/g = \alpha(\Delta G/G)$, which have periods of 12 h sidereal time. By comparing with gravimeter data (measurements of "Earth tides"), he was able to rule out Whitehead's theory, which predicted an effect 200-times larger than the experimental limit, as well as putting an upper limit on the parameter combination $(\Delta_2 + \xi - 1)$ to within three percent.

2.1. Chronometric Geodesy

The next application of relativistic geodesy to be explored, and probably the most interesting and promising, is the use of clocks to determine the spacetime metric. Indeed, the gravitational redshift effect discovered by Einstein must be taken into account when comparing the frequencies of distant clocks. Instead of using our knowledge of the Earth gravitational field to predict frequency shifts between distant clocks, one can revert the problem and ask if the measurement of frequency shifts between distant clocks can improve our knowledge of the gravitational field. To do simple orders of magnitude estimates, it is good to have in mind some correspondences:

$$1 \text{ meter} \leftrightarrow \frac{\Delta \nu}{\nu} \sim 10^{-16} \leftrightarrow \Delta W \sim 10 \text{ m}^2 \cdot \text{s}^{-2} \tag{1}$$

where one meter is the height difference between two clocks, $\Delta \nu$ is the frequency difference in a frequency transfer between the same two clocks and ΔW is the gravity potential difference between the locations of these clocks.

From this correspondence, we can already recognize two direct applications of clocks in geodesy: if we are capable to compare clocks to 10^{-16} accuracy, we can determine height differences between clocks with one-meter accuracy (leveling) or determine geopotential differences with 10-m$^2 \cdot$s^{-2} accuracy.

The first article to explore seriously this possibility was written in 1983 [12]. The article is named "chronometric leveling". The term "chronometric" seems well suited for qualifying the method of using clocks to determine directly gravitational potential differences, as "chronometry" is the science of the measurement of time. However, the term "leveling" seems too restrictive with respect to all of the applications one could think of for using the results of clock comparisons. Therefore, we will use the term "chronometric geodesy" to name of the scientific discipline that deals with the measurement and representation of the Earth, including its gravitational field, with the help of atomic clocks. It is sometimes named "clock-based geodesy" or "relativistic geodesy". However, this last designation is improper as relativistic geodesy aims at describing all possible techniques (including, e.g., gravimetry and gradiometry) in a relativistic framework. The natural arena of chronometric geodesy is the four-dimensional space-time. At the lowest order, there is proportionality between relative frequency shift measurements, corrected from the first order Doppler effect, and (Newtonian) gravity potential differences. To calculate this relation, one does not need the theory of general relativity, but only to postulate local position invariance. Therefore, if the measurement accuracy does not reach the magnitude of the higher order terms, it is perfectly possible to use clock comparison measurements, corrected for the first order Doppler effect, as a direct measurement of (the differences of) the gravity potential that is considered in classical geodesy. Comparisons between two clocks on the ground generally use a third clock in space. In this article, we calculate explicitly the higher order terms in the PPN formalism.

In his article, Martin Vermeer explores the "possibilities for technical realization of a system for measuring potential differences over intercontinental distances" using clock comparisons [12]. The two main ingredients are of course accurate clocks and a means to compare them. He considers hydrogen maser clocks. For the links, he considers a two-way satellite link over a geostationary satellite, or GPS receivers in interferometric mode. He has also to consider a means to compare the proper frequencies of the different hydrogen maser clocks. Today, this can be overcome by comparing primary frequency standards (PFS), which have a well-defined proper frequency based on a transition of cesium 133, used for the definition of the second. Secondary frequency standards, i.e., standards based on a transition other than the defining one, may nevertheless be used if the uncertainty in systematic effects has been fully evaluated, in the same way as for a PFS. It often happens that this evaluation can be done more accurately than for the defining transition. This is one of the purposes of the European project "International timescales with optical clocks" [13] (projects.npl.co.uk/itoc), where optical clocks based on different atoms are compared to each other locally and to the PFS. It is planned also to do a proof-of-principle experiment of chronometric geodesy, by comparing two optical clocks separated by a height difference of around 1 km using an optical fiber link. For more information about atomic clock relativistic time and frequency transfer, see [3,14].

Few authors have seriously considered chronometric geodesy. Following the Vermeer idea, the possibility of using GPS observations to solve the problem of the determination of geoid heights has been explored in [15]. They consider two techniques based on frequency comparisons and direct clock readings. However, they leave aside the practical feasibility of such techniques. The value and future applicability of chronometric geodesy has been discussed in [16], including direct geoid mapping on continents and joint gravity-geopotential surveying to invert for subsurface density anomalies. They find that a geoid perturbation caused by a 1.5-km radius sphere with a 20 percent density anomaly buried at a 2-km depth in the Earth's crust is already detectable by atomic clocks of achievable accuracy. The potentiality of the new generation of atomic clocks has been shown in [17], based on optical transitions, to measure heights with a resolution of around 30 cm.

The possibility of determining the geopotential at high spatial resolution thanks to chronometric geodesy is thoroughly explored and evaluated in [18]. The authors consider the Alps-Mediterranean area, which comprises high reliefs and a land/sea transition, leading to variations of the gravitational field over a range of spatial scales. In such type of region, the scarcity of gravity data is an important limitation in deriving accurate high resolution geopotential models. Through numerical simulations, the contribution of clocks comparisons data in the geopotential recovery is assessed in combination with ground gravity measurements. It is shown that adding only a few clock data (around 30 comparisons) reduces the geopotential recovery bias significantly and improves the standard deviation by a factor of three. The effect of the data coverage and data quality are explored, as well as the trade-off between the measurement noise level and the number of data.

2.2. The Chronometric Geoid

Arne Bjerhammar in 1985 gave a precise definition of the "relativistic geoid" [19,20]:

"The relativistic geoid is the surface where precise clocks run with the same speed and the surface is nearest to mean sea level"

This is an operational definition, which has been translated in the context of post-Newtonian theory [21,22]. A different operational definition of the relativistic geoid has been introduced based on gravimetric measurements: a surface orthogonal everywhere to the direction of the plumb-line and closest to mean sea level. They call the two surfaces obtained with clocks and gravimetric measurements respectively the "u-geoid" and the "a-geoid". They prove that these two surfaces coincide in the case of a stationary metric. In order to distinguish the operational definition of the geoid from its theoretical description, it is less ambiguous to give a name based on the particular technique to measure it. The term "relativistic geoid" is too vague, as Soffel et al. have defined two different ones. The names chosen by Soffel et al. are not particularly explicit, so instead of "u-geoid" and "a-geoid", one can call them "chronometric geoid" and "gravimetric geoid", respectively. There can be no confusion with the geoid derived from satellite measurements, as this is a quasi-geoid that does not coincide with the geoid on the continents [23]. Other considerations on the chronometric geoid can be found in [6,24,25].

We notice that the problem of defining a reference isochronometric surface is closely related to the problem of realizing terrestrial time (TT). This is developed in more detail in Section 3.4.

Recently, extensive work has been done aiming at developing an exact relativistic theory of Earth's geoid undulation [26], as well as developing a theory of the reference level surface in the context of post-Newtonian gravity [27,28]. This goes beyond the problem of the realization of a reference isochronometric surface and tackles the tough work of extending all concepts of classical physical geodesy (see, e.g., [23]) in the framework of general relativity.

2.3. Gravimetry and Gradiometry

Following Will's work, a PPN theory of gravimetric measurements was developed [21] taking into account only PPN parameters γ and β [11], with an accuracy of 10^{-11} g. In particular, they take into account the influence of all bodies in the Solar System and show that the relative second order corrections to gravimetric measurements (of order c^{-2}) are of the form $(\gamma + 2\beta - 2)U_\oplus/c^2$ and $(\gamma - 4\beta + 3)U_*/c^2$, where $U_\oplus$ and U_* are respectively potentials related to the Earth and to all other Solar System bodies. It is claimed in [29] that it is impossible to measure the second order corrections to gravimetric measurements with two measurements, one at the South Pole and another one at the Equator, because of the errors induced by the uncertainty in the Earth's flatness and mean equatorial radius. However, the study could go further and consider using more points at different latitudes.

In parallel, a theory of gradiometry measurements was developed, with a particular emphasis on measurements on-board a satellite and the feasibility of such measurements with superconducting gradiometers [30–35]. Recently, a test of the Chern–Simons modified gravity has been proposed with such an experiment [36], as well as a test of post-Newtonian physics of semi-conservative metric theories [37].

3. Theoretical Tools of Relativistic Geodesy

The theoretical background for relativistic geodesy is general relativity. We consider the spacetime as a Lorentzian manifold $(\mathcal{M}, g)$ of dimension four. We consider the components of the metric $g_{\mu\nu}$ to be given in an initial coordinate system (x^μ), defined in an open subset $\mathcal{U}$. The infinitesimal interval $ds^2 = g_{\mu\nu}dx^\mu dx^\nu$ between two neighboring events is invariant under coordinate transformation.

All of the theoretical tools introduced in this section can be applied to any initial metric g, which is an exact or approximate solution of the Einstein equations and the components of which are given in any coordinate system. For applications in the vicinity of the Earth or in the Solar System, the International Astronomical Union (IAU) recommends to use respectively the GCRSor the BCRS, which both use harmonic coordinates. Explicit expressions of the metric components in these coordinate systems are given in [38]. Other approaches exist in this context based on generalized Fermi coordinates [39–41], or a perturbed Schwarzschild metric [42]. In the different context of a slowly-rotating astronomical object, the Kerr metric is used in [35].

The goal of this section is to describe a local experiment, such as a gravimeter, a gradiometer or a clock. To do so, we need to introduce a local frame and local coordinates adapted to the apparatus. There are several ways of introducing a local frame and coordinates [43]. From the principle of general covariance, any coordinate system can be used to describe the local measurements. Here, we use Fermi normal coordinates, which have the advantage of displaying "beautiful ties" to the Riemann curvature tensor, as well as to the physical acceleration and rotation of the observer. Indeed, the metric components in the Fermi normal coordinates can be expressed with the help of the Riemann tensor, as well as the accelerations and rotations measured by the observer. The metric components are the same as the ones of special relativity up to the first order in the local coordinates, such that they are "as Minkowskian as possible" [44,45]. As a consequence, the Fermi normal frame can be spatially fixed w.r.t. to an observer in any kinematical state.

The use of Fermi coordinates is not adapted for a self-gravitating body, the mass-energy of which contributes to the determination of the initial metric *g* when solving the Einstein equations. For this reason, harmonic coordinates are preferred and recommended for the definition and realization of relativistic celestial reference systems [25,38,46,47], where the frame origin can be centered on the center-of-mass of a massive body. However, a local apparatus is a test body that does not contribute to the background metric. Therefore, the definition of the Fermi coordinates is not a problem in this context. On the contrary, they have the advantage of separating the problem of the determination of the background metric and the observer trajectory in the initial coordinate system, on the one hand, and from the definition, description and modelization of the observables of the local experiment, on the other hand.

When using Fermi normal coordinates, unlike the harmonic coordinates approach, no matching procedure between the initial frame and the local frame is required in order to obtain the metric in the local frame, and the explicit coordinate transformations from the initial coordinate system to the local one are not required. Moreover, all frames obtained from a spatial rotation of the Fermi normal frame are still Fermi normal frames. This is not the case for the harmonic frame, as the harmonic gauge condition does not admit the rigidly rotating frames of [25] (Chapter 8). Therefore, obtaining a harmonic frame with a spatially-fixed axis w.r.t. the apparatus is a priori not possible. For all these reasons, we believe that using a Fermi frame and corresponding Fermi coordinates is a better choice in order to describe a local experiment. When possible, we will compare the results from both approaches: the Fermi normal frame and the harmonic frame.

3.1. Notations and Conventions

In this work, the signature of the Lorentzian metric g is $(+, -, -, -)$. Greek indices run from zero to three, and Latin indices run from one to three. The partial derivative of A will be noted $A_{,\alpha} = \partial A / \partial x^{\alpha}$. We use the summation rule on repeated indices (one up and one down). $\eta_{\alpha\beta}$ are the components of the Minkowski metric. The convention for the Riemann tensor is:

$$R^{\mu}{}_{\alpha\nu\beta} = \Gamma^{\mu}{}_{\alpha\nu,\beta} - \Gamma^{\mu}{}_{\alpha\beta,\nu} - \Gamma^{\mu}{}_{\nu\sigma}\Gamma^{\sigma}{}_{\alpha\beta} + \Gamma^{\mu}{}_{\beta\sigma}\Gamma^{\sigma}{}_{\alpha\nu}$$

In this section, the indices for tensor components in the proper reference frame and the Fermi frame are denoted with a hat, i.e., $A^{\hat{\alpha}} \equiv (A^{\hat{0}}, A^{\hat{i}})$, as well as the partial derivative in the Fermi frame, i.e., $A^{\hat{\alpha}}{}_{,\hat{j}} = \partial A^{\hat{\alpha}} / \partial X^{\hat{j}}$.

3.2. The Local Frame

Let $\mathcal{C}$ be the observer world line; this world line is a timelike path $(ds^2 > 0)$. We call τ, the proper time, that is the integral value $\int d\tau \equiv \int \sqrt{ds^2/c^2}$ along $\mathcal{C}$ between a chosen origin O and an arbitrary event P along $\mathcal{C}$. The observer world line is parameterized with the proper time:

$$\mathcal{C} : x^{\mu} = f^{\mu}(\tau) \tag{2}$$

The four-velocity is $u^{\mu} = df^{\mu}/d\tau$, and the four-acceleration is $\gamma^{\mu} = Du^{\mu}/D\tau$, where $D/D\tau$ is the covariant differentiation along the world line $\mathcal{C}$.

The Proper Reference Frame

We define the proper reference frame with coordinates $(X^{\hat{a}})$ as in [45]. It is entirely determined by these two conditions:

1. On the observer world line, the temporal coordinate $X^{\hat{0}}/c$ of the proper reference frame is equal to the proper time τ of the observer, and the spatial coordinates $X^{\hat{i}}$ are constant.
2. At first order in the new coordinates $X^{\hat{a}}$, we want to recover the metric of an accelerated and rotating observer in special relativity [44].

The new coordinate system $(X^{\hat{a}})$ is defined in an ad-hoc subset $\mathcal{U_C} \subset \mathcal{U}$, so that $\mathcal{C}$ is included in $\mathcal{U_C}$. We select P an event along $\mathcal{C}$, so that $x_P^{\mu} = f^{\mu}(\tau)$. From Condition (1) we infer:

$$X_P^{\hat{0}} = c\tau \tag{3}$$

where c is the velocity of light in vacuum.

The origin O is defined so that $x_O^{\mu} = f^{\mu}(0)$, without loss of generality. The coordinate transformation from $(X^{\hat{a}})$ to (x^{μ}) is a diffeomorphism $\mathcal{Y} : X^{\hat{a}}(\mathcal{U_C}) \to x^{\mu}(\mathcal{U_C})$. The partial derivatives of $\mathcal{Y}$ at point P are defined by the components of the Jacobian matrix:

$$e_{\hat{a}}^{\beta} = \left\{ x_{,\hat{a}}^{\beta} \right\}_P \equiv \bar{x}_{,\hat{a}}^{\beta} \tag{4}$$

where $x^{\beta} = x^{\beta}(X^{\hat{a}})$ are the components of $\mathcal{Y}$, and the bar stands for the value of a function at point P (as P is arbitrary along the world line $\mathcal{C}$, then the bar stands for the value of the function all along $\mathcal{C}$, which means that all quantities with the bar over them are functions of the proper time τ). The inverse transformations follow:

$$e_{\mu}^{\hat{\beta}} e_{\hat{a}}^{\mu} = \delta_{\hat{a}}^{\hat{\beta}} \tag{5}$$

$$e_{\mu}^{\hat{\beta}} e_{\hat{\beta}}^{\nu} = \delta_{\mu}^{\nu} \tag{6}$$

where δ is the Kronecker delta. We note that $e_{\hat{0}}^{\mu} = u^{\mu}/c$.

For the sake of simplicity, $X^{\hat{i}}(\mathcal{C}) = 0$, i.e., the world line constitutes the spatial origin of the proper reference frame. The vector $e_{\hat{0}}^{\mu}$ is determined by the observer world line, while the vectors $e_{\hat{j}}^{\mu}$ are chosen, such that $\left(e_{\hat{a}}^{\mu} \right)$ constitutes a basis of the tangent space for each event along $\mathcal{C}$. $(e_{\hat{j}}^{\mu})$ is the spatial frame of the observer at event P. The transformation relations of the metric tensor are:

$$g_{\hat{a}\hat{\beta}} = g_{\mu\nu} x_{,\hat{a}}^{\mu} x_{,\hat{\beta}}^{\nu} \tag{7}$$

where $g_{\hat{a}\hat{\beta}}$ are the components of the metric tensor in the proper reference frame. For simplicity, we choose the vectors $e_{\hat{a}}^{\mu}$, so that they form an orthonormal basis, so-called a tetrad, such that:

$$g_{\hat{a}\hat{\beta}}(\mathcal{C}) \equiv \eta_{\hat{a}\hat{\beta}} = g_{\mu\nu}(\mathcal{C}) e_{\hat{a}}^{\mu} e_{\hat{\beta}}^{\nu} \tag{8}$$

where $\eta_{\hat{a}\hat{\beta}}$ is the Minkowski metric.

Then, it is shown in [45] that the metric in the proper reference frame can be written, up to first order in the new coordinates $(X^{\hat{a}})$:

$$ds^2 = \left[1 - 2\gamma_{\hat{j}} X^{\hat{j}} + \mathcal{O}\left(X^2 \right) \right] c^2 d\tau^2 + \left[2\Omega_{\hat{m}\hat{j}} X^{\hat{j}} + \mathcal{O}\left(X^2 \right) \right] dX^{\hat{m}} c d\tau + \left[\eta_{\hat{i}\hat{m}} + \mathcal{O}\left(X^2 \right) \right] dX^{\hat{i}} dX^{\hat{m}} \tag{9}$$

where $\Omega_{\hat{\alpha}\hat{\beta}}$ is the antisymmetric rotation matrix defined with:

$$\Omega_{\hat{\alpha}\hat{\beta}} = \frac{1}{2}\bar{g}_{\mu\nu}\left(e_{\hat{\alpha}}^{\mu}\frac{De_{\hat{\beta}}^{\nu}}{D\tau} - e_{\hat{\beta}}^{\mu}\frac{De_{\hat{\alpha}}^{\nu}}{D\tau}\right) \tag{10}$$

and $\gamma_{\hat{j}} = \Omega_{\hat{j}\hat{0}} = g_{\mu\nu}(\mathcal{C})e_{\hat{j}}^{\mu}\gamma^{\nu}$. The function $\Omega_{\hat{\alpha}\hat{\beta}}(\tau)$ defines the tetrad transport along the observer trajectory:

$$\frac{De_{\hat{\beta}}^{\mu}}{D\tau} = \Omega^{\hat{\alpha}}{}_{\hat{\beta}}e_{\hat{\alpha}}^{\mu} \tag{11}$$

where $\Omega^{\hat{\alpha}}{}_{\hat{\beta}} = \bar{g}^{\hat{\alpha}\hat{\alpha}}\Omega_{\hat{\alpha}\hat{\beta}}$.

Moreover, we define the vector $\Omega^{\hat{k}}$, so that $\Omega_{\hat{i}\hat{j}} = \varepsilon_{\hat{i}\hat{j}\hat{k}}\Omega^{\hat{k}}$, with $\varepsilon_{\hat{i}\hat{j}\hat{k}}$ the Levi–Civita symbol. $\Omega^{\hat{k}}$ is the rotation of the observer spatial frame $(e_{\hat{j}}^{\mu})$, as it can be measured with three gyroscopes. $\gamma^{\hat{k}}$ is the acceleration vector of the observer, as it can be measured with accelerometers. If $\Omega^{\hat{k}} = 0$, the frame is Fermi–Walker transported; if $\Omega^{\hat{k}} = 0$ and $\gamma^{\hat{k}} = 0$, the frame is parallel transported (i.e., $\mathcal{C}$ is a geodesic).

The Fermi Normal Frame

Up to the second order in the coordinates, there is a certain choice of freedom to prolongate the coordinates lines of the proper reference frame. For the sake of mathematical simplicity, what is usually done is to define the Fermi normal frame, where the coordinate lines are taken as geodesics [45]. In the Fermi normal frame, the metric can be written:

$$\begin{aligned}
ds^2 = &\left[1 - 2\gamma_{\hat{j}}X^{\hat{j}} + \left(\Omega_{\hat{\alpha}\hat{j}}\Omega^{\hat{\alpha}}{}_{\hat{k}} + \bar{R}_{\hat{0}\hat{j}\hat{0}\hat{k}}\right)X^{\hat{j}}X^{\hat{k}} + \mathcal{O}(X^3)\right]c^2d\tau^2\\
&+ \left[2\Omega_{\hat{m}\hat{j}}X^{\hat{j}} + \frac{4}{3}\bar{R}_{\hat{0}\hat{j}\hat{m}\hat{k}}X^{\hat{j}}X^{\hat{k}} + \mathcal{O}(X^3)\right]dX^{\hat{m}}cd\tau\\
&+ \left[\eta_{\hat{l}\hat{m}} + \frac{1}{3}\bar{R}_{\hat{l}\hat{j}\hat{m}\hat{k}}X^{\hat{j}}X^{\hat{k}} + \mathcal{O}(X^3)\right]dX^{\hat{l}}dX^{\hat{m}}
\end{aligned} \tag{12}$$

A different approach is used in [24,25], where harmonic coordinates are used to build the local frame, named the topocentric reference frame. We prefer to name it the harmonic frame here as a reference to the corresponding harmonic coordinates. As a result, the metric components are very different when using harmonic coordinates (HC) [25] (see Equations (8.40) to (8.42)) from the ones using Fermi normal coordinates (FNC). The cross-component of the metric has no first order term in the HC, while in the FNC, the first order term depends on the observer rotation as in special relativity. Indeed, the harmonic frame is dynamically non-rotating [24,48], and therefore, it cannot be adapted to rotating observers. As in special relativity, there is no first order term in the spatial component of the metric in the FNC, while there is one when using the HC. The metric calculated in the HC in [24,25] is an expansion in both the local coordinates and c^{-1} (post-Newtonian approach), while there in no post-Newtonian expansion in the approach presented here. Finally, as a consequence of the equivalence principle, the metric components in the FNC do not depend on the chosen initial coordinate system, while the metric components in the HC do through the matching procedure. Moreover, the metric components in the FNC are expressed in terms of the Riemann tensor, acceleration and rotation of the observer, while in the HC, the metric components are expressed with non-tensorial quantities to be determined through the matching procedure.

3.3. Geodesic Equation in the Local Frame

The gravimetric and gradiometric observables can be deduced from the general dynamical equation of a test body written in the local frame:

$$\frac{d^2X^{\hat{\alpha}}}{d\lambda^2} + c^2\Gamma^{\hat{\alpha}}{}_{\hat{0}\hat{0}}\left(\frac{d\tau}{d\lambda}\right)^2 + 2c\Gamma^{\hat{\alpha}}{}_{\hat{0}\hat{i}}\frac{d\tau}{d\lambda}\frac{dX^{\hat{i}}}{d\lambda} + \Gamma^{\hat{\alpha}}{}_{\hat{i}\hat{j}}\frac{dX^{\hat{i}}}{d\lambda}\frac{dX^{\hat{j}}}{d\lambda} = \frac{F^{\hat{\alpha}}}{m} \equiv \Gamma^{\hat{\alpha}} \tag{13}$$

where λ is an affine parameter along the test body trajectory.

From [49], we can write it in all generality. We will consider the gravimeter/gradiometer to be made of components that are at rest with respect to the local frame, i.e., $dX^{\hat{i}}/d\lambda = d^2X^{\hat{i}}/d\lambda^2 = 0$, thanks to some local forces voluntarily applied to the apparatus components. In the local frame, we decompose a four-vector as $(V^{\hat{0}}, V^{\hat{i}}) \equiv (V^{\hat{0}}, V)$. Therefore, we deduce from [49] (Equation (25)):

$$
\begin{aligned}
\Gamma^{\hat{i}} &= \gamma^{\hat{i}} + [\boldsymbol{\Omega} \times (\boldsymbol{\Omega} \times \boldsymbol{X})]^{\hat{i}} + (\boldsymbol{\eta} \times \boldsymbol{X})^{\hat{i}} + c^2 X^{\hat{l}} R_{\hat{0}\hat{i}\hat{0}\hat{l}} \\
&\quad + \tfrac{1}{c^2}\left[(\boldsymbol{\gamma}\cdot\boldsymbol{X})\gamma^{\hat{i}} - (\boldsymbol{b}\cdot\boldsymbol{X})(\boldsymbol{\Omega} \times \boldsymbol{X})^{\hat{i}} - 2\boldsymbol{\gamma}\cdot(\boldsymbol{\Omega} \times \boldsymbol{X})(\boldsymbol{\Omega} \times \boldsymbol{X})^{\hat{i}}\right] \\
&\quad - \tfrac{1}{3}\gamma^{\hat{k}} R_{\hat{i}\hat{l}\hat{k}\hat{m}} X^{\hat{l}} X^{\hat{m}} - 2c X^{\hat{l}} (\boldsymbol{\Omega} \times \boldsymbol{X})^{\hat{k}} R_{\hat{0}\hat{l}\hat{i}\hat{k}} + 2X^{\hat{l}}(\boldsymbol{\gamma}\cdot\boldsymbol{X})R_{\hat{0}\hat{i}\hat{0}\hat{l}} \\
&\quad + \tfrac{c^2}{2}\left(R_{\hat{i}\hat{l}\hat{m}\hat{0};\hat{0}} + R_{\hat{i}\hat{0}\hat{l}\hat{0};\hat{m}}\right) X^{\hat{l}} X^{\hat{m}} + \mathcal{O}(X^3)
\end{aligned}
\tag{14}
$$

where $\boldsymbol{b} = \frac{d\boldsymbol{\gamma}}{d\tau} + \boldsymbol{\Omega} \times \boldsymbol{\gamma}$ and $\boldsymbol{\eta} = \frac{d\boldsymbol{\Omega}}{d\tau}$.

The geodesic equation using harmonic coordinates can be found in [24,25]. As the harmonic frame is dynamically non-rotating, all terms depending on rotation are absent. Moreover, the coefficients of the equation are found through the matching procedure; therefore, they depend on the choice of the initial metric and coordinate system. Then, it is not possible to compare the terms with Equation (14) here, which is more general.

3.3.1. Gravimetric Observables

We suppose that we apply a force to the gravimeter mass to keep it at the center of the local frame. Therefore, Equation (14) reduces to:

$$
\Gamma^{\hat{i}} = \gamma^{\hat{i}}
\tag{15}
$$

which is simply the physical acceleration of the local frame. We emphasize that here we suppose that the mass of the gravimeter is kept fixed at the center of the local frame. Therefore, the measured quantity is the force vector $\boldsymbol{F}$ applied to the mass in order to be still, such that $\Gamma^{\hat{i}} = F^{\hat{i}}/m$.

3.3.2. Gradiometric Observables

Suppose now that we have a two masses located in the direction $\boldsymbol{e}_{\hat{j}}$ of the spatial part of the local frame basis, at a distance $l/2$ and $-l/2$ from the center of the frame. The local distance l is supposed to be constant here, e.g., by putting both accelerometers on a rigid structure. We define the local distance as the Euclidean distance calculated in terms of the coordinates of the local frame: $l = \sqrt{\sum_{\hat{i}}(X^{\hat{i}})^2}$. Then, we define the quantities measured by the differential accelerometer, or gradiometer, with:

$$
\Gamma^{\hat{i}}_{\ \hat{j}} = \frac{1}{l}\left[\Gamma^{\hat{i}}\left(\tau, \frac{l}{2}\boldsymbol{e}_{\hat{j}}\right) - \Gamma^{\hat{i}}\left(\tau, -\frac{l}{2}\boldsymbol{e}_{\hat{j}}\right)\right]
\tag{16}
$$

By doing this, we suppose that the geometrical center of the gradiometer is at the origin of the local frame. From Equation (14), we deduce:

$$
\Gamma^{\hat{i}}_{\ \hat{j}} = \left[\boldsymbol{\Omega} \times (\boldsymbol{\Omega} \times \boldsymbol{e}_{\hat{j}})\right]^{\hat{i}} + (\boldsymbol{\eta} \times \boldsymbol{e}_{\hat{j}})^{\hat{i}} + c^2 R_{\hat{0}\hat{i}\hat{0}\hat{j}} + \frac{1}{c^2}\gamma^{\hat{i}}\gamma^{\hat{j}} + \mathcal{O}(l^2)
\tag{17}
$$

In case of a free-falling and non-rotating gradiometer, one simply has:

$$
\Gamma^{\hat{i}}_{\ \hat{j}} = c^2 R_{\hat{0}\hat{i}\hat{0}\hat{j}} + \mathcal{O}(l^2)
\tag{18}
$$

3.4. Clock Frequency Comparisons and Syntonization

The principle of clock frequency comparison is to measure the frequency of an electromagnetic signal with the help of the emitting clock, A, and then with the receiving clock, B. We obtain respectively two measurements ν_A and ν_B. However, in general, one measures the time of flight of the electromagnetic signal between emission and reception. Then, the ratio ν_A/ν_B can be obtained by deriving the time of flight measurements with respect to the time of reception.

Let $S(x^\alpha)$ be the phase of the electromagnetic signal emitted by clock A. It can be shown that light rays are contained in hypersurfaces of constant phase. The frequency measured by A/B is:

$$\nu_{A/B} = \frac{1}{2\pi}\frac{dS}{d\tau_{A/B}} \tag{19}$$

where $\tau_{A/B}$ is the proper time along the world line of clock A/B. We introduce the wave vector $k_\alpha^{A/B} = (\partial_\alpha S)_{A/B}$ to obtain:

$$\nu_{A/B} = \frac{1}{2\pi}k_\alpha^{A/B}u_{A/B}^\alpha \tag{20}$$

where $u_{A/B}^\alpha = dx_{A/B}^\alpha/d\tau$ is the four-velocity of clock A/B. Finally, we obtain a fundamental relation for frequency transfer:

$$\frac{\nu_A}{\nu_B} = \frac{k_\alpha^A u_A^\alpha}{k_\alpha^B u_B^\alpha} \tag{21}$$

This formula does not depend on a particular theory and then can be used to perform tests of general relativity. It is needed in the context of relativistic geodesy, in order to calculate the gravitational potential difference between two clocks from the measurement of the ratio of the frequencies ν_A/ν_B.

Introducing $v^i = dx^i/dt$ and $\hat{k}_i = k_i/k_0$, it is usually written as:

$$\frac{\nu_A}{\nu_B} = \frac{u_A^0}{u_B^0}\frac{k_0^A}{k_0^B}\frac{1+\frac{\hat{k}_i^A v_A^i}{c}}{1+\frac{\hat{k}_i^B v_B^i}{c}} \tag{22}$$

From Equation (19), we deduce that:

$$\frac{\nu_A}{\nu_B} = \frac{d\tau_B}{d\tau_A} = \left(\frac{dt}{d\tau}\right)_A \frac{dt_B}{dt_A}\left(\frac{d\tau}{dt}\right)_B \tag{23}$$

In the case of propagation in free space, if we suppose that the space-time is stationary, i.e., $\partial_0 g_{\alpha\beta} = 0$, then it can be shown that k_0 is constant along the light ray, meaning that $k_0^A = k_0^B$. Then, from Equations (22) and (23), we deduce that:

$$\frac{dt_B}{dt_A} = \frac{1+\frac{\hat{k}_i^A v_A^i}{c}}{1+\frac{\hat{k}_i^B v_B^i}{c}} \tag{24}$$

This term depends on how the signal propagates from A to B. For a free propagation in a vacuum, it is calculated up to order c^{-3} in [50], for the more general metric GCRS. Up to second order, it does not depend on the gravitational field, but only on the relative motion of the two clocks. It is simply the first order Doppler effect of order $\frac{v}{c}$. At third order, there is a term of order $\frac{Gm}{rc^2}\frac{v}{c}$. It is less than 3.6×10^{-14} for a satellite and around 2.2×10^{-15} on the ground. In [51], the term (24) is calculated for the metric given later in Equations (27) to (29) up to the fourth order. It is stressed that the J_2 term of the expansion (30) in the third order term can amount to 1.3×10^{-16} for a satellite in low orbit.

If the signal propagates in an optical fiber, the term (24) has been calculated up to order c^{-3} in [52]. Up to second order, it does not depend on the gravitational field, as for the free propagation in vacuum. The first order term is due to the variation of the fiber length (e.g., due to thermal expansion) and of its refractive index. For a 1000-km fiber with refractive index $n = 1.5$, this term is equal to 3.6×10^{-14}. The second order term is the derivative of the Sagnac effect, which is of order 10^{-19} or less for a 1000-km fiber. Finally, the third order term is of the order of 10^{-22} for a 1000-km fiber.

Syntonization is a different problem and is needed for the realization of coordinate time scales (such as TAI (Temps Atomique International)). It depends on the particular coordinate system chosen as a reference and is given by the derivative of proper time with respect to coordinate time. In a metric theory, one has $c d\tau = \sqrt{g_{\alpha\beta} dx^\alpha dx^\beta}$, so that:

$$\frac{d\tau}{dt} \equiv \left(u^0\right)^{-1} = \left[g_{00} + 2g_{0i}\frac{v^i}{c} + g_{ij}\frac{v^i v^j}{c^2}\right]^{-1/2} \tag{25}$$

where we defined the coordinate velocity $v^i = dx^i / dt$.

In the context of relativistic geodesy, this quantity is needed for the realization of the chronometric geoid. An isochronometric surface is a surface where all clocks beat at the same rate with respect to a reference coordinate time, such that the quantity (25) is constant. This reference coordinate time is usually taken as TT (terrestrial time), for which TAI is a realization, or TCG (Temps Coordonné géocentrique). The chronometric geoid is a reference isochronometric surface that should coincide up to some level with the classical geoid—a level surface of the gravity potential closest to the topographic mean sea level—so that a possible definition is:

$$\frac{d\tau}{d(TT)} = 1 \tag{26}$$

TT is itself defined with respect to TCG with $d(TT)/d(TCG) = 1 - L_G$, where L_G is a defining constant [38], chosen such that the reference isochronometric surface defined from TT (26) coincides with some level with the classical geoid.

An interesting problem is that the chronometric geoid will differ in the future from the classical geoid. Indeed, the value of the potential on the geoid, W_0, depends on the global ocean level, which changes with time. In addition, there are several methods to realize that the geoid is "closest to the mean sea level", so that there is yet no adopted standard to define a reference geoid and W_0 value (see, e.g., the discussion in [53]). Several authors have considered the time variation of W_0 (see, e.g., [54,55]), but there is some uncertainty in what is accounted for in such a linear model. A recent estimate over 1993 to 2009 is $dW_0/dt = -2.7 \times 10^{-2} \text{ m}^2 \cdot \text{s}^{-2} \cdot \text{year}^{-1}$, mostly driven by the sea level change of +2.9 mm/year [55]. However, the rate of change of the global ocean level could vary during the next few decades, and predictions are highly model dependent [56]. Nevertheless, to state an order of magnitude, considering a systematic variation in the sea level of order 2 mm/year, different definitions of a reference surface for the gravity potential could yield differences in the frequency of order 2×10^{-18} in a decade. Comparisons of accurate clocks could therefore help in the future to establish a worldwide vertical datum.

4. Application to a Stationary PPN Metric Tensor

4.1. PPN Metric of an Isolated, Axisymmetric Rotating Body

In order to calculate the observables of relativistic geodesy with respect to the initial coordinate system and evaluate the higher order terms, we simplify the Earth metric. We consider that the Earth is a body in uniform rotation, isolated and axisymmetric. Moreover, in order to assess the potential of relativistic geodesy for general relativity tests, we generalize the metric to the so-called Will–Nordtvedt formalism [10,11]. This formalism contains ten parameters $\gamma, \beta, \xi, \alpha_1, \ldots, \alpha_3, \zeta_1, \ldots,$ ζ_4. The parameters α and γ are the usual Eddington–Robertson–Schiff parameters used to describe the classical tests of general relativity (=1 in GR), while other parameters measure preferred-location and preferred-frame effects and the violation of the conservation of total momentum. They are all zero in GR. Theories that possess conservation laws for total momentum, called "semi-conservative", have five free PPN parameters ($\gamma, \beta, \xi, \alpha_1, \alpha_2$). The PPN parameter γ has been constrained to $|\gamma - 1| \leq 2.3 \times 10^{-5}$ using the Cassini spacecraft [57], β to $|\beta - 1| \leq 3 \times 10^{-5}$ thanks to planetary ephemeris [58] and α_1 and α_2 to $|\alpha_1| \leq 6 \times 10^{-6}$ and $|\alpha_2| \leq 3.5 \times 10^{-5}$ thanks to refined values of Solar System planetary precessions [59]. Many other constraints have been put on PPN parameters, and a summary of these constraints can be found in, e.g., [11,60].

The assumptions done in this article to write the metric are simplistic, and for the analysis of a particular experiment, one should use a complete description of the metric around the Earth (see, e.g., [38,61]). This has been done in particular in the context of the detection of the Lense–Thirring effect in the Solar System (see, e.g., [62,63] and the references therein).

For the sake of simplicity, the only PPN parameters used here are γ, β and α_1. $\vec{w}$ is the speed of the Earth center of mass with respect to a preferred rest frame, if one exists. We use a non-rotating geocentric reference system with initial coordinate system $(ct; x^i) \equiv \left(ct; \vec{x}\right)$. This case has been rigorously studied in [51]; the metric is stationary, and it is given by:

$$g_{00}\left(\vec{x}\right) = 1 - \frac{2}{c^2}W\left(\vec{x}\right) + \frac{2\beta}{c^4}W^2\left(\vec{x}\right) + \tilde{O}_6 \tag{27}$$

$$g_{0j}\left(\vec{x}\right) = \frac{2}{c^3}\left[\left(1 + \gamma + \frac{\alpha_1}{4}\right)\vec{W}\left(\vec{x}\right) + \frac{1}{4}\alpha_1 W\left(\vec{x}\right)\vec{w}\right] + \tilde{O}_5 \tag{28}$$

$$g_{ij}\left(\vec{x}\right) = -\delta_{ij}\left(1 + \frac{2\gamma}{c^2}W\left(\vec{x}\right)\right) + \tilde{O}_4 \tag{29}$$

where:

$$W\left(\vec{x}\right) = \frac{GM}{r}\left[1 - \sum_{n=2}^{\infty} J_n\left(\frac{R}{r}\right)^n P_n(\cos\theta)\right], \tag{30}$$

$$\vec{W}\left(\vec{x}\right) = \frac{GI\vec{\omega}\times\vec{x}}{2r^3}\left[1 - \sum_{n=1}^{\infty} K_n\left(\frac{R}{r}\right)^n P'_{n+1}(\cos\theta)\right], \tag{31}$$

$r = \|\vec{x}\|$, θ is the angle between $\vec{x}$ and the axis of rotation of the Earth, $\vec{\omega}$ its angular velocity, assumed constant, R its equatorial radius, the P_n the Legendre polynomials, and the coefficients M, $J_2, \ldots, J_n, \ldots$ and $I, K_1, \ldots, K_n, \ldots$ are the multipole moments of the multipolar expansion of the potentials W and $\vec{W}$, for which convergence is assumed. The angular momentum of the central body is $\vec{J} = I\vec{\omega}$. The order of the metric expansion is:

$$\tilde{O}_n = \left(\frac{GM}{rc^2}\right)^{n/2} \tag{32}$$

This is a practical notation: the term $\tilde{O}_n$, when it is at the end of a sum, means $\mathcal{O}(\tilde{O}_n)$.

4.2. Clock Observables

From Equations (25) and the given metric (27) to (29), we calculate:

$$\begin{aligned}
\left(u^0\right)^{-1} \equiv \frac{d\tau}{dt} = 1 &- \frac{1}{c^2}\left(W + \frac{1}{2}v^2\right) \\
&+ \frac{1}{c^4}\left[\left(\beta - \frac{1}{2}\right)W^2 - \left(\gamma + \frac{1}{2}\right)Wv^2 - \frac{1}{8}v^4\right. \\
&\left. + 2\left(\gamma + 1 + \frac{\alpha_1}{4}\right)\vec{W}\cdot\vec{v} + \frac{1}{2}\alpha_1 W\vec{v}\cdot\vec{w}\right] + \mathcal{O}\left(\frac{1}{c^6}\right)
\end{aligned} \tag{33}$$

where $v = |d\vec{x}/dt|$ is the coordinate velocity; from which we deduce that:

$$
\begin{aligned}
\frac{(u^0)_A}{(u^0)_B} &= 1 + \tfrac{1}{c^2}\left(W_A - W_B + \tfrac{1}{2}v_A^2 - \tfrac{1}{2}v_B^2\right) \\
&\quad + \tfrac{1}{c^4}\left[(\gamma+1)\left(W_A v_A^2 - W_B v_B^2\right) + \tfrac{3}{8}v_A^4 - \tfrac{1}{8}v_B^4 - \tfrac{1}{4}v_A^2 v_B^2\right. \\
&\quad + 2\left(\gamma + 1 + \tfrac{\alpha_1}{4}\right)\left(\vec{W}_B \cdot \vec{v}_B - \vec{W}_A \cdot \vec{v}_A\right) \\
&\quad + \tfrac{1}{2}\alpha_1 \vec{w} \cdot \left(\vec{v}_B W_B - \vec{v}_A W_A\right) \\
&\quad \left. \tfrac{1}{2}(W_A - W_B)(W_A - W_B + 2(1-\beta)(W_A + W_B) + v_A^2 - v_B^2)\right] \\
&\quad + \mathcal{O}\left(\tfrac{1}{c^6}\right)
\end{aligned}
\tag{34}
$$

This result coincides with the one of [51] when putting $\alpha_1 = 0$.

4.3. Gravimetry Observables

We find a tetrad that satisfies $e_0^\alpha = u^\alpha/c$ and Equation (8) for the metric (27) to (29), valid for a general trajectory with coordinate velocity v^i:

$$
\begin{aligned}
e_0^0 &= 1 + \tfrac{1}{c^2}\left(W + \tfrac{1}{2}v^2\right) \\
&\quad + \tfrac{1}{c^4}\left[-\left(\beta - \tfrac{3}{2}\right)W^2 + \left(\gamma + \tfrac{3}{2}\right)Wv^2 + \tfrac{3}{8}v^4\right. \\
&\quad \left. -2\left(\gamma + 1 + \tfrac{\alpha_1}{4}\right)\vec{W}\cdot\vec{v} - \tfrac{1}{2}\alpha_1 W\vec{v}\cdot\vec{w}\right] + \mathcal{O}\left(\tfrac{1}{c^6}\right)
\end{aligned}
\tag{35}
$$

$$
e_0^i = \frac{1}{c}v^i e_0^0
\tag{36}
$$

$$
e_j^0 = \frac{1}{c}v^j + \frac{1}{c^3}v^j\left((\gamma+2)W + \frac{1}{2}v^2\right) - g_{0j} + \mathcal{O}\left(\frac{1}{c^5}\right)
\tag{37}
$$

$$
e_j^i = \delta_{ij} + \frac{1}{c^2}\left(\frac{1}{2}v^i v^j - \delta_{ij}\gamma W\right) + \mathcal{O}\left(\frac{1}{c^4}\right)
\tag{38}
$$

where δ_{ij} is the Kronecker symbol and g_{0j} is given by (28). This tetrad coincides with the results found elsewhere for the the GR case, e.g., [47] (Equation (5.21)). Here, the tetrad is chosen such that it is identity at zeroth order: the spatial part is non-rotating with respect to distant stars as for the initial coordinate system. The case of a rotating tetrad does not change the gravimetry observable, as it is defined as a local quantity. However, it will be studied for the gradiometry observable.

As in Section 3.3, a four-vector is decomposed in the local frame defined in (35) to (38) as $(V^0, V^i) \equiv (V^0, V)$, while it is decomposed as $(V^0, V^i) \equiv (V^0, \vec{V})$ in the initial coordinate system.

Then, we calculate the antisymmetric rotation matrix with Formula (10). We deduce from this matrix the physical acceleration experienced by the observer:

$$
\begin{aligned}
\gamma &= \vec{a} - \vec{\nabla}W \\
&\quad + \tfrac{1}{c^2}\left[(\gamma + 2\beta - 2)W\vec{\nabla}W + 2\left(\gamma + 1 + \tfrac{\alpha_1}{4}\right)\vec{v}\times(\vec{\nabla}\times\vec{W})\right. \\
&\quad -(\gamma+1)v^2\vec{\nabla}W + 2\left(\gamma + \tfrac{3}{2}\right)(\vec{v}\cdot\vec{\nabla}W)\vec{v} + (v^2 + (\gamma+2)W)\vec{a} \\
&\quad \left. + \tfrac{1}{2}\alpha_1\left(\vec{\nabla}W(\vec{v}\cdot\vec{w}) - \vec{w}(\vec{v}\cdot\vec{\nabla}W)\right) + \tfrac{1}{2}(\vec{v}\cdot\vec{a})\vec{v}\right] + \mathcal{O}\left(\tfrac{1}{c^4}\right)
\end{aligned}
\tag{39}
$$

where $\vec{a} = d\vec{v}/dt$, and its rotation as measured with gyroscopes:

$$
\Omega = \Omega_{\mathrm{LT}} + \Omega_{\mathrm{dS}} + \Omega_{\mathrm{Th}} + \mathcal{O}\left(\frac{1}{c^4}\right)
\tag{40}
$$

where:

$$\Omega_{\text{LT}} = -\frac{1}{2c^2} \vec{\nabla} \times \vec{g} = -\frac{1}{c^2}\left[\left(1 + \gamma + \frac{\alpha_1}{4}\right)\vec{\nabla} \times \vec{W} + \frac{\alpha_1}{4}\vec{\nabla}W \times \vec{w}\right] \tag{41}$$

$$\Omega_{\text{dS}} = -\frac{1}{c^2}\left(\gamma + \frac{1}{2}\right)\vec{v} \times \vec{\nabla}W \tag{42}$$

$$\Omega_{\text{Th}} = \frac{1}{2c^2}\vec{v} \times \gamma \tag{43}$$

where $\left(\vec{g}\right)^j \equiv c^3 g_{0j}$. This result coincides with the results found elsewhere (see, e.g., [64] (Equation (3.4.38)) for $\alpha_1 = 0$ and [65] (Equations (5) to (7))).

4.4. Gradiometry Observables

Now, we consider a gradiometer fixed to the surface of the Earth and rotating together with the Earth. In this paper, we suppose a uniform rotation of the Earth and a stationary metric. Irregularities in the Earth rotation are not included here. We define a tetrad $e^\mu_{(\alpha)}$ fixed to the gradiometer, such that the axes of the gradiometer are in the direction of the tetrad axes. This tetrad is related to the tetrad (35) to (38) aligned with the GCRS system by the following relations:

$$e^\mu_{(0)} = e^\mu_{\hat{0}} = u^\mu/c \tag{44}$$

$$e^\mu_{(i)} = \Lambda^{\hat{j}}_{(i)} e^\mu_{\hat{j}} \tag{45}$$

i.e., the spatial part is rotated by a matrix $\Lambda^{\hat{j}}_{(i)} \in SO(3)$. This matrix is a function of a parameter along the gradiometer path, e.g., $\Lambda^{\hat{j}}_{(i)}(t)$ with t being the GCRS coordinate time. The gradiometer rotates together with the Earth surface, i.e., in cylindrical coordinates (ct, ρ, ϕ, z) related to the GCRS coordinates $\left(ct, \vec{x}\right) = (ct, x, y, z)$ as $x = \rho \cos\phi$, $y = \rho \sin\phi$; its points move along orbits of the vector field $\partial_t + \omega\partial_\phi$ where $\omega = d\phi/dt$ is the angular velocity of the Earth surface. The tetrad $e^\mu_{(\alpha)}$ is then Lie transported (to the order of interest) by this vector field. We obtain:

$$\Lambda^{\hat{j}}_{(i)}(t) = B^{(k)}_{(i)} A^{\hat{j}}_{(k)}(t) \tag{46}$$

with:

$$A^{\hat{j}}_{(k)}(t) = \begin{pmatrix} \cos\omega(t-t_0) & -\sin\omega(t-t_0) & 0 \\ \sin\omega(t-t_0) & \cos\omega(t-t_0) & 0 \\ 0 & 0 & 1 \end{pmatrix}, \tag{47}$$

This leads to:

$$\frac{d\Lambda^{\hat{j}}_{(i)}}{dt} = -\Lambda^{\hat{k}}_{(i)} \varepsilon^{\hat{j}}_{\ \hat{k}\hat{l}}\omega^{\hat{l}} \tag{48}$$

with $\omega^{\hat{l}} = (0, 0, \omega)$.

We can calculate the vector $\Omega^{(j)}$ based on Formula (10) for the rotating tetrad $e^\mu_{(\alpha)}$. We obtain $\Omega^{(j)} = \left(\Lambda^{-1}\right)^{(j)}_{\hat{i}}\Omega^{\hat{i}} = \Lambda^{\hat{i}}_{(j)}\Omega^{\hat{i}}$ with:

$$\Omega = \omega + \frac{1}{c^2}\left(W + \frac{1}{2}v^2\right)\omega + \Omega_{\text{LT}} + \Omega_{\text{dS}} + \Omega_{\text{Th}} + O\left(\frac{1}{c^4}\right) \tag{49}$$

where Ω_{LT}, Ω_{dS} and Ω_{Th} are defined respectively in (41) to (43).

Here, a four-vector is decomposed in the rotating local frame defined in (44) to (45) as $(V^{(0)}, V^{(i)}) \equiv (V^{(0)}, \tilde{V})$. For the first term of (17), we then obtain:

$$\left[\tilde{\Omega} \times \left(\tilde{\Omega} \times \tilde{e}_{(n)}\right)\right]^{(m)} = \left(\Lambda^{-1}\right)_{\hat{\imath}}^{(m)} \Lambda_{(n)}^{\hat{\jmath}} [\Omega \times (\Omega \times e_{\hat{\jmath}})]^{\hat{\imath}} = \Lambda_{(m)}^{\hat{\imath}} \Lambda_{(n)}^{\hat{\jmath}} [\Omega \times (\Omega \times e_{\hat{\jmath}})]^{\hat{\imath}} \tag{50}$$

with:

$$
\begin{aligned}
[\Omega \times (\Omega \times e_{\hat{\jmath}})]^{\hat{\imath}} =\;& [\omega \times (\omega \times e_{\hat{\jmath}})]^{\hat{\imath}}\left(1 + \tfrac{1}{c^2}(2W + v^2)\right) \\
& + 2\omega^{(\hat{\imath}}\Omega^{\hat{\jmath})}{}_{\mathrm{LT}} - 2\delta_{\hat{\jmath}}^{\hat{\imath}} \\
& + 2\omega^{(\hat{\imath}}\Omega^{\hat{\jmath})}{}_{\mathrm{dS}} - 2\delta_{\hat{\jmath}}^{\hat{\imath}} \\
& + 2\omega^{(\hat{\imath}}\Omega^{\hat{\jmath})}{}_{\mathrm{Th}} - 2\delta_{\hat{\jmath}}^{\hat{\imath}} \\
& + O\left(\tfrac{1}{c^4}\right)
\end{aligned}
\tag{51}
$$

$$
\begin{aligned}
=\;& [\omega \times (\omega \times e_{\hat{\jmath}})]^{\hat{\imath}}\left(1 + \tfrac{1}{c^2}(2W + v^2)\right) \\
& + \tfrac{1}{c^2}(\omega^{(\hat{\imath}}e^{\hat{\jmath})mn} - \delta_{\hat{\jmath}}^{\hat{\imath}}\omega_{\hat{k}}\varepsilon^{kmn})\left((2\gamma + 2 + \tfrac{\alpha_1}{2})W_{m,n} + \tfrac{\alpha_1}{2}w_m W_{,n}\right) \\
& + \tfrac{1}{c^2}(\omega^{(\hat{\imath}}e^{\hat{\jmath})mn} - \delta_{\hat{\jmath}}^{\hat{\imath}}\omega_{\hat{k}}\varepsilon^{kmn})v_m(a_n - (2\gamma + 2)W_{,n}) \\
& + O\left(\tfrac{1}{c^4}\right)
\end{aligned}
\tag{52}
$$

The second term of (17) containing the angular acceleration is non-vanishing due to the term with vector $\vec{w}$ in the Lense–Thirring part of the angular velocity, which is not axially symmetric. The angular acceleration $\tilde{\eta} = \frac{d\tilde{\Omega}}{d\tau}$ is given as $\eta^{(j)} = \left(\Lambda^{-1}\right)_{\hat{\imath}}^{(j)}\eta^{\hat{\imath}} = \Lambda_{(j)}^{\hat{\imath}}\eta^{\hat{\imath}}$ with the components $\eta^{\hat{\imath}}$ given by:

$$\eta = \frac{1}{c^2}\frac{\alpha_1}{4}\left((\vec{w}\cdot\vec{\nabla}W)\omega - (\omega\cdot\vec{\nabla}W)\vec{w}\right) + O\left(\frac{1}{c^4}\right) \tag{53}$$

For the second term of (17), we then obtain:

$$\left(\tilde{\eta} \times \tilde{e}_{(n)}\right)^{(m)} = \left(\Lambda^{-1}\right)_{\hat{\imath}}^{(m)}\Lambda_{(n)}^{\hat{\jmath}}(\eta \times e_{\hat{\jmath}})^{\hat{\imath}} = \Lambda_{(m)}^{\hat{\imath}}\Lambda_{(n)}^{\hat{\jmath}}(\eta \times e_{\hat{\jmath}})^{\hat{\imath}}. \tag{54}$$

The third term of (17) can be obtained based on Formula (2.26) of [32]. This formula was derived for the metric (27) to (29) with $\vec{w} = 0$ and general functions $W\left(\vec{x}\right)$ and $\vec{W}\left(\vec{x}\right)$ ($U^*(x^i)$ and $V_j(x^i)$ in the notation of [32]) and using the tetrad (35) to (38). The terms with $\vec{w}$ can be added by substitution of a corresponding function for V_j. Thus, we get $R_{(0)(m)(0)(n)} = \Lambda_{(m)}^{\hat{\imath}}\Lambda_{(n)}^{\hat{\jmath}}R_{\hat{0}\hat{\imath}\hat{0}\hat{\jmath}}$ with:

$$
\begin{aligned}
c^2 R_{\hat{0}\hat{\imath}\hat{0}\hat{\jmath}} =\;& -W_{,ij} + \tfrac{1}{c^2}\big[(2(\beta + \gamma - 1)W - (\gamma + 1)v^2)W_{,ij} \\
& + (2\beta + 2\gamma - 1)W_{,i}W_{,j} - \gamma\delta_{ij}|\nabla W|^2 \\
& + (2\gamma + 1)v^k v_{(i}W_{,j)k} - \gamma\delta_{ij}v^m v^n W_{,mn} \\
& + (2\gamma + 2 + \tfrac{\alpha_1}{2})\left(v^k W_{k,ij} - v^k W_{(i,j)k}\right) \\
& + \tfrac{\alpha_1}{2}\left(v^k w_k W_{,ij} - v^k w_{(i}W_{,j)k}\right)\big] + O\left(\tfrac{1}{c^4}\right)
\end{aligned}
\tag{55}
$$

The last term of (17) can be expressed as $\gamma^{(m)}\gamma^{(n)} = \left(\Lambda^{-1}\right)_{\hat{\imath}}^{(m)}\left(\Lambda^{-1}\right)_{\hat{\jmath}}^{(n)}\gamma^{\hat{\imath}}\gamma^{\hat{\jmath}} = \Lambda_{(m)}^{\hat{\imath}}\Lambda_{(n)}^{\hat{\jmath}}\gamma^{\hat{\imath}}\gamma^{\hat{\jmath}}$, where using (39), we obtain:

$$\frac{1}{c^2}\gamma^{\hat{\imath}}\gamma^{\hat{\jmath}} = \frac{1}{c^2}\left(a^i - W_{,i}\right)\left(a^j - W_{,j}\right) + O\left(\frac{1}{c^4}\right) \tag{56}$$

with a^i being the centrifugal acceleration (circular motion is considered) of the center of the gradiometer with GCRS coordinates x_c, y_c given as $a^i = -\omega^2(x_c, y_c, 0)$.

5. Orders of Magnitudes

Let us take a clock that is on the surface of the Earth, at rest in the rotating Earth frame. Then, in usual spherical coordinates (r, θ, ϕ), one has:

$$\vec{v} = v_\phi \vec{u}_\phi = R\omega \sin\theta \vec{u}_\phi \tag{57}$$

$$\vec{a} = a_r \vec{u}_r + a_\theta \vec{u}_\theta = -R\omega^2 \sin\theta \left(\sin\theta \vec{u}_r + \cos\theta \vec{u}_\theta \right) \tag{58}$$

where R and ω are the Earth radius and angular velocity. Assuming that $W = W(r,\theta)$, $\vec{W} = W_\phi(r,\theta)\vec{u}_\phi$, $\gamma = \beta = 1$ and $\alpha_1 = 0$, we deduce from (39):

$$\gamma_r = a_r \left[1 + \frac{1}{c^2}\left(v_\phi^2 + 3W\right) \right] - W_{,r}\left[1 + \frac{1}{c^2}\left(2v_\phi^2 - W\right) \right] + \frac{4}{c^2}\frac{v_\phi}{r}\frac{\partial(rW_\phi)}{\partial r} \tag{59}$$

$$\gamma_\theta = a_\theta \left[1 + \frac{1}{c^2}\left(v_\phi^2 + 3W\right) \right] - \frac{W_{,\theta}}{r}\left[1 + \frac{1}{c^2}\left(2v_\phi^2 - W\right) \right] + \frac{4}{c^2}\frac{v_\phi}{r\sin\theta}\frac{\partial(W_\phi \sin\theta)}{\partial\theta} \tag{60}$$

$$\gamma_\phi = 0 \tag{61}$$

We take into account the gravitational potential up to second order:

$$W(r,\theta) = \frac{GM}{r}\left(1 - J_2\left(\frac{R}{r}\right)^2 P_2(\theta) \right) \tag{62}$$

where $P_2(\theta) = \frac{1}{2}(3\cos^2\theta - 1)$ is the Legendre function and $J_2 \sim 1.083 \times 10^{-3}$ is the Earth's flatness. Moreover, we take into account only the monopole of the gravitational potential vector:

$$\vec{W}(r,\theta) = \frac{G\vec{J} \times \vec{r}}{2r^3} \tag{63}$$

where $\vec{J} = I\vec{\omega}$ is the Earth angular momentum. Attempts to measure the monopole of the gravitational potential vector with orbiting gyroscopes and satellites were performed in recent years (see, e.g., [66–68] and the references therein).

The order of magnitude of the fourth order relativistic effect in the absolute clock observable (33) is 2×10^{-19}, which is below the current clock accuracy. In terms of geoid height, it corresponds to 2 mm, which is also below the actual accuracy of the geoid determination (which is around 1 to 10 cm). The different contributions of non-linear terms to the geoid height are given in more detail in [6].

The second order contribution to the local acceleration γ is shown in Figure 1. One can see that the relativistic effects are below or just at the μGal level (1 Gal = 10^{-2} m·s^{-2}), which is the accuracy of the best absolute gravimeters nowadays.

For gradiometry, the contribution of the relativistic effect is of the order of a few μE (1 E = 1 eotvos = 10^{-9} s^{-2}), when the accuracy of the best gradiometers to date is of the order of a few 100 μE. However, it has been claimed that by integrating the effect over one year, it could become observable [32,69].

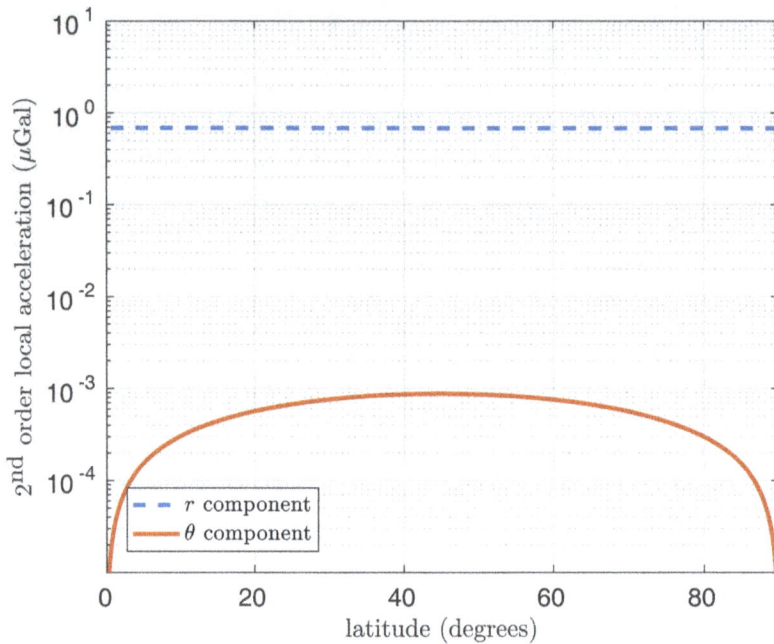

Figure 1. Second order (c^{-2}) contributions to the local acceleration γ.

6. Conclusions

We have reviewed the literature of relativistic geodesy. We introduced the theoretical tools of relativistic geodesy and applied them for a stationary PPN metric. We applied the calculation to the case of a stationary clock on the Earth. Some interesting conclusions concerning post-Newtonian corrections are that:

- differences between the chronometric geoid and the Newtonian geoid are of order 2 mm;
- post-Newtonian corrections for gravimeters are below or just at the level of current accuracy of the best absolute gravimeters, which is about 1 μGal;
- post-Newtonian corrections for gradiometers are below current accuracy, a few μE, but could be measurable by integrating for a long time.

Acknowledgments: This work was partly funded by the European Metrology Research Programme (EMRP) within the SIB-55 ITOCproject. The EMRP is jointly funded by the EMRP participating countries within EURAMET and the European Union.

Author Contributions: P.D. and J.G. conducted all calculations and wrote the article together.

Conflicts of Interest: The authors declare no conflict of interest.

References

1. Cacciapuoti, L.; Salomon, C. Space Clocks and Fundamental Tests: The ACES Experiment. *Eur. Phys. J. Spec. Top.* **2009**, *172*, 57–68. [CrossRef]
2. Koller, S.B.; Grotti, J.; Vogt, S.; Al-Masoudi, A.; Dörscher, S.; Häfner, S.; Sterr, U.; Lisdat, C. Transportable Optical Lattice Clock with 7×10^{-17} Uncertainty. *Phys. Rev. Lett.* **2017**, *118*, 073601. [CrossRef] [PubMed]
3. Delva, P.; Lodewyck, J. Atomic Clocks: New Prospects in Metrology and Geodesy. *Acta Futura* **2013**, *7*, 67–78.
4. Flury, J. Relativistic Geodesy. *J. Phys. Conf. Ser.* **2016**, *723*, 012051. [CrossRef]

5. Denker, H.; Timmen, L.; Voigt, C.; Weyers, S.; Peik, E.; Margolis, H.S.; Delva, P.; Wolf, P.; Petit, G. Geodetic methods to determine the relativistic redshift at the level of 10^{-18} in the context of international timescales—A review and practical results. *J. Geod.* **2017**, in press.
6. Müller, J.; Soffel, M.; Klioner, S.A. Geodesy and Relativity. *J. Geod.* **2007**, *82*, 133–145. [CrossRef]
7. Soffel, M.; Kopeikin, S.; Han, W.B. Advanced Relativistic VLBI Model for Geodesy. *J. Geod.* **2016**, 1–19. [CrossRef]
8. Will, C.M. Relativistic Gravity in the Solar System. II. Anisotropy in the Newtonian Gravitational Constant. *Astrophys. J.* **1971**, *169*, 141–155. [CrossRef]
9. Will, C.M. Theoretical Frameworks for Testing Relativistic Gravity. II. Parametrized Post-Newtonian Hydrodynamics, and the Nordtvedt Effect. *Astrophys. J.* **1971**, *163*, 611–628. [CrossRef]
10. Will, C.M. *Theory and Experiment in Gravitational Physics*; Cambridge University Press: Cambridge, UK, 1993.
11. Will, C.M. The Confrontation between General Relativity and Experiment. *Living Rev. Relativ.* **2014**, *17*, 4. [CrossRef] [PubMed]
12. Vermeer, M. *Chronometric Levelling*; Technical Report; Finnish Geodetic Institute: Helsinki, Finland, 1983.
13. Margolis, H.; Godun, R.; Gill, P.; Johnson, L.; Shemar, S.; Whibberley, P.; Calonico, D.; Levi, F.; Lorini, L.; Pizzocaro, M.; et al. International Timescales with Optical Clocks (ITOC). In Proceedings of the 2013 Joint European Frequency and Time Forum International Frequency Control Symposium (EFTF/IFC), Prague, Czech Republic, 21–25 July 2013; pp. 908–911.
14. Petit, G.; Wolf, P.; Delva, P. Atomic Time, Clocks, and Clock Comparisons in Relativistic Spacetime: A Review. In *Frontiers in Relativistic Celestial Mechanics—Volume 2: Applications and Experiments*; Kopeikin, S.M., Ed.; De Gruyter Studies in Mathematical Physics; De Gruyter: Berlin, Germany, 2014; pp. 249–279.
15. Brumberg, V.A.; Groten, E. On Determination of Heights by Using Terrestrial Clocks and GPS Signals. *J. Geod.* **2002**, *76*, 49–54. [CrossRef]
16. Bondarescu, R.; Bondarescu, M.; Hetényi, G.; Boschi, L.; Jetzer, P.; Balakrishna, J. Geophysical Applicability of Atomic Clocks: Direct Continental Geoid Mapping. *Geophys. J. Int.* **2012**, *191*, 78–82. [CrossRef]
17. Chou, C.W.; Hume, D.B.; Rosenband, T.; Wineland, D.J. Optical Clocks and Relativity. *Science* **2010**, *329*, 1630–1633. [CrossRef] [PubMed]
18. Lion, G.; Panet, I.; Wolf, P.; Guerlin, C.; Bize, S.; Delva, P. Determination of a High Spatial Resolution Geopotential Model Using Atomic Clock Comparisons. *J. Geod.* **2017**, 1–15. [CrossRef]
19. Bjerhammar, A. On a Relativistic Geodesy. *Bull. Geod.* **1985**, *59*, 207–220. [CrossRef]
20. Bjerhammar, A. *Relativistic Geodesy*; Technical Report NON118 NGS36; National Oceanic and Atmospheric Administration (NOAA): Silver Spring, MD, USA, 1986.
21. Soffel, M.; Herold, H.; Ruder, H.; Schneider, M. Relativistic Theory of Gravimetric Measurements and Definition of the Geoid. *Manuscr. Geod.* **1988**, *13*, 143–146.
22. Soffel, M.H. *Relativity in Astrometry, Celestial Mechanics, and Geodesy*; Springer: New York, NY, USA, 1989.
23. Hofmann-Wellenhof, B.; Moritz, H. *Physical Geodesy*; Springer Science & Business Media: New York, NY, USA, 2006.
24. Kopejkin, S.M. Relativistic Manifestations of Gravitational Fields in Gravimetry and Geodesy. *Manuscr. Geod.* **1991**, *16*, 301–312.
25. Kopeikin, S.M.; Efroimsky, M.; Kaplan, G. *Relativistic Celestial Mechanics of the Solar System*; John Wiley & Sons: New York, NY, USA, 2011.
26. Kopeikin, S.M.; Mazurova, E.M.; Karpik, A.P. Towards an Exact Relativistic Theory of Earth's Geoid Undulation. *Phys. Lett. A* **2015**, *379*, 1555–1562. [CrossRef]
27. Kopeikin, S.M.; Han, W.; Mazurova, E. Post-Newtonian Reference Ellipsoid for Relativistic Geodesy. *Phys. Rev. D* **2016**, *93*, 044069. [CrossRef]
28. Kopeikin, S.M. Reference Ellipsoid and Geoid in Chronometric Geodesy. *Front. Astron. Space Sci.* **2016**, *3*, 5. [CrossRef]
29. Iorio, L. On the Impossibility of Measuring the General Relativistic Part of the Terrestrial Acceleration of Gravity with Superconducting Gravimeters. *Geophys. J. Int.* **2006**, *167*, 567–569. [CrossRef]
30. Mashhoon, B.; Theiss, D.S. Relativistic Tidal Forces and the Possibility of Measuring Them. *Phys. Rev. Lett.* **1982**, *49*, 1542–1545. [CrossRef]
31. Theiss, D.S. A General Relativistic Effect of a Rotating Spherical Mass and the Possibility of Measuring It in a Space Experiment. *Phys. Lett. A* **1985**, *109*, 19–22. [CrossRef]

32. Mashhoon, B.; Paik, H.J.; Will, C.M. Detection of the Gravitomagnetic Field Using an Orbiting Superconducting Gravity Gradiometer. Theoretical Principles. *Phys. Rev. D* **1989**, *39*, 2825–2838. [CrossRef]
33. Li, X.Q.; Shao, M.X.; Paik, H.J.; Huang, Y.C.; Song, T.X.; Bian, X. Effects of Satellite Positioning Errors and Earth's Multipole Moments in the Detection of the Gravitomagnetic Field with an Orbiting Gravity Gradiometer. *Gen. Relativ. Gravit.* **2014**, *46*, 1–14. [CrossRef]
34. Xu, P.; Paik, H.J. First-Order Post-Newtonian Analysis of the Relativistic Tidal Effects for Satellite Gradiometry and the Mashhoon-Theiss Anomaly. *Phys. Rev. D* **2016**, *93*, 044057. [CrossRef]
35. Bini, D.; Mashhoon, B. Relativistic Gravity Gradiometry. *Phys. Rev. D* **2016**, *94*, 124009. [CrossRef]
36. Qiang, L.E.; Xu, P. Testing Chern-Simons Modified Gravity with Orbiting Superconductive Gravity Gradiometers: The Non-Dynamical Formulation. *Gen. Relativ. Gravit.* **2015**, *47*, 1–15. [CrossRef]
37. Qiang, L.E.; Xu, P. Probing the Post-Newtonian Physics of Semi-Conservative Metric Theories through Secular Tidal Effects in Satellite Gradiometry Missions. *Int. J. Mod. Phys. D* **2016**, *25*, 1650070. [CrossRef]
38. Soffel, M.; Klioner, S.A.; Petit, G.; Wolf, P.; Kopeikin, S.M.; Bretagnon, P.; Brumberg, V.A.; Capitaine, N.; Damour, T.; Fukushima, T.; et al. The IAU 2000 Resolutions for Astrometry, Celestial Mechanics, and Metrology in the Relativistic Framework: Explanatory Supplement. *Astron. J.* **2003**, *126*, 2687–2706. [CrossRef]
39. Fukushima, T. The Fermi Coordinate System in the Post-Newtonian Framework. *Celest. Mech.* **1988**, *44*, 61–75. [CrossRef]
40. Ashby, N.; Bertotti, B. Relativistic Perturbations of an Earth Satellite. *Phys. Rev. Lett.* **1984**, *52*, 485–488. [CrossRef]
41. Ashby, N.; Bertotti, B. Relativistic Effects in Local Inertial Frames. *Phys. Rev. D* **1986**, *34*, 2246–2259. [CrossRef]
42. Kostić, U.; Horvat, M.; Gomboc, A. Relativistic Positioning System in Perturbed Spacetime. *Class. Quantum Gravity* **2015**, *32*, 215004. [CrossRef]
43. De Felice, F.; Bini, D. *Classical Measurements in Curved Space-Times*; Cambridge University Press: Cambridge, UK, 2010.
44. Misner, C.W.; Thorne, K.S.; Wheeler, J.A. *Gravitation*; W. H. Freeman: New York, NY, USA, 1973.
45. Delva, P.; Angonin, M.C. Extended Fermi Coordinates. *Gen. Relativ. Gravit.* **2012**, *44*, 1–19. [CrossRef]
46. Kopejkin, S.M. Celestial Coordinate Reference Systems in Curved Space-Time. *Celest. Mech.* **1988**, *44*, 87–115. [CrossRef]
47. Damour, T.; Soffel, M.; Xu, C. General-Relativistic Celestial Mechanics. I. Method and Definition of Reference Systems. *Phys. Rev. D* **1991**, *43*, 3273–3307. [CrossRef]
48. Brumberg, V.A.; Kopejkin, S.M. Relativistic Reference Systems and Motion of Test Bodies in the Vicinity of the Earth. *Nuovo Cimento* **1989**, *103*, 63–98. [CrossRef]
49. Li, W.Q.; Ni, W.T. Coupled Inertial and Gravitational Effects in the Proper Reference Frame of an Accelerated, Rotating Observer. *J. Math. Phys.* **1979**, *20*, 1473–1480. [CrossRef]
50. Blanchet, L.; Salomon, C.; Teyssandier, P.; Wolf, P. Relativistic Theory for Time and Frequency Transfer to Order. *Astron. Astrophys.* **2001**, *370*, 320–329. [CrossRef]
51. Linet, B.; Teyssandier, P. Time transfer and frequency shift to the order $1/c^4$ in the field of an axisymmetric rotating body. *Phys. Rev. D* **2002**, *66*, 024045. [CrossRef]
52. Geršl, J.; Delva, P.; Wolf, P. Relativistic Corrections for Time and Frequency Transfer in Optical Fibres. *Metrologia* **2015**, *52*, 552. [CrossRef]
53. Sánchez, L. Towards a Vertical Datum Standardisation under the Umbrella of Global Geodetic Observing System. *J. Geod. Sci.* **2012**, *2*, 325–342. [CrossRef]
54. Burša, M.; Kenyon, S.; Kouba, J.; Šíma, Z.; Vatrt, V.; Vítek, V.; Vojtíšková, M. The Geopotential Value W_0 for Specifying the Relativistic Atomic Time Scale and a Global Vertical Reference System. *J. Geod.* **2006**, *81*, 103–110. [CrossRef]
55. Dayoub, N.; Edwards, S.J.; Moore, P. The Gauss–Listing Geopotential Value W_0 and Its Rate from Altimetric Mean Sea Level and GRACE. *J. Geod.* **2012**, *86*, 681–694. [CrossRef]
56. Jevrejeva, S.; Moore, J.C.; Grinsted, A. Sea Level Projections to AD2500 with a New Generation of Climate Change Scenarios. *Glob. Planet. Chang.* **2012**, *80–81*, 14–20. [CrossRef]
57. Bertotti, B.; Iess, L.; Tortora, P. A Test of General Relativity Using Radio Links with the Cassini Spacecraft. *Nature* **2003**, *425*, 374–376. [CrossRef] [PubMed]

58. Pitjeva, E.V.; Pitjev, N.P. Relativistic Effects and Dark Matter in the Solar System from Observations of Planets and Spacecraft. *Mon. Not. R. Astron. Soc.* **2013**, *432*, 3431–3437. [CrossRef]

59. Iorio, L. Constraining the preferred frame α_1, α_2 parameters from Solar System planetary precessions. *Int. J. Mod. Phys. D* **2013**, *23*, 1450006. [CrossRef]

60. Debono, I.; Smoot, G.F. General Relativity and Cosmology: Unsolved Questions and Future Directions. *Universe* **2016**, *2*, 23. [CrossRef]

61. Soffel, M.; Frutos, F. On the Usefulness of Relativistic Space-Times for the Description of the Earth's Gravitational Field. *J. Geod.* **2016**, *90*, 1345–1357. [CrossRef]

62. Iorio, L.; Lichtenegger, H.I.M.; Ruggiero, M.L.; Corda, C. Phenomenology of the Lense-Thirring Effect in the Solar System. *Astrophys. Space Sci.* **2010**, *331*, 351–395. [CrossRef]

63. Renzetti, G. History of the Attempts to Measure Orbital Frame-Dragging with Artificial Satellites. *Open Phys.* **2013**, *11*, 531–544. [CrossRef]

64. Ciufolini, I.; Wheeler, J.A. *Gravitation and Inertia*; Princeton University Press: Princeton, NJ, USA, 1995.

65. Angonin, M.C.; Tourrenc, P.; Delva, P. Cold Atom Interferometer in a Satellite: Orders of Magnitude of the Tidal Effect. *Appl. Phys. B* **2006**, *84*, 579–584. [CrossRef]

66. Everitt, C.W.F.; Muhlfelder, B.; DeBra, D.B.; Parkinson, B.W.; Turneaure, J.P.; Silbergleit, A.S.; Acworth, E.B.; Adams, M.; Adler, R.; Bencze, W.J.; et al. The Gravity Probe B Test of General Relativity. *Class. Quantum Gravity* **2015**, *32*, 224001. [CrossRef]

67. Iorio, L. An Assessment of the Systematic Uncertainty in Present and Future Tests of the Lense-Thirring Effect with Satellite Laser Ranging. *Space Sci. Rev.* **2009**, *148*, 363–381. [CrossRef]

68. Renzetti, G. First Results from LARES: An Analysis. *New Astron.* **2013**, *23–24*, 63–66. [CrossRef]

69. Paik, H.J. Detection of the Gravitomagnetic Field Using an Orbiting Superconducting Gravity Gradiometer: Principle and Experimental Considerations. *Gen. Relativ. Gravit.* **2008**, *40*, 907–919. [CrossRef]

universe

MDPI

Review

Testing General Relativity with the Radio Science Experiment of the BepiColombo mission to Mercury

Giulia Schettino [1,*] and Giacomo Tommei [2]

[1] IFAC-CNR, Via Madonna del Piano 10, 50019 Sesto Fiorentino (FI), Italy
[2] Department of Mathematics, University of Pisa, Largo Bruno Pontecorvo 5, 56127 Pisa, Italy; giacomo.tommei@unipi.it
*   Correspondence: g.schettino@ifac.cnr.it; Tel.: +39-055-522-6315

Academic Editors: Lorenzo Iorio and Elias C. Vagenas
Received: 29 June 2016; Accepted: 1 September 2016; Published: 12 September 2016

Abstract: The relativity experiment is part of the Mercury Orbiter Radio science Experiment (MORE) on-board the ESA/JAXA BepiColombo mission to Mercury. Thanks to very precise radio tracking from the Earth and accelerometer, it will be possible to perform an accurate test of General Relativity, by constraining a number of post-Newtonian and related parameters with an unprecedented level of accuracy. The Celestial Mechanics Group of the University of Pisa developed a new dedicated software, ORBIT14, to perform the simulations and to determine simultaneously all the parameters of interest within a global least squares fit. After highlighting some critical issues, we report on the results of a full set of simulations, carried out in the most up-to-date mission scenario. For each parameter we discuss the achievable accuracy, in terms of a formal analysis through the covariance matrix and, furthermore, by the introduction of an alternative, more representative, estimation of the errors. We show that, for example, an accuracy of some parts in 10^{-6} for the Eddington parameter β and of 10^{-5} for the Nordtvedt parameter η can be attained, while accuracies at the level of 5×10^{-7} and 1×10^{-7} can be achieved for the preferred frames parameters α_1 and α_2, respectively.

Keywords: general relativity and gravitation; experimental studies of gravity; Mercury; BepiColombo mission

1. Introduction

BepiColombo is a mission for the exploration of the planet Mercury, jointly developed by the European Space Agency (ESA) and the Japan Aerospace eXploration Agency (JAXA). The mission is scheduled for launch in April 2018 and for orbit insertion around Mercury at the end of 2024. The science mission consists of two separated spacecraft, which will be inserted in two different orbits around the planet: the Mercury Planetary Orbiter (MPO), devoted to the study of the surface and internal composition [1], and the Mercury Magnetospheric Orbiter (MMO), designed for the study of the planet's magnetosphere [2]. In particular, the Mercury Orbiter Radio science Experiment (MORE) is one of the experiments on-board the MPO spacecraft.

Thanks to the state-of-the-art on-board and on-ground instrumentation [3], MORE will enable a better understanding of both Mercury geophysics and fundamental physics. The main goals of the MORE radio science experiment are concerned with the gravity of Mercury [4–8], the rotation of Mercury [9–11] and General Relativity (GR) tests [12–16]. The global experiment consists in determining the value and the formal uncertainty (as defined in Section 2.2) of a number of parameters of general interest, with the addition of further parameters characterizing each specific goal of the experiment. The quantities to be determined can be partitioned as follows:

(a)   spacecraft state vector (position and velocity) at given times (*Mercurycentric orbit determination*);
(b)   spherical harmonics of the gravity field of Mercury [17] and tidal Love number k_2 [18], in order to constrain physical models of the interior of Mercury (*gravimetry experiment*);
(c)   parameters defining the model of the Mercury's rotation (*rotation experiment*);
(d)   digital calibrations for the Italian Spring Accelerometer (ISA) [19,20];
(e)   state vector of Mercury and Earth-Moon Barycenter (EMB) orbits at some reference epoch, in order to improve the ephemerides (*Mercury and EMB orbit determination*);
(f)   post-Newtonian (PN) parameters [12,13,21,22], together with some related parameters, like the solar oblateness factor $J_{2\odot}$, the solar gravitational factor $\mu_\odot = GM_\odot$, where G is the gravitational constant and $M_\odot$ the Sun's mass, and its time variation, $\zeta = (1/\mu_\odot)\,d\mu_\odot/dt$, in order to test gravitational theories (*relativity experiment*).

A good initial guess for each of the above parameters will also be necessary: for example, for the PN parameters we use the GR values, while for gravimetry we refer to the most recent MESSENGER results as nominal values in simulations [23].

To cope with the extreme complexity of MORE and its challenging goals, the Celestial Mechanics Group of the University of Pisa developed (under an Italian Space Agency commission) a dedicated software, ORBIT14, which is now ready for use. The software enables the generation of the simulated observables and the determination of the solve-for parameters by means of a global least squares fit.

In this paper, the results of a full set of simulations are discussed, carried out in the most up-to-date mission scenario [24], focusing on the parameters of interest for the relativity experiment. The paper is organized as follows: in Section 2 we describe in details the structure of the ORBIT14 software, while in Section 3 we introduce the mathematical models adopted to perform a full relativistic and coherent analysis of the observations. The simulation scenario and the adopted assumptions are detailed in Section 4, while the results are presented in Section 5, together with a comprehensive discussion on the achievable accuracies. Finally, conclusions are drawn in Section 6.

2. The ORBIT14 Software

The ORBIT14 software system has been developed by the Celestial Mechanics Group of the University of Pisa starting since 2007 as a new dedicated software for the MORE experiment. In Section 2.1 we describe the global structure of the software, while in Section 2.2 we briefly recall the non-linear least squares method. Details on the adopted multi-arc strategy are given in Section 2.3.

2.1. Global Structure

The global structure of the code is outlined in Figure 1. The main programs belong to two categories: **data simulator** (short: simulator) and **differential corrector** (short: corrector). Code is written in Fortran90 language. The simulator is needed to predict possible scientific results of the experiment. It generates simulated observables (range and range-rate, accelerometer readings) and the nominal value for orbital elements of the Mercurycentric orbit of the spacecraft, of Mercury and of the EMB orbits. The program structure of the simulator is quite simple if compared with the differential corrector, the most demanding part being the implementation of the dynamical, observational and error models.

The actual core of the code is the corrector, which solves for all the parameters **u** which can be determined by a least squares fit (possibly constrained and/or decomposed in a multi-arc structure). The corrector structure has been designed in order to exploit parallel computing, especially for the most computationally expensive portion of the processing. An outline of the steps involved in a single corrector's iteration is shown in Figure 2.

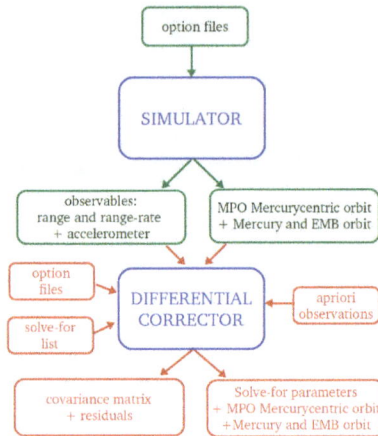

Figure 1. Block diagram of the ORBIT14 code: simulation and differential corrections stages. Green arrows refer to simulator inputs/outputs and orange arrows to corrector inputs/outputs. The input option files for simulator and corrector are similar and include, for example, the state vector of the spacecraft at the initial epoch, the number of considered arcs, the time steps for the orbit propagation of the spacecraft, Mercury and EMB, the time sampling for range, range-rate and accelerometer data.

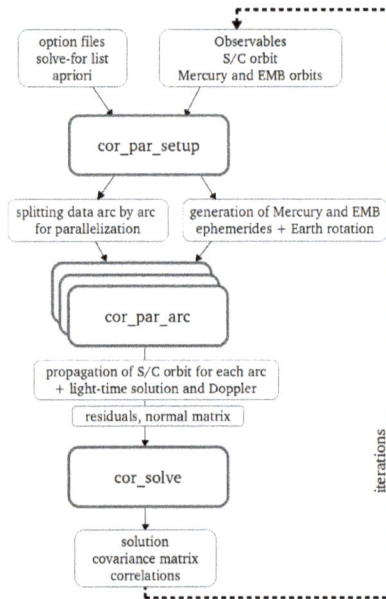

Figure 2. Block diagram of a differential corrector decomposed in three steps: (1) in "cor_par_setup" all the input options are read, data are split for the following parallel computation and the orbits of Mercury and EMB are propagated; (2) "cor_par_arc" contains most of the computationally expensive processing and is parallelized, by executing multiple copies of the same code, without need for interprocess communication; at this stage, the orbit of the spacecraft is propagated at each arc, the light-time computation is performed and residuals and normal matrix are given as output for the next step; (3) in "cor_solve" the covariance matrix and the LS solution are computed.

2.2. Non-Linear Least Squares Fit

Following a classical approach (see, for instance, [25]—Chapter 5), the non-linear least squares (LS) fit leads to compute a set of parameters **u** which minimizes the following target function:

$$Q(\mathbf{u}) = \frac{1}{m}\boldsymbol{\xi}^T(\mathbf{u})W\boldsymbol{\xi}(\mathbf{u}) = \frac{1}{m}\sum_{i=1}^{m} w_i \xi_i^2(\mathbf{u}), \tag{1}$$

where m is the number of observations and $\boldsymbol{\xi}(\mathbf{u}) = \mathcal{O} - \mathcal{C}(\mathbf{u})$ is the vector of *residuals*, i.e., the difference between the observed $\mathcal{O}$ and the predicted quantities $\mathcal{C}(\mathbf{u})$, computed following suitable mathematical models and assumptions. In our case, $\mathcal{O}$ are tracking data (range, range-rate and non-gravitational accelerations from the accelerometer), while $\mathcal{C}(\mathbf{u})$ are the results of the light-time computation (see [22] for details) as a function of all the parameters **u**. Finally, w_i is the weight associated to the i-th observation. Among the parameters **u**, the ones introduced in Section 1 in (a), (b), (c) and (d) occur in the equation of motion of the Mercurycentric orbit of the spacecraft, while those in e) and f) occur in the equations of the orbits of Mercury and the Earth-Moon barycenter with respect to the Solar System Barycenter (SSB). Other information required for such orbit propagations are supposed to be known: positions and velocities of the other planets of the Solar System are obtained from the JPL ephemerides DE421 [26], while the rotation of the Earth is provided by the interpolation table made public by the International Earth Rotation Service (IERS: http://www.iers.org.) and the coordinates associated with the ground stations are expected to be available.

The procedure to compute $\mathbf{u}^*$, the set of parameters which minimizes $Q(\mathbf{u})$, is based on a modified Newton's method known in the literature as *differential corrections method*. All the details can be found in [25]—Chapter 5. Let us define:

$$B = \frac{\partial \boldsymbol{\xi}}{\partial \mathbf{u}}(\mathbf{u}), \qquad C = B^T W B,$$

which are called *design matrix* and *normal matrix*, respectively. Then, the correction

$$\Delta \mathbf{u} = C^{-1}D \quad \text{with } D = -B^T W \boldsymbol{\xi}$$

is applied iteratively until either Q does not change meaningfully from one iteration to the other or $\Delta \mathbf{u}$ becomes smaller than a given tolerance. Introducing the inverse of the normal matrix, $\Gamma = C^{-1}$, we always adopt the probabilistic interpretation of Γ as the *covariance matrix* of the vector **u**, considered as a multivariate Gaussian distribution with mean $\mathbf{u}^*$ in the space of parameters.

2.3. Pure and Constrained Multi-Arc Strategy

The tracking measurements from the Earth to the spacecraft are not continuous because of the mutual geometric configuration between the observing station and the antenna on the probe. The following visibility conditions are defined to account for: (i) the occultation of the spacecraft behind Mercury as seen from the Earth; (ii) the elevation of Mercury above the horizon at the observing station (the data received when Mercury is below a minimum elevation of 15° from the horizon are discarded because they are too noisy and could degrade the results); (iii) the angle between Mercury and the Sun as seen from the Earth. As a result, the observations are split in arcs, with a duration of ~24 h. Considering two observing stations (see Section 4), the adopted visibility conditions provide tracking sessions with an average duration of 15–16 h, called *observed arcs*, followed by a period of some hours without observations.

In orbit determination, the estimation approach consists of a combined solution called **multi-arc strategy** (see, e.g., [25]). According to this method, every single arc of observations has its own set of initial conditions (position and velocity at the reference central epoch of the considered time interval), as it belongs to a different object. In this way, due to lack of knowledge in the dynamical models, the actual errors in the orbit propagation can be reduced by an over-parameterization of the initial conditions. A different choice has been made in ORBIT14, implementing the so called **constrained multi-arc strategy** [10,14,27]. The method is based on the idea that each observed arc belongs to the same object (the spacecraft). First of all, an *extended arc* is defined as the observed arc broadened to half the preceding and to half the following periods without tracking, as shown in Figure 3. The orbits of two consecutive extended arcs should coincide at the connection time in the middle of the non-observed interval. We refer to [10,27] for a complete description of the constrained multi-arc strategy.

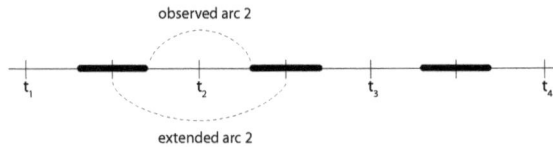

Figure 3. Schematic representation of observed and extended arc. The times t_i $(i = 1, .., 4)$ are the central epoch of each arc; the black bars correspond to dark intervals, without tracking from Earth. See the text for more explanation.

In the constrained multi-arc approach, the parameters **u** can be classified, depending on the arc they refer to, as:

- **Global Parameters (g):** parameters that affect the dynamical equations of every observed (and extended) arc. The PN parameters and the spherical harmonic coefficients of Mercury are an example.
- **Local Parameters (l^k):** parameters that affect the dynamical equations of a single observed arc k. The state vector of the Mercurycentric orbit associated with the arc and the desaturation manoeuvres applied during the tracking are few examples.
- **Local External Parameters ($le^{k,k+1}$):** parameters that affect only the dynamical equations in the period without tracking between two subsequent observed arcs k and $k + 1$. These are the desaturation maneuvers taking place out of the observed arcs.

To implement the constrained multi-arc strategy in the framework of the LS fit described in Section 2.2, we define the *discrepancy vector* between the k and $k + 1$ arcs, $\mathbf{d}^{k,k+1}$, as:

$$\mathbf{d}^{k,k+1} = \Phi(t_c^k; t_0^{k+1}, \mathbf{X}_0^{k+1}) - \Phi(t_c^k; t_0^k, \mathbf{X}_0^k),$$

where t_0^k and t_0^{k+1} denote the central time of the k and $k + 1$ arc, respectively, t_c^k is the connection time between the two extended k and $k + 1$ arcs, $\mathbf{X}_0^k$ and $\mathbf{X}_0^{k+1}$ are the state vector at t_0^k and t_0^{k+1}, respectively, and $\Phi(t_c^k; t_0^{k+1}, \mathbf{X}_0^{k+1})$ is the image of $(t_0^{k+1}, \mathbf{X}_0^{k+1})$ under the flow of the vector field associated with the Mercurycentric orbit at time $t = t_c^k$ (analogously for $\Phi(t_c^k; t_0^k, \mathbf{X}_0^k)$). Thus, we have:

$$\mathbf{d}^{k,k+1} = \mathbf{d}^{k,k+1}(g, l^k, le^{k,k+1}, l^{k+1}),$$

where $g, l^k, le^{k,k+1}$ are the parameters **u** included in the LS fit, classified, respectively, as global (g), local (l^k) for the k-th arc and local external ($le^{k,k+1}$) between the k-th and $(k + 1)$-th arcs. The constrained multi-arc strategy consists in minimizing the target function:

$$Q(\mathbf{u}) = \frac{1}{m + 6(n - 1)} \sum_{i=1}^{m} w_i \varsigma_i^2 +$$
$$+ \frac{1}{\mu} \frac{1}{m + 6(n - 1)} \sum_{k=1}^{n-1} \mathbf{d}^{k,k+1} \cdot C^{k,k+1} \mathbf{d}^{k,k+1},$$

where n is the total number of extended arcs, μ is a *penalty parameter* and $C^{k,k+1}$ is a weight matrix for the discrepancy vectors. Two possible approaches can be followed. The first is called **internally constrained multi-arc strategy**. In this case, we consider the confidence ellipsoids associated with $\mathbf{X}_0^k$ and $\mathbf{X}_0^{k+1}$ at t_0^k and t_0^{k+1}, respectively, and we propagate them to t_c^k through the corresponding state transition matrices. This means that we expect $\mathbf{d}^{k,k+1}$ to be normally distributed with mean $\Phi(t_c^k; t_0^{k+1}, \mathbf{X}_0^{k+1}) - \Phi(t_c^k; t_0^k, \mathbf{X}_0^k)$ and covariance

$$\Gamma_c := (C^k)^{-1} + (C^{k+1})^{-1},$$

where C^k and C^{k+1} are the 6×6 normal matrices associated with $\mathbf{X}_0^k$ and $\mathbf{X}_0^{k+1}$ at t_0^k and t_0^{k+1} and propagated to t_c^k, respectively. It follows that

$$C^{k,k+1} = \Gamma_c^{-1} \qquad \text{and} \qquad \mu = 1.$$

The second approach is called **apriori constrained multi-arc strategy** and it takes care of the degeneracy in orbit determination due to the orbit geometry (details can be found in [28]). In particular, we deal with an approximated version of the exact symmetry described in [28], where the small parameter of the perturbation is the angle of displacement of the Earth-Mercury vector in an inertial frame. In this case, the normal matrix has one eigenvalue significantly smaller than the others. As a consequence of this weakness, the confidence ellipsoid associated with the discrepancy and defined by $C^{k,k+1}$ could be very elongated. The basic idea of this approach is to constrain the discrepancy $\mathbf{d}^{k,k+1}$ inside a sphere of given radius, that can be suitably shrunk by varying μ. This can be interpreted as adding apriori observations. On the contrary, in the internally constrained multi-arc strategy, the discrepancy is constrained inside the intersection of the two ellipsoids propagated from t_0^k and t_0^{k+1}. All the details are extensively explained in [27]. For the results presented in this review, we will always adopt an apriori constrained multi-arc strategy. Finally, it can be noted that in the multi-arc method the residuals ξ depend only on global and local parameters, and this applies to the target function Q defined in Equation (1) as well.

3. Mathematical Models

The purpose of the MORE relativity experiment is to perform a test of General Relativity comparing theory with experiment. The majority of the Solar System tests of gravitation can be set in the context of the slow-motion, weak field limit [29], usually known as the *post-Newtonian* approximation. In this limit, the space-time metric can be written as an expansion about the Minkowski metric in terms of dimensionless gravitational potentials. In the *parametrized* PN formalism, each potential term in the metric is expressed by a specific parameter, which measures a general property of the metric. The basic idea of the MORE relativity experiment is to investigate the dependence of the equation of motion from the PN parameters. By isolating the effects of each parameter on the motion, it is possible to constrain the parameters values within some accuracy threshold, testing the validity of GR predictions.

The PN parameters of interest for our analysis are the following:

- the *Eddington parameters* β and γ. β accounts for the modification of the non-linear three-body general relativistic interaction and γ parametrizes the velocity-dependent modification of the two-body interaction and accounts also for the space-time curvature through the Shapiro effect [30]. These are the only non-zero PN parameters in GR (they are both equal to unity);
- the *Nordtvedt parameter* η. The effect of η in the equations of motion is to produce a polarization of the Mercury and Earth orbits in the gravitational field of the other planets and it is related to possible violations of the Strong Equivalence Principle (see, e.g., the discussion in [31]);
- the *preferred frame effects parameters* α_1 and α_2. They phenomenologically describe the effects due to the presence of a gravitationally preferred frame; we follow the standard assumption to identify the preferred frame with the rest frame of the cosmic microwave background [32].

Beside the five PN parameters mentioned above, we consider a few additional parameters. These do not properly produce relativistic effects, but the uncertainty in their knowledge generates orbital effects at least comparable with the perturbations expected from the tested relativistic parameters. We consider two kinds of "Newtonian" orbital effects: a small change in the Sun's gravity oblateness $J_{2\odot}$ and a small change in the Sun's gravitational factor $\mu_\odot = GM_\odot$, where G is the gravitational constant and $M_\odot$ is the Sun's mass. Regarding the second quantity, since many alternative theories of gravitation allow for a possible time variation of the gravitational constant, we included in the solve-for parameters the time derivative of $\mu_\odot$ described by the parameter $\zeta = \frac{1}{\mu_\odot} \frac{d\mu_\odot}{dt}$. Indeed, within the MORE experiment we cannot discriminate between a time variation of $M_\odot$ or G, but assuming an independent estimate for the rate of $M_\odot$, it could be possible to draw information on the time rate of G.

The effects linked to each PN parameter, corresponding to modifications of the space-time metric, affect both the propagation of the tracking signal (range and range-rate) and the equations of motion. In the following we describe the mathematical models adopted in our analysis: in Section 3.1 we define the computed tracking observables in a coherent relativistic background and in Section 3.2 we present the Langrangian formulation of the planetary dynamics in the context of the first-order PN approximation. Finally, in Section 3.3 we define the Mercurycentric dynamical model.

3.1. Computation of Observables

In a radio science experiment, the observational technique is complicated by many factors (for example plasma reduction) but in simulations it can be merely considered as a tracking from an Earth-based station, giving range and range-rate information (see, e.g., [3]). In order to compute the range distance from the ground station on the Earth to the spacecraft around Mercury (or in an interplanetary trajectory), and the corresponding range-rate, we introduce the following state vectors, each one evolving according to a specific dynamical model (see Figure 4):

- the Mercurycentric position of the spacecraft, $\mathbf{x}_{sat}$;
- the SSB positions of Mercury and of the EMB, $\mathbf{x}_M$ and $\mathbf{x}_{EM}$;
- the geocentric position of the ground antenna, $\mathbf{x}_{ant}$;
- the position of the Earth barycenter with respect to the EMB, $\mathbf{x}_E$.

They can be combined to define the range distance using the following formula, as a first approximation:

$$r = |\mathbf{r}| = |(\mathbf{x}_{sat} + \mathbf{x}_M) - (\mathbf{x}_{EM} + \mathbf{x}_E + \mathbf{x}_{ant})| . \tag{2}$$

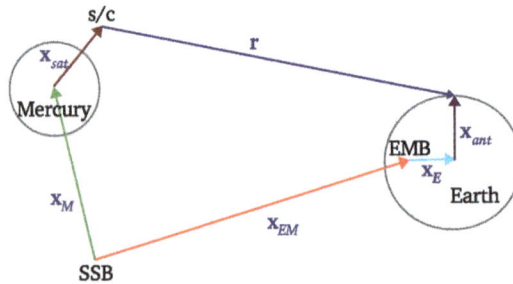

Figure 4. Vectors involved in the multiple dynamics for the tracking of the spacecraft from the Earth.

As explained in [22], Equation (2) corresponds to model the space as a flat arena (r is the Euclidean distance) and the time as an absolute parameter. Obviously, this is not a suitable assumption, since it is clear that beyond some threshold of accuracy, as expected for the BepiColombo radio science experiment, space and time must be formulated in the framework of General Relativity. Moreover, we have to take into account the different times at which the events have to be computed: the transmission of the signal at the transmitting time (t_t), the signal at the Mercury orbiter at the time of bounce (t_b) and the reception of the signal at the receiving time (t_r). Equation (2) can be used as a starting point to construct a correct relativistic formulation, containing not all the possible relativistic effects, but the ones which are measurable in the experiment.

The five vectors in Equation (2) have to be computed at their own time, which corresponds to the epoch of different events: e.g., $\mathbf{x}_{ant}$, $\mathbf{x}_{EM}$ and $\mathbf{x}_E$ are computed at both the antenna transmission time t_t and receiving time t_r of the signal, while $\mathbf{x}_M$ and $\mathbf{x}_{sat}$ are computed at the bounce time t_b (when the signal has reached the orbiter and is sent back, with correction for the delay of the transponder). To perform the vectorial sums and differences, these vectors must be converted to a common space-time reference system, the only possible choice being some realization of the BCRS (Barycentric Celestial Reference System). We adopt a realization of the BCRS that we call SSB (Solar System Barycentric) reference frame in which the time is a re-definition of the TDB (Barycentric Dynamic Time), according to the IAU 2006 Resolution B3 (https://www.iau.org/static/resolutions/IAU2006_Resol3.pdf). Other possible choices, such as TCB (Barycentric Coordinate Time), can only differ by linear scaling. The TDB choice of the SSB time scale entails also the appropriate linear scaling of space-coordinates and planetary masses as described, for instance, in [33,34].

The vectors $\mathbf{x}_M$, $\mathbf{x}_E$, and $\mathbf{x}_{EM}$ are already in the SSB reference frame as provided by numerical integration and external ephemerides, while the vectors $\mathbf{x}_{ant}$ and $\mathbf{x}_{sat}$ must be converted to SSB from the geocentric and Mercurycentric systems, respectively. Of course, the conversion of reference systems implies also the conversion of the time coordinate. There are three different time coordinates to be considered. The currently published planetary ephemerides are provided in TDB. The observations from the Earth are based on averages of clock and frequency measurements on the Earth surface: this defines another time coordinate called TT (Terrestrial Time). Thus for each observation the times of transmission t_t and reception t_r need to be converted from TT to TDB to find the corresponding positions of the planets. This time conversion step is necessary for the accurate processing of each set of interplanetary tracking data. The main term in the difference TT-TDB is periodic, with period 1 year and amplitude $\simeq 1.6 \times 10^{-3}$ s, while there is essentially no linear trend, as a result of a suitable definition of the TDB. Finally, the equation of motion of the spacecraft orbiting Mercury has been approximated, to the required level of accuracy, following what done in [35] for the case of a near-Earth spacecraft in a geocentric frame of reference. We consider the Newtonian dynamics in a local Mercurycentric frame assuming as independent variable a suitably defined time coordinate. Moreover, we add the relativistic perturbative acceleration from the one-body Schwarzschild isotropic metric for Mercury and the acceleration due to the geodesic precession, as explained in [10]. Thus, for MORE we have defined a new time coordinate TDM (Mercury Dynamic Time), as described in [13], containing terms of 1-PN order depending mostly upon the distance from the Sun and velocity of Mercury.

In general, the differential equation giving the local time T (in our case TT or TDM) as a function of the SSB time t, which we are currently assuming to be TDB, is the following:

$$\frac{dT}{dt} = 1 - \frac{1}{c^2} \left[U + \frac{v^2}{2} - L \right] , \qquad (3)$$

where U is the gravitational potential (the list of contributing bodies depends upon the required accuracy: in our implementation we use Sun, Mercury to Neptune, Moon) at the planet center and v is the velocity with respect to the SSB of the same planet. The constant term L is used to perform the conventional rescaling motivated by removal of secular terms [36].

The space-time transformations needed to coherently compute the vector $\mathbf{x}_{ant}$ involve essentially the position of the antenna and of the orbiter. The geocentric coordinates of the antenna should be transformed into TDB-compatible coordinates [34]. The transformation is expressed by:

$$\mathbf{x}_{ant}^{TDB} = \mathbf{x}_{ant}^{TT} \left(1 - \frac{U}{c^2} - L_C \right) - \frac{1}{2} \left(\frac{\mathbf{v}_E^{TDB} \cdot \mathbf{x}_{ant}^{TT}}{c^2} \right) \mathbf{v}_E^{TDB},$$

where U is the gravitational potential at the geocenter (excluding the Earth mass), $L_C = 1.48082686741 \times 10^{-8}$ is a scaling factor given as definition [37], supposed to be a good approximation for removing secular terms from the transformation and $\mathbf{v}_E^{TDB}$ is the barycentric velocity of the Earth. The following equation contains the effect on the velocities of the time coordinate change, which should be consistently used together with the coordinate change:

$$\mathbf{v}_{ant}^{TDB} = \left[\mathbf{v}_{ant}^{TT} \left(1 - \frac{U}{c^2} - L_C \right) - \frac{1}{2} \left(\frac{\mathbf{v}_E^{TDB} \cdot \mathbf{v}_{ant}^{TT}}{c^2} \right) \mathbf{v}_E^{TDB} \right] \left[\frac{dT}{dt} \right].$$

Note that the previous formula contains the factor dT/dt (expressed by Equation (3)) that deals with a time transformation: T is the local time for Earth, that is TT, and t is the corresponding TDB time. To compute the coordinates of the orbiter (vector $\mathbf{x}_{sat}$) we adopt similar equations, as discussed in [22], where we neglected the terms of the SSB acceleration of the planet center [38], because they contain, beside $1/c^2$, the additional small parameter *distance from planet center divided by planet distance to the Sun*, which is of the order of 10^{-4} even for a Mercury orbiter.

In Figure 5 the behaviour of the range and range-rate observable, computed with and without the just explained relativistic corrections, is shown. It is quite evident that the differences are significant, at a signal-to-noise ratio $S/N \simeq 1$ for range, much larger for range-rate, with an especially strong signature from the orbital velocity of the Mercurycentric orbit (with $S/N > 50$).

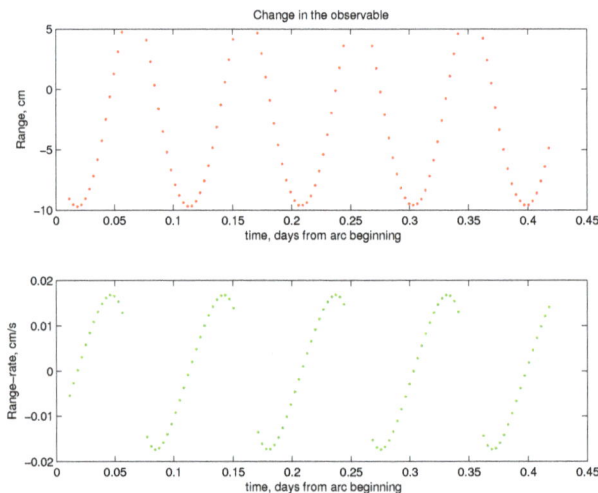

Figure 5. Cont.

Figure 5. The difference in the observables range and range-rate for one pass of Mercury above the horizon for a ground station, by using an hybrid model in which the position and velocity of the orbiter have not been transformed to TDB-compatible quantities and a correct model in which all quantities are TDB-compatible. Gaps of the signal are due to spacecraft passage behind Mercury as seen for the Earth station. (**Top**): for a hybrid model with the satellite position and velocity not transformed to TDB-compatible; (**Bottom**): for a hybrid model with the position and velocity of the antenna not transformed to TDB-compatible.

3.2. Dynamical Relativistic Models

To constrain the PN and related parameters, we need to determine the orbit of Mercury. A relativistic model for the motion of the planet is necessary. We choose to start from a Lagrangian formulation (see, e.g., [21]) in order to compute the terms of accelerations to be included in the right-hand side of the differential equations for Mercury and EMB (the dynamics of the other planets, as well as the relative EMB-Earth position, are given by the JPL ephemerides). In particular, let us assume that the motion of the considered planets is described by the sum of different Lagrangians:

$$L = L_{New} + L_{GR} + (\gamma - 1)L_{\gamma} + (\beta - 1)L_{\beta} + \zeta\,L_{\zeta} + J_{2\odot}L_{J2\odot} + \alpha_1 L_{\alpha_1} + \alpha_2 L_{\alpha_2} + \eta L_{\eta}\,, \qquad (4)$$

where L_{New} is the Lagrangian of the Newtonian N-body problem, L_{GR} is the corrective term taking into account General Relativity in the post-Newtonian approximation, L_{γ} and L_{β} are the terms taking into account the PN parameters γ and β, respectively, L_{ζ} is the Lagrangian for the time variation of the gravitational parameter of the Sun $\mu_{\odot} = GM_{\odot}$, $L_{J2\odot}$ takes into account the effect of the oblateness of the Sun, L_{α_1}, L_{α_2} describe the preferred-frame effects through the parameters α_1, α_2, and, finally, L_{η} checks for possible violations of the strong equivalence principle (see [12,13,31]).

To express the Lagrangian terms, we follow the notation of [35] and we introduce two parameters: the total mass of EMB, μ_{EMB}, which is the sum of the gravitational parameter of the Earth and the Moon, and the mass ratio, $\bar{\mu}$, defined as the ratio of the gravitational parameter of the Earth over that of the Moon; both quantities μ_{EMB} and $\bar{\mu}$ are assumed to be fixed, hence we do not solve for them. Defining $\mu_E = \bar{\mu}/(1 + \bar{\mu})$ and $\mu_{Moon} = 1/(1 + \bar{\mu})$, we have:

$$\mathbf{r}_{EM} = \mu_E\,\mathbf{r}_{Earth} + \mu_{Moon}\,\mathbf{r}_{Moon}$$

with analogous expressions for velocity and acceleration.

Notice that the usual Lagrangians are multiplied by G, so that only $\mu_i = G M_i$ appear in the overall Lagrangian. Indeed, the gravitational constant cannot be determined from any form of orbit determination (apart artificial systems). In the following we give the explicit expressions for each Lagrangian term in Equation (4).

- **N-body Newtonian Lagrangian:**

$$L_{New} = \frac{1}{2} \sum_i \mu_i v_i^2 + \frac{1}{2} \sum_i \sum_{j \neq i} \frac{\mu_i \mu_j}{r_{ij}}.$$

- **Post-Newtonian General Relativistic Lagrangian:**

$$
\begin{aligned}
L_{GR} &= \frac{1}{8 c^2} \sum_i \mu_i v_i^4 - \frac{1}{2 c^2} \sum_i \sum_{j \neq i} \sum_{k \neq i} \frac{\mu_i \mu_j \mu_k}{r_{ij} r_{ik}} + \\
&+ \frac{1}{2} \sum_i \sum_{j \neq i} \frac{\mu_i \mu_j}{r_{ij}} \left[\frac{3}{2 c^2} (v_i^2 + v_j^2) - \frac{7}{2 c^2} (\mathbf{v_i} \cdot \mathbf{v_j}) - \frac{1}{2 c^2} (\mathbf{n_{ij}} \cdot \mathbf{v_i})(\mathbf{n_{ij}} \cdot \mathbf{v_j}) \right].
\end{aligned}
$$

- **Lagrangian for PN parameter γ:**

$$L_\gamma = \frac{1}{2 c^2} \sum_i \sum_{j \neq i} \frac{\mu_i \mu_j}{r_{ij}} (\mathbf{v_i} - \mathbf{v_j})^2.$$

- **Lagrangian for PN parameter β:**

$$L_\beta = -\frac{1}{c^2} \sum_i \sum_{j \neq i} \sum_{k \neq i} \frac{\mu_i \mu_j \mu_k}{r_{ij} r_{ik}}.$$

- **Lagrangian for parameter ζ:**
  L_ζ describes the effect of a time variation of the gravitational parameter of the Sun, $\mu_\odot$:

$$\mu_\odot = \mu_\odot(t_0) + \dot{\mu}_\odot(t_0)(t - t_0) + \dots;$$

defining

$$\zeta = \frac{\dot{\mu}_\odot(t_0)}{\mu_\odot(t_0)} = \frac{d}{dt} \ln \mu_\odot(t_0),$$

we have:

$$L_\zeta = (t - t_0) \sum_{i \neq 0} \frac{\mu_\odot \mu_i}{r_{0i}}.$$

- **Lagrangian for $J_{2\odot}$ effect:**

$$L_{J_{2\odot}} = -\frac{1}{2} \sum_{i \neq 0} \frac{\mu_0 \mu_i}{r_{0i}} \left(\frac{R_\odot}{r_{0i}} \right)^2 [3(\mathbf{n}_{0i} \cdot \mathbf{e}_0)^2 - 1],$$

where $R_\odot$ is the radius of the Sun, $\mathbf{n}_{0i} = \mathbf{r}_{0i}/r_{0i}$ is the heliocentric position of body i and $\mathbf{e}_0$ is the unit vector along the rotation axis of the Sun. The unit vector $\mathbf{e}_0$ is given in standard equatorial coordinates with equinox J2000 at epoch J2000.0 (JD 2451545.0 TCB): $\alpha_0 = 286.13°$, $\delta_0 = 63.87°$ [39].

- **Lagrangian for preferred frame effects, PN α_1 and α_2:**

$$L_{\alpha_1} = -\frac{1}{4 c^2} \sum_j \sum_{i \neq j} \frac{\mu_i \mu_j}{r_{ij}} (\mathbf{z_i} \cdot \mathbf{z_j}),$$

$$L_{\alpha_2} = \frac{1}{4\,c^2} \sum_j \sum_{i \neq j} \frac{\mu_i\,\mu_j}{r_{ij}} \left[(\mathbf{z}_i \cdot \mathbf{z}_j) - (\mathbf{n_{ij}} \cdot \mathbf{z}_i)\,(\mathbf{n_{ij}} \cdot \mathbf{z}_j) \right],$$

where $\mathbf{z}_i = \mathbf{w} + \mathbf{v_i}$ and $\mathbf{w}$ is the velocity of the considered reference system with respect to the PN preferred reference frame, which is a reference frame whose outer regions are at rest with respect to the universe rest frame (see [21]). In the case of the SSB reference frame, that could be the one of cosmic microwave background, $|\mathbf{w}| = 370 \pm 10$ km/s, in the direction $(\alpha, \delta) = (168°, 7°)$ in the Equatorial J2000 reference frame (see [12]). Notice that we can combine the two previous Lagrangians and the parameters α_1 and α_2 obtaining an unique Lagrangian for the preferred frame effects:

$$\begin{aligned}
L_\alpha = \alpha_1\,L_{\alpha_1} + \alpha_2\,L_{\alpha_2} = \ & \frac{\alpha_2 - \alpha_1}{4\,c^2} \sum_j \sum_{i \neq j} \frac{\mu_i\,\mu_j}{r_{ij}} (\mathbf{v_i} + \mathbf{w}) \cdot (\mathbf{v_j} + \mathbf{w}) + \\
& - \frac{\alpha_2}{4\,c^2} \sum_j \sum_{i \neq j} (\mathbf{r_{ji}} \cdot (\mathbf{v_j} + \mathbf{w}))\,(\mathbf{r_{ji}} \cdot (\mathbf{v_i} + \mathbf{w})) \frac{\mu_i\,\mu_j}{r_{ij}^3}.
\end{aligned}$$

- **Lagrangian for possible violation of the equivalence principle, PN η:**
  With the Lagrangian multiplied by G, the Newtonian kinetic energy is:

$$T = \frac{1}{2} \sum_i \mu_i\,v_i^2,$$

where we assume that the inertial mass and the gravitational mass are the same (or, at least, exactly proportional). If some form of mass has a different gravitational coupling, there are, for each body i, two quantities μ_i and μ_i^I, one appearing in the gravitational potential (including the relativistic part) and the other appearing in the kinetic energy. If there is a violation of the strong equivalence principle involving body i, with a fraction Ω_i of its mass due to gravitational self-energy (for the moment we are using the approximation of constant density: $\Omega_i = -3\mu_i/5Rc^2$; notice that Ω_i is $\mathcal{O}(c^{-2})$):

$$\mu_i = (1 + \eta\Omega_i)\,\mu_i^I \iff \mu_i^I = (1 - \eta\Omega_i)\,\mu_i + \mathcal{O}(\eta^2)$$

with η the PN parameter for this violation. Neglecting $\mathcal{O}(\eta^2)$ terms (and also $\mathcal{O}[\eta\,(\gamma - 1)],..)$ this is expressed by a Lagrangian term $\eta\,L_\eta$, where:

$$L_\eta = -\frac{1}{2} \sum_i \Omega_i\,\mu_i\,v_i^2.$$

Considering an inertial reference system, the equations of motion for the i-th body are described by the Lagrangian equations:

$$\frac{d}{d\,t} \frac{\partial L}{\partial \mathbf{v}_i} = \frac{\partial L}{\partial \mathbf{r}_i},$$

which in general give an implicit expression for the acceleration of the form $f(\mathbf{a}_i) = g(\mathbf{r}_i, \mathbf{v}_i)$. However, since the main term is the N-body Newtonian acceleration $\mathbf{a}_i^{New}$ and the other terms are small perturbations, we can use the following approximation for the total acceleration of the i-th body:

$$\mu_i \mathbf{a}_i = \mu_i \mathbf{a}_i^{New} + \frac{\partial (L - L_{New})}{\partial \mathbf{r}_i} - \left[\frac{d}{d\,t} \frac{\partial (L - L_{New})}{\partial \mathbf{v}_i} \right] \Big|_{\mathbf{a}_i = \mathbf{a}_i^{New}}.$$

If we call $Y = [\mathbf{r}_1, \mathbf{r}_{EM}, \mathbf{v}_1, \mathbf{v}_{EM}]^T$ the 12-dimensional state vector (for Mercury and EMB) we want to propagate, we can write the equations of motion in the more complete form:

$$\frac{d}{dt}Y = \begin{pmatrix} \mathbf{v}_1 \\ \mathbf{v}_{EM} \\ \mathbf{a}_1 \\ \mathbf{a}_{EM} \end{pmatrix} = F(\mathbf{r}_1, \mathbf{r}_{EM}, \mathbf{v}_1, \mathbf{v}_{EM}, \dots).$$

The reference system for these dynamics is centered in the SSB, and it is inertial in the PN approximation. On the other hand, if we consider possible violations, as it happens in the case of parameterized PN formalism, we need to reassess the total linear momentum conservation theorem. Using Noether's theorem we can compute the integral of the total linear momentum of the system:

$$\frac{d}{dt}\mathbf{P} = 0, \qquad \text{where} \qquad \mathbf{P} = \sum_i \frac{\partial L}{\partial \mathbf{v}_i}.$$

Since L_β, L_ζ, $L_{J2\odot}$ do not depend on velocities and, because of the antisymmetry, we have that $\sum_i \frac{\partial L_\gamma}{\partial \mathbf{v}_i} = 0$ and L_γ does not contribute; thus, the total linear momentum of the system reads:

$$\mathbf{P} = \sum_i \frac{\partial(L_{New} + L_{GR} + L_\alpha + \eta L_\eta)}{\partial \mathbf{v}_i}.$$

In the PN approximation the total linear momentum is simply:

$$\begin{aligned}
\mathbf{P} &= \sum_i \frac{\partial(L_{New} + L_{GR})}{\partial \mathbf{v}_i} = \\
&= \sum_i \mu_i \mathbf{v}_i \left[1 + \frac{1}{2c^2}v_i^2 - \frac{1}{2c^2}\sum_{k \neq i}\frac{\mu_k}{r_{ik}} \right] - \frac{1}{2c^2}\sum_i \sum_{k \neq i}\frac{\mu_i \mu_k}{r_{ik}}(\mathbf{n}_{ik} \cdot \mathbf{v}_k)\,\mathbf{n}_{ik},
\end{aligned}$$

and the vector:

$$\mathbf{R} = \sum_i \mu_i \mathbf{r}_i \left(1 + \frac{1}{2c^2}v_i^2 - \frac{1}{2c^2}\sum_{k \neq i}\frac{\mu_k}{r_{ik}} \right)$$

is such that:

$$\frac{d}{dt}\mathbf{R} = \mathbf{P},$$

to the $O(c^{-2})$ level of accuracy. Thus $\mathbf{R}$ (rescaled by the total mass) plays the role of the barycenter of the Solar System and can be used to eliminate the Sun from the equations of motion:

$$\mathbf{r}_0 = -\frac{\sum_{i \neq 0}\mu_i \mathbf{r}_i \left(1 + \frac{v_i^2}{2c^2} - \frac{U_i}{2c^2} \right)}{\mu_\odot \left(1 + \frac{v_0^2}{2c^2} - \frac{U_0}{2c^2} \right)}, \qquad U_i = \sum_{k \neq i}\frac{\mu_k}{r_{ik}}.$$

If we now take into account the PN parameters effects, we can write the linear momentum as:

$$\mathbf{P} = \mathbf{P}_0 + \mathbf{P}_\alpha, \qquad \mathbf{P}_0 = \sum_j \frac{\partial(L_{New} + L_{GR} + \eta L_\eta)}{\partial \mathbf{v}_j}, \qquad \mathbf{P}_\alpha = \sum_j \frac{\partial L_\alpha}{\partial \mathbf{v}_j}.$$

In this way, defining the center of mass of the system (rescaled by the total mass) as:

$$\mathbf{R} = \sum_i \mu_i(1 - \eta\Omega_i)\mathbf{r}_i \left(1 + \frac{1}{2c^2}v_i^2 - \frac{1}{2c^2}\sum_{k \neq i}\frac{\mu_k}{r_{ik}} \right),$$

we have

$$\frac{d}{dt}\mathbf{R} = \mathbf{P}_0,$$

and the position of the Sun in this barycentric system is now:

$$\mathbf{r}_0 = -\frac{\sum_{i\neq 0}\mu_i(1-\eta\Omega_i)\mathbf{r}_i\left(1+\frac{v_i^2}{2c^2}-\frac{U_i}{2c^2}\right)}{\mu_\odot(1-\eta\Omega_0)\left(1+\frac{v_0^2}{2c^2}-\frac{U_0}{2c^2}\right)}.$$

Finally, since we have:

$$\dot{\mathbf{P}} = 0 \Longrightarrow \dot{\mathbf{P}}_0 = -\dot{\mathbf{P}}_\alpha \Longrightarrow \ddot{\mathbf{R}} = -\dot{\mathbf{P}}_\alpha,$$

it means that the barycentric reference frame is accelerated. Thus, the equations of motion for the *i*-th body in this reference frame need to be corrected by the acceleration of the barycenter **B**, keeping the $O(c^{-2})$ level of accuracy:

$$\mathbf{a}_i = \mathbf{a}_i^{New} + \frac{1}{\mu_i}\frac{\partial(L-L_{New})}{\partial\mathbf{r}_i} - \left[\frac{d}{dt}\frac{\partial(L-L_{New})}{\partial\mathbf{v}_i}\right]\Big|_{\mathbf{a}_i=\mathbf{a}_i^{New}} - \ddot{\mathbf{B}},$$

where:

$$\mathbf{B} = \frac{\mathbf{R}}{\sum_i\mu_i\left(1-\eta\,\Omega_i\right)\left(1+\frac{v_i^2}{2c^2}-\frac{U_i}{2c^2}\right)}.$$

3.3. Mercurycentric Dynamical Model

In this Section we briefly describe the models adopted to compute the Mercurycentric position of the spacecraft $\mathbf{x}_{sat}$, introduced in Section 3.1 in the expression for the range distance r.

3.3.1. Mercury Gravity Field (Static Part)

The motion of the satellite around Mercury is dominated by the gravity field of the planet. In a Mercurycentric reference frame and using spherical coordinates (r, θ, λ), the gravitational potential of the planet, intended as a static rigid mass, can be expanded in a spherical harmonics series as (see, e.g., [25]—Chapter 13):

$$V(r,\theta,\lambda) = \frac{GM_M}{r} + \sum_{\ell=2}^{+\infty}\frac{GM_M R_M^\ell}{r^{\ell+1}}\sum_{m=0}^{\ell}P_{\ell m}(\sin\theta)[C_{\ell m}\cos m\lambda + S_{\ell m}\sin m\lambda], \qquad (5)$$

where $r > 0$ is the distance from the center of the planet, $-\pi/2 < \theta < \pi/2$ the latitude and $0 \leq \lambda < 2\pi$ the longitude, M_M and R_M are Mercury's mass and mean radius, respectively, $P_{\ell m}$ the Legendre associated functions, $C_{\ell m}$, $S_{\ell m}$ the spherical harmonics coefficients and the summation starts from $\ell = 2$ because the potential is referred to the center of Mercury.

3.3.2. Tidal Perturbations

Mercury cannot be exhaustively described as a rigid body. The gravitational field of the Sun exerts solid tides on Mercury with the tidal bulge oriented in the direction of the Sun. This deformation can be described by adding to the Newtonian potential of Equation (5) a quantity V_L called *Love potential* [18,40,41]:

$$V_L = \frac{GM_\odot k_2 R_M^5}{r_S^3 r^3}\left(\frac{3}{2}\cos^2\psi - \frac{1}{2}\right),$$

where r_S is the Mercury-Sun distance and ψ is the angle between the Mercurycentric position of the spacecraft, $\mathbf{r}$, and the Sun Mercurycentric position. The *Love number* k_2 is the elastic constant characterizing the effect.

3.3.3. Sun and Planetary Perturbations

The solar and planetary gravitational effects on the spacecraft that orbits around Mercury can be computed as a "third-body" perturbative acceleration $\mathbf{a}_{third-body}$ in a local Mercurycentric reference frame. The N bodies acting in the perturbation are: Sun, Venus, Earth-Moon, Mars, Jupiter, Saturn, Uranus and Neptune:

$$\mathbf{a}_{third-body} = \sum_{i=0}^{N-1} GM_i \left(\frac{\mathbf{d}_i}{d_i^3} - \frac{\mathbf{r}_i}{r_i^3} \right),$$

where $\mathbf{d}_i$ is the position of the i-th body of mass M_i with respect to the spacecraft and $\mathbf{r}_i$ is its position with respect to Mercury (see, e.g., [35,42])

3.3.4. Rotational Dynamics

The gravity field development given by Equation (5) is valid in a body-fixed reference frame, like the Mercury body-fixed frame of reference, Ψ_{BF}, defined by the principal inertia axes, with the x-axis along the minimum inertia axis, assumed as rotational reference meridian (see [10] for details). If we define the space-fixed Mercurycentric frame, Ψ_{MC}, in which writing the equation of motion of the spacecraft, then we need to compute the rotation matrix $\mathcal{R}$ to convert the probe coordinates from Ψ_{BF} to Ψ_{MC}. To this aim, we adopt the semi-empirical model defined in [9]. Referring to that paper for an exhaustive discussion, we recall that the rotation matrix can be decomposed as $\mathcal{R} = \mathcal{R}_3(\phi)\mathcal{R}_1(\delta_2)\mathcal{R}_2(\delta_1)$, where $\mathcal{R}_i(\alpha)$ is the matrix associated with the rotation by an angle α about the i-th axis ($i = 1,2,3$), (δ_1, δ_2) define the space-fixed direction of the rotation axis in the Ψ_{MC} frame and ϕ is the rotation angle around the rotation axis, assuming the unit vector along the longest axis of the equator of Mercury (minimum momentum of inertia) as the rotational reference meridian.

The fundamental assumptions to describe the rotational state of Mercury in the adopted semi-empirical model are the following, as defined in [43,44]: (i) the *Cassini state theory*, defining the obliquity η with respect to the orbit normal as $\cos \eta = \cos \delta_2 \cos \delta_1$, assumed to be constant over the mission time span; (ii) addition in the description of two librations in longitude terms, the amplitude ε_1 of 88 days forced librations and the amplitude ε_2 of the Jupiter forced librations, possibly near-resonant with the free libration frequency (see, e.g., [45,46]).

3.3.5. Non-Gravitational Perturbations

The spacecraft around Mercury is perturbed significantly by non-gravitational forces such as the direct radiation pressure from the Sun, the indirect emission from the planet surface, the thermal re-emission from the spacecraft itself. The non-gravitational effects on the Mercurycentric orbit of the spacecraft are so intense that, if not properly taken into account, they would lead to a significantly biased orbit determination. Due to the general difficulty of modeling these effects, an accelerometer (ISA—Italian Spring Accelerometer) will be placed on board the spacecraft [20]. This instrument is able to measure differential accelerations between a sensitive element and its rigid frame (cage) and thus to give accurate information on the non-gravitational accelerations. During the scientific phase of the mission, the accelerometer readings will be available nearly continuously at the rate of 1 Hz.

For the purpose of simulations, we introduce a simplified model of non-gravitational perturbations in order to include the accelerometer readings among the observables. We account for the effect of direct solar radiation pressure $\mathbf{a}_{rad}$ assuming a spherical satellite with coefficient 1 (i.e., we neglect diffusive terms). The shadow of the planet is computed accurately, taking into account the penumbra effects. Moreover, we include the acceleration due to the thermal radiation from the planet, $\mathbf{a}_{th}$, assuming a zero relaxation time for the thermal re-emission of Mercury (details on the model, supplied by D.Vokrouhlicky, Charles University of Prague, can be found in [10]). The whole non-gravitational perturbations experienced by the spacecraft are, then, $\mathbf{a}_{ng} = \mathbf{a}_{rad} + \mathbf{a}_{th}$. We need to stress out that this model, although simplified, is accurate enough for the purpose of simulations. As will be detailed in Section 4.1.2, the key issue is that the accelerometer readings suffer from both random and systematic errors, which are the critical terms to deal with. We can write the accelerometer contribution to the equation of motion as: $\mathbf{a}_{ISA} = -\mathbf{a}_{ng} + \varepsilon$, where ε represents the contribution of all the error sources in the ISA readings. As already highlighted, one of the main goals of the radio science experiment is to perform a very accurate orbit determination of the Mercurycentric motion of the spacecraft. To this aim, what really matters is to remove in the most suitable way any bias introduced in the accelerometer readings by instrumental errors. For this reason, in our analysis we mainly focus on the techniques to handle these error terms instead of accurately modeling the non-gravitational perturbations themselves.

4. Simulation Scenario and Assumptions

In the following section, we outline the observational and dynamical scenario of the numerical simulations of the relativity experiment. The latest mission scenario provides for a one year orbital phase starting from 28 March 2025. The initial Mercurycentric orbit is polar and near-circular (480×1500 km) with the pericenter located at $\sim 15°$ N. The orbital period of the spacecraft is about 2.3 h. We assume that two ground stations are available for tracking, one at the Goldstone Deep Space Communications Complex in California (USA), providing observations in the *Ka* band, and one located at the Cebreros station in Spain, supplying only *X* band observations. An average of 15–16 h of tracking per day is expected, with an average of 8 h in the *Ka* band. Range and range-rate measurements are simulated every 120 and 30 s, respectively. The propagation of the Mercurycentric dynamics in simulation stage is based on the gravity field of Mercury measured by MESSENGER [23], up to degree and order 25, with the addition of the Sun tidal effects described by the Love number k_2 and on the semi-empirical model for the planet rotation outlined in Section 3.3.4. For the Love number we adopted the value $k_2 = 0.45$ measured by MESSENGER [23]. For the rotational parameters we used the following values: the orientation of the rotation axis is defined, in our semi-empirical model, by the arbitrary angles $\delta_1 = 3$ arcmin and $\delta_2 = 1$ arcmin; the amplitudes of the librations in longitude are $\varepsilon_1 = 38.9$ arcsec, as measured by MESSENGER [47], and $\varepsilon_2 = 40$ arcsec [45]. Concerning the relativity parameters, we adopted the GR values for the PN parameters: $\beta = \gamma = 1$, $\eta = 0$, $\alpha_1 = \alpha_2 = 0$. The values of $\mu_\odot$ and $J_{2\odot}$ are taken from the DE421 ephemerides and we assume $\zeta = 0$. In the case of γ, we added the apriori $\gamma = 1 \pm 5 \times 10^{-6}$ in differential correction stage. In fact, the PN parameter γ appears both in the equations of motion for Mercury and EMB and in the equations for radio waves propagation. The delay of light propagation due to the space-time curvature, called Shapiro effect [30], is enhanced during a solar superior conjunction. Thus, a Superior Conjunction Experiment (SCE) is devised during BepiColombo cruise phase [12], with the aim of updating the constraint provided by the Cassini-Huygens mission [48]. The adopted apriori value on γ has been obtained from dedicated SCE simulations [49].

4.1. Observables Error Models

The observables we are dealing with are the tracking data (range and range-rate) and the non-gravitational accelerations, measured by the on-board accelerometer. To perform simulations in a realistic scenario, we need to properly add some measurement error to each observable.

4.1.1. Range and Range-Rate

According to [3], a nominal white noise can be associated to each tracking observation. Defining the simulated one-way range and range-rate observables as two-way measurements divided by 2 and assuming top accuracy performances of the transponder, we add a Gaussian error of $\sigma_r = 23.7$ cm to the 120 s range observables and $\sigma_{\dot{r}} = 8.7 \times 10^{-4}$ cm/s to the 30 s range-rate measurements [6].

These values represent the optimal performances in the *Ka* band, while for the *X* band we assume 10 times larger errors.

From a comparison of the accuracies in the range and range-rate, it turns out that $\sigma_r/\sigma_{\dot{r}} \sim 10^5$ s (according to Gaussian statistics, the standard deviations can be rescaled in order to be compared over the same integration time). Range measurements are, hence, more accurate than range-rate when we are observing phenomena with a period longer than 10^5 s, while the opposite is true for range-rate. We can conclude that the relativity experiment, which involves long-term periodicity phenomena, is mainly performed through the range tracking data, while gravimetry and rotation experiments mainly with the range-rate (we recall that the Mercurycentric orbital period is less than 10^4 s).

At the level of accuracy provided by the MORE relativity experiment, it could be necessary to account for additional sources of uncertainty in the range measurements. Indeed, instrumental related effects, such as residual signatures from the calibrator or residual biases after ground system calibration, can affect the observations in a non-negligible way. To account for these spurious effects, we add to the range Gaussian noise a generic systematic term, described by a bias of 3 cm and a sinusoidal trend (as already done in [12]) which reaches an amplitude of 3 cm after one year of observations. The choice of this functional behavior can be replaced by other assumptions, as done in [50]; it merely accounts for a possible scenario, which is the purpose of our simulations.

4.1.2. Accelerometer Readings and Calibration Strategy

As outlined in Section 3.3.5, we write the accelerometer contribution to the Mercurycentric motion of the spacecraft as $\mathbf{a}_{ISA} = -\mathbf{a}_{ng} + \boldsymbol{\varepsilon}$, where $\mathbf{a}_{ng}$ is computed according to our simplified model. Concerning the error term $\boldsymbol{\varepsilon}$, we assume the model provided by ISA team (private communications). It consists of a random background with some periodic terms superimposed: the main ones are a thermal term, resulting in a sinusoid at Mercury sidereal period (7.6×10^6 s) and a resonant term, resulting in a sinusoid at the orbital period of the spacecraft (8.3×10^3 s). All the details on the adopted model and the effects of the main components on the Mercurycentric orbit determination are described in [10].

The key issue in dealing with the accelerometer error term is that if we simply add it to the right hand side of the equation of motion, its detrimental effect causes a downgrading of the orbit determination of the spacecraft by orders of magnitude, vanifying the radio science experiment. In fact, this spurious instrumental effect is absorbed by the solve for parameters (like the state vector of the spacecraft at each arc) just like any other physical effect, resulting in a totally biased solution. To overcome this problem, the basic idea is to add to the right hand side of the equation of motion an additional term $c(\boldsymbol{\psi};t)$, function of a further set of parameters $\boldsymbol{\psi}$, to be added in the solve for list, and of time, such that $\boldsymbol{\varepsilon}(t) - c(\boldsymbol{\psi};t) \simeq 0$. In such a way, the *calibration function* $c(\boldsymbol{\psi};t)$ absorbs most of the accelerometer error and the physical parameters of interest for the radio science experiment are, in principle, not anymore biased. In the ORBIT14 software we implemented a novel calibration strategy, in which the calibration function is represented by a $\mathcal{C}^1$ cubic spline. All the details can be found in [19]. As a consequence, six additional parameters per arc (two per direction) are determined. We point out, as extensively discussed in [10], that this calibration strategy is able to absorb the low frequencies (i.e., longer than one day) error terms and the random component; in fact, the coefficients of the spline polynomials are computed once per arc, hence features with a periodicity lower than one day cannot be accounted for. This means that the resonant term, which shows a periodicity significantly lower than one day (about 2.3 h), is not absorbed by calibration at all. While this term results highly critical for what concerns the gravimetry and rotation experiments, we will see that its amplitude is not significantly detrimental for the relativity experiment.

4.2. Desaturation Maneuvres

Additional sources of perturbation on the orbit of the spacecraft around Mercury are the reaction wheels desaturation manoeuvres. We will assume as a general scenario to have one dump maneuvre during tracking and one dump maneuvre in the periods without tracking, hence a maximum amount of two dump manoeuvres per arc, as specified by the mission requirements. Each desaturation maneuvre, needed to maintain the desired attitude of the spacecraft, affects the precise Mercurycentric orbit determination. The result is a significant velocity change in the radial and out-of-plane directions and a linear momentum transfer in the transversal direction. To guarantee the expected level of accuracy in the orbit determination, these effects need to be modeled and removed from the estimation of the parameters. Each maneuvre appears in the spacecraft equation of motion as an additional acceleration acting on the probe. The downgrading effect on the spacecraft orbit determination can be significant, up to tens of meters in the range observations (see, e.g., [27]). For this reason, the velocity change $\Delta \mathbf{v}$ due to each maneuvre is added to the list of solve-for parameters, removing most of the downgrading effect from the orbit determination. The values for $\Delta \mathbf{v}$ adopted in simulations, along with all the details on the modelization and implementation of the maneuvres scenario, are given in [27]. The presence of orbital maneuvres, which are in general much larger than the desaturation maneuvres, is not considered here.

4.3. Metric Theories of Gravitation

A critical issue of the MORE relativity experiment, already discussed in [12], is that the Eddington parameter β and the Sun oblateness $J_{2\odot}$ show a near 1 correlation, as it appears from the covariance matrix obtained through the LS fit. This effect can be interpreted from a geometrical point of view considering that the main orbital effect of β is a precession of the argument of perihelion, which is a displacement taking place in the plane of the orbit of Mercury, while $J_{2\odot}$ affects the precession of the longitude of the node, thus producing a displacement in the plane of the solar equator. The angle between these two planes is only $\theta = 3.3°$, hence, being $\cos\theta \simeq 1$, we can expect such a high correlation between the two parameters. The consequence is a significant deterioration of the formal accuracies of both parameters. Since, unavoidably, the geometrical configuration cannot be changed, a possible solution to determine both parameters without a significant loss in accuracy is to add a suitable constraint on one of the involved parameter. A meaningful possibility is to link the PN parameters through the Nordtvedt equation [51]:

$$\eta = 4(\beta - 1) - (\gamma - 1) - \alpha_1 - \frac{2}{3}\alpha_2 .$$

In such a way, the knowledge on β is determined from the value of η: the correlation between β and η becomes almost 1, but that between β and $J_{2\odot}$ is greatly reduced. The introduction of the Nordtvedt equation is justified if we assume that gravitation must be described by a metric theory. In the following this becomes a basic assumption of our scenario.

4.4. Rank Deficiencies in the Mercury and EMB Orbit Determination Problem

As already stated, to perform the MORE relativity experiment we need to determine the orbit of Mercury and the EMB, that is to compute their state vector (position and velocity) at a given reference epoch. In practice, we find that we cannot solve for all the 12 components of the 2 state vectors without running into a significant deterioration of the results. The issue arises from the fact that this orbit determination problem, including simultaneously Mercury and the Earth, shows an approximate rank deficiency of order 4 (see, e.g., [12] for details).

An approximate rank deficiency of order 3 results from the breaking of an exact symmetry of the problem with respect to the full rotation group $SO(3)$. If there were only the Sun, the Earth and Mercury and if the Sun was exactly spherically symmetric (i.e., $J_{2\odot} = 0$), there would be an exact symmetry for rotation in determining both the orbits of Mercury and the Earth and therefore an exact rank deficiency of order 3. Due to the coupling with the other planets and to the asphericity of the Sun, the exact symmetry is broken but only by a small parameter (of the order of the relative size of the mutual perturbations by the other planets on the orbits of Mercury and the Earth and of the size of $J_{2\odot}$), bringing a residual approximate rank deficiency of order 3 in the problem.

A further exact symmetry would be present, if there were only the Sun, the Earth and Mercury. Changing all the lengths by a factor L, all the masses by a factor M and all the time intervals by a factor T, provided that the scaling factors are related by $L^3 = T^2 M$ (Kepler's third law), the equation of motion of the gravitational 3-body problem would remain unchanged. Again, this symmetry is broken by a small parameter, and an approximate rank deficiency of order 1 remains, leading to a total rank deficiency of order 4.

The standard solution already adopted in [12] is to solve for only 8 of the 12 components of the state vectors, assuming the remaining 4 as consider parameters. In the following we adopt the same assumption and we do not solve for the three position components of the EMB orbit (x_{EM}, y_{EM}, z_{EM}) and for the EMB velocity component perpendicular to the EMB orbital plane ($\dot{z}_{EM}$). The adopted technique is called *descoping*. A different approach to a problem of rank deficiency of order d is to add d constraint equations as apriori observations, instead of removing d parameters from the solve for list (see [25]—Chapter 6 for details). This is the technique we apply, for example, assuming the validity of the Nordtvedt equation in order to remove the degeneracy between β and $J_{2\odot}$. In ORBIT14 we have implemented also the possibility of determining all the 12 state vectors components, by adding 4 apriori constraints between the state vectors components and the Sun's mass $\mu_\odot$ in order to remove the degeneracy. A detailed discussion on this topic will be presented in a future paper by our group. In the following we assume to determine only 8 out of the 12 components.

5. Results

In this Section we will present and discuss the results of our simulations. In this review, we are mainly interested in the MORE relativity experiment: the results concerning PN and related parameters will be given in Section 5.1. For completeness, we will discuss the results concerning gravimetry and rotation in Section 5.2.

At each iteration of the differential correction process we solve for the following parameters:

- Global dynamical:

    - PN parameters: β, γ, η, α_1, α_2;
    - other parameters of interest for the relativity experiment: $\mu_\odot$, ζ, $J_{2\odot}$;
    - the state vectors of Mercury and EMB (8 components): $(x_M, y_M, z_M; \dot{x}_M, \dot{y}_M, \dot{z}_M)$; $(\dot{x}_{EM}, \dot{y}_{EM})$;
    - normalized harmonic coefficients of the gravity field of Mercury up to degree and order 25 and the Love number k_2;
    - rotational parameters: δ_1, δ_2, ε_1, ε_2;
    - six accelerometer calibration coefficients for each arc, plus 6+6 boundary conditions;

- Local dynamical:

    - state vector of the Mercurycentric orbit of the spacecraft, in the Ecliptic J2000 inertial reference frame, at the central time of each observed arc;
    - three dump manoeuvre components, $\Delta \mathbf{v}$, taking place during tracking, for each observed arc;

- External local dynamical:

    - three dump manoeuvre components, $\Delta \mathbf{v}$, taking place in the period without tracking between each pair of consecutive observed arcs.

5.1. The Relativity Experiment Results

The results for the PN and related parameters of interest for the relativity experiment are shown in Table 1. For each parameter we report the following quantities: (i) the formal uncertainty; (ii) the true error; (iii) the true-to-formal (T/F) error ratio; (iv) the current accuracy with which the parameter is presently known.

Table 1. Simulation results for the parameters of interest in the MORE relativity experiment (errors on $\mu_\odot$ are in cm^3/s^2, on ζ in y^{-1}).

Parameter	Formal Error	True Error	T/F Error Ratio	Current Accuracy
β	7.3×10^{-7}	2.6×10^{-6}	3.6	7×10^{-5} [52]
γ	9.3×10^{-7}	1.1×10^{-6}	1.2	2.3×10^{-5} [48]
η	2.2×10^{-6}	1.1×10^{-5}	4.9	4.5×10^{-4} [53]
α_1	4.9×10^{-7}	4.9×10^{-7}	1.0	6.0×10^{-6} [54]
α_2	8.3×10^{-8}	1.0×10^{-7}	1.2	3.5×10^{-5} [54]
$\mu_\odot$	4.2×10^{13}	4.2×10^{13}	1.0	$10^{16}, 8 \times 10^{15}$ [55,56]
ζ	2.3×10^{-14}	3.6×10^{-14}	1.5	4.3×10^{-14} [57]
$J_{2\odot}$	4.1×10^{-10}	4.1×10^{-10}	1.0	1.2×10^{-8} [52]

The formal error is obtained from the diagonal terms of the covariance matrix. The main limitation of formal analysis is that it does not account at all for any error that is non-Gaussian, like systematic errors, or time-correlated, unless they are in some way calibrated introducing further parameters in the solve-for list. Besides the formal analysis, we introduce a second quantity, which we call "true" error, to assess the expected accuracies in a more realistic way. This quantity is defined for each parameter as the difference between the nominal value of the parameter (used in simulations) and the value determined at convergence of the differential correction process. In such a way, the systematic effects are included in the computation of the accuracies. The true errors shown in Table 1 have been obtained as rms values over a number of runs carried out by changing the seed of the random numbers generator. We found that ~10 runs are adequately representative to quantify systematic errors. The ideal situation would occur when T/F error ratios follow Gaussian statistics, which means either that no systematic effects are present at all or that they are accounted for, through calibration parameters, in the formal analysis. In practice, this ratio is almost always greater than 1, but what does matter is that it is limited within a maximum of T/F ~ 3. Any higher value would be representative of a wrong or lacking modelization of some effects.

Analyzing the two sources of systematic effects included in simulation, i.e., the error model for accelerometer readings and the spurious effects from the ranging system, we found that T/F values higher than 1 for the relativity parameters can be only partially ascribed to non-perfectly calibrated long term components in the accelerometer error model, which are not fully absorbed by the C^1 spline calibration. The main downgrading effect turns out to be the presence of systematic terms in the range error model, which are not calibrated at all. The effect is particularly detrimental for the determination of β and η. We remind that the range error model includes a bias term of 3 cm and a sinusoidal trend up to 3 cm after one year. Analyzing individually the two error terms, we found that the bias term is responsible for most of the deterioration in estimating β and η. Adding the linear term does not further deteriorate true errors in a significant way. Moreover, as extensively discussed in [50], increasing the adopted value for the bias term results in a corresponding increase of the true errors of all the PN parameters.

A possible approach to the problem would be to introduce in the solve-for list an additional global parameter, that is a bias in modeling the range observables. Estimating the bias, it could be possible in principle to absorb most of the spurious effect in range, leading to a better estimate of the PN parameters in terms of T/F error ratios. This approach has a disadvantage that immediately appears when we take correlations into account. The correlations between PN and related parameters are shown in Table 2. As expected, the correlation between β and η is almost one. In fact we adopted the Nordvedt relation as an apriori constraint between PN parameters and we included an apriori constraint on γ from the SCE simulations during cruise phase. As a consequence, η is deduced from β

and their correlation is very high. If we add a bias in range to the solve-for list, the correlation between the bias term and, especially, β and η turns out to be almost one. This would lead to a worsening in the formal error of both β and η by more than one order of magnitude. This result would not be compliant with the scientific goals expected from MORE in terms of accuracies. In fact, the goal is to determine η at a level of, at least, 10^{-5} and β at a level of some parts in 10^{-6} [3,6,12]. In conclusion, if the systematic effects due to the ranging system remain at the level of few cm, the downgrading effect on the accuracies is still acceptable, as can be envisaged comparing the present results with current accuracies. Conversely, if the systematic terms, especially a spurious bias in the range measurements, become more significant, some suitable calibration strategy would be mandatory. As sketched in Section 4.4, the different approach of estimating all the 12, instead of only 8, components of Mercury and the EMB state vectors by adding some apriori constraints is under analysis. Preliminary attempts have shown that in such a case the correlation between the bias term in range and η and β would significantly decrease. We will report our conclusions in a future work.

Table 2. Correlations between PN and related parameters (values higher than 0.7 are highlighted in bold).

	β	γ	η	α_1	α_2	$\mu_\odot$	ζ	$J_{2\odot}$
$J_{2\odot}$	0.15	0.21	0.11	**0.90**	0.29	**0.89**	0.10	–
ζ	<0.1	0.28	<0.1	<0.1	0.17	<0.1	–	
$\mu_\odot$	0.20	0.14	0.14	**0.84**	<0.1	–		
α_2	0.35	0.28	0.36	0.27	–			
α_1	0.35	0.12	0.22	–				
η	**0.96**	0.60	–					
γ	**0.77**	–						
β	–							

A critical issue in the MORE relativity experiment, already remarked in [12] (p. 17), concerns the effects of a lacking knowledge of the Solar System model. In our simulations, we assumed that all the parameters of the SS model not included in the solve-for list (for example, the masses of the planets) are known from the ephemerides well enough that no spurious effects are introduced in the parameters estimation. An extensive discussion on this approximation has been carried out in [31] and the issue is still controversial.

Finally, we point out that the Lense-Thirring effect on the Mercury's orbit due to the Sun's angular momentum has been neglected. This choice has presently been made in order to simplify the development and implementation of the dynamical models. However, the effect is expected to be relevant [58], hence in future work we will investigate on its possible impact on the relativity parameters determination.

5.2. Results for Gravimetry and Rotation

In Section 4.1.1, we introduced one of the basic issues of the BepiColombo radio science experiment. Comparing the expected accuracies of range and range-rate, it turns out that range-rate measurements are more accurate than range data when observing phenomena with periodicity shorter than $\sim 10^5$ s, while the opposite is true for long-term periodicity phenomena. As a consequence, gravimetry and rotation experiments are mainly performed by means of range-rate data, while the relativity experiment by means of range. MORE is a comprehensive experiment in which all the parameters are solved simultaneously in the non-linear LS fit, but the expected independence between gravimetry/rotation on one side and relativity on the other suggests that, for the purpose of simulations, we can perform the experiments individually or all together and achieve the same results. We checked the validity of this statement by performing additional simulations. Referring to the solve-for list in Section 4, we ran the following simulations:

- *relativity simulations:* we removed from the solve-for list the gravimetry and rotational parameters, i.e. the gravity field spherical harmonic coefficients, Love number k_2, the angles (δ_1, δ_2), the libration amplitudes $\varepsilon_1, \varepsilon_2$;
- *gravimetry and rotation simulations:* we removed from the solve-for list the PN and related parameters.

These are mandatory tests since the chance that any further unforeseen rank deficiency between relativity and gravimetry/rotation parameters appears in the global fit needs to be verified. The results confirmed our expectations. The accuracies of PN and related parameters achieved in the global simulation, discussed in Section 5.1, and in the relativity simulations are equivalent, and the same is true for gravimetry and rotation. We have already extensively reported in [10] on the results for the MORE gravimetry and rotation experiments, together with a discussion on the achievable accuracy in the orbit determination. Therefor, we do not duplicate here the same results and we refer to that paper for a discussion on these topics. We point out that in the simulations described here we did not include among the observables the optical data from the high resolution camera HRIC, part of the SIMBIO-SYS payload [59]. In fact, camera observations significantly support range-rate measurements in the determination of the rotational parameters, while they do not contribute at all to the relativity experiment.

6. Discussion and Conclusions

In this review, we summarized all the issues concerning the BepiColombo relativity experiment. After recalling the global structure of the ORBIT14 software and the techniques to determine the parameters of interest, we detailed the essential mathematical models on which the experiment is based and the fundamental assumptions adopted. We finally presented the results of a full cycle of simulations carried out in the latest mission scenario.

At the beginning of 2000's our group performed a similar set of simulations, whose results are reported in [12], with the specific aim of dictating the mission and instrumentation requirements in order to make the BepiColombo relativity experiment feasible. Several underlying issues concerning the experiment have since been reconsidered and updated and the software has undergone significant revision. The formal results of the present paper are compared with the formal errors obtained in 2002 in [12] in Table 3, where the results reported in [6], representing the goal accuracies required for the MORE relativity experiment, have also been included. We refer to Experiment D in [12], where Nordtvedt equation has been assumed to link PN parameters. In the comparison we did not consider the ζ parameter because in [12] the quantity $\dot{G}/G$ was included instead of ζ.

Table 3. Comparison between the results in Schettino & Tommei (2016) (this paper) and previous results of the relativity experiment ($\mu_\odot$ in cm^3/s^2).

Parameter	Schettino & Tommei (2016)	Milani et al. (2002) [12]	Iess et al. (2009) [6]
β	7.3×10^{-7}	9.2×10^{-7}	2×10^{-6}
γ	9.3×10^{-7}	2×10^{-6} (SCE)	2×10^{-6}
η	2.2×10^{-6}	3.3×10^{-6}	8×10^{-6}
α_1	4.9×10^{-7}	7.1×10^{-7}	–
α_2	8.3×10^{-8}	1.9×10^{-7}	–
$\mu_\odot$	4.2×10^{13}	4.1×10^{13}	–
$J_{2\odot}$	4.1×10^{-10}	6.2×10^{-10}	2×10^{-9}

It can be seen that a slight improvement of the 2002 expectations [12] has been achieved. It can be remarked that in both cases formal errors turn out to be significantly lower than the goal accuracies of MORE. A quantitative comparison with the 2015 results in [15] is difficult because the mission time span scenario is different and, furthermore, the assumption of a metric theory of gravitation was not included in [15].

At this stage, the spacecraft is almost ready for launch and no significant modification of the mission can be addressed anymore. The aim of the set of simulations described in this review is, hence, clear: assuming the performances expected and tested for each instrument and the revised launch scenario, we want to establish the feasibility of the relativity experiment and provide the results that can be achieved in terms of accuracies. Two key issues were pointed out already during the past years: the impact on the solution from the errors in the accelerometer readings and from the aging of the transponder. Concerning the first issue, the main downgrading source was found in the thermal effects which produce periodic spurious signatures, with the periodicity of both the orbital period of the spacecraft around Mercury and the sidereal period of Mercury around the Sun. These signatures mainly affect the Mercurycentric orbit determination. An extensive discussion on the potentially downgrading effects for the gravimetry and rotation experiments have been recently discussed by our group in [10]. The results shown in Table 1 and the discussions presented in [14,50] lead us to the conclusion that, if the accelerometer error model is compliant to the one adopted, the effects on the relativity experiment are not detrimental. More critical is the question on how the ranging system affects the results. In [12] it was shown that, describing the transponder aging with a sinusoidal trend up to some tens of cm after one year, the effect was highly detrimental for the relativity parameters estimation. This issue has been tackled by introducing an on-board calibrator to account for the aging of the transponder. Nevertheless, residual spurious effects due, e.g., to the calibrator itself or to the on-ground instrumentation can still lead to a systematic error of a few cm. In our simulations, we assumed a bias of 3 cm and a sinusoidal trend up to 3 cm after one year. In such a case, the detrimental effects on the parameters are restrained, but in an unfavorable scenario in which they exceed the value of 5 cm on the one-way range, the solution would be significantly downgraded, as shown in [50]. In such a case, we envisage the need of a calibration strategy within the LS fit. In any case, in the realistic scenario presented here, we can conclude that the accuracies achievable by the BepiColombo relativity experiment for each of the PN parameter, compared with current accuracies, would represent a significant improvement of our knowledge of gravitational theories.

Acknowledgments: The results of the research presented in this paper have been performed within the scope of the Addendum n. I/080/09/1 of the contract n. I/080/09/1 with the Italian Space Agency. The authors would like to thank the three anonymous reviewers and the editors for the valuable comments and the significant improvements to the earlier version of the manuscript.

Author Contributions: Giulia Schettino and Giacomo Tommei conceived and designed the simulations; Giulia Schettino run the simulations; Giulia Schettino and Giacomo Tommei analyzed the output of the simulations; Giulia Schettino and Giacomo Tommei wrote the paper.

Conflicts of Interest: The authors declare no conflict of interest. The founding sponsors had no role in the design of the study; in the collection, analyses, or interpretation of data; in the writing of the manuscript, and in the decision to publish the results.

References

1. Benkhoff, J.; van Casteren, J.; Hayakawa, H.; Fujimoto, M.; Laakso, H.; Novara, M.; Ferri, P.; Middleton, H.R.; Ziethe, R. BepiColombo-Comprehensive exploration of Mercury: Mission overview and science goals. *Planet. Space Sic.* **2010**, *58*, 2–20.
2. Mukai, T.; Yamakawa, H.; Hayakawa, H.; Kasaba, Y.; Ogawa, H. Present status of the BepiColombo/Mercury magnetospheric orbiter. *Adv. Space Res.* **2006**, *38*, 578–582.
3. Iess, L.; Boscagli, G. Advanced radio science instrumentation for the mission BepiColombo to Mercury. *Planet. Space Sci.* **2001**, *49*, 1597–1608.
4. Milani, A.; Rossi, A.; Vokrouhlický, D.; Villani, D.; Bonanno, C. Gravity field and rotation state of Mercury from the BepiColombo Radio Science Experiments. *Planet. Space Sci.* **2001**, *49*, 1579–1596.
5. Sanchez Ortiz, N.; Belló Mora, M.; Jehn, R. BepiColombo mission: Estimation of Mercury gravity field and rotation parameters. *Acta Astronaut.* **2006**, *58*, 236–242.
6. Iess, L.; Asmar, S.; Tortora, P. MORE: An advanced tracking experiment for the exploration of Mercury with the mission BepiColombo. *Acta Astronaut.* **2009**, *65*, 666–675.

7. Genova, A.; Marabucci, M.; Iess, L. Mercury radio science experiment of the mission BepiColombo. *Mem. Soc. Astron. Ital. Suppl.* **2012**, *20*, 127–132.

8. Schettino, G.; Di Ruzza, S.; De Marchi, F.; Cicalò, S.; Tommei, G.; Milani, A. The radio science experiment with BepiColombo mission to Mercury. *Mem. Soc. Astron. Ital.* **2016**, *87*, 24–29.

9. Cicalò, S.; Milani, A. Determination of the rotation of Mercury from satellite gravimetry. *Mon. Not. R. Astron. Soc.* **2012**, *427*, 468–482.

10. Cicalò, S.; Schettino, G.; Di Ruzza, S.; Alessi, E.M.; Tommei, G.; Milani, A. The BepiColombo MORE gravimetry and rotation experiments with the ORBIT14 software. *Mon. Not. R. Astron. Soc.* **2016**, *457*, 1507–1521.

11. Palli, A.; Bevilacqua, A.; Genova, A.; Gherardi, A.; Iess, L.; Meriggiola, R.; Tortora, P. Implementation of an End to End Simulator for the BepiColombo Rotation Experiment. In Proceedings of the European Planetary Space Congress 2012, Madrid, Spain, 23–28 September 2012.

12. Milani, A.; Vokrouhlicky, D.; Villani, D.; Bonanno, C.; Rossi, A. Testing general relativity with the BepiColombo Radio Science Experiment. *Phys. Rev. D* **2002**, *66*, doi:10.1103/PhysRevD.66.082001.

13. Milani, A.; Tommei, G.; Vokrouhlicky, D.; Latorre, E.; Cicalò, S. Relativistic models for the BepiColombo radioscience experiment. In *Relativity in Fundamental Astronomy: Dynamics, Reference Frames, and Data Analysis*; Cambridge University Press: Cambridge, UK, 2010.

14. Schettino, G.; Cicalò, S.; Di Ruzza, S.; Tommei, G. The relativity experiment of MORE: Global full-cycle simulation and results. In Proceedings of the IEEE Metrology for Aerospace (MetroAeroSpace), Benevento, Italy, 4–5 June 2015; pp. 141–145.

15. Schuster, A.K.; Jehn, R.; Montagnon, E. Spacecraft design impacts on the post-Newtonian parameter estimation. In Proceedings of the IEEE Metrology for Aerospace (MetroAeroSpace), Benevento, Italy, 4–5 June 2015; pp. 82–87.

16. Tommei, G.; De Marchi, F.; Serra, D.; Schettino, G. On the Bepicolombo and Juno Radio Science Experiments: Precise models and critical estimates. In Proceedings of the IEEE Metrology for Aerospace (MetroAeroSpace), Benevento, Italy, 4–5 June 2015; pp. 323–328.

17. Kaula, W.M. *Theory of Satellite Geodesy: Applications of Satellites to Geodesy*; Blaisdell: Waltham, MA, USA, 1996.

18. Kozai, Y. Effects of the tidal deformation of the Earth on the motion of close Earth satellites. *Publ. Astron. Soc. Jpn.* **1965**, *17*, 395–402.

19. Alessi, E.M.; Cicalò, S.; Milani, A. Accelerometer data handling for the BepiColombo orbit determination. In *Advances in the Astronautical Science*, Proccedings of the 1st IAA Conference on Dynamics and Control of Space Systems, Porto, Portugal, 19–21 March 2012.

20. Iafolla, V.; Nozzoli S. Italian Spring Accelerometer (ISA): A high sensitive accelerometer for BepiColombo ESA Cornerstone. *Planet. Space Sci.* **2001**, *49*, 1609–1617.

21. Will, C.M. *Theory and Experiment in Gravitational Physics*; Cambridge University Press: Cambridge, UK, 1993.

22. Tommei, G.; Milani, A.; Vokrouhlicky, D. Light-time computations for the BepiColombo Radio Science Experiment. *Celest. Mech. Dyn. Astron.* **2010**, *107*, 285–298.

23. Mazarico, E.; Genova, A.; Goossens, S.; Lemoine, F.G.; Neumann, G.A.; Zuber, M.T.; Smith, D.E.; Solomon, S.C. The gravity field, orientation, and ephemeris of Mercury from MESSENGER observations after three years in orbit. *J. Geophys. Res.* **2014**, *119*, 2417–2436.

24. Jehn, R.; Rocchi, A. *BepiColombo Mercury Cornerstone Mission Analysis: The April 2018 Launch Option*; MAS Working Paper No. 608; BC-ESC-RP-50013; 24 March 2016; Issue 2.1. Available online: http://issfd.org/ISSFD_2014/ISSFD24_Paper_S6-5_jehn.pdf (accessed on 5 August 2016).

25. Milani, A.; Gronchi, G.F. *Theory of Orbit Determination*; Cambridge University Press: Cambridge, UK, 2010.

26. Folkner, W.M.; Williams, J.G.; Boggs, D.H. *The Planetary and Lunar Ephemeris DE 421*; JPL Publication: Pasadena, CA, USA, 2008.

27. Alessi, E.M.; Cicalò, S.; Milani, A.; Tommei, G. Desaturation Manoeuvres and Precise Orbit Determination for the BepiColombo Mission. *Mon. Not. R. Astron. Soc.* **2012**, *423*, 2270–2278.

28. Bonanno, C.; Milani, A. Symmetries and rank deficiency in the orbit determination around another planet. *Celest. Mech. Dyn. Astron.* **2002**, *83*, 17–33.

29. Will, C.M. The Confrontation between General Relativity and Experiment. *Living Rev. Relativ.* **2014**, *17*, 1–117.

30. Shapiro, I.I. Fourth Test of General Relativity. *Phys. Rev. Lett.* **1964**, *13*, 789–791.

31. De Marchi, F.; Tommei, G.; Milani, A.; Schettino, G. Constraining the Nordtvedt parameter with the BepiColombo Radioscience experiment. *Phys. Rev. D* **2016**, *93*, 123014.
32. Peebles P.J.E. *Principles of Physical Cosmology*; Princeton University Press: Princeton, NJ, USA, 1993.
33. Klioner, S.A. Relativistic scaling of astronomical quantities and the system of astronomical units. *Astron. Astrophys.* **2008**, *478*, 951–958.
34. Klioner, S.A.; Capitaine, N.; Folkner, W.; Guinot, B.; Huang, T.Y.; Kopeikin, S.; Petit, G.; Pitjeva, E.; Seidelmann, P.K.; Soffel, M. Units of Relativistic Time Scales and Associated Quantities. In *Relativity in Fundamental Astronomy: Dynamics, Reference Frames, and Data Analysis*; Cambridge University Press: Cambridge, UK, 2010.
35. Moyer, T.D. *Formulation for Observed and Computed Values of Deep Space Network Data Types for Navigation, NASA-JPL, Deep Space Communication and Navigation Series, Monograph 2*; John Wiley & Sons: Hoboken, NJ, USA, 2000.
36. Soffel, M.; Klioner, S.A.; Petit, G.; Wolf, P.; Kopeikin, S.M.; Bretagnon, P.; Brumberg, V.A.; Capitaine, N.; Damour, T.; Fukushima, T. The IAU Resolutions for Astrometry, Celestial Mechanics, and Metrology in the Relativistic Framework: Explanatory Supplement. *Astron. J.* **2003**, *126*, 2687–2706.
37. Irwin, A.W.; Fukushima, T. A numerical time ephemeris of the Earth. *Astron. Astrophys.* **1999**, *348*, 642–652.
38. Damour, T.; Soffel, M.; Hu, C. General-relativistic celestial mechanics. IV. Theory of satellite motion. *Phys. Rev. D* **1994**, *49*, 618–635.
39. Archinal, B.A.; A'Hearn, M.F.; Bowell, E.; Conrad, A.; Consolmagno, G.J.; Courtin, R.; Fukushima, T.; Hestroffer, D.; Hilton, J.L.; Krasinskyet, G.A.; et al. Report of the IAU Working Group on Cartographic Coordinates and Rotational Elements: 2009. *Celest. Mech. Dyn. Astron.* **2011**, *2*, 101–135.
40. Montenbruck, O.; Gill, E. *Satellite Orbits: Models, Methods and Applications*; Springer: Berlin, Germany, 2005.
41. Padovan, S.; Margot, J.-L.; Hauck, S.A., II; Moore, W.B.; Solomon, S.C. The tides of Mercury and possible implications for its interior structure. *J. Geophys. Res.* **2014**, *119*, 850–866.
42. Roy, A.E. *Orbital Motion*, 4th ed.; CRC Press: Boca Raton, FL, USA, 2005.
43. Peale, S.J. *The Rotational Dynamics of Mercury and the State of Its Core*; University Arizona press: Tucson, AZ, USA, 1988; pp.461–493.
44. Peale, S.J. The proximity of Mercury's spin to Cassini state 1 from adiabatic invariance. *Icarus* **2006**, *181*, 338–347.
45. Yseboodt, M.; Margot, J.L.; Peale, S.J. Analytical model of the long-period forced longitude librations of Mercury. *Icarus* **2010**, *207*, 536–544.
46. Yseboodt, M.; Rivoldini, A.; Van Hoolst, T.; Dumberry, M. Influence of an inner core on the long-period forced librations of Mercury. *Icarus* **2013**, *226*, 41–51.
47. Stark, A.; Oberst, J.; Preusker, F.; Peale, S.J.; Margot, J.-L.; Phillips, R.J.; Neumann, G.A.; Smith, D.E.; Zuber, M.T.; Solomon, S.C. First MESSENGER orbital observations of Mercury's librations. *Geophys. Res. Lett.* **2015**, *42*, 7881–7889.
48. Bertotti, B.; Iess, L.; Tortora, P. A test of general relativity using radio links with the Cassini spacecraft. *Nature* **2003**, *425*, 374–376.
49. Imperi, L.; Iess, L. Testing general relativity during the cruise phase of the BepiColombo mission to Mercury. In Proceedings of the IEEE Metrology for Aerospace (MetroAeroSpace), Benevento, Italy, 4–5 June 2015.
50. Schettino, G.; Imperi, L.; Iess, L.; Tommei, G. Sensitivity study of systematic errors in the BepiColombo relativity experiment. In Proceedings of the Metrology for Aerospace (MetroAeroSpace), Florence, Italy, 22–23 June 2016, in press.
51. Nordtvedt, K.J. Post-Newtonian Metric for a General Class of Scalar-Tensor Gravitational Theories and Observational Consequences. *Astrophys. J.* **1970**, *161*, 1059–1067.
52. Fienga, A.; Laskar, J.; Exertier, P.; Manche, H.; Gastineau, M. Numerical estimation of the sensitivity of INPOP planetary ephemerides to general relativity parameters. *Celest. Mech. Dyn. Astron.* **2015**, *123*, 325–349.

53. Williams, J.G.; Turyshev, S.G.; Boggs, D.H. Lunar Laser Ranging Tests of the Equivalence Principle with the Earth and Moon. *Int. J. Mod. Phys. D* **2009**, *18*, 1129–1175.
54. Iorio, L. Constraining the Preferred-Frame α_1, α_2 Parameters from Solar System Planetary Precessions. *Int. J. Mod. Phys. D* **2014**, *23*, 1450006.

55. Pitjeva, E.V. Determination of the Value of the Heliocentric Gravitational Constant ($GM_\odot$) from Modern Observations of Planets and Spacecraft. *J. Phys. Chem. Ref. Data* **2015**, *44*, 031210.

56. The Value has Been Obtained from Latest JPL Ephemerides and Has Been Published on the NASA HORIZONS Web-Interface. Available online: http://ssd.jpl.nasa.gov/?constants (accessed on 5 August 2016).

57. Pitjeva, E.V.; Pitjev, N.P. Relativistic effects and dark matter in the Solar system from observations of planets and spacecraft. *Mon. Not. R. Astron. Soc.* **2013**, *432*, 3431–3437.

58. Iorio, L.; Lichtenegger, H.I.M.; Ruggiero, M.L.; Corda, C. Phenomenology of the Lense-Thirring effect in the solar system. *Astrophys. Space Sci.* **2011**, *331*, 351–395.

59. Flamini, E.; Capaccioni, F.; Colangeli, L.; Cremonese, G.; Doressoundiram, A.; Josset, J.L.; Langevin, L.; Debei, S.; Capria, M.T.; De Sanctis, M.C.; et al. SIMBIO-SYS: The spectrometer and imagers integrated observatory system for the BepiColombo planetary orbiter. *Planet. Space Sci.* **2010**, *58*, 125–143.

![universe logo] *universe*

MDPI

Article

Strategies to Ascertain the Sign of the Spatial Curvature

Pedro C. Ferreira [1] and Diego Pavón [2,*]

[1]   Escola de Ciências e Tecnologia, Universidade Federal do Rio Grande do Norte, Natal 59072-970, Rio Grande do Norte, Brazil; pedro.ferreira@ect.ufrn.br
[2]   Departamento de Física, Universidad Autónoma de Barcelona, Bellaterra, Barcelona 08193, Spain
*   Correspondence: diego.pavon@uab.es

Academic Editors: Lorenzo Iorio and Elias C. Vagenas
Received: 14 October 2016; Accepted: 15 November 2016; Published: 24 November 2016

Abstract: The second law of thermodynamics, in the presence of gravity, is known to hold at small scales, as in the case of black holes and self-gravitating radiation spheres. Using the Friedmann–Lemaître–Robertson–Walker metric and the history of the Hubble factor, we argue that this law also holds at cosmological scales. Based on this, we study the connection between the deceleration parameter and the spatial curvature of the metric, Ω_k, and set limits on the latter, valid for any homogeneous and isotropic cosmological model. Likewise, we devise strategies to determine the sign of the spatial curvature index k. Finally, assuming the lambda cold dark matter model is correct, we find that the acceleration of the cosmic expansion is increasing today.

Keywords: mathematical cosmology; spatial curvatur; thermodynamics

1. Introduction

The validity of the second law of thermodynamics for systems dominated by gravity should not be taken for granted. Gravity is a long-ranged interaction while the formulation of the second is based on the observation of ordinary systems, i.e., those dominated by short-ranged interactions. In actual fact, its validity for the former systems was studied only recently, notably in the case of black holes and self-gravitating radiation spheres. In the former case, Bekenstein demonstrated that the black-hole entropy, in addition to the entropy of the black-hole exterior, never decreases [1,2]. In the latter, it was shown that the static stable configurations of a sphere of self-gravitating radiation are those that maximize the radiation entropy [3,4]. Both instances correspond to small scale systems. Although different authors assumed it to be in order to constrain the evolution of cosmological models (see, e.g., [5] and references therein), as far as we know, the validity of the said law at large (i.e., cosmic) scales has not been explored as yet. The main purpose of this work is to fill this gap. Our study analysis rests on the simplest realistic large-scale space-time metric, namely, the Friedmann–Lemaître–Robertson–Walker (FLRW) one alongside a selected set of observational data about the history of cosmic expansion.

Homogeneous and isotropic universe models are usually described by the FLRW metric

$$ds^2 = -c^2 dt^2 + a^2(t) \left\{ \frac{dr^2}{1 - kr^2} + r^2 \left(d\theta^2 + \sin^2 \theta \, d\phi^2 \right) \right\},$$ (1)

coupled to the sources of the gravitational field. This metric relies on the cosmological principle [6–8] whose validity, at large scales, has not been contradicted thus far [9] and it looks rather robust [10–12]. The curvature index, k, is either $0, +1$, or -1 depending on whether the spatial part of the metric is flat, positively curved (closed), or negatively curved (hyperbolic), respectively.

This constant index, like the scale factor $a(t)$, is not a directly observable quantity. In principle, however, it can be determined through the knowledge of the dimensionless, fractional curvature density, $\Omega_k \equiv -k/(a^2 H^2)$, which is accessible to observation, albeit indirectly. As usual, $H = c\, d \ln a/dt$ denotes the Hubble factor. Current measurements of Ω_k only indicate that its present absolute value is small ($|\Omega_{k0}| \lesssim 10^{-3}$ [13,14]). Note that this constraint was obtained under the assumption that the universe is accurately described by the ΛCDM model. Thus the sign of k remains unknown.

The aim of this research is fourfold: (i) To determine whether the second law of thermodynamics is fulfilled at cosmological scales and; if so, (ii) constrain Ω_k as much as possible and (iii) determine the sign of k; finally, (iv) to derive a thermodynamic constraint relating the present value of the deceleration and jerk parameters. For the first three objectives, neither a cosmological model nor theory of gravity will be assumed. We shall just use the FLRW metric, the history $H(z)$ of the Hubble factor and the second law of thermodynamics. For the fourth objective, we will assume Einstein gravity and the ΛCDM model. As is customary, a subindex zero attached to any quantity means that the latter should be evaluated at present time.

2. Cosmological Consequences of the Second Law

Given the strong connection between gravity and thermodynamics [1,2,15–17], it is natural to expect that the universe behaves as a normal thermodynamic system; it therefore must tend to a state of maximum entropy in the long run [18,19].

For comoving observers, FLRW models entail "normal", "trapped" and "anti-trapped" regions. In the first one, the expansion of outgoing null geodesic congruences, normal to the spatial two-sphere of radius $\tilde{r}(= ra(t))$ centered at the origin (i.e., at the position of the observer), is positive, and negative for ingoing null geodesic congruences. In the trapped region, both kind of geodesic congruences have negative expansion. By contrast, in the anti-trapped region the expansion of both congruences is positive. The boundary hyper-surface of the space-time anti-trapped region is called the apparent horizon; its radius is $\tilde{r}_A = [(H/c)^2 + ka^{-2}]^{-1/2}$. Since the observer has no information about what might be going on beyond the horizon, the latter has an entropy, namely: $S_A = k_B \pi \tilde{r}_A^2 / \ell_{\rm pl}^2$, where $\ell_{\rm pl}$ is Planck's length. For details, see [20]. (Bear in mind that $\tilde{r}$ and H have dimensions of length and length^{-1}, respectively, k of length^{-2}, and a is dimensionless.)

A rather reasonable assumption concerning the entropy of the observable universe is that it is dominated by the entropy of the cosmic horizon. In the current universe, the entropy of the horizon exceeds that of supermassive black holes, stellar black holes, relic neutrinos and CMB photons by 18, 25, 33 and 33 orders of magnitude, respectively [21]. There are several possible choices for the cosmic horizon: the particle horizon, the event horizon, the apparent horizon and the Hubble horizon. Given that the first one does not exist for accelerating universes and the second only exists if the universe accelerates forever in the future, we take the apparent horizon, which, on the one hand, always exists, both for ever-expanding and ever-contracting universes, and, on the other hand, by contrast to the other mentioned possibilities, the laws of thermodynamics are fulfilled on it [22]. The Hubble horizon is a particular case of the apparent horizon when $k = 0$.

To support the above claim that the entropy of the horizon dominates over the entropy of any form of energy inside the horizon, especially at late times, we shall consider the entropy of pressureless matter. The latter is given by $S_m = k_B\, n\, V_k$ [23], with $n = n_0 a^{-3}$, being n_0 the present number density of matter particles, and

$$V_k = 2\pi a^2 \left[\sqrt{|k|}\, a \sin^{-1}(\sqrt{|k|}\, a^{-1} \tilde{r}_A) - k \tilde{r}_A^2 H \right] \tag{2}$$

the volume enclosed by the apparent horizon for $k = +1$ and -1 (for the flat case, $V_{k=0} = (4\pi/3)\, \tilde{r}_A^3$). For $k = -1$ one follows $S_m(a \gg 1) \to 2k_B\, n_0\, \pi a^{-1} \tilde{r}_A^2 H$. Hence, when $a \gg 1$ the ratio S_A/S_m results proportional to a/H. For $k = +1$ one has $S_m(a \gg 1) \to 2k_B\, n_0\, a^{-1}\left(1 - \sqrt{1 - \tilde{r}_A^2 a^{-1}}\right)$, hence

$$\frac{S_A}{S_m} \propto \frac{a\, \tilde{r}_A^2}{1 - \sqrt{1 - \frac{\tilde{r}_A^2}{a}}}.$$

Accordingly, in all three cases ($k = 0, +1, -1$) the entropy of the horizon overwhelms that of the matter inside it, especially at late times.

Recalling that $S_A \propto \mathcal{A}$ with $\mathcal{A} = 4\pi(H^2 + ka^{-2})^{-1}$ the area of the horizon (henceforward we set $c = 1$), the second law of thermodynamics $S'_A \geq 0$ leads to

$$\mathcal{A}' = -\frac{\mathcal{A}^2}{2\pi}\left(HH' - \frac{k}{a^3}\right) \geq 0 \quad \Rightarrow \quad HH' \leq \frac{k}{a^3}, \tag{3}$$

where the prime means d/da.

The second inequality tells us that if H' is or has been positive at any stage of cosmic expansion (excluding, possibly, the pre-Planckian era), then $k = +1$ and that, in principle, any sign of k is compatible with $H' < 0$. Multiplying the said inequality by $-aH^{-2}$ produces $-aH'/H \geq \Omega_k$, which can be recast in terms of the redshift as

$$(1+z)\frac{d\ln H}{dz} \geq \Omega_k. \tag{4}$$

Thus, if $dH/dz > 0$ for all $z \geq 0$, then both $k = +1$ and $k = 0$ are consistent with the second law of thermodynamics at large scales. However, given the present ample uncertainties in the observational data regarding the Hubble history, if k were -1, then the said law could break down at cosmic scales. To explore this, we set $k = -1$ in Equation (4) and integrate the resulting expression in the interval $z_1 \leq z \leq z_2$ to get

$$H_2^2 - H_1^2 \geq 2(z_2 - z_1) + (z_2^2 - z_1^2). \tag{5}$$

Therefore, if this relationship failed for whatever pair of points (z_i, H_i), with $i = 1, 2$, it should mean that the choice $k = -1$ would not be consistent with the second law at the said scales.

We use Equation (5) alongside the 28 experimental data H vs. z, in the interval $0.1 \leq z \leq 2.36$, with their 1σ error bars, compiled by Farook et al. [24] and listed in Table 1 (see also Figure 1) for the reader convenience, to draw Figure 2. The latter suggests that, given the experimental uncertainties, the possibility $k = -1$ also appears compatible with the inequality $S'_A \geq 0$. While wider compilations of $H(z)$ are available, we believe this one is preferable because it does not include any obviously correlated data, nor does it contain older, less reliable data, some with much weight from anomalously small error bars.

Equation (4) can alternatively be written as

$$1 + q \geq \Omega_k, \tag{6}$$

where $q = -\ddot{a}/(aH^2)$ is the dimensionless deceleration parameter. The last equation, like (4), imposes an upper bound (that depends on redshift) on Ω_k. In the radiation dominated era q was close to 1; a result that, in spite of having been derived for spatially flat universes described by general relativity, should hold irrespective of the sign of the curvature and the gravity theory employed. Notice that even a mild deviation of $q \simeq 1$ at that time would conflict with the observational results about the primordial nucleosynthesis of light elements [25]. This suggests an easily verifiable test on modified gravity theories, namely, that they should be consistent with the bound $\Omega_k \leq 2$ at the radiation era. However, if general relativity is the right theory of gravity, the first Friedmann equation implies the stronger bound $\Omega_k < 1$ at all epochs. Nevertheless, even if one uses general relativity, Equation (6) might provide a useful bound when $q < 0$.

Table 1. Hubble Parameter vs. Redshift Data.

z	$H(z)$ (km·s^{-1}·Mpc^{-1})	Reference
0.100	69 ± 12	[26]
0.170	83 ± 8	[26]
0.179	75 ± 4	[27]
0.199	75 ± 5	[27]
0.270	77 ± 14	[26]
0.320	79.2 ± 5.6	[28]
0.352	83 ± 14	[27]
0.400	95 ± 17	[26]
0.440	82.6 ± 7.8	[29]
0.480	97 ± 62	[30]
0.570	100.3 ± 3.7	[28]
0.593	104 ± 13	[27]
0.600	87.9 ± 6.1	[29]
0.680	92 ± 8	[27]
0.730	97.3 ± 7	[29]
0.781	105 ± 12	[27]
0.875	125 ± 17	[27]
0.880	90 ± 40	[30]
0.900	117 ± 23	[26]
1.037	154 ± 20	[27]
1.300	168 ± 17	[26]
1.363	160 ± 33.6	[31]
1.430	177 ± 18	[26]
1.530	140 ± 14	[26]
1.750	202 ± 40	[26]
1.965	186.5 ± 50.4	[31]
2.340	222 ± 7	[32]
2.360	226 ± 8	[33]

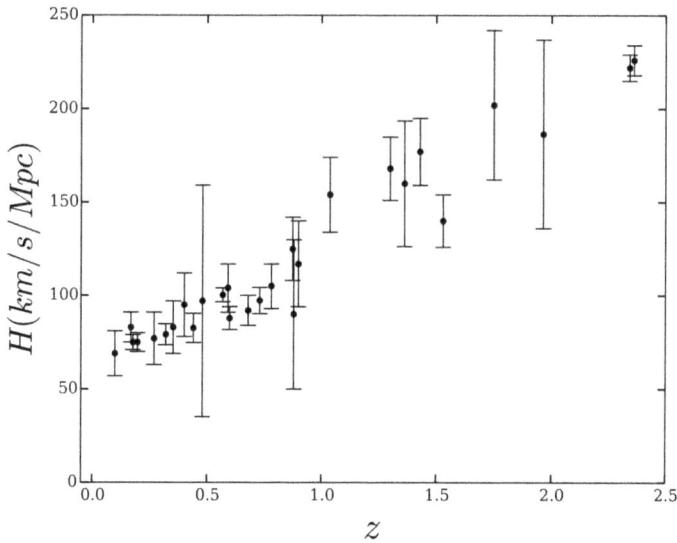

Figure 1. 28 $H(z)$ data points with their 1σ uncertainty.

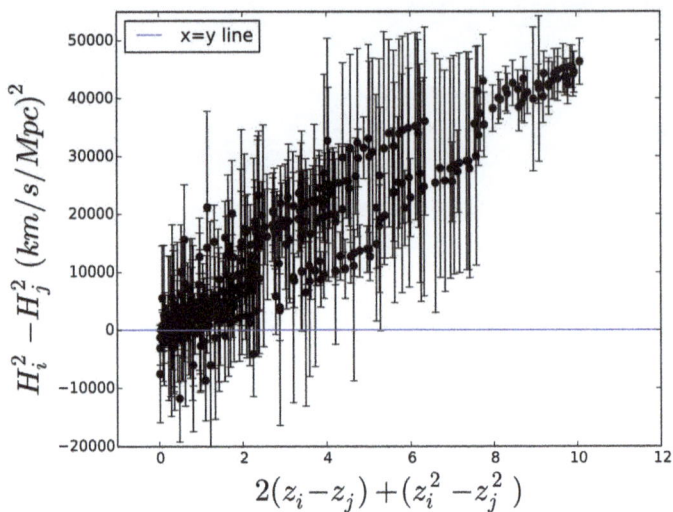

Figure 2. Left-hand side vs. right-hand side of Equation (5) for all possible $i > j$ combinations of the data shown in Table 1. The error bars denote 1σ confidence level.

We can draw further consequences from the thermodynamic bound (6). To this end, we first apply the model independent Gaussian process (GP) introduced by Seikel et al. [34] to smooth the 28 observational $H(z)$ data depicted in Figure 1. Figure 3 shows the outcome.

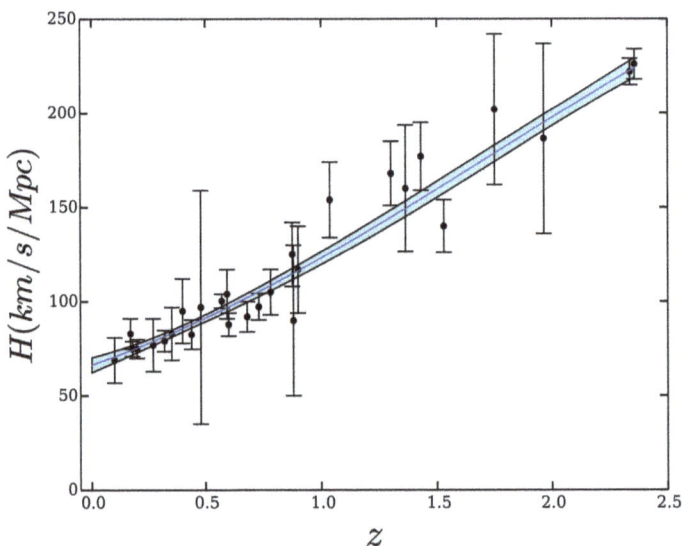

Figure 3. Gaussian process reconstruction of the history of the Hubble factor from the raw $H(z)$ data depicted in Figure 1, as well as here for convenience of the reader. The blue shaded region shows the 1σ uncertainty.

Inspection of the latter suggests that $dH/dz \geq 0$ in the redshift range be considered. If this gets confirmed by future $H(z)$ data of much higher quality, any sign of the curvature scalar index k will be consistent with the second law of thermodynamics. The following analysis, based on the smoothed data shown in Figure 3, allows the quantification of the gap between $1 + q$ and Ω_k.

The quantity $1 + q$ alongside its 1σ, uncertainty is obtained by computing the quantity in the left-hand side of (4) using the smoothed $H(z)$ data, and similarly Ω_k by computing $-k(1+z)^2/H^2(z)$ using the same data. Figures 4 and 5 summarize the results for $k = +1$ and -1, respectively. It is apparent that, whatever the sign of k, the second law is fulfilled by a generous margin. Likewise, inspection of the left panels of the aforesaid figures indicates that $\Omega_{k0} \leq 0.64$. Obviously, this upper bound is much more loose than the one obtained in [14] ($6.5 \times 10^{-3} \leq \Omega_{k0} \leq -6.6 \times 10^{-3}$), but the latter is based on a particular (though so far successful) cosmological model—the ΛCDM—that rests on a number of assumptions, some of which can be justified only a posteriori. By contrast, this other rests just on the FLRW metric and the second law of thermodynamics. Combining the readings on the vertical axes of the right panels of the same figures yields the constraint $2 \times 10^{-4} \leq \Omega_{k0} \leq -2.6 \times 10^{-4}$.

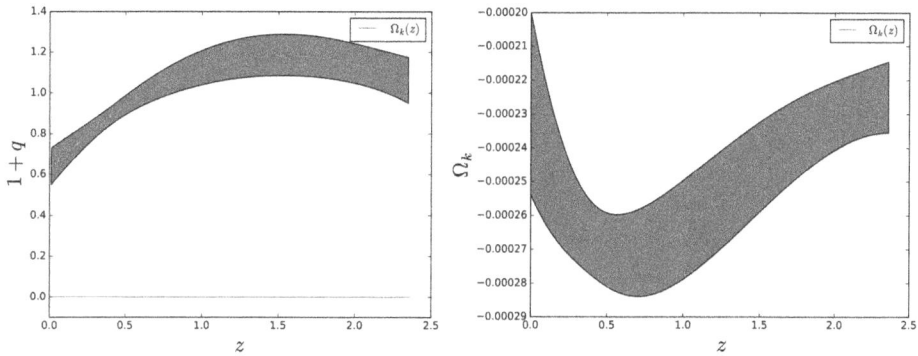

Figure 4. Left panel: $1 + q$ vs. redshift after smoothing the 28 $H(z)$ data as depicted in Figure 3. Also shown is Ω_k for $k = +1$. Clearly, the latter is practically zero. Right panel: Zoom of Ω_k and its 1σ uncertainty interval.

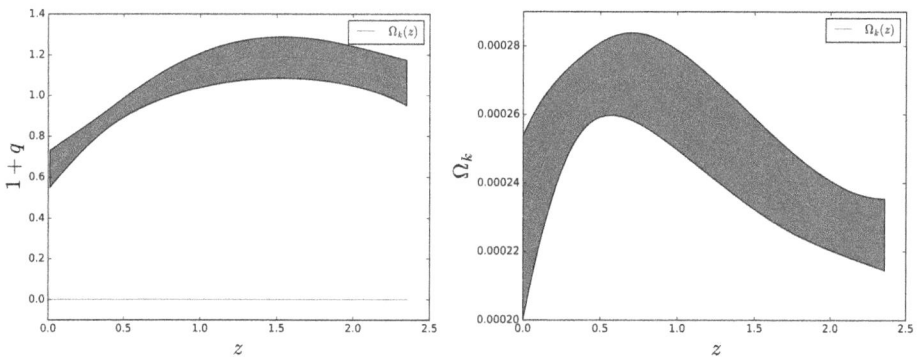

Figure 5. Same as Figure 4 but for $k = -1$.

Regrettably, as hinted above, the quality of the available sets of $H(z)$ data is not good enough to directly constrain Ω_k into a small range, much less to discriminate the sign of k. One has to apply some smoothing procedure to the data of the Hubble history (the GP process in our case) to downsize the error bars and thus obtain a tighter constraint. However, one should not be fully confident about the outcome since the said procedure, though efficient, is not exempt of potential shortcomings.

Nevertheless, the situation is expected to improve greatly in the not so distant future thanks to the Sandage–Loeb (SL) test [35,36] based on the Mc-Vittie formula [37]

$$H(z_s) = H_0[1 + z_s(t_0)] - \frac{\Delta z_s}{\Delta t_0} \tag{7}$$

that governs the drift of the redshift. Here, z_s stands for the redshift of the source (e.g., quasar, globular cluster, HI region, ...). With the use of high precision spectrographs, such as CODEX [38], and extremely large telescopes, as the ELT [39], the SL test will provide us accurate $H(z)$ data sets at different redshift intervals. These data will be free of any assumption whatsoever about the spatial curvature, gravity theory or cosmological model.

Observational data in the $0 < z < 1.0$ interval will be provided by the square kilometer array (SKA) radio-telescope [40], likewise the wide radio-sky survey PARKES will scan 21-cm radio-sources [41] as well as the experiment CHIME in the $0.8 < z < 2.5$ interval [42]. To collect a useful sample of $H(z)$ data will take between one and four decades, approximately. Details can be found in References [43,44].

If the data revealed that, in some redshift, interval H decreased with increasing z, it would immediately imply $k = +1$. On the contrary, if H always increased in every z interval, the application of (4) would require more effort, but in any case it will (hopefully) permit one to discern the sign of k.

If the above strategy would fail, for instance if the data would indicate different signs for Ω_k in separate intervals, it would mean either that the second law of thermodynamics does fail at large scales or that the FLRW metric should not be trusted after all.

3. The Jerk Parameter

By expanding the scale factor in terms of its successive derivatives we can write

$$a(t) = a_0 \left\{ 1 + H_0 (t - t_0) - \tfrac{1}{2} q_0 H_0^2 (t - t_0)^2 + \tfrac{1}{6} j_0 H_0^3 (t - t_0)^3 + \tfrac{1}{24} s_0 H_0^4 (t - t_0)^4 + \mathcal{O}([t - t_0]^5) \right\}, \tag{8}$$

where $j = \dddot{a}/(aH^3)$ and $s = (aH^4)^{-1} d^4 a/dt^4$ are the dimensionless jerk and snap parameters, respectively.

Here, we shall focus on the current value of the jerk parameter of a universe dominated by pressureless matter and the cosmological constant (subindexes m and Λ, respectively). Thus far, we did not specialize to any cosmological model nor theory of gravity. In what follows, to constrain the theoretical value of j_0, we adopt general relativity and the ΛCDM model because they are the simplest theory and model, respectively, that comply, at least at the background level, with the observational data [14]. In this model, the Hubble factor, as well as the deceleration and jerk parameters, read in terms of the redshift

$$H(z) = H_0 \sqrt{\Omega_{m0}(1+z)^3 + \Omega_{\Lambda 0} + \Omega_{k0}(1+z)^2}, \tag{9}$$

$$q(z) = \frac{1}{2} \frac{\Omega_{m0}(1+z)^3 - 2\Omega_{\Lambda 0}}{\Omega_{m0}(1+z)^3 + \Omega_{\Lambda 0} + \Omega_{k0}(1+z)^2}, \tag{10}$$

$$j(z) = 1 - \frac{\Omega_{k0}(1+z)^2}{\Omega_{m0}(1+z)^3 + \Omega_{\Lambda 0} + \Omega_{k0}(1+z)^2}, \tag{11}$$

where the various Ω_{i0}, with $i = m, \Lambda$, and k, stand for the current values of the fractional energy densities.

Bearing in mind the Friedmann constraint $\Omega_m + \Omega_\Lambda + \Omega_k = 1$ we readily get

$$j_0 = 1 - \Omega_{k0} \tag{12}$$

from Equation (11). Thereby if future accurate measurements show that j_0 deviates from unity, we will know that our universe (modulo the FLRW metric and the ΛCDM are correct) is not spatially flat, and the deviation will coincide with minus the present value of the spatial curvature. Unfortunately, current measurements of j_0 come along only with great latitude, $-7.6 \leq j_0 \leq 8.5$ [45]. However, this wide observational uncertainty gets substantially reduced after combining (12) with Equation (6), specialized to the ΛCDM model. It readily yields $q_0 + j_0 \geq 0$. For instance, using the experimental constraint on q_0 of Daly et al. [46], $q_0 = -0.48 \pm 0.11$, we find (within 1σ)

$$j_0 \geq 0.37. \tag{13}$$

The simple fact that, observationally, q_0 is negative [24,46–48], renders j_0 positive in the said model; i.e., cosmic acceleration should be increasing nowadays.

4. Concluding Remarks

The validity of the second law, in the presence of gravity, is well supported at small scales by the thermodynamics of astrophysical-sized collapsed objects, in particular of black holes [1,2], and of self-gravitating radiation spheres [3,4] but, to the best of our knowledge, this law had not been tested at cosmological scales thus far. Here, assuming the correctness of the FLRW metric at large scales and using the history of the Hubble factor—see Equations (4) and (5) and Figure 2—we found that the second law likely holds at these scales as well. However, due to the sizable error bars of the $H(z)$ data, the thermodynamic constraint on $\mid \Omega_{k0} \mid$ is rather loose. As we have shown, the situation greatly improves by applying the GP procedure of Reference [34] to these data. Then, $\mid \Omega_{k0} \mid \sim 10^{-4}$—see the right-hand panel of Figures 4 and 5. However, although the procedure rests on very reasonable assumptions, these are hard to test. On the other hand, we could not determine the sign of k. Nevertheless, we suggested that by means of Mc Vittie formula, Equation (7) of the drift of the redshift [37] and the use of advanced telescopes and spectrographs that will be in service soon, it will be possible to obtain accurate $H(z)$ data capable of discerning it. Further, in the context of the ΛCDM model, we demonstrated a very simple relationship, Equation (12), between the present value of the jerk parameter and Ω_{k0}. Finally, we showed that the second law drastically reduces the ample uncertainty about the current value of the jerk and using current constraints on q_0 sets a lower bound on it.

Author Contributions: The authors contributed equally to this paper.

Conflicts of Interest: The authors declare no conflict of interest.

References

1. Bekenstein, J.D. Generalized second law of thermodynamics in black-hole physics. *Phys. Rev. D* **1974**, *9*, 3292–3300.
2. Bekenstein, J.D. Statistical Black Hole Thermodynamics. *Phys. Rev. D* **1975**, *12*, 3077–3085.
3. Sorkin, R.D.; Wald, R.M.; Jiu, Z.Z. Entropy of Self-Gravitating Radiation. *Gen. Relativ. Gravit.* **1981**, *13*, 1127–1146.
4. Pavón, D.; Landsberg, P.T. Heat capacity of a self-gravitating radiation sphere. *Gen. Relativ. Gravit.* **1988**, *20*, 457–461.
5. Ferreira, P.C.; Pavón, D. Thermodynamics of nonsingular bouncing universes. *Eur. Phys. J. C* **2016**, *76*, 37.
6. Robertson, H.P. Kinematics and World-Structure. *Astrophys. J.* **1935**, *82*, 284–301.
7. Robertson, H.P. Kinematics and World-Structure III. *Astrophys. J.* **1936**, *83*, 257–271.
8. Walker, A.G. On Milne's theory of world-structure. *Proc. Lond. Math. Soc.* **1936**, *s2-42*, 90–127.
9. Clarkson, C.; Basset, B.; Lu, T.H.-C. A general test of the Copernican principle *Phys. Rev. Lett.* **2008**, *101*, 011301.
10. Zhang, P.; Stebbins, A. Confirmation of the Copernican principle through the anisotropic kinetic Sunyaev Zel'dovich effect. *Phil. Trans. R. Soc. A* **2011**, *369*, 5138–5145.
11. Bentivegna, E.; Bruni, M. Effects of Nonlinear Inhomogeneity on the Cosmic Expansion with Numerical Relativity. *Phys. Rev. Lett.* **2016**, *116*, 251302.

12. Saadeh, D.; Feeney, S.M.; Pontzen, A.; Peiris, H.V.; McEwen, J.D. How Isotropic is the Universe? *Phys. Rev. Lett.* **2016**, *117*, 131302.
13. Komatsu, E.; Smith, K.M.; Dunkley, J.; Bennett, C.L.; Gold, B.; Hinshaw, G.; Jarosik, N.; Larson, D.; Nolta, M.R.; Page, L.; et al. Seven-year Wilkinson Microwave Anisotropy Probe (WMAP) Observations: Cosmological Interpretation. *Astrophys. J. Suppl. Ser.* **2011**, *192*, 18.
14. Ade, P.R.; Aghanim, N.; Armitage-Caplan, C.; Arnaud, M.; Ashdown, M.; Atrio-Barandela, F.; Aumont, J.; Baccigalupi, C.; Banday, A.J.; Barreiro, R.B.; et al. Planck 2013 results. XVI. Cosmological parameters. *Astron. Astrophys.* **2014**, *571*, A16.
15. Hawking, S.W. Black hole explosions? *Nature* **1974**, *248*, 30–31.
16. Jacobson, T. Thermodynamics of Spacetime: The Einstein Equation of State. *Phys. Rev. Lett.* **1995**, *75*, 1260–1263.
17. Padmanabhan, T. Gravity and the thermodynamics of horizons. *Phys. Rep.* **2005**, *406*, 49–125.
18. Radicella, N.; Pavón, D. A thermodynamic motivation for dark energy. *Gen. Relativ. Grav.* **2012**, *44*, 685–702.
19. Pavón, D.; Radicella, N. Does the entropy of the Universe tend to a maximum? *Gen. Relativ. Grav.* **2013**, *45*, 63–68.
20. Bak, D.; Rey, S.-J. Cosmic holography. *Class. Quantum Grav.* **2000**, *17*, L83.
21. Egan, C.; Lineweaver, C.L. A Larger Estimate of the Entropy of the Universe. *Astrophys. J.* **2010**, *710*, 1825–1834.
22. Wang, B.; Gong, Y.; Abdalla, E. Thermodynamics of an accelerated expanding universe. *Phys. Rev. D* **2006**, *74*, 083520.
23. Frautschi, S. Entropy in an Expanding Universe. *Science* **1982**, *217*, 593–599.
24. Farook, O.; Madiyar, F.R.; Crandall, S.; Ratra, B. Hubble Parameter Measurement Constraints on the Redshift of the Deceleration-Acceleration Transition, Dynamical Dark Energy, and Space Curvature. 2016, arXiv:1607.03537.
25. Cyburt, R.H.; Fields, B.D.; Olive, K.; Yeh, T.H. Big bang nucleosynthesis: Present status. *Rev. Mod. Phys.* **2016**, *88*, 015004.
26. Simon, J.; Verde, L.; Jiménez, R. Constraints on the redshift dependence of the dark energy potential. *Phys. Rev. D* **2005**, *71*, 123001.
27. Moresco, M.; Cimatti, A.; Jimenez, R.; Pozzetti, L.; Zamorani, G.; Bolzonella, M.; Dunlop, J.; Lamareille, F.; Mignoli, M.; Pearce, H.; et al. Improved constraints on the expansion rate of the Universe up to $z \sim 1.1$ from the spectroscopic evolution of cosmic chronometers. *J. Cosmol. Astropart. Phys.* **2012**, *2012*, 006.
28. Cuesta, A.J.; Vargas-Magaña, M.; Beutler, F.; Bolton, A.S.; Brownstein, J.R.; Eisenstein, D.J.; Gil-Marín, H.; Ho, S.; McBride, C.K.; Maraston, C.; et al. The clustering of galaxies in the SDSS-III Baryon Oscillation Spectroscopic Survey: Baryon acoustic oscillations in the correlation function of LOWZ and CMASS galaxies in Data Release 12. *Mont. Not. R. Astron. Soc.* **2016**, *457*, 1770–1785.
29. Blake, C.; Brough, S.; Colless, M.; Contreras, C.; Couch, W.; Croom, S.; Croton, D.; Davis, T.M.; Drinkwater, M.J.; Forster, K.; et al. The WiggleZ Dark Energy Survey: Joint measurements of the expansion and growth history at $z < 1$. *Mon. Not. R. Astron. Soc.* **2012**, *425*, 405–141.
30. Stern, D.; Jimenez, R.; Verde, L.; Kamionkowski, M.; Adam, S. Cosmic Chronometers: Constraining the Equation of State of Dark Energy. I: H(z) Measurements. *J. Cosmol. Astropart. Phys.* **2010**, *2010*, 008.
31. Moresco, M. Raising the bar: New constraints on the Hubble parameter with cosmic chronometers at $z \sim 2$. *Mon. Not. R. Astron. Soc.* **2015**, *450*, L16–L20.
32. Delubac, T.; Bautista, J.E.; Busca, N.G.; Rich, J.; Kirkby, D.; Bailey, S.; Font-Ribera, A.; Slosar, A.; Lee, K.-G.; Pieri, M.M.; et al. aryon acoustic oscillations in the Lyα forest of BOSS DR11 quasars. *Astron. Astrophys.* **2015**, *574*, A59.
33. Font-Ribera, A.; Kirkby, D.; Busca, N.; Miralda-Escudé, J.; Ross, N.P.; Slosar, A.; Rich, J.; Aubourg, E.; Bailey, S.; Bhardwaj, V.; et al. Quasar-Lyman α forest cross-correlation from BOSS DR11: Baryon Acoustic Oscillations. *J. Cosmol. Astropart. Phys.* **2014**, *2014*, 027.
34. Seikel, M.; Clarkson, C.; Smith, M. Reconstruction of dark energy and expansion dynamics using Gaussian processes. *J. Cosmol. Astropart. Phys.* **2012**, *2012*, 036.
35. Sandage, A. The Change of Redshift and Apparent Luminosity of Galaxies due to the Deceleration of Selected Expanding Universes. *Astrophys. J.* **1962**, *136*, 319–333.

36. Loeb, A. The Change of Redshift and Apparent Luminosity of Galaxies due to the Deceleration of Selected Expanding Universes. *Astrophys. J.* **1998**, *499*, L111–L114.
37. Vittie, G.C.M. *Cosmological Theory*, 2nd ed.; Wiley: New York, NY, USA, 1949.
38. Spectrograph CODEX. Available online: http://www.iac.es/proyecto/codex/ (accessed on 22 November 2016).
39. The European Extremely Large Telescope. Available online: http://www.eso.org/public/teles-instr/e-elt/ (accessed on 22 November 2016).
40. Klockner, H.R.; Obreschkow, D.; Martins, C.; Raccanelli, A.; Champion, D.; Roy, A.; Lobanov, A.; Wagner, J.; Keller, R. Real time cosmology-A direct measure of the expansion rate of the Universe with the SKA. *Proc. Sci.* **2015**, *AASKA14*, 027.
41. Parkers 21 cm Multibeam Project. Available online:http://www.atnf.csiro.au/research/multibeam/(accessed on 22 November 2016).
42. The Canadian Hydrogen Intensity Mapping Experiment. Available online: chime.phas.ubc.ca/ (accessed on 22 November 2016).
43. Yu, H.R.; Zhang, T.J.; Pen, U.L. Method for Direct Measurement of Cosmic Acceleration by 21-cm Absorption Systems. *Phys. Rev. Lett.* **2014**, *113*, 041303.
44. Liske, J.; Grazian, A.; Vanzella, E.; Dessauges, M.; Viel, M.; Pasquini, L.; Haehnelt, M.; Cristiani, S.; Pepe, F.; Avila, G.; et al. Cosmic dynamics in the era of Extremely Large Telescopes. *Mon. Not. R. Astron. Soc.* **2008**, *386*, 1192–1218.
45. Bochner, B.; Pappas, D.; Dong, M. Testing Lambda and the Limits of. Cosmography with the Union2.1. Supernova Compilation. *Astrophys. J.* **2015**, *814*, 7.
46. Daly, R.; Djorgovski, S.G.; Freeman, K.A.; Mory, M.P.; O'Dea, C.P.; Kharb, P.; Baum, S. Improved Constraints on the Acceleration History of the Universe and the Properties of the Dark Energ. *Astrophys. J.* **2008**, *677*, 1–11.
47. Perlmutter, S.; Aldering, G.; Della Valle, M.; Deustua, S.; Ellis, R.S.; Fabbro, S.; Fruchter, A.; Goldhaber, G; Groom, D.E.; Hook, I.M.; et al. Discovery of a supernova explosion at half the age of the Universe. *Nature* **1998**, *391*, 51–54.
48. Riess, A.G.; Kirshner, R.P.; Schmidt, B.P.; Jha, S.; Challis, P.; Garnavich, P.M.; Esin, A.A.; Carpenter, C.; Grashius, R.; Schild, R.E.; et al. BV RI light curves for 22 type Ia supernovae. *Astron. J.* **1999**, *117*, 707–724.

universe

MDPI

Article

Warm Inflation

Øyvind Grøn

Art and Design, Faculty of Technology, Oslo and Akershus University College of Applied Sciences,
P.O. Box 4 St., Olavs Plass, NO-0130 Oslo, Norway; Oyvind.Gron@hioa.no

Academic Editors: Elias C. Vagenas and Lorenzo Iorio
Received: 25 July 2016; Accepted: 27 August 2016; Published: 6 September 2016

Abstract: I show here that there are some interesting differences between the predictions of warm and cold inflation models focusing in particular upon the scalar spectral index n_s and the tensor-to-scalar ratio r. The first thing to be noted is that the warm inflation models in general predict a vanishingly small value of r. Cold inflationary models with the potential $V = M^4 (\phi/M_P)^p$ and a number of e-folds $N = 60$ predict $\delta_{nsC} \equiv 1 - n_s \approx (p+2)/120$, where n_s is the scalar spectral index, while the corresponding warm inflation models with constant value of the dissipation parameter Γ predict $\delta_{nsW} = [(20 + p)/(4 + p)]/120$. For example, for $p = 2$ this gives $\delta_{nsW} = 1.1\delta_{nsC}$. The warm polynomial model with $\Gamma = V$ seems to be in conflict with the Planck data. However, the warm natural inflation model can be adjusted to be in agreement with the Planck data. It has, however, more adjustable parameters in the expressions for the spectral parameters than the corresponding cold inflation model, and is hence a weaker model with less predictive force. However, it should be noted that the warm inflation models take into account physical processes such as dissipation of inflaton energy to radiation energy, which is neglected in the cold inflationary models.

Keywords: General relativity; Cosmology; The inflationary era

---

1. Introduction

In the usual (cold) inflationary models, dissipative effects with decay of inflaton energy into radiation energy are neglected. However, during the evolution of warm inflation dissipative effects are important, and inflaton field energy is transformed to radiation energy. This produces heat and viscosity, which make the inflationary phase last longer. Warm inflation models were introduced and developed by Berera and coworkers [1–14]. However, even earlier inflation models with dissipation of inflaton energy to radiation and particles had been considered [15–22]. Introductions to warm inflation models and references to works prior to 2009 on warm inflation are found in [8] and [23]. For later works, see [9] and [24] and references in these articles. Further developments are found in the articles [25–43].

In this scenario, there is no need for a reheating at the end of the inflationary era. The universe heats up and becomes radiation dominated during the inflationary era, so there is a smooth transition to a radiation dominated phase (Figure 1).

In the present work, I will review the foundations of warm inflation and some of the most recent phenomenological models of this type, focusing in particular on the comparison with the experimental measurements of the scalar spectral index n_s and the tensor to scalar ratio r by the Planck observatory.

The article is organized as follows. In Section 2, the definition and current measurements of these quantities are given. Then, the optical parameters in the warm inflation scenario are considered. We go on and study some phenomenological models in the subsequent sections: monomial-, natural- and viscous inflation. The models are compared in Section 7, and the results are summarized in the final section.

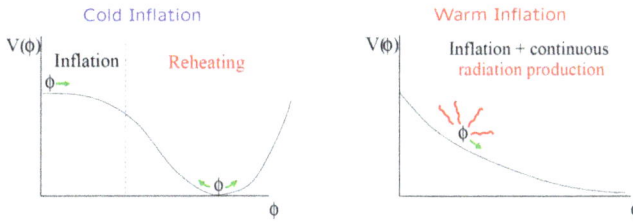

Figure 1. Illustration of the difference between cold inflation and warm inflation (Berera et al. (2009)).

2. Definition and Measured Values of the Optical Parameters

We shall here briefly review a few of the mathematical quantities that are used to describe the temperature fluctuations in the CMB. The power spectra of scalar and tensor fluctuations are represented by [44]

$$
P_S = A_S\left(k_*\right)\left(\frac{k}{k_*}\right)^{n_S - 1 + (1/2)\alpha_S \ln(k/k_*) + \cdots}, \quad P_T = A_T\left(k_*\right)\left(\frac{k}{k_*}\right)^{n_T + (1/2)\alpha_T \ln(k/k_*) + \cdots},
$$
$$
A_S = \frac{V}{24\pi^2 \varepsilon M_P^4} = \left(\frac{H^2}{2\pi\dot{\phi}}\right)^2, \qquad A_T = \frac{2V}{3\pi^2 M_P^4} = \varepsilon\left(\frac{2H^2}{\pi\dot{\phi}}\right)^2 \tag{2.1}
$$

Here, k is the wave number of the perturbation which is a measure of the average spatial extension for a perturbation with a given power, and k_* is the value of k at a reference scale usually chosen as the scale at horizon crossing, called the pivot scale. One often writes $k = \dot{a} = aH$, where a is the scale factor representing the ratio of the physical distance between reference particles in the universe relative to their present distance. The quantities A_S and A_T are amplitudes at the pivot scale of the scalar- and tensor fluctuations, and n_S and n_T are the *spectral indices* of the corresponding fluctuations. We shall represent the scalar spectral index by the quantity $\delta_{ns} \equiv 1 - n_S$. The quantities n_S and n_T are called the *tilt* of the power spectrum of curvature perturbations and tensor modes, respectively, because they represent the deviation of the values $\delta_{ns} = n_t = 0$ that represent a scale invariant spectrum.

The quantities α_S and α_T are factors representing the k-dependence of the spectral indices. They are called the *running of the spectral indices* and are defined by

$$
\alpha_S = \frac{dn_S}{d\ln k}, \quad \alpha_T = \frac{dn_T}{d\ln k} \tag{2.2}
$$

They will, however, not be further considered in this article.

As mentioned above, if $n_S = 1$ the spectrum of the scalar fluctuations is said to be *scale invariant*. An invariant mass-density power spectrum is called a *Harrison-Zel'dovich spectrum*. One of the predictions of the inflationary universe models is that the cosmic mass distribution has a spectrum that is *nearly* scale invariant, but not exactly. The observations and analysis of the Planck team [45] have given the result $n_S = 0.968 \pm 0.006$. Hence, we shall use $n_S = 0.968$ as the preferred value of n_S. Different inflationary models will be evaluated against the Planck 2015 value of the tilt of the scalar curvature fluctuations, $\delta_{ns} = 0.032$.

The tensor-to-scalar ratio r is defined by

$$
r \equiv \frac{P_T\left(k_*\right)}{P_S\left(k_*\right)} = \frac{A_T}{A_S} \tag{2.3}
$$

As noted by [46], the tensor-to-scalar ratio is a measure of the energy scale of inflation, $V^{1/4} = (100r)^{1/4} 10^{16} GeV$. From Equations (2.1) and (2.3), we have

$$
r = 16\varepsilon \tag{2.4}
$$

The Planck observational data have given $r < 0.11$.

3. Optical Parameters in Warm Inflation

During the warm inflation era, both the inflaton field energy with density ρ_ϕ and the electromagnetic radiation with energy density ρ_r are important for the evolution of the universe. The first Friedmann equation takes the form

$$H^2 = \frac{\kappa}{3}\left(\rho_\phi + \rho_r\right) \tag{3.1}$$

We shall here use units so that $\kappa = 1/M_P^2$ where M_P is the reduced Planck mass. In these models, the continuity equations for the inflaton field and the radiation take the form

$$\dot{\rho}_\phi + 3H\left(\rho_\phi + p_\phi\right) = -\Gamma\dot{\phi}^2 \quad , \quad \dot{\rho}_r + 4H\rho_r = \Gamma\dot{\phi}^2 \tag{3.2}$$

respectively, where the dot denotes differentiation with respect to cosmic time, and Γ is a dissipation coefficient of a process which transforms inflaton energy into radiation. In general, Γ is temperature dependent. The density and pressure of the inflaton field are given in terms of the kinetic and potential energy of the inflaton field as

$$\rho_\phi = \frac{\dot{\phi}^2}{2} + V \quad , \quad p_\phi = \frac{\dot{\phi}^2}{2} - V \tag{3.3}$$

During warm inflation, the dark energy predominates over radiation, i.e., $\rho_\phi \gg \rho_r$, and H, ϕ and Γ are slowly varying so that the production of radiation is quasi-static, $\ddot{\phi} \ll H\dot{\phi}$, $\dot{\rho}_r \ll 4H\rho_r$ and $\dot{\rho}_r \ll \Gamma\dot{\phi}^2$. Note that in the slow roll era the kinetic energy of the inflaton field energy can be neglected compared to its potential energy. Then, the inflaton field obeys the equation of state $p_\phi \approx -\rho_\phi$. Also, in this era, the second of Equation (3.2) gives $\rho_r = 0$ in the case of vanishing dissipation, $\Gamma = 0$, i.e., in the warm inflation model all of the radiation is produced by dissipation of the inflaton energy. Then, the first Friedmann equation and the equation for the evolution of the inflaton field take the form

$$3H^2 = \kappa\rho_\phi = \kappa V \quad , \quad (3H + \Gamma)\dot{\phi} = -V' \tag{3.4}$$

respectively. Here, a prime denotes differentiation with respect to the inflaton field ϕ.

Defining the so-called dissipative ratio by

$$Q \equiv \Gamma/3H \tag{3.5}$$

the last of Equation (3.4) may be written as

$$3H(1+Q)\dot{\phi} = -V' \tag{3.6}$$

The quantity Q represents the effectiveness at which inflaton energy is transformed to radiation energy. If $Q \gg 1$ one says that there is a strong, dissipative regime, and if $Q \ll 1$ there is a weak dissipative regime.

During warm inflation, the second of the Equation (3.2) reduces to

$$\rho_r = (3/4)Q\dot{\phi}^2 \tag{3.7}$$

In the warm inflation scenario, a thermalized radiation component is present with $T > H$, where both T and H are expressed in units of energy. Then, the tensor-to-scalar ratio defined in Equation (2.3), is modified with respect to standard cold inflation, so that [12]

$$r_W = \frac{H/T}{(1+Q)^{5/2}}r \tag{3.8}$$

Hence, the tensor-to-scalar ratio is suppressed by the factor $(T/H)(1+Q)^{5/2}$ compared with the standard cold inflation.

Hall, Moss and Berera [9] have calculated the spectral index in warm inflation for the strong dissipative regime with $Q \gg 1$ or $\Gamma \gg 3H$. We shall here follow Visinelli [47] and permit arbitrary values of Q. Differentiating the first of the Equation (3.4) and using Equation (3.6) gives

$$\dot{H} = -(\kappa/2)(1+Q)\dot{\phi}^2 \tag{3.9}$$

Hence $\dot{H} < 0$.

We define the potential slow roll parameters ε and η by

$$\varepsilon \equiv \frac{1}{2\kappa}\left(\frac{V'}{V}\right)^2 \quad , \quad \eta \equiv \frac{1}{\kappa}\frac{V''}{V} \tag{3.10}$$

These expressions are to be evaluated at the beginning of the slow roll era. Using Equations (3.4), (3.6) and (3.9) and the first of Equation (3.10) we get

$$\varepsilon = -(1+Q)\frac{\dot{H}}{H^2} \tag{3.11}$$

Differentiation of Equation (3.6) and using that $\left(\dot{\phi}\right)' = \ddot{\phi}/\dot{\phi}$ gives

$$V'' = \frac{\Gamma'V'}{\Gamma + 3H} - 3H(1+Q)\frac{\ddot{\phi}}{\dot{\phi}} - 3\dot{H} \tag{3.12}$$

Dividing by κV and using the first of Equation (3.4) in the two last terms leads to

$$\eta = \frac{Q}{1+Q}\frac{1}{\kappa}\frac{\Gamma'V'}{\Gamma V} - \frac{1+Q}{H}\frac{\ddot{\phi}}{\dot{\phi}} - \frac{\dot{H}}{H^2} \tag{3.13}$$

Defining

$$\beta \equiv \frac{1}{\kappa}\frac{\Gamma'V'}{\Gamma V} \tag{3.14}$$

and using Equation (3.12) we get

$$\frac{\ddot{\phi}}{H\dot{\phi}} = -\frac{1}{1+Q}\left(\eta - \beta + \frac{\beta - \eta}{1+Q}\right) \tag{3.15}$$

in agreement with Equation (3.14) of Visinelli [47].

It follows from Equation (3.6) that

$$\frac{d}{d\phi} = -\frac{3H(1+Q)}{V'}\frac{d}{dt} \tag{3.16}$$

From Equation (3.5) and the first of Equation (3.4) we have

$$H\Gamma = \kappa VQ \tag{3.17}$$

Using Equations (3.14), (3.16) and (3.17) can be written as

$$\frac{\dot{\Gamma}}{H\Gamma} = -\frac{\beta}{1+Q} \tag{3.18}$$

During slow roll the second of the Equation (3.2) reduces to

$$4H\rho_r = \Gamma\dot{\phi}^2 \qquad (3.19)$$

Differentiation gives

$$\frac{\dot{\rho}_r}{H\rho_r} = \frac{\dot{\Gamma}}{H\Gamma} + 2\frac{\ddot{\phi}}{H\dot{\phi}} - \frac{\dot{H}}{H^2} \qquad (3.20)$$

Inserting Equations (3.11), (3.15) and (3.18) into Equation (3.20) gives

$$\frac{\dot{\rho}_r}{H\rho_r} = -\frac{1}{1+Q}\left(2\eta - \beta - \varepsilon + 2\frac{\beta - \varepsilon}{1+Q}\right) \qquad (3.21)$$

We now define $\delta_{ns} \equiv 1 - n_s$, where n_s is the scalar spectral index. Visinelli [48] has deduced

$$\delta_{ns} = 4\frac{\dot{H}}{H^2} - 2\frac{\ddot{\phi}}{H\dot{\phi}} - \frac{\dot{\omega}}{H(1+\omega)} \qquad (3.22)$$

where

$$\omega = \frac{T}{H}\frac{2\sqrt{3}\pi Q}{\sqrt{3+4\pi Q}} \qquad (3.23)$$

Since $\rho_r \propto T^4$ we have that

$$\omega \propto \frac{\rho_r^{1/4}Q}{H\sqrt{3+4\pi Q}} \qquad (3.24)$$

Differentiating this we get

$$\frac{\dot{\omega}}{H\omega} = \frac{1}{4}\frac{\dot{\rho}_r}{H\rho_r} - \frac{\dot{H}}{H^2} + \frac{3+2\pi Q}{3+4\pi Q}\frac{\dot{Q}}{HQ} \qquad (3.25)$$

Differentiating Equation (3.5) gives

$$\frac{\dot{Q}}{HQ} = \frac{\dot{\Gamma}}{H\Gamma} - \frac{\dot{H}}{H^2} \qquad (3.26)$$

Using Equations (3.11) and (3.18) then leads to

$$\frac{\dot{Q}}{HQ} = \frac{\varepsilon - \beta}{1+Q} \qquad (3.27)$$

Inserting Equations (3.11), (3.21) and (3.27) into Equation (3.25) gives

$$\dot{\omega} = -\frac{H\omega}{1+Q}\left[\frac{2\eta - \beta - 5\varepsilon}{4} + \frac{1}{2}\frac{\beta - \varepsilon}{1+Q} + \frac{3+2\pi Q}{3+4\pi Q}(\beta - \varepsilon)\right] \qquad (3.28)$$

Visinelli has rewritten this as follows

$$\dot{\omega} = -\frac{H\omega}{1+Q}\left[\frac{2\eta + \beta - 7\varepsilon}{4} + \frac{6+(3+4\pi)Q}{(1+Q)(3+4\pi Q)}(\beta - \varepsilon)\right] \qquad (3.29)$$

Inserting the expressions (3.11), (3.15) and (3.29) into Equation (3.22) gives

$$\delta_{ns} = \frac{1}{1+Q}\left[4\varepsilon - 2\left(\eta - \beta + \frac{\beta-\varepsilon}{1+Q}\right) + \frac{\omega}{1+\omega}\left(\frac{2\eta+\beta-7\varepsilon}{4} + \frac{6+(3+4\pi)Q}{(1+Q)(3+4\pi Q)}(\beta - \varepsilon)\right)\right] \qquad (3.30)$$

The usual cold inflation is found in the limit $Q \to 0$ and $T << H$, i.e., $\omega \to 0$. Then,

$$\delta_{ns} \to 2\,(3\varepsilon - \eta) \tag{3.31}$$

In the strong regime of warm inflation, $Q >> 1$, $\omega >> 1$ we get

$$\delta_{ns} = \frac{3}{2Q}\left[\frac{3}{2}\,(\varepsilon + \beta) - \eta\right] \tag{3.32}$$

In the weak regime, $Q << 1$, Equation (3.16) leads to

$$\delta_{ns} = 2\,(3\varepsilon - \eta) - \frac{\omega/4}{1+\omega}\,(15\varepsilon - 2\eta - 9\beta) \tag{3.33}$$

It may be noted that in warm inflation the condition for slow roll is that the absolute values of ε, η and β are much smaller than $1 + Q$.

Visinelli has found that the tensor-to-scalar ratio in warm inflation is

$$r = \frac{16\varepsilon}{(1+Q)^2\,(1+\omega)} \tag{3.34}$$

In the cold inflation limit, this reduces to

$$r \to 16\varepsilon \tag{3.35}$$

In the strong dissipation regime warm inflation gives in general

$$r \to \frac{16}{Q^2\omega}\varepsilon << \varepsilon \tag{3.36}$$

Hence, all the warm inflation models predict an extremely small tensor-to-scalar-ratio in the strong dissipation regime with $Q >> 1$ and $\omega >> 1$.

4. Warm Monomial Inflation

Visinelli [48] has investigated warm inflation with a polynomial potential which we write in the form

$$V = M^4\,(\phi/M_P)^p \tag{4.1}$$

since the potential and the inflaton field have dimensions equal to the fourth and first power of energy, respectively. Here, M represents the energy scale of the potential when the inflaton field has Planck mass. Furthermore he assumes that the dissipative term is also monomial

$$\Gamma = \Gamma_0\,(\phi/M_P)^{q/2} \tag{4.2}$$

He considered models with $p > 0$ and $q > p$. However, in the present article, we shall also consider polynomial models with $p < 0$. From Equations (3.3) and (3.4) we have

$$Q = Q_0\left(\frac{\phi}{M_P}\right)^{\frac{q-p}{2}}\quad,\quad Q_0 = \frac{\Gamma_0 M_P}{\sqrt{3}M^2} \tag{4.3}$$

The constant Q_0 represents the strength of the dissipation. For $q = p$ the dissipative ratio is constant, $Q = Q_0$. We shall here consider the strong dissipative regime where $Q >> 1$. Then, the second of Equation (3.3) reduces to

$$\dot{\phi} = -\frac{V'}{\Gamma} \tag{4.4}$$

Inserting Equations (4.1) and (4.2) gives

$$\dot{\phi} = -\frac{pM^4}{\Gamma_0 M_P}\left(\frac{\phi}{M_P}\right)^{p-\frac{q}{2}-1} \tag{4.5}$$

Integration leads to

$$\phi(t) = \left[\frac{4+q-2p}{2}\left(K - \frac{pM^4}{\Gamma_0 M^{p-\frac{q}{2}}}t\right)\right]^{\frac{2}{4+q-2p}} \quad , \quad q > 2\,(p-2) \tag{4.6}$$

where K is a constant of integration. The initial condition $\phi\,(0) = 0$ gives $K = 0$.

The special cases (i) $\Gamma = V/M_P^3$, i.e., $\Gamma_0 = M^4/M_P^3$, $q = 2p$ and (ii) $\Gamma = \Gamma_0$, i.e., $q = 0$, both with the initial condition $\phi\,(0) = 0$, i.e., $K = 0$, have been considered by Sharif and Saleem (2015). For these cases, the condition $\phi\,(t) > 0$ requires $p < 0$. In the first case, Equation (3.6) reduces to

$$\phi = M_P\sqrt{-2pM_Pt} \tag{4.7}$$

Note that the time has dimension inverse mass with the present units, so that M_Pt is dimensionless.

Visinelli, however, has considered polynomial models with $p > 0$. Then, we have to change the initial condition. The corresponding solution of Equation (4.5) with $q = 2p$ and the inflaton field equal to the Planck mass at the Planck time gives

$$\phi = M_P\sqrt{1 - 2pM_P\,(t - t_P)} \tag{4.8}$$

It may be noted that $q = 2\,(p-2)$ gives a different time evolution of the inflaton field. Then, Equation (3.5) with the boundary condition $\phi\,(t_P) = M_P$ has the solution

$$\phi = M_P\exp\left[-\frac{pM^4}{\Gamma_0 M_P^2}\,(t - t_P)\right] \tag{4.9}$$

In this case, the inflaton field decreases or increases exponentially, depending upon the sign of p. Inserting Equations (4.1) and (4.2) into Equations (3.9) and (3.13), the slow-roll parameters are

$$\varepsilon = \frac{p^2}{2}\left(\frac{M_P}{\phi}\right)^2 \quad , \quad \eta = \frac{2\,(p-1)}{p}\varepsilon \quad , \quad \beta = \frac{q}{p}\varepsilon \tag{4.10}$$

With these expressions Equation (3.32) valid in the regime of strong dissipation, $Q \gg 1$, gives

$$\delta_{ns} = \frac{3\,(4+3q-p)}{4p}\frac{\varepsilon}{Q} \tag{4.11}$$

The slow-roll regime ends when at least one of the parameters (4.10) is not much smaller than $1 + Q$. In the strong dissipative regime $Q \gg 1$ and $\varepsilon_f = Q_f$. Using Equations (4.3) and (4.10) we then get

$$\phi_f = M_P\left(\frac{p^2}{2Q_0}\right)^{\frac{2}{4+q-p}} \tag{4.12}$$

The number of e-folds, N, in the slow roll era for this model has been calculated by Visinelli [48] . It is defined by

$$N = \ln\frac{a_f}{a} = \int_t^{t_f} H dt = \int_\phi^{\phi_f} \frac{H}{\dot{\phi}}d\phi \tag{4.13}$$

Using Equations (3.3) and (3.5) we get

$$N = \frac{1}{M_P^2} \int_{\phi_f}^{\phi} (1+Q) \frac{V}{V'} d\phi \qquad (4.14)$$

Inserting the potential (4.1), performing the integration and considering the strong dissipative regime gives

$$N \approx \frac{2Q_0}{p\,(4+q-p)} \left[\left(\frac{\phi}{M_P} \right)^{\frac{4+q-p}{2}} - \left(\frac{\phi_f}{M_P} \right)^{\frac{4+q-p}{2}} \right] \qquad (4.15)$$

The time dependence of the inflaton field is given by Equation (4.6) when $p < 0$ showing that $\phi_f > \phi$ in this case, and by Equation (4.8) when $p > 0$ implying $\phi_f < \phi$ in that case, showing that $N > 0$ in both cases (not dot here)

$$\frac{\phi}{M_P} \approx \left(\frac{p\,(4+q-p)\,N}{2Q_0} \right)^{\frac{2}{4+q-p}} \qquad (4.16)$$

Inserting this into the first of Equations (4.10) and (4.3) gives

$$\varepsilon \approx \frac{p^2}{2} \left[\frac{2Q_0}{p\,(4+q-p)\,N} \right]^{\frac{4}{4+q-p}} , \quad Q \approx Q_0 \left[\frac{p\,(4+q-p)\,N}{2Q_0} \right]^{\frac{q-p}{4+q-p}} \qquad (4.17)$$

Inserting these expressions into Equation (4.11) gives

$$\delta_{ns} \approx \frac{3\,(4+3q-p)}{4\,(4+q-p)} \frac{1}{N} \qquad (4.18)$$

Note that with $q = 0$, i.e., a constant value of the dissipation parameter Γ, Equation (4.18) reduces to

$$\delta_{ns} = \frac{3}{4N} \qquad (4.19)$$

for all values of p. Then $N = 60$ gives $\delta_{ns} = 0.012$ which is smaller than the preferred value from the Planck data, $\delta_{ns} = 0.032$. Inserting $q = 2p$ in Equation (4.18) and solving the equation with respect to p gives,

$$p = \frac{4\,(4N\delta_{ns} - 3)}{15 - 4N\delta_{ns}} \qquad (4.20)$$

The Planck values $\delta_{ns} = 0.032$, $N = 60$ give $p = 2.56$ and $q = 5.11$.

Panotopoulos and Videla [24] have investigated the tensor-to-scalar ratio in warm in inflation for inflationary models with an inflaton field given by the potential

$$V = (M/M_P)^4 \phi^4 \qquad (4.21)$$

where M is the energy scale of the potential when the inflaton field has Planck mass, M_P. Let us choose $p = q = 4$ in the monomial models above. Inserting this in Equation (3.18) gives $\delta_{ns} = 9/4N$. With $\delta_{ns} = 0.032$ we get $N = 70$.

In this case $\delta_{ns} = 2/N$ for cold inflation. For $\delta_{ns} = 0.032$ this corresponds to $N \approx 62$ which is an acceptable number of e-folds. Then, the tensor-to-scalar ratio is $r = 0.32$, which is much larger than allowed by the Planck observations [45]. Panotopoulos and Videla found the corresponding $\delta_{ns}, r-$ relation in warm inflation with $\Gamma = aT$, where a is a dimensionless parameter. They considered two cases.

(A) The weak dissipative regime. In this case $Q << 1$ and Equation (3.7) reduces to $r_W = (H/T)r$. They then found

$$r_W \approx \frac{0.01}{\sqrt{a}} \delta_{ns} \tag{4.22}$$

With the Planck values $\delta_{ns} = 0.032$ and $r_W < 0.12$ this requires $a > 7 \cdot 10^{-6}$. However, they also found that in this case $\delta_{ns} = 1/N$ giving $N = 31$ which is too small to be compatible with the standard inflationary scenario.

(B) The strong dissipative regime. Then, $R >> 1$ and $r_W \approx \left(H/TR^{5/2} \right) r$. They then found

$$\delta_{ns} = \frac{45}{28N} \quad , \quad r_W = \frac{3.8 \cdot 10^{-7}}{a^4} \delta_{ns} \tag{4.23}$$

Then $N = 50$ and $a > 1.8 \cdot 10^{-2}$, so this is a promising model.

5. Warm Natural Inflation

Visinelli [47] has also investigated warm natural inflation with the potential

$$V(\phi) = V_0 \left(1 + \cos\tilde{\phi} \right) = 2V_0\cos^2 \left(\tilde{\phi}/2 \right) \tag{5.1}$$

where $\tilde{\phi} = \phi/M$, and M is the spontaneous symmetry breaking scale, and $M > M_P$ in order for inflation to occur. The constant V_0 is a characteristic energy scale for the model. The potential V has a minimum at $\tilde{\phi} = \pi$. Inserting the potential (5.1) into the expressions (3.9) we get

$$\varepsilon = \frac{b}{2} \frac{1 - \cos\tilde{\phi}_i}{1 + \cos\tilde{\phi}_i} \quad , \quad \eta = \varepsilon - \frac{b}{2} \quad , \quad b = \left(\frac{M_P}{M} \right)^2 \tag{5.2}$$

From Equation (3.3) with the potential (5.1) we have

$$H = \sqrt{(\kappa/3) V_0 \left(1 + \cos\tilde{\phi} \right)} \tag{5.3}$$

Equations (3.4) and (5.3) then give

$$Q = \frac{\Gamma M_P}{\sqrt{3V_0 \left(1 + \cos\tilde{\phi} \right)}} \tag{5.4}$$

During the slow roll era we must have $\varepsilon << R$. Using the expressions (5.2) and (5.4) we find that this corresponds to

$$\frac{1 - \cos\tilde{\phi}}{\sqrt{1 + \cos\tilde{\phi}}} << 1/\beta \quad , \quad \beta = \frac{\sqrt{6V_0}}{\Gamma M_P} b \tag{5.5}$$

Inserting Equations (5.2) and (5.4) into Equation (3.31) with $\beta = 0$ gives in the strong dissipative regime

$$\delta_{ns} = \frac{3}{4\alpha} \frac{3 + \cos\tilde{\phi}_i}{\sqrt{1 + \cos\tilde{\phi}_i}} \tag{5.6}$$

We shall now express the δ_{ns} in terms of the number of e-folds of expansion during the slow roll era for this inflationary universe model, again following Visinelli. Assuming that the dissipation parameter Γ is independent of ϕ, i.e., that $\beta = 0$, the number of e-folds is given by

$$N = -\Gamma \int_{\phi_i}^{\phi_f} \frac{H(\phi)}{V'(\phi)} d\phi \tag{5.7}$$

Differentiating the potential (5.1) and inserting Equation (5.3) we get

$$N = \frac{\alpha}{2} \int_{\tilde{\phi}_i}^{\tilde{\phi}_f} \frac{\sqrt{1+\cos x}}{\sin x} dx = \frac{\alpha}{\sqrt{2}} \ln \frac{\tan \left(\tilde{\phi}_f / 4 \right)}{\tan \left(\tilde{\phi}_i / 4 \right)} \tag{5.8}$$

Hence,

$$\tan \frac{\tilde{\phi}_i}{4} = \tan \frac{\tilde{\phi}_f}{4} exp \left(-\frac{\beta N}{2} \right) \tag{5.9}$$

Visinelli has argued that

$$\tilde{\phi}_f = \pi - \beta \tag{5.10}$$

giving

$$\tan \frac{\tilde{\phi}_f}{4} = \frac{1 - \tan (\beta/4)}{1 + \tan (\beta/4)} \tag{5.11}$$

Inserting this into Equation (5.9) gives

$$\tan \frac{\tilde{\phi}_i}{4} = \gamma exp \left(-\frac{\beta N}{2} \right) \quad , \quad \gamma = \frac{1 - \tan (\beta/4)}{1 + \tan (\beta/4)} \tag{5.12}$$

Applying the trigonometric identity

$$\sqrt{1 + \cos \theta} = \sqrt{2} \frac{1 - \tan^2 (\theta/4)}{1 + \tan^2 (\theta/4)} \tag{5.13}$$

in the expression (5.12) and inserting the result into Equation (5.6) we finally arrive at

$$\delta_{ns} = \frac{3}{8} \beta \frac{exp (2\beta N) + \gamma^4}{exp (2\beta N) - \gamma^4} \tag{5.14}$$

Here, we must have $\beta \ll 1$ in order to give the Planck value $\delta_{ns} = 0.032$ for $N = 60$. Hence, Equation (5.12) gives $\gamma \approx 1$. A good approximation for δ_{ns} is therefore

$$\delta_{ns} \approx (3/8) \beta \coth (\beta N) \tag{5.15}$$

Inserting $\delta_{ns} = 0.032$ and $N = 60$ gives $\beta = 0.08$.
Visinelli (2011) further found that the tensor-to-scalar ratio for this inflationary model is

$$r = 128\kappa \sqrt{\frac{\pi}{\Gamma} \frac{\dot{\phi}^2}{T\sqrt{H}}} \tag{5.16}$$

Differentiating the expression (5.3) gives

$$\dot{H} = -\frac{\kappa V_0}{6M} \frac{s_\phi \dot{\phi}}{H} \quad , \quad s_\phi \equiv \sin \tilde{\phi} \tag{5.17}$$

Combining this with Equation (3.8) in the strong dissipative regime and using Equation (3.4) gives

$$\dot{\phi} = \frac{3V_0 s_\phi}{M\Gamma} \tag{5.18}$$

The energy density of the radiation is

$$\rho_\gamma = aT^4 \tag{5.19}$$

where $a = 7.5657 \times 10^{-16}$ J·m^{-3}·K$^{-4} = 4.69 \times 10^{-6}$ GeV·m^{-3}·K^{-4} is the radiation constant. Combining with Equation (3.6) we get

$$T = \left(\frac{\Gamma}{4aH}\right)^{1/4} \dot{\phi}^{1/2} \tag{5.20}$$

Equations (5.15), (5.18) and (5.19) give

$$r = B\frac{s_\phi^{3/2}}{(1+\cos\widetilde{\phi})^{1/8}} \quad , \quad B = \frac{384 \cdot 3^{5/8} \kappa^{7/8} \sqrt{6\pi} V^{11/8} a^{1/4}}{M^{3/2}\Gamma^{9/4}} \tag{5.21}$$

Visinelli [47] has evaluated the constant B and concluded that for this type of inflationary universe model the expected value of r is extremely low. If observations give a value $r > 10^{-14}$ this model has to be abandoned. On the other hand, the predictions of this model are in accordance with the observations so far.

6. Warm Viscous Inflation

As noted by del Campo, Herrera and Pavón [29], it has been usual, for the sake of simplicity, to study warm inflation models containing an inflaton field and radiation, only, (comma here) ignoring the existence of particles with mass that will appear due to the decay of the inflaton field. However, these particles modify the fluid pressure in two ways: (i) The relationship between pressure and energy density is no longer $p = (1/3)\rho$ as it is for radiation. A simple generalization is to use the equation of state $p = w\rho$, where w is a constant with value $0 \leq w \leq 1$; (ii) Due to interactions between the particles and the radiation there will appear a bulk viscosity so that the effective pressure takes the form

$$p_{eff} = p - 3\varsigma H \tag{6.1}$$

where ς is a coefficient of bulk viscosity.

We shall now consider isotropic universe models corresponding to the anisotropic models considered by Sharif and Saleem [37]. Equation (3.8) can be written

$$\dot{\phi} = \pm M_P\sqrt{-2\dot{H}/(1+Q)} \tag{6.2}$$

For these models, the time dependence of the scale factor during the inflationary era may be written

$$a(t) = a_0\exp\left(\frac{t}{t_1}\right)^\beta \quad , \quad 0 < \beta \leq 1, \tag{6.3}$$

where a_0 is the value of the scale factor at $t = 0$ before the slow roll era has started, and t_1 is the Hubble time of the corresponding De Sitter model having $\beta = 1$. The Hubble parameter and its rate of change with time is

$$H = \frac{\beta}{t_1}\left(\frac{t}{t_1}\right)^{\beta-1} \quad , \quad \dot{H} = \frac{\beta(\beta-1)}{t_1^2}\left(\frac{t}{t_1}\right)^{\beta-2} \tag{6.4}$$

Note that $\dot{H} < 0$ for $\beta < 1$. Inserting the second expression into Equation (6.2) gives

$$\dot{\phi} = \pm\frac{M_P}{t_1}\sqrt{\frac{2\beta(1-\beta)}{1+Q}}\left(\frac{t}{t_1}\right)^{\frac{\beta}{2}-1} \tag{6.5}$$

Sharif and Saleem considered two cases. In the first one $\Gamma = \Gamma(\phi) = \kappa V(\phi)/M_P$. Equations (3.3) and (3.4) then gives $Q = H/M_P$. Furthermore, for several reasons, they restricted their analysis to the strong dissipative regime where $Q >> 1$. Equation (6.5) then reduces to

$$\dot{\phi} = \pm M_P\sqrt{2M_P(1-\beta)}t^{-1/2} \tag{6.6}$$

Integrating with the initial condition $\phi(0) = 0$ and assuming that $\phi(t) > 0$ we get

$$\phi(t) = 2M_P \sqrt{2M_P(1-\beta)t} \tag{6.7}$$

Hence, ϕ is an increasing function of time. Inserting the first of the expressions (6.4) into the first of the Equation (3.3) gives

$$V(t) = 3\left(\frac{\beta M_P}{t_1}\right)^2 \left(\frac{t}{t_1}\right)^{2(\beta-1)} \tag{6.8}$$

Combining this with Equation (6.7) leads to

$$V(\phi) = 3\left(\frac{\beta M_P}{t_1}\right)^2 \left(\frac{\phi}{2M_P\sqrt{2(1-\beta)}\,M_P t_1}\right)^{4(\beta-1)} \tag{6.9}$$

Sharif and Saleem used the Hubble slow roll parameters,

$$\varepsilon_H \equiv -\frac{\dot{H}}{H^2} = \frac{1}{2(1+Q)}\left(\frac{V'}{V}\right)^2 \quad, \quad \eta_H \equiv -\frac{\ddot{H}}{2H\dot{H}} = \frac{1}{1+Q}\left[\frac{V''}{V} - \frac{1}{2}\left(\frac{V'}{V}\right)^2\right] \tag{6.10}$$

Note that $\varepsilon_H = 1 + q$, where q is the deceleration parameter. In the present case and in the strong dissipative regime, we can replace $1 + Q$ by $H = \sqrt{\kappa V/3}$. Then $\varepsilon_H = (1/Q)\varepsilon$ and $\eta_H = (1/Q)(\eta - \varepsilon)$. Differentiating the expression (6.9) then gives

$$\varepsilon_H = \frac{1-\beta}{\beta}\left(\frac{\phi}{2M_P\sqrt{2(1-\beta)}M_P t_1}\right)^{-2\beta} \quad, \quad \eta_H = \frac{3-2\beta}{2\beta}\left(\frac{\phi}{2M_P\sqrt{2(1-\beta)}M_P t_1}\right)^{-2\beta} = \frac{3-2\beta}{2(1-\beta)}\varepsilon_H \tag{6.11}$$

The slow roll era ends when the inflaton field has a value ϕ_f so that $\varepsilon_H(\phi_f) = 1$, corresponding to $\varepsilon(\phi_f) = Q$, which gives

$$\left(\frac{\phi_f}{2M_P\sqrt{2(1-\beta)}\,M_P t_1}\right)^{2\beta} = \frac{1-\beta}{\beta} \tag{6.12}$$

The number of e-folds is given by Equation (4.15), which in the present case takes the form

$$N = \frac{1}{\sqrt{3}M_P}\int_{\phi_f}^{\phi}\frac{V^{3/2}}{V'}d\phi \tag{6.13}$$

Inserting the potential (6.9) and integrating gives

$$N = \left(\frac{\phi_f}{2M_P\sqrt{2(1-\beta)}M_P t_1}\right)^{2\beta} - \left(\frac{\phi}{2M_P\sqrt{2(1-\beta)}M_P t_1}\right)^{2\beta} = \frac{1-\beta}{\beta} - \left(\frac{\phi}{2M_P\sqrt{2(1-\beta)}M_P t_1}\right)^{2\beta} \tag{6.14}$$

Hence

$$\left(\frac{\phi}{2M_P\sqrt{2(1-\beta)}\,M_P t_1}\right)^{2\beta} = \frac{1-\beta}{\beta} - N \tag{6.15}$$

Since the left hand side is positive, this requires that $N < (1-\beta)/\beta$ or $\beta < 1/(N+1)$. For $N > 50$ this means that $0 < \beta < 0.02$.

Sharif and Saleem have calculated the scalar spectral index with the result

$$\delta_{ns} = \frac{3\beta - 2}{\beta}\left(\frac{\phi}{2M_P\sqrt{2(1-\beta)}\,M_P t_1}\right)^{-2\beta} \tag{6.16}$$

Using Equation (6.15) we get

$$\delta_{ns} = \frac{3\beta - 2}{1 - \beta - \beta N} \approx \frac{2 - 3\beta}{\beta} \frac{1}{N} \tag{6.17}$$

This equation can be written

$$\beta \approx \frac{2}{3 + N\delta_{ns}} \tag{6.18}$$

Inserting the Planck value $\delta_{ns} = 0.032$ and $N = 60$, give $\beta = 0.41$ corresponding to $p = -2.36$. This value of β is not allowed by Equation (6.15).

In the second case, Sharif and Saleem assumed that $\Gamma = \Gamma_0$. Equations (3.3) and (3.4) then give $Q = \Gamma_0/3H$. Using Equations (6.2) and (6.4) and integrating with the initial condition $\phi(0) = 0$, leads to

$$\phi(t) = \lambda \left(\frac{t}{t_1}\right)^{\beta - 1/2} \quad , \quad V(\phi) = 3\left(\frac{\beta M_P}{t_1}\right)^2 \left(\frac{\phi}{\lambda}\right)^{\frac{4(1-\beta)}{2\beta-1}} \quad , \quad \lambda = \frac{2\beta M_P}{2\beta - 1}\sqrt{\frac{6(1-\beta)}{t_1 \Gamma_0}} \tag{6.19}$$

In this case ε_H and η_H becomes

$$\varepsilon_H = \frac{1-\beta}{\beta}\left(\frac{\phi}{\lambda}\right)^{-\frac{2\beta}{2\beta-1}} \quad , \quad \eta_H = \frac{2-\beta}{\beta}\left(\frac{\phi}{\lambda}\right)^{-\frac{2\beta}{2\beta-1}} = \frac{2-\beta}{1-\beta}\varepsilon_H \tag{6.20}$$

The final value of ϕ_f is given by

$$\left(\frac{\phi_f}{\lambda}\right)^{\frac{2\beta}{2\beta-1}} = \frac{1-\beta}{\beta} \tag{6.21}$$

The number of e-folds is

$$N = \left(\frac{\phi}{\lambda}\right)^{\frac{2\beta}{2\beta-1}} - \left(\frac{\phi_f}{\lambda}\right)^{\frac{2\beta}{2\beta-1}} = \left(\frac{\phi}{\lambda}\right)^{2\beta} - \frac{1-\beta}{\beta} \tag{6.22}$$

Hence

$$\left(\frac{\phi}{\lambda}\right)^{\frac{2\beta}{2\beta-1}} = N + \frac{1-\beta}{\beta} \tag{6.23}$$

The scalar spectral index is

$$\delta_{ns} = \frac{4+\beta}{2\beta}\left(\frac{\phi}{\lambda}\right)^{-\frac{2\beta}{2\beta-1}} = \frac{4+\beta}{2(\beta N + 1 - \beta)} \approx \frac{4+\beta}{2\beta}\frac{1}{N} \tag{6.24}$$

which can be written

$$\beta = \frac{4}{2N\delta_{ns} - 1} \tag{6.25}$$

Inserting the Planck value $\delta_{ns} = 0.032$ and $N = 60$ gives $\beta = 1.4$ outside the range $\beta < 1$ which requires $N > 78$. However, in the anisotropic case considered by Sharif and Saleem, one may obtain agreement with the Planck data for $\beta < 1$. As noted above, the tensor to scalar ratio has a very small value in these models. The time evolution of the inflaton field is given by Equation (6.7).

7. Comparison of Models

The models of Sharif and Saleem are a class of the monomial models. Comparing Equations (4.1) and (6.9) we have $p = 4(\beta - 1)$ or $\beta = 1 + p/4$. Hence, for $\beta < 1$ we must have $p < 0$ while Visinelli considered models with $p > 0$. Furthermore, in the first case of Sharif and Saleem with $\Gamma = V$ we have

$q = 2p$ and in the case with $\Gamma = \Gamma_0$ we have $q = 0$. Also, it should be noted that Visinelly has deduced the expression for the spectral parameters from the potential slow roll parameters, while Sharif and Saleem have used the Hubble slow roll parameters, and they have got slightly different expressions.

Let us consider an isotropic monomial model with scale as given in Equation (6.3). Then, we have two formulae for the potential—Equations (4.1) and (6.9). Hence

$$t_1 = \left(\sqrt{3}\beta\right)^{\frac{1}{\beta}} [8(1-\beta)]^{\frac{1-\beta}{\beta}} \left(\frac{M_P}{M}\right)^{2/\beta} t_P \tag{7.1}$$

where $t_P = 1/M_P$ is the Planck time. As mentioned above in Sharif and Saleem's first case $\Gamma = \Gamma(\phi) = \kappa V(\phi)/M_P$. Combining this with the first Equation (3.3) we get $\Gamma = 3H^2/M_P$. Furthermore they considered the strong dissipative regime with $\Gamma \gg 3H$. Hence $H \gg M_P$. The slow roll era begins at a point of time, t_i, when the inflaton field is given by Equation (6.23). This leads to

$$t_i = \left(N + \frac{1-\beta}{\beta}\right)^{1/\beta} t_1 \tag{7.2}$$

The Hubble parameter is given by the first equation in (6.4) with a maximal value at the beginning of the inflationary era. Hence, the condition $H \gg M_P$ requires that

$$t_i = \left(\frac{\beta}{M_P t_1}\right)^{\frac{1}{1-\beta}} t_1 \tag{7.3}$$

Inserting the expression (7.2) for t_1 we arrive at

$$t_i \ll (\beta N + 1 - \beta) t_P \tag{7.4}$$

Hence in this model with for example $\beta = 1/2$ and $N = 60$ the inflationary era begins much earlier than at around 30 Planck times. Inserting the inequality (7.4) into Equation (7.1) we get

$$M \gg \sqrt{\sqrt{3}[8(1-\beta)]^{1-\beta}(\beta N + 1 - \beta) M_P} \tag{7.5}$$

Hence $M \gg M_P$, so these models are large field inflation models.

V. Kamali and M. R. Setare [49] have considered warm viscous inflation models in the context of brane cosmology using the so-called chaotic potential (3.1) with $p = 2$, i.e., $\beta = 3/2$. We have considered the corresponding models in ordinary (not brane) spacetime which corresponds to taking the limit that the brane tension $\lambda \to \infty$ in their equations. They first considered the case $\Gamma = \Gamma_0$, i.e., $q = 0$. Then, the time evolution of the inflaton field is given by Equation (4.9) with $p = 2$. As noted above, in this case $\delta_{ns} = 0.012$ which is smaller than the preferred value from the Planck data. It may be noted that Kamali and M. R. Setare got a different result. Letting $\lambda \to \infty$ in their Equation (68) gives $\delta_{ns} = 0$, i.e., a scale invariant spectrum.

Next, they considered the case $\Gamma = \Gamma(\phi) = \alpha V(\phi)$. With $\alpha = 1$ this corresponds to the first case considered by Sharif and Saleem [37].

8. Conclusions

Warm inflation is a promising model of inflation, taking account of dissipative processes that are neglected in the usual, cold inflationary models. In warm inflation, radiation is produced by dissipation of the inflaton field, and reheating is not necessary. This type of inflationary model was introduced and developed initially by Berera and coworkers. Also, interactions between the inflaton field and the radiation provide a mechanism for producing viscosity.

In this article, I have given a review of some recent models with particular emphasis on their predictions of optical parameters, making it possible to evaluate the models against the observational data obtained by the Planck team. In particular, power law potential inflation, PI, and natural inflation, NI, in the warm inflation scenario have been considered.

I have emphasized that there are some interesting differences between the predictions of these models and the corresponding cold inflation models. The first thing to be noted is that the warm inflation models in general predict a vanishingly small value of the tensor-to-scalar ratio, r. I the present paper I have parametrized the scalar spectral index n_s by $\delta_{ns} = 1 - n_s$. The Planck data favor the value $\delta_{ns} = 0.032$, $r < 0.11$ and a number of e-folds $N = 60$.

Cold PI with the potential (4.1) predicts $\delta_{ns} = \frac{2(p+2)}{p+4N}$ and $r = \frac{16p}{p+4N}$. Inserting $\delta_{ns} = 0.032$ and $N = 60$ gives $p = 1.8$ and $r = 0.12$. The corresponding warm PI model with constant value of the dissipation parameter Γ predicts, according to Equation (6.24), $\delta_{ns} = \frac{20+p}{4+p}\frac{1}{2N}$ giving $p = 2.8$. The corresponding model with $\Gamma = \Gamma(\phi) = V(\phi)$ predicts $\delta_{ns} = -\frac{4+3p}{4+p}\frac{1}{N}$ giving $p = -2.36$. However, according to Equation (6.15), this model is only consistent for $-4 < p < -3.92$. Hence, this model is in conflict with the Planck data.

Cold natural inflation predicts

$$\delta_{ns} = b\frac{(2+b)\,e^{bN} + b}{(2+b)\,e^{bN} - b} \quad , \quad r = \frac{8b^2}{(2+b)\,e^{bN} - b} \quad , \quad b = \left(\frac{M_P}{M}\right)^2 \tag{8.1}$$

Inserting $\delta_{ns} = 0.032$ and $N = 60$ gives $b = 0.032$ or $M = 5.5M_P$, giving $r = 0.0006$. Since $M > M_P$ this is large field inflation according to the standard definition of this classification (Lyth [50], Dine and Pack [51]). The corresponding warm natural inflation model has two parameters, Γ and V_0, contained in β in the expression for δ_{ns}. Hence, some assumption concerning the relationship between Γ and V_0, is needed to make a prediction of the value of δ_{ns} in this model.

Acknowledgments: I would like to thank Luca Visinelli for useful correspondence concerning this work and the referees for valuable suggestions and for providing several references to old articles describing inflation models with dissipation of inflaton energy.

Conflicts of Interest: The author declare no conflict of interest.

References

1. Berera, A. Warm inflation. *Phys. Rev. Lett.* **1995**, *75*, 3218–3221. [CrossRef] [PubMed]
2. Berera, A. Thermal properties of an inflationary universe. *Phys. Rev. D* **1996**, *54*, 2519–2534. [CrossRef]
3. Berera, A. Interpolating the stage of exponential expansion in the early universe: Possible alternative with no reheating. *Phys. Rev. D* **1997**, *55*, 3346–3357. [CrossRef]
4. Berera, A. Warm inflation in the adiabatic regime—A model, an existence proof for inflationary dynamics in quantum field theory. *Nucl. Phys. B* **2000**, *585*, 666–714. [CrossRef]
5. Berera, A. The warm inflationary universe. *Contemp. Phys.* **2006**, *47*, 33–49. [CrossRef]
6. Berera, A. Developments in inflationary cosmology. *Pramana* **2009**, *72*, 169–182. [CrossRef]
7. Berera, A.; Gleiser, M.; Ramos, R.O. Strong Dissipative Behavior in Quantum Field Theory. *Phys. Rev. D* **1998**, *58*, 123508. [CrossRef]
8. Hall, L.; Moss, I.G.; Berera, A. Constraining warm inflation with the cosmic microwave background. *Phys. Lett. B* **2004**, *589*, 1–6. [CrossRef]
9. Hall, L.; Moss, I.G.; Berera, A. Scalar perturbation spectra from warm inflation. *Phys. Rev. D* **2004**, *69*, 083525. [CrossRef]
10. Berera, A.; Moss, I.G.; Ramos, R.O. Warm Inflation and its Microphysical Basis. *Rep. Prog. Phys.* **2009**, *72*, 026901. [CrossRef]
11. Bartrum, S.; Berera, A.; Rosa, J.G. Warming up for Planck. *J. Cosmol. Astropart. Phys.* **2013**, *2013*, 025. [CrossRef]
12. Bastero-Gil, M.; Berera, A. Warm Inflation model building. *Int. J. Mod. Phys.* **2009**, *A24*, 2207–2240. [CrossRef]
13. Bastero-Gil, M.; Berera, A.; Ramos, R.O.; Rosa, J.G. General dissipation coefficient in low-temperature warm inflation. *J. Cosmol. Astropart. Phys.* **2013**, *2013*, 016. [CrossRef]

14. Bastero-Gil, M.; Berera, A.; Kronberg, N. Exploring parameter space of warm-inflation models. *J. Cosmol. Astropart. Phys.* **2015**, *2015*, 046. [CrossRef]

15. Abbott, L.F.; Farhi, E.; Wise, M.B. Particle production in the new inflationary cosmology. *Phys. Lett. B* **1982**, *117*, 29–33. [CrossRef]

16. Albrecht, A.; Steinhardt, P.J.; Turner, M.S.; Wilczek, F. Reheating an Inflationary Universe. *Phys. Rev. Lett.* **1982**, *48*, 1437–1440. [CrossRef]

17. Morikawa, M.; Sasaki, M. Entropy Production in the Inflationary Universe. *Prog. Theor. Phys.* **1984**, *72*, 782–798. [CrossRef]

18. Hosoya, A.; Sakagami, M. Time development of Higgs field at finite temperature. *Phys. Rev. D* **1984**, *29*, 2228–2239. [CrossRef]

19. Moss, I.G. Primordial inflation with spontaneous symmetry breaking. *Phys. Lett. B* **1985**, *154*, 120–124. [CrossRef]

20. Lonsdale, S.R.; Moss, I.G. A superstring cosmological model. *Phys. Lett. B* **1987**, *189*, 12–16. [CrossRef]

21. Yokoyama, J.; Maeda, K. On the Dynamics of the Power Law Inflation Due to an Exponential Potential. *Phys. Lett. B* **1988**, *207*, 31–35. [CrossRef]

22. Liddle, A.R. Power Law Inflation with Exponential Potentials. *Phys. Lett. B* **1989**, *220*, 502–508. [CrossRef]

23. Del Campo, S. Warm Inflationary Universe Models. In *Aspects of Today's Cosmology*; Alfonso-Faus, A., Ed.; InTech: Rijeka, Croatia, 2011.

24. Panotopoulos, G.; Videla, N. Warm $(\lambda/4)\,\phi^4$ inflationary universe model in light of Planck 2015 results. 2015, arXiv:1510.0698.

25. Bellini, M. Warm inflation and classicality conditions. *Phys. Lett. B* **1998**, *428*, 31–36. [CrossRef]

26. Lee, W.; Fang, L.-Z. Mass density perturbations from ination with thermal dissipation. *Phys. Rev. D* **1999**, *59*, 083503. [CrossRef]

27. Maia, J.M.F.; Lima, J.A.S. Extended warm inflation. *Phys. Rev. D* **1999**, *60*, 101301. [CrossRef]

28. Herrera, R.; del Campo, S.; Campuzano, C. Tachyon warm inflationary universe models. *J. Cosmol. Astropart. Phys.* **2006**, *2006*, 9. [CrossRef]

29. Del Campo, S.; Herrera, R.; Pavón, D. Cosmological perturbations in warm inflationary models with viscous pressure. *Phys. Rev. D* **2007**, *75*, 083518. [CrossRef]

30. Hall, L.M.H.; Peiris, H.V. Cosmological Constraints on Dissipative Models of Ination. *J. Cosmol. Astropart. Phys.* **2008**, *2008*, 027. [CrossRef]

31. Moss, I.G.; Xiong, C. On the consistency of warm inflation. 2008, arXiv:0808.0261. [CrossRef]

32. Deshamukhya, A.; Panda, S. Warm tachyonic inflation in warped background. *Int. J. Mod. Phys. D* **2009**, *18*, 2093–2106. [CrossRef]

33. Nozari, K.; Fazlpour, B. Non-Minimal Warm Ination and Perturbations on the Warped DGP Brane with Modified Induced Gravity. *Gen. Relativ. Gravit.* **2011**, *43*, 207–234. [CrossRef]

34. Cai, Y.F.; Dent, J.B.; Easson, D.A. Warm DBI Inflation. *Phys. Rev. D* **2011**, *83*, 101301. [CrossRef]

35. Cerezo, R.; Rosa, J.G. Warm inflection. *High Energy Phys.* **2013**, *2013*, 24. [CrossRef]

36. Sharif, M.; Saleem, R. Warm Anisotropic Inflationary Universe Model. *Eur. Phys. J. C* **2014**, *74*, 2738. [CrossRef]

37. Sharif, M.; Saleem, R. Warm anisotropic inflation with bulk viscous pressure in intermediate era. *Astropart. Phys.* **2015**, *62*, 241–248. [CrossRef]

38. Setare, M.R.; Kamali, V. Warm-intermediate inflationary model with viscous pressure in high dissipative regime. *Gen. Relativ. Gravit.* **2014**, *46*, 1698. [CrossRef]

39. Chimento, L.P.; Jacubi, A.S.; Zuccala, N.A.; Pavon, D. Synergistic warm ination. *Phys. Rev. D* **2002**, *65*, 083510. [CrossRef]

40. Kinney, W.H.; Kolb, E.W.; Melchiorri, A.; Riotto, A. Inflation model constraints from the Wilkinson Microwave Anisotropy Probe three-year data. *Phys. Rev. D* **2006**, *74*, 023502. [CrossRef]

41. Mishra, H.; Mohanty, S.; Nautiyal, A. Warm natural inflation. *Phys. Lett. B* **2012**, *710*, 245–250. [CrossRef]

42. Sánchez, J.C.; Bastero-Gill, B.M.; Berera, A.; Dimoupoulos, K. Warm hilltop inflation. *Phys. Rev. D* **2008**, *77*, 123527. [CrossRef]

43. Setare, M.R.; Sepehri, A.; Kamali, V. Constructing warm inflationary model in brane-antibrane system. *Phys. Lett. B* **2014**, *735*, 84–89. [CrossRef]

44. Kinney, W.H. Cosmology, inflation, and the physics of nothing. In *Techniques and Concepts of High-Energy Physics XII*; NATO Science Series; Springer: Berlin, Germany, 2003; Volume 123, pp. 189–243.

45. Ade, P.A.R.; Aghanim, N.; Arnaud, M.; Arroja, F.; Ashdown, M.; Aumont, J.; Baccigalupi, C.; Ballardini, M.; Banday, A.J.; Barreiro, R.B.; et al. Planck 2015 results. XIII. Constraints on inflation. 2015, arXiv:1502.01589.

46. Baumann, D. TASI Lectures on Inflation. 2012, arXiv:0907.5424.

47. Visinelli, L. Natural Warm Inflation. 2011, arXiv:1107.3523. [CrossRef]

48. Visinelli, L. Observational constraints on Monomial Warm Inflation. *JCAP07* **2016**, 054. [CrossRef]

49. Kamali, V.; Setare, M.R. Warm-viscous inflation model on the brane in light of Planck data. *Class. Quantum Gravity* **2015**, *32*, 235005. [CrossRef]

50. Lyth, D.H. Particle physics models of inflation. *Lect. Notes Phys.* **2008**, *738*, 81–118.

51. Dine, M.; Pack, L. Studies in small field inflation. *J. Cosmol. Astropart. Phys.* **2012**, *2012*, 033. [CrossRef]

universe

MDPI

Article

A Solution of the Mitra Paradox

Øyvind Grøn

Oslo and Akershus University College of Applied Sciences, Faculty of Technology, Art and Sciences, PB 4 St. Olavs. Pl., NO-0130 Oslo, Norway; oyvind.gron@hioa.no

Academic Editors: Lorenzo Iorio and Elias C. Vagenas
Received: 7 September 2016; Accepted: 1 November 2016; Published: 4 November 2016

Abstract: The "Mitra paradox" refers to the fact that while the de Sitter spacetime appears non-static in a freely falling reference frame, it looks static with reference to a fixed reference frame. The coordinate-independent nature of the paradox may be gauged from the fact that the relevant expansion scalar, $\theta = \sqrt{3\Lambda}$, is finite if $\Lambda > 0$. The trivial resolution of the paradox would obviously be to set $\Lambda = 0$. However, here it is assumed that $\Lambda > 0$, and the paradox is resolved by invoking the concept of "expansion of space". This is a reference-dependent concept, and it is pointed out that the solution of the Mitra paradox is obtained by taking into account the properties of the reference frame in which the coordinates are co-moving.

Keywords: general theory of relativity; exact solutions; spherical symmetry; physical interpretation

PACS: 04.20-q; 04.20.Cv; 04.20.Jb

1. Introduction

Abbas Mitra [1] has recently discussed an interesting problem concerning the physical interpretation of the de Sitter spacetime. He has pointed out that seemingly there is a contradiction between the static form of the de Sitter metric and the non-static, expanding representation of this spacetime as a Friedmann-Lemaitre-Robertson-Walker universe model of the steady state type. This will here be called the Mitra paradox.

Mitra writes that there is a physical or at least interpretational self-contradiction between the original static interpretation of the de Sitter metric and the present day non-static de Sitter view. Furthermore he writes that a metric represents a physical point of view, and due to the principle of covariance, the essential physical picture should not depend on the choice of coordinates. He also points out that that there has not been any attempt for physical resolution to reconcile the static and non-static versions of for example the de Sitter metric. In this paper I will provide such a reconciliation.

Since there is a similar conflict between the Minkowski spacetime and the Milne universe model [2,3], I will start the present discussion by considering the corresponding Mitra paradox for these metrics. Then the de Sitter spacetime will be considered and finally the Schwarzschild and the Schwarschild-de Sitter spacetime.

Write Schutz [4] writes that we define a *static* spacetime to be one in which we can find a time coordinate t with two properties: (i) all metric components are independent of t; and (ii) the geometry is unchanged by time reversal, $t \rightarrow -t$. A spacetime with the property (i) but not (ii) is said to be *stationary*. This definition can be formulated in terms of Killing vectors. A static spacetime is a spacetime, which has a time-like Killing vector field that is hypersurface orthogonal. This is a coordinate-independent characterization of a static spacetime. If any coordinate system exists in which none of the metric components depend upon time, there exists a time-like Killing vector in the spacetime. In this case the actual physical geometry of the spacetime is unchanging with time. Although the geometry of a stationary spacetime does not change in time, it can rotate. If the spacetime does not permit a time coordinate so that all the metric components are independent of *t* it is non-static and non-stationary. In this case the spacetime has no time-like Killing vector field.

The somewhat surprising fact is that even *a static spacetime can have a time-dependent metric*. The Mitra paradox is concerned with finding the proper physical meaning of this fact.

2. The Connection between the Global Minkowski Metric and the Milne Universe

It is well known that the Minkowski and Milne metrics are connected by a change of reference frame [5–8]. This will here be utilized to shed some light upon the Mitra paradox. We consider Minkowski spacetime with spherical coordinates (R, θ, φ) and a time coordinate T so that the line element takes the form

$$ds^2 = -dT^2 + dR^2 + R^2 d\Omega^2 \quad , \quad d\Omega^2 = d\theta^2 + \sin^2\theta \, d\varphi^2 \tag{1}$$

where we have used units so that $c = 1$. Then we introduce new coordinates (t, r) by the transformation

$$t = \sqrt{T^2 - R^2} \quad , \quad r = \frac{t_0 R}{\sqrt{T^2 - R^2}} \tag{2}$$

The inverse transformation is

$$T = t\sqrt{1 + \frac{r^2}{t_0^2}} \quad , \quad R = \frac{r\,t}{t_0} \tag{3}$$

where t_0 is the present age of the universe. It follows from this transformation that

$$R = \frac{r}{\sqrt{1 + r^2/t_0^2}} \frac{T}{t_0} \quad , \quad T^2 - R^2 = t^2 \tag{4}$$

We see that the world-lines of the reference particles defining the reference frame in which (t, r) are co-moving, i.e., $r = $ constant, are straight lines, and the simultaneity curves $t = $ constant are hyperbolae.

It is seen that while the coordinates (T, R) are co-moving in a static reference frame, the coordinates (t, r) are co-moving in an expanding reference frame. A reference particle with $r = $ constant in the expanding frame has a coordinate velocity

$$v_R = \frac{r/t_0}{\sqrt{1 + r^2/t_0^2}} = \frac{R}{T} \tag{5}$$

in the rigid frame. In the expanding frame the line element of the Minkowski spacetime takes the form

$$ds^2 = -dt^2 + \left(\frac{t}{t_0}\right)^2 \left(\frac{dr^2}{1 + r^2/t_0^2} + r^2 d\Omega^2\right) \tag{6}$$

This is the line element of an empty, expanding universe model with negative spatial curvature—the Milne universe.

The Minkowski coordinates (T, R) are the co-moving coordinates of a rigid inertial reference frame of an arbitrarily chosen reference particle P in the expanding cloud of particles defining the Milne universe model. The time T is the *private time* of P. The time t is measured on clocks following all of the reference particles. As seen from the first of Equation (4) the space $T = $ constant has a finite extent, $R_{\max} = \lim_{r \to \infty} R = T$. This space is the *private space* of an observer following the particle P. The space $t = $ constant is represented by a hyperbola given in the second of Equation (4) as shown in the Minkowski diagram of the P observer. It is defined by simultaneity of the clocks carried by all the reference particles, and is called the *public space* or simply the space of the universe model. It has infinite extension in spite of the fact that the Big Bang has the character of a point event in the Milne universe model.

In the inertial and rigid Minkowski coordinate system the velocity of a reference particle with co-moving coordinates is less than *c* for all values of. However, in the expanding cosmic frame it is different. Here the velocity of the reference particles as defined by an observer at the origin is given by Hubble's law. Hence the reference particles have superluminal velocity at sufficiently great distances from the observer. According to special relativistic kinematics, superluminal velocity is problematic because the particles cannot move through space with a velocity greater than *c*. However, according to the general relativistic interpretation, the reference particles define the public space of the universe model, and there is no limit to how fast space itself can expand.

The metric (1) is static and the metric (6) not. The Mitra paradox is concerned with a reconciliation of these properties of two metrics that are connected by a coordinate transformation, and hence that represent one and the same spacetime.

The Mitra paradox makes it clear that one cannot define a static spacetime as a spacetime where the metric is independent of time. The metric is coordinate-dependent, and may be independent of time in one coordinate system, but dependent on it in another, while the static property of a spacetime is invariant.

3. Proposal for a Solution of the Mitra Paradox

An important point when we try to solve the Mitra paradox is to distinguish between coordinate-dependent quantities and coordinate-independent physical quantities. The term *metric* is usually taken to mean the functions that appear in the line element multiplied by the coordinate differentials. Hence the metric is understood to mean the *components* of the metric tensor. This means that *the metric is a coordinate-dependent quantity*. It is natural, therefore, that at least in some cases, one and the same spacetime can be represented by both a static and a non-static metric.

Another important distinction is the difference between a coordinate system and a reference frame. In four-dimensional spacetime, a *coordinate system* provides a region of spacetime with a continuum of 4-tuples so that each event in spacetime is marked with a 4-tuple, different events with different 4-tuples. A coordinate system is a mathematical quantity. A *reference frame* is a continuum of world-lines representing reference particles with specified motions. This is a physical quantity. *Co-moving coordinates* in a reference frame are coordinates so that the reference particles of the frame have constant spatial coordinates.

The Mitra paradox is not only concerned with the metric, but also with 3-space. Hence it is important to distinguish between a coordinate 3-space and a coordinate-independent physical 3-space. Here we meet an important difficulty of the Mitra paradox. The 3-space has two very different qualities. On the one hand, it is a set of simultaneous events measured by clocks at rest in the chosen reference frame. Again this is coordinate-dependent or better, reference-dependent, due to the relativity of simultaneity.

It follows from Friedmann's 1. equation for a flat universe, $H = \sqrt{8\pi G\rho/3}$, that if the Minkowski spacetime is perceived as the limit of the Friedmann-Lemaitre-Robertson-Walker universe model with empty space, the Hubble parameter vanishes, and the 3-space is static. In a similar way, the De Sitter spacetime is then the limit of empty space with a cosmological constant, Λ, having a positive, constant Hubble parameter, $H = \sqrt{\Lambda/3}$, and the 3-space of the de Sitter spacetime is non-static and expanding. However, both of these spacetimes have maximal symmetry and have a time-like hyper surface orthogonal Killing vector, meaning that these spacetimes are static. Hence it is important to note the difference between a static 3-space and a static spacetime. The Mitra paradox is concerned with 3-space.

We considered the globally flat spacetime above. In the standard coordinates co-moving with a static reference frame with time-independent distances between the reference points, the 3-space is static. But as described in terms of coordinates co-moving in an expanding reference frame, the 3-space is not static, but expands. The flat, static spacetime then looks like an expanding universe—the Milne universe model.

This seems strange. So far we have defined 3-space as a set of simultaneous events. There is no motion involved in this definition. Hence the definition of a 3-space should be supplied by a second quality permitting space to expand. We can then supply the definition of a 3-space: A 3-space is made up of a set of reference particles at a given point of time. The 3-space of a reference frame is defined by identifying the reference particles of the 3-space with the reference particles of the frame. The most simple mathematical description of the 3-space is obtained by using coordinates co-moving with the reference frame of the 3-space. *The 3-space is said to be stationary if the physical distances between the reference points does not change.* In this case the reference frame can be said to be *rigid*. If these physical distances change, the 3-space is non-stationary. If the rotation of the velocity field of the reference particles of a stationary 3-space vanishes, the 3-space is said to be *static*.

It is then clear that whether a 3-space appears static or non-static depends upon the reference frame it is associated with. This means that *the static or non-static character of a 3-space is a coordinate-dependent quality of the spacetime.* It will be made clear below that this is an important ingredient in the solution of the Mitra paradox.

One may wonder whether this means that physical 3-space does not exist. It is then important to make one more distinction: between *physical* and *invariant*. A quantity is said to be invariant if it has the same value in every reference frame or coordinate system. A physical property need not be invariant. For example a 3-velocity of a particle is a physical property, but it is not invariant. It can even be transformed away by going into the rest frame of the particle.

We should not talk about *the* 3-space of Minkowski spacetime. We should talk about *a* 3-space. Minkowki spacetime can have a static 3-space and equally well a non-static 3-space. Although the property of a 3-space of being static is a physical property, its static character is not invariant. This is the proposed solution of the Mitra paradox, which will be further worked out below.

4. Static and Expanding 3-Space

Let us first consider a static spacetime as described in the co-moving coordinates of a rigid reference frame, RF, so that the metric is static and has the form

$$ds^2 = -f(R)\,dt^2 + \frac{dR^2}{f(R)} + R^2 d\Omega^2 \qquad (7)$$

Here the radial coordinate is chosen so that the invariant area of a spherical surface with radius R is equal to $4\pi R^2$. This radial coordinate is sometimes called the curvature radius or alternatively the area coordinate. The 3-space of simultaneous events as measured by clocks carried by the reference particles of RF, is static. This is the preferred 3-space of spacetimes with a localized mass distribution, such as the Schwarzschild spacetime.

Assume that there exists a surface with coordinate radius $R = R_0$ so that a particle permanently at rest on this surface has vanishing 4-acceleration, i.e., a free particle instantaneously at rest at this surface will remain at rest on the surface. The radial component of the 4-acceleration of a particle at rest in the coordinate system is according to the geodesic equation given in terms of certain Christoffel symbols and the time component of the 4-velocity as

$$a^R = \Gamma^R_{TT}\left(u^T\right)^2 = (1/2)\,f\prime(R) \qquad (8)$$

Hence the radius R_0 is given by $f\prime(R_0) = 0$. Let us now consider the 3-space of simultaneous events as shown by clocks carried by *free particles* starting their movements from a state of rest at $R = R_0$. These particles make up a locally inertial frame, IF. Hence this 3-space may be called an *inertial 3-space*. This is the preferred 3-space of the relativistic universe models. In [9] it was shown that the 3-velocity of the inertial 3-space as given with respect to the orthonormal basis of an observer at rest in RF is

$$v_{3-space} = \left(\frac{d\hat{x}^R}{d\tau}\right)_{3-space} = \pm\sqrt{1 - \frac{f(R)}{f(R_0)}} \tag{9}$$

In the case of the Minkowski metric (1) this gives $v_{3-space} = 0$, and the inertial 3-space is then at rest in an arbitrary rigid frame in flat spacetime.

Let us now describe the 3-space with reference to an expanding reference frame in which the metric is of the Friedmann-Lemaitre-Robertson-Walker type. Then the reference frame consists of a set of freely moving particles expanding together with the cosmic fluid. Let t be the proper time of clocks co-moving with the reference particles of this frame, and r a co-moving radial coordinate following the cosmic fluid. Then the line element has the form

$$ds^2 = -dt^2 + [a(t)]^2\left(\frac{dr^2}{1 - kr^2} + r^2d\Omega^2\right) \tag{10}$$

where k is a constant, which can have the values $k = \{-1/t_0^2, 0, 1/t_0^2\}$, and $a(t)$ is the scale factor. If it is normalized to $a(t_0) = 1$ at the present time t_0, it represents the ratio of the cosmic distances between the reference particles at an arbitrary point of time and their present distances. In this coordinate system the 3-space has an expansion, $\theta = 3\dot{a}/a$. Hence the 3-space is not static in this frame if $\dot{a} \neq 0$.

So the 3-space of the Minkowski spacetime may be pictured as either static or non-static depending upon the reference frame that is used. This freedom of point of view is due to the Lorentz invariance of this solution of Einstein's field equations. It is not typical of the solutions of the field equations in general. But there are two other solutions that share this property of Lorentz invariance with the Minkowski spacetime, and those are the de Sitter and anti-de Sitter spacetimes. Let us consider the de Sitter spacetime.

5. The de Sitter Spacetime

We consider the de Sitter spacetime with spherical coordinates (R, θ, φ) and a time coordinate T, so that the line element takes the form

$$ds^2 = -\left(1 - H^2R^2\right)dT^2 + \frac{dR^2}{1 - H^2R^2} + R^2d\Omega^2 \tag{11}$$

for $R < R_H = 1/H$, where $H = \sqrt{\Lambda/3}$, and the cosmological constant $\Lambda = 8\pi G\rho_\Lambda$ represents the constant density ρ_Λ of a Lorentz Invariant Vacuum Energy, LIVE, with stress $p_\Lambda = -\rho_\Lambda$. It should be noted that the coordinate clocks showing T go with a position-independent rate equal to that of a standard clock at $R = 0$.

We introduce new coordinates (t, r) by the transformation

$$t = T + \frac{1}{2H}\ln\sqrt{1 - H^2R^2} \quad , \quad r = \frac{R}{\sqrt{1 - H^2R^2}}e^{-HT} \tag{12}$$

The inverse transformation is

$$T = t - (1/2H)\ln\sqrt{1 - H^2r^2e^{2Ht}} \quad , \quad R = re^{Ht} \tag{13}$$

Differentiating the second of Equation (12) with respect to T with constant r gives the coordinate velocity of a reference particle in the (t, r)-system with respect to the (T, R)-system

$$v_R = He^{Ht} = \sqrt{1 - H^2R^2}\,e^{HT} \tag{14}$$

Hence the (t, r) coordinates are co-moving in an expanding reference frame relative to the rigid (T, R)-system. In the coordinates (t, r) the line element has the form

$$ds^2 = -dt^2 + e^{2Ht}\left(dr^2 + r^2 d\Omega^2\right) \tag{15}$$

The (t, r)-coordinates are co-moving with free particles, as is the case for all the FLRW-universe models. Equation (14) shows that the free particles in this spacetime move outwards with an accelerated motion.

It follows from Equation (8) that for the metric (11) the 4-acceleration of a particle fixed in the rigid reference frame, is

$$a^R = H^2 R \tag{16}$$

Hence a free particle at rest in the rigid reference frame must be at the position $R_0 = 0$. The 3-space made up of simultaneous events as measured by clocks carried by these particles is the inertial 3-space. In this case the velocity of inertial 3-space as given by Equation (9) is

$$v_{3-space} = HR \tag{17}$$

Even if the metric (11) is static, the velocity of the inertial 3-space obeys the Hubble lav. There is an expansion equal to $3H$. Hence the 3-space expands in accordance with the Hubble law. This is often called the *Hubble flow*. The inertial 3-space flows with the velocity of light at $R = R_H = 1/H$ and with superluminal velocity for $R > R_H$. There is a horizon at $R = R_H$.

The coordinate time t is shown on standard clocks following the freely falling reference particles of the 3-space, and the coordinate r is co-moving with those particles. That is the reason for the time dependence of the metric (15) in this coordinate system. Hence there is no contradiction between the static form of the line element (11) and the non-static form (15). The first form reflects the rigidity of the reference frame in which the coordinates T, R are co-moving, and the second reflects the expansion of the reference frame in which the coordinates t, r are co-moving. This solves the Mitra paradox.

However, there is something strange about the metric (11). There is a coordinate singularity at $R_H = 1/H$. Note that $\lim_{R \to 1/H} r = \infty$, and that $v_{3-space}(R_H) = 1$. Hence the 3-space is flowing with the velocity of light at this surface.

Consider radially moving light in the metric (11). The coordinate velocity of light is the same in the inwards and outwards direction, and is equal to

$$(dR/dT)_L = \pm\left(1 - H^2 R^2\right) \tag{18}$$

which vanishes at $R = R_H$. Hence in this coordinate system the light cone collapses at $R = R_H$.

In order to have open light cones at $R = R_H$ one may introduce a new time coordinate. There are several related such coordinates, and it may be useful to compare the description of the light cones in three of them. All of them are given by an *internal coordinate transformation* in the sense that the coordinate clocks are at rest in the same reference frame as those showing T.

We first consider a light cone coordinate, $\overline{T}$, used by Spradlin, Strominger and Volovich [10], given by the coordinate transformation

$$\overline{T} = T - \frac{1}{2H}\ln\frac{1 + HR}{1 - HR} \tag{19}$$

By using L'Hopital's rule we get $\lim_{H \to 0} \overline{T} = T - R$ showing that $\overline{T}$ reduces to an ordinary light cone coordinate in Minkowski spacetime. Differentiating we have

$$d\overline{T} = dT - \frac{dR}{1 - H^2 R^2} \tag{20}$$

Hence the coordinate clocks showing $\overline{T}$ have another simultaneity than those showing T. With the new time coordinate the line element of the de Sitter spacetime takes the form

$$ds^2 = -\left(1 - H^2 R^2\right) d\overline{T}^2 - 2d\overline{T} \, dR + R^2 d\Omega^2 \tag{21}$$

For light moving radially we have $ds^2 = d\Omega^2 = 0$ and hence,

$$2d\overline{T} \, dR = -\left(1 - H^2 R^2\right) d\overline{T}^2 \tag{22}$$

For light moving outwards $dR > 0$, which is not permitted by Equation (22). However, for light moving outwards, Equations (16) and (18) give $d\overline{T} = 0$, which *is* permitted. Hence $\overline{T}$ is a light cone coordinate for outgoing light. For light moving inwards, Equation (22) gives the coordinate velocity

$$\left(\frac{dR}{d\overline{T}}\right)_{L-} = -\frac{1}{2}\left(1 - H^2 R^2\right) \tag{23}$$

which vanishes at $R = R_H$. The "inwards directed" velocity of light changes sign at $R = R_H$ and becomes outwards directed for $R > R_H$.

Another time coordinate T_P called the Painlevé-de Sitter coordinate, was used by Parikh [11] and is given by the transformation

$$T_P = T + (1/2\,H) \ln\left(1 - H^2 R^2\right) \tag{24}$$

Comparing with equation the first of the Equation (12) we see that $T_P = t$. Hence the Paainlevé-de Sitter time is the same as the cosmic time, which is measured by standard clocks following freely moving particles. Differentiating gives

$$dT_P = dT - \frac{HR}{1 - H^2 R^2} dR \tag{25}$$

Inserting this into Equation (11) we find that the line element takes the form

$$ds^2 = -\left(1 - H^2 R^2\right) dT_P^2 - 2HR \, dT_P \, dR + dR^2 + R^2 d\Omega^2 \tag{26}$$

The coordinate velocity of outgoing and ingoing light is

$$(dR/dT_P)_+ = 1 + HR \quad , \quad (dR/dT_P)_- = -(1 - HR) \tag{27}$$

At the horizon $(dR/dT_P)_+ = 2, (dR/dT_P)_- = 0$. The velocity of the ingoing light changes sign at the horizon, and moves outwards outside the horizon. This is an effect of the repulsive gravity due to the LIVE, which fills this spacetime and causes an accelerated expansion of the inertial 3-space. Note that in this context "accelerated" means non-vanishing 3-acceleration. The 4-acceleration of the reference particles of the inertial 3-space vanishes, since the particles are freely falling.

Finally a time coordinate corresponding to the ingoing Eddington-Finkelstein coordinate in the Schwarzshild spacetime is defined by the condition that the coordinate velocity of outgoing light is equal to 1. This was used by Braeck and Grøn [9] and is given by

$$\widetilde{T} = T + \frac{1}{2H} \ln \frac{1 - HR}{1 + HR} + R \tag{28}$$

Differentiation gives

$$d\widetilde{T} = dT - \frac{H^2 R^2}{1 - H^2 R^2} dR \tag{29}$$

With this time coordinate the line element takes the form

$$ds^2 = -\left(1 - H^2 R^2\right) d\tilde{T}^2 - 2H^2 R^2\, d\tilde{T}\, dR + \left(1 + H^2 R^2\right) dR^2 + R^2 d\Omega^2 \tag{30}$$

The coordinate velocity of outgoing and ingoing light is

$$\left(dR/d\tilde{T}\right)_+ = 1, \quad \left(dR/d\tilde{T}\right)_- = -\frac{1 - H^2 R^2}{1 + H^2 R^2} \tag{31}$$

At the horizon $\left(dR/d\tilde{T}\right)_+ = 1, \left(dR/d\tilde{T}\right)_- = 0$. Again the velocity of the ingoing light changes sign at the horizon, i.e., the light cones turn outwards, implying that nothing can enter the horizon from the outside region.

All of the line elements (21), (26) and (30) are stationary, although they are not static. The stationary character shows that the coordinate R is co-moving in a rigid reference frame. The reason that they are not static is that the coordinate clocks are not Einstein synchronized. Their simultaneity is not that of Einstein synchronized clocks at rest in the rigid reference frame.

The de Sitter spacetime is static since there exists a coordinate system where the metric is static and the time-like basis vector is a Killing vector.

Nevertheless this spacetime is filled with vacuum energy that expands. This energy causes repulsive gravity, which acts back upon the energy itself and makes the expansion accelerate. It should be noted that in a homogeneous universe there is no pressure gradient, so the accelerated expansion is not a pressure effect, but a gravitational effect. The negative pressure, $p = -\rho c^2$, contributes to the effective gravitational mass density, $\rho_{grav} = \rho + 3p/c^2$, making it negative, which means that gravity is repulsive [12]. Hence there is energy with accelerated expansion in this spacetime. Is it then reasonable to say that it is static?

Compare 3-space with a river, and consider the river now and an hour later, assuming that there is the same amount of water in the river at these points of time. In this situation, the river has not changed. The river is static. But the water is not static. It flows. Similarly spacetime is static, but 3-space is expanding.

In spacetime the river corresponds to the geometry of space at a certain position, and the flowing water corresponds to the flowing reference particles constituting the 3-space. In the de Sitter spacetime the geometry of space is unchanged at a fixed position in a rigid reference frame. Hence it is a static spacetime; but the 3-space is flowing. It is not static. The metric in a coordinate system co-moving with the reference particles of the 3-space is not static, but depends upon time as in the metric of Equation (15).

6. The Schwarzschild Spacetime

Outside the Schwarzschild radius the Schwarzschild spacetime has a time-like Killing vector field that is hypersurface orthogonal. Hence it is static and there exists a coordinate system in which the metric is independent of time and the line element has no product terms where a spatial differential is multiplied by a time differential. One such coordinate system is the standard so-called curvature coordinates where the invariant area of a surface with coordinate radius R around the origin is $4\pi R^2$. In this coordinate system the line element takes the form

$$ds^2 = -\left(1 - R_S/R\right) dT^2 + \frac{dR^2}{1 - R_S/R} + R^2 d\Omega^2 \tag{32}$$

where $R_S = 2GM$ is the Schwarzschild radius of the central mass. Inside the Schwarzschild radius the Killing vector field is spacelike, and in this region the Schwarzschild spacetime is not static.

Consider now an observer moving with the inertial 3-space in this spacetime, i.e., he is falling freely from a state of rest infinitely far away from the central mass. The co-moving radius of this observer is r and he carries with him a standard clock showing t. The new coordinates are given by transformation [13–15]

$$T = t + R_S \ln \frac{\sqrt{R/R_S} + 1}{\sqrt{R/R_S} - 1} - 2\sqrt{R_S R} \quad , \quad R^{3/2} = -(3/2)\sqrt{R_S}(t + r) \tag{33}$$

In terms of the co-moving coordinates of the inertial 3-space the line element of the Schwarzschild spacetime takes the form

$$ds^2 = -dt^2 + \left(\frac{2}{3}\frac{R_S}{t + r}\right)^{2/3} dr^2 + \left(\frac{3}{2}\sqrt{R_S}(t + r)\right)^{4/3} d\Omega^2 \tag{34}$$

An observer with $r = r_0$ has initially a large negative value of t, which increases towards $-r_0 - (2/3)R_S$ as the observer passes the Schwarzschild horizon. Hence the line element (34) corresponds to that of an inhomogeneous universe with an anisotropic and position-dependent scale factor, and the inertial 3-space expands in the radial direction and contracts in the tangential direction. In these coordinates the metric of the Schwarzschild spacetime is not static. These geometrical changes with time of the inertial 3-space are due to tidal forces becoming stronger at the position of the reference particles co-moving with the inertial 3-space, as they approach the central mass.

The coordinate transformation is well defined only for $R > R_S$. This is due to the rigid character of the reference frame in which the coordinates (T, R) are co-moving, which is physically possible only outside the Schwarzschild horizon. However the line element (34) has no coordinate singularity at the Schwarzschild horizon. The coordinates t and r are well defined in all of spacetime outside the central singularity, also inside the horizon, and the line element gives a singularity-free description of the Schwarzschild spacetime in the whole of this region. This illustration shows that a static spacetime, which is usually expressed so that 3-space is static, may also be expressed so that the 3-space is non-static.

7. The Schwarzschild-de Sitter Spacetime

This is a static spacetime in which the line element may be written

$$ds^2 = -\left(1 - \frac{R_S}{R} - H^2 R^2\right) dT^2 + \frac{dR^2}{1 - \frac{R_S}{R} - H^2 R^2} + R^2 d\Omega^2 \tag{35}$$

In this spacetime the inertial 3-space has a rather interesting behavior. At the surface with

$$R_0 = \left(R_S R_H^2 / 2\right)^{1/3} \tag{36}$$

the 4-acceleration of a particle permanently at rest vanishes [9]. Hence the reference particles of the inertial 3-space are at rest at this surface. But the inertial 3-space diverges at this surface. It expands outside this surface and contracts inside it.

8. Static Form of the FLRW Metric

Mitra has recently deduced an interesting form of the Friedmann-Lemaitre-Robertson-Walker metric in curvature coordinates [16] and used this to investigate when an expanding universe can look static [17]. The FLRW-metric is first written in the usual form (10). Mitra then found that the metric can be written in curvature coordinates as follows

$$ds^2 = -\left(\frac{\partial t}{\partial T}\right)^2 \frac{1 - k(R/a)^2 - (\dot{a}/a)^2 R^2}{1 - k(R/a)^2} dT^2 + \frac{dR^2}{1 - k(R/a)^2 - (\dot{a}/a)^2 R^2} + R^2 d\Omega^2 \tag{37}$$

where

$$R = r\,a\,(t) \tag{38}$$

He then showed that essentially only the Milne universe and the de Sitter and anti-de Sitter universe models can be written in static form using curvature coordinates.

As an illustrating example we will here consider only the first model studied by Mitra. It is the de Sitter universe model with negative spatial curvature, $k = -1$ and $\Lambda > 0$. For this model the solution of the Friedmann equations gives the scale factor

$$a\,(t) = (1/H)\sinh\,(H\,t) \tag{39}$$

where $H = \sqrt{\Lambda/3}$. This universe model is filled by vacuum energy with constant density and stress given by

$$p_\Lambda = -\rho_\Lambda = -\Lambda/8\pi G \tag{40}$$

Inserting Equations (38) and (39) into Equation (37), the line element takes the form

$$ds^2 = -\left(\frac{\partial t}{\partial T}\right)^2 \frac{1 - H^2 R^2}{1 + r^2} dT^2 + \frac{dR^2}{1 - H^2 R^2} + R^2 d\Omega^2 \tag{41}$$

Comparing with Equation (11) we obtain

$$\frac{\partial t}{\partial T} = \sqrt{1 + r^2} \tag{42}$$

Mitra has shown that Equations (38), (39) and (42) lead to the transformation

$$HR = r\sinh\,(Ht), \quad \tanh\,(HT) = \sqrt{1 + r^2}\tanh\,(Ht) \tag{43}$$

The inverse transformation may be written

$$\sinh\,(H\,t) = \sqrt{\sinh^2\,(HT) - H^2 R^2 \cosh^2\,(HT)}, \quad r = \frac{HR}{\sqrt{\sinh^2\,(HT) - H^2 R^2 \cosh^2\,(HT)}} \tag{44}$$

Differentiating the first of these equations partially with respect to T and using the second equation, we get

$$\frac{\partial t}{\partial T} = \sqrt{\frac{1 - H^2 R^2}{\tanh^2\,(HT) - H^2 R^2}}\,\tanh\,(HT) = \sqrt{1 + r^2} \tag{45}$$

in agreement with Equation (42). Furthermore, by taking the differentials of the transformation (43) and inserting the expressions into Equation (11) one finds that (43) transforms the static metric (11) to the line element of the expanding de Sitter universe model with a negatively curved 3-space,

$$ds^2 = -dt^2 + \sinh^2\,(Ht)\left(\frac{dr^2}{1 + r^2} + r^2 d\Omega^2\right) \tag{46}$$

Equation (38) shows that the value of R increases with time for a fixed value of r. Again we see that the reconciliation of the static and non-static forms of the line element for one and the same spacetime is in recognizing the motion of the reference frames in which the radial coordinates are co-moving. The radial coordinate of the time-dependent metric is co-moving with an expanding reference frame, and the radial coordinate of the static metric is co-moving with a rigid reference frame. This is the solution of the Mitra paradox as applied to the present spacetime.

9. Energy Conservation

Writing the line element of the de Sitter spacetime in terms of coordinates co-moving with free particles, i.e., in the form (15), Mitra deduced that the vacuum energy inside a radius r is

$$U = (\Lambda/6\,G)\, r^3 e^{3Ht} \tag{47}$$

Mitra concluded: "Thus the total energy of the de Sitter model increases in an exponential manner. Such a bad violation of the "Principle of Conservation of Energy" in the co-moving frame is in sharp contrast with the corresponding nice behavior in the Schwarzschild frame". (He uses units in which the gravitational constant is $G = 1$, but I have kept G in the formulae.)

The solution of this seeming paradox is as follows. The Friedmann equations lead to

$$dU + p_\Lambda dV = 0 \tag{48}$$

where $V = (4\pi/3)\,a^3$, $U = \rho_\Lambda V$, and a is the scale factor of Equation (37). Equation (49) is the 1. Law of thermodynamics for adiabatic expansion as applied to a co-moving region with radius $r = 1$ around an observer. It expresses the law of energy conservation. Heat is defined as transport of energy due to temperature difference. In a homogeneous universe there are no large scale temperature differences, and this is the reason that the universe expands adiabatically.

Using $\rho_\Lambda = \Lambda/8\pi G$ and $a = e^{Ht}$, we get $dU = \rho_\Lambda 4\pi a^2 da = (\Lambda H/2G)\, e^{3Ht} dt$, which is the same as we get by taking the differential of U in Equation (47). The volume work performed at the boundary of the region is $dW = p_\Lambda dV = -\rho_\Lambda 4\pi a^2 da$. Hence the energy conservation equation is obeyed in spite of the fact that the amount of vacuum energy is increasing inside the co-moving surface. The reason is that there is a negative work at the boundary, which transfers energy from the outside region to the inside region. Imagining that the region is extended so that the boundary is infinitely far from the observer, one may say that the density of the vacuum energy is kept constant in spite of the expansion by extracting energy from an infinitely far region. This shows that global energy conservation is indeed a problematic concept at least for a universe with infinitely great spatial extension.

10. Results and Discussion

Mitra has pointed out that there seems to be an interpretational self-contradiction between the static interpretation of the de Sitter metric and the non-static de Sitter universe model. This has here been called the Mitra paradox. He also writes that there has not been any attempt for a physical resolution to reconcile the static and non-static versions of for example the de Sitter metric.

Both the problem and the resolution are of a conceptual nature of great significance for a proper way of teaching the general theory of relativity. A theory is much more than some rules for calculating physical effects, making it possible to falsify the theory. The theory also provides us with concepts representing the foundations of our world picture. As said by Einstein: It is the theory that tells what we observe.

Hence it is extremely important to obtain a proper physical interpretation of the general theory of relativity, free of contradictions. This also means that interpretational problems such as that formulated by Mitra, should not be neglected. The present article has been an effort to give a constructive discussion of this problem—and to solve it. I have here provided a resolution by focusing upon the difference between 3-space and spacetime and pointing out the significance of the motion of the reference frames in which different coordinate systems are co-moving.

11. Conclusions

The Mitra paradox is concerned with the physical reconciliation of two metrics, where one is static and the other time-dependent, that are connected by a coordinate transformation, and hence that represent one and the same spacetime.

The Mitra paradox makes it clear that one cannot define a static spacetime as a spacetime where the metric is independent of time. The metric is coordinate-dependent, and may be independent of time in one coordinate system, but dependent on it in another, while the static property of a spacetime is invariant and characterized by the existence of a hypersurface orthogonal Killing vector.

The solution of the Mitra paradox lies in recognizing that the metric is not determined by the geometric properties of the spacetime. In general there are ten independent components of the metric tensor and only six independent field equations, leaving the freedom of choosing the coordinate system. By choosing coordinates in a given spacetime as co-moving with a rigid reference frame one obtains a time-independent metric—otherwise a time-dependent one.

These general properties of the solution to the Mitra paradox have been illustrated in the present paper by considering several cases, the first of which being flat spacetime. With coordinates co-moving in a rigid inertial frame one obtains the usual Minkowski metric, and with coordinates co-moving in an expanding reference frame one obtains the time-dependent Milne metric. Secondly, the de Sitter spacetime has been considered. Again, by using coordinates co-moving in a rigid frame one obtains a static metric and using coordinates co-moving with freely moving particles that make up a system that expands due to repulsive gravity in this spacetime, one obtains a time-dependent metric. Thirdly, we have discussed the Schwarzschild spacetime. Again the metric is static in a rigid frame. But using coordinates co-moving with a system of freely falling particles one obtains a time-dependent metric, still representing the Schwarzschild spacetime.

Finally, as shown by Mitra, and interpreted physically here, a similar result is obtained for just three different Friedmann-Lemaitre Robertson-Walker universe models, the Milne universe and the de Sitter and anti-de Sitter universe models. All these models are solutions of Einstein's equations for empty space, the first one without a cosmological constant and the two latter ones with a positive and negative cosmological constant, respectively.

The universe models with matter or radiation energy are solutions of the field equations with a time-dependent energy-momentum tensor. Hence Mitra's result implies that universe models with a time-dependent energy-momentum tensor cannot be represented globally by a line element with a time-independent metric. This should be formulated in a coordinate-independent way.

For a perfect fluid with energy-momentum tensor

$$T^{\mu\nu} = (\rho + p)\, u^\mu u^\nu + p g^{\mu\nu} \tag{49}$$

we may define the energy-momentum scalar

$$T^{\mu\nu} T_{\mu\nu} = \rho^2 + 3p^2 \tag{50}$$

Hence we may conclude by formulating Mitra's result for the FLRW-universe models in the following way: The line element cannot be written in a globally time-independent way for a universe model with a time-dependent energy-momentum scalar.

Acknowledgments: I would like to thank the referees for useful suggestions that contributed to improvements of the article.

Conflicts of Interest: The author declares no conflict of interest.

References

1.	Mitra, A. Interpretational conflicts between the static and non-static forms of the de Sitter metric. *Sci. Rep.* **2012**, *2*, 923. [CrossRef] [PubMed]
2.	Rindler, V. Finite foliations of open FRW universes and the point-like big bang. *Phys. Lett. A* **2000**, *276*, 52–58. [CrossRef]
3.	Grøn, Ø.; Elgarøy, Ø. Is space expanding in the Friedmann universe models. *Am. J. Phys.* **2007**, *75*, 151–157. [CrossRef]

4. Schutz, B.F. *A First Course in General Rkkelativity*; Cambridge University Press: Cambridge, UK, 1985.
5. Grøn, Ø.; Hervik, S. *Einstein's General Theory of Relativity*; Springer: New York, NY, USA, 2007.
6. Carroll, S.M. *Spacetime and Geometry*, 1st ed.; Addison-Wesley: Boston, MA, USA, 2004.
7. Mukhanov, V. *Physical Foundations of Cosmology*; Cambridge University Press: Cambridge, UK, 2005.
8. Misner, C.; Thorne, K.; Wheeler, J.A. *Gravitation*; W. H. Freeman and Company: New York, NY, USA, 1971.
9. Bræck, S.; Grøn, Ø. A river model of space. *Eur. Phys. J. Plus* **2013**, *128*, 24. [CrossRef]
10. Spradlin, M.; Strominger, A.; Volovich, A. Unity from Duality: Gravity, Gauge Theory and Strings. In *Les Houches—Ecole d'Ete de Physique Theorique*; Springer: Berlin/Heidelberg, Germany, 2002.
11. Parikh, M.K. New coordinates for de Sitter space and de Sitter radiation. *Phys. Lett. B* **2002**, *546*, 189–195. [CrossRef]
12. Grøn, Ø. Repulsive gravitation and electron models. *Phys. Rev. D* **1985**, *31*, 2129–2131. [CrossRef]
13. Landau, L.; Lifshitz, E.M. *The Classical Theory of Fields*; Reed Educational and Professional Publishing Ltd.: Oxford, UK, 2002.
14. Lightman, A.P.; Press, W.H.; Price, R.H.; Teukolsky, S.A. *Problem Book in Relativity and Gravitation*; Princeton University Press: Princeton, NJ, USA, 1975.
15. Dai, S.; Guan, C.B. Maximally symmetric subspace decomposition of the Schwarzschild black hole. *arXiv* **2004**, arXiv:gr-qc/0406109.
16. Mitra, A. Friedmann-Robertson-Walker metric in curvature coordinates and its applications. *Gravit. Cosmol.* **2013**, *19*, 134–137. [CrossRef]
17. Mitra, A. When can an "Expanding Universe" look "Static" and vice versa: A comprehensive study. *Int. J. Mod. Phys. D* **2015**, *24*, 1550032. [CrossRef]

universe

MDPI

Article

What Is the Validity Domain of Einstein's Equations? Distributional Solutions over Singularities and Topological Links in Geometrodynamics

Elias Zafiris

Parmenides Foundation, Center for the Conceptual Foundations of Science, Kirchplatz 1, Pullach, 82049 Munich, Germany; elias.zafiris@parmenides-foundation.org

Academic Editors: Stephon Alexander, Jean-Michel Alimi, Elias C. Vagenas and Lorenzo Iorio
Received: 13 July 2016; Accepted: 22 August 2016; Published: 29 August 2016

Abstract: The existence of singularities alerts that one of the highest priorities of a centennial perspective on general relativity should be a careful re-thinking of the validity domain of Einstein's field equations. We address the problem of constructing distinguishable extensions of the smooth spacetime manifold model, which can incorporate singularities, while retaining the form of the field equations. The sheaf-theoretic formulation of this problem is tantamount to extending the algebra sheaf of smooth functions to a distribution-like algebra sheaf in which the former may be embedded, satisfying the pertinent cohomological conditions required for the coordinatization of all of the tensorial physical quantities, such that the form of the field equations is preserved. We present in detail the construction of these distribution-like algebra sheaves in terms of residue classes of sequences of smooth functions modulo the information of singular loci encoded in suitable ideals. Finally, we consider the application of these distribution-like solution sheaves in geometrodynamics by modeling topologically-circular boundaries of singular loci in three-dimensional space in terms of topological links. It turns out that the Borromean link represents higher order wormhole solutions.

Keywords: general relativity; sheaf cohomology; abstract differential geometry; singularities; geometrodynamics; distributions; generalized functions; nowhere dense algebras; algebra sheaves; topological links; wormholes; Borromean rings

---

1. Introduction

One hundred years after Einstein's initial conception and formulation of the general theory of relativity, it still remains a vibrant subject of intense research and formidable depth. In this way, during all of these years, our understanding of gravitation in differential geometric terms is being continuously refined. We believe that one of the highest priorities of a centennial perspective on general relativity should be a careful re-examination of the validity domain of Einstein's field equations. These equations constitute the irreducible kernel of general relativity and the possibility of retaining the form of Einstein's equations, while concurrently extending their domain of validity is promising for shedding new light on old problems and guiding toward their effective resolution. These problems are primarily related to the following perennial issues: (a) the smooth manifold background of the theory; (b) the existence of singular loci in spacetime where the metric breaks down or the curvature blows up; and (c) the non-geometric nature of the second part of Einstein's equations involving the energy-momentum tensor. It turns out that these problems are intrinsically related to each other and require a critical re-thinking of the initial assumptions referring to the domain of validity of Einstein's equations.

In this communication, first of all, we would like to consider the problem of constructing distinguishable extensions of the smooth spacetime manifold solution space of Einstein's equations incorporating singularities by taking into account recent developments in differential geometry. These developments pertain to the possible generalization of the technical framework of differential manifolds, on which the formalism and interpretation of general relativity is based on, to non-smooth or singular topological spaces by applying concepts and methods of sheaf theory and sheaf cohomology. In a nutshell, it turns out that all of the usual local constructions of differential geometry, re-interpreted sheaf-theoretically, do not require the notion of a global smooth manifold, but are based on much weaker conditions of an essentially cohomological nature. The physical interpretation of these findings, referring to appropriate extensions of Einstein's equations over singular domains, is tantamount to the viable possibility of extending the covariant formulation of Einstein's equations using continuous distribution-like or even non-smooth sheaves of coefficients for all of the involved tensorial physical quantities.

Second, we would like to show explicitly how certain generalized distribution-like solutions of partial differential equations, which fit appropriately in the above-mentioned sheaf-theoretic framework of differential geometry, bear significance in relation to obtaining singularity-free solutions of Einstein's equations in extended domains. We scrutinize the generation of these distribution-like algebra sheaves of coefficients from a physical perspective and explain the means of their construction in terms of residue classes of sequences of smooth functions modulo the information of singular loci encoded in suitable ideals.

Finally, we consider the application of these distribution-like solution sheaves in geometrodynamics. The geometrodynamical formalism is very instructive in relation to the proposed extensions because it leads to the conclusion that active positive gravitational mass may emerge from purely topological considerations taking into account the constraints imposed by Einstein's field equations in the vacuum. In this manner, we may re-assessfruitfully Wheeler's insights referring to "mass without mass" and "charge without charge", as well as re-evaluate the notion of wormhole solutions from a cohomological point of view. In this context, we propose to model topologically-circular boundaries of singular loci in three-dimensional space in terms of topological links. It turns out that there exists a universal topological link bearing the connectivity property of the Borromean rings. The cohomological expression of the Borromean link points to its physical interpretation as a higher order wormhole solution of the field equations.

2. General Relativity from the Perspective of Sheaf Theory

In the standard formulation of general relativity, the spacetime event structure is represented by means of a connected, four-dimensional real smooth manifold X. The chronogeometric relations on the event manifold X are expressed in terms of a pseudo-Riemannian metric of the Lorentzian signature, called the spacetime metric. The chronogeometric relations are not fixed kinematically a priori, like in all predecessor classical field theories, but they should be obtained dynamically in terms of the metric as a solution of Einstein's field equations depending on the energy-momentum matter field distributions. In this manner, all of the pertinent chronogeometric relations defined on a four-dimensional smooth manifold, endowing it with the structure of a spacetime manifold, become variable. The dynamical constitution of these relations by means of the field equations requires the imposition of a compatibility requirement relating the metric tensor, which represents the spacetime geometry, with the affine connection, which represents the differential evolution of the gravitational field. A spacetime manifold is considered to be without singularities if the coefficients of the metric tensor field are smooth and the manifold X is geodesically complete with respect to the metric. In this case, all timelike geodesic curves can be extended to arbitrary length in the smooth spacetime manifold X. From a physical viewpoint, according to the above requirements, the notion of localization at a spacetime point-event is sensible only if the coefficients of the metric tensor field are smooth in an open neighborhood of this point.

Algebraically speaking, a real smooth manifold X can be reconstructed entirely from the $\mathbb{R}$-algebra $\mathbb{C}^{\infty}(X)$ of smooth real-valued functions on it, and in particular, the points of X are derived from the algebra $C^\infty(X)$ as the $\mathbb{R}$-algebra homomorphisms $C^\infty(X) \to \mathbb{R}$. This important observation in relation to general relativity has been first proposed and explicated by Geroch in the form of Einstein algebras [1]. From a modern mathematical perspective, it is a consequence of the Gelfand representation theorem applied to the case of smooth manifolds [2,3]. In this way, manifold points constitute the $\mathbb{R}$-spectrum of the algebra of smooth functions $C^\infty(X)$, being isomorphic with the maximal ideals of this algebra. Notice that the $\mathbb{R}$-algebra $C^\infty(X)$ is a commutative topological algebra that contains the field of real numbers $\mathbb{R}$ as a distinguished subalgebra, encapsulating the predominant physical assumption that our means of characterizing events is conducted by evaluations in the field of real numbers $\mathbb{R}$.

The algebraic viewpoint is instructive because it makes clear that in the standard differential geometric setting of general relativity, all of the tensorial physical quantities are coordinatized by means of the commutative $\mathbb{R}$-algebra of globally-defined smooth real-valued functions $C^\infty(X)$. Hence, the background of the theory remains fixed as the $\mathbb{R}$-spectrum of the commutative topological algebra $C^\infty(X)$, supplying smooth coefficients for the coordinatization of physical quantities. The points of the manifold X, although not dynamically localizable degrees of freedom in general relativity, serve as the semantic information carriers of spacetime events. More precisely, the points are marked on a smooth manifold in terms of global evaluations of the smooth algebra $C^\infty(X)$ in the field of real numbers. The subtlety of general relativity is exactly that manifold points are not dynamically localizable entities in the theory. More precisely, manifold points assume an indirect reference as indicators of spacetime events, only after the dynamical specification of chronogeometrical relations among them, as particular solutions of the generally covariant field equations. Clearly, the existence of singular loci in spacetime where the metric breaks down in terms of smooth function coefficients forbids the association of smooth manifold points with spacetime events. What remains is an emergent notion of an event horizon of a singular locus where spacetime information may be encoded appropriately.

The dynamical variability of the coefficients coordinatizing all tensorial physical quantities requires the action of a covariant differential operator to be applied upon them. This takes place via the notion of an affine connection, which is expressed as a covariant derivative acting on these smooth coefficients. The result of differentiation is encoded in $C^\infty(X)$-modules over the algebra $C^\infty(X)$, called modules of differential forms Ω and their duals $\Xi = Hom(\Omega, C^\infty(X))$, as well as their higher powers constructed by means of exterior algebra.

In the same algebraic context, the role of a metric geometry on a smooth manifold, as related to the above modules of differential forms and their duals in general relativity, pertains to the representability of spacetime events by points of the manifold, which in turn necessitates their coordinatization in terms of real numbers. This is tantamount to the requirement that all types of differentially-variable quantities should possess uniquely-defined dual types, such that their point-event representability can be made possible by means of real numbers. This is precisely the role of a geometry induced by a metric. Concretely, the spacetime metric assigns a unique dual to each differentially-variable quantity, by effecting an isomorphism between the modules Ω and $\Xi := Hom(\Omega, C^\infty(X))$, that is $g : \Omega \simeq \Xi$, such that $df \mapsto v_f := g(df)$.

The important thing to notice is that all of these constructions can be performed strictly locally, that is by using only sections defined in the neighborhood of points. This is an implication that differential geometric constructions should be expressed not in terms of global algebra coefficients, but in terms of sheaves of coefficients defined locally. Then, the task is to study the maximal extendibility of these constructions from the local to the global level, which is technically expressed via the theory of sheaf cohomology.

In the context of general relativity, the modeling of the dynamical variability, caused by the gravitational field by means of the Levi–Civita connection, from a local sheaf-theoretic perspective, is becoming even more relevant in view of the spacetime metric compatibility of this connection and the associated solution space of the theory. Einstein's equations are formulated in terms of non-linear partial differential equations involving smooth functions, playing the role of local coefficients coordinatizing the metric tensor, the Ricci tensor and the scalar curvature. The solution of these equations in terms of the spacetime metric determines the local metrical properties of the spacetime manifold around any point, depending on the energy-momentum tensor. Notwithstanding this, all of the global cosmological predictions of the theory are obtained not from these local solutions of the field equations per se, but from the possibility of the continuation of some local solution to an extended region. The method of the continuation or extension of some solution from the local to the global level is mathematically of a sheaf-theoretic nature.

In view of the problem of singularities in general relativity, this is a clear warning that distribution-like sheaves of coefficients may be more appropriate for the continuation of some local solution over extended regions when the smooth ones become ill-defined over singular loci. It is a natural requirement that these sheaves of coefficients contain the standard smooth ones as a subalgebra, or equivalently, there is an algebra sheaf embedding of the smooth coefficients into the generalized ones. It is expected that distribution-like sheaves of coefficients can prevent the breaking down of the metric at singularities and, therefore, provide the means to extend the domain of validity of the field equations, under the proviso that the same tensorial equations can be re-expressed covariantly in terms of these generalized sheaves of coefficients.

3. Cohomological Conditions for Extending the Smooth Sheaf of Coefficients in General Relativity

Cohomology theory constitutes a sophisticated algebraic-topological method of assigning global invariants to a topological space in a homotopy-invariant way. The cohomology groups measure the global obstructions for extending sections from the local to the global level, for instance extending local solutions of a differential equation to a global solution. The differential geometric mechanism of smooth manifolds is essentially based on the set-up of the de Rham complex in terms of locally-defined smooth coefficients. In particular, de Rham cohomology measures the extent that closed differential forms fail to be exact and, thus, the obstruction to integrability. In this context, the central role is played by the lemma of Poincaré, according to which every closed differential form is locally exact in terms of smooth coefficients. The de Rham theorem asserts that the homomorphism from the de Rham cohomology ring to the differentiable singular cohomology ring, given by the integration of closed forms over differentiable singular cycles, is a ring isomorphism. The sheaf-theoretic understanding of this deep result came after the realization that both the de Rham cohomology and the differentiable singular cohomology are actually special isomorphic cases of sheaf cohomology. In particular, it has been also crystallized that the de Rham cohomology of a differential manifold depends only on the property of paracompactness of the underlying topological space. In turn, the paracompactness property, which is required in the definition of a differential manifold, can be also characterized cohomologically via the acyclic behavior of soft sheaves, like the sheaf of smooth functions. In other words, soft sheaves, namely sheaves whose sections over any closed subset can be extended to a global section, are acyclic over a paracompact topological space.

The re-interpretation and generalization of the standard de Rham cohomology theory on manifolds in sheaf-cohomological terms is physically significant, because it provides an intrinsic way to set up and solve differential equations expressing the dynamical variability of physical quantities. The concepts and technical tools of sheaf cohomology have been developed through the ground-breaking work of Grothendieck in geometry [4,5]. What should be initially kept in mind for physical applications is that the natural argument of a cohomology theory is a pair consisting of a topological space together with a sheaf of commutative algebras defined over it, rather than just a space.

It is instructive to include the basic definition characterizing the notion of a sheaf of sets on a topological space X, which also gives rise in a direct way to the notion of a sheaf of commutative algebras over X that we will employ in the sequel:

A presheaf $\mathbb{F}$ of sets on a topological space X, consists of the following information:

(I)   For every open set U of X, a set denoted by $\mathbb{F}(U)$, and

(II)   For every inclusion $V \hookrightarrow U$ of open sets of X, a restriction morphism of sets in the opposite direction:

$$r(U|V) : \mathbb{F}(U) \to \mathbb{F}(V) \tag{1}$$

such that:

(a) $r(U|U) =$ identity at $\mathbb{F}(U)$ for all open sets U of X.
(b) $r(V|W) \circ r(U|V) = r(U|W)$ for all open sets $W \hookrightarrow V \hookrightarrow U$. Usually, the following simplifying notation is used: $r(U|V)(s) := s|_V$.

A presheaf $\mathbb{F}$ of sets on a topological space X is defined to be a sheaf if it satisfies the following two conditions, for every family $V_a, a \in I$, of local open covers of V, where V open set in X, such that $V = \cup_a V_a$:

(1)   Local identity axiom of sheaf:

Given $s, t \in \mathbb{F}(V)$ with $s|_{V_a} = t|_{V_a}$ for all $a \in I$, then $s = t$.
(2)   Gluing axiom of sheaf:

Given $s_a \in \mathbb{F}(V_a), s_b \in \mathbb{F}(V_b), a, b \in I$, such that:

$$s_a|_{V_a \cap V_b} = s_b|_{V_a \cap V_b}, \tag{2}$$

for all $a, b \in I$, then there exists a unique $s \in \mathbb{F}(V)$, such that: $s|_{V_a} = s_a \in F(V_a)$ and $s|_{V_b} = s_b \in F(V_b)$.

As a basic example, if $\mathbb{F}$ denotes the presheaf that assigns to each open set $U \subset X$ the commutative algebra of all real-valued continuous functions on U, then $\mathbb{F}$ is actually a sheaf. This is intuitively clear since the specification of a topology on X is solely used for the definition of the continuous functions on X. Thus, the continuity of each function can be determined locally. This means that continuity respects the operation of restriction to open sets and, moreover, that continuous functions can be amalgamated together in a unique manner, as is required for the satisfaction of the sheaf condition.

The realization that the natural argument of a cohomology theory is not only a space, but it is actually a pair consisting of a topological space together with a sheaf of commutative algebras localized over it, has given rise to the notion of a commutative locally $\mathbb{R}$-algebraized space, defined by means of a pair $(X, \mathbb{A})$ consisting of a topological space X and a sheaf of commutative $\mathbb{R}$-algebras $\mathbb{A}$ on X, such that the restriction $\mathbb{A}_x$ is a local commutative $\mathbb{R}$-algebra for any point $x \in X$. Regarding the possibility of extending consistently all of the standard local differential geometric constructions in the context of smooth manifolds to singular spaces, in terms of locally $\mathbb{R}$-algebraized spaces, where a suitable sheaf of commutative $\mathbb{R}$-algebras $\mathbb{A}$ on X substitutes the smooth sheaf of $\mathbb{R}$-algebras $\mathbb{C}^\infty(X)$), a full-grown theory has been recently developed, called Abstract Differential Geometry (ADG). This theory has shown that the standard differential-analytic tools of locally-Euclidean spaces and smooth manifolds leading to the formulation and solution of differential equations can be actually re-produced and generalized to non-smooth or singular topological spaces by means of sheaf cohomology. Equivalently, the suitability of a sheaf of commutative $\mathbb{R}$-algebras $\mathbb{A}$ on an abstract topological space X for expressing the differential geometric mechanism in terms of these coefficients instead of the smooth ones is entirely determined only by the satisfaction of precise cohomological conditions pertaining to the characterization of the algebra sheaf $\mathbb{A}$. We note, in passing, that for the economy of symbols, we denote algebra sheaves by the same symbols as we used for the algebras before, since the difference is clear from the context.

The mathematical theory of ADG has been built rigorously by Mallios [6,7] (see also [8]), based on critical prior work of Selesnick [9]. The significance of ADG for physics has been also shown by an explicit reconstruction and generalization of the framework of the Maxwell and Yang–Mills gauge field theories in sheaf cohomological terms [10,11]; see also [12–14]. An exposition of the basic didactics of ADG in relation to its physical applications has been presented by Raptis in [15]. The basic method introduced for the generalization of the standard analytic tools of Classical Differential Geometry (CDG) consists of the following: Initially, a concept of CDG is suitable for the extension to a broader differential context (beyond the context of smooth manifolds) if it is liable to a process of sheaf-theoretic localization [16]. In CDG, all of the differential geometric constructions require that the base space is a smooth manifold. The underlying reason is that the means of differentiation are lifted locally from the structure of a Euclidean space. In this way, the de Rham complex is fixed with respect to smooth coefficients, and all tensorial quantities are coordinatized in smooth terms. In ADG, the base space provides merely a topological basis of sheaf-theoretic localization, such that all of the pertinent differential geometric constructions can take place locally, whereas the latter are not subordinate to this topological basis, meaning that they are not dependent on any particular localization basis. Thus, the object of primary significance in ADG is not the base space itself, but the algebra sheaf of coefficients localized over it. The differentiation structure is built in the algebra sheaf of coefficients by means of the notion of a connection defined independently of any locally-Euclidean considerations. In this way, the associated de Rham complex can be satisfied by various possible algebra sheaves of coefficients modulo some well-defined cohomological conditions. We emphasize that the prominent role in the context of ADG is played by the algebra sheaf of coefficients, interpreted as a "functional coordinate arithmetic" [14,17] (see also [18–22]), meaning that all geometric objects involved in the formalism are locally expressed in terms of its sections. In this way, an algebra sheaf of coefficients is not constrained ab initio to be a smooth one, restricting the geometric solution space within the spectrum of a smooth manifold. More generally, a suitable algebra sheaf of coefficients turns out to be an algebra sheaf of generalized functions, including distributions, defined by Rosinger in the context of solutions to non-linear partial differential equations [23,24].

Concerning general relativity, which is formulated using the CDG of smooth manifolds, the possibilities offered by ADG bear a remarkable physical significance. In particular, there arises the possibility of re-assessing the global problems of general relativity related to the existence of singularities, where the metric breaks down, from the perspective of appropriate generalized algebra sheaves of coefficients. In this manner, the validity of Einstein's equations may be extended beyond differential manifolds, under the condition that the covariance properties of all tensorial physical quantities are maintained under these extensions, expressed in terms of the new sheaves of coefficients. From a physical viewpoint, this approach would allow one to obtain solutions in terms of distribution-like sheaves corresponding to non-punctual localization properties, which would nevertheless still satisfy the field equations. This clearly vindicates the following critical remark of Weyl [25]: "While topology has succeeded fairly well in mastering continuity, we do not yet understand the inner meaning of the restriction to differential manifolds. Perhaps one day physics will be able to discard it".

The possibility of obtaining extended admissible solution spaces in terms of generalized algebra sheaves of coefficients is based on the fact that the validity of the de Rham complex, in its sheaf-theoretic guise, is not restricted exclusively to the coordinatization of the tensorial physical quantities by smooth coefficients $\mathbb{C}^\infty$, as is actually the case when the $\mathbb{R}$-spectrum of the coefficients is a smooth manifold. Thus, we may consider distribution-like sheaves of coefficients satisfying the validity of the de Rham complex and, therefore, formulate and solve the field equations in terms of these distribution coefficients instead of the smooth ones. More precisely, this is the case if the following sequence of $\mathbb{R}$-linear sheaf morphisms:

$$\mathbb{A} \rightarrow \Omega^1(\mathbb{A}) \rightarrow \ldots \rightarrow \Omega^n(\mathbb{A}) \rightarrow \ldots \tag{3}$$

is a complex of $\mathbb{R}$-vector space sheaves, identified as the sheaf-theoretic de Rham complex of $\mathbb{A}$.

In this case, if the cohomological condition expressing the Poincaré Lemma, $Ker(d^0) = \mathbb{R}$ is satisfied with respect to the algebra sheaf $\mathbb{A}$ and requiring that $\mathbb{A}$ is a soft algebra sheaf, viz. any section over any closed subset of X can be extended to a global section, we obtain that the sequence:

$$0 \to \mathbb{R} \to \mathbb{A} \to \Omega^1(\mathbb{A}) \to \ldots \to \Omega^n(\mathbb{A}) \to \ldots \tag{4}$$

is an exact sequence of $\mathbb{R}$-vector space sheaves. Thus, the sheaf-theoretic de Rham complex of the algebra sheaf $\mathbb{A}$ constitutes an acyclic resolution of the constant sheaf $\mathbb{R}$.

The physical interpretation of this fact is the following: First of all, the essential feature of the localization method, utilizing coefficients from algebra sheaves instead of global algebras, is that the sheaf-theoretic de Rham complex is actually an acyclic resolution of the constant sheaf of the reals coordinatizing the events. For instance, referring to the CDG of smooth manifolds, the de Rham complex, expressed in terms of local smooth coefficients and their differential forms of higher orders, provides such an acyclic resolution of the constant sheaf $\mathbb{R}$. What has been uncovered by ADG is that the smooth algebra sheaf $\mathbb{C}^\infty(X))$ is not unique in this respect. More concretely, any other soft algebra sheaf $\mathbb{A}$ constituting an acyclic resolution of the constant sheaf $\mathbb{R}$ is a viable source of coefficients for the coordinatization of the tensors, maintaining at the same time all of their covariance properties in terms of the new local coefficients. This crucial fact essentially questions the uniqueness of the role of local smooth coefficients for formulating the means of dynamical variability. In other words, it questions the unique role of smooth manifold geometric spectrums as domains of validity of the field equations.

The idea to address the problem of singularities from the perspective of ADG has been proposed already, for instance in [11,12]. More concretely, in particular relation to the issue of spacetime singularities, Mallios and Rosinger [23,24] have applied ADG using as an algebra sheaf of coefficients, a variety of the so-called "spacetime foam algebras", and by Raptis [26], building up on prior work by Mallios and Raptis [27], using as a sheaf of coefficients "differential incidence algebras" defined over a locally-finite poset substitute of a continuous manifold.

Our present proposal constitutes a twist of perspective in comparison to these works, which is actually implemented by physical criteria of suitability going beyond the satisfaction of the cohomological conditions. Our quest is related to the possibility of using a particular type of a "spacetime foam algebra" as a kind of a distribution-like sheaf of coefficients, distinguished on physical grounds, for extending the domain of validity of Einstein's field equations. For this purpose, from the whole variety of "spacetime foam algebras", we distinguish only the "nowhere dense generalized function algebra" as bearing physical significance in relation to the field equations of general relativity. This is based on a physical criterion determining which properties should be characterized as intrinsic to the gravitational field and eventually deciding what should be generic with respect to its function or not. This physical criterion refers to the viable possibility of expressing the gravitational field sources via the instantiation of these generalized algebra sheaves of coefficients. Our rationale is based on the idea that in an intrinsically dynamically-variable theory, like general relativity, it should be the pertinent physical conditions or the sources of the field themselves that determine the type of these extensions as solutions to the field equations.

4. Coping with Spacetime Singularities: Conceptual and Technical Aspects

In the classical differential geometric formulation of general relativity, spacetime is represented as a connected, paracompact and Hausdorff four-dimensional $\mathbb{C}^\infty$ manifold X, endowed with a pseudo-Riemannian metric of the Lorentzian signature, which is obtained as a solution of Einstein's field equations. A spacetime manifold is considered to be without singularities if the coefficients of the metric tensor field are at least of class $\mathbb{C}^2$ and X is geodesically complete with respect to the metric, meaning that all timelike geodesic curves can be extended to arbitrary length [28,29]. Consequently, a spacetime manifold is considered to be singular if there exist incomplete geodesic curves, or equivalently finite affine length geodesics that cannot be extended. A spacetime singularity delimits a locus where the behavior of the metric tensor coefficients become ill-defined with respect to the smooth characterization of the manifold. Usually, the singular locus is identified as a locus where the spacetime curvature blows up. We note that the localization at a spacetime point-event is meaningful if the metric coefficients are smooth, or at least of class $\mathbb{C}^2$ in a neighborhood of this point.

The usual way to cope with a spacetime singularity is to consider it as a singular spacetime boundary rather than a locus within spacetime. For instance, a spacetime boundary may be defined in terms of a set of incomplete curves S. This takes place by the imposition of an appropriate equivalence relation $\sim$ on the set S, such that the quotient set $S/\sim := \partial X$ is interpreted as the singular boundary of X. The criterion of equivalence is determined by the choice of those equivalence classes, which are forced to play the role of ideal points in the extension of X by ∂X. There have been proposed various possible choices, for example Geroch's "g-boundary" or Schmidt's "b-boundary", but it is always assumed that X is topologically dense in $X \sqcup \partial X$ [30,31]. Following this approach, Heller and Sasin have shown that Einstein's field equations can be formulated in the extension of X by ∂X, that is on $X \sqcup \partial X$ defined as an "Einstein structured space" [32]. Actually, this is the Gelfand spectrum of a sheaf of Einstein algebras, which constitutes the sheaf-theoretic localization of an Einstein algebra, a notion proposed initially by Geroch in his attempt to re-formulate general relativity in algebraic terms without invoking directly a spacetime manifold background [1]. In particular, it has been proven that the closed Friedmann world model and the Schwarzschild solution, combined with Schmidt's "b-boundary" construction, fit nicely in the sheaf-theoretic context of an "Einstein structured space". In turn, this has been a first indication that the validity of Einstein's equations may be extended to bigger domains incorporating singular loci, which are not smooth manifolds anymore. It has been also pointed out that some sorts of singularities can also appear when there exists a transition to the quantum gravity regime. More concretely, the smooth manifold structure of spacetime can break down, and the possible validity of Einstein's equations should be sought for in further extended and generalized non-smooth spectrums of appropriate sheaves of algebras, where the singularities are not necessarily forced to some type of spacetime boundary.

From a broader conceptual perspective, the issue of singularities in general relativity as impossibilities of extending smooth metric solutions of Einstein's equations necessitates the coordinatization of all of the tensorial quantities by distributional coefficients effecting a type of topological coarse-graining over singular loci and, thus, localizing the point-event stratum in their terms. Under the proviso that these distributional coefficients form algebra sheaves fulfilling all of the required cohomological conditions, the means of extending local distributional solutions generalizes the standard method of extending timelike geodesic curves in a smooth manifold. The physical significance of this generalization is that the domain of validity of the field equations can be extended beyond the notion of a smooth manifold. Not only this, but additionally, these distinguishable extensions may be associated intrinsically with the gravitational field, under the constraint that sources of the field itself giving rise to singularities can be expressed topologically in the terms of distribution-like algebra sheaves.

In this state of affairs, the smoothness assumption can be retained, at best, only locally and certainly far from singular loci. Mathematically, there should exist an embedding of the algebra sheaf of smooth functions into a distribution-like algebra sheaf of coefficients qualified as a solution of the extended field equations. An illuminating way to think of the proposed approach in non-technical terms is that coping efficiently with singularities requires a process of folding out of the smooth point-event manifold background. This viewpoint has been emphasized by von Müller [33], according to whom the process of folding out into a "statu-nascendi" level should be considered in the context of a whole new categorical apparatus qualifying its intrinsic characteristics in contradistinction to the event stratum. In this manner, we suggest that the existence of a distribution-like sheaf of coefficients as a solution of the field equations within an appropriately-extended domain characterized by some generic gravitational criterion paves the way for understanding the precise nature of this folding out of the smooth point-event stratum.

The possibility of extending the formulation of Einstein's equations in the case of non-smooth spectrums using the sheaf-theoretic technique of localization in the context of ADG is of major significance. We note that non-smooth spectrums of algebra sheaves do not require the consideration of singularities as ideal points on the boundary of a smooth manifold. In other words, singular loci are allowed to be located, according to specific topological criteria, within a manifold. Of course, a natural requirement should be that the exclusion of singular loci would recast Einstein's equation in the familiar form in terms of smooth coefficients. However, clearly in the case that Einstein's equations become meaningfully extended over singular loci, then the coefficients of the metric and curvature tensors cannot be smooth any more. Therefore, from a smooth perspective, a singularity functions as an obstruction to the extension of a local solution to the field equations expressed in terms of smooth coefficients. Thus, more precisely, a singular locus plays the role of a cohomological obstruction to the extendibility of a local smooth solution. This criterion incorporates and generalizes sheaf-cohomologically the initial definition of singular behavior in terms of non-extendibility of geodesics. Essentially, the reason is that the notion of extendibility of local solutions is of a sheaf-theoretic nature, recalling for instance the well-known procedure of analytic continuation.

There are two important physical consequences emanating from the possibility of formulating Einstein's equations in terms of generalized non-smooth sheaves of coefficients. The first is related to the natural question concerning the criterion of depicting a particular sheaf of algebras for this purpose. The second is related to a possible re-evaluation of the status of the energy-momentum source term in Einstein's equations, which currently is not implemented by any process of geometrization.

Regarding the first, the required physical condition is the following: Since the formulation of Einstein's equations can be extended over singular loci, it should precisely be the nature and specification of these singular loci that would determine the appropriate sheaf of coefficients, such that a solution can be expressed eventually in terms of these coefficients. In the non-singular case, we know already that a solution can be expressed in terms of smooth coefficients. In other words, we already know that if no singularity is present, the spacetime metric, obtained as a particular solution of the vacuum Einstein equations, for example, is always expressible in terms of smooth coefficients, i.e., in terms of the sheaf of algebras $\mathbb{C}^\infty(X)$. Hence, we expect that in the presence of a particular type of a singular locus over which Einstein's equations hold in terms of a distribution-like sheaf of coefficients, there exists a metric solution expressed in terms of these coefficients. Not only this, but additionally, since the knowledge of the metric solution is completely expressible in terms of these coefficients, considered as unknowns when plugged into the equations, the specification of a singular locus should force a corresponding algebra sheaf as the solution. In other words, the nature of a singular locus should determine the differentiability properties of a metric solution in the case that Einstein's equations can be extended over this locus. As we stressed previously, the physical association of singular loci with sources of the gravitational field itself, giving rise to distinguishable extensions of the standard smooth manifold spacetime model of general relativity, implies that sources can be expressed topologically after all, if solutions of the field equations are expressed in terms of appropriate distribution-like algebra sheaves.

In this context, the physical significance of ADG is that it determines rigorously the criteria that these algebra sheaves of coefficients have to satisfy, such that Einstein's equations can be satisfied over various sorts of singular loci, expressed in terms of these coefficients. Not surprisingly, these criteria are of a cohomological nature. Essentially, they determine viable algebra sheaves of coefficients by the requirement that they are soft, and thus acyclic, such that the validity of the de Rham complex remains intact. In turn, the basic idea is that the Poincaré lemma should remain in force, viz. closed differential forms expressed in these generalized coefficients should be locally exact as in the smooth case, so that the differential geometric mechanism can be extended over singularities without breaking down. We will present a general form of these algebra sheaves consisting of distribution-like coefficients in the sequel. According to Clarke, the answer to many of the problems related to singularities "involve detailed considerations of distributional solutions to Einstein's equations, leading into an area that is only starting to be explored ..." [28]. We propose that the extension of the validity of Einstein's equations over singular loci in terms of appropriate sheaves of algebras, which are generally non-smooth, sheds new light on the problem of singular behavior in general relativity.

Regarding the second physical consequence, it is instructive to remind that the energy-momentum source term in the smooth formulation of Einstein's equations is not of any geometric nature. The energy-momentum tensor attributes the source of curvature entirely to matter (including the cosmological dark energy), as it does not incorporate the stress-energy associated with the gravitational field itself. There is an underlying assumption that spacetime is somehow empty unless it is filled in by matter, expressed in terms of the smooth coefficients of the energy-momentum tensor. This is the reason that when the energy-momentum part is zero, then the equations are called vacuum equations. Now, the validity of Einstein's equations over singular domains in terms of generalized non-smooth algebra sheaves casts serious doubts on this assumption. Namely, the form of Einstein's equations with the vanishing non-geometric second part may turn out to be the fundamental form of these equations. The reason is that sources of the gravitational field itself might be implemented in terms of non-smooth algebra sheaves, and thus, what is called a vacuum is not empty at all, precisely because it engulfs these sources. This idea is not actually as controversial as it sounds, if we take seriously into account that all classical experimental tests of general relativity involve a vanishing energy-momentum tensor, and thus, what they really verify is the equation $\mathbf{R}_{\mu\nu} = 0$. This issue has been also pointed out and argued for extensively, from a non sheaf-theoretic point of view, by Vishwakarma [34], who conducted a careful analysis based on the observational tests of the theory. In the sequel, we will discuss this issue in more detail from a geometrodynamical perspective in light of the particular form of distribution-like algebra sheaves.

5. Spacetime Extensions in Terms of Singularity-Free Distributional Algebra Sheaves

It is physically reasonable to expect that an admissible commutative algebra sheaf of coefficients in terms of which Einstein's equations may be extended over a singular locus should be distribution-like. For example, we may think of a matter distribution confined to a submanifold of spacetime whose density is integrable over this submanifold. In the context of a linear field theory, this should be naturally modeled in terms of a linear distribution. Unfortunately, this is not possible in the context of general relativity, which is a non-linear theory. In other words, Schwarz's linear distributions are not suitable candidates for expressing the information of singular loci.

The unsuitability of linear distributions rests on the fact that the space $\mathbb{D}'$ they form is only a linear space, but it is not an algebra. This is characterized as the "Schwarz impossibility" and may be formulated as follows: There is no symmetric bilinear morphism:

$$\circ : \mathbb{D}'(V) \times \mathbb{D}'(V) \ni (S, T) \to S \circ T \in \mathbb{D}'(V)$$

so that $S \circ T$ is the usual point-wise product of continuous functions, when $S, T \in \mathbb{C}^0(V)$. Equivalently, $\mathbb{D}'(V)$ is not closed under any multiplication that extends the usual multiplication of continuous functions, where V is an open subset X. Since all of the involved arguments are of a local character, without loss of generality, we may simply consider V as an open subset of $\mathbb{R}^4$.

A physically natural way to bypass "Schwarz impossibility" is to assume the existence of an embedding morphism $\mathbb{D}'(V) \hookrightarrow \mathbb{A}(V)$, which embeds the vector space of distributions $\mathbb{D}'(V)$ as a vector subspace in $\mathbb{A}(V)$, where $\mathbb{A}(V)$ is the quotient algebra:

$$\mathbb{A}(V) = \mathbb{K}(V)/\mathbb{I}, \tag{5}$$

and $\mathbb{K}(V)$ is a subalgebra in $\mathbb{C}^\infty(V)^\Lambda$, for some index set Λ, whereas $\mathbb{I}$ is an ideal in $\mathbb{K}(V)$. This approach was initiated by Rosinger [35,36] and developed further in [37–40].

We will restrict ourselves to a certain subclass of this type of algebras, namely the unital, associative and commutative algebras of generalized functions, whose suitably-defined ideals can engulf algebraically the information of singular loci. These algebras, introduced by Rosinger [36], have been formed in such a way as to express generalized solutions of non-linear partial differential equations. We may describe the generation of these algebras locally as follows:

Let $V \subseteq \mathbb{R}^4$ be an open set and $L = (\Lambda, \leq)$ be a right-directed partial order on some specified index set Λ. That is, for all $\lambda, \mu \in \Lambda$, there exists $\nu \in \Lambda$, such that $\lambda, \mu \leq \nu$. With respect to the usual componentwise operations, $\mathbb{C}^\infty(V)^\Lambda$ is a unital and commutative algebra over the reals. We define the following ideal $\mathbb{I}_L$ in $\mathbb{C}^\infty(V)^\Lambda$, whose physical meaning will be described in the sequel:

$$\mathbb{I}_L(V) = \left\{ \phi = (\phi_\lambda)_{\lambda \in \Lambda} \middle| \begin{array}{l} \exists \quad \Gamma \subset V \text{ closed nowhere dense} : \\ \forall \quad x \in [V \setminus \Gamma] \text{ being dense} : \\ \exists \quad \lambda \in \Lambda : \\ \forall \quad \mu \in \Lambda, \mu \geq \lambda : \\ \qquad \phi_\mu(x) = 0, \partial^p \phi_\mu(x) = 0 \end{array} \right\} \tag{6}$$

In the above definition, we think of Γ as a singular locus in $\mathbb{R}^4$, characterized as a closed and nowhere dense subset relative to the open set $V \subseteq \mathbb{R}^4$, such that its complement $V \setminus \Gamma$ in V is dense. The unital and commutative algebra $\mathbb{C}^\infty(V)^\Lambda$ contains smooth functions ϕ_λ indexed by the set Λ and defined over V, to be thought of as diagrams or sequences of Λ-indexed smooth functions. The requirement of the right-directed partial order on the specified index set Λ, which is denoted by $L = (\Lambda, \leq)$, is technically necessary in order that the above set forms actually an ideal in $\mathbb{C}^\infty(V)^\Lambda$. Now, the ideal $\mathbb{I}_L(V)$ in $\mathbb{C}^\infty(V)^\Lambda$ includes all of these sequences of smooth functions ϕ_λ that vanish asymptotically outside the singular locus Γ together with all of their partial derivatives. Therefore, intuitively speaking, the ideal of the form $\mathbb{I}_L(V)$ incorporates all of these sequences of smooth functions indexed by Λ whose support covers the singular locus Γ, whereas they vanish outside it. In this manner, the information of the singular locus Γ is encoded in the ideal $\mathbb{I}_L(V)$ in $\mathbb{C}^\infty(V)^\Lambda$. Hence, the quotient commutative algebra $\mathbb{A}_L(V) = \mathbb{C}^\infty(V)^\Lambda / \mathbb{I}_L(V)$ is an algebra of residues of sequences of smooth functions modulo the singular information ideal $\mathbb{I}_L(V)$.

A natural question in the above context refers to the requirement that the complement $V \setminus \Gamma$ of the singular locus Γ in V should be dense. The necessity of this requirement can be understood by the fact that we wish to obtain an embedding ι of the algebra of smooth functions $\mathbb{C}^\infty(V)$ into the algebra of generalized functions $\mathbb{A}_L(V)$:

$$\iota : \mathbb{C}^\infty(V) \hookrightarrow \mathbb{A}_L(V) = \frac{\mathbb{C}^\infty(V)^\Lambda}{\mathbb{I}_L(V)} \tag{7}$$

such that:

$$\varphi \hookrightarrow \iota(\varphi) = \Delta(\varphi) + [\mathbb{I}_L(V)] \tag{8}$$

where $\Delta_\Lambda |_V : \mathbb{C}^\infty(V) \to \mathbb{C}^\infty(V)^\Lambda$ is the diagonal morphism with respect to Λ, defined for an open set V as follows:

$$\Delta_\Lambda(\varphi) |_V = \{\Delta(\varphi) = (\varphi_\lambda)_{\lambda \in \Lambda} \mid \varphi_\lambda = \varphi, \forall \lambda \in \Lambda, \varphi \in \mathbb{C}^\infty(V)\}.$$

Hence, for every smooth function φ in $\mathbb{C}^\infty(V)$, the diagonal image $\Delta(\varphi)$ of φ in $\mathbb{C}^\infty(V)^\Lambda$ is a sequence of smooth functions all identical to φ, indexed by Λ. The embedding ι is feasible according to the above, if and only if the ideal $\mathbb{I}_L(V)$ satisfies the off diagonality condition:

$$\mathbb{I}_L(V) \cap \Delta_\Lambda |_V = \{0\}. \tag{9}$$

Therefore, it remains to show that if the complement $V \setminus \Gamma$ of the singular locus Γ in V is dense, according to the specification in (6), then the ideal $\mathbb{I}_L(V)$ actually satisfies the above off diagonality condition. Therefore, we suppose that $V \setminus \Gamma$ is dense in V and consider a smooth function χ in $\mathbb{C}^\infty(V)$. If $\Delta_\Lambda(\chi)|_V := \Delta(\chi)$ belongs to the ideal $\mathbb{I}_L(V)$, then the asymptotic vanishing condition in (6) implies that $\chi = 0$ in $V \setminus \Gamma$, and therefore, we must have $\chi = 0$ in V because $V \setminus \Gamma$ is dense in V by hypothesis. Thus, it follows that the ideal $\mathbb{I}_L(V)$ satisfies the off diagonality condition (9), as required.

Conclusively, there exists a canonical injective homomorphism of commutative algebras, or equivalently, an embedding ι of the algebra of smooth functions $\mathbb{C}^\infty(V)$ into the algebra of generalized functions $\mathbb{A}_L(V)$:

$$\iota : \mathbb{C}^\infty(V) \hookrightarrow \mathbb{A}_L(V) = \frac{\mathbb{C}^\infty(V)^\Lambda}{\mathbb{I}_L(V)} \tag{10}$$

Furthermore, in view of (6), it follows immediately that the partial differential operators:

$$\partial^p : \mathbb{C}^\infty(V)^\Lambda \ni \phi = (\phi_\lambda) \mapsto \partial^p \phi = (\partial^p \phi_\lambda) \in \mathbb{C}^\infty(V)^\Lambda$$

satisfy the inclusion:

$$\partial^p(\mathbb{I}_L(V)) \subseteq \mathbb{I}_L(V). \tag{11}$$

Thus, the standard partial derivative operators on $\mathbb{C}^\infty(V)$ extend to $\mathbb{A}_L(V)$:

$$\partial^p : \mathbb{A}_L(V) \ni [\phi + \mathbb{I}_L(V)] \mapsto [\partial^p \phi + \mathbb{I}_L(V)] \in \mathbb{A}_L(V), \tag{12}$$

We conclude that the embedding of commutative algebras (10) extends to an embedding of differential algebras. Therefore, the following diagram commutes:

$$
\begin{array}{ccc}
\mathbb{C}^\infty(V) & \xrightarrow{\ \partial^p\ } & \mathbb{C}^\infty(V) \\
\downarrow & & \downarrow \\
\mathbb{A}_L(V) & \xrightarrow{\ \partial^p\ } & \mathbb{A}_L(V)
\end{array}
$$

We emphasize that the embedding (10) preserves not only the algebraic structure of $\mathbb{C}^\infty(V)$, but also its differential structure. The off diagonality condition (9) implies also the existence of an injective, linear morphism:

$$\mathbb{D}'(V) \hookrightarrow \mathbb{A}_L(V). \tag{13}$$

Therefore, the differential algebra $\mathbb{A}_L(V)$ contains the space of distributions as a linear subspace; see [38] (pp. 234–244), where those algebras that admit linear embeddings of distributions are characterized in terms of such off diagonality conditions. However, in contradistinction with (10), the embedding (13) does not commute with partial derivatives, and thus, the partial derivatives on $\mathbb{A}_L(V)$ do not, in general, coincide with distributional derivatives, when restricted to $\mathbb{D}'(V)$.

Finally, it is crucial to observe that a subset of a topological space is closed and nowhere dense if and only if it satisfies this condition locally. This is the key idea used to prove that the algebras of generalized functions $\mathbb{A}_L(V)$ form actually sheaves of commutative algebras, which additionally, are soft and flasque or flabby [23,24]. Thus, they are characterized as cohomologically-appropriate sheaves of coefficients according to ADG. More precisely, the distribution-like soft algebra sheaves of the form $\mathbb{A}_L$ constitute an acyclic resolution of the constant sheaf of the reals coordinatizing the events. Thus, we conclude that the de Rham complex can be rigorously expressed in terms of these coefficients instead of the smooth ones, and consequently, Einstein's equations can be formulated with respect to coefficients from the algebra sheaf $\mathbb{A}_L$ instead of the smooth ones from $\mathbb{C}^\infty$. Consequently, the validity of Einstein's equations can be extended over singular loci in a covariant manner by utilizing coefficients from the sheaf $\mathbb{A}_L$ for expressing all involved differential geometric tensorial quantities. Reciprocally, according to the intended physical interpretation of these algebra sheaves, pertaining to expressing sources of the gravitational field in terms of closed and nowhere dense subsets, the presence of a singular locus forces an algebra sheaf of the form $\mathbb{A}_L$ as coefficients with respect to which Einstein's equations retain their validity over this locus and do not break down, like in the case of insisting to use indiscriminately-smooth coefficients.

For the sake of completeness, it is instructive to remind that the softness property of the sheaves of the form $\mathbb{A}_L$ means that any section over any closed subset can be extended to a global section. Thus, these types of sheaves characterize cohomologically the topological property of paracompactness by means of acyclicity. Equivalently, soft sheaves are acyclic over a paracompact topological space. Moreover, sheaves of the form $\mathbb{A}_L$ are not only soft, but they are flasque or flabby, as well, which is a local property. This means that the restriction morphism of sections in the sheaf definition is an epimorphism. Hence, in this case, we can always extend any local section by zero to obtain a global section of $\mathbb{A}_L$.

We may recapitulate by pointing out that the first basic idea involved in the construction of distribution-like algebra sheaves of coefficients, in their role to coordinatize solutions of non-linear partial differential equations, is to model a singular locus Γ in $\mathbb{R}^4$ as a closed and nowhere dense subset relative to an open set $V \subseteq \mathbb{R}^4$, such that its complement $V \setminus \Gamma$ in V is dense. The second basic idea is to express such a closed and nowhere dense singular locus as an ideal in an algebra sheaf constructed as an extension of the smooth one over a partially-ordered set. In this manner, the ideal expressing algebraically a singular locus contains diagrams of locally-defined smooth functions indexed by Λ whose support covers the singular locus Γ, whereas they vanish outside it. Then, it can be shown that the quotient commutative algebra sheaf $\mathbb{A}_L(V) = \mathbb{C}^\infty(V)^\Lambda / \mathbb{I}_L(V)$ is an algebra sheaf of residues of diagrams of smooth functions modulo the closed nowhere dense singular ideal $\mathbb{I}_L(V)$.

It is instructive to emphasize that the algebra of global sections of the sheaf $\mathbb{A}_L(V)$ contains the space of Schwarz distributions $\mathbb{D}'(V)$ only as a linear subspace and not as a commutative subalgebra. For example, Dirac's delta, considered as a distribution, is represented in terms of a generalized function whose pertinent closed and nowhere dense set is an one-point set. It is well known that the square of the delta distribution is not a distribution itself, since the operation of point-wise multiplication of distributions is not well-defined in $\mathbb{D}'(V)$. Notwithstanding this fact, the representative generalized function may be unproblematically squared providing a legitimate generalized function without being a linear distribution itself. Clearly, by the rules of the construction of these commutative algebras of generalized functions, arbitrary nonlinear continuous operations may be applied to a generalized function giving another generalized function in the same algebra. In passing, it is also worth pointing out that the linear space of Schwarz distributions does not give rise to a flasque vector sheaf in contradistinction to the case of the embedding sheaf $\mathbb{A}_L(V)$, a property that is crucial for the global extendibility of all standard local differential geometric constructions.

In the sequel, we are going to propose a concrete class of closed and nowhere dense sets modeling the boundaries of singular loci and forming a topological link in 3D space. Conceptually, this essentially means that the semantics of folding out a local smooth event stratum into a singular domain may be associated with the formation of some topological link configuration and its concomitant algebraic expression in terms of an algebra sheaf of the type $\mathbb{A}_L$. At the final stage, we have to examine if this algebra sheaf satisfies the cohomological conditions necessary for expressing the differential geometric mechanism of general relativity in these terms instead of the globally-smooth ones. This turns out to be actually the case, and therefore, algebra sheaves of the type $\mathbb{A}_L$ can be used legitimately to express the metric solution of Einstein's field equations extended now over singularities.

The important consequence is that we can retain not only the validity, but not the form and covariance property of Einstein's equations even over singular loci. The reason is that all physical quantities can be still transformed according to a tensor law for any arbitrary admissible coordinate transformation. The difference in comparison to the smooth case is that the coordinates are allowed to be non-standard or non-smooth, while at the same time, all of the machinery of differential geometry can be applied with respect to them. In particular, while the coefficients of the tensorial physical quantities are non-smooth, all of the usual differential-geometric constructions can be carried out as in the smooth case. The only price to be paid for this generalization is the rejection of the fixed absolute smooth manifold background of the theory. We consider this fact as physically non-disturbing, since the essence of general relativity is in the covariant formulation and validity of Einstein's equations and not on the existence of a smooth background manifold. In particular, what we gain from such a generalization is not only that Einstein's equations can be extended covariantly over singular loci, but also that the solution of these equations in terms of coefficients from a sheaf of the form $\mathbb{A}_L$ are free of singularities.

6. Topological Links in Geometrodynamics

According to the paradigm of geometrodynamics [41], we may foliate a spacetime manifold X into three-dimensional spacelike leaves Σ_t by utilizing an one-parameter family of embeddings $\varepsilon_t : \Sigma \hookrightarrow X$, such that $\varepsilon_t(\Sigma) = \Sigma_t$. In the geometrodynamical formulation, the three-dimensional Riemannian manifold (Σ, h) is thought of as dynamically evolving, where the corresponding metric at time t, $h_t = \varepsilon_t{}^* g$, is derived by pulling back the spacetime metric g via ε_t. It is implicitly assumed that all three-dimensional spacelike leaves Σ_t are mutually disjoint, such that the Lorentzian manifold $(\mathbb{R} \times \Sigma, \varepsilon^* g)$ represents X, where the leaves of the considered foliation correspond to the constant time hypersurfaces.

The geometrodynamical picture is instructive for our purposes because it shows that active gravitational mass may emerge from purely topological considerations taking into account the constraints imposed by Einstein's field equations in the vacuum [42]. From a physical perspective, this may be interpreted in a novel way according to Wheeler's insight referring to "mass without mass" [43,44] as follows: Localized configurations of topologically-singular loci in open sets of a spacetime manifold restricted to closed nowhere dense subsets amount to active gravitational mass/energy in their complementary open dense subsets. In particular, if we consider that the Lorentzian manifold $(\mathbb{R} \times \Sigma, \varepsilon^* g)$ represents X, the singular loci may be localized within the three-dimensional manifold Σ. In this context, if Σ has a non-trivial topology, Gannon's theorem [45] implies that spacetime is geodesically incomplete and, thus, singular. The simplest way to implement a non-trivial topology on Σ is via the hypothesis of non-simple connectivity. More precisely, the existence of singular loci in Σ, localized in closed nowhere dense subsets makes Σ a multiple-connected topological space and, thus, topologically different from $\mathbb{R}^3$. We may recapitulate our conclusion up to now by asserting that the existence of singular loci in closed nowhere dense subsets of Σ, making it a multiply-connected topological space, implies active gravitational mass/energy in the complementary open dense subsets. Moreover, according to the "positive mass theorem" considered in the vacuum case, this gravitational mass/energy is non-zero and strictly positive. In passing, we would like to stress that Gannon's theorem should be conceived of as a significant generalization of the Penrose–Hawking singularity theorems [29], in the sense of replacing the usual geometric hypothesis of closed trapped surfaces in Σ by the more general applicable topological hypothesis of the non-simple connectivity of Σ.

In the same vein of ideas, we may also consider the system of Einstein–Maxwell equations without sources for the Maxwell field and, in this way, address from our perspective the alternative Wheeler's insight referring to "charge without charge" [43,44]. This has been originally tied to the assumption that Σ is orientable and bears the standard wormhole topology, that is homotopically equivalent to $S^1 \times S^2 - \{point\}$, such that the magnetic flux lines thread through the wormhole. In this case, the homology class of all two-spheres containing both of the wormhole mouths has zero charge, whereas the two individual wormhole mouths may be considered as having equal and opposite charges. In this context, a wormhole may be thought of in terms of a one-dimensional homology class in spacetime. From the general results of low-dimensional geometric topology [46], we know that every homology class of a four-dimensional spacetime can be represented by an embedded submanifold. Using the geometrodynamic foliation, we may restrict this representation to Σ. In this manner, we can instantiate a higher-order wormhole solution, for example by considering an appropriate two-dimensional homology class.

We are going to outline a general method of generating these types of solutions guided by the form of the algebra sheaves $\mathbb{A}_L$ incorporating gravitational properties defined on dense open sets of X and by restriction to dense open sets of Σ. For this purpose, we may consider a singular locus with boundary in $\mathbb{R}^3$ or in its compactification S^3, which is excised from $\mathbb{R}^3$ or S^3. We consider a singular locus as a singular disk cut off from S^3, which may be visualized in terms of a cone whose apex is at infinity and whose base lies at the boundary of the singular locus. A singular disk of this form excised from S^3 gives rise to a two-dimensional relative homology class of S^3, which may be interpreted according to the above as a two-dimensional embedded compact submanifold. The circular boundary of this singular disk is a closed and nowhere dense subset with respect to an open set of S^3. Analogously, we may consider the excision of more than one singular disks from S^3, such that their circular boundaries collectively define a closed and nowhere dense subset of an open set of S^3. We propose to think of these circular singular boundaries as giving rise to topological links.

The notion of a topological link is based on the underlying idea of connectivity among a collection of topological circles, called simply loops [47]. We consider that a loop is a tame closed curve. The property of tameness means that a closed curve can be deformed continuously and without self-intersections into a polygonal one, that is a closed curve formed by a finite collection of straight-line segments. Given this qualification, a loop is characterized by the following properties: First, it is a one-dimensional object. Second, it is bounded, meaning that it is contained in some sphere of sufficiently large radius. Third, a single cut at a point cannot separate a loop into two pieces, whereas any set of two cuts at two different points does separate a loop into two pieces. Moreover, a loop is called knotted if it cannot be continuously deformed into a circle without self-intersection. We only consider unknotted tame closed curves. A topological N-link is a collection of N loops, where N is a natural number. Regarding the connectivity of a collection of N loops, the crucial property is the property of the splittability of the corresponding N-link. We say that a topological N-link is splittable if it can be deformed continuously, such that part of the link lies within B and the rest of the link lies within C, where B, C denote mutually-exclusive solid spheres (balls). Intuitively, the property of splittability of an N-link means that the link can come at least partly apart without cutting. Complete splittability means that the link can come completely apart without cutting. On the other side, non-splittability means that not even one of the involved loops, or any pair of them, or any combination of them, can be separated from the rest without cutting.

According to our hypothesis, a collection of circular singular boundaries defining a closed and nowhere dense subset of an open set of S^3 gives rise to a topological link in S^3. We may now replace the loop components of such a topological link by open non-intersecting tubular neighborhoods such that the complement of the link in S^3 can be given by the structure of a three-dimensional compact and oriented manifold with a boundary. Clearly, this space is homologically equivalent to the original one since it is just its deformation retracted. Next, we may consider an ordering of the loops $l_1, l_2, \ldots l_N$ constituting the link, or equivalently, an ordering of their tubular neighborhoods $\lambda_1, \lambda_2, \ldots \lambda_N$. Then, if we take λ_i, λ_j, together with their ordering, we define the relative homology class σ_{ij} that is represented by the compact oriented embedded submanifold whose two boundary components lie on the total boundary, that is the first one in $\partial \lambda_i$ and the second in $\partial \lambda_j$. The orientation is defined as being negative on the first boundary component and positive on the second, so that we have a path from λ_i to λ_j in this case.

7. The Borromean Rings as a Universal Nowhere Dense Singular Link

According to the formalism of geometrodynamics, we consider the Lorentzian manifold $(\mathbb{R} \times \Sigma, \varepsilon^* g)$ as a representative of X, where the singularities are localized within the three-dimensional manifold Σ. We remind that, according to Gannon's theorem, if Σ is multiple-connected as a topological space, then spacetime is geodesically incomplete. According to our previous analysis, a collection of circular singular boundaries defining a closed and nowhere dense subset of an open set of S^3 gives rise to a topological link in S^3. Moreover, this implies the existence of active gravitational mass/energy in the complementary open dense subsets, which is non-zero and strictly positive.

In this context, it is important to examine if there exists a universal way via which we can obtain the three-dimensional manifold Σ by the information incorporated in a topological link in S^3 representing the singular boundaries, forming collectively a closed and nowhere dense subset. This sheds more light on the role of the algebra sheaves $\mathbb{A}_L$ utilized to express gravitational properties defined on dense open sets of X and by restriction to dense open sets of Σ and is guiding in our quest of exploring generalized wormhole-types of solutions based on topological links and their associated homology classes.

It turns out that a universal way to obtain Σ by using a topological link in S^3 representing the singular boundaries, according to the above, actually exists and is based on the notion of a universal topological link. In view of the type of solutions we are interested in, such a universal link is defined by the Borromean rings. In particular, using methods of geometric topology, it can be shown that any compact oriented three-dimensional manifold Σ without boundary can be obtained as the branched covering space of the three-sphere S^3 with the branch set the Borromean rings [48]. In this manner, the Borromean rings constitute a universal topological link.

The notion of a branched covering space is a generalization of the standard notion of a covering space, characterized as a local homeomorphism bearing the unique path lifting and homotopy lifting property [49]. More precisely, a branched covering space of the three-sphere S^3 is considered as a map from Σ to S^3 such that this map is a covering space after we delete or exclude a locus of points, called the branched locus. The universality property says that Σ can be obtained in this way if the branched locus is formed by the Borromean rings, considered as a closed and nowhere dense set with respect to an open set in S^3 in our setting. In a well-defined sense, this branched covering space provides the geometric representation of an algebra sheaf of the form $\mathbb{A}_L$ restricted to the three spatial dimensions, where the closed and nowhere dense subset formed by the Borromean rings is localized. We may extend this closed and nowhere dense subset to four dimensions by considering a timelike axis perpendicular to the Borromean rings, which plays the role of a three-fold symmetry axis of rotation.

The Borromean rings consist of three rings localized in S^3, which are linked together in such a way that each of the rings lies completely over one of the other two, and completely under the other, as is shown in the pictures below:

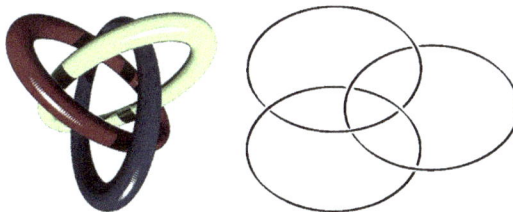

This particular type of topological linking displayed by the Borromean rings is called the Borromean link and is characterized by the following distinguishing property: if any one of the rings is removed from the Borromean link, the remaining two come completely apart. It is important to emphasize that the rings should be modeled in terms of unknotted tame closed curves and not as perfectly circular geometric circles. The adjective topological means that they can be deformed continuously under the constraint that the particular type of linkage forming the Borromean configuration is preserved.

From the viewpoint of the theory of topological links, the Borromean link constitutes an interlocking family of three loops, such that if any one of them is cut at a point and removed, then the remaining two loops become completely unlinked [47,49–52]. In more precise terms, the Borromean link is characterized topologically by the property of splittability as follows: The Borromean link is a non-splittable three-link (because it consists of three loops), such that every two-sub-link of this three-link is completely splittable. It is clear that it is a non-splittable three-link because not even one of the three loops, or any pair of them, can be separated from the rest without cutting. A two-sub-link is simply any sub-collection of two loops obtained by erasing the loop that does not belong to this sub-collection. Since the Borromean link is characterized by the property that if we erase any one of the three interlocking loops, then the remaining two loops become unlinked, it is clear that every two-sub-link of the non-splittable three-link is completely splittable.

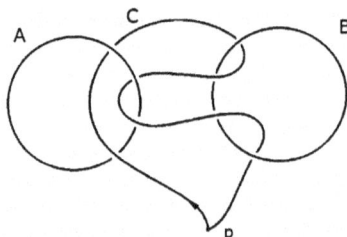

In our context, we conclude that if a triad of circular singular boundaries defining a closed and nowhere dense subset of an open set of S^3 are connected in the form of the Borromean topological link, then Σ as a compact oriented three-dimensional manifold can be obtained as the branched covering space of the three-sphere S^3 with the branch set these Borromean-linked boundaries. Based on these findings, we would like to explore their semantics in relation to the instantiation of a higher-order wormhole solution. For this purpose, we remind that the standard wormhole solution is thought of in terms of a one-dimensional homology class in a space homotopically equivalent to $S^1 \times S^2 - \{point\}$. In our framework, we do not need to impose a particular topology on Σ ab initio, since it can now be derived universally as the branched covering space of S^3 over the branch nowhere dense subset of singular boundaries forming a Borromean link. The fact that the Borromean link is a non-splittable three-link, such that every two-sub-link of this three-link is completely splittable, is characterized in homology theory by a non-vanishing triple Massey product, where all pairwise intersection products of one-dimensional homology classes vanish, reflecting the fact that the components of the Borromean link are not pairwise linked. If we denote the components of the Borromean link $\mathcal{B}$ by $\lambda_1, \lambda_2, \lambda_3$, the triple Massey product [49] is expressed as a two-dimensional cohomology class in the dense complement of $\mathcal{B}$ in S^3, that is it defines a non-trivial class in $H^2(S^3 \backslash (\lambda_1 \sqcup \lambda_2 \sqcup \lambda_3))$.

8. Conclusions

The main purpose of this communication, one hundred years after Einstein's formulation of the general theory of relativity, has been an invitation to re-think the validity domain of the field equations. The primary motivations emanate from three distinct sources: The first comes from Clarke's assertion concerning the problem of singularities, according to which, the answers "involve detailed considerations of distributional solutions to Einstein's equations, leading into an area that is only starting to be explored …". The second comes from Weyl's critical remark regarding the role of a background differential manifold, according to which, "while topology has succeeded fairly well in mastering continuity, we do not yet understand the inner meaning of the restriction to differential manifolds. Perhaps one day physics will be able to discard it". The third comes from Wheeler's ideas regarding the notions of "mass without mass" and "charge without charge" in the vacuum, which can be given a more precise mathematical formulation in topological terms.

The sheaf-theoretic re-formulation of the usual differential geometric framework of smooth manifolds points to the conclusion that there exist distinguishable extensions of the standard smooth manifold spacetime model of general relativity, which are utilized by appropriate extensions of the sheaf of coefficients parameterizing all tensorial physical quantities of the theory. The criteria of the suitability of these extensions are determined by sheaf-cohomological means and maintain the standard covariance properties of the theory in domains, including singular loci. We have presented and discussed in detail a concrete distribution-like sheaf of coefficients incorporating singularities in closed and nowhere dense subsets of an open set of a four-dimensional spacetime. An instructive way to think of these generalized algebra sheaves of coefficients refers to the role of a singularity as an obstruction to the existence of a solution to the field equations, expressed in terms of smooth coefficients. Thus, more generally, a singular locus may be thought of as a cohomological obstruction to the extendibility of a local smooth solution. This criterion incorporates and generalizes sheaf-cohomologically the initial definition of singular behavior in terms of the non-extendibility of geodesics. Essentially, the reason is that the notion of the extendibility of some local solution is of a sheaf-theoretic nature.

At a further stage involving the formalism of geometrodynamics, the existence of singular loci in closed and nowhere dense subsets of a spatial hypersurface, making it a multiply-connected topological space, implies active gravitational mass/energy in the complementary open dense subsets. Moreover, according to the "positive mass theorem" considered in the vacuum case, this gravitational mass/energy is non-zero and strictly positive. We show that it is enough for this purpose to consider singular boundaries forming closed and nowhere dense subsets and forcing a multiple-connected topology, which in turn implies that spacetime is geodesically incomplete. In view of expressing generalized wormhole solutions in this context, we propose that closed singular boundaries may form topological links. In this manner, using the results of geometric topology, we point out that the Borromean topological link is characterized as a universal link. Since this link is characterized cohomologically by a higher order invariant, it may be associated with a generalized wormhole model, which reinforces Wheeler's ideas in geometrodynamics.

Finally, we express the hope that the proposed approach paves the way for a further technical and semantical refinement of the following two of Einstein's fundamental insights in building up general relativity, which have not been addressed in satisfactory completeness up to the present:

> "Under the influence of the ideas of Faraday and Maxwell the notion developed that the whole of physical reality could perhaps be represented as a field whose components depend on four space-time parameters. If the laws of this field are in general covariant, that is, are not dependent on a particular choice of coordinate system, then the introduction of an independent (absolute) space is no longer necessary. That which constitutes the spatial character of reality is then simply the four-dimensionality of the field. There is then no "empty" space, that is, there is no space without a field." [53]

> "A field theory is not yet completely determined by the system of field equations. Should one admit the appearance of singularities? ... It is my opinion that singularities must be excluded. It does not seem reasonable to me to introduce into a continuum theory points (or lines etc.) for which the field equations do not hold ..." [54]

In a nutshell, regarding the first, the utilization of distribution-like sheaves of coefficients extending the smooth one over singularities, and thus, extending the domain of validity of the field equations beyond globally-smooth manifolds, shows in agreement with geometrodynamics that active gravitational mass/energy may emerge from purely topological considerations taking into account the constraints imposed by the field equations in the vacuum. These topological considerations pertain to the modeling of singularities in terms of closed and nowhere dense sets, such that their complements who bear the induced active gravitational mass/energy are open and dense. In this manner, the vacuum can be legitimately considered as a structural quality of the field itself. Regarding the second, it is indeed unreasonable to consider singular loci in a continuum theory, where the field equations do not hold. The existence of distribution-like sheaves of coefficients provides precisely the means to bypass this problem by coordinatizing all of the tensorial quantities in their terms, extending the smooth ones and, therefore, extending the domain of validity of the field equations.

Conflicts of Interest: The authors declare no conflict of interest.

References

1. Geroch, R. Einstein algebras. *Commun. Math. Phys.* **1972**, *26*, 271–275.
2. Mallios, A. On geometric topological algebras. *J. Math. Anal. Appl.* **1993**, *172*, 301–322.
3. Mallios, A. The de Rham-Kähler complex of the Gelfand sheaf of a topological algebra. *J. Math. Anal. Appl.* **1993**, *175*, 143–168.
4. Grothendieck, A. Sur quelques points d' algèbre homologique. *Tôhoku Math. J.* **1957**, *9*, 119–221.
5. Grothendieck, A. *A General Theory of Fiber Spaces with Structure Sheaf*; Univ. Kansas: Lawrence, KS, USA, 1958.
6. Mallios, A. *Geometry of Vector Sheaves: An Axiomatic Approach to Differential Geometry, Vol I: Vector Sheaves, General Theory*; Kluwer Academic Publishers: Dordrecht, The Netherlands, 1998.
7. Mallios, A. *Geometry of Vector Sheaves: An Axiomatic Approach to Differential Geometry, Vol II: Geometry Examples and Applications*; Kluwer Academic Publishers: Dordrecht, The Netherlands, 1998.
8. Vassiliou, E. *Geometry of Principal Sheaves*; Kluwer Academic Publishers: Dordrecht, The Netherlands, 2004.
9. Selesnick, S.A. Line bundles and harmonic analysis on compact groups. *Math. Z.* **1976**, *146*, 53–67.
10. Mallios, A. *Modern Differential Geometry in Gauge Theories: Vol. 1. Maxwell Fields*; Birkhäuser: Boston, MA, USA, 2006.
11. Mallios, A. *Modern Differential Geometry in Gauge Theories: Vol. 2. Yang-Mills Fields*; Birkhäuser: Boston, MA, USA, 2009.
12. Mallios, A. Quantum gravity and "singularities". *Note Mat.* **2006**, *25*, 57.
13. Mallios, A. Geometry and physics today. *Int. J. Theor. Phys.* **2006**, *45*, 1552–1588.
14. Mallios, A. A-invariance: An axiomatic approach to quantum relativity. *Int. J. Theor. Phys.* **2008**, *47*, 1929–1948.
15. Raptis, I. A dodecalogue of basic didactics from applications of abstract differential geometry to quantum gravity. *Int. J. Theor. Phys.* **2007**, *46*, 3009–3021.
16. Mallios, A. On localizing topological algebras. *Contemp. Math.* **2004**, *341*, 79.
17. Mallios, A. On algebra spaces. *Contemp. Math.* **2007**, *427*, 263.
18. Epperson, M.; Zafiris, E. *Foundations of Relational Realism: A Topological Approach to Quantum Mechanics and the Philosophy of Nature*; Lexington Books: Lanham, MD, USA, 2013.
19. Mallios, A.; Zafiris, E. *Differential Sheaves and Connections: A Natural Approach to Physical Geometry*; World Scientific: Singapore, 2016.
20. Zafiris, E. Boolean coverings of quantum observable structure: A setting for an abstract differential geometric mechanism. *J. Geom. Phys.* **2004**, *50*, 99–114.
21. Zafiris, E. Interpreting observables in a quantum world from the categorical standpoint. *Int. J. Theor. Phys.* **2004**, *43*, 265–298.
22. Zafiris, E. Quantum observables algebras and abstract differential geometry: The topos-theoretic dynamics of diagrams of commutative algebraic localizations. *Int. J. Theor. Phys.* **2007**, *46*, 319–382.
23. Mallios, A.; Rosinger, E.E. Abstract differential geometry, differential algebras of generalized functions, and de Rham cohomology. *Acta Appl. Math.* **1999**, *55*, 231.
24. Mallios, A.; Rosinger, E.E. Space-time foam dense singularities and de Rham cohomology. *Acta Appl. Math.* **2001**, *67*, 59–89.
25. Weyl, H. *Philosophy of Mathematics and Natural Science*; Princeton Univ. Press: Princeton, NJ, USA, 2009.
26. Raptis, I. Finitary-algebraic "resolution" of the inner Schwartzschild singularity. *Int. J. Theor. Phys.* **2006**, *45*, 79–128.

27. Mallios, A.; Raptis, I. Finitary, causal and quantal vacuum einstein gravity. *Int. J. Theor. Phys.* **2003**, *42*, 1479–1619.

28. Clarke, C.J.S. *The Analysis of Space-Time Singularities*; Cambridge University Press: Cambridge, UK, 1993.

29. Hawking, S.W.; Ellis, G.F.R. *The Large Scale Structure of Space-Time*; Cambridge University Press: Cambridge, UK, 1973.

30. Bosshard, B. On the b-boundary of the closed friedmann models. *Commun. Math. Phys.* **1976**, *46*, 263–268.

31. Schmidt, B.G. A new definition of singular points in general relativity. *Gen. Relat. Grav.* **1971**, *1*, 269–280.

32. Heller, M.; Sasin, W. Structured spaces and their application to relativistic physics. *J. Math. Phys.* **1995**, *36*, 3644–3662.

33. Von Müller, A. *The Forgotten Present. In: Re-Thinking Time at the Interface of Physics and Philosophy*; von Müller, A., Filk, T., Eds.; Springer: Heidelberg, Germany, 2015; pp. 1–46.

34. Vishwakarma, R.G. Mysteries of $R_{ik} = 0$: A novel paradigm in Einstein's theory of gravitation. *Front. Phys.* **2014**, *9*, 98–112.

35. Rosinger, E.E. *Distributions and Nonlinear Partial Differential Equations*; Springer: Berlin, Germany, 1978.

36. Rosinger, E.E. Nonlinear Partial Differential Equations, Sequential and Weak Solutions. In *North Holland Mathematics Studies*; Elsevier: Amsterdam, The Netherlands, 1980.

37. Rosinger, E.E. Generalized Solutions of Nonlinear Partial Differential Equations. In *North Holland Mathematics Studies*; Elsevier: Amsterdam, The Netherlands, 1987.

38. Rosinger, E.E. Nonlinear Partial Differential Equations, an Algebraic View of Generalized Solutions. In *North Holland Mathematics Studies*; Elsevier: Amsterdam, Netherlands, 1990.

39. Rosinger, E. How to solve smooth nonlinear PDEs in algebras of generalized functions with dense singularities. *Appl. Anal.* **2001**, *78*, 355–378.

40. Rosinger, E.E. Differential algebras with dense singularities on manifolds. *Acta Appl. Math.* **2007**, *95*, 233–256.

41. Misner, C.W.; Thorne, K.S.; Wheeler, J.A. *Gravitation*; W. H. Freeman and Company: New York, NY, USA, 1970.

42. Arnowitt, R.; Deser, S.; Misner, C.W. The dynamics of general relativity. In *Gravitation: An Introduction to Current Research*; Witten, L., Ed.; John Wiley and Sons: New York, NY, USA; London, UK, 1962; pp. 227–265.

43. Misner, C.W.; Wheeler, J.A. Classical physics as geometry: Gravitation, electromagnetism, unquantized charge, and mass as properties of empty space. *Ann. Phys.* **1957**, *2*, 525–603.

44. Wheeler, J.A. On the nature of quantum geometrodynamics. *Ann. Phys.* **1957**, *2*, 604–614.

45. Gannon, D. Singularities in nonsimply connected space-times. *J. Math. Phys.* **1975**, *16*, 2364–2367.

46. Scorpan, A. *The Wild World of 4-Manifolds*; Americal Mathematical Society: Providence, RI, USA, 2005.

47. Kawauchi, A. *A Survey of Knot Theory*; Springer: Berlin, Germany, 1996.

48. Hilden, H.M.; Losano, M.T.; Montesinos, J.M.; Whitten, W. On universal groups and 3-manifolds. *Invent. Math.* **1987**, *87*, 441–456.

49. Hatcher, A. *Algebraic Topology*; Cambridge University Press: Cambridge, UK, 2002.

50. Cromwell, P.; Beltrami, E.; Rampichini, M. The borromean rings. *Math. Intell.* **1998**, *20*, 53–62.

51. Debrunner, H. Links of brunnian type. *Duke Math. J.* **1961**, *28*, 17–23.

52. Lindström, B.; Zetterström, H.-O. Borromean circles are impossible. *Am. Math. Mon.* **1991**, *98*, 340–341.

53. Jammer, M. *Concepts of Space: The History of Theories of Space in Physics; Foreword by Albert Einstein*, 3rd ed.; Dover: Mineola, NY, USA, 1993.

54. Einstein, A. *The Meaning of Relativity*, 5th ed.; Princeton University Press: Princeton, NJ, USA, 1956.

universe

|MDPI|

Review

Einstein and Beyond: A Critical Perspective on General Relativity

Ram Gopal Vishwakarma

Unidad Académica de Matemáticas, Universidad Autónoma de Zacatecas, Zacatecas, ZAC C.P. 98000, Mexico; vishwa@uaz.edu.mx

Academic Editors: Lorenzo Iorio and Elias C. Vagenas
Received: 4 January 2016; Accepted: 11 May 2016; Published: 30 May 2016

Abstract: An alternative approach to Einstein's theory of General Relativity (GR) is reviewed, which is motivated by a range of serious theoretical issues inflicting the theory, such as the cosmological constant problem, presence of non-Machian solutions, problems related with the energy-stress tensor T^{ik} and unphysical solutions. The new approach emanates from a critical analysis of these problems, providing a novel insight that the matter fields, together with the ensuing gravitational field, are already present inherently in the spacetime without taking recourse to T^{ik}. Supported by lots of evidence, the new insight revolutionizes our views on the representation of the source of gravitation and establishes the spacetime itself as the source, which becomes crucial for understanding the unresolved issues in a unified manner. This leads to a new paradigm in GR by establishing equation $R^{ik} = 0$ as the field equation of gravitation plus inertia in the very presence of matter.

Keywords: gravitation; general relativity; fundamental problems and general formalism; Mach's principle

1. Introduction

The year 2015 marks the centenary of the advent of Albert Einstein's theory of General Relativity (GR), which constitutes the current description of gravitation in modern physics. It is undoubtedly one of the towering theoretical achievements of 20th-century physics, which is recognized as an intellectual achievement par excellence.

Einstein first revolutionized, in 1905, the concepts of absolute space and absolute time by superseding them with a single four-dimensional spacetime fabric, which only had an absolute meaning. He discovered this in his theory of Special Relativity (SR), which he formulated by postulating that the laws of physics are the same in all non-accelerating reference frames and the speed of light in vacuum never changes. He then made a great leap from SR to GR through his penetrating insight that the gravitational field in a small neighborhood of spacetime is indistinguishable from an appropriate acceleration of the reference frame (principle of equivalence), and hence gravitation can be added to SR (which is valid only in the absence of gravitation) by generalizing it for the accelerating observers. This leads to a curved spacetime.

This dramatically revolutionized the Newtonian notion of gravitation as a force by heralding that gravitation is a manifestation of the dynamically curved spacetime created by the presence of matter. The principle of general covariance (the laws of physics should be the same in all coordinate systems, including the accelerating ones) then suggests that the theory must be formulated by using the language of tensors. This leads to the famous Einstein equation:

$$G^{ik} \equiv R^{ik} - \frac{1}{2}g^{ik}R = -\frac{8\pi G}{c^4}T^{ik} \tag{1}$$

which represents how geometry, encoded in the left-hand side (which is a function of the spacetime curvature), behaves in response to matter encoded in the energy-momentum-stress tensor T^{ik}. [Here, as usual, g^{ik} is the contravariant form of the metric tensor g_{ik} representing the spacetime geometry, which is defined by $ds^2 = g_{ik}dx^i dx^k$. R^{ik} is the Ricci tensor defined by $R^{ik} = g_{hj}R^{hijk}$ in terms of the Riemann tensor R^{hijk}. $R = g_{ik}R^{ik}$ is the Ricci scalar and G^{ik} the Einstein tensor. T^{ik} is the energy-stress tensor of matter (which can very well absorb the cosmological constant or any other candidate of dark energy). G is the Newtonian constant of gravitation and c the speed of light in vacuum. The Latin indices range and sum over the values 0, 1, 2, 3 unless stated otherwise.] This, in a sense, completes the identification of gravitation with geometry. It turns out that the spacetime geometry is no longer a fixed inert background, rather it is a key player in physics, which acts on matter and can be acted upon. This constitutes a profound paradigm shift.

The theory has made remarkable progress on both theoretical and observational fronts [1–5]. It is remarkable that, born a century ago out of almost pure thought, the theory has managed to survive extensive experimental/observational scrutiny and describes accurately all gravitational phenomena ranging from the solar system to the largest scale—the Universe itself. Nevertheless, a number of questions remain open. On the one hand, the theory requires the dark matter and dark energy—two of the largest contributions to T^{ik}—which have entirely mysterious physical origins and do not have any non-gravitational or laboratory evidence. On the other hand, the theory suffers from profound theoretical difficulties, some of which are reviewed in the following. Nonetheless, if a theory requires more than 95% of "dark entities" in order to describe the observations, it is an alarming signal for us to turn back to the very foundations of the theory, rather than just keep adding epicycles to it.

Although Einstein, and then others, were mesmerized by the "inner consistency" and elegance of the theory, many theoretical issues were discovered even during the lifetime of Einstein which were not consistent with the founding principles of GR. In the following, we provide a critical review of the historical development of GR and some ensuing problems, most of which are generally ignored or not given the proper attention they deserve. This review will differ from the conventional reviews in the sense that, unlike most of the traditional reviews, it will not recount a well-documented story of the discovery of GR, rather it will focus on some key problems which insinuate an underlying new insight on a geometric theory of gravitation, thereby providing a possible way out in the framework of GR itself.

2. Issues Warranting Attention: Mysteries of the Present with Roots in the Past

Mach's Principle: Mach's principle, akin to the equivalence principle, was the primary motivation and guiding principle for Einstein in the formulation of GR. (The name "Mach's principle" was coined by Einstein for the general inspiration that he found in Mach's works on mechanics [6], even though the principle itself was never formulated succinctly by Mach himself.) Though in the absence of a clear statement from Ernst Mach, there exist a number of formulations of Mach's principle, in essence the principle advocates to shun all vestiges of the unobservable absolute space and time of Newton in favor of the directly observable background matter in the Universe, which determines its geometry and the inertia of an object.

As the principle of general covariance (non-existence of a privileged reference frame) emerges as a consequence of Mach's denial of absolute space, Einstein expected that his theory would automatically obey Mach's principle. However, it turned out not to be so, as there appear several anti-Machian features in GR. According to Mach's principle, the presence of a material background is essential for defining motion and a meaningful spacetime geometry. This means that an isolated object in an otherwise empty Universe should not possess any inertial properties. However, this is clearly violated by the Minkowski solution, which possesses timelike geodesics and a well-defined notion of inertia in the total absence of T^{ik}. Similarly, the cosmological constant also violates Mach's principle (if it does not represent the vacuum energy, but just a constant of nature—as is believed by some authors) in the sense in which the geometry should be determined completely by the mass distribution. In the same vein, there exists a class of singularity-free curved solutions, which admit Einstein's equations in the absence of T^{ik}. Furthermore, a global rotation, which is not allowed by Mach's principle (in the absence of an absolute frame of reference), is revealed in the Gödel solution [7], which describes a Universe with a uniform rotation in the whole spacetime.

After failing to formulate GR in a fully Machian sense, Einstein himself moved away from Mach's principle in his later years. Nevertheless, the principle continued to attract a lot of sympathy due to its aesthetic appeal and enormous impact, and it is widely believed that a viable theory of gravitation must be Machian. Moreover, the consistency of GR with SR, which abolishes the absolute space akin to Mach's principle, also persuades us that GR must be Machian. This characterization has however remained just wishful thinking.

Equivalence Principle: The equivalence principle—the physical foundation of any metric theory of gravitation—first expressed by Galileo and later reformulated by Newton, was assumed by Einstein as one of the defining principles of GR. According to the principle, one can choose a locally inertial coordinate system (LICS) (*i.e.*, a freely-falling one) at any spacetime point in an arbitrary gravitational field such that within a sufficiently small region of the point in question, the laws of nature take the same form as in unaccelerated Cartesian coordinate systems in the absence of gravitation [8]. As has been mentioned earlier, this equivalence of gravitation and accelerated reference frames paved the way for the formulation of GR. Since the principle rests on the conviction that the equality of the gravitational and inertial mass is exact [8,9], one expects the same to hold in GR solutions. However, the inertial and the (active) gravitational mass have remained unequal in general. For instance, for the case of T^{ik} representing a perfect fluid:

$$T^{ik} = (\rho + p)u^i u^k - pg^{ik} \qquad (2)$$

various solutions of Equation (1) indicate that the inertial mass density (=passive gravitational mass density) $= (\rho + p)/c^2$, while the active gravitational mass density $= (\rho + 3p)/c^2$, where ρ is the energy density of the fluid (which includes all the sources of energy of the fluid except the gravitational field energy) and p is its pressure. The binding energy of the gravitational field is believed to be responsible for this discrepancy. However, why the contributions from the gravitational energy to the different masses are not equal, has remained a mystery.

T^{ik} **and Gravitational Energy:** Appearing as the source term in Equation (1), T^{ik} is expected to include all the inertial and gravitational aspects of matter, *i.e.*, all the possible sources of gravitation. However, this requirement does not seem to be met on at least two counts. Firstly, T^{ik} fails to support, in a general spacetime with no symmetries, an unambiguous definition of angular momentum, which is a fundamental and unavoidable characteristic of matter, as is witnessed from the subatomic to the galactic scales. While a meaningful notion of the angular momentum in GR always needs the introduction of some additional structure in the form of symmetries, quasi-symmetries, or some other background structure, it can be unambiguously defined only for isolated systems [10,11].

Secondly, T^{ik} fails to include the energy of the gravitational field, which also gravitates. Einstein and Grossmann emphasized that, *akin to all other fields, the gravitational field must also have an energy-momentum tensor which should be included in the "source term"* [9]. However, after failing to find a tensor representation of the gravitational field, Einstein then commented that *"there may very well be gravitational fields without stress and energy density"* [12] and finally admitted that *"the energy tensor can be regarded only as a provisional means of representing matter"* [13]. Alas, a century-long dedicated effort to discover a unanimous formulation of the energy- stress tensor of the gravitational field, has failed concluding that a proper energy-stress tensor of the gravitational field does not exist. [It can be safely said that despite the century-long dedicated efforts of many luminaries, like Einstein, Tolman, Papapetrou, Landau-Lifshitz, Möller and Weinberg, the attempts to discover a unanimous formulation of the gravitational field energy has failed due to the following three reasons: (i) the non-tensorial character of the energy-stress 'complexes' (pseudo tensors) of the gravitational field; (ii) the lack of a unique agreed-upon formula for the gravitational field pseudo tensor in view of various formulations thereof, which may lead to different distributions even in the same spacetime background. Moreover, a pseudo tensor, unlike a true tensor, can be made to vanish at any pre-assigned point by an appropriate transformation of coordinates, rendering its status rather nebulous; (iii) according to the equivalence principle, the gravitational energy cannot be localized.] Since then, neither Einstein nor anyone else has been able to discover the true form of T^{ik}, although it is at the heart of the current efforts to reconcile GR with quantum mechanics.

It is an undeniable fact that the standards of T^{ik}, in terms of elegance, consistency and mathematical completeness, do not match the vibrant geometrical side of Equation (1), which is determined almost uniquely by pure mathematical requirements. Einstein himself conceded this fact when he famously remarked: *"GR is similar to a building, one wing of which is made of fine marble, but the other wing of which is built of low grade wood"*. It was his obsession that attempts should be directed to convert the "wood" into "marble".

The doubt envisioned by Einstein about representing matter by T^{ik}, is further strengthened by a recent study which discovers some surprising inconsistencies and paradoxes in the formulation of the energy-stress tensor of the matter fields, concluding that the formulation of T^{ik} does not seem consistent with the geometric description of gravitation [14]. This is reminiscent of the view expressed about four decades ago by J. L. Synge, one of the most distinguished mathematical physicists of the 20th Century: *"the concept of energy-momentum* (tensor) *is simply incompatible with general relativity"* [15] (which may seem radical from today's mainstream perspective).

Unphysical Solutions: Since its very inception, GR started having observational support which substantiated the theory. Its predictions have been well-tested in the limit of the weak gravitational field in the solar system, and in the stronger fields present in the systems of binary pulsars. This has been done through two solutions—the Schwarzschild and Kerr solutions.

However, there exist many other 'vacuum' solutions of Equation (1) which are considered *unphysical*, since they represent curvature in the absence of any conventional source. The solutions falling in this category are the de Sitter solution, Taub-NUT Solution, Ozsváth–Schücking solution and two newly discovered [16,17] solutions (given by Equations (6) and (7) in the following). (Another solution, which falls in this category, is the Gödel solution which admits closed timelike-curves and hence permits a possibility to travel in the past, violating the concepts of causality and creating paradoxes: "what happens if you go back in the past and kill your father when he was a baby!") Hence the theory has been supplemented by additional "physical grounds" that are used to exclude otherwise exact solutions of Einstein's equation.

This situation is very reminiscent of what Kinnersley wrote about the GR solutions, *"most of the known exact solutions describe situations which are frankly unphysical"* [18].

This is however misleading because not only does it reject *a priori* the majority of the exact solutions claiming "unphysical" and "extraneous", but also mars the general validity of the theory and introduces an element of subjectivity in it. Perhaps we fail to interpret a solution correctly and pronounce it unphysical because the interpretation is done in the framework of the conventional wisdom, which may not be correct [14,19].

Interior Solutions: As mentioned earlier, GR successfully describes the gravitational field outside the Sun in terms of the Schwarzschild (exterior) and Kerr solutions. Nevertheless, the theory has not been that successful in describing the interior of a massive body.

Soon after discovering his famous and successful (exterior) solution (with $T^{ik} = 0$), Schwarzschild discovered another solution of Equation (1) (with a non-zero T^{ik}) representing the interior of a static, spherically symmetric non-rotating massive body, generally called the Schwarzschild interior solution. SInce then, many other, similar interior solutions have been discovered with different matter distributions. It appears, however, that the picture the conventional interiors provide is not conceptually satisfying. For example, the Schwarzschild-interior solution assumes a static sphere of matter consisting of an incompressible perfect fluid of constant density (in order to obtain a mathematically simple solution). Hence, the solution turns out to be unphysical, since the speed of sound $= c\sqrt{dp/d\rho}$ becomes infinite in the fluid with a constant density ρ and a variable pressure p.

The Kerr solution, representing the exterior of a rotating mass, has remained unmatched to any known non-vacuum solution that could represent the interior of a rotating mass. It seems that we have been searching for the interior solutions in the wrong place [17].

Dark Matter and Dark Energy: Soon after formulating GR, Einstein applied his theory to model the Universe. At that time, Einstein believed in a static Universe, perhaps guided by his religious conviction that the Universe must be eternal and unchanging. As Equation (1) in its original form does not permit a static Universe, he inserted a term—the famous 'cosmological constant Λ' to force the equation to predict a static Universe. However, it was realized later that this gave an unstable Universe. It was then realized that a naive prediction of Equation (1) was an expanding Universe, which subsequently found consistent with the observations. Realizing this, Einstein retracted the introduction of Λ terming it his *"biggest blunder"*.

The cosmological constant has however reentered the theory in the guise of dark energy. As has been mentioned earlier, in order to explain various observations, the theory requires two mysterious, invisible, and as yet unidentified ingredients—dark matter and dark energy—and Λ is the principal candidate of dark energy.

One the one hand, the theory predicts that about 27% of the total content of the Universe is made of non-baryonic dark matter particles, which should certainly be predicted by some extension of the Standard Model of particles physics. However, there is no indication of any new physics beyond the Standard Model which has been successfully verified at the Large Hadron Collider. Curious discrepancies also appear to exist between the predicted clustering properties of dark matter on small scales and observations. Obviously, the dark matter has eluded our every effort to bring it out of the shadows.

On the other hand, the dark energy is believed to constitute about 68% of the total content of the Universe. The biggest mystery is not that the majority of the content of T^{ik} cannot be seen, but that it cannot be comprehended. Moreover, the most favored candidate of dark energy—the cosmological constant Λ—poses serious conceptual issues, including the cosmological constant problem—"why does Λ appear to take such an unnatural value?" That is, "why is the observed value of the energy associated with Λ so small (by a factor of $\approx 10^{-120}$!) compared to its value (Planck mass) predicted by the quantum field theory?" and the coincidence problem—"why is this observed value so close to the present matter density?".

The cosmological constant problem in fact arises from a structural defect of the field Equation (1). While in all non-gravitational physics, the dynamical equations describing a system do not change if we shift the "zero point" of energy, this symmetry is not respected by Equation (1) wherein all sources of energy and stress appear through T^{ik} and hence gravitate (*i.e.*, affect the curvature). As the Λ-term can very well be assimilated in T^{ik}, adding this constant to Equation (1) changes the solution. It may be noted that no dynamical solution of the cosmological constant problem is possible within the existing framework of GR [20].

Horizon Problem: Why does the cosmic microwave background (CMB) radiation look the same in all directions despite being emitted from regions of space failing to be causally connected? The size of the largest coherent region on the last scattering surface, in which the homogenizing signals passed at sound speed, can be measured in terms of the sound horizon. In the standard cosmology, this implies, however, that the CMB ought to exhibit large anisotropies (*not isotropy*) for angular scales of theorder of 1° or larger—a result contrary to what is observed [8]. Hence, it seems that the isotropy of the CMB cannot be explained in terms of some physical process operating under the principle of causality in the standard paradigm.

Inflation comes to the rescue. It is generally believed that inflation made the Universe smooth and left the seeds of structures, on the surface of the last scatter, of the order of the Hubble distance at that time. However, inflation has its own problems either unsolved or fundamentally unresolvable. There is no consensus on which (if any) the inflation model is correct, given that there are many different inflation models. A physical mechanism that could cause inflation is not known, though there are many speculations. There are also difficulties on how to turn off the inflation once it starts—the "graceful exit" problem.

Flatness Problem: In the standard cosmology, the total energy density ρ in the early Universe appears to be extremely fine-tuned to its critical value $\rho_c = 3H^2/(8\pi G)$, which corresponds to a flat spatial geometry of the Universe, where H is the Hubble parameter. Since ρ departs rapidly from ρ_c over cosmic time, even a small deviation from this value would have had massive effects on the nature of the present Universe. For instance, the theory requires ρ at the Planck time to be within one part in 10^{57} of ρ_c in order to meet the observed uncertainties in ρ at present! That is, the Universe was almost flat just after the Big Bang—but how?

If a theory predicts a fine-tuned value for some parameter, there should be some underlying physical symmetry in the theory. In the present case however, this appears just an unnatural and *ad hoc* assumption in order to reproduce observation. Inflation comes to the rescue again. Irregularities in the geometry were evened out by inflation's rapid accelerated expansion causing space to become flatter and hence forcing ρ toward its critical value, no matter what its initial value was.

However, it should also be mentioned that flatness and horizon problems are not problems of GR. Rather, they are problems concerned with the cosmologist's conception of the Universe, very much in the same vein as was Einstein's conception of a static Universe.

Scale Invariance: It is well-known that GR, unlike the rest of physics, is not scale invariant in the field Equation (1) [21]. As scale invariance is one of the most fundamental symmetries of physics, any physical theory, including GR, is desired to be scale invariant.

3. A New Perspective on Gravity

Hence, with a substantial amount of anomalies, paradoxes and unexplained phenomena, one would question whether the pursued approach to GR is correct. It appears that we have misunderstood the true nature of a geometric theory of gravitation because of the way the theory has evolved. Taken at face value, these problems insinuate that our understanding of gravitation in terms of the conventional GR is grossly incomplete (if not incorrect) and we need yet another paradigm shift.

Science advances more from what we do not understand than by what we do understand. From a careful re-examination of the above-mentioned problems, a new insight with deeper vision of a geometric theory of gravitation emerges, which appears as the missing piece of the theory. It may appear surprising at first sight though that these seemingly disconnected problems can lead to any coherent, meaningful solution. Nevertheless, as we shall see in the following, the analysis develops drastic revolutionary changes in our conventional views of GR and offers an enlightened view wherein all the above-mentioned difficulties disappear.

3.1. Revisiting Mach's Principle

Guided by the principle of covariance, GR has been formulated in the language of tensors. As the principle of covariance results as a consequence from Mach's principle, one naturally expects the theory to be perfectly Machian, as Einstein did. Then, why do some of the solutions of GR contradict Mach's philosophy? Perhaps we have missed the real message these solutions want to convey. Particularly, the curious presence of the timelike geodesics and a well-defined notion of inertia in the solutions of Equation (1) obtained in the absence of T^{ik} must not be just coincidental and there must be some source.

In order to witness this, let us try to impose the philosophy of Mach on the existing framework of GR by quantifying Mach's principle with a precise formulation in which matter and geometry appear to be in one-to-one correspondence. The key insight is the observation that not only inertia, but also space and time emerge from the interaction of matter. As space is an abstraction from the totality of distance-relations between matter, it follows that the existence of matter (fields) is a necessary and sufficient condition for the existence of spacetime. This idea can be formulated in terms of the following postulate:

Postulate: *Spacetime cannot exist in the absence of fields.*

The postulate posits that spacetime is not something to which one can ascribe a separate existence, independently of the matter fields, and the very existence of spacetime signifies the presence of the matter (fields). This is very much in the spirit of Mach's principle which implies that the existence of a spacetime structure has any meaning only in the presence of matter, which is bound so tightly to the former that one can not exist without the other.

Inspired by this, Einstein had envisioned that *"space as opposed to 'what fills space', has no separate existence"* [22] thought he could not implement it in his field Equation (1), wherein the "space" (represented in the left-hand side of the equation) and "what fills space" (represented by its right-hand side) do have separate existence: as has been mentioned earlier, there exist various meaningful spacetime solutions of Equation (1) in the total absence of T^{ik}. The adopted postulate, on the other hand, emphasizes that spacetime has no independent existence without a material background, which is present universally regardless of the geometry of the spacetime.

As the matter field is always accompanied by the ensuing gravitational field and since the latter also gravitates, an important consequence of the adopted postulate is that the geometry of the resulting spacetime should be determined by the net contribution from the two fields. Thus, the metric field is entirely governed by considered matter fields, as one should expect from a Machian theory.

3.2. Fields without T^{ik}: An Inescapable Consequence of Mach's Principle

The theoretical appeal of the above-described hypothesis is that it is naive, self evident and plausible. However, more than that, it has potential to shape a theory and gives rise to a new vision of GR with novel, dramatic implications. For instance, it makes a powerful prediction that the resulting theory should not have any bearing on the energy-stress tensor T^{ik} in order to represent the source fields. [The source of curvature in a solution of Einstein's field Equation (1), in the absence of T^{ik}, is conventionally attributed to a singularity. This prescription is however rendered nebulous by the presence of various singularity-free curved solutions of Equation (1) in the absence of T^{ik}]. Let us recall that Equation (1) does admit various meaningful spacetime solutions in the absence of the "source" term T^{ik}.

According to the postulate, as fields are present universally in all spacetimes irrespective of their geometry, the flat Minkowskian spacetime should not be an exception, and it must also be endowed with the matter fields and the ensuing gravitational field. Now let us recall that the Minkowski spacetime appears as a solution of Einstein's field Equation (1) only in the absence of T^{ik}, in which case the effective field equation yields

$$R^{ik} = 0 \tag{3}$$

However, if the fields can exist in the Minkowski spacetime (as asserted by the founding postulate) in the absence of T^{ik}, they can also exist in other spacetimes in the absence of T^{ik}. Hence, the requirement of uniqueness of the field equation of a viable theory dictates that T^{ik} must not be the carrier of the source fields in a theory resulting from the adopted postulate, and, thus, the canonical Equation (3) emerges as the field equation of the resulting theory in the very presence of matter. In fact, this is what happens if we accept, at their face value, the implications of Mach's principle applied to GR.

This novel feature that GR would acquire—that the spacetime solutions of Equation (3), including the Minkowskian one, are not devoid of fields—provides an appealing first principle approach and a linchpin to understand various unsolved issues in a unified scheme. It becomes remarkably decisive for the theory on Machianity. It was the earlier-mentioned characteristic of the Minkowski and other solutions of Equation (3) to possess timelike geodesics and a well-defined notion of inertia, that pronounced these solutions non-Machian, as they are conventionally regarded to represent empty spacetimes. The new insight, however, renders them perfectly Machian and physically meaningful by bestowing a matter-full dignity on them. Moreover, this novel feature of the Minkowski solution also explains another so-far unexplained issue: It has been noticed that the Noether current associated with an arbitrary vector field in the Minkowski solution is non-zero in general [23], which remains unexplained in the conventional 'empty' Minkowskian spacetime.

Though the proposed scheme of having matter fields in the absence of T^{ik} may sound surprising and orthogonal to the prevailing perspective, it seems to have many advantages over the conventional approach, as we shall see in the following. The issue is whether it can be made realistic. That is, if Equation (3) is claimed to constitute the field equation of a viable theory of gravitation in the very presence of matter, its solutions must possess some imprint of this matter. Thus, do we have any evidence of such imprints in the solutions of Equation (3)? The answer is, yes.

3.3. Evidence of the Presence of Fields in the Absence of T^{ik}

As Mach's principle denies unobservable absolute spacetime in favor of the observable quantities (the background matter) which determine its geometry, the principle would expect the source of curvature in a solution to be attributable entirely to some directly observable quantity, such as mass-energy, momentum, and angular momentum or their densities. Thus, if GR is correct and it must be Machian, these quantities are expected to be supported by some dimension-full parameters appearing in the curved spacetime solutions in such a way that the parameters vanish as the observable quantities vanish, reducing the solutions to the Minkowskian form.

Interestingly, it has been shown recently [16,17] that it is always possible to write a curved solution of Equation (3) in a form containing some dimension-full parameters, which appear in the Riemann tensor generatively and can be attributed to the source of curvature. The study further shows that these parameters can support physical observable quantities such as the mass-energy, momentum or angular momentum or their densities. For instance, the source of curvature in the Schwarzschild solution

$$ds^2 = \left(1 + \frac{K}{r}\right) c^2 dt^2 - \frac{dr^2}{(1 + K/r)} - r^2 d\theta^2 - r^2 \sin^2\theta \, d\phi^2 \qquad (4)$$

can be attributed to the mass m (of the isotropic matter situated at $r = 0$) through the parameter $K = -2Gm/c^2$. Similarly, the dimension-full parameters present in the Kerr solution can be attributed to the mass and the angular momentum of the source mass; those in the Taub-NUT solution to the mass and the momentum of the source; and the parameters in the Kerr-NUT solution to the mass, momentum and angular momentum [16,17].

A remarkable piece of evidence of the presence of fields in the absence of T^{ik} is provided by the Kasner solution, which exemplifies that even in the standard paradigm, all the well-known curved solutions of Equation (3) do not represent space outside a gravitating mass in an empty space. [It is conventionally believed that only those curved solutions of Equation (3) are meaningful which represent space outside some source matter, otherwise the solutions represent an empty spacetime. However, Equation (3) cannot decipher just from the symmetry of a solution that it necessarily belongs to a spacetime structure in an empty space outside a mass, since the same symmetry can also be shared by a spacetime structure inside a matter distribution.] Although the Kasner solution in its standard form does not contain any dimension-full parameter that can be attributed to its curvature, the solution can however be transformed to the form

$$ds^2 = c^2 dt^2 - (1 + nt)^{2p_1} dx^2 - (1 + nt)^{2p_2} dy^2 - (1 + nt)^{2p_3} dz^2 \qquad (5)$$

where n is an arbitrary constant parameter (which is dimension-full) and the dimensionless parameters p_1, p_2, p_3 satisfy $p_1 + p_2 + p_3 = 1 = p_1^2 + p_2^2 + p_3^2$.

A dimensional analysis suggests that, in order to meet its natural dimension (which is of the dimension of the inverse of time), the parameter n can support only the densities of the observables energy, momentum or angular momentum and *not* the energy, momentum or angular momentum themselves [such that Equation (5) becomes Minkowskian when the observables vanish]. However, the energy density and the angular momentum density vanish here: while the symmetries of Equation (5) discard any possibility for the angular momentum density, the energy density disappears as it is canceled by the negative gravitational energy [17,24]. That is, the parameter n in Equation (5) can be expressed in terms of the momentum density $\mathcal{P}$ as $n = \gamma\sqrt{G\mathcal{P}/c}$, where γ is a dimensionless constant. This indicates that Equation (5) results from a (uniform) matter distribution (throughout space) and not from a spacetime outside a point mass as in the cases of the Schwarzschild and Kerr solutions. Thus, the Kasner solution represents a homogeneous distribution of matter expanding and contracting anisotropically (at different rates in different directions), which can give rise to a net non-zero momentum density represented through the parameter n serving as the source of curvature, thus demystifying the solution.

This new insight on the source of curvature is authenticated by two new solutions of Equation (3) discovered in [16,17] whose discovery is facilitated by the new insight. The first solution, whose source of curvature cannot be explained with the conventional wisdom (as it is singularity-free), provides a powerful support to the Machian strategy of representing the source in terms of the dimension-full source-carrier parameters (here ℓ). The solution is given by

$$ds^2 = \left(1 - \frac{\ell^2 x^2}{8}\right)c^2 dt^2 - dx^2 - dy^2 - \left(1 + \frac{\ell^2 x^2}{8}\right)dz^2 + \ell x(cdt - dz)dy + \frac{\ell^2 x^2}{4}cdt\,dz \qquad (6)$$

which has been derived by defining the parameter ℓ in terms of the angular momentum density $\mathcal{J}$ via $\ell = G\mathcal{J}/c^3$ [16]. The fact that the parameter ℓ can support only the density of angular momentum and *not* the angular momentum itself asserts that Equation (6) results from a rotating *matter distribution* (confined to $-\frac{2\sqrt{2}}{|\ell|} < x < \frac{2\sqrt{2}}{|\ell|}$) and not from a spacetime outside a point mass as are the cases of the Schwarzschild and Kerr solutions. This is in perfect agreement with the founding postulate that the fields are not different from the spacetime.

Equation (6) as a new solution of field Equation (3) is important in its own right. Moreover, it illuminates the so far obscure source of curvature in the well-known Ozsváth–Schücking solution, which would otherwise be in stark contrast with the new strategy in the absence of any free parameter. It has been shown in [16] that the Ozsváth–Schücking solution results from Equation (6) by assigning a particular value to the parameter ℓ.

Following the new insight, another new solution of Equation (3) has been discovered recently in [17], whose curvature is supported by the energy density (the author recently came to know that solution Equation (7) has also been reported in [25]). The solution is given by:

$$ds^2 = \frac{(1 + 4\mu z^2)}{(1 + \mu r^2)^2}c^2 dt^2 - \frac{dr^2}{(1 + \mu r^2)^4} - r^2 d\phi^2 - \frac{dz^2}{(1 + 4\mu z^2)(1 + \mu r^2)^2} \qquad (7)$$

which represents a inhomogeneous axisymmetric distribution of matter, with the parameter μ given in terms of the energy density $\mathcal{E}$ as $\mu = G\mathcal{E}/c^4$. As Equation (7) is curved but singularity-free for all finite values of the coordinates, it provides, in the absence of any conventional source there, a strong support to the new strategy of source representation.

3.4. A New Vision of Gravity in the Framework of GR: Spacetime Becomes a Physical Entity

What does the presence of these dimension-full parameters we witness in the solutions of the field Equation (3) signify? As the physical observable quantities sustained by the parameters—*i.e.*, energy, momentum, angular momentum and their densities have any meaning only in the presence of matter, the presence of such parameters in the solutions of Equation (3) must not be just a big coincidence, and, at face value, their ubiquitous presence in the solutions of Equation (3) insinuates that fields are universally present in the spacetime in Equation (3).

Not only does this provide a strong support to the founding postulate establishing GR as a Machian theory, but also establishes, on firm grounds, Equation (3) as the field equation of a feasible theory of gravitation in the very presence of fields. More than that, there emerges a radically new vision of a geometric theory of gravitation through drastic revolutionary changes in our views on the representation of the source of gravitation, which must be through the geometry and *not* through T^{ik}. By reconceptualizing our previous notions of spacetime, this constitutes a paradigm shift in GR wherein the spacetime itself becomes a physical entity, we may call it the "emergent matter" in a relativistic/geometric theory of gravitation. From the ubiquitous presence of fields in all geometries, it becomes clear that there is no empty space solution in the new paradigm, as one should expect from a Machian theory. The same was also envisioned by Einstein (though could not be achieved).

One may wonder how the properties of matter can be incorporated into the dynamical equations of the new theory without taking recourse to T^{ik}. This can be achieved by applying the conservation laws and symmetry principles to the new conviction that all spacetimes harbor fields, inertial and gravitational, whose net contribution determines their geometry. For instance, by assuming that the sum of the gravitational and inertial energies in a uniform matter distribution should be vanishing [17,24], it has been shown recently that the homogeneous, isotropic Universe in the new paradigm leads to the Friedmann equation of the standard "concordance" cosmology [17]. This should not be a surprise, as the Friedmann equation for dust can also be derived in Newtonian cosmology or in a kinematic theory (like the Milne model) by using the continuity equation and the Navier–Stokes equation of fluid dynamics [26,27].

3.5. Equivalence Principle in the New Perspective

The perfect equivalence between gravitational and inertial masses, first noted by Galileo and Newton, was more or less accidental. For Einstein, however, this served as a key to a deeper understanding of inertia and gravitation. From his valuable insight that the kinematic acceleration and the acceleration due to gravity are intrinsically identical, he was able to unearth a hitherto unknown mystery of nature—that gravitation is a geometric phenomenon.

It however seems that the full implications of the equivalence principle have not yet been appreciated. If gravitation is a geometric phenomenon, then through the (local) equivalence of gravitation and inertia, the inertia of matter should also be considered geometrical in nature, at least when it appears in a geometric theory of gravitation. A purely geometrical interpretation of gravitation would be impossible unless the gravitational as well as the inertial properties of matter are intrinsically geometrical. This would, however, have revolutionary implications. Considering T^{ik} (which represents the inertial fields) of a purely geometric origin, Equation (1) would imply

$$G^{ik} + \frac{8\pi G}{c^4} T^{ik} \equiv \chi^{ik} = 0 \qquad (8)$$

where χ^{ik} appears a tensor of purely geometric origin. This would however be nothing else but the Ricci tensor R^{ik} (with a suitable g_{ik}), since the only tensor of rank two having a purely geometric origin (emerging from the Riemann tensor), is the Ricci tensor. That is, Equation (8) would reduce to the field Equation (3)! In this way, the consequences of the equivalence principle would be in perfect agreement with the adopted Machian postulate—that spacetime has no separate existence from matter, *i.e.*, the parameters of the spacetime geometry determine entirely the combined effects of gravitation and inertia.

Therefore, the consequence of the equivalence principle—that the gravitational and inertial fields are entirely geometrical by nature—takes GR to its logical extreme in that the spacetime emerges from the interaction of matter. This reconceptualizes the previous notion of spacetime by establishing it as the very source of gravitation. The matter is in fact more intrinsically related to the geometry than is believed in the conventional GR and all the aspects of matter fields (including the ensuing gravitational field) are already present inherently in the spacetime geometry. This establishes Equation (3) as a competent field equation of gravitation plus inertia. This is well-supported by our observation that while the gravitational field is present in the Schwarzschild, Kerr and Taub-NUT solutions (as these represent the spacetimes outside the source mass), the inertial as well as the gravitational fields are present in Equations (5)–(7) including the Minkowskian one, which represent matter distribution.

A precise specification of the fields, which are being claimed to be present in the spacetime, is possible only when a precise formulation thereof is available. Nevertheless, in view of the newly gained insight, at least this much can be declared that the matter fields present in the geometry of Equation (3) are those which are attempted to be introduced in Equation (1) or (8) via T^{ik} (which has now been absorbed in Equation (3)).

4. A Closer Look at the Conventional Four-Dimensional Formulation of Matter

Modeling matter by T^{ik} in Equation (1) has modified at the deepest level the way we used to think about the source of gravitation. As mass density is the source of gravitation in Newtonian theory, the energy density was expected to take over this role in the relativistic generalization of Poisson's equation. To our surprise, however, all ten (independent) components of T^{ik} become contributing sources of gravitation. We need not doubt this novelty, as new theories originated from innovative ideas are expected to have innovative features. However, the way the non-conventional sources appear in the dynamical equations, appears to create inconsistencies and paradoxes, which warrants a second look at the relativist formulation of matter given by T^{ik}.

Everyone will agree that, like the conservation of momentum, the conservation of energy of an isolated system is an absolute symmetry of nature and this fundamental principle is expected to be respected by any physical theory. Nonetheless, the principle is violated in GR in many different situations including the cosmological scenarios (see, for example, [28]). The blame rests with the energy of the gravitational field, which has been of an obscure nature and a controversial history, as has been mentioned earlier. We shall, however, see that the gravitational energy is not to be blamed for the trouble. This is ascertained beyond a doubt in the following analysis by filtering out the gravitational energy from the equations.

4.1. Problems with T^{ik}

As is well-known, the formulation of the energy-stress tensor T^{ik} given by Equation (2) is obtained by first deriving it in the absence of gravity in SR, by considering a fluid element in a small neighborhood of an LICS, which exists admittedly at all points of spacetime (by courtesy of the principle of equivalence). Then, the expression for the tensor in the presence of gravity is imported, from SR to GR, through a coordinate transformation. It would be insightful to reconsider the same LICS to understand the mysterious implications of T^{ik}, since the subtleties of gravitation and the gravitational energy disappear locally in this coordinate system. Let us then study the divergence of T^{ik} in the considered LICS, which is known for describing the mechanical behavior of the fluid. Through the vanishing divergence of G^{ik}, Equation (1) implies that $T^{ij}{}_{;j} = 0$, which, in the chosen coordinates, reduces to

$$\frac{\partial T^{ij}}{\partial x^j} = 0 \tag{9}$$

For the case of a perfect fluid given by Equation (2), it is easy to show that Equation (9), in the chosen LICS, yields [29]

$$\frac{\partial p}{\partial x} + \frac{(\rho + p)}{c^2} \frac{du_x}{dt} = 0 \tag{10}$$

for the case $i = 1$, where du_x/dt is the acceleration of the considered fluid element in the x-direction. As any role of gravity and gravitational energy is absent in this equation, it can be interpreted as the relativistic analogue of the Newtonian law of motion: the fluid element of unit volume, which moves under the action of the force applied by the pressure gradient $\partial p / \partial x$, has got the inertial mass $(\rho + p)/c^2$. Let us, however, recall that the term ρ in Equation (2) includes in it, by definition, not only the rest mass of the individual particles of the fluid but also their kinetic energy, internal energy (for example, the energy of compression, energy of nuclear binding, *etc.*) and *all other sources of mass-energy* [11]. Therefore, the additional contribution to the inertial mass entering through the term p, appears to violate the celebrated law of the conservation of energy. Though Equation (10) is usually interpreted as a momentum conservation equation, an alternative (but viable) interpretation is not expected to defy the energy conservation.

Similar problems seem to afflict the temporal component of Equation (9) for $i = 0$, which can be written as the following [29]:

$$\frac{d}{dt}(\rho \delta v) + p \frac{d}{dt}(\delta v) = 0 \tag{11}$$

where δv is the proper volume of the fluid element. The usual interpretation to this equation says: the rate of change in the energy of the fluid element is given in terms of the work done against the external pressure. This seems reasonable at first sight, but cracks seem to appear in it after a little reflection. The concern, as also noticed by Tolman [29], is that the fluid of a finite size can be divided into similar fluid elements and the same Equation (11) can be applied to each of these elements, meaning that the proper energy $(\rho \delta v)$ of every element is decreasing when the fluid is expanding or increasing when the fluid is contracting. This leads to a paradoxical situation that the sum of the proper energies of the fluid elements which make up an isolated system, is not constant. Tolman overlooked this problem by assuming a possible role of the gravitational energy in it. We note, however, that no such possibility exists as Equation (11) has been derived in an LICS.

The total energy E, including the gravitational energy, of an isolated time-independent fluid sphere comprised of perfect fluid given by Equation (2) and occupying volume V of the three-space $x^0 = $ constant, is given by the Tolman formula [29]:

$$E = \int_V (\rho + 3p) \sqrt{|g_{00}|} \, dV \tag{12}$$

which measures the strength of the gravitational field produced by the fluid sphere. The formula is believed to be consistent, for the case of the disordered radiation ($p = \rho/3$), with the observed deflection of starlight (twice as much as predicted by a heuristic argument made in Newtonian gravity), when it passes the Sun. Ironically, this expectation is contradicted by the weak-field approximation of the same Equation (12). In a weak field, like that of the Sun, where Newtonian gravitation can be regarded as a satisfactory approximation, Equation (12) can be written, following Tolman (see page 250 of [29]), as $E = \int \rho dV + (1/2c^2) \int \rho \psi dV$, where ψ is the Newtonian gravitational potential. As ψ is negative, we note that the general relativistic active gravitational mass E/c^2 of the gravitating body, here the Sun, is obviously less than its Newtonian value $(1/c^2) \int \rho dV$ and is expected to give a lower value for the gravitational deflection of light than the corresponding Newtonian value. (Let us recall that the correct interpretation of the observations of the bending of starlight, when it passes past the Sun, comes from the correct geometry around Sun resulting from the Schwarzschild solution.

As has been mentioned earlier, the (active) gravitational and inertial mass are in general unequal in GR solutions (the discrepancy thereof is supposed to be accounted by the gravitational energy). Thus, in an LICS, which nullifies gravitation and hence gravitational mass locally, we expect a unique value for the mass in the equations. To check this, let us calculate the Tolman integral Equation (12) (density) in the considered LICS wherein it reduces to

$$E = \int (\rho + 3p) \, dV \tag{13}$$

which may now be valid for a sufficiently small volume of the fluid. Surprisingly, we still encounter different unequal values of mass (density) in Equations (10), (11) and (13). [Equation (11) can be written alternatively as $\delta v \, d\rho/dt + (\rho + p)d(\delta v)/dt = 0$.] While Equations (10) and (11) give this value as $(\rho + p)/c^2$, Equation (13) provides a different value $(\rho + 3p)/c^2$. Perhaps the origin of the problem is not in the gravitational energy but in T^{ik} itself.

Given this backdrop, it thus appears that the relativistic formulation of matter given by T^{ik} suffers from some subtle inherent problems. The point to note is that there is no role of the notorious (pseudo) energy of the gravitational field in these problems. It would not be correct to conclude that the above-analysis advocates denial of fluid pressure in GR (as the problems are evaded in the absence of pressure). Rather it insinuates that the four-dimensional description of matter in terms of T^{ik} is not compatible with the geometric description of gravitation. It is perhaps not correct to patchwork a four-dimensional tensor from two basically distinct kinds of three-dimensional quantities—(i) the energy density, a non-directional quantity and (ii) the momenta and stresses, directional quantities. The tensor, however, treats them on equal footing by recognizing a component T^{ik} as a scalar (irrespective of the values of i and k) linked with the surface specified by i and k in the *hypothetical* four-dimensional fluid, in the same way as the component G^{ik} is linked with the curvature of the same surface. This leads to sound mathematics, and we do not notice any inconsistency until we relate the tensor T^{ik} with the real fluid, which is three-dimensional and not four-dimensional.

Does it then mean that Einstein's "wood" is not only low grade compared to the standards of his "marble" but it is also infested? It should be noted that the relativistic formulation of the matter, in terms of the tensor T^{ik}, has never been tested in any direct experiment. It may be recalled that the crucial tests of GR, which have substantiated the theory beyond doubt, are based on the solutions of Equation (3) only, *viz.* the Schwarzschild and Kerr solutions.

It thus becomes increasingly clear that the development of GR was led astray by formulating matter in terms of T^{ik}. This is corroborated by the fact that whenever the theory takes recourse to T^{ik} in Equation (1), trouble shows up in the form of either the dark energy or the inviabilities of Godel's solution and Schwarzschild's interior solution, *etc.* In view of the new finding, this assertion acquires a new meaning—we have been searching for the matter in the wrong place. The correct place to search for it is the geometry. We have seen in innumerable examples that matter is already present in the geometry of Equation (3) without taking recourse to T^{ik}. That is, the "wood" is already included into the 'marble', dramatically fulfilling Einstein's obsession.

5. Successes of the Novel Gravity Formulation

5.1. Observational Support for the New Paradigm

The last words on a putative theory have to be spoken by observations and experiments. The consistency of the field Equation (3) with the local observations in the solar system and binary pulsars, has already been established in the standard tests of GR—the only satisfactory testimonial of the theory among the conventional tests, which do not require any epicycle of the dark sectors.

Interestingly, as has been shown recently [27], all the cosmological observations can also be explained successfully in terms of a homogeneous, isotropic solution of Equation (3). This solution can be obtained by solving Equation (3) for the Robertson–Walker metric, yielding

$$ds^2 = c^2dt^2 - c^2t^2 \left(\frac{dr^2}{1+r^2} + r^2d\theta^2 + r^2\sin^2\theta \, d\phi^2 \right) \tag{14}$$

which represents the homogeneous, isotropic Universe in the new paradigm. It may be mentioned that solution Equation (14) (which is generally recognized as the Milne model), wherein the Universe appears dynamic in terms of the comoving coordinates and the cosmic time t, can be reduced to the Minkowskian form by using the locally defined measures of space and time [27].

The observational tests considered in [27] include the observations of the high-redshift supernovae (SNe) Ia, the observations of high-redshift radio sources, observations of starburst galaxies, the CMB observations and compatibility of the age of the Universe with the oldest objects in it (for instance, the globular clusters) for the currently measured values of the Hubble parameter. It may also be mentioned that, by preforming a rigorous statistical test on a much bigger sample of SNe Ia, (by taking account of the empirical procedure by which corrections are made to their absolute magnitudes), a recent study has found only marginal evidence for an accelerated expansion, and the data are quite consistent with the Milne model [30].

One may wonder how the new model, which does not possess dark energy (and hence is not an accelerated expansion), manages to reconcile with the observations. The mystery lies in the special expansion dynamics of the model at a constant rate throughout the evolution, as is clear from Equation (14), wherein the Robertson–Walker scale factor $S = ct$. We note that, unlike the standard cosmology, Equation (14) provides efficiently different measures of distances without requiring any input from the matter fields. For instance, the luminosity distance d_L of a source of redshift z, in the present case, is given by

$$d_L = cH_0^{-1}(1+z)\sinh[\ln(1+z)] \tag{15}$$

where H_0 represents the present value of the Hubble parameter $H = \dot{S}/S$. As has been shown in Figure 1, the luminosity distance of an object of redshift z in the new cosmology is almost the same as that in the standard cosmology for $z \lesssim 1.3$. This explains why both models are equally consistent with the SNe Ia data wherein the majority of the SNe belong to this range of redshift. However, for $z > 1.3$, the new model departs significantly from the standard cosmology, as is clear from the figure. Hence, observations of more SNe Ia at higher redshifts will be decisive for both paradigms.

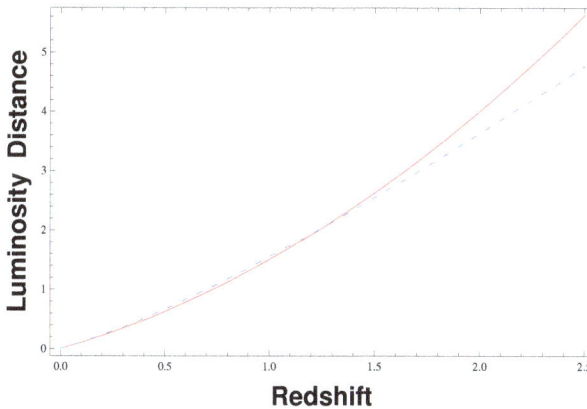

Figure 1. Luminosity distance in the new model (continuous curve) is compared with that in the Λ CDM concordance model $\Omega_m = 1 - \Omega_\Lambda = 0.3$ (broken curve). Distances shown on the vertical axis are measured in units of cH_0^{-1}. The two models significantly depart for $z \gtrsim 1.3$.

5.2. Different Pieces Fit Together

As the dark energy can be assimilated in the energy-stress tensor, and since the latter is absent from the dynamical equations in the new paradigm (wherein the fields appear through the geometry), the dark energy and its associated problems, for instance the cosmological constant problem (which appears due to a conflict between the energy-stress tensor T^{ik} in Equation (1) and the energy density of vacuum in the quantum field theory) and the coincidence problem, are evaded in the new paradigm.

For the same reason, the flatness problem is circumvented due to the absence of the energy-stress tensor in the new paradigm.

As has been mentioned earlier, the observed isotropy of CMB cannot be explained in the standard paradigm in terms of some homogenization process that has taken place in the baryon-photon plasma operating under the principle of causality, since a finite value for the particle horizon $d_{PH}(t) = cS(t) \int_0^t dt'/S(t')$ (the largest distance from which light could have reached the present observer) exists in the theory. As $d_{PH} = \infty$ always for $S = ct$, no horizon exists in the new paradigm, and the whole Universe is always causally connected, which explains the observed overall uniformity of CMB without invoking inflation [19].

As the Big Bang singularity is a breakdown of the laws of physics and the geometrical structure of spacetime, there have been attempts to discover singularity-free cosmological solutions of Einstein equations, which are usually achieved by violating the energy conditions.

Although Equation (14), which represents the cosmological model in the new paradigm, has well-behaved metric potentials at $t = 0$, the volume of the spatial slices vanishes there, resulting in a blowup in the accompanied matter density. However, this is just a coordinate effect which can be removed in the Minkowskian form of solution Equation (14) by considering the locally defined coordinates of space and time.

Moreover, as the locally defined time scale τ is related with the cosmic time t through the transformation $\tau = t_0 \ln(t/t_0)$ [27], the epoch corresponding to the Big Bang, is pushed back to the infinite past giving an infinite age to the Universe which can accommodate even older objects than the standard cosmology can. Interestingly, even in terms of the cosmic time t, wherein the Universe appears dynamic, the age of the Universe appears higher than that in the standard paradigm [27].

As has been mentioned earlier, the conventional 'source' term T^{ik} in Equation (1) fails to include the energy, momentum or angular momentum of the gravitational field. Remarkably, these quantities, akin to the matter fields, are inherently present in the geometry of Equation (3), substantiating the new strategy of the new paradigm to represent the source through geometry. For instance, the term $K/r = -2Gm/(c^2r)$ in the Schwarzschild solution Equation (4) contains the gravitational energy at the point r. It perfectly agrees with the Newtonian estimate of the gravitational energy given by $-Gm/r$, indicating that the term $-2Gm/(c^2r)$ is just its relativistic analogue. Assigning the gravitational energy to K/r is also supported by the locality of GR, which becomes an intrinsic characteristic of the theory as soon as the Newtonian concept of gravitation as a force (action-at-a-distance) is superseded by the curvature. Being a local theory, GR then assigns the curvature present at a particular point, to the source present at that very point. Thus, the agent responsible for the curvature in Equation (4) must be the gravitational energy, since matter exists only at $r = 0$, whereas Equation (4) is curved at all finite values of r. Hence, the presence of curvature in the Schwarzschild solution implies that the gravitational energy does gravitate just as does every other form of energy, and the gravitational field is obviously present in the geometry of Equation (3).

Similarly, the angular momentum of the gravitational field, arising from the rotation of the mass m, is revealed through the geometry of the Kerr solution, and its momentum in the Taub-NUT solution. Thus, the long-sought-after gravitational field energy-momentum-angular momentum of GR is already present in the geometry.

It may be interesting to note that new interior solutions, based on the solutions of Equation (3), have been formulated in the new paradigm that forms the Schwarzschild interior and the Kerr interior [17]. The new interiors are conceptually satisfying and free from the earlier mentioned problems.

As the Newtonian theory of gravitation provides excellent approximations under a wide range of astrophysical cases, the first crucial test of any theory of gravitation is that it reduces to the Newtonian gravitation in the limit of a weak gravitational field. In this context, it has been shown recently [17] that the new paradigm consistently admits the Poisson equation in the case of a slowly varying weak gravitational field when the concerned velocities are considered much less than c, provided we take into account the inertial as well as the gravitational properties of matter, as should correctly be expected in a true Machian theory. The standard paradigm on the other hand fails to fulfill this requirement as Equation (1), in the limit of the weak field, does not reduce to the Poisson equation in the presence of a non-zero Λ (or any other candidate of dark energy), which becomes unavoidable in the standard paradigm. In addition, it would not be correct to argue that a Λ as small as $\approx 10^{-56}$ cm^{-2} (as inferred from the cosmological observations) cannot contribute to the physics appreciably in the local problems. It has been shown recently that even this value of Λ does indeed contribute to the bending of light and to the advance of the perihelion of planets [31].

Interestingly, the new paradigm becomes scale invariant, since the new field Equation (3) is manifestly scale invariant. This becomes a remarkable achievement in the sense that one of the most common ways for a theory with continuous field to be renormalizable is for it to be scale-invariant.

Since the Universe in the new paradigm is flat, the symmetries of its Minkowskian form make it possible to validate the conservation of energy, solving the long-standing problems associated with the conservation of energy. As has been shown by Noether, it is the symmetry of the Minkowskian space that is the cause of the conservation of the energy momentum of a physical field [32,33].

5.3. Geometrization of Electromagnetism in the New Paradigm

How can the electromagnetic field be added to the new paradigm? While the equivalence principle renders the gravitational and inertial fields essentially geometrical (owing to the fact that the ratio of the gravitational and inertial mass is strictly unity for all matter), this is not so in the case of the electromagnetic field (since the ratio of electric charge to mass varies from particle to particle). Hence, the addition of the electromagnetic energy tensor E^{ik} to Equation (1), results in

$$R^{ik} = -\frac{8\pi G}{c^4} E^{ik} \tag{16}$$

since T^{ik} is absorbed in the geometry (as we have noted earlier), and $g_{ik}E^{ik} = 0$ reduces $R = 0$ identically. The tensor E_k^i is given, in terms of the skew-symmetric electromagnetic field tensor F_{ik}, as usual:

$$E^{ik} = \nu\left[-g^{k\ell}F^{ij}F_{\ell j} + \frac{1}{4}g^{ik}F_{\ell j}F^{\ell j}\right] \tag{17}$$

where ν is a constant. It has already been shown that Equation (16), taken together with the 'source-free' Maxwell equations,

$$\left.\begin{array}{c}\dfrac{\partial F_{ik}}{\partial x^\ell} + \dfrac{\partial F_{k\ell}}{\partial x^i} + \dfrac{\partial F_{\ell i}}{\partial x^k} = 0 \\[2mm] \dfrac{\partial}{\partial x^k}(\sqrt{-g}F^{ik}) = 0\end{array}\right\} \quad , \quad g = \det((g_{ik})) \tag{18}$$

consistently represents the electromagnetic field in the presence of gravitation [17]. As the existence of charge is intimately related with the existence of the charge-carrier matter, and since the new paradigm claims the inherent presence of matter in the geometry, it is reasonable to expect the charge also to appear through the geometry. This view is indeed supported not only by the Reissner–Nordstrom and Kerr–Newman solutions, but also by the cosmological solutions—the so-called "electrovac universes" [17], wherein the charge does appear through the geometry. [Let us note that unlike the Reissner–Nordstrom and Kerr–Newman solutions (which represent the field outside the charged matter), the electrovac solutions are not expected to contain any "outside" where the charge-carrier matter can exist.]

Thus, Equations (16)–(18) of restricted validity in the standard paradigm [wherein they are believed to represent the electromagnetic field in vacuum, very much in the same vein as Equation (3) is believed to represent the gravitational field in vacuum] get full validity and represent a unified theory of gravitation, inertia and electromagnetism.

Interestingly, Misner and Wheeler also expressed similar views long ago and advocated to represent *"gravitation, electromagnetism, unquantized charge and unquantized mass as properties of curved empty space"* [34]. Although they failed to realize the presence of fields in the flat spacetimes; nonetheless, they also realized that Equations (16)–(18) provide a unified theory of electromagnetism and gravitation. [The removal of charge (by switching off E^{ik} from Equation (16), in which case the "electrovac universes" become flat) does not mean that mass (which was carrying charge) must necessarily disappear from these solutions.]

6. Summary and Conclusions: What Next?

GR is undoubtedly a theory of unrivaled elegance. The theory indoctrinates that gravitation is a manifestation of the spacetime geometry—one of the most precious insights in the history of science. It has emerged as a highly successful theory of gravitation and cosmology, predicting several new phenomena, most of them have already been confirmed by observations. The theory has passed every observational test ranging from the solar system to the largest scale, the Universe itself.

Nevertheless, GR ceases to be the ultimate description of gravitation, an epitome of a perfect theory, despite all these feathers in its cap. Besides its much-talked-about incompatibility with quantum mechanics, the theory suffers from many other conceptual problems, most of which are generally ignored. If in a Universe where, according to the standard paradigm, some 95% of the total content is still missing, it is an alarming signal for us to turn back to the very foundations of the theory. In view of these problems (discussed in the paper), we are led to believe that the historical development of GR was indeed on the wrong track, and the theory requires modification or at least reformulation.

By a critical analysis of Mach's principle and the equivalence principle, a new insight with a deeper vision of a geometric theory of gravitation emerges: matter, in its entirety of gravitational, inertial and electromagnetic properties, can be fashioned out of spacetime itself. This revolutionizes our views on the representation of the source of curvature/gravitation by dismissing the conventional source representation through T^{ik} and establishing spacetime itself as the source.

This appears as the missing link of the theory and posits that spacetime does not exist without matter, the former is just an offshoot of the latter. The conventional assumption that matter only fills the already existing spacetime, does not seem correct. This establishes the canonical equation $R^{ik} = 0$ as the field equation of gravitation plus inertia in the very presence of matter, giving rise to a new paradigm in the framework of GR. Though there seems to exist some emotional resistance in the community to tinkering with the elegance of GR, the new paradigm dramatically enhances the beauty of the theory in terms of the deceptively simple new field equation $R^{ik} = 0$. Remarkably, the new paradigm explains the observations at all scales without requiring the epicycle of dark energy.

This review provides an increasingly clear picture that the new paradigm is a viable possibility in the framework of GR, which is valid at all scales, avoids the fallacies, dilemmas and paradoxes, and answers the questions that the old framework could not address.

Though we have witnessed numerous evidences of the presence of fields in the solutions of the field Equation (3), however, the challenge to discover, from more fundamental considerations, a concrete mathematical formulation of the fields in purely geometric terms is still to be met. This formulation is expected to use the gravito-electromagnetic features of GR in the new paradigm and is expected to achieve the following:

1. It should explain the observed flat rotation curves of galaxies without requiring the ad-hoc dark matter.
2. The net field in a homogeneous and isotropic background must be vanishing.

Acknowledgments: The author thanks the IUCAA for hospitality where part of this work was done during a visit. Thanks are also due to two anonymous referees: to one for making critical, constructive comments which helped improving the manuscript; and to other for pointing out an old work of Misner and Wheeler [34], which is deeply connected with and strongly supporting the present work.

Conflicts of Interest: The author declares no conflict of interest.

References

1. Ni, W.-T. Empirical Foundations of the Relativistic Gravity. *Int. J. Mod. Phys. D* **2005**, *14*, 901–921.
2. Will, C.M. The confrontation between general relativity and experiment. *Living Rev. Relativ.* **2006**, *9*, 3.
3. Turyshev, S.G. Experimental tests of general relativity: Recent progress and future directions. *Phys. Usp.* **2009**, *52*, 1–27.
4. Ashtekar, A. (Ed.) *100 Years of Relativity. Space-Time Structure: Einstein and Beyond*; World Scientific: Singapore, 2005.
5. Pdmanabhan, T. One hundred years of General Relativity: Summary, status and prospects. *Curr. Sci.* **2015**, *109*, 1215–1219.
6. Mach, E. *The Science of Mechanics: A Critical and Historical Account of Its Development*; Open Court Publishing: London, UK, 1919.
7. Gödel, K. An example of a new type of cosmological solution of Einstein's field equations of gravitation. *Rev. Mod. Phys.* **1949**, *21*, 447–450.
8. Weinberg, S. *Gravitation and Cosmology: Principles and Applications of the General Theory of Relativity*; John Wiley & Sons: New York, NY, USA, 1972.
9. Einstein, A.; Grossmann, M. Outline of a generalized theory of relativity and of a theory of gravitation. *Z. Math. Phys.* **1913**, *62*, 225–261.
10. Jaramillo, J.L.; Gourgoulhon, E. Mass and angular momentum in general relativity. In *Mass and Motion in General Relativity*; Springer: Dordrecht, The Netherlands, 2011.
11. Misner, C.W., Thorn, K.S.; Wheeler, J.A. *Gravitation*; W. H. Freeman and Company: New York, NY, USA, 1970.
12. Einstein, A. Note on E. Schrödinger's Paper: The energy components of the gravitational field. *Phys. Z.* **1918**, *19*, 115–116.
13. Einstein, A. *The Meaning of Relativity*; Princeton University Press: Princeton, NJ, USA, 1922.
14. Vishwakarma, R.G. On the relativistic formulation of matter. *Astrophys. Space Sci.* **2012**, *340*, 373–379.
15. Cooperstock, F.I.; Dupre, M.J. Covariant energy-momentum and an uncertainty principle for general relativity. *Ann. Phys.* **2013**, *339*, 531–541.
16. Vishwakarma, R.G. A new solution of Einstein's vacuum field equations. *Pramana J. Phys.* **2015**, *85*, 1101–1110.
17. Vishwakarma, R.G. A Machian approach to General Relativity. *Int. J. Geom. Methods Mod. Phys.* **2015**, *12*, 1550116.
18. Kinnersley, W. Recent progress in exact solutions. In Proceedings of the 7th International Conference on General Relativity and Gravitation (GR7), Tel-Aviv, Israel, 23–28 June 1974.
19. Vishwakarma, R.G. Mysteries of $R^{ik} = 0$: A novel paradigm in Einstein's theory of gravitation. *Front. Phys.* **2014**, *9*, 98–112.
20. Weinberg, S. The cosmological constant problem. *Rev. Mod. Phys.* **1989**, *61*, 1–23.
21. Hoyle, F.; Burbidge, G.; Narlikar, J.V. The basic theory underlying the Quasi-Steady-State cosmology. *Proc. R. Soc. Lond. A* **1995**, *448*, 191–212.
22. Einstein, A. *Relativity: The Special and the General Theory*; Create Space Independent Publishing Platform: London, UK, 2015.
23. Padmanabhan, T. Momentum density of spacetime and the gravitational dynamics. *Gen. Relativ. Grav.* **2016**, *48*, 4.
24. Hawking, S.; Milodinow, L. *The Grand Design*; Bantam Books: New York, NY, USA, 2010.
25. Giardino, S. Axisymmmetric empty space: Light propagation, orbits and dark matter. *J. Mod. Phys.* **2014**, *5*, 1402–1411.
26. Narlikar, J.V. *An Introduction to Cosmology*; Cambridge University Press: Cambridge, UK, 2002.
27. Vishwakarma, R.G. A curious explanation of some cosmological phenomena. *Phys. Scripta* **2013**, *87*, 5.
28. Harrison, E.R. Mining Energy in an Expanding Universe. *Astrophys. J.* **1995**, *446*, 63.
29. Tolman, R. C. *Relativity, Thermodynamics and Cosmology*; Oxford University Press: Oxford, UK, 1934.
30. Nielsen, J.T.; Guffanti, A.; Sarkar, S. Marginal evidence for cosmic acceleration from Type Ia supernovae. 2015, arXiv:1506.01354.
31. Ishak, M.; Rindler, W.; Dossett, J.; Moldenhauer, J.; Allison, C. A new independent limit on the cosmological constant/dark energy from the relativistic bending of light by galaxies and clusters of galaxies. *Mom. Not. R. Astron. Soc.* **2008**, *388*, 1279–1283.

32. Noether, E. Invariant variation problems. *Transp. Theory Stat. Phys.* **1971**, *1*, 186–207;
33. Baryshev, Y.V. Energy-momentum of the gravitational field: Crucial point for gravitation physics and cosmology. *Pract. Cosmol.* **2008**, *1*, 276–286.
34. Misner, C.W.; Wheeler, J.A. Classical physics as geometry: Gravitation, electromagnetism, unquantized charge, and mass as properties of empty space. *Ann. Phys.* **1957**, *2*, 525–603.

universe

MDPI

Article
Virial Theorem in Nonlocal Newtonian Gravity

Bahram Mashhoon

Department of Physics and Astronomy, University of Missouri, Columbia, MO 65211, USA;
mashhoonb@missouri.edu

Academic Editors: Lorenzo Iorio and Elias C. Vagenas
Received: 6 April 2016; Accepted: 11 May 2016; Published: 30 May 2016

Abstract: Nonlocal gravity is the recent classical nonlocal generalization of Einstein's theory of gravitation in which the past history of the gravitational field is taken into account. In this theory, nonlocality appears to simulate dark matter. The virial theorem for the Newtonian regime of nonlocal gravity theory is derived and its consequences for "isolated" astronomical systems in virial equilibrium at the present epoch are investigated. In particular, for a sufficiently isolated nearby *galaxy* in virial equilibrium, the galaxy's baryonic diameter $\mathcal{D}_0$—namely, the diameter of the smallest sphere that completely surrounds the baryonic system at the present time—is predicted to be larger than the effective dark matter fraction f_{DM} times a universal length that is the basic nonlocality length scale $\lambda_0 \approx 3 \pm 2$ kpc.

Keywords: nonlocal gravity; celestial mechanics; dark matter

1. Introduction

In the standard theory of relativity, physics is local in the sense that a postulate of locality permeates through the special and general theories of relativity. First, Lorentz invariance is extended in a pointwise manner to actual, namely, accelerated, observers in Minkowski spacetime. This *hypothesis of locality* is then employed crucially in Einstein's local principle of equivalence to render observers pointwise inertial in a gravitational field [1]. Field measurements are intrinsically nonlocal, however. To go beyond the locality postulate in Minkowski spacetime, the past history of the accelerated observer must be taken into account. The observer in general carries the memory of its past acceleration. The deep connection between inertia and gravitation suggests that gravity could be nonlocal as well, and, in nonlocal gravity, the gravitational memory of past events must then be taken into account. Along this line of thought, a classical nonlocal generalization of Einstein's theory of gravitation has recently been developed [2–13]. In this theory, the gravitational field is local but satisfies partial integro-differential field equations. Moreover, a significant observational consequence of this theory is that the nonlocal aspect of gravity appears to simulate dark matter. The physical foundations of this classical theory, from nonlocal special relativity theory to nonlocal general relativity, sets it completely apart from purely phenomenological and *ad hoc* approaches to the problem of dark matter.

Dark matter is currently required in astrophysics for explaining the gravitational dynamics of galaxies as well as clusters of galaxies [9], gravitational lensing observations [10] and structure formation in cosmology [13]. We emphasize that only some of the implications of nonlocal gravity theory have thus far been confronted with observation [9,12]. It is also important to mention here that many other approaches to nonlocal gravitation theory exist that are, however, inspired by developments in quantum field theory. The consideration of such theories is well beyond the scope of this purely classical work.

In this paper, we are concerned with the Newtonian regime of nonlocal gravity, where Poisson's equation of Newtonian gravity is modified by the addition of a certain average over the gravitational field. This nonlocal term involves a kernel function q whose functional form can perhaps be derived from a future more complete theory, but, at the present stage of the development of nonlocal gravity, must be determined using observational data. It is necessary that a unique kernel be eventually chosen in this way, but kernel q at the present time could be either q_1 or q_2 [6]. Each of these kernels is spherically symmetric in space and contains three length scales a_0, λ_0, and μ_0^{-1} such that $a_0 < \lambda_0 < \mu_0^{-1}$. The basic scale of nonlocality is a galactic length λ_0 of order 1 kpc, while a_0 is a short-range parameter that controls the behavior of $q(r)$ as $r \to 0$. At the other extreme, $r \to \infty$, $q(r)$ decays exponentially as $\exp(-\mu_0 r)$, indicating the fading of spatial memory with distance. The short-range parameter a_0 is necessary in dealing with the gravitational physics of the Solar System, globular clusters and isolated dwarf galaxies; however, it may be safely neglected in dealing with larger systems such as clusters of galaxies. When $a_0 = 0$, q_1 and q_2 reduce to a single kernel q_0, $q_1 = q_2 = q_0$, and the remaining parameters (λ_0 and μ_0) have been determined from a comparison of the theory with the astronomical data regarding a sample of 12 spiral galaxies from the THINGS catalog—see reference [9] for a detailed treatment. The results can be expressed, for the sake of convenience, as $\lambda_0 \approx 3$ kpc and $\mu_0^{-1} \approx 17$ kpc. Moreover, lower limits have been placed on a_0 from the study of the precession of perihelia of planetary orbits in the Solar System [12,14,15].

It is interesting to explore the implications of the virial theorem for nonlocal gravity. In general, the virial theorem of Newtonian physics establishes a simple linear relation between the time averages of the kinetic and potential energies of an isolated material system for which the potential energy is a homogeneous function of spatial coordinates. For an isolated *gravitational* N-body system, the significance of the virial theorem has to do with the circumstance that the kinetic energy is a sum of terms each proportional to the mass of a body in the system, while the potential energy is a sum of terms each proportional to the product of two masses in the system. Thus, under favorable conditions, the virial theorem can be used to connect the total dynamic mass of an isolated relaxed gravitational system with its average internal motion.

The main purpose of the present paper is to discuss, within the Newtonian regime of nonlocal gravity, the consequences of the extension of the virial theorem to nonlocal gravity. Though such an extension is technically straightforward, it is nevertheless physically quite significant as it allows the possibility of making *predictions* regarding the effective dark-matter content of cosmologically nearby isolated N-body gravitational systems in virial equilibrium.

2. Modification of the Inverse Square Force Law

It can be shown [12] that, in the Newtonian regime of nonlocal gravity, the force of gravity on point mass m due to point mass m' is given by:

$$\mathbf{F}(\mathbf{r}) = -Gmm'\frac{\hat{\mathbf{r}}}{r^2}\left\{[1 - \mathcal{E}(r) + \alpha_0] - \alpha_0\left(1 + \frac{1}{2}\mu_0 r\right)e^{-\mu_0 r}\right\} \tag{1}$$

where $\mathbf{r} = \mathbf{x}_m - \mathbf{x}_{m'}$, $r = |\mathbf{r}|$ and $\hat{\mathbf{r}} = \mathbf{r}/r$. The quantity in curly brackets is henceforth denoted by $1 + \mathbb{N}$, where $\mathbb{N}$ is the contribution of nonlocality to the force law and depends upon three parameters, namely, α_0, μ_0 and a short-range parameter a_0 that is contained in $\mathcal{E}$; in fact, $\mathcal{E} = 0$ when $a_0 = 0$. We will show in the next section that $\mathbb{N}$ starts out from zero at $r = 0$ with vanishing slope and monotonically increases toward an asymptotic value of about 10 as $r \to \infty$. Thus, the gravitational force in Equation (1) is *always attractive*; moreover, this force is central, conservative and satisfies Newton's third law of motion.

Nonlocal gravity is in the early stages of development and, depending on whether we choose kernel q_1 or kernel q_2, $\mathcal{E}(r)$ at the present time can be either

$$\mathcal{E}_1(r) = \frac{a_0}{\lambda_0}e^p\left[E_1(p) - E_1(p + \mu_0 r)\right] \tag{2}$$

or

$$\mathcal{E}_2(r) = \frac{a_0}{\lambda_0}\left\{-\frac{r}{r+a_0}e^{-\mu_0 r} + 2e^p\left[E_1(p) - E_1(p+\mu_0 r)\right]\right\} \tag{3}$$

respectively, where $p = \mu_0 a_0$, $\lambda_0 = 2/(\alpha_0\,\mu_0)$ and $E_1(u)$ is the *exponential integral function* [16]:

$$E_1(u) = \int_u^\infty \frac{e^{-t}}{t}dt \tag{4}$$

For $u : 0 \to \infty$, $E_1(u) > 0$ monotonically decreases from infinity to zero. In fact, near $u = 0$, $E_1(u)$ behaves like $-\ln u$ and as $u \to \infty$, $E_1(u)$ vanishes exponentially. Furthermore,

$$E_1(x) = -C - \ln x - \sum_{n=1}^\infty \frac{(-x)^n}{n\,n!} \tag{5}$$

where $C = 0.577\ldots$ is Euler's constant. It is useful to note that

$$\frac{e^{-u}}{u+1} < E_1(u) \leq \frac{e^{-u}}{u} \tag{6}$$

(see Equation 5.1.19 in reference [16]).

It is clear from Equation (1) that α_0 is dimensionless, while μ_0^{-1}, λ_0 and a_0 have dimensions of length. In fact, we expect that $a_0 < \lambda_0 < \mu_0^{-1}$; moreover, the short-range parameter a_0 and $\mathcal{E}$ may be neglected in Equation (1) when dealing with the rotation curves of spiral galaxies and the internal gravitational physics of clusters of galaxies. In this way, α_0 and μ_0 have been *tentatively* determined from a detailed comparison of nonlocal gravity with observational data [9]:

$$\alpha_0 = 10.94 \pm 2.56\,, \qquad \mu_0 = 0.059 \pm 0.028 \text{ kpc}^{-1} \tag{7}$$

Hence, we find $\lambda_0 = 2/(\alpha_0\,\mu_0) \approx 3 \pm 2$ kpc. It is important to mention here that λ_0 *is the fundamental length scale of nonlocal gravity at the present epoch*; indeed, for $\lambda_0 \to \infty$, $\mathbb{N} \to 0$ and Equation (1) reduces to Newton's inverse square force law. In what follows, we usually assume $\alpha_0 \approx 11$ and $\mu_0^{-1} \approx 17$ kpc for the sake of convenience. Furthermore, we expect that $p = \mu_0 a_0$ is such that $0 < p < \frac{1}{5}$. In reference [12], preliminary lower limits have been placed on a_0 on the basis of current data regarding planetary orbits in the Solar System. For instance, using the data for the orbit of Saturn, a preliminary lower limit of $a_0 \gtrsim 2 \times 10^{15}$ cm can be established if we use $\mathcal{E}_1$, while $a_0 \gtrsim 5.5 \times 10^{14}$ cm if we use $\mathcal{E}_2$.

Let us note that

$$\frac{d\mathcal{E}_1}{dr} = \frac{a_0}{\lambda_0}\frac{1}{a_0+r}e^{-\mu_0 r} \tag{8}$$

and

$$\frac{d\mathcal{E}_2}{dr} = \frac{a_0}{\lambda_0}\frac{a_0+2r+\mu_0 r(a_0+r)}{(a_0+r)^2}e^{-\mu_0 r} \tag{9}$$

Therefore, $\mathcal{E}_1(r)$ and $\mathcal{E}_2(r)$ start from zero at $r = 0$ and monotonically increase as $r \to \infty$; furthermore, they asymptotically approach $\mathcal{E}_1(\infty) = \mathcal{E}_\infty$ and $\mathcal{E}_2(\infty) = 2\,\mathcal{E}_\infty$, respectively, where

$$\mathcal{E}_\infty = \frac{1}{2}\,\alpha_0\,p\,e^p\,E_1(p) \tag{10}$$

It is a consequence of (6) that $\mathcal{E}_\infty < \alpha_0/2$, so that, in the gravitational force in Equation (1),

$$\alpha_0 - \mathcal{E}(r) > 0 \tag{11}$$

In the Newtonian regime, where we formally let the speed of light $c \to \infty$, retardation effects vanish and gravitational memory is purely spatial. The resulting gravitational force in Equation (1) thus consists of two parts: an enhanced attractive "Newtonian" part and a repulsive fading spatial memory ("Yukawa") part with an exponential decay length of $\mu_0^{-1} \approx 17$ kpc. Equation (1) is such that it reduces to Newton's inverse square force law for $r \to 0$, as it should [17–21], and on galactic scales, it is a generalization of the phenomenological Tohline-Kuhn modified gravity approach to the flat rotation curves of spiral galaxies [22–25]. An excellent review of the Tohline-Kuhn work is contained in the paper of Bekenstein [26].

For $r \gg \mu_0^{-1}$, the exponentially decaying ("fading memory") part of Equation (1) can be neglected and

$$\mathbf{F}(\mathbf{r}) \approx -\frac{Gmm' \left[1 + \alpha_0 - \mathcal{E}(\infty) \right]}{r^2} \, \hat{\mathbf{r}} \tag{12}$$

so that $m' \left[\alpha_0 - \mathcal{E}(\infty) \right]$ has the interpretation of the *total effective dark mass* associated with m'. For $a_0 = 0$, the net effective dark mass associated with point mass m' is simply $\alpha_0 m'$, where $\alpha_0 \approx 11$ [9]. On the other hand, for $a_0 \neq 0$, the corresponding result is $\alpha_0 \, \epsilon(p) \, m'$, where

$$\epsilon_1(p) = 1 - \frac{1}{2} \, p \, e^p \, E_1(p) \,, \qquad \epsilon_2(p) = 1 - p \, e^p \, E_1(p) \tag{13}$$

depending on whether we use $\mathcal{E}_1$ or $\mathcal{E}_2$, respectively. The functions in Equation (13) start from unity at $p = 0$ and decrease monotonically to $\epsilon_1(0.2) \approx 0.85$ and $\epsilon_2(0.2) \approx 0.70$ at $p = 0.2$; they are plotted in Figure 1 of reference [12] for $p : 0 \to 0.2$. If a_0 turns out to be just a few parsecs or smaller, for instance, then $\epsilon_1 \approx \epsilon_2 \approx 1$.

A detailed investigation reveals that it is possible to approximate the exterior gravitational force due to a star or a planet by assuming that its mass is concentrated at its center [12]. In this connection, we note that the radius of a star or a planet is generally much smaller than the length scales a_0, λ_0 and μ_0^{-1} that appear in the nonlocal contribution to the gravitational force. Therefore, one can employ Equation (1) in the approximate treatment of the two-body problem in astronomical systems such as binary pulsars and the Solar System, where possible deviations from general relativity may become measurable in the future.

Consider, for instance, the deviation from the Newtonian inverse square force law, namely,

$$\delta \mathbf{F}(\mathbf{r}) = -\frac{Gmm' \, \hat{\mathbf{r}}}{r^2} \, \mathbb{N}(r) \tag{14}$$

For $r < a_0$, it is possible to show via an expansion in powers of r/a_0 that [12]

$$\delta \mathbf{F}_1(\mathbf{r}) = -\frac{1}{2} \frac{Gmm'}{\lambda_0 \, a_0} \, (1 + p) \, \hat{\mathbf{r}} + \frac{1}{3} \frac{Gmm'}{\lambda_0 \, a_0} \, (1 + p + p^2) \, \frac{r}{a_0} \, \hat{\mathbf{r}} + \cdots \tag{15}$$

if $\mathcal{E}_1$ is employed, or

$$\delta \mathbf{F}_2(\mathbf{r}) = -\frac{1}{3} \frac{Gmm'}{\lambda_0 \, a_0} \, (1 + p) \, \frac{r}{a_0} \, \hat{\mathbf{r}} + \cdots \tag{16}$$

if $\mathcal{E}_2$ is employed. Perhaps dedicated missions, such as ESA's Gaia mission that was launched in 2013, can measure the imprint of nonlocal gravity in the Solar System [27,28]. In this connection, we note that

$$\frac{1}{2} \frac{G \, M_\odot}{\lambda_0 \, a_0} \, (1 + p) \approx \left(\frac{10^{18} \, \text{cm}}{a_0} \right) 10^{-14} \, \text{cm s}^{-2} \tag{17}$$

which, combined with lower limits on a_0 established in reference [12], is at least three orders of magnitude smaller than the acceleration involved in the Pioneer anomaly ($\sim 10^{-7} \, \text{cm s}^{-2}$). It follows from these results that nonlocal gravity is consistent with the gravitational physics of the Solar System.

3. Virial Theorem

Consider an idealized isolated system of N Newtonian point particles with fixed masses m_i, $i = 1, 2, \ldots, N$. We assume that the particles occupy a finite region of space and interact with each other only gravitationally such that the center of mass of the isolated system is at rest in a global inertial frame and the isolated system permanently occupies a compact region of space. The equation of motion of the particle with mass m_i and state $(\mathbf{x}_i, \mathbf{v}_i)$ is then

$$m_i \frac{d\,\mathbf{v}_i}{dt} = -\sum_j{}' \frac{G\,m_i\,m_j\,(\mathbf{x}_i - \mathbf{x}_j)}{|\mathbf{x}_i - \mathbf{x}_j|^3} \left[1 + \mathbb{N}(|\mathbf{x}_i - \mathbf{x}_j|)\right] \tag{18}$$

for $j = 1, 2, \ldots, N$, but the case $j = i$ is excluded in the sum by convention. In fact, a prime over the summation sign indicates that in the sum $j \neq i$. Here, $1 + \mathbb{N}(r)$ is a *universal* function that is inside the curly brackets in Equation (1) and the contribution of nonlocality, $\mathbb{N}(r)$, is given by

$$\mathbb{N}(r) = \alpha_0 \left[1 - \left(1 + \frac{1}{2}\mu_0 r\right) e^{-\mu_0 r}\right] - \mathcal{E}(r) \tag{19}$$

Consider next the quantities

$$\mathbb{I} = \frac{1}{2}\sum_i m_i\,x_i^2, \qquad \frac{d\,\mathbb{I}}{dt} = \sum_i m_i\,\mathbf{x}_i \cdot \mathbf{v}_i \tag{20}$$

where $x_i = |\mathbf{x}_i|$ and

$$\frac{d^2\,\mathbb{I}}{dt^2} = \sum_i m_i\,v_i^2 + \sum_i m_i\,\mathbf{x}_i \cdot \frac{d\,\mathbf{v}_i}{dt} \tag{21}$$

It follows from Equation (18) that

$$\sum_i m_i\,\mathbf{x}_i \cdot \frac{d\,\mathbf{v}_i}{dt} = -\sum_{i,j}{}' \frac{G\,m_i\,m_j\,(\mathbf{x}_i - \mathbf{x}_j)\cdot\mathbf{x}_i}{|\mathbf{x}_i - \mathbf{x}_j|^3} \left[1 + \mathbb{N}(|\mathbf{x}_i - \mathbf{x}_j|)\right] \tag{22}$$

Exchanging i and j in the expression on the right-hand side of Equation (22), we get

$$\sum_i m_i\,\mathbf{x}_i \cdot \frac{d\,\mathbf{v}_i}{dt} = \sum_{i,j}{}' \frac{G\,m_i\,m_j\,(\mathbf{x}_i - \mathbf{x}_j)\cdot\mathbf{x}_j}{|\mathbf{x}_i - \mathbf{x}_j|^3} \left[1 + \mathbb{N}(|\mathbf{x}_i - \mathbf{x}_j|)\right] \tag{23}$$

Adding Equations (22) and (23) results in

$$\sum_i m_i\,\mathbf{x}_i \cdot \frac{d\,\mathbf{v}_i}{dt} = -\frac{1}{2}\sum_{i,j}{}' \frac{G\,m_i\,m_j}{|\mathbf{x}_i - \mathbf{x}_j|} \left[1 + \mathbb{N}(|\mathbf{x}_i - \mathbf{x}_j|)\right] \tag{24}$$

Using this result, Equation (21) takes the form

$$\frac{d^2\,\mathbb{I}}{dt^2} = \sum_i m_i\,v_i^2 - \frac{1}{2}\sum_{i,j}{}' \frac{G\,m_i\,m_j}{|\mathbf{x}_i - \mathbf{x}_j|} \left[1 + \mathbb{N}(|\mathbf{x}_i - \mathbf{x}_j|)\right] \tag{25}$$

Let us recall that the net kinetic energy and the Newtonian gravitational potential energy of the system are given by

$$\mathbb{T} = \frac{1}{2}\sum_i m_i\,v_i^2, \qquad \mathbb{W}_N = -\frac{1}{2}\sum_{i,j}{}' \frac{G\,m_i\,m_j}{|\mathbf{x}_i - \mathbf{x}_j|} \tag{26}$$

Hence,

$$\frac{d^2\,\mathbb{I}}{dt^2} = 2\,\mathbb{T} + \mathbb{W}_N + \mathbb{D} \tag{27}$$

where

$$\mathbb{D} = -\frac{1}{2}\sum_{i,j}' \frac{G\,m_i\,m_j}{|\mathbf{x}_i - \mathbf{x}_j|}\,\mathbb{N}(|\mathbf{x}_i - \mathbf{x}_j|) \tag{28}$$

and $\mathbb{N}$ is given by Equation (19).

Finally, we are interested in the average of Equation (27) over time. Let $<f>$ denote the time average of f, where

$$<f> = \lim_{\tau \to \infty}\frac{1}{\tau}\int_0^\tau f(t)\,dt \tag{29}$$

Then, it follows from averaging Equation (27) over time that

$$2 <\mathbb{T}> = -<\mathbb{W}_N> - <\mathbb{D}> \tag{30}$$

since $d\,\mathbb{I}/dt$, which is the sum of $m\,\mathbf{x}\cdot\mathbf{v}$ over all particles in the system, is a bounded function of time and hence the time average of $d^2\,\mathbb{I}/dt^2$ vanishes. This is clearly based on the assumption that the spatial coordinates and velocities of all particles indeed remain finite for all time. Equation (30) expresses the *virial theorem* in nonlocal Newtonian gravity.

It is important to digress here and re-examine some of the assumptions involved in our derivation of the virial theorem. In general, any consequence of the gravitational interaction involves the whole mass-energy content of the universe due to the universality of the gravitational interaction; therefore, an astronomical system may be considered isolated only to the extent that the tidal influence of the rest of the universe on the internal dynamics of the system can be neglected. Moreover, the parameters of the force law in Equation (1) refer to the present epoch and hence the virial theorem in Equation (30) ignores cosmological evolution. Thus, the temporal average over an infinite period of time in Equation (30) must be reinterpreted here to mean that the relatively isolated system under consideration has evolved under its own gravity such that it is at the present epoch in a steady equilibrium state. That is, the system is currently in virial equilibrium. Finally, we recall that a point particle of mass m in Equation (30) could reasonably represent a star of mass m as well, where the mass of the star is assumed to be concentrated at its center.

The deviation of the virial theorem in Equation (30) from the Newtonian result is contained in $<\mathbb{D}>$, where $\mathbb{D}$ is given by Equation (28). More explicitly, we have

$$\mathbb{D} = -\frac{1}{2}\sum_{i,j}' \frac{G\,m_i\,m_j}{|\mathbf{x}_i - \mathbf{x}_j|}\left[\alpha_0 - \alpha_0\left(1 + \frac{1}{2}\mu_0\,|\mathbf{x}_i - \mathbf{x}_j|\right)e^{-\mu_0\,|\mathbf{x}_i - \mathbf{x}_j|} - \mathcal{E}(|\mathbf{x}_i - \mathbf{x}_j|)\right] \tag{31}$$

It proves useful at this point to study some of the properties of the function $\mathbb{N}$, which is the contribution of nonlocality that is inside the square brackets in Equation (31). The argument of this function is $|\mathbf{x}_i - \mathbf{x}_j| > 0$ for $i \ne j$; therefore, $|\mathbf{x}_i - \mathbf{x}_j|$ varies over the interval $(0, \mathcal{D}_0]$, where $\mathcal{D}_0$ is the largest possible distance between any two baryonic point masses in the system. Thus, $\mathbb{N}(r)$, in the context of the virial theorem, is defined for the interval $0 < r \le \mathcal{D}_0$, where $\mathcal{D}_0$ is the diameter of the smallest sphere that completely encloses the *baryonic* system for all time. In general, however, $\mathbb{N}(0) = 0$ and $\mathbb{N}(\infty) = \alpha_0 - \mathcal{E}(\infty) > 0$, where $\mathcal{E}(\infty) = \mathcal{E}_\infty$ or $2\,\mathcal{E}_\infty$, depending on whether we use $\mathcal{E}_1$ or $\mathcal{E}_2$, respectively. Moreover, $d\,\mathbb{N}(r)/dr$ is given by

$$\frac{d}{dr}\,\mathbb{N}_1(r) = \frac{1}{2}\alpha_0\,\mu_0\,\frac{r\left[1 + \mu_0\,(a_0 + r)\right]}{a_0 + r}\,e^{-\mu_0\,r} \tag{32}$$

if we use $\mathcal{E}_1$ or

$$\frac{d}{dr}\,\mathbb{N}_2(r) = \frac{1}{2}\alpha_0\,\mu_0\,\frac{r^2\left[1 + \mu_0\,(a_0 + r)\right]}{(a_0 + r)^2}\,e^{-\mu_0\,r} \tag{33}$$

if we use $\mathcal{E}_2$. Writing $\exp(\mu_0\,r) = 1 + \mu_0\,r + \mathcal{R}$, where $\mathcal{R} > 0$ represents the remainder of the power series, it is straightforward to see that for $r \ge 0$ and $n = 1, 2, \ldots,$

$$e^{\mu_0 r} (a_0 + r)^n > r^n \left[1 + \mu_0 (a_0 + r) \right] \qquad (34)$$

This result, for $n = 1$ and $n = 2$, implies that the right-hand sides of Equations (32) and (33), respectively, are less than $\alpha_0 \mu_0 / 2$. Therefore, it follows that, in general,

$$\frac{d}{dr} \mathbb{N}(r) < \frac{1}{2} \alpha_0 \mu_0 \qquad (35)$$

Moreover, for $r > 0$, (35) implies:

$$\mathbb{N}(r) = \int_0^r \frac{d \mathbb{N}(x)}{dx} \, dx < \frac{1}{2} \alpha_0 \mu_0 r \qquad (36)$$

We conclude that $\mathbb{N}$ is a monotonically increasing function of r that is zero at $r = 0$ with a slope that vanishes at $r = 0$. For $r \gg \mu_0^{-1}$, $\mathbb{N}(r)$ asymptotically approaches a constant $\alpha_0 \epsilon := \alpha_0 - \mathcal{E}(\infty)$. Here, $\epsilon(p)$ is either $\epsilon_1(p)$ or $\epsilon_2(p)$ depending on whether we use $\mathcal{E}_1$ or $\mathcal{E}_2$, respectively. The functions $\epsilon_1(p)$ and $\epsilon_2(p)$ are defined in Equation (13).

4. Dark Matter

Most of the matter in the universe is currently thought to be in the form of certain elusive particles that have not been directly detected [29–32]. The existence and properties of this *dark matter* have thus far been deduced only through its gravity. We are interested here in dark matter only as it pertains to stellar systems such as galaxies and clusters of galaxies [33–39]. We mention that dark matter is also essential in the explanation of gravitational lensing observations [40,41] and in the solution of the problem of structure formation in cosmology [13,42]; however, these topics are beyond the scope of this work.

Actual (mainly baryonic) mass is observationally estimated for astronomical systems using the mass-to-light ratio M/L. However, it turns out that the dynamic mass of the system is usually larger and this observational fact is normally attributed to the possible existence of nonbaryonic dark matter. Let M be the baryonic mass and M_{DM} be the mass of the nonbaryonic dark matter needed to explain the gravitational dynamics of the system. Then,

$$f_{DM} = \frac{M_{DM}}{M} \qquad (37)$$

is the dark matter fraction and $M + M_{DM} = M (1 + f_{DM})$ is the dynamic mass of the system.

In observational astrophysics, the virial theorem of Newtonian gravity is interpreted to be a relationship between the dynamic (virial) mass of the entire system and its average internal motion deduced from the rotation curve or velocity dispersion of the bound collection of masses in virial equilibrium. Therefore, regardless of how the net amount of dark matter in galaxies and clusters of galaxies is operationally estimated and the corresponding f_{DM} is thereby determined, for sufficiently isolated self-gravitating astronomical systems in virial equilibrium, we must have

$$2 < \mathbb{T} > \; = -(1 + f_{DM}) < \mathbb{W}_N > \qquad (38)$$

That is, virial theorem Equation (38) is employed in astronomy to infer in some way the total dynamic mass of the system. Indeed, Zwicky first noted the need for dark matter in his application of the standard virial theorem of Newtonian gravity to the Coma cluster of galaxies [33,34].

5. Effective Dark Matter

A significant physical consequence of nonlocal gravity theory is that it appears to simulate dark matter [9]. In particular, in the Newtonian regime of nonlocal gravity, the Poisson equation is modified such that the density of ordinary matter ρ is accompanied by a term ρ_D that is obtained from the folding (convolution) of ρ with the reciprocal kernel of nonlocal gravity. Thus, ρ_D has the interpretation of the density of *effective dark matter* and $\rho + \rho_D$ is the density of the *effective dynamic mass*.

The virial theorem makes it possible to elucidate in a simple way the manner in which nonlocality can simulate dark matter. It follows from a comparison of Equations (30) and (38) that nonlocal gravity can account for this "excess mass" if

$$< \mathbb{D} > \; = f_{DM} < \mathbb{W}_N > \tag{39}$$

where $\mathbb{W}_N$ and $\mathbb{D}$ are given in Equations (26) and (28), respectively.

It is interesting to apply the virial theorem of nonlocal gravity to sufficiently isolated astronomical N-body systems. The configurations that we briefly consider below consist of clusters of galaxies with diameters $\mathcal{D}_0 \gg \mu_0^{-1} \approx 17$ kpc, galaxies with $\mathcal{D}_0 \sim \mu_0^{-1}$ and globular star clusters with $\mathcal{D}_0 \ll \mu_0^{-1}$. The results presented in this section follow from certain general properties of the function $\mathbb{N}(r)$ and are completely independent of how the baryonic matter is distributed within the astronomical system under consideration.

We emphasize that, after setting the short-range parameter $a_0 = 0$, the parameters α_0 and μ_0, and hence λ_0, were originally determined from the combined observational data for the rotation curves of a sample of 12 nearby spiral galaxies from the THINGS catalog [9]. These tentative values are given in Equation (7). These parameter values were then found to be in reasonable agreement with the internal dynamics of a sample of 10 rich nearby clusters of galaxies from the Chandra X-ray catalog [9]. In the present paper, we use these parameter values to make predictions about *all* nearby isolated N-body gravitational systems that are in virial equilibrium.

5.1. Clusters of Galaxies: $f_{DM} \approx \alpha_0 \, \epsilon(p)$

Consider, for example, a cluster of galaxies, where nearly all of the relevant distances are much larger than $\mu_0^{-1} \approx 17$ kpc. In this case, $\mu_0 \, r \gg 1$ and hence $\mathbb{N}$ approaches its asymptotic value, namely,

$$\mathbb{N} \approx \alpha_0 \, \epsilon(p) \tag{40}$$

where $\epsilon = \epsilon_1$ or ϵ_2, defined in Equation (13), depending on whether we use $\mathcal{E}_1$ or $\mathcal{E}_2$, respectively. Hence, Equation (28) can be written as:

$$< \mathbb{D} > \approx \alpha_0 \, \epsilon(p) < \mathbb{W}_N > \tag{41}$$

It then follows from Equation (39) that, for galaxy clusters,

$$f_{DM} \approx \alpha_0 \, \epsilon(p) \tag{42}$$

in nonlocal gravity. We recall that ϵ is only weakly sensitive to the magnitude of a_0. It follows from $\alpha_0 \approx 11$ that f_{DM} for galaxy clusters is about 10, in general agreement with observational data [9]. This theoretical result is essentially equivalent to the work on galaxy clusters contained in reference [9], except that Equation (42) takes into account the existence of the short-range parameter a_0.

Nonlocal gravity thus predicts that the effective dark matter fraction f_{DM} has approximately the same constant value of about 10 for all isolated nearby clusters of galaxies that are in equilibrium.

5.2. Galaxies: $f_{DM} < \mathcal{D}_0/\lambda_0$

Consider next a sufficiently isolated galaxy of diameter $\mathcal{D}_0$ in virial equilibrium. In this case, we recall that $\mathbb{N}(r)$ is a monotonically increasing function of r, so that for $0 < r \le \mathcal{D}_0$, Equation (36) implies

$$\mathbb{N}(r) \le \mathbb{N}(\mathcal{D}_0) < \frac{1}{2} \alpha_0 \, \mu_0 \, \mathcal{D}_0 \tag{43}$$

Therefore, it follows from Equation (28) that, in this case,

$$\mathbb{D} > \left(\frac{1}{2}\,\alpha_0\,\mu_0\,\mathcal{D}_0\right)\mathbb{W}_N \tag{44}$$

The virial theorem for nonlocal gravity in the case of an isolated galaxy is then

$$2 < \mathbb{T} > + < \mathbb{W}_N > \ < \ -\left(\frac{1}{2}\,\alpha_0\,\mu_0\,\mathcal{D}_0\right) < \mathbb{W}_N > \tag{45}$$

which means, when compared with Equation (38), that

$$f_{DM} < \frac{1}{2}\,\alpha_0\,\mu_0\,\mathcal{D}_0 \tag{46}$$

Let us note that

$$\frac{1}{2}\,\alpha_0\,\mu_0 = \frac{1}{\lambda_0} \tag{47}$$

where λ_0 is the basic nonlocality length scale. Its exact value is not known; however, from the results of reference [9], we have $\lambda_0 \approx 3 \pm 2$ kpc. If we formally let $\lambda_0 \to \infty$, then (46), namely, $f_{DM} < \mathcal{D}_0/\lambda_0$, implies that in this case nonlocality and the effective dark matter both disappear, as expected. Therefore, for a sufficiently isolated galaxy in virial equilibrium, the ratio of its baryonic diameter to dark matter fraction f_{DM} must always be above a fixed length λ_0 of about 3 ± 2 kpc; that is,

$$\frac{\mathcal{D}_0}{f_{DM}} > \lambda_0 \tag{48}$$

To illustrate (48), consider, for instance, the Andromeda Galaxy (M31) with a diameter $\mathcal{D}_0$ of about 67 kpc. In this case, we have $f_{DM} \approx 12.7$ [43,44], so that for this spiral galaxy

$$\frac{\mathcal{D}_0}{f_{DM}} \,(\text{Andromeda Galaxy}) \approx 5.3 \text{ kpc} \tag{49}$$

More recently, the distribution of dark matter in M31 has been further studied in reference [45]. Similarly, for the Triangulum Galaxy (M33), we have $\mathcal{D}_0 \approx 34$ kpc and $f_{DM} \approx 5$ [46], so that

$$\frac{\mathcal{D}_0}{f_{DM}} \,(\text{Triangulum Galaxy}) \approx 6.8 \text{ kpc} \tag{50}$$

Turning next to an elliptical galaxy, namely, the massive E0 galaxy NGC 1407, we have $\mathcal{D}_0 \approx 160$ kpc and $f_{DM} \approx 31$ [47], so that

$$\frac{\mathcal{D}_0}{f_{DM}} \,(\text{NGC 1407}) \approx 5.2 \text{ kpc} \tag{51}$$

Moreover, for the intermediate-luminosity elliptical galaxy NGC 4494, which has a half-light radius of $R_e \approx 3.77$ kpc, the dark matter fraction has been found to be $f_{DM} = 0.6 \pm 0.1$ [48]. Assuming that the baryonic system has a radius of $2\,R_e$, we have $\mathcal{D}_0 = 4\,R_e \approx 15$ kpc and $f_{DM} \approx 0.6$; hence,

$$\frac{\mathcal{D}_0}{f_{DM}} \,(\text{NGC 4494}) \approx 25 \text{ kpc} \tag{52}$$

Let us note that the results presented here are essentially for the present epoch in the expansion of the universe. Observations indicate, however, that the diameters of massive galaxies can increase with decreasing redshift z. For a discussion of such *massive compact galaxies*, see reference [49].

Finally, it is interesting to consider f_{DM} at the other extreme, namely, for the case of globular star clusters and isolated dwarf galaxies. The diameter of a globular star cluster is about 40 pc. We can therefore conclude from (48) with $\lambda_0 \approx 3$ kpc that for globular star clusters:

$$f_{DM} \text{ (globular star cluster)} \lesssim 10^{-2} \tag{53}$$

Thus, according to the virial theorem of nonlocal gravity, less than about one percent of the mass of a globular star cluster must appear as effective dark matter if the system is sufficiently isolated and is in virial equilibrium. It is not clear to what extent such systems can be considered isolated. It is usually assumed that observational data are consistent with the existence of almost no dark matter in globular star clusters. However, a recent investigation of six galactic globular clusters has led to the conclusion that $f_{DM} \approx 0.4$ [50]. The resolution of this discrepancy is beyond the scope of the present work.

Isolated dwarf galaxies with diameters $\mathcal{D}_0 \ll \mu_0^{-1}$ would similarly be expected to contain a relatively small percentage of effective dark matter. There is a significant discrepancy here as well, see reference [51]; again, the resolution of this difficulty is beyond the scope of this paper. In dwarf systems that are not isolated, the tidal influence of a much larger neighboring galaxy on the dynamics of the dwarf spheroidal galaxy cannot be ignored [52–54].

6. Discussion

Nonlocal gravity theory predicts that the amount of effective dark matter in a sufficiently isolated nearby galaxy in virial equilibrium is such that f_{DM} has an upper bound, $\mathcal{D}_0/\lambda_0$, that is completely independent of the distribution of baryonic matter in the galaxy. However, it is possible to derive an *improved* upper bound for f_{DM}, which does depend on how baryons are distributed within the galaxy. To this end, we note that Equation (28) for $\mathbb{D}$ and $\mathbb{N}(r) < r/\lambda_0$ imply:

$$\mathbb{D} > -\frac{1}{2} \sum_{i,j}' \frac{G m_i m_j}{\lambda_0} \tag{54}$$

If follows from this result together with Equation (39) that

$$< \mathbb{W}_N > f_{DM} > -\frac{1}{2} \sum_{i,j}' \frac{G m_i m_j}{\lambda_0} \tag{55}$$

Let us define a characteristic length, R_{av}, for the average extent of the distribution of baryons in the galaxy via

$$R_{av} < \mathbb{W}_N > = -\frac{1}{2} \sum_{i,j}' G m_i m_j \tag{56}$$

Then, it follows from (55) and Equation (56), that

$$f_{DM} < \frac{R_{av}}{\lambda_0} \tag{57}$$

Clearly, R_{av} depends upon the density of baryons in the galaxy. In the Newtonian gravitational potential energy in Equation (56), $0 < |\mathbf{x}_i - \mathbf{x}_j| \leq \mathcal{D}_0$; therefore, in general, $R_{av} \leq \mathcal{D}_0$; hence, we recover from the new inequality, namely, $f_{DM} < R_{av}/\lambda_0$, our previous less tight but more general result $f_{DM} < \mathcal{D}_0/\lambda_0$.

Acknowledgments: I am grateful to Jeffrey Kuhn, Sohrab Rahvar and Haojing Yan for valuable discussions.

Conflicts of Interest: The author declares no conflict of interest.

References

1. Einstein, A. *The Meaning of Relativity*; Princeton University Press: Princeton, NJ, USA, 1955.
2. Hehl, F.W.; Mashhoon, B. Nonlocal gravity simulates dark matter. *Phys. Lett. B* **2009**, *673*, 279–282.

3.   Hehl, F.W.; Mashhoon, B. Formal framework for a nonlocal generalization of Einstein's theory of gravitation. *Phys. Rev. D* **2009**, *79*, 064028.

4.   Blome, H.-J.; Chicone, C.; Hehl, F.W.; Mashhoon, B. Nonlocal modification of Newtonian gravity. *Phys. Rev. D* **2010**, *81*, 065020.

5.   Mashhoon, B. Nonlocal gravity. In *Cosmology and Gravitation*; Novello, M., Begliaffa, S.E.P., Eds.; Cambridge Scientific Publishers: Cambridge, UK, 2011; pp. 1–9.

6.   Chicone, C.; Mashhoon, B. Nonlocal gravity: Modified Poisson's equation. *J. Math. Phys.* **2012**, *53*, 042501.

7.   Chicone, C.; Mashhoon, B. Linearized gravitational waves in nonlocal general relativity. *Phys. Rev. D* **2013**, *87*, 064015.

8.   Mashhoon, B. Nonlocal gravity: Damping of linearized gravitational waves. *Class. Quantum Gravity* **2013**, *30*, 155008.

9.   Rahvar, S.; Mashhoon, B. Observational tests of nonlocal gravity: Galaxy rotation curves and clusters of galaxies. *Phys. Rev. D* **2014**, *89*, 104011.

10.   Mashhoon, B. Nonlocal gravity: The general linear approximation. *Phys. Rev. D* **2014**, *90*, 124031.

11.   Mashhoon, B. Nonlocal general relativity. *Galaxies* **2015**, *3*, 1–17.

12.   Chicone, C.; Mashhoon, B. Nonlocal gravity in the Solar System. *Class. Quantum Gravity* **2016**, *33*, 075005.

13.   Chicone, C.; Mashhoon, B. Nonlocal Newtonian cosmology. 2015, arXiv:1510.07316 [gr-qc].

14.   Iorio, L. Gravitational Anomalies in the Solar System? *Int. J. Mod. Phys. D* **2015**, *24*, 1530015.

15.   Deng, X.-M.; Xie, Y. Solar System test of the nonlocal gravity and the necessity for a screening mechanism. *Ann. Phys.* **2015**, *361*, 62–71.

16.   Abramowitz, M.; Stegun, I.A. *Handbook of Mathematical Functions*; National Bureau of Standards: Washington, DC, USA, 1964.

17.   Adelberger, E.G.; Heckel, B.R.; Nelson, A.E. Tests of the Gravitational Inverse-Square Law. *Ann. Rev. Nucl. Part. Sci.* **2003**, *53*, 77–121.

18.   Hoyle, C.D.; Kapner, D.J.; Heckel, B.R.; Adelberger, E.G.; Gundlach, J.H.; Schmidt, U.; Swanson, H.E. Sub-millimeter tests of the gravitational inverse-square law. *Phys. Rev. D* **2004**, *70*, 042004.

19.   Adelberger, E.G.; Heckel, B.R.; Hoedl, S.A.; Hoyle, C.D.; Kapner, D.J.; Upadhye, A. Particle-Physics Implications of a Recent Test of the Gravitational Inverse-Square Law. *Phys. Rev. Lett.* **2007**, *98*, 131104.

20.   Kapner, D.J.; Cook, T.S.; Adelberger, E.G.; Gundlach, J.H.; Heckel, B.R.; Hoyle, C.D.; Swanson, H.E. Tests of the Gravitational Inverse-Square Law below the Dark-Energy Length Scale. *Phys. Rev. Lett.* **2007**, *98*, 021101.

21.   Little, S.; Little, M. Laboratory test of Newton's law of gravity for small accelerations. *Class. Quantum Gravity* **2014**, *31*, 195008.

22.   Tohline, J.E. Stabilizing a Cold Disk with a 1/r Force Law. In *IAU Symposium 100, Internal Kinematics and Dynamics of Galaxies*; Athanassoula, E., Ed.; Reidel: Dordrecht, The Netherlands, 1983; pp. 205–206.

23.   Tohline, J.E. Does Gravity Exhibit a 1/r Force on the Scale of Galaxies? *Ann. N. Y. Acad. Sci.* **1984**, *422*, 390–390.

24.   Kuhn, J.R.; Burns, C.A.; Schorr, A.J. Numerical Coincidences, Fictional Forces, and the Galactic Dark Matter Distribution. 1986, unpublished work.

25.   Kuhn, J.R.; Kruglyak, L. Non-Newtonian forces and the invisible mass problem. *Astrophys. J.* **1987**, *313*, 1–12.

26.   Bekenstein, J.D. *Second Canadian Conference on General Relativity and Relativistic Astrophysics*; Coley, A., Dyer, C., Tupper, T., Eds.; World Scientific: Singapore, 1988; p. 68.

27.   Hees, A.; Hestroffer, D.; Le Poncin-Lafitte, C.; David, P. Tests of gravitation with Gaia observations of Solar System Objects. 2015, arXiv: 1509.06868.

28.   Buscaino, B.; DeBra, D.; Graham, P.W.; Gratta, G.; Wiser, T.D. Testing long-distance modifications of gravity to 100 astronomical units. *Phys. Rev. D* **2015**, *92*, 104048.

29.   Aprile, E.; Alfonsi, M.; Arisaka, K.; Arneodo, F.; Balan, C.; Baudis, L.; Bauermeister, B.; Behrens, A.; Beltrame, P.; Bokeloh, K.; *et al.* Dark Matter Results from 225 Live Days of XENON100 Data. *Phys. Rev. Lett.* **2012**, *109*, 181301.

30.   Akerib, D.S.; Araújo, H.M.; Bai, X.; Bailey, A.J.; Balajthy, J.; Bedikian, S.; Bernard, E.; Bernstein, A.; Bolozdynya, A.; Bradley, A.; *et al.* First Results from the LUX Dark Matter Experiment at the Sanford Underground Research Facility. *Phys. Rev. Lett.* **2014**, *112*, 091303.

31.   Agnese,R.; Anderson, A.J.; Asai, M.; Balakishiyeva, D.; Basu Thakur, R.; Bauer, D.A.; Beaty, J.; Billard, J.; Borgland, A.; Bowles, M.A.; *et al.* Search for Low-Mass Weakly Interacting Massive Particles with SuperCDMS. *Phys. Rev. Lett.* **2014**, *112*, 241302.

32. Baudis, L. Dark matter searches. *Ann. Phys.* **2016**, *528*, 74–83.
33. Zwicky, F. Die Rotverschiebung von extragalaktischen Nebeln. *Helv. Phys. Acta* **1933**, *6*, 110–127.
34. Zwicky, F. On the Masses of Nebulae and of Clusters of Nebulae. *Astrophys. J.* **1937**, *86*, 217–246.
35. Rubin, V.C.; Ford, W.K. Rotation of the Andromeda Nebula from a Spectroscopic Survey of Emission Regions. *Astrophys. J.* **1970**, *159*, 379–403.
36. Roberts, M.S.; Whitehurst, R.N. The rotation curve and geometry of M31 at large galactocentric distances. *Astrophys. J.* **1975**, *201*, 327–346.
37. Sofue, Y.; Rubin, V. Rotation Curves of Spiral Galaxies. *Annu. Rev. Astron. Astrophys.* **2001**, *39*, 137–174.
38. Seigar, M.S. *Dark Matter in the Universe*; Morgan and Claypool: San Rafael, CA, USA, 2015.
39. Harvey, D.; Massey, R.; Kitching, T.; Taylor, A.; Tittley, E. The nongravitational interactions of dark matter in colliding galaxy clusters. *Science* **2015**, *347*, 1462–1465.
40. Clowe, D.; Bradač, M.; Gonzalez, A.H.; Markevitch, M.; Randall, S.W.; Jones, C.; Zaritsky, D. A direct empirical proof of the existence of dark matter. *Astrophys. J. Lett.* **2006**, *648*, L109–L113.
41. Clowe, D.; Randall, S.W.; Markevitch, M. Catching a bullet: direct evidence for the existence of dark matter. *Nucl. Phys. B Proc. Suppl.* **2007**, *173*, 28–31.
42. Bini, D.; Mashhoon, B. Nonlocal gravity: Conformally flat spacetimes. *Int. J. Geom. Methods Mod. Phys.* **2016**, *13*, 1650081.
43. Barmby, P.; Ashby, M.L.N.; Bianchi, L.; Engelbracht, C.W.; Gehrz, R.D.; Gordon, K.D.; Hinz, J.L.; Huchra, J.P.; Humphreys, R.M.; Pahre, M.A.; *et al.* Dusty waves on a starry sea: The mid-infrared view of M31 *Astrophys. J.* **2006**, *650*, L45–L49.
44. Barmby, P.; Ashby, M.L.N.; Bianchi, L.; Engelbracht, C.W.; Gehrz, R.D.; Gordon, K.D.; Hinz, J.L.; Huchra, J.P.; Humphreys, R.M.; Pahre, M.A.; *et al.* Erratum: "Dusty Waves on a Starry Sea: The Mid-Infrared View of M31". *Astrophys. J.* **2007**, *655*, L61–L61.
45. Tamm, A.; Tempel, E.; Tenjes, P.; Tihhonova, O.; Tuvikene, T. Stellar mass map and dark matter distribution in M31. *Astron. Astrophys.* **2012**, *546*, A4.
46. Corbelli, E. Dark matter and visible baryons in M33. *Mon. Not. R. Astron. Soc.* **2003**, *342*, 199–207.
47. Pota, V.; Romanowsky, A.J.; Brodie, J.P.; Peñarrubia, J.; Forbes, D.A.; Napolitano, N.R.; Foster, C.; Walker, M.G.; Strader, J.; Roediger, J.C. The SLUGGS survey: Multipopulation dynamical modelling of the elliptical galaxy NGC 1407 from stars and globular clusters. *Mon. Not. R. Astron. Soc.* **2015**, *450*, 3345–3358.
48. Morganti, L.; Gerhard, O.; Coccato, L.; Martinez-Valpuesta, I.; Arnaboldi, M. Elliptical galaxies with rapidly decreasing velocity dispersion profiles: NMAGIC models and dark halo parameter estimates for NGC 4494 *Mon. Not. R. Astron. Soc.* **2013**, *431*, 3570–3588.
49. De Arriba, L.P.; Balcells, M.; Falcón-Barroso, J.; Trujillo, I. The discrepancy between dynamical and stellar masses in massive compact galaxies traces non-homology. *Mon. Not. R. Astron. Soc.* **2014**, *440*, 1634–1648.
50. Sollima, A.; Bellazzini, M.; Lee, J.-W. A comparison between the stellar and dynamical masses of six globular clusters. *Astrophys. J.* **2012**, *755*, 156.
51. Oh, S.-H.; Hunter, D.A.; Brinks, E.; Elmegreen, B.G.; Schruba, A.; Walter, F.; Rupen, M.P.; Young, L.M.; Simpson, C.E.; Johnson, M.C. High-resolution mass models of dwarf galaxies from LITTLE THINGS. *Astron. J.* **2015**, *149*, 180.
52. Kuhn, J.R.; Miller, R.H. Dwarf spheroidal galaxies and resonant orbital coupling. *Astrophys. J. Lett.* **1989**, *341*, L41–L45.
53. Fleck, J.-J.; Kuhn, J.R. Parametric dwarf spheroidal tidal interaction. *Astrophys. J.* **2003**, *592*, 147–160.
54. Muñoz, R.R.; Frinchaboy, P.M.; Majewski, S.R.; Kuhn, J.R.; Chou, M-Y.; Palma, C.; Sohn, S.T.; Patterson, R.J.; Siegel, M.H. Exploring Halo Substructure with Giant Stars: The Velocity Dispersion Profiles of the Ursa Minor and Draco Dwarf Spheroidal Galaxies at Large Angular Separations. *Astrophys. J. Lett.* **2005**, *631*, L137–L141.

universe

MDPI

Review

The Scales of Gravitational Lensing

Francesco De Paolis [1,2,*,†], Mosè Giordano [1,2,†], Gabriele Ingrosso [1,2,†], Luigi Manni [1,2,†], Achille Nucita [1,2,†] and Francesco Strafella [1,2,†]

1   Department of Mathematics and Physics "Ennio De Giorgi", University of Salento, CP 193, I-73100 Lecce, Italy; giordano@le.infn.it (M.G.); ingrosso@le.infn.it (G.I.); luigi.manni@le.infn.it (L.M.); nucita@le.infn.it (A.N.); strafella@le.infn.it (F.S.)
2   INFN (Istituto Nazionale di Fisica Nucleare), Sezione di Lecce CP-193, Italy
*   Correspondence: francesco.depaolis@le.infn.it; Tel.: +39-0832-297493; Fax: +39-0832-297505
†   These authors contributed equally to this work.

Academic Editor: Lorenzo Iorio
Received: 5 January 2016; Accepted: 7 March 2016; Published: 14 March 2016

Abstract: After exactly a century since the formulation of the general theory of relativity, the phenomenon of gravitational lensing is still an extremely powerful method for investigating in astrophysics and cosmology. Indeed, it is adopted to study the distribution of the stellar component in the Milky Way, to study dark matter and dark energy on very large scales and even to discover exoplanets. Moreover, thanks to technological developments, it will allow the measure of the physical parameters (mass, angular momentum and electric charge) of supermassive black holes in the center of ours and nearby galaxies.

Keywords: gravitational lensing

1. Introduction

In 1911, while he was still involved in the development of the general theory of relativity (subsequently published in 1916), Einstein made the first calculation of light deflection by the Sun [1]. He correctly understood that a massive body may act as a gravitational lens deflecting light rays passing close to the body surface. However, his calculation, based on Newtonian mechanics, gave a deflection angle wrong by a factor of two. On 14 October 1913, Einstein wrote to Hale, the renowned astronomer, inquiring whether it was possible to measure a deflection angle of about $0.84''$ toward the Sun. The answer was negative, but Einstein did not give up, and when, in 1915, he made the calculation again using the general theory of relativity, he found the right value $\phi = 2r_s/b$ (where $r_s = 2GM/c^2$ is the Schwarzschild radius and b is the light rays' impact parameter) that corresponds to an angle of about $1.75''$ in the case of the Sun. That result was resoundingly confirmed during the Solar eclipse of 1919 [2].

In 1924, Chwolson [3] considered the particular case when the source, the lens and the observer are aligned and noticed the possibility of observing a luminous ring when a far source undergoes the lensing effect by a massive star. In 1936, after the insistence of Rudi Mandl, Einstein published a paper on science [4] describing the gravitational lensing effect of one star on another, the formation of the luminous ring, today called the Einstein ring, and giving the expression for the source amplification. However, Einstein considered this effect exceedingly curious and useless, since in his opinion, there was no hope to actually observe it.

On this issue, however, Einstein was wrong: he underestimated technological progress and did not foresee the motivations that today induce one to widely use the gravitational lensing phenomenon. Indeed, Zwicky promptly understood that galaxies were gravitational lenses more powerful than stars and might give rise to images with a detectable angular separation. In two letters published in 1937 [5,6], Zwicky noticed that the observation of galaxy lensing, in addition to giving a further proof of the general theory of relativity, might allow observing sources otherwise invisible, thanks to the light gravitational amplification, thereby obtaining a more direct and accurate estimate of the lens galaxy dynamical mass. He also found that the probability to observe lensed galaxies was much larger than that of star on star. This shows the foresight of this eclectic scientist, since the first strong lensing event was discovered only in 1979: the double quasar QSO 0957+561 a/b [7], shortly followed by the observation of tens of other gravitational lenses, Einstein rings and gravitational arcs. All of that plays today an extremely relevant role for the comprehension of the evolution of the structures and the measure of the parameters of the so-called cosmological standard model.

Actually, there are different scales in gravitational lensing, on which we shall briefly concentrate in the next sections, after a short introduction to the basics of the theory of gravitational lensing (Section 2). Generally speaking, gravitational lens images separated by more than a few tenths of arcsecs are clearly seen as distinct images by the observer. This was the case considered by Zwicky, and the gravitational lensing in this regime is called *strong (or macro) lensing* (see Section 3), which also includes distorted galaxy images, like Einstein rings or arcs. If instead, the distortions induced by the gravitational fields on background objects are much smaller, we have the *weak lensing* effect (see Section 4). On the other side, if one considers the phenomenology of the star-on-star lensing (as Einstein did), the resulting angular distance between the images is of the order of a few μas, generally not separable by telescopes. Gravitational lensing in this regime is called, following Paczyński [8], *microlensing*, and the observable is an achromatic change in the brightness of the source star over time, due to the relative motion of the lens and the source with respect to the line of sight of the source (see Section 5). In all of these regimes, the gravitational field can be treated in the weak field approximation. Another scale on which gravitational lensing applies is that involving black holes. In particular, when light rays come very close to the event horizon, they are subject to strong gravitational field effects, and thereby, the deflection angles are large. This effect is called *retro-lensing*, and we shall discuss it in Section 6. The observation of retro-lensing events is of great importance also because the general theory of relativity still stands practically untested in the strong gravitational field regime (see [9] for a very recent review), apart from gravitational waves [10]. A short final discussion is then offered in Section 7.

2. Basics of Gravitational Lensing

In the general theory of relativity, light rays follow null geodesics, *i.e.*, the minimum distance paths in a curved space-time. Therefore, when a light ray from a far source interacts with the gravitational field due to a massive body, it is bent by an angle approximately equal to $\alpha_S(b) = 2r_s/b$. By looking at Figure 1, assuming the ideal case of a thin lens and noting that $\alpha_S D_{LS} = (\theta - \theta_S)D_S$, one can easily derive the so-called lens equation:

$$\theta_S = \theta - \frac{\theta_E^2}{\theta}, \tag{1}$$

where θ_S indicates the source position and:

$$\theta_E = \left(\frac{4GM}{c^2} \frac{D_{LS}}{D_S D_L} \right)^{1/2} \tag{2}$$

is the Einstein ring radius, which is the angular radius of the image when lens and source are perfectly aligned ($\theta_S = 0$). Therefore, one can see that two images appear in the source plane, whose positions can be obtained by solving Equation (1).

More generally (see for details [11]), the light deflection between the two-dimensional position of the source θ_S and the position of the image θ is given by the lens mapping equation:

$$\theta_S = \theta - \nabla\phi(\theta), \tag{3}$$

where $\phi = 2D_{LS}\Phi_N^{2D}/(D_Sc^2)$ is the so-called lensing potential and Φ_N^{2D} is the two-dimensional Newtonian projected gravitational potential of the lens. We also note, in turn, that the ratio D_{LS}/D_S depends on the redshift of the source and the lens, as well as on the cosmological parameters $\Omega_M = \rho_M/\rho_c$ and $\Omega_\Lambda = \rho_\Lambda/\rho_c$, being $\rho_c = 3H_0^2/(8\pi G)$, ρ_M and ρ_Λ the critical, the matter and the dark energy densities, respectively. The transformation above is thus a mapping from the source plane to the image plane, and the Jacobian $\mathcal{J}$ of the transformation is given by:

$$\mathcal{J} = \frac{d\boldsymbol{\theta}_S}{d\boldsymbol{\theta}} = \mathcal{A}^{-1} = \begin{pmatrix} 1-\phi_{,11} & -\phi_{,12} \\ -\phi_{,12} & 1-\phi_{,22} \end{pmatrix} = \begin{pmatrix} 1-\kappa-\gamma_1 & -\gamma_2 \\ -\gamma_2 & 1-\kappa+\gamma_1 \end{pmatrix}, \tag{4}$$

where the commas are the partial derivatives with respect to the two components of $\boldsymbol{\theta}$. Here, κ is the convergence, which turns out to be equal to $\Sigma/2\Sigma_{cr}$, where:

$$\Sigma_{cr} = c^2 D_S/(4\pi G D_L D_{LS}) \tag{5}$$

is the critical surface density, $\gamma = (\gamma_1, \gamma_2)$ is the shear and $\mathcal{A}$ is the magnification matrix. Thus, the previous equations define the convergence and shear as second derivatives of the potential, *i.e.*,

$$\kappa = \frac{1}{2}(\partial_1\partial_1 + \partial_2\partial_2)\phi = \frac{1}{2}\nabla^2\phi, \quad \gamma_1 = (\partial_1\partial_1 - \partial_2\partial_2)\phi, \quad \gamma_2 = \partial_1\partial_2\phi. \tag{6}$$

From the above discussion, it is clear that gravitational lensing may allow one to probe the total mass distribution within the lens system, which reproduces the observed image configurations and distortions. This, in turn, may allow one to constrain the cosmological parameters, although this is a second order effect.

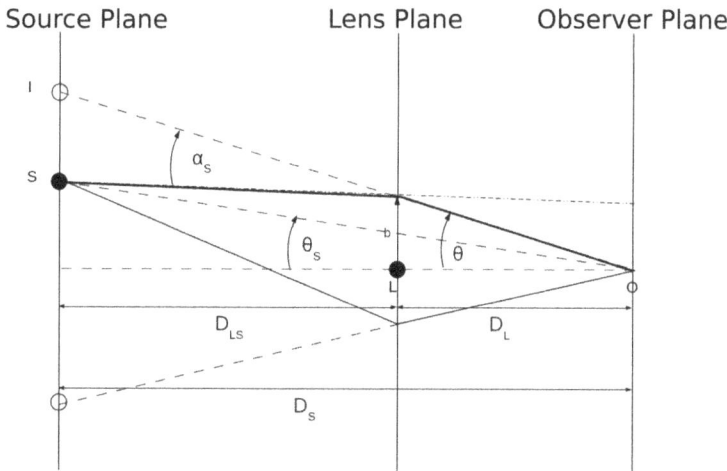

Figure 1. Schematics of the lensing phenomenon.

3. Strong Lensing

Quasars are the brightest astronomical objects, visible even at a distance of billions of parsecs. After the identification of the first quasar in 1963 [12], these objects remained a mystery for quite a long time, but today, we know that they are powered by mass accretion on a supermassive black hole, with a mass billions of times that of the Sun. The first strong gravitational lens, discovered in 1979, was indeed linked to a quasar (QSO 0957+561 [7]), and although the phenomenon was expected on theoretical grounds, it left the astronomers astonished. The existence of two objects separated by about 6″ and characterized by an identical spectrum led to the conclusion that they were the doubled image of the same quasar, clearly showing that Zwicky was perfectly right and that galaxies may act as gravitational lenses. Afterwards, also the lens galaxy was identified, and it was established that its dynamical mass, responsible for the light deflection, was at least ten-times larger than the visible mass. This double quasar was also the first object for which the time delay (about 420 days) between the two images [13], due to the different paths of the photons forming the two images, has been measured. This has also allowed obtaining an independent estimate of the lens galaxy dynamical mass. Observations can also show four images of the same quasar, as in the case of the so-called Einstein Cross, or when the lens and the source are closely aligned, one can observe the Einstein ring, e.g., in the case of MG1654-1346 [14]. The macroscopic effect of multiple images' formation is generally called strong lensing, which also consists of the formation of arcs, as those clearly visible in the deep sky field images by the Sloan Digital Sky Survey (SDSS; see, e.g., [15]). The sources of strong lensing events are often quasars, galaxies, galaxy clusters and supernovae, whereas the lenses are usually galaxies or galaxy clusters. The image separation is generally larger than a few tenths of an arcsec, often up to a few arcsecs.

Over the years, many strong lensing events have been found in deep surveys of the sky, such as the CLASS [16], the Sloan ACS [17], the SDSS, one of the most successful surveys in the history of astronomy (see, e.g., [18] and references therein), the SQLS (the Sloan Digital Sky Survey for Quasar Lens Search) [19], and so on.

Strong gravitational lensing is nowadays a powerful tool for investigation in astrophysics and cosmology (see, e.g. [20,21]). As already mentioned in the previous section, strong lensing gives a unique opportunity to measure the dynamical mass of the lens object using, for example, the mass estimator $M(< R_E) = \pi \Sigma_{cr} \theta_E^2$, which directly gives the mass within R_E, using Equation (5) in this regime. The result is that masses obtained in this way are almost always larger than the visible mass of the lensing object, showing that galaxy and galaxy cluster masses are dominated by dark matter. In any case, accurately constraining the mass distribution of the lens system (e.g., a galaxy cluster) is a generally degenerate problem, in the sense that there are several mass distributions that can fit the observables; thus, the best way to solve it is to use multiple images (see, e.g., [22]).

Another important application of strong lensing is the study of dark matter halo substructures. Indeed, sometimes flux ratio anomalies in the lensed quasar images are detected (see, e.g., [23,24]), and while smooth mass models of the lensing galaxy may generally explain the observed image positions, the prediction of such models of the corresponding fluxes is frequently violated. Especially in the radio band observations, since the quasar radio emitting region is quite large, the observed radio flux anomalies are explained as being due to the presence of substructures of about $10^6 - 10^8$ M$_\odot$ along the line of sight. After some controversy regarding whether ΛCDM (cold dark matter plus Cosmological Constant) simulations predict enough dark matter substructures to account for the observations (for example, in [25], some indication is found of an excess of massive galaxy satellites), more recent analysis, taking also into account the uncertainty in the lens system ellipticity, finds results consistent with those predicted by the standard cosmological model [26,27]. However, at present, the list of multiply-imaged quasars observed in the radio and mid-IR bands is quite short, and further observational and theoretical work would be very helpful in this respect. Another indication of dark matter halo substructures comes from detailed analysis of galaxy-galaxy lensing. Although the results obtained are generally consistent with ΛCDM simulations, more data should be analyzed in order to get strong constraints [28,29]. Strongly lensed quasars have been observed to show a certain variability of one image with respect to the others. This can be often attributed to microlensing (see Section 5) by the stars throughout the lens galaxy. This effect, and in particular its variation with respect to the wavelength, has provided an opportunity to study in detail the central engine of the source quasar, and the magnitude of the microlensing variability has allowed astrophysicists to constrain the stellar density in the lens galaxy [30–32].

Strong gravitational lensing may be used as a natural telescope that magnifies dim galaxies, making them easier to be studied in detail. For this reason, mass concentrations, like galaxies and clusters of galaxies, can be effectively used as cosmic telescopes to study faint sources that would not be possible to detect in the absence of gravitational lensing (see, e.g., [33,34]). At present, there is also an event of a high magnified supernova multiply imaged and also seen exploding again, being lensed by a galaxy in the cluster MACS J1149.6+2223 [35,36].

The ultimate goal of strong lensing is not only to get information on the large-scale structure of the Universe, but also to constrain the cosmological parameters. For instance, analyzing the time delay among the lensed source images, it is possible to estimate also the value of the Hubble constant H_0. Indeed, the time delay is given by the difference of the light paths from the images and is inversely proportional to H_0, as first understood by Refsdal [37] (see also the review in [38]). At present, one of the most accurate measurement of the Hubble constant using a gravitational lens is provided in [39]. There is also a project (COSMOGRAIL) particularly devoted to the time delay measurements of doubly- or multiply-lensed quasars (see [40] and the references therein). Moreover, the measure of both the frequency of occurrence and the redshift of multiple images in deep sky surveys may allow one to constrain the values of Ω_M and Ω_Λ in an independent way with respect to other methods, such as those coming from SN Ia or the CMB (Cosmic Microwave Background) power spectrum.

4. Weak Lensing

In addition to the macroscopic deformations discussed in the previous section, in the deep field surveys of the sky, also arclets (*i.e.*, single distorted images with an elliptical shape) and weakly distorted images of galaxies, with an almost invisible individual elongation, have been detected. This effect is known as weak lensing and is playing an increasingly important role in cosmology.

The weak lensing's main feature is the shape deformation of background galaxies, whose light crosses a mass distribution (e.g., a galaxy or a galaxy cluster) that acts as a gravitational lens. Actually, as discussed in Section 2, gravitational lensing gives rise to two distinct effects on a source image: convergence, which is isotropic, and shear, which is anisotropic. In the weak lensing regime, the observer makes use of the shear, that is the image deformation (sometimes related to the galaxy orientation), while the convergence effect is not used, since the intrinsic luminosity and the size of the lensed objects are unknown. For a complete and in-depth review on the basics of weak gravitational lensing, with full mathematical details of all the most important concepts, we refer the reader to [41].

The first weak lensing event was detected in 1990 as statistical tangential alignment of galaxies behind massive clusters [42], but only in 2000, coherent galaxy distortions were measured in blind fields, showing the existence of the cosmic shear (see, e.g., [43,44]). Here, we remark that the weak lensing cannot be measured by a single galaxy, but its observation relies on the statistical analysis of the shape and alignment of a large number of galaxies in a certain direction.

Therefore, the game is to measure the galaxy ellipticities and orientations and to relate them to the surface mass density distribution of the lens system (generally a galaxy cluster placed in between). There are at least two major issues in weak lensing studies, one mainly relying on the theory, the other one on observations: the former concerns finding the best way to reconstruct the intervening mass distribution from the shear field $\gamma = (\gamma_1, \gamma_2)$, the latter with looking for the best way to determine the *true* ellipticity of a faint galaxy, which is smeared out by the instrumental point spread function (PSF). To solve these issues, several approaches have been proposed, which can be distinguished into two broad families: direct and inverse methods. On the theoretical side, the direct approaches are: the integral method, which consists of expressing the projected mass density distribution as the convolution of γ by a kernel (see, e.g., [45]), and the local inversion method, which instead starts from the gradient of γ (see, e.g., [46] and the references therein). The inverse approaches work on the lensing potential ϕ (see Equation 3), and they include the use of the maximum likelihood [47,48] or the maximum entropy methods [49] to determine the most likely projected mass distribution that reproduces the shear field. The inverse methods are particularly useful since they make it possible to quantify the errors in the resultant lensing mass estimates, as, for instance, errors deriving from the assumption of a spherical mass model when fitting a non-spherical system [50,51].

The inverse methods allow one also to derive constraints from external observations, such as X-ray data on galaxy clusters' strong lensing or CMB lensing. In particular, one can compare mass measurements from weak lensing and X-ray observations for large samples of galaxy clusters [52]. In this respect, [53] used a large sample of nearby clusters with good weak lensing and X-ray measurements to investigate the agreement between mass estimates based on weak lensing and X-ray data, as well as studied the potential sources of errors in both methods. Moreover, a combination of weak lensing and CMB data may provide powerful constraints on the cosmological parameters, especially on the Hubble constant H_0, the amplitude of fluctuations σ_8 and the matter cosmic density Ω_m [54,55]. We also mention, in this respect, that one way to determine the fluid-mechanical properties of dark energy, characterized by its sound speed and its viscosity apart from its equation of state, is to combine Planck data with galaxy clustering and weak lensing observations by Euclid, yielding one percent sensitivity on the dark energy sound speed and viscosity [56] (see the end of this section).

On the observational side, the first priority is to use a telescope with a wide field of view, appropriate to probe the large-scale structure distribution at least of a galaxy cluster. On the other hand, it is also necessary to minimize the source of noise in the determination of the ellipticity of very faint galaxies, so that the best-seeing conditions for a ground-based telescope or, better, a space-based instrument, are extremely useful.

Very promising results have been obtained with the weak lensing technique so far, as, for example, the best measure, until today, of the existence and distribution of dark matter within the famous Bullet cluster [57] (actually constituted by a pair of galaxy clusters observed in the act of colliding). Astronomers found that the shocked plasma was almost entirely in the region between the two clusters, separated from the galaxies. However, weak lensing observations showed that the mass was largely concentrated around the galaxies themselves, and this enabled a clear, independent measurement of the amount of dark matter.

With the major aim to map, through the weak lensing effect, the mass distribution in the Universe and the dark energy contribution by measuring the shape and redshift of billions of very far away galaxies (for a review, see [58]), the European Space Agency (ESA) is planning to launch the Euclid satellite in the near future. Also ground-based telescopes will allow one to detect an enormous number of weak and strong lensing events. An example is given by the LSST (Large Synoptic Survey Telescope) project, located on the Cerro Pachón ridge in north-central Chile, which will become operative in 2022. Its 8.4-meter telescope uses a special three-mirror design, creating an exceptionally wide field of view, and has the ability to survey the entire sky in only three nights. The effective number density n_{eff} of weak lensing galaxies (which is a measure of the statistical power of a weak lensing survey) that will be discovered by LSST is, conservatively, in the range of 18–24 arcmin^{-2} (see Table 4 in [59]). The very large (about 1.5×10^4 square degrees) and deep survey of the sky that will be performed by Euclid will allow astrophysicists to address fundamental questions in physics and cosmology about the nature and the properties of dark matter and dark energy, as well as in the physics of the early Universe and the initial conditions that provided the seeds for the formation of cosmic structure. Before closing this section, we also mention that strong systematics may be present in weak lensing surveys. For example, the intrinsic alignment of background sources may mimic to an extent the effects of shear and may contaminate the weak lensing signal. However, these systematics may be controlled if also the galaxy redshifts are acquired, and this fully removes the unknown intrinsic alignment errors from weak lensing detections (for further details, see [60,61]).

5. Microlensing

Let us consider now the microlensing scale of the lensing phenomenon that occurs when θ_E is smaller than the typical telescope angular resolution, as in the case of stars lensing the light from background stars (for a review on gravitational microlensing and its astrophysical applications, we refer to, e.g., [62]). As is clear from the discussion in Section 2, by solving the lens Equation (1), one can determine the angular positions of the primary (I_1) and secondary (I_2) images. In Figure 2, these positions are shown for four different values of the impact parameter θ_S in the case of a point-like source. If the source and the lens are aligned (first panel on the left), the circular symmetry of the problem leads to the formation of a luminous annulus having radius θ_E around the lens position. Otherwise, increasing the θ_S value, the secondary image gets closer to the lens position, while the primary image drifts apart from it, and in the limit of $\theta_S \gg \theta_E$, the microlensing phenomenon tends to disappear. However, observing multiple images during a microlensing event is practically impossible with the present technology. For instance, in the case in which the phenomenon is maximized, corresponding to the perfect alignment, for a star in the galactic bulge (about 8 kpc away), one has $\Delta\theta = 2\theta_E \simeq 1\ \mu\mathrm{as}$, which is well below the angular resolving power, even of the Hubble Space Telescope (about 43 mas at 500 nm); see, e.g., http://www.coseti.org/9008-065.htm.

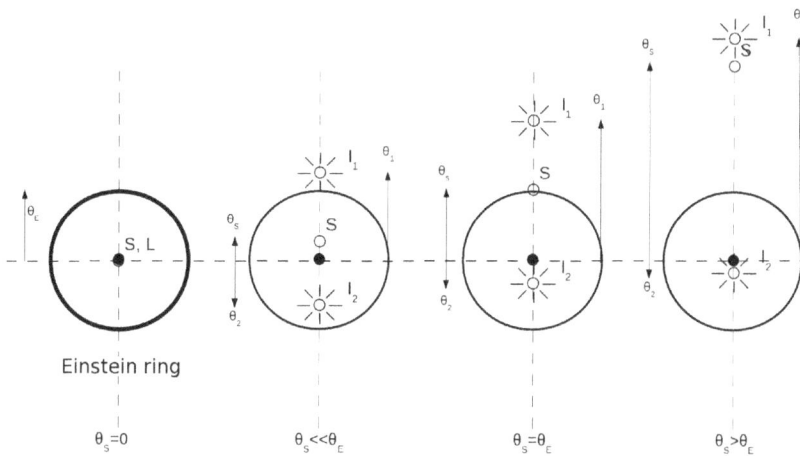

Figure 2. Angular positions of the primary (I_1) and secondary (I_2) images for four different values of the source impact parameter θ_S in the case of a point-like source.

When a source is microlensed, its images do not have the same luminosity; therefore, the observer receives a total flux (or magnitude) different from that of the unlensed source. The flux difference can be described very simply in terms of the light magnification and the law of the conservation of the specific intensity I, which represents the energy, with frequency in the range $d\nu$ crossing the surface dA during the time interval dt in the solid angle $d\Omega$ around the direction orthogonal to the surface. Indeed, the light specific intensity turns out to be conserved in the absence of absorption phenomena, interstellar scattering or Doppler shifts. This is also a consequence of Liouville's theorem, which claims that the density of states in the phase space remains constant if the interacting forces are non-collisional (and gravitation fulfills this condition due to its weak coupling constant), and the propagating medium is approximately transparent (as is the case for interstellar space). This effect can produce a magnification or a de-magnification of the images of an extended light source (see Figure 3). If the image is magnified, it means that it certainly subtends a wider angle with respect to that subtended by the source in the absence of the lens. In microlensing, the source disk size should not be neglected in general. Within the framework of the finite source approximation for a source with flux F_S and assuming $\theta_E \leq \theta_S$, one can show that the magnification A of an image at angular position θ is given by $(1 - \theta_E^4/\theta^4)^{-1}$. As a consequence, the observed flux corresponds to $F = AF_S$. Of course, when the source star disk gradually moves away from the line of sight, the magnification decreases, and the unlensed F_S flux is then recovered.

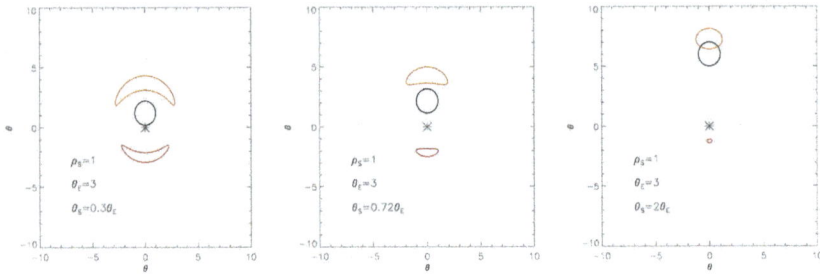

Figure 3. As in Figure 2, but with a conformal transformation of the source boundary, considered extended and with radius ρ_S, by a point-like lens. Each point of the source disk behaves as a point-like source. The black circle represents the source disk, while the red and yellow arcs are the deformed primary and secondary images.

As already anticipated, the observer cannot see, in the microlensing case, well-separated images, but, instead, detects a single image made by the overlapping of the primary and the secondary images. In this case, one can easily obtain the classical magnification factor A by summing up the individual magnifications, *i.e.*, $A = (u^2 + 2)/\sqrt{u^2(u^2 + 4)}$, where $u = \theta_S/\theta_E$ is the impact factor. If there is a relative movement between the lens and the source, u changes with time, and a standard Paczyński curve [8] does emerge.

An important role in gravitational microlensing is played by the *caustics*, the geometric loci of the points belonging to the lens plane where the light magnification of a point-like source becomes infinite, and by the corresponding *critical* curves in the source plane. In the case of a single lens, the caustic is a point coinciding with the lens position; therefore, the magnification diverges when the impact parameter approaches zero. However, real sources are not point-like, so we always have finite magnifications that can be calculated by an average procedure:

$$\langle A \rangle = \frac{\int A(\mathbf{y})I(\mathbf{y})d^2y}{\int I(\mathbf{y})d^2y}, \tag{7}$$

where $A(\mathbf{y})$ is the point-like source magnification, $I(\mathbf{y})$ is the brightness profile of the stellar disk (the limb darkening profile) and the integral is extended over the source star disk.

Observations show that about half of all stars are in binary systems, and moreover, thousands of exoplanets are being discovered around their host stars by different techniques and instruments. Therefore, it is worth considering binary and multiple systems as lenses in microlensing observations. In this case, the lens equation, obviously, becomes more complicated, but it can still be solved by numerical methods in order to obtain the magnification map where caustics take on distinctive shapes depending on the specific geometry of the system. In Figure 4, we show the magnification map and the resulting light curve for a simulated microlensing event due to a binary lens with mass ratio $q = M_1/M_2 \simeq 0.01$ (e.g., a solar mass as the primary component and a Jupiter-like planet as the secondary one). In these cases, the resulting light curve may be rather different with respect to the typical Paczyński one, depending on the system parameters. The study of these anomalies in the microlensing light curves behavior is becoming more and more important nowadays, since it allows one to estimate some of the parameters of the lensing system (see, e.g., [63]). The main advantage of this technique, compared to the other methods adopted by the exoplanets hunters (e.g., radial velocity, direct imaging, transits), is the possibility to detect even very small planets orbiting their own star at enormous distances from Earth. It also allows one to discover the so-called free-floating planets (FFPs), otherwise hardly detectable [64]. By studying the PA99-N2 microlensing event, detected in 1999 by the French-British collaboration POINT-AGAPE [66], Ingrosso *et al.* [65] revealed in 2009 that the anomaly observed was compatible with the presence of a super-Jupiter with a mass of $\simeq 5M_J$ around a star lying in the Andromeda galaxy (see Figure 5), thus finding the first putative exoplanet in another galaxy.

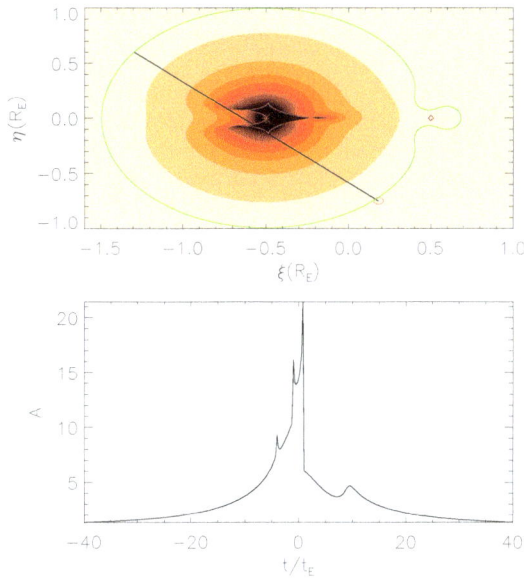

Figure 4. Magnification map for a binary lens system characterized by two objects separated by a projected distance of $1R_E$ and mass ratio $q = 0.01$. The green and red closed lines indicate the critical and caustic curves obtained by solving the lens equation in Equation (1). The black line indicates the trajectory of the source star, which has a radius of $0.03R_E$. The simulated light curve is shown in the lower panel.

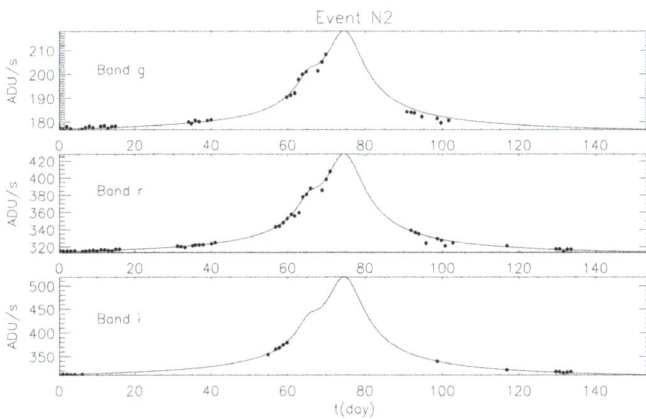

Figure 5. Light curve in three different bands (g, r and i) of the PA99-N2 event detected in 1999 toward the Andromeda galaxy.

5.1. Astrometric Microlensing

During an ongoing microlensing event, the centroid of the multiple images and the source star positions move in the lens plane giving rise to a phenomenon known as astrometric microlensing (see, e.g., [67,68] and the references therein). In the simplest case of a point-like lens, for a source at angular distance θ_S, the position θ of the images with respect to the lens can be obtained by solving the lens Equation (1). Since the Einstein radius $R_E = D_L \theta_E$ defines the scale length on the lens plane, the lens equation reads:

$$d^2 - dd_S - R_E^2 = 0, \tag{8}$$

where d_S and d are the linear distances, in the lens plane, of the source and images from the gravitational lens, respectively. Moreover, using the dimensionless source-lens distances $u = \theta_S / \theta_E$ and $\tilde{u} = \theta / \theta_E$, the previous relation can be further simplified as:

$$\tilde{u}^2 - u\tilde{u} - 1 = 0. \tag{9}$$

Denoting with u_+ and u_- the solutions of this equation, one notes that, in the lens plane, the $+$ image resides always outside the circular ring centered on the lens position with radius equal to the Einstein angle, while the $-$ image is always within the ring. As the source-lens distance increases, the $+$ image approaches the source position, while the $-$ one (becoming fainter) moves towards the lens location. For a source moving in the lens plane with transverse velocity $v_\perp$ directed along the ξ axis (η is perpendicular to it), the projected coordinates of the source result in being:

$$\xi(t) = \frac{t - t_0}{t_E}, \quad \eta(t) = u_0. \tag{10}$$

where $t_E = R_E / v_\perp$ and u_0 is the impact parameter (in this case, lying on the η axis). Since $u^2 = \xi^2 + \eta^2$ is time dependent, the two images move in the lens plane during the gravitational lensing event.

By weighting the $+$ and $-$ image position with the associated magnification [69], one gets:

$$\tilde{u} \equiv \frac{\tilde{u}_+ \mu_+ + \tilde{u}_- \mu_-}{\mu_+ + \mu_-} = \frac{u(u^2 + 3)}{u^2 + 2}. \tag{11}$$

Finally, the observable is defined as the displacement of the centroid with respect to the source,

$$\Delta \equiv \tilde{u} - u = \frac{u}{2 + u^2}. \tag{12}$$

Note that the centroid shift may be viewed as a vector:

$$\Delta = \frac{\mathbf{u}}{2 + u^2} \tag{13}$$

with components along the axes:

$$\Delta_\xi = \frac{\xi(t)}{2 + u^2}, \quad \Delta_\eta = \frac{u_0}{2 + u^2}. \tag{14}$$

Here, we remind that all of the angular quantities are given in units of the Einstein angle θ_E, which, for a source at distance $D_S \gg D_L$, results in being:

$$\theta_E \simeq 2 \left(\frac{M}{0.5 \, M_\odot} \right)^{1/2} \left(\frac{D_L}{\text{kpc}} \right)^{-1/2} \text{mas}, \tag{15}$$

which fixes the scale of the phenomenon.

It is straightforward to show (see [69]) that during a microlensing event, the centroid shift Δ traces (in the Δ_ξ, Δ_η plane) an ellipse centered in the point $(0, b)$. The ellipse semi-major axis a (along Δ_η) and semi-minor axis b (along Δ_ξ) are:

$$a = \frac{1}{2}\frac{1}{\sqrt{u_0^2 + 2}}, \quad b = \frac{1}{2}\frac{u_0}{u_0^2 + 2}. \tag{16}$$

Then, for $u_0 \to \infty$, the ellipse becomes a circle with radius $1/(2u_0)$, while it degenerates into a straight line of length $1/\sqrt{2}$ for u_0 approaching zero. Note also that Equation (16) implies:

$$u_0^2 = 2(b/a)^2 \left[1 - (b/a)^2\right]^{-1}, \tag{17}$$

so that by measuring a and b, one can determine the event impact parameter u_0.

As observed in [67], Δ falls more slowly than the magnification, implying that the centroid shift may be an interesting observable also for large source-lens distances, *i.e.*, far from the light curve peak. In fact, in astrometric microlensing, the threshold impact parameter u_{th} (*i.e.*, the value of the impact parameter that gives an astrometric centroid signal larger than a certain quantity δ_{th}) is given by $u_{th} = \sqrt{T_{obs}v_\perp/(\delta_{th}D_L)}$, where T_{obs} is the observing time and $v_\perp$ the relative velocity of the source with respect to the lens. For example, the Gaia satellite should reach an astrometric precision $\sigma_G \simeq 300$ μas (for objects with visual magnitude $\simeq 20$) in five years of observation [70]. Then, assuming a threshold centroid shift $\delta_{th} \simeq \sigma_G$, one has $u_{th} \simeq 60$ for a lens at a distance of 0.1 kpc and transverse velocity $v_\perp \simeq 100$ km s^{-1}. For comparison, the threshold impact parameter for a ground-based photometric observation is $\simeq 1$. Consequently, the cross-section for an astrometric microlensing measurement is much larger than the photometric one, since it scales as u_{th}^2. Hence, in the absence of finite-source and blending effects, by measuring a and b, one can directly estimate the impact parameter u_0.

A further advantage of the astrometric microlensing is that some events can be predicted in advance [71]. In fact, by studying in detail the characteristics of stars with large proper motions, Proft *et al.* [72] identified tens of candidates to measure astrometric microlensing by the Gaia satellite, an European Space Agency (ESA) mission that will perform photometry, spectroscopy and high precision astrometry (see [70]).

5.2. Polarization and Orbital Motion Effects in Microlensing Events

Gravitational microlensing observations may also offer a unique tool to study the atmospheres of far away stars by detecting a characteristic polarization signal [73]. In fact, it is well known that the light received from stars is linearly polarized by the photon scattering occurring in the stellar atmospheres. The mechanism is particularly effective for the hot stars (of A or B type) that have a free electron atmosphere giving rise to a polarization degree increasing from the center to the stellar limb [74]. By a minor extent, polarization may be also induced in main sequence F or G stars by the scattering of star light off atoms/molecules and in evolved, cool giant stars by photon scattering on dust grains contained in their extended envelopes.

Following the approach in [74], the polarization P in the direction making an angle $\chi = \arccos(\mu)$ with the normal to the star surface is $P(\mu) = [I_r(\mu) - I_l(\mu)]/[I_r(\mu) + I_l(\mu)]$, where $I_l(\mu)$ is the intensity in the plane containing the line of sight and the normal, and $I_r(\mu)$ is the intensity in the direction perpendicular to this plane. Here, $\mu = \sqrt{1 - (r/R)^2}$, where r is the distance of a star disk element from the center and R the star radius, and we are assuming that light propagates in the direction $\mathbf{r} \times \mathbf{l}$.

For isolated stars, a polarization signal has been measured only for the Sun for which, due to the distance, the projected disk is spatially resolved. Instead, when a star is significantly far away and can be considered as point-like, only the polarization $\langle P \rangle$ averaged over the stellar disk can be measured, and usually $\langle P \rangle = 0$, since the flux from each stellar disk element is the same. A net polarization of the light appears if a suitable asymmetry in the stellar disk is present (caused by, e.g., eclipses, tidal distortions, stellar spots, fast rotation, magnetic fields). In the microlensing context, the polarization arises since different regions of the source star disk are magnified differently during the event. Indeed, during an ongoing microlensing event, the gravitational lens scans the disk of the background star, giving rise not only to a time-dependent light magnification, but also to a time-dependent polarization.

This effect (see also [75]) is particularly relevant in the microlensing events where: (1) the magnification turns out to be significant; (2) the source star radius and the lens impact parameter are comparable; (3) the source star is a red giant, characterized by a rather low surface temperature ($T \leq 3000\,K$), around which the formation of dust grains is possible. This occurs beyond the distance R_h from the star center at which the gas temperature in the stellar wind becomes lower than the grain sublimation temperature ($\simeq 1400\,$K). The intensity of the expected polarization signal relies on the dust grain optical depth τ and can reach values of 0.1%–1%, which could be reasonably observed using, for example, the ESO VLTtelescope (see [76]). In Figure 6, we show some typical polarization curves, expected in *bypass* (continuous curves) and *transit* events (dashed curves), in which the lens trajectory approaches or passes through the source regions where the dust grains are present. In Figure 7, the distribution of the peak polarization values (given in percent) as a function of the intrinsic source star color index $(V - I)_{int}$ (*i.e.*, the de-reddened color of the unlensed source star) is shown for a sample of OGLE-type microlensing events generated by a synthetic stellar catalog simulating the bulge stellar population. As one can see, red giants with $(V - I)_{int} \leq 3$, which corresponds to the events inside the regions delimited by dashed lines, have $P_{max} \leq 1$ percent values. These are the typical events observed by the OGLE-III microlensing campaign. There are, however, a few events with $1 \leq P_{max} \leq 10$ percent, characterized by $(V - I)_{int} \geq 3$, corresponding to source stars in the AGB phase. These stars, which are rather rare in the galactic bulge, have not been sources of microlensing events observed in the OGLE-III campaign, but they are expected to exist in the galactic bulge. In this respect, the significant increase in the event rate by the forthcoming generation of microlensing surveys towards the galactic bulge, both ground-based, like KMTNet [77], and space-based, like EUCLID [78] and WFIRST [79], opens the possibility to develop an alert system able to trigger polarization measurements in ongoing microlensing events.

Another way to study the atmosphere of the source star is to analyze the amplification curve and look for dips and peaks, typically due to the presence of stellar spots on the photosphere of the star [80,81]. These features may be easily confused, however, with the signatures of a binary lensing system. When the source star has a relevant rotation motion during the lensing event, there is the possibility to really detect the stellar spots on the source's surface and to estimate the rotation period of the star [82]. A new generation of networks of telescopes dedicated to microlensing surveys, like KMTNet [83], will provide high-precision and high-cadence photometry that will enable us to observe spots on the source's surface. We remark that also multicolor observations of the event would help to disentangle the aforementioned degeneracy, as the ratio between the brightness of the spot and the surrounding photosphere strongly depends on the frequency of the observation. It has been shown that stellar spots can be detected also through polarimetric observations of microlensing caustic-crossing events [84].

Under certain circumstances, binary lens systems are characterized by the *close-wide degeneracy*: if the two objects are separated by a projected distance s or $1/s$, the resulting caustics have the same structure, and also, the observed light curves will appear the same [85,86]. This happens, for example, in systems with small mass ratio q, like planetary systems [87]. It is possible to resolve this degeneracy in the case of short-period binary lenses, the so-called *rapidly rotating lenses*, as the orbital motion induces repeating features in the amplification curve that can be exploited to estimate important physical parameters of the lensing systems, including the orbital period, the projected separation and the mass [88,89].

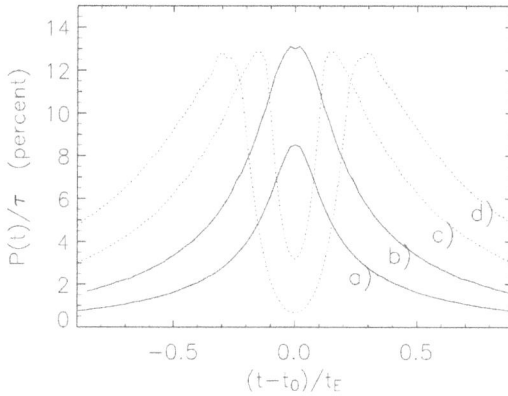

Figure 6. Assuming parameter values $u_0 = 0.09$, $t_E = 60$ day and $R_h/R_S = 5$, the $P(t)/\tau$ polarization curves are shown as a function of $(t - t_0)/t_E$, for increasing values of $R_h/u_0 = 0.35$, 0.75, 1.5, 2.5, corresponding to the curves labeled (a), (b) (c) and (d), respectively. Continuous curves, (a) and (b), are *bypass* events; dotted lines, (c) and (d), are *transit* events. Here, R_h is the minimum distance for the formation of dust grains, and R_S is the source star radius.

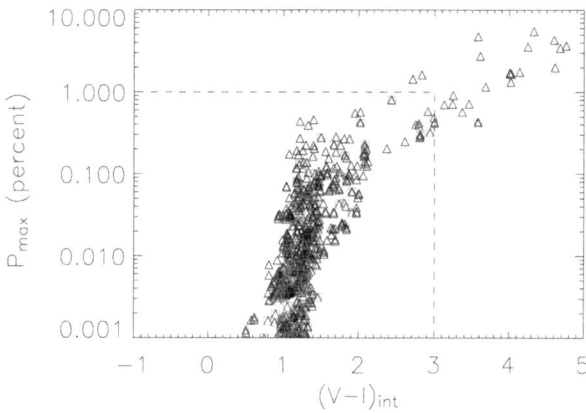

Figure 7. Distribution of the peak value of the polarization signal as a function of the intrinsic color index $(V - I)_{int}$ of the source star for simulated *transit* (triangles) and *bypass* (purple squares) events. The dashed lines indicate the region in which the events observed by the OGLECollaboration are expected to lie.

6. Retro-Lensing: Measuring the Black Hole Features

Gravitational lensing at the scales considered in the previous sections can be treated in the weak gravitational field approximation of the general theory of relativity, since in those cases, photons are deflected by very small angles. This is not the case when one considers black holes, for which it may happen that photons get very close to the event horizon of these compact objects.

Black holes are relatively simple objects. The *no-hair theorem* postulates that they are completely described by only three parameters: mass, angular momentum (generally indicated by the spin parameter *a*) and electric charge; any other information (for which *hair* is a metaphor) disappears behind the event horizon, and it is therefore inaccessible to external observers. Depending on the values of these parameters, black holes can be classified into Schwarzschild black holes (non-rotating and non-charged), Kerr black holes (rotating and non-charged), Reissner–Nordström black holes (non-rotating and charged) and Kerr–Newman black holes (rotating and charged).

Even though they appear so simple, black holes are mathematically complicated to describe (see, e.g., [90]). Nowadays, we know that black holes are placed at the center of the majority of galaxies, active or not, and in many binary systems emitting X-rays. Moreover, they are the engine of gamma-ray bursts (GRBs) and play an essential role in better understanding stellar evolution, galaxy formation and evolution, jets and, in the end, the nature of space and time. One goal astrophysicists have been pursuing for a long time is to probe the immediate vicinity of a black hole with an angular resolution as close as possible to the size of the event horizon. This kind of observations would give a new opportunity to study strong gravitational fields, and as we will see at the end of this section, we think we are very close to reaching this goal.

How do we measure the mass, angular momentum and electric charge of a black hole? One possibility, rich with interesting consequences, was suggested by Holz and Wheeler [91], who considered a phenomenon that was already known to be possible around black holes. They used the Sun as the source of light rays and a black hole far from the solar system. As shown in Figure 8, some photons would have the right impact parameter to turn around the black hole and come back to Earth. Other photons, with a slightly smaller impact parameter, can even rotate twice around the black hole, and so on. A series of concentric rings should then appear if the observer, the Sun and the black hole are perfectly aligned. The two authors also suggested to do a survey and look for concentric rings in the sky in order to discover black holes. Unfortunately, there are two problems with this idea. First, it is unlikely that the Sun, Earth and a black hole are perfectly aligned, and in any case, Earth moves around the Sun, so that the alignment can occur only for a short time interval. The second and most important problem is that the retro-image of the Sun is so dim, that even using the Hubble Space Telescope (HST), only a black hole with a mass larger than 10 $M_\odot$ within 0.01 pc from the Earth could be observed with the proposed technique. Moreover, we already know that such an object cannot be so close to the solar system without causing observable perturbations in the planet orbits.

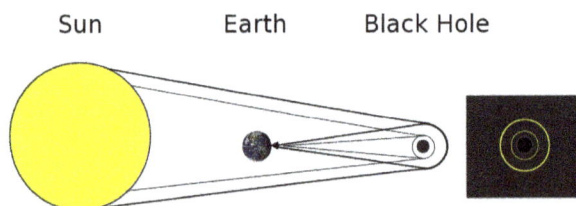

Figure 8. Retro-lensing of the Sun light by a black hole as seen from Earth. On the right-hand side, the series of rings around the black hole's event horizon an observer would observe in the case of perfect alignment. For clarity, only two rings are shown.

A better approach to test the idea proposed by Holz and Wheeler is to consider a well-known supermassive black hole and a bright star around it. Of course, the brighter the source star, the brighter will be the retro-image. Some of us [92] soon proposed to consider retro-lensing around the black hole at the galactic center, and in particular, the retro-lensing image of the closest star orbiting around it. Indeed, it is known that at the center of our galaxy, there is a supermassive black hole, with mass about $(4.2 \pm 0.2) \times 10^6$ $M_\odot$, identified by studying the orbits of several bright stars orbiting around it (see [93,94] and the references therein). A method to determine the mass and the angular momentum of this black hole could then be to measure the periastron or apoastron shifts of some of the stars orbiting around it. Another method to estimate the black hole spin a is based on the analysis of the quasi-periodical oscillations towards Sgr A*. Recently, the analysis of the data in the X-ray and IR bands have allowed some astrophysicists to find that $a = 0.65 \pm 0.05$ [94,95]. However, there is a drawback in this approach: periastron and apoastron shift of orbits depend not only on the black hole parameters, but also on how stars are distributed around the black hole and on the mass density profile of the dark matter possibly present in the region surrounding the black hole. It is possible to understand the difficulty of the measure by noting that the difference of the periastron shift of the S2 star (the closest one to the black hole at the center of our galaxy) induced by a Schwarzschild black hole or a Kerr black hole with spin parameter $a = 1$ (and the same mass of the Schwarzschild one) is only of $\simeq 10$ μas (for the dependance of the periastron shift on the black hole spin orientation see [96]). Then, even if one had succeeded in measuring the periastron shift of the closest star to the central black hole, it would be unlikely to derive the amount of the black hole angular momentum. Our goal could be achieved anyway by measuring the periastron shift of many stars orbiting around the center of the Galaxy. The measure of the periastron shift could give, in turn, also an estimate of the parameters of the dark matter concentration expected to lie towards the Sgr A^* region [97], as well as to test different modifications of the general theory of relativity [98–100] (see also [101] for the constraints on R^n theories by Solar System data)). However, this is anything but easy [102]. An important step forward in this direction has been provided recently by near-infrared astrometric observations of many stars around Sgr A^* with a precision of about 170 μas in position and $\simeq 0.07$ mas$\cdot$ yr^{-1} in velocity [103]. A further improvement, hopefully in the near future, would make possible the direct detection of relativistic effects in the orbits of stars orbiting the central black hole.

Retro-lensing images of bright stars retro-lensed by the black hole at the galactic center might give an alternative method to estimate the Sgr A* black hole parameters. Even though in general it is difficult to calculate the retro-lensing images, since this requires integrating with high precision the trajectories followed by the light, it is possible to numerically do these calculations not only for a Schwarzschild black hole, but also for Kerr and Reissner–Nordström black holes. As discussed in several papers (see, e.g., [104] and the references therein), one finds that the shape of the retro-lensing image depends on the black hole spin (see Figure 9), and then, in principle, a single precise enough observation of the retro-lensing image of a star could allow one to unambiguously estimate the parameters of the black hole in Sgr A*. It is possible to show that also the electric charge of a Reissner–Nordström black hole can be obtained [105]. In fact, although the formation of a Reissner–Nordström black hole may be problematic, charged black holes are objects of intensive investigations, and the black hole charge can be estimated by using the size of the retro-lensing images that can be revealed by future astrometrical missions. The shape of the retro-lensing (or shadow) image depends in fact also on the electric charge of the black hole, and it becomes smaller as the electric charge increases. The mirage size difference between the extreme charged black hole and the Schwarzschild black hole case is about 30%, and in the case of the black hole in Sgr A*, the shadow typical angular sizes are about 52 μas for the Schwarzschild case and about 40 μas for a maximally charged Reissner–Nordström black hole. Therefore, a charged black hole could be, in principle, distinguished by a Schwarzschild black hole with RADIOASTRON, at least if its charge is close to the maximal value. We also mention that the black hole spin gives rise also to chromatic effects (while for non-rotating lenses, the gravitational lensing effect is always achromatic), making one side of the image bluer than the other side [104].

Can we really hope to observe these retro-lensing images towards Sgr A*? Despite what one could think, we are not so far from this goal. The successor of the Hubble Space Telescope, the James Webb Space Telescope (JWST), scheduled for launch in October 2018, has the sensitivity to observe the retro-lensing image of the S2 star produced by the black hole at the galactic center with an exposure time of about thirty hours. In Figure 10, we show the magnification (upper panel) and the magnitude (bottom panel) light curves (in K band) of the retro-lensing image of the S2 star produced by the black hole at the galactic center (see also [106]). Unfortunately, JWST has not the angular resolution necessary to provide information about the shape of the retro-lensing image. The right angular resolution could be gained with the next generation of radio interferometers. In fact, the diameter of the retro-lensing image around the central black hole should be of about 30 μas, and already in 2008, Doeleman and his collaborators [107] managed to achieve an angular resolution of about 37 μas, very close to the required one, by using interferometrically different radio telescopes with a baseline of about 4500 km. Progress in this field is so fast, that it is not hard to think we can eventually reach this aim in the near future by, e.g., the EHT (Event Horizon Telescope) project, or by the planned Russian space observatory, Millimetron (the spectrum-M project), or by combined observations with different interferometers, such as the Very Large Array (VLA) and ALMA (Atacama Large Millimeter Array).

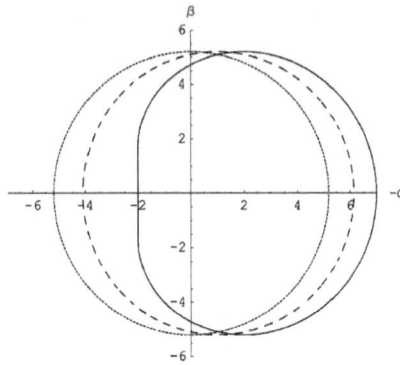

Figure 9. Retro-lensing images of a source by a Schwarzschild black hole (dotted circle), a Kerr black hole with spin parameter $a = 0.5$ (dashed line) and a maximally spinning black hole with $a = 1$ (continuous line). The line of sight of the observer is perpendicular to the spin axis of the black hole.

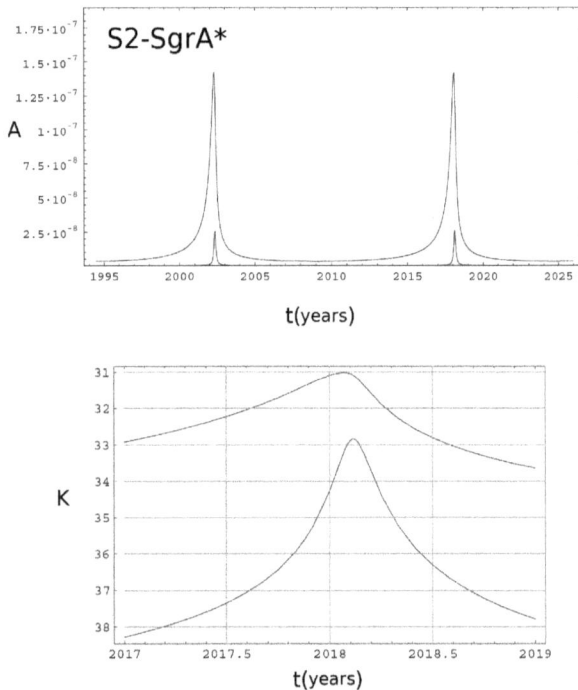

Figure 10. Upper panel: amplification as a function of time for the primary (upper curve) and secondary (lower curve) retro-lensing images of the S2 star by the black hole in the Galaxy center. Lower panel: light curve in K-band magnitude of the two retro-lensing images (adapted from [106]). The standard interstellar absorption coefficient towards the Galaxy center has been assumed.

7. Conclusions

In the paper, we have discussed the various scales in which gravitational lensing manifests itself and that may lead us to obtain valuable information about a great variety of astronomical issues ranging from the star distribution in the Milky Way, the study of stellar atmospheres, the discovery of exoplanets in the Milky Way and also in nearby galaxies, the study of far away galaxies, galaxy clusters and black holes. Gravitational lensing, in particular in the strong and weak lensing regime, may also allow scientists to answer, in the near future, fundamental questions in cosmology related to the nature of dark matter, why the Universe is accelerating and what is the nature of the source responsible for the acceleration, which physicists refer to as dark energy.

Acknowledgments: The authors acknowledge the support of the TAsP (Theoretical Astroparticle Physics) Project funded by INFN. Mosè Giordano acknowledges the support of the Max Planck Institute for Astronomy, Heidelberg, where part of this work has been done. We thank Asghar Qadir and Alexander Zakharov for the valuable discussions on the paper subject during many years.

Author Contributions: All authors have equally contributed to this paper and have read and approved the final version.

Conflicts of Interest: The authors declare no conflict of interest.

References

1. Einstein, A. On the Influence of Gravitation on the Propagation of Light. *Ann. Phys.* **1911**, *340*, 898–908.
2. Dyson, F.W.; Eddington, A.; Davidson, C. A determination of the deflection of light by the sun's gravitational field from observations made at the total eclipse of May 29, 1919. *Phil. Trans. Roy. Soc. A* **1920**, *220*, 291–333.
3. Chwolson, O. Über eine mögliche Form fiktiver Doppelsterne. *Astron. Nachr.* **1924**, *221*, 329–330.
4. Einstein, A. Lens-Like Action of a Star by the Deviation of Light in the Gravitational Field. *Science* **1936**, *84*, 506–507.
5. Zwicky, F. Nebulae as Gravitational Lenses. *Phys. Rev.* **1937**, *51*, 290.
6. Zwicky, F. On the Probability of Detecting Nebulae which Act as Gravitational lenses. *Phys. Rev.* **1937**, *51*, 679.
7. Walsh, D.; Carswell, R.F.; Weymann, R.J. 0957 + 561 A,B: Twin quasistellar objects or gravitational lens? *Nature* **1979**, *279*, 381–384.
8. Paczyński, B. Gravitational microlensing by the galactic halo. *Astrophys. J.* **1986**, *304*, 1–5.
9. Johannsen, T. Sgr A* and general relativity. **2015**, arXiv:1512.03818.
10. Abbott, B.P.; Abbott, R.; Abbott, T.D.; Abernathy, M.R.; Acernese, F.; Ackley, K.; Adams, C.; Adams, T.; Addesso, P.; Adhikari, R.X.; *et al.* Observation of Gravitational Waves from a Binary Black Hole Merger. *Phys. Rev. Lett.* **2016**, *116*, 061102.
11. Schneider, P.; Ehlers, G.; Falco, E.E. Gravitational Lenses; Springer Verlag: Berlin, Germany, 1992.
12. Schmidt, M. 3C 273: A star-like object with large red-shift. *Nature* **1963**, *197*, 1040.
13. Pelt, J.; Kayser, R.; Refsdal, S.; Schramm, T. The light curve and the time delay of QSO 0957+561. *Astron. Astrophys.* **1996**, *306*, 97–106.
14. Chen, G.H.; Kochanek, C.S.; Hewitt, J.N. The Mass Distribution of the Lens Galaxy in MG 1131+0456. *Astrophys. J.* **1995**, *447*, 62–81.
15. Hennawi, J.F.; Gladders, M.D.; Oguri, M.; Dalal, N.; Koester, B.; Natarajan, P.; Strauss, M.A.; Inada, N.; Kayo, I.; Lin, H. A New Survey for Giant Arcs. *Astron. J.* **2008**, *135*, 664–681.
16. Browne, I.W.A.; Wilkinson, P.N.; Jackson, N.J.F.; Myers, S.T.; Fassnacht, C.D.; Koopmans, L.V.E.; Marlow, D.R.; Norbury, M.; Rusin, D.; Sykes, C.M.; *et al.* The Cosmic Lens All-Sky Survey - II. Gravitational lens candidate selection and follow-up. *Mon. Not. R. Astron. Soc.* **2003**, *341*, 13–32.
17. Bolton, A.S.; Burles, S.; Koopmans, L.V.E.; Treu, T.; Moustakas, L.A. The Sloan Lens ACS Survey. I. A Large Spectroscopically Selected Sample of Massive Early-Type Lens Galaxies. *Astrophys. J.* **2006**, *638*, 703–724.
18. Alam, S.; Albareti, F.D.; Allende, P.; Anders, F.; Anderson, S.F.; Anderton, T.; Andrews, B.H.; Armengaud, E.; Aubourg, E.; Bailey, S.; *et al.* The Eleventh and Twelfth Data Releases of the Sloan Digital Sky Survey: Final Data from SDSS-III. *Astrophys. J. Suppl. Ser.* **2015**, *219*, 12.

19. Oguri, M.; Inada, N.; Pindor, B.; Strauss, M.A.; Richards, G.T.; Hennawi, J.F.; Turner, E.L.; Lupton, R.H.; Schneider, D.P.; Fukugita, M. The Sloan Digital Sky Survey Quasar Lens Search. I. Candidate Selection Algorithm. *Astrophys. J.* **2006**, *132*, 999–1013.

20. Blandford, R.D.; Narajan, R. Cosmological applications of gravitational lensing. *Annu. Rev. Astron. Astrophys.* **1992**, *30*, 311–358.

21. Treu, T. Strong lensing by galaxies. *Annu. Rev. Astron. Astrophys.* **2010**, *48*, 87–125.

22. Kneib, J.-P.; Ellis, R.S.; Smail, I.; Couch, W. J.; Sharples, R.M. Hubble Space Telescope Observations of the Lensing Cluster Abell 2218. *Astrophys. J.* **1996**, *471*, 643–656.

23. Mao, S.; Schneider, P. Evidence for substructure in lens galaxies? *Mon. Not. R. Astron. Soc.* **1998**, *95*, 587–594.

24. Metcalf, R.B.; Madau, P. Compound Gravitational Lensing as a Probe of Dark Matter Substructure within Galaxy Halos. *Astrophys. J.* **2001**, *563*, 9–20.

25. Xu, D.D.; Mao, S.; Wang, J.; Springel, V.; Gao, L.; White, S.D.M.; Frenk, C.S.; Jenkins, A.; Li, G.; Navarro, J.F. Effects of dark matter substructures on gravitational lensing: results from the Aquarius simulations. *Mon. Not. R. Astron. Soc.* **2009**, *408*, 1235–1253.

26. Metcalf, R.B.; Amara, A. Small-scale structures of dark matter and flux anomalies in quasar gravitational lenses. *Mon. Not. R. Astron. Soc.* **2012**, *419*, 3414–3425.

27. Xu, D.D.; Sluse, D.; Gao, L.; Wang, J.; Frenk, C.; Mao, S.; Schneider, P.; Springel, V. How well can cold dark matter substructures account for the observed radio flux-ratio anomalies. *Mon. Not. R. Astron. Soc.* **2015**, *447*, 3189–3206.

28. Vegetti, S.; Lagattuta, D.J.; McKean, J.P.; Auger, M.W.; Fassnacht, C.D.; Koopmans, L.V.E. Gravitational detection of a low-mass dark satellite galaxy at cosmological distance. *Nature* **2012**, *481*, 341–343.

29. Vegetti, S.; Koopmans, L.V.E.; Auger, M.W.; Treu, T.; Bolton, A.S. Inference of the cold dark matter substructure mass function at z = 0.2 using strong gravitational lenses. *Mon. Not. R. Astron. Soc.* **2014**, *442*, 2017–2035.

30. Kochanek, C.S. Quantitative Interpretation of Quasar Microlensing Light Curves. *Astrophys. J.* **2004**, *605*, 58–77.

31. Mosquera, A.M.; Kochanek, C.S. The Microlensing Properties of a Sample of 87 Lensed Quasars. *Astrophys. J.* **2011**, *738*, 96.

32. Pooley, D.; Rappaport, S.; Blackburne, J.A.; Schechter, P.L.; Wambsganss, J. X-Ray and Optical Flux Ratio Anomalies in Quadruply Lensed Quasars. II. Mapping the Dark Matter Content in Elliptical Galaxies. *Astrophys. J.* **2012**, *744*, 111.

33. Soucail, G.; Fort, B.; Mellier, Y.; Picat, J.P. A blue ring-like structure, in the center of the A 370 cluster of galaxies. *Astron. Astrophys.* **1987**, *172*, L14–L16.

34. Stark, D.P.; Swinbank, A.M.; Ellis, R.S. Dye, S.; Smail, I.R; Richard, J. The formation and assembly of a typical star-forming galaxy at redshift z~3. *Nature* **2008**, *455*, 775–777.

35. Kelly, P.L.; Rodney, S.A.; Treu, T.; Foley, R.J.; Brammer, G.; Schmidt, K.B.; Zitrin, A.; Sonnenfeld, A.; Strolger, L.-G.; Graur, O.; *et al.* Multiple images of a highly magnified supernova formed by an early-type cluster galaxy lens. *Science* **2015**, *347*, 1123–1126.

36. Treu, T.; Brammer, G.; Diego, J.M.; Grillo, C.; Kelly, P.L.; Oguri, M.; Rodney, A.; Rosati, P.; Sharon, K.; Zitrin, A. "Refsdal" Meets Popper: Comparing Predictions of the Re-appearance of the Multiply Imaged Supernova Behind MACSJ1149.5+2223. *Astrophys. J.* **2016**, *817*, 60.

37. Refsdal, S. On the possibility of determining Hubble's parameter and the masses of galaxies from the gravitational lens effect. *Mon. Not. R. Astron. Soc.* **1964**, *128*, 307–310.

38. Jackson, N. The Hubble Constant. *Living Rev. Relativ.* **2007**, *10*, 4.

39. Suyu, S.H.; Marshall, P.J.; Auger, M.W.; Hilbert, S.; Blandford, R.D.; Koopmans, L.V.E.; Fassnacht, C.D.; Treu, T.; *et al.* Dissecting the Gravitational lens B1608+656. II. Precision Measurements of the Hubble Constant, Spatial Curvature, and the Dark Energy Equation of State. *Astrophys. J.* **2010**, *711*, 201–221.

40. Bonvin, V.; Tewes, M.; Courbin, F.; Kuntzer, T.; Sluse, D.; Meylan, G. COSMOGRAIL: the COSmological MOnitoring of GRAvItational Lenses. XV. Assessing the achievability and precision of time-delay measurements. *Astron. Astrophys.* **2016**, *585*, A88.

41. Bartelmann, M.; Schneider, P. Weak gravitational lensing. *Phys. Rep.* **2001**, *340*, 291–472.

42. Tyson, J.A.; Wenk, R.A.; Valdes, F. Detection of systematic gravitational lens galaxy image alignments—Mapping dark matter in galaxy clusters. *Astrophys. J.* **1990**, *349*, L1–L4.

43. Bacon, D.J.; Refregier, A.R.; Ellis, R.S. Detection of weak gravitational lensing by large-scale structure. *Mon. Not. R. Astron. Soc.* **2000**, *318*, 625–640.
44. Kaiser, N. A New Shear Estimator for Weak-Lensing Observations. *Astrophys. J.* **2000**, *537*, 555–577.
45. Seitz, S.; Schneider, P. Cluster lens reconstruction using only observed local data: An improved finite-field inversion technique. *Astron. Astrophys.* **1996**, *305*, 383–401.
46. Lombardi, M.; Bertin, G. A fast direct method of mass reconstruction for gravitational lenses. *Astron. Astrophys.* **1999**, *348*, 38–42.
47. Schneider, P.; King, L.; Erben, T. Cluster mass profiles from weak lensing: Constraints from shear and magnification information. *Astron. Astrophys.* **2000**, *353*, 41–56.
48. Han, J.; Eke, V.R.; Frenk, C.S.; Mandelbaum, R.; Norberg, P.; Schneider, M.D.; Peacock, J.A.; Jing, Y.; Baldry, I.; Bland-Hawthorn, J.; *et al.* Galaxy And Mass Assembly (GAMA): the halo mass of galaxy groups from maximum-likelihood weak lensing. *Mon. Not. R. Astron. Soc.* **2015**, *446*, 1356–1379.
49. Marshall, P.J.; Hobson, M.P.; Gull, S.F.; Bridle, S.L. Maximum-entropy weak lens reconstruction: improved methods and application to data. *Mon. Not. R. Astron. Soc.* **2002**, *335*, 1037–1048.
50. Clowe, D.; De Lucia, G.; King, L. Effects of asphericity and substructure on the determination of cluster mass with weak gravitational lensing. *Mon. Not. R. Astron. Soc.* **2004**, *350*, 1038–1048.
51. Corless, V.L.; King, L.J. A statistical study of weak lensing by triaxial dark matter haloes: consequences for parameter estimation. *Mon. Not. R. Astron. Soc.* **2007**, *380*, 149–161.
52. Hoekstra, H. A comparison of weak-lensing masses and X-ray properties of galaxy clusters. *Mon. Not. R. Astron. Soc.* **2007**, *379*, 317–330.
53. Zhang, Y.-Y.; Finoguenov, A.; Böhringer, H.; Kneib, J.-P.; Smith, G.P.; Kneissl, R.; Okabe, N.; Dahle, H. LoCuSS: Comparison of observed X-ray and lensing galaxy cluster scaling relations with simulations. *Astron. Astrophys.* **2008**, *482*, 451–472.
54. Contaldi, C.R.; Hoekstra, H.; Lewis, A. Joint Cosmic Microwave Background and Weak Lensing Analysis: Constraints on Cosmological Parameters. *Phys. Rev. Lett.* **2003**, *22*, 221303.
55. Hollenstein, L.; Sapone, D.; Crittenden, R.; Schäfer, B.M. Constraints on early dark energy from CMB lensing and weak lensing tomography. *J. Cosmol. Astropart. Physucs* **2009**, *04*, id. 012.
56. Majerotto, E.; Sapone, D.; Schäfer, B,M. Combined constraints on deviations of dark energy from an ideal fluid from Euclid and Planck. *Mon. Not. R. Astron. Soc.* **2016**, *456*, 109–118.
57. Clowe, D.; Bradaĉ, M.; Gonzalez, A.H.; Markevitch, M.; Randall, S.W.; Jones, C.; Zaritsky, D. A Direct Empirical Proof of the Existence of Dark Matte. *Astrophys. J.* **2006**, *648*, L109–L113.
58. Amendola, L.; Appleby, S.; Bacon, D. Baker, T. Baldi, M. Bartolo, N. Blanchard, A. Bonvin, C. Borgani, S. Branchini, E.; *et al.* Cosmology and Fundamental Physics with the Euclid Satellite. *Living Rev. Relat.* **2013**, *16*, 6.
59. Chang, C.; Jarvis, M.; Jain, B. Kahn, S.M.; Kirkby, D.; Connolly, A.; Krughoff, S.; Peng, E.-H.; Peterson, J.R. The effective number density of galaxies for weak lensing measurements in the LSST project. *Mon. Not. R. Astron. Soc.* **2013**, *434*, 2121–2135.
60. Heymans, C.; Heavens, A. Weak gravitational lensing: reducing the contamination by intrinsic alignments. *Mon. Not. R. Astron. Soc.* **2003**, *339*, 711–720.
61. Valageas, P. Source-lens clustering and intrinsic-alignment bias of weak-lensing estimators. *Astron. Astrophys.* **2014**, *561*, A53.
62. Mao, S. Astrophysical applications of gravitational microlensing. *Res. Astron. Astrophys.* **2012**, *12*, 947–972.
63. Perryman, M. *The Exoplanet Handbook*; University Press: Cambridge, UK, 2014.
64. Sumi, T.; Kamiya, K.; Udalski, A.; Bennett, D.P.; Bond, I.A.; Abe, F.; Botzler, C.S.; Fukui, A.; Furusawa, K.; Hearnshaw, J.B.; Itow, Y.; *et al.* Unbound or distant planetary mass population detected by gravitational microlensing. *Nature* **2011**, *473*, 349–352.
65. Ingrosso, G.; Calchi Novati, S.; De Paolis, F.; Jetzer, P.; Nucita, A.A.; Strafella, F. Pixel lensing as a way to detect extrasolar planets in M31. *Mon. Not. R. Astron. Soc.* **2009**, *399*, 219–228.
66. An, J.H.; Evans, N.W.; Kerins, E.; Baillon, P.; Calchi-Novati, S.; Carr, B.J.; Creze, M.; Giraud-Heraud, Y.; Gould, A.; Hewett, P.; *et al.* The Anomaly in the Candidate Microlensing Event PA-99-N2. *Astrophys. J.* **2004**, *601*, 845–857.
67. Dominik, M.; Sahu, K.C. Astrometric Microlensing of Stars. *Astrophys. J.* **2000**, *534*, 213–226.

68. Lee, C.-H.; Seitz, S.; Riffeser, A.; Bender, R. Finite-source and finite-lens effects in astrometric microlensing. *Mon. Not. R. Astron. Soc.* **2010**, *407*, 1597–1608.
69. Walker, M.A. Microlensed Image Motions. *Astrophys. J.* **1995**, *453*, 37–39.
70. Eyer, L.; Holl, B.; Pourbaix, D.; Mowlavi, N.; Siopis, C.; Barblan, F.; Evans, D.W.; North, P. The Gaia Mission. *Cent. Eur. Astrophys. Bull. (CEAB)* **2013**, *37*, 115–126.
71. Paczyński, B. The Masses of Nearby Dwarfs and Brown Dwarfs with the HST. *Acta Astron.* **1996**, *46*, 291–296.
72. Proft, S.; Demleitner, M.; Wambsganss, J. Prediction of astrometric microlensing events during the Gaia mission. *Astron. Astrophys.* **2011**, *536*, A50.
73. Ingrosso, G.; Calchi Novati, S.; De Paolis, F.; Jetzer, P.; Nucita, A.A.; Strafella, F.; Zakharov, A.F. Polarization in microlensing events towards the Galactic bulge. *Mon. Not. R. Astron. Soc.* **2012**, *426*, 1496–1506.
74. Chandrasekhar, S. *Radiative Transfer*; Clarendon Press: Oxford, UK, 1950.
75. Simmons, J.F.L.; Bjorkman, J.E.; Ignace, R.; Coleman, I.J. Polarization from microlensing of spherical circumstellar envelopes by a point lens. *Mon. Not. R. Astron. Soc.* **2002**, *336*, 501–510.
76. Ingrosso, G.; Calchi Novati, S.; De Paolis, F.; Jetzer, P.; Nucita, A.A.; Strafella, F. Measuring polarization in microlensing events. *Mon. Not. R. Astron. Soc.* **2015**, *446*, 1090–1097.
77. Henderson, C.B.; Gaudi, B.S.; Han, C.; Skowron, J.; Penny, M.T.; Nataf, D.; Gould, A.P. Optimal Survey Strategies and Predicted Planet Yields for the Korean Microlensing Telescope Network. *Astrophys. J.* **2014**, *794*, 52.
78. Penny, M.T.; Kerins, E.; Rattenbury, N.; Beaulieu, J.-P.; Robin, A.C.; Mao, S.; Batista, V.; Calchi Novati, S.; Cassan, A.; Fouque, P.; *et al.* Zapatero Osorio ExELS: An exoplanet legacy science proposal for the ESA Euclid mission - I. Cold exoplanets. *Mon. Not. R. Astron. Soc.* **2013**, *434*, 2–22.
79. Yee, J.C.; Albrow, M.; Barry, R.K.; Bennett, D.; Bryden, G.; Chung, S.-J.; Gaudi, B.S.; Gehrels, N.; Gould, A.; Penny, M.T.; *et al.* Takahiro SumiNASA ExoPAG Study Analysis Group 11: Preparing for the WFIRST Microlensing Survey. **2014**, arXiv:1409.2759.
80. Heyrovský, D.; Sasselov, D. Detecting Stellar Spots by Gravitational Microlensing. *Astrophys. J.* **2000**, *529*, 69–76.
81. Hendry, M.A.; Bryce, H.M.; Valls-Gabaud, D. The microlensing signatures of photospheric starspots. *Mon. Not. R. Astron. Soc.* **2002**, *335*, 539–549.
82. Giordano, M.; Nucita, A.A.; De Paolis, F.; Ingrosso, G. Starspot induced effects in microlensing events with rotating source star. *Mon. Not. R. Astron. Soc.* **2015**, *453*, 2017–2021.
83. Kim, S.-L.; Park, B.-G.; Lee, C.-U.; Yuk, I.-S.; Han, C.; O'Brien, T.; Gould, A.; Lee, J.W.; Kimet, D.-J. Technical specifications of the KMTNet observation system. *Proc. SPIE* **2010**, doi:10.1117/12.856833.
84. Sajadian, S. Detecting stellar spots through polarimetric observations of microlensing events in caustic-crossing. *Mon. Not. R. Astron. Soc.* **2015**, *452*, 2587–2596.
85. Dominik, M. The binary gravitational lens and its extreme cases. *Astron. Astrophys.* **1999**, *349*, 108–125.
86. An, J.H. Gravitational lens under perturbations: symmetry of perturbing potentials with invariant caustics. *Mon. Not. R. Astron. Soc.* **2005**, *356*, 1409–1428.
87. Griest, K.; Safizadeh, N. The Use of High-Magnification Microlensing Events in Discovering Extrasolar Planets. *Astrophys. J.* **1998**, *500*, 37–50.
88. Penny, M.T.; Kerins, E.; Mao, S. Rapidly rotating lenses: Repeating features in the light curves of short-period binary microlenses. *Mon. Not. R. Astron. Soc.* **2011**, *417*, 2216–2229.
89. Nucita, A.A.; Giordano, M.; De Paolis, F.; Ingrosso, G. Signatures of rotating binaries in microlensing experiments. *Mon. Not. R. Astron. Soc.* **2014**, *438*, 2466–2473.
90. Chandrasekhar, S. *The Mathematical Theory of Black Holes*; Oxford University Press: Oxford, UK, 1983.
91. Holz, D.E.; Wheeler, J.A. Retro-MACHOs: pi in the Sky. *Astrophys. J.* **2002**, *57*, 330–334.
92. De Paolis, F.; Geralico, A.; Ingrosso, G.; Nucita, A.A. The black hole at the galactic center as a possible retro-lens for the S2 orbiting star. *Astron. Astrophys.* **2003**, *409*, 809–812.
93. Gillessen, S.; Eisenhauer, F.; Trippe, S.; Alexander, T.; Genzel, R.; Martins, F.; Ott, T. Monitoring Stellar Orbits Around the Massive Black Hole in the Galactic Center. *Astrophys. J.* **2009**, *692*, 1075–1109.
94. Dokuchaev, V.I. Spin and mass of the nearest supermassive black hole. *Gen. Rel. Gravit.* **2014**, *46*, 1832.
95. Dokuchaev, V.I.; Eroshenko, Y.N. Physical Laboratory at the center of the Galaxy. *Phys. Uspekhi* **2015**, *58*, 772–784.

96. Iorio, L. Perturbed stellar motions around the rotating black hole in Sgr A* for a generic orientation of its spin axis. *Phys. Rev. D* **2012**, *84*, 124001.
97. Zakharov, A.F.; Nucita, A.A.; De Paolis, F.; Ingrosso, G. Apoastron shift constraints on dark matter distribution at the Galactic Center. *Phys. Rev. D* **2007**, *76*, 062001.
98. Zakharov, A.F.; de Paolis, F.; Ingrosso, G.; Nucita, A.A. Shadows as a tool to evaluate the black hole parameters and a dimension of spacetime. *New Astron. Rev.* **2012**, *56*, 64–73.
99. Falcke, H.; Markoff, S.B. Toward the event horizon—The supermassive black hole in the galactic center. *Classicala Quantum Gravit.* **2013**, *30*, 244003.
100. Zakharov, A.F.; Borka, D.; Borka Jovanović, V.; Jovanović, P. Constraints on R^n gravity from precession of orbits of S2-like stars: A case of a bulk distribution of mass. *Adv. Space Res.* **2014**, *54*, 1108–1112.
101. Zakharov, A.F.; Nucita, A.A.; De Paolis, F.; Ingrosso, G. Solar system constraints on R^n gravity. *Phys. Rev. D* **2006**, *74*, 107101.
102. Nucita, A.A.; De Paolis, F.; Ingrosso, G.; Qadir, A.; Zakharov, A.F. Sgr A*: A Laboratory to Measure the Central Black Hole and Stellar Cluster Parameters. *Publ. Astron. Soc. Pac.* **2007**, *119*, 349–359.
103. Plewa, P.M. Gillessen, S.; Eisenhauer, F.; Ott, T.; Pfuhl, O.; George, E.; Dexter, J.; Habibi, M.; Genzel, R.; Reid, M.J.; *et al.* Pinpointing the near-infrared location of Sgr A* by correcting optical distortion in the NACO imager. *Mon. Not. R. Astron. Soc.* **2015**, *453*, 3234–3244.
104. De Paolis, F.; Ingrosso, G.; Nucita, A.A.; Qadir, A.; Zakharov, A.F. Estimating the parameters of the Sgr A* black hole. *Gen. Rel. Gravit.* **2011**, *43*, 977–988.
105. Zakharov, A.F.; De Paolis, F.; Ingrosso, G.; Nucita, A. Direct measurements of black hole charge with future astrometrical missions. *Astron. Astrophys.* **2005**, *442*, 795–799.
106. Bozza, V.; Mancini, L. Gravitational Lensing by Black Holes: A Comprehensive Treatment and the Case of the Star S2. *Astrophys. J.* **2005**, *611*, 1045–1053.
107. Doeleman, S.S.; Weintroub, J.; Rogers, A.E.E.; Plambeck, R.; Freund, R.; Tilanus, R.P.J.; Friberg, P.; Ziurys, L.M.; Moran, J.M.; Corey, B.; *et al.* Event-horizon-scale structure in the supermassive black hole candidate at the Galactic Centre. *Nature* **2008**, *455*, 78–80.

![universe logo] *universe*

MDPI

Article

On the Effect of the Cosmological Expansion on the Gravitational Lensing by a Point Mass

Oliver F. Piattella

Physics Department, Universidade Federal do Espírito Santo, Vitória 29075-910, Brazil;
oliver.piattella@pq.cnpq.br

Academic Editors: Lorenzo Iorio and Elias C. Vagenas
Received: 2 September 2016; Accepted: 9 October 2016; Published: 18 October 2016

Abstract: We analyse the effect of the cosmological expansion on the deflection of light caused by a point mass, adopting the McVittie metric as the geometrical description of a point-like lens embedded in an expanding universe. In the case of a generic, non-constant Hubble parameter, H, we derive and approximately solve the null geodesic equations, finding an expression for the bending angle δ, which we expand in powers of the mass-to-closest approach distance ratio and of the impact parameter-to-lens distance ratio. It turns out that the leading order of the aforementioned expansion is the same as the one calculated for the Schwarzschild metric and that cosmological corrections contribute to δ only at sub-dominant orders. We explicitly calculate these cosmological corrections for the case of the H constant and find that they provide a correction of order 10^{-11} on the lens mass estimate.

Keywords: McVittie metric; gravitational lensing; cosmology

PACS: 95.30.Sf; 04.70.Bw; 95.36.+x

1. Introduction

The effect of the cosmological constant Λ (and thus, by extension, of cosmology) on the bending of light is an issue which has raised interest since a pioneering work by Rindler and Ishak in 2007 [1]. The common intuition is that Λ cannot have any local effect on the deflection of light because it is homogeneously distributed in the universe, thus not forming lumps which may act as lenses. Moreover, General Relativity (GR) is assumed as the fundamental theory of gravity, and the Kottler metric is considered [2] (We use throughout this paper $G = c = 1$ units):

$$ds^2 = f(r)dt^2 - f^{-1}(r)dr^2 - r^2 d\Omega^2 , \tag{1}$$

where $d\Omega^2 = d\theta^2 + \sin^2\theta d\phi^2$ and:

$$f(r) \equiv 1 - \frac{2M}{r} - \frac{\Lambda r^2}{3} , \tag{2}$$

as the description of a point mass M embedded in a de Sitter space. It turns out that Λ does not appear in the null geodesic trajectory equation, cf. e.g., Equation (17) of Ref. [3]. Indeed:

$$\frac{d^2u}{d\phi^2} + u = 3Mu^2 , \tag{3}$$

where $u = 1/r$ is the inverse of the radial distance from the lens. Therefore, one may conclude that Λ does not affect the bending of light, which is then entirely due to the presence of the point mass M.

On the other hand, in Ref. [1], the authors point out that the bending angle cannot be calculated as the angle between the asymptotic directions of the light ray, since these do not exist. Indeed, from Equation (2), one sees that $r \leq \sqrt{3/\Lambda}$, i.e., a cosmological horizon exists. In other words, the effect of Λ enters into the boundary conditions that we choose when solving Equation (3).

Thus, the authors of Ref. [1] find that Λ enters the definition of the deflection angle in the following way (cf. their Equation (17)):

$$\psi_0 \approx \frac{2M}{R} \left(1 - \frac{2M^2}{R^2} - \frac{\Lambda R^4}{24M^2} \right) ,$$ (4)

where R is the closest approach distance. Twice ψ_0 is the deflection angle. Therefore, one identifies the well-known Schwarzschild contribution $4M/R$, weighed by Λ, which tends to thwart the deflection.

After Ref. [1], many authors confirmed with their calculations that Λ does enter the formula for the deflection angle, although sometimes in a way different from the one in Equation (4). See e.g., Refs. [4–12].

On the other hand, there are a few works that do not agree with the above-mentioned results (see e.g., [13–16]). The main criticism is that the **Hubble flow** is not properly taken into account, i.e., the relative motion among source, lens and observer is neglected. In particular, the authors of Ref. [15] argue that the Λ contribution in Equation (4) is cancelled by the aberration effect due to the cosmological relative motion. Another interesting remark made in Ref. [15] is that the contribution of Λ to the deflection angle does not vanish for $M \to 0$ in Equation (4). In this respect, consider also e.g., Equation (25) of Ref. [10]:

$$\delta = \frac{4M}{b} - Mb \left(\frac{1}{r_S^2} + \frac{1}{r_O^2} \right) + \frac{2Mb\Lambda}{3} - \frac{b\Lambda}{6}(r_S + r_O) - \frac{b^3\Lambda}{12} \left(\frac{1}{r_S} + \frac{1}{r_O} \right) + \frac{Mb^3\Lambda}{6} \left(\frac{1}{r_S^2} + \frac{1}{r_O^2} \right) + \cdots .$$ (5)

Here, b is the impact parameter and r_S and r_O are the radial distances from the lens to the source and to the observer, respectively. Taking the limit $M \to 0$ in the above equation does not imply $\delta \to 0$.

This seems to be odd since we do not expect lensing without a lens. However, this is the result that one obtains when the **Hubble flow** is not taken into account. Indeed, the author of Ref. [16] constructs "by hand" cosmological observers in the Kottler metric and finds that Λ has no observable effect on the deflection of light.

Therefore, according to the results of Refs. [13–16], the standard approach to gravitational lensing does not need modifications. For the sake of clarity, the standard approach to gravitational lensing consists of using the result on the deflection angle obtained from the Schwarzschild metric (which models the lens) together with the cosmological angular diameter distances calculated from the Friedmann–Lemaître-Robertson–Walker (FLRW) metric. See e.g., Ref. [17].

In Ref. [18], we also tackled the investigation of whether a cosmological constant might affect the gravitational lensing by adopting the McVittie metric [19] as the description of the lens. The McVittie metric is an exact **spherically symmetric solution** of Einstein equations in presence of a point mass and a cosmic perfect fluid. See e.g., Refs. [20–28] for mathematical investigations of the geometrical properties of McVittie metric. In Ref. [18], we considered a constant Hubble factor, thus the geometry involved is the very Kottler one considered by most of the authors cited in this paper, but written in a different reference frame. Our results corroborate those of Refs. [13–16].

In the present paper, we generalise the results of Ref. [18] to the case of a generic time-dependent H. See also Ref. [29]. This is necessary in order to make contact with the current standard model of cosmology, the ΛCDM model, in which pressure-less matter prevents H to be a constant. In particular, Friedmann equation for the **spatially-flat** ΛCDM model reads:

$$\frac{H^2}{H_0^2} = \Omega_m a^{-3} + \Omega_\Lambda ,$$ (6)

where H_0 is the Hubble constant, a is the scale factor, Ω_m is the present density parameter of pressure-less matter and $\Omega_\Lambda = 1 - \Omega_m$ is the **present** density parameter of the cosmological constant. The time-derivative of H can be easily computed as:

$$\frac{\dot{H}}{H_0^2} = -\frac{3}{2}\Omega_m a^{-3} . \tag{7}$$

Since $\Omega_m \approx 0.3$, one can see that $\dot{H}_0$ and H_0^2 are of the same order at present time. Moreover, $|\dot{H}| \geq H^2$ for $1 + z \geq \sqrt[3]{2\Omega_\Lambda/\Omega_m}$. This gives a redshift $z > 1.67$. Many of the observed sources and lenses have redshifts larger than this limit (see e.g., the CASTLES survey, https://www.cfa.harvard.edu/castles/); therefore, the above calculation shows that assuming H constant is a very bad approximation and if one plans to make contact with observation, it is necessary to go beyond the static case of the Kottler metric. The McVittie metric with a generic, non-constant H provides an opportunity to do this.

Very recently, the deflection of light in a cosmological context with a generic $H(t)$ has been considered in Ref. [30]. The main result is the following, cf. Equation (11) of Ref. [30]:

$$\Delta\tilde{\varphi} = \frac{4\tilde{M}_{MSH}}{\tilde{R}} - 2H^2\tilde{R}^2 , \tag{8}$$

where $\tilde{M}_{MSH}$ is the Misner–Sharp–Hernandez mass [31,32] contained in a radius $\tilde{R} = a(t)r$. Hence, it turns out that the effect of cosmology on the gravitational lensing depends on whether one takes into account the total contribution (local plus cosmological) to the mass or just the local one. However, in both cases, $\tilde{M}_{MSH}$ can be decomposed in the local m contribution plus the cosmological one, which is cancelled by the $-2H^2\tilde{R}^2$ in the above Equation (8). Thus, it appears that the net result is that the cosmic fluid does not contribute directly to the gravitational lensing.

As a final remark, we must stress that the McVittie metric is an oversimplified model of an actual lens, which has a more complicated structure than that of a point. However, the results of our investigation may shed an important light and give valuable insights for future research that takes into account a more complex structure of the lens. As far as we know, lenses with structures different from a point have been considered only by Ref. [10].

The present paper is structured as follows. In Section 2, we present the McVittie metric and its principal features. In Section 3, we obtain the null geodesic equations and calculate the deflection angle. In Section 4, we focus on the case of a constant H and calculate exactly the subdominant contribution to the deflection angle, estimating a relative correction on the mass determination of about 10^{-11}, due to the **Hubble flow**. In Section 5, we present our conclusions. Throughout the paper, we use $G = c = 1$ units.

2. The McVittie Metric

The McVittie metric [19] can be written in the following form:

$$ds^2 = -\left(\frac{1-\mu}{1+\mu}\right)^2 dt^2 + (1+\mu)^4 a(t)^2(d\rho^2 + \rho^2 d\Omega^2) , \tag{9}$$

where $a(t)$ is the scale factor, $d\Omega^2 = d\theta^2 + \sin^2\theta d\phi^2$ and

$$\mu \equiv \frac{M}{2a(t)\rho} , \tag{10}$$

where M is the mass of the point. One can check that for $a = $ constant, the Schwarzschild metric in isotropic coordinates is recovered, whereas for $M = 0$, the FLRW metric is recovered.

When $\mu \ll 1$, the McVittie metric (9) can be approximated by:

$$ds^2 = -(1 - 4\mu)\, dt^2 + (1 + 4\mu)a(t)^2(d\rho^2 + \rho^2 d\Omega^2) , \tag{11}$$

i.e., it takes the form of a perturbed FLRW metric in the Newtonian gauge with gravitational potential 2μ. Since there is only a single gravitational potential, then no anisotropic pressure is present [17,33].

Since null geodesics are conformally invariant, the scale factor $a(t)$ can be written as a conformal factor in front of Equation (11), replacing the cosmic time with the conformal time. The time dependence of the photon trajectory appears then only in μ which, from Equation (10), represents an effective mass decreasing inversely proportional to the cosmic expansion. We are going to show that most of the deflection takes place at the closest approach distance to the lens. Therefore, we can already estimate that at this moment the effective mass of the lens would be M/a_L, where a_L is the scale factor evaluated at the moment of closest approach. Therefore, the bending angle shall be proportional to M/a_L, and this is indeed our result from Equation (47).

Note also that the gravitational potential $2\mu = M/(a\rho)$ resembles the Newtonian one but with an effective mass that reduces with time or a gravitational radius that increases with time. On the other hand, we know from cosmological linear perturbation theory that, in the matter-dominated epoch, the gravitational potential is time-independent. Therefore, the McVittie metric cannot offer a realistic model of a gravitational lens.

Calculating the Einstein tensor from the McVittie line element (9), one gets:

$$G^t{}_t = 3H^2 , \quad G^r{}_r = G^\theta{}_\theta = G^\phi{}_\phi = 3H^2 + \frac{2\dot{H}(1+\mu)}{1-\mu} , \tag{12}$$

from which one deduces that the pressure of the cosmological medium has the following form:

$$P = -\frac{1}{8\pi} \left[3H^2 + \frac{2\dot{H}(1+\mu)}{1-\mu} \right] , \tag{13}$$

i.e., it is not homogeneous and diverging when $\mu = 1$. If $H = \text{constant}$, then there is no divergence and the pressure is also a constant. This is the case of the Schwarzschild–de Sitter space, described by the Kottler metric [2]. When $\dot{H} \neq 0$, **far away** from the point mass, i.e., for $\mu \ll 1$, one gets the usual result of cosmology:

$$P = -\frac{1}{8\pi} \left(3H^2 + 2\dot{H} \right) + \mathcal{O}(\mu) = -\rho - \frac{\dot{H}}{4\pi} + \mathcal{O}(\mu) , \tag{14}$$

i.e., the acceleration equation. Isotropy is preserved since $G^r{}_r = G^\theta{}_\theta = G^\phi{}_\phi$, i.e., there is no anisotropic pressure, as we already mentioned.

Following Faraoni [28], but also Park [13], the McVittie metric (9) can be reformulated in terms of the areal radius

$$R = a\rho(1+\mu)^2 , \tag{15}$$

and gets the following form:

$$ds^2 = -\left(1 - \frac{2M}{R} - H^2 R^2 \right) dt^2 + \frac{dR^2}{1 - \frac{2M}{R}} - \frac{2HR}{\sqrt{1 - \frac{2M}{R}}} dt dR + R^2 d\Omega^2 . \tag{16}$$

Changing the time coordinate to

$$F(r,t)dT = dt + \frac{HR}{\sqrt{1 - \frac{2M}{R}}\left(1 - \frac{2M}{R} - H^2 R^2\right)} dR , \tag{17}$$

the above line element (16) can be finally cast as

$$ds^2 = -\left(1 - \frac{2M}{R} - H^2R^2\right)F^2dT^2 + \frac{dR^2}{1 - \frac{2M}{R} - H^2R^2} + R^2d\Omega^2 \,. \tag{18}$$

If H is constant, then F can be set to unity and we recover the Kottler metric. Using Equation (18) and calculating the Misner–Sharp–Hernandez mass [31,32] of a sphere of proper radius R, one finds [28]:

$$m_{\text{MSH}} = M + \frac{H^2R^3}{2} = M + \frac{4\pi}{3}\rho R^3 \,, \tag{19}$$

which contains the time-independent contribution M from the point mass plus the mass of the cosmic fluid contained in the sphere. Therefore, M has indeed the physical meaning of the mass of the point. In the Kottler case, one can also define a Komar integral, or Komar mass, and verify that it is indeed equal to M [34].

3. The Bending of Light in the McVittie Metric

We now revisit the calculation for the bending angle performed in Ref. [18], but take into account a general non-constant Hubble factor $H(t)$. We perform the calculations in two different ways: the one in this section is also used in Ref. [18] and is based on the approach usually adopted to study weak lensing (see e.g., Ref. [33,35]). In this approach, the origin of the coordinate system is occupied by the observer. The second way is the one in which the lens is put at the origin of the coordinate system and it is employed in Appendix A.

We adopt μ as perturbative parameter and work at the first order approximation in μ. The observed angle of a lensed source is of the order of the arc second, which corresponds to $\theta_O \approx 10^{-6}$ radians. See e.g., the CASTLES survey lens database (https://www.cfa.harvard.edu/castles/). At least for Einstein ring systems, the bending angle is of order $\delta \approx 10^{-6}$, i.e., it is of the same order as the observed deflection angle. However, at the same time, the bending angle is of the same order of μ. Therefore, we draw the conclusion that $\mu \approx 10^{-6}$ and the truncation error, when working at first order in μ, is $\mathcal{O}(10^{-12})$.

The geometry of the lensing process is depicted in Figure 1. In this scheme, the observer stays at the origin of the spatial coordinate system and x is the comoving coordinate along the observer-lens axis.

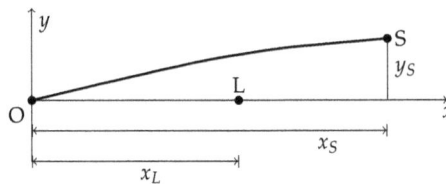

Figure 1. Scheme of lensing.

The observer has spatial position $(0,0,0)$ and the lens has spatial position $(x_L,0,0)$. The McVittie metric (9) written in Cartesian spatial coordinates is the following:

$$ds^2 = -\left(\frac{1-\mu}{1+\mu}\right)^2 dt^2 + (1+\mu)^4 a(t)^2 \delta_{ij} dx^i dx^j \,. \tag{20}$$

Note that the spherical symmetry of the McVittie metric implies rotational symmetry about the observer-lens axis. Therefore, we set $z = 0$ without losing generality and the source has thus spatial position $(x_S, y_S, 0)$.

Since in metric (9) the lens lays at the origin of the coordinate system, we have to perform a translation along the x axis in the Cartesian coordinates of metric (20), so that μ gets the following form:

$$\mu = \frac{M}{2a(t)\sqrt{(x - x_L)^2 + y^2 + z^2}} . \tag{21}$$

Introducing an affine parameter λ and the four-momentum $P^\mu = dx^\mu / d\lambda$, we can derive from metric (20) the following relation:

$$g_{\mu\nu} P^\mu P^\nu = 0 \quad \Rightarrow \quad P^0 = \frac{1+\mu}{1-\mu} p \sim (1 + 2\mu) p , \tag{22}$$

where $p^2 = g_{ij} P^i P^j$ is the proper momentum. The above equation represents the usual gravitational redshift experienced by a photon passing through the potential well generated by the point mass. Note that this potential well is not static since μ is time-dependent.

Now, we calculate the geodesic equations for the photon propagating in the McVittie metric (20):

$$\frac{d^2 x^\nu}{d\lambda^2} + \Gamma^\nu_{\alpha\beta} \frac{dx^\alpha}{d\lambda} \frac{dx^\beta}{d\lambda} = 0 . \tag{23}$$

The geodesic equation for $\nu = 0$ has the following form:

$$\frac{dP^0}{d\lambda} = 2\dot\mu (P^0)^2 + 4\mu_{,i} P^0 P^i - p^2 [H(1 + 4\mu) + 2\dot\mu] . \tag{24}$$

Using Equation (22) and the fact that, from Equation (10), $\dot\mu = -H\mu$, one finds:

$$p\frac{dp}{dt} = -Hp^2 + 2H\mu p^2 + 4\mu_{,i} P^i p . \tag{25}$$

The zeroth-order term Hp^2 represents the usual cosmological redshift term. The spatial geodesics equations have the form:

$$\frac{dP^i}{d\lambda} = \frac{4\delta^{il} \mu_{,l} p^2}{a^2} - 2HpP^i - 4P^i P^k \mu_{,k} . \tag{26}$$

We now look for an equation for the quantity $dy/dx \equiv \tan\theta$, which represents the slope of the line tangent to the photon trajectory. Note that θ is a physical angle because of the isotropic form of metric (20).

Since we can invert $x(\lambda)$ to $\lambda(x)$, being it a monotonic function, we can rewrite Equation (26) for y and change the variable to x:

$$P^x \frac{d}{dx}\left(P^x \frac{dy}{dx}\right) = 4\mu_{,y}[(P^x)^2 + (P^y)^2] - 2Ha(1 + 2\mu)\sqrt{(P^x)^2 + (P^y)^2} P^x \frac{dy}{dx} - 4\frac{dx}{dz} P^x [P^y \mu_{,y} + P^x \mu_{,x}] , \tag{27}$$

where we used the fact that

$$p^2 = a^2(1 + 4\mu)[(P^x)^2 + (P^y)^2] . \tag{28}$$

Expanding the left-hand side and using $P^y / P^x = dy/dx$, we obtain:

$$\frac{d^2 y}{dx^2} + \frac{1}{P^x}\frac{dP^x}{dx}\frac{dy}{dx} = 4\mu_{,y}\left[1 + \left(\frac{dy}{dx}\right)^2\right] - 2Ha(1 + 2\mu)\sqrt{1 + \left(\frac{dy}{dx}\right)^2}\frac{dy}{dx} - 4\frac{dy}{dx}\left(\frac{dy}{dx}\mu_{,y} + \mu_{,z}\right) . \tag{29}$$

In order to determine the second term on the left-hand side, we use Equation (26) for x:

$$\frac{1}{P^x}\frac{dP^x}{dx} = 4\mu_{,x}\left[1 + \left(\frac{dy}{dx}\right)^2\right] - 2Ha(1 + 2\mu)\sqrt{1 + \left(\frac{dy}{dx}\right)^2} - 4\left(\frac{dy}{dx}\mu_{,y} + \mu_{,z}\right) . \tag{30}$$

Combining the two Equations (29) and (30), we finally find:

$$\frac{d^2y}{dx^2} = 4\mu_{,y}\left[1+\left(\frac{dy}{dx}\right)^2\right] - 4\mu_{,x}\left[1+\left(\frac{dy}{dx}\right)^2\right]\frac{dy}{dx} .$$

(31)

Let's discuss a little about the **spatial derivative** of μ. From Equation (21), we get:

$$\mu_{,y} = -\mu\frac{y}{\sqrt{(x-x_L)^2+y^2}} ,$$

(32)

and

$$\mu_{,x} = -\mu\frac{(x-x_L)}{\sqrt{(x-x_L)^2+y^2}} .$$

(33)

Notice that, when $\mu = 0$, Equation (31) becomes:

$$\frac{d^2y}{dx^2} = 0 ,$$

(34)

i.e., the zeroth-order trajectory is, as expected, a straight line in comoving coordinates.

We now make a second approximation: we assume $dy/dx = \tan\theta$ to be small. From the CASTLES survey we know that the observed angle θ_O is of the order of the arc second, which corresponds to $\theta_O \approx 10^{-6}$ radians. The latter is larger that the actual angular position of the source, say θ_S, because of the lensing geometry, see e.g., Figure 2. For this reason, we can assume dy/dx to be small along all the trajectory. Since $dy/dx = \tan\theta$, then $dy/dx = \theta + \theta^3/3 + \cdots$. The truncation error is then of order $\mathcal{O}(\theta^3) \sim 10^{-18}$.

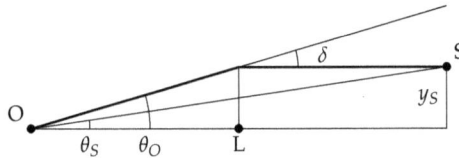

Figure 2. The thin-lens approximation.

We consider Equation (31) up to the lowest order term, i.e.,

$$\frac{d^2y}{dx^2} = 4\mu_{,y} + \mathcal{O}(\mu\theta) ,$$

(35)

where $\mathcal{O}(\mu\theta) \sim 10^{-12}$. Using Equation (32) for the derivative $\mu_{,y}$, the above equation becomes:

$$\frac{d^2y}{dx^2} = -\frac{2My}{a(x)\left[(x-x_L)^2+y^2\right]^{3/2}} .$$

(36)

Note that a is a function of time, but inverting $x(t)$, we can write a as a function of x. For simplicity, we normalise x, y and $2M$ to x_L, thus obtaining:

$$\frac{d^2Y}{dX^2} = -\alpha\frac{Y}{a(X)\left[(X-1)^2+Y^2\right]^{3/2}} ,$$

(37)

where $Y \equiv y/x_L$, $X \equiv x/x_L$ and $\alpha \equiv 2M/x_L$. The above equation was already found in Ref. [18]. Since $dY/dX = \tan\theta$ and $\tan\theta \sim \theta$, we can cast the above equation in the following form:

$$\frac{d\theta}{dX} = -\alpha \frac{Y}{a(X)\left[(X-1)^2+Y^2\right]^{3/2}} . \tag{38}$$

We *define* the bending angle as follows:

$$\delta \equiv \int_{\theta_S}^{\theta_O} d\theta = -\alpha \int_{X_S}^{0} \frac{Y(X)dX}{a(X)\left[(X-1)^2+Y(X)^2\right]^{3/2}} , \tag{39}$$

i.e., as the variation of the slope of the trajectory between the source and the observer.

We shall solve the above equation keeping the first order in α. The order of magnitude of α can be estimated as follows:

$$\alpha \equiv \frac{2M}{x_L} \approx 2MH_0/z_L , \tag{40}$$

where we assumed a small redshift z_L. The above approximation becomes an exact result in the case of a constant Hubble parameter (see e.g., (51)).

Therefore, α is proportional to the ratio between the Schwarzschild radius of the lens and the Hubble radius. This is $H_0M \sim 10^{-12}$ for a galaxy of 10^{10} $M_\odot$ and $H_0M \sim 10^{-9}$ for a cluster of 10^3 galaxies each of mass 10^{10} $M_\odot$.

We now devote a small paragraph to the zeroth-order solution.

3.1. The Zeroth-Order Solution

The zeroth-order solution (i.e., the one for $\alpha = 0$) of Equation (37) is a straight line in comoving coordinates:

$$y = \theta_S(x - x_S) + y_S , \tag{41}$$

where $\theta_S \ll 1$ is the slope of the trajectory and (x_S, y_S) are the comoving coordinates of the source. See Figure 3.

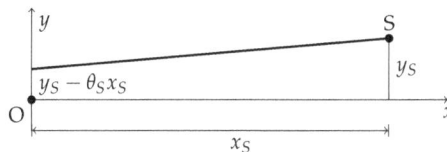

Figure 3. Zeroth-order solution.

When we pass from comoving to proper distances by multiplying by the scale factor $a(x)$ we obtain:

$$y_p = \theta_S x_p + \frac{a(x_p)}{a_S}(y_{pS} - \theta_S x_{pS}) , \tag{42}$$

where we used a subscript p to indicate the proper distance. The above is not a straight line trajectory, as also noticed by the authors of Ref. [15]. It is bent because of the $a(x_p)$ factor on the right-hand side, whose effect vanishes only for $y_{pS} = \theta_S x_{pS}$. The latter condition, when substituted in Equation (41), represents the ray which gets to $y = 0$ when $x = 0$, i.e., the observed ray. See Figure 4.

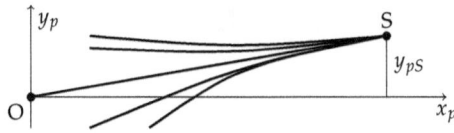

Figure 4. Zeroth-order solution, using proper distances.

The **Hubble flow** seems to bend away the trajectories such that we cannot detect any light. This happens isotropically, i.e., no observer could ever detect a bent ray but just the straight one coming directly from the source.

On the other hand, let's speculate about the following. If a cosmologically bent ray passes sufficiently close to a lens, then its trajectory could be bent back by the gravitational field of the lens, possibly allowing us to detect it. See Figure 5.

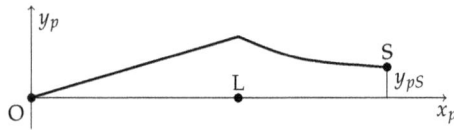

Figure 5. A cosmologically bent ray, bent back. The "back-bending".

This "back-bending" seems to suggest that the bending angle must increase and therefore cosmology must somehow enter the gravitational lensing phenomenon.

3.2. Calculation of the Bending Angle

We now integrate Equation (39) retaining the first order only in α. For this reason, the $Y(X)$ entering the integral is the zeroth-order solution, which we discussed in the previous subsection.

Since we are working at the first order in α, we can assume without losing generality that the zeroth-order trajectory is horizontal, i.e., $Y^{(0)} = Y_S \equiv y_S / x_L$.

The equation for the slope, i.e., Equation (38), becomes

$$\frac{d\theta}{dX} = -\frac{\alpha Y_S}{a(X)\left[(X-1)^2 + Y_S^2\right]^{3/2}} \, . \tag{43}$$

In order to determine $a(X)$, we take advantage of metric (20) and write:

$$dx^2 = \frac{1 - 8\mu}{1 + (dy/dx)^2} \frac{dt^2}{a^2} \, , \tag{44}$$

which is a very complicated integration to perform, since it includes the very trajectory we want to determine. On the other hand, we are staying at the lowest possible order of approximation; therefore:

$$dx = -\frac{dt}{a} \, , \tag{45}$$

i.e., all the contributions coming from μ and θ of Equation (44) are of negligible order in Equation (43). Now, write Equation (43) as follows:

$$d\theta = -\frac{\alpha \, dX}{a(X)Y_S^2\left[1 + \frac{(X-1)^2}{Y_S^2}\right]^{3/2}} \, . \tag{46}$$

When $(X-1)^2 \gg Y_S^2$, the above integration, whatever function a might be of X, is $\mathcal{O}(Y_S)$. On the other hand, when $(X-1)^2 \ll Y_S^2$, the above integration is $\mathcal{O}(1/Y_S^2)$.

Therefore, the main contribution comes from $X = 1$ and spans the interval $1 - Y_S < X < 1 + Y_S$. That is, most of the deflection takes place very close to the lens, as it happens for the case of the Schwarzschild metric. For this reason, we also approximate $a(X)$ with a_L, which is the scale factor when $x = x_L$.

Therefore, we end up with the following bending angle:

$$\delta = -\frac{\alpha}{a_L Y_S^2}(-2Y_S) = \frac{2\alpha}{a_L Y_S} = \frac{4M}{a_L y_S} . \tag{47}$$

The above formula is general, valid for any kind of **Hubble flow**. We derive it using another method in Appendix A and prove its validity in the case of a dust-dominated universe, for which an exact calculation is possible, in Appendix B.

Now we apply Equation (47) in the lens equation. Let us refer to Figure 2. The geometry of this figure is justified by the fact that, as we showed earlier, the bending happens predominantly at the closest approach distance to the lens.

In Figure 2, θ_S is the angular position of the source, so that $\theta_S D_S$ is the proper transversal position of the source, where D_S is the angular-diameter distance from the observer to the source. The angle θ_O is the angular apparent position of the source, so that $\theta_O D_S$ is the transversal apparent position of the source.

Therefore, the lens equation in the thin-lens approximation can be written as:

$$\theta_O D_S = \theta_S D_S + \delta D_{LS} , \tag{48}$$

where D_{LS} is the angular-diameter distance between lens and source. Using the result of Equation (47), we get

$$\theta_O - \theta_S = \frac{4M}{a_L y_S} \frac{D_{LS}}{D_S} . \tag{49}$$

In the standard lens equation, one has the closest approach distance to the lens, let's call it R, in place of $a_L y_S$. One then writes $R = \theta_O D_L$ and thus finds the usual formula, see e.g., [17].

Now, since we found that the deflection occurs almost completely at the closest position to the lens, we can approximate $y_S \approx y_L$. Moreover, one also has $y_L = \theta_O x_L$ and $D_L = a_L x_L$, from the definition of the angular-diameter distance to the lens. Thus, $a_L y_S \approx a_L \theta_O x_L = \theta_O D_L$, and we recover the usual well-known formula:

$$\theta_O(\theta_O - \theta_S) = \frac{4M}{D_L} \frac{D_{LS}}{D_S} . \tag{50}$$

Therefore, we can conclude that cosmology does not modify the bending angle at the leading order of the expansion in powers of μ and θ. The cosmological "drift" discussed earlier for the zeroth-order solution is already taken into account when using angular-diameter distances so that the final result does not change.

However, sub-dominant terms do carry a cosmological signature, as we show in the next section. Here, we address the simple case of a cosmological constant-dominated universe, where analytical calculations are possible.

4. Next-to-Leading Order Contributions to the Bending Angle in the Case of a Cosmological Constant-Dominated Universe

As we saw in Equation (47), the leading contribution in the expansion for the bending angle calculated in the McVittie metric is the same as the one calculated for the Schwarzschild metric. Therefore, it is interesting to check if next-to-leading orders do carry a signature of the cosmological embedding of the point lens. We tackle this issue here in the case of a cosmological constant-dominated universe, for which exact calculations are possible, and leave a more general treatment as a future work.

When $H = H_0 = $ constant, one can find an analytic expression for $a(x)$:

$$x = \int_a^1 \frac{da'}{H_0 a'^2} = \frac{1}{H_0} \left(\frac{1}{a} - 1 \right) = \frac{z}{H_0} , \qquad (51)$$

where in the last equality we introduced the redshift. The scale factor as function of the comoving distance is thus:

$$\frac{1}{a(x)} = H_0 x + 1 = H_0 X x_L + 1 = z_L X + 1 , \qquad (52)$$

and Equation (43) becomes:

$$\frac{d\theta}{dX} = -\frac{\alpha Y_S (z_L X + 1)}{[(X-1)^2 + Y_S^2]^{3/2}} . \qquad (53)$$

As we anticipated, this equation can be solved exactly and the bending angle, as we defined it in Equation (39), is the following:

$$\delta = \frac{\alpha}{Y_S} \left[\frac{1 + z_L + z_L Y_S^2}{\sqrt{1 + Y_S^2}} + \frac{(X_S - 1)(1 + z_L) - z_L Y_S^2}{\sqrt{(X_S - 1)^2 + Y_S^2}} \right] . \qquad (54)$$

Expanding this solution for a small impact parameter Y_S, one gets:

$$\delta = \frac{2\alpha(1 + z_L)}{Y_S} \left[1 + Y_S^2 \frac{2(z_L - 1) + X_S[2 + X_S(z_L - 1) - 4z_L]}{4(z_L + 1)(X_S - 1)^2} \right] , \qquad (55)$$

where we have already truncated $\mathcal{O}(Y_S^4)$ terms and put in evidence the leading order contribution $2\alpha(1 + z_L)/Y_S$ (see Equation (47)).

Recovering the physical quantities $Y_S = y_S/x_L$, $X_S = x_S/x_L$, $\alpha = 2M/x_L$ and using Equation (51) in order to express x as the redshift, we get:

$$\delta = \frac{4M(1 + z_L)}{y_S} \left[1 + \frac{y_S^2}{x_L^2} \frac{2z_L^2(z_L - 1) + z_S[2z_L + z_S(z_L - 1) - 4z_L^2]}{4(z_L + 1)(z_S - z_L)^2} \right] . \qquad (56)$$

We already showed in the discussion leading to Equation (50) that $y_S \approx \theta_O x_L$, so that:

$$\delta = \frac{4M}{\theta_O D_L} \left[1 + \theta_O^2 \frac{2z_L^2(z_L - 1) + z_S[2z_L + z_S(z_L - 1) - 4z_L^2]}{4(z_L + 1)(z_S - z_L)^2} \right] , \qquad (57)$$

and in the lens equation:

$$\theta_O(\theta_O - \theta_S) = \frac{4MD_{LS}}{D_L D_S} \left[1 + \theta_O^2 \frac{2z_L^2(z_L - 1) + z_S[2z_L + z_S(z_L - 1) - 4z_L^2]}{4(z_L + 1)(z_S - z_L)^2} \right] . \qquad (58)$$

Let's focus on Einstein ring systems, i.e., $\theta_S = 0$. We have in this case the mass estimate (it is actually an estimate on the product $H_0 M$, due to the presence of the angular-diameter distances, see Ref. [17]):

$$\frac{4MD_{LS}}{D_L D_S} = \theta_O^2 \left[1 - \theta_O^2 \frac{2z_L^2(z_L - 1) + z_S[2z_L + z_S(z_L - 1) - 4z_L^2]}{4(z_L + 1)(z_S - z_L)^2} \right] . \qquad (59)$$

The next-to-leading order correction is $\mathcal{O}(\theta_O^4)$ and depends on the redshifts of the lens and of the source.

Consider, for example, the Einstein ring Q0047-2808 of the CASTLES survey, for which $\theta_O = 2.7''$, $z_S = 3.60$ and $z_L = 0.48$. Substituting these numbers in Equation (59), the correction on the mass estimate is therefore

$$\frac{4MD_{LS}}{\theta_O^2 D_L D_S} = 1 + 0.12\,\theta_O^2 = 1 + 2.03 \cdot 10^{-11}\,. \tag{60}$$

This is an extremely small correction which nonetheless depends on cosmology. Note that it is only one order of magnitude larger than the terms $\mathcal{O}(\mu^2)$ that we have neglected in our calculations.

5. Conclusions

We investigated whether cosmology affects the gravitational lensing caused by a point mass. To this purpose, we used McVittie metric as the description of the point-like lens embedded in an expanding universe. The reason for this choice is to use a metric which properly takes into account the **Hubble flow** to which source, lens and observer are subject. We considered the general case in which the Hubble factor is a generic function of time and find that no contribution coming from cosmology enters the bending angle at the leading order (see Equation (47)), thus strengthening the results obtained by [13–16].

We addressed the sub-dominant contributions to the bending angle in the special case of a constant Hubble factor $H = H_0$, for which exact calculations are possible. We found that in this case cosmology does affect the bending of light, through a combination of the lens and source redshifts, given in Equation (58). This correction is of order 10^{-11} for the Einstein ring Q0047-2808.

We conclude that the standard approach to gravitational lensing on cosmological distances, which consists of patching together the results coming from the Schwarzschild metric (which models the lens) and Friedmann–Lemaître-Robertson–Walker (FLRW) metric (which serves to calculate the cosmological angular-diameter distances), does not require modifications.

Future developments of this investigation should address the entity subdominant orders of the expansion for the bending angle in a model-independent way. We expect the latter to depend on H_L, i.e., the Hubble parameter evaluated at the lens redshift. If these corrections were measurable, they might provide a new cosmological probe for determining the value of the Hubble parameter at different redshifts.

Finally, we must stress that McVittie metric is a particular and oversimplified description of the geometry of a lens, this being a galaxy or a cluster of galaxies. Therefore, another improvement would be that of tackling the analysis of the gravitational lensing and of the bending angle by constructing for the lens a density profile more realistic than a Dirac delta (i.e., the one used here for a point mass). We expect that different lens density profiles would lead to different results in the mass estimates also from the point of view of the cosmological corrections, as shown in Ref. [10] for the case of the Kottler metric.

Acknowledgments: The author thanks CNPq (Brasilia, Brazil) for partial financial support. He is also indebted to D. Bacon, V. Marra, H. Velten and the anonymous referees for stimulating discussions and suggestions.

Conflicts of Interest: The author declares no conflict of interest.

Appendix A. Standard Approach to Gravitational Lensing

We now place the lens at the origin of the reference frame and use polar coordinates, as in Figure A1.

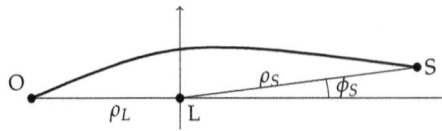

Figure A1. Scheme of lensing, with the lens at the origin of the coordinate system.

Again, we work at first order in μ. For a photon, metric (11) with $\theta = \pi/2$ gives:

$$0 = -(1 - 4\mu)\dot{t}^2 + (1 + 4\mu)a^2\dot{\rho}^2 + (1 + 4\mu)a^2\rho^2\dot{\phi}^2 , \tag{A1}$$

where the dot denotes derivation with respect to the affine parameter λ. Since $\zeta_{(\phi)} = \delta^\mu_\phi \partial_\mu$ is a Killing vector, there exists the following conserved quantity:

$$L = g_{\mu\nu}P^\mu \zeta^\nu_{(\phi)} = (1 + 4\mu)a^2\rho^2\dot{\phi} , \tag{A2}$$

where P^μ is the photon four-momentum. The geodesic equation for t, cf. Equation (24), can be cast as follows:

$$(1 - 4\mu)\ddot{t} - 4\frac{\partial\mu}{\partial\rho}\dot{t}\dot{\rho} + \frac{1}{a}\frac{da}{dt}(1 - 4\mu)\dot{t}^2 = 0 , \tag{A3}$$

and written in the following compact form:

$$\frac{d}{d\lambda}\left[(1 - 4\mu)a\dot{t}\right] + 4a\dot{\mu}\dot{t}^2 = 0 . \tag{A4}$$

Recalling the definition of μ in Equation (10), i.e., $\mu = M/(2a\rho)$, one can easily determine that $\dot{\mu} = -H\mu$. We neglect this contribution since indeed the ratio between the gravitational radius of the lens and the Hubble radius must be very small, as we discussed after Equation (40).

Therefore, neglecting HM, Equation (A4) can be exactly integrated, giving the following result:

$$\left(1 - \frac{2M}{a\rho}\right)\dot{t} = \frac{E}{a} + \mathcal{O}(HM) , \tag{A5}$$

where E is an integration constant. We found a mixture of the known results for the Schwarzschild metric and for the FLRW one. Indeed, if $H = 0$, then a is an unimportant constant that we can incorporate into the definitions of ρ and E, and we recover the result for the Schwarzschild metric. On the other hand, with $M = 0$, we recover the usual cosmological decay of the energy of a photon, which is inversely proportional to the scale factor.

Combining Equation (A1) with Equation (A5), we can write the following equation for ρ:

$$\frac{a^4}{E^2}\dot{\rho}^2 = 1 - \frac{D^2}{\rho^2}(1 - 8\mu) , \tag{A6}$$

where $D \equiv L/E$ is a parameter associated to the closest approach distance ρ_L, defined as the one for which $\dot{\rho}_L = 0$, i.e.,

$$\rho_L = D(1 - 4\mu_L) , \tag{A7}$$

where μ_L is μ evaluated at the closest approach distance, i.e., $\mu_L = M/(2a_L\rho_L)$.

We use now the definition of the bending angle proposed by Rindler and Ishak in Ref. [1], based on the following formula:

$$\tan\psi = \frac{\sqrt{g_{\phi\phi}}}{\sqrt{g_{\rho\rho}}}\left|\frac{d\phi}{d\rho}\right| , \tag{A8}$$

which represents the angle between the radial and the tangential directions of the photon trajectory (see Figure A2). Using Equations (A2) and (A6) we find:

$$\tan \psi = \frac{D}{\rho} \frac{1 - 4\mu}{\sqrt{1 - \frac{D^2}{\rho^2}(1 - 8\mu)}} \, . \tag{A9}$$

This expression can be rewritten in terms of the closest approach radius ρ_L as follows:

$$\tan \psi = \frac{\rho_L/\rho}{\sqrt{1 - \rho_L^2/\rho^2}} \left(1 - \frac{2M}{a\rho} + \frac{2M}{a_L\rho_L} \right) \, . \tag{A10}$$

For $M = 0$, we obtain from Equation (A10) that

$$\tan \psi = \frac{\rho_L/\rho}{\sqrt{1 - \rho_L^2/\rho^2}} = \tan \phi \, , \tag{A11}$$

i.e., we recover the straight trajectory. Therefore, at any given position along the trajectory, $\psi - \phi$ gives the local bending angle, i.e., the deviation from the straight-line trajectory. See Figure A2.

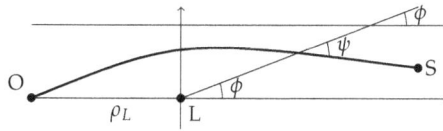

Figure A2. Schematic definition of the angle ψ, defined in Equation (A8). See also Figure 2 of Ref. [1].

The total bending angle is given by:

$$\delta = \psi_S + \psi_O - \phi_S - \phi_O \, . \tag{A12}$$

If we assume ρ_S and ρ_O are much larger than ρ_L, then the contributions from ψ_S and ψ_O are very small and practically negligible. Therefore, the dominant contribution to δ comes from $\phi_S + \phi_O$. In order to determine this sum, we must analyse the equation for the trajectory, i.e.,

$$\frac{d\rho}{d\phi} = \pm \rho \left(1 + \frac{4\mu - 4\mu_L}{1 - \rho_L^2/\rho^2} \right) \sqrt{\frac{\rho^2}{\rho_L^2} - 1} \, . \tag{A13}$$

For $\rho \gg \rho_L$, one can simplify this equation as follows:

$$\frac{d\rho}{d\phi} = \pm \frac{\rho^2}{\rho_L} (1 - 4\mu_L) \, , \tag{A14}$$

where we have considered only the leading-order correction to the equation for the straight line. The above equation tells us that the trajectory still is a straight line, far away from the lens, but tilted of an angle $4\mu_L$ from each side with respect to the horizontal. Therefore, the bending angle is

$$\delta = 8\mu_L = \frac{4M}{a_L\rho_L} \, , \tag{A15}$$

which is identical to the result of Equation (47) and also valid for a time-dependent H.

Appendix B. Bending Angle in a Matter-Dominated Universe

We check here Formula (47) in the case of a matter-dominated universe, described by the Friedmann equation $H^2 = H_0^2/a^3$. The scale factor as a function of the comoving distance x can be calculated as follows:

$$x = \int_a^1 \frac{da'}{H(a')a^2} = \frac{2}{H_0}(1 - \sqrt{a}) , \tag{B1}$$

which implies $a(x) = (1 - H_0 x/2)^2$. With this $a(x)$, Equation (43) can be solved exactly and the bending angle is the following:

$$\delta = \frac{2\alpha}{Y_S}\frac{4}{(H_0 x_L - 2)^2} + \mathcal{O}(Y_S) = \frac{2\alpha}{a_L Y_S} + \mathcal{O}(Y_S) , \tag{B2}$$

i.e., the same result found in Equation (47).

References

1. Rindler, W.; Ishak, M. Contribution of the cosmological constant to the relativistic bending of light revisited. *Phys. Rev. D* **2007**, *76*, 043006.
2. Kottler, F. Über die physikalischen grundlagen der Einsteinschen gravitationstheorie. *Ann. Phys.* **1918**, *361*, 401–462.
3. Islam, J. The cosmological constant and classical tests of general relativity. *Phys. Lett. A* **1983**, *97*, 239–241.
4. Schucker, T. Cosmological constant and lensing. *Gen. Relativ. Gravit.* **2009**, *41*, 67–75.
5. Ishak, M. Light Deflection, Lensing, and Time Delays from Gravitational Potentials and Fermat's Principle in the Presence of a Cosmological Constant. *Phys. Rev. D* **2008**, *78*, 103006.
6. Ishak, M.; Rindler, W.; Dossett, J. More on Lensing by a Cosmological Constant. *Mon. Not. Roy. Astron. Soc.* **2010**, *403*, 2152–2156.
7. Sereno, M. The role of Lambda in the cosmological lens equation. *Phys. Rev. Lett.* **2009**, *102*, 021301.
8. Kantowski, R.; Chen, B.; Dai, X. Gravitational Lensing Corrections in Flat ΛCDM Cosmology. *Astrophys. J.* **2010**, *718*, 913–919.
9. Ishak, M.; Rindler, W. The Relevance of the Cosmological Constant for Lensing. *Gen. Relativ. Gravit.* **2010**, *42*, 2247–2268.
10. Biressa, T.; de Freitas Pacheco, J. The Cosmological Constant and the Gravitational Light Bending. *Gen. Relativ. Gravit.* **2011**, *43*, 2649–2659.
11. Hammad, F. A note on the effect of the cosmological constant on the bending of light. *Mod. Phys. Lett.* **2013**, *A28*, 1350181.
12. Arakida, H. Effect of the Cosmological Constant on Light Deflection: Time Transfer Function Approach. *Universe* **2016**, *2*, 5.
13. Park, M. Rigorous Approach to the Gravitational Lensing. *Phys. Rev. D* **2008**, *78*, 023014.
14. Khriplovich, I.B.; Pomeransky, A.A. Does Cosmological Term Influence Gravitational Lensing? *Int. J. Mod. Phys. D* **2008**, *17*, 2255–2259.
15. Simpson, F.; Peacock, J.A.; Heavens, A.F. On lensing by a cosmological constant. *Mon. Not. Roy. Astron. Soc.* **2010**, *402*, 2009.
16. Butcher, L.M. Lambda does not Lens: Deflection of Light in the Schwarzschild-de Sitter Spacetime. 2016, arXiv:gr-qc/1602.02751.
17. Weinberg, S. *Cosmology*; Oxford University Press: Oxford, UK, 2008.
18. Piattella, O.F. Lensing in the McVittie metric. *Phys. Rev. D* **2016**, *93*, 024020.
19. McVittie, G. The mass-particle in an expanding universe. *Mon. Not. Roy. Astron. Soc.* **1933**, *93*, 325–339.
20. Nolan, B.C. A Point mass in an isotropic universe: Existence, uniqueness and basic properties. *Phys. Rev. D* **1998**, *58*, 064006.
21. Nolan, B. A Point mass in an isotropic universe. 2. Global properties. *Class. Quantum Gravity* **1999**, *16*, 1227–1254.
22. Nolan, B.C. A Point mass in an isotropic universe. 3. The region R less than or = to 2m. *Class. Quantum Gravity* **1999**, *16*, 3183–3191.

23. Kaloper, N.; Kleban, M.; Martin, D. McVittie's Legacy: Black Holes in an Expanding Universe. *Phys. Rev. D* **2010**, *81*, 104044.
24. Lake, K.; Abdelqader, M. More on McVittie's Legacy: A Schwarzschild—De Sitter black and white hole embedded in an asymptotically ΛCDM cosmology. *Phys. Rev. D* **2011**, *84*, 044045.
25. Nandra, R.; Lasenby, A.N.; Hobson, M.P. The effect of a massive object on an expanding universe. *Mon. Not. Roy. Astron. Soc.* **2012**, *422*, 2931–2944.
26. Nandra, R.; Lasenby, A.N.; Hobson, M.P. The effect of an expanding universe on massive objects. *Mon. Not. Roy. Astron. Soc.* **2012**, *422*, 2945–2959.
27. Nolan, B.C. Particle and photon orbits in McVittie spacetimes. *Class. Quantum Gravity* **2014**, *31*, 235008.
28. Faraoni, V. *Cosmological and Black Hole Apparent Horizons*; Springer: Cham, Switzerland, 2015.
29. Aghili, M.E.; Bolen, B.; Bombelli, L. Effect of Accelerated Global Expansion on Bending of Light. 2014, arXiv:gr-qc/1408.0786.
30. Faraoni, V.; Lapierre-Leonard, M. Beyond Lensing by the Cosmological Constant. 2016, arXiv:gr-qc/1608.03164.
31. Misner, C.W.; Sharp, D.H. Relativistic equations for adiabatic, spherically symmetric gravitational collapse. *Phys. Rev.* **1964**, *136*, B571–B576.
32. Hernandez, W.C.; Misner, C.W. Observer Time as a Coordinate in Relativistic Spherical Hydrodynamics. *Astrophys. J.* **1966**, *143*, 452.
33. Dodelson, S. *Modern Cosmology*; Academic Press: Cambridge, MA, USA, 2003.
34. Kastor, D. Komar Integrals in Higher (and Lower) Derivative Gravity. *Class. Quantum Gravity* **2008**, *25*, 175007.
35. Bartelmann, M.; Schneider, P. Weak gravitational lensing. *Phys. Rep.* **2001**, *340*, 291–472.

universe

Article

Effect of the Cosmological Constant on Light Deflection: Time Transfer Function Approach

Hideyoshi Arakida

College of Engineering, Nihon University, Koriyama, Fukushima 963-8642, Japan; arakida@ge.ce.nihon-u.ac.jp

Academic Editor: Lorenzo Iorio
Received: 9 February 2016; Accepted: 8 March 2016; Published: 14 March 2015

Abstract: We revisit the role of the cosmological constant Λ in the deflection of light by means of the Schwarzschild–de Sitter/Kottler metric. In order to obtain the total deflection angle α, the time transfer function approach is adopted, instead of the commonly used approach of solving the geodesic equation of photon. We show that the cosmological constant does appear in expression of the deflection angle, and it diminishes light bending due to the mass of the central body M. However, in contrast to previous results, for instance, that by Rindler and Ishak (Phys. Rev. D. 2007), the leading order effect due to the cosmological constant does not couple with the mass of the central body M.

Keywords: light deflection; cosmological constant; time transfer function; relativity

PACS: 95.30.Sf; 98.62.Sb; 98.80.Es

1. Introduction

The cosmological constant problem is the old one concerning closely the general theory of relativity (See reviews by, e.g., [1,2]). After establishing the general theory of relativity by Einstein in 1915–1916, he introduced the cosmological constant Λ to describe the static universe since the original Einstein equation cannot represent the picture of static universe. Though the discovery of the cosmic expansion by Hubble made a denial of Einstein's first purpose, the cosmological constant is realized again because of the find of accelerating expansion of the Universe [3–5], and it is popularly considered that the cosmological constant Λ or dark energy generally has the highest potential for explaining the observed accelerating expansion of the Universe. However, its details are still far from clear; therefore, this hypothesis must be verified through not only cosmological observations but also other astronomical/astrophysical measurements.

Among such efforts, the most straightforward approach is to investigate the role of Λ in the classical tests of general relativity, such as the perihelion advance of planetary orbits and the bending of light rays. Thus far, it was found that the cosmological constant Λ contributes to the perihelion shift in principle even though this contribution is presently difficult to detect because of its very small effect (See [6–8] and the references therein, and corresponding topic to perihelion advance [9–12]).

While in the case of bending of light under the Schwarzschild–de Sitter/Kottler spacetime (see Equation (12)), contrary to the expectation, the second-order geodesic equation of a photon does not contain Λ,

$$\frac{d^2u}{d\phi^2} = -u + \frac{3}{2}r_g u^2, \quad r_g = \frac{2GM}{c^2}, \quad u = \frac{1}{r} \tag{1}$$

then, as a consequence, it is considered that the deflection angle in the Schwarzschild–de Sitter or Kottler metric coincides with that of Schwarzschild case. However, recently, Rindler and Ishak [13] reported that Λ does affect the bending of light by means of the Schwarzschild–de Sitter or Kottler metric and the invariant formula for the cosine. Subsequently, many authors have argued its appearance in many different ways and the generality of these arguments advocated the appearance of Λ in the deflection angle α. Nevertheless, presently, it seems that a conclusion has not yet been reached, for instance, on whether the leading order effect due to Λ is coupled with the mass of the central body M or not. See [14] for a review and also [15–25]. In addition, for the cosmological constant and cosmological lensing equation, see, e.g., [17,18,26,27].

As we assess the circumstances, the origin of confusion, e.g., the appearance/disappearance of Λ or the coupling/uncoupling with the mass of the central body M, is essentially attributable to the use of the standard geodesic equation of a photon to obtain light deflection due to Λ, because Λ does not appear. Therefore, it is worthy to revisit this problem using another theoretical approach.

In this paper, we will revisit the role of the cosmological constant Λ in terms of the time transfer function recently proposed in [28,29], which is originally related to Synge's world function $\Omega(x_A, x_B)$ and which enables us to circumvent the integration of the null geodesic equation. In Section 2, we will briefly summarize the time transfer function method. In Section 3, the effect of Λ on light deflection will be re-investigated. Section 4 is devoted to a short summary of this paper.

2. Outline of the Time Transfer Function

Before calculating the light deflection due to the cosmological constant Λ, let us briefly summarize the time transfer function method presented in [28,29].

Synge's world function is defined by [30]

$$\Omega(x_A, x_B) \equiv \frac{1}{2}(\lambda_B - \lambda_A) \int_{\lambda_A}^{\lambda_B} g_{\mu\nu} \frac{dx^\mu}{d\lambda} \frac{dx^\nu}{d\lambda} d\lambda \tag{2}$$

where $g_{\mu\nu}$ is a metric tensor of spacetime; $x_A = (x_A^0 = ct_A, x_A^i = \vec{x}_A)$ and $x_B = (x_B^0 = ct_B, x_B^i = \vec{x}_B)$ are the coordinates of the two end-points A and B, respectively, on the geodesic world-line; and λ is the affine parameter. Then, the world function $\Omega(x_A, x_B)$ is defined as the half length of the world-line between A and B.

It is generally difficult to acquire the form of the world function concretely. Nonetheless, in the case of the Minkowskian flat spacetime, the world function is easily obtained using the parameter equation $x(\lambda) = (x_B - x_A)\lambda + x_A$ and by setting $\lambda_A = 0$ and $\lambda_B = 1$ [28,30],

$$\Omega^{(0)}(x_A, x_B) = \frac{1}{2}\eta_{\mu\nu}(x_B^\mu - x_A^\mu)(x_B^\nu - x_A^\nu) \tag{3}$$

where x^μ ($\mu = 0, 1, 2, 3$) are the Minkowskian coordinates with respect to the Minkowski metric $\eta_{\mu\nu} = \mathrm{diag}(-1, 1, 1, 1)$.

For the null geodesic, the world function $\Omega(x_A, x_B)$ satisfies the condition

$$\Omega(x_A, x_B) = 0 \tag{4}$$

because $ds^2 = 0$. Hence, from Equations (3) and (4), the travel time between A and B, namely $t_B - t_A$, in the Minkowskian flat spacetime becomes

$$c^2(t_B - t_A)^2 = \delta_{ij}(x_B^i - x_A^i)(x_B^j - x_A^j) = R_{AB}^2 \tag{5}$$

where δ_{ij} is Kronecker's delta, and c is the speed of light in vacuum. The time transfer function starts from Equation (5), and the weak-field approximation is developed recursively with respect to the gravitational constant G.

If the metric has the form

$$g_{\mu\nu} = \eta_{\mu\nu} + h_{\mu\nu}, \quad |h_{\mu\nu}| \ll 1 \tag{6}$$

where $h_{\mu\nu}$ is a perturbation to $\eta_{\mu\nu}$, the time transfer functions that yield the travel time of the light ray are formally expressed as follows:

$$t_B - t_A = \mathcal{T}_e(t_A, \vec{x}_A, \vec{x}_B) = \frac{1}{c}[R_{AB} + \Delta_e(t_A, \vec{x}_A, \vec{x}_B)] \tag{7}$$

$$= \mathcal{T}_r(\vec{x}_A, t_B, \vec{x}_B) = \frac{1}{c}[R_{AB} + \Delta_r(\vec{x}_A, t_B, \vec{x}_B)] \tag{8}$$

where $\mathcal{T}_e(t_A, \vec{x}_A, \vec{x}_B)$ is the emission time transfer function for the spatial coordinates $\vec{x}_A, \vec{x}_B$ and signal emission time t_A; $\mathcal{T}_r(\vec{x}_A, t_B, \vec{x}_B)$ is the reception time transfer function for the spatial coordinates $\vec{x}_A, \vec{x}_B$ and signal reception time t_B; $R_{AB} = |\vec{x}_B - \vec{x}_A|$; and Δ_e and Δ_r are called the emission time delay function and reception time delay function, respectively. Δ_e and Δ_r characterize the gravitational time delay. R_{AB} in Equations (7) and (8) comes from Equation (5). Henceforth, A corresponds to the emission and B corresponds to the reception.

In general, the time transfer function depends on either the emission time t_A or reception time t_B, and this dependence feature is applied to obtain the gravitational time delay in the McVittie spacetime [31]. However, if the spacetime is static, the first order formulae reduce to

$$\Delta^{(1)}(\vec{x}_A, \vec{x}_B) = -\frac{R_{AB}}{2}\int_0^1 \left[g_{(1)}^{00} - 2N_{AB}^i g_{(1)}^{0i} + N_{AB}^i N_{AB}^j g_{(1)}^{ij} \right] d\mu \tag{9}$$

where $\vec{N}_{AB} = N_{AB}^i = (x_B^i - x_A^i)/R_{AB}$. The above equation is integrated along the parameter equation $\vec{x}(\mu) = \vec{x}_A + \mu(\vec{x}_B - \vec{x}_A)$ on the Minkowskian spacetime. From Equation (9), the time delay is calculated with the remaining form of the metric $g_{\mu\nu}$, though the weak-field approximation is presumed.

Once the time transfer function $\mathcal{T}$ is determined, the direction of the light ray can be obtained by

$$(k_0)_A = -1, \quad (k_i)_A = -c\frac{\partial \mathcal{T}}{\partial x_A^i} \tag{10}$$

$$(k_0)_B = -1, \quad (k_i)_B = c\frac{\partial \mathcal{T}}{\partial x_B^i} \tag{11}$$

Equations (10) and (11) enable us to calculate light deflection directly from the time transfer function $\mathcal{T}$.

We note that Equations (9)–(11) have an opposite sign with respect to the corresponding equations given in [28,29], as we now adopt the signature of Minkowski metric as $(-,+,+,+)$ and because of which, the time transfer function should essentially be a positive value, $\mathcal{T} > 0$.

3. Effect of the Cosmological Constant on Light Deflection

Now, let us revisit the contribution of Λ to the light deflection with consideration of the time transfer function $\mathcal{T}$. To this end, we adopt the Schwarzschild–de Sitter or Kottler metric [32];

$$ds^2 = -\left(1 - \frac{r_g}{r} - \frac{\Lambda}{3}r^2\right)c^2 dt^2 + \left(1 - \frac{r_g}{r} - \frac{\Lambda}{3}r^2\right)^{-1} dr^2 + r^2 d\Omega^2$$

$$= -\left(1 - \frac{r_g}{r} - \frac{\Lambda}{3}r^2\right)c^2 dt^2$$

$$+ \left(1 + \frac{r_g}{r} + \frac{\Lambda}{3}r^2 + \mathcal{O}(r_g^2, \Lambda^2)\right) dr^2 + r^2 d\Omega^2 \tag{12}$$

where $r_g = 2GM/c^2$ is the Schwarzschild radius, $d\Omega^2 = d\theta^2 + \sin^2\theta d\phi^2$, and the dr^2 component is linearized from the first line to the second.

Here, we consider the validity and limitation of weak-field approximation supposed in Equation (12), The bending of light due to the point mass M is characterized by r_g/r; on the other hand, the bending due to the cosmological constant Λ is derived from $\Lambda r^2/3$ term. Therefore, it may be suitable to estimate the validity of approximation by

$$\frac{\Lambda}{3} r^2 < \frac{r_g}{b}, \quad \frac{r_g}{b} \ll 1 \tag{13}$$

where b is the impact parameter. Then, r should range $b < r < d$ and d is estimated from the relation $r_g/b \sim \Lambda d^2/3$. As an example of a deflector or lens object, let us choose the Sun ($M \approx M_\odot = 2.0 \times 10^{30}$ [kg], $b \approx R_\odot = 7.0 \times 10^8$ [m]) and the galaxy ($M \approx 10^{12} M_\odot$, $b \approx R_{\text{galaxy}} \approx 10^5$ [ly]); it is found that $d \sim 10^{23}$ [m] ~ 10 [Mpc] in both cases (we assumed $\Lambda \approx 10^{-52}$ [m^{-2}]). This value is comparable with the distance from our galaxy to the Virgo Cluster but one or two orders of magnitude smaller than the distance from our galaxy to quasars, the typical range of which is from 100 [Mpc] to 1000 [Mpc].

It is beneficial to transform the spherical coordinates into rectangular ones since it is easy to set up the rectilinear line as the first approximation of the light path (straight line in flat spacetime). However, it is difficult to transform the standard Schwarzschild–de Sitter/Kottler metric into the isotropic form; hence, employing the approach used in [33], we recast Equation (12) in rectangular form. By the coordinate transformation,

$$x = r\sin\theta\cos\phi, \quad y = r\sin\theta\sin\phi, \quad z = r\cos\theta \tag{14}$$

Equation (12) is rewritten as

$$
\begin{aligned}
ds^2 \;=\; & -\left(1 - \frac{r_g}{r} - \frac{\Lambda}{3}r^2\right) c^2 dt^2 \\
& + \left[\delta_{ij} + \left(\frac{r_g}{r} + \frac{\Lambda}{3}r^2\right)\frac{x^i x^j}{r^2} + \mathcal{O}(r_g^2, \Lambda^2)\right] dx^i dx^j
\end{aligned}
\tag{15}
$$

in which indices i, j run from 1 to 3 (spatial coordinates). We presume that the light travels in x-y plane; that is, $\vec{x}_A = (x_A, y_A)$, $\vec{x}_B = (x_B, y_B)$, and from Equations (9) and (15), the time transfer function $\mathcal{T}(\vec{x}_A, \vec{x}_B)$ can be obtained as

$$\mathcal{T} = \frac{1}{c}(R_{AB} + \Delta\mathcal{T}) \tag{16}$$

$$
\begin{aligned}
\Delta\mathcal{T} =\; & \frac{r_g}{2} \ln \frac{R_B + \vec{x}_B \cdot \vec{N}_{AB}}{R_A + \vec{x}_A \cdot \vec{N}_{AB}} \\
& + \frac{1}{2}\left[(N^x_{AB})^2(x_B - x_A)^2 + 2N^x_{AB}N^y_{AB}(x_B - x_A)(y_B - y_A) + (N^y_{AB})^2(y_B - y_A)^2\right] \\
& \times \left\{ r_g \left[\frac{1}{R^2_{AB}} \ln \frac{R_B + \vec{x}_B \cdot \vec{N}_{AB}}{R_A + \vec{x}_A \cdot \vec{N}_{AB}} \right.\right. \\
& \left.\left. - \frac{(\vec{x}_A \cdot \vec{N}_{AB})[2\vec{x}_A \cdot \vec{x}_B - R_A(R_A + R_B)] - R_{AB} R^2_A}{R_B\{[\vec{x}_A \cdot (\vec{x}_B - \vec{x}_A)]^2 - R^2_{AB}R^2_A\}} \right] + \frac{\Lambda}{9} \right\} \\
& + \left[(N^x_{AB})^2 x_A(x_B - x_A) + 2N^x_{AB}N^y_{AB}(x_B y_A + x_A y_B - 2x_A y_A) + (N^y_{AB})^2 y_A(y_B - y_A)\right] \\
& \times R_{AB}\left[r_g \frac{\vec{x}_A \cdot (\vec{x}_B - \vec{x}_A) R_A(R_B - R_A)}{R_B\{[\vec{x}_A \cdot (\vec{x}_B - \vec{x}_A)]^2 - R^2_{AB}R^2_A\}} + \frac{\Lambda}{6} \right] \\
& + \left[(N^x_{AB})^2 x_A^2 + 2N^x_{AB}N^y_{AB}x_A y_A + (N^y_{AB})^2 y_A^2\right] \\
& \times \frac{R_{AB}}{2}\left[r_g \frac{(R_B - R_A)[\vec{x}_A \cdot (\vec{x}_B - \vec{x}_A)] - R^2_{AB}R_A}{R_A R_B\{[\vec{x}_A \cdot (\vec{x}_B - \vec{x}_A)]^2 - R^2_{AB}R^2_A\}} + \frac{\Lambda}{3} \right]
\end{aligned}
\tag{17}
$$

in which $R_A = |\vec{x}_A|$, $R_B = |\vec{x}_B|$, $\vec{N}_{AB} = (N_{AB}^x, N_{AB}^y)$. The slightly complicated expression in Equation (17) originates from $g_{(1)}^{ij}$ in Equation (15). We are interested in how Λ modulates the total deflection angle in the Schwarzschild case, $\alpha_{GR} = 4GM/(c^2 b)$, where b is the impact parameter. Then, in order to extract the influence of Λ on the bending angle of light rays, let us re-define the coordinate system in such a way that the emission point A and the reception point B have the same value of the y coordinate, namely, $y_A = y_B = b$. Further, let us assume that the source and the observer are at rest with respect to the lens (deflector), the light is emitted at x_A and received at x_B and that $x_A < x_B$ holds, then $N_{AB}^x = 1$, $|\vec{x}_B - \vec{x}_A| = x_B - x_A > 0$, and so on. Then, Equation (17) reduces to a simple form,

$$\mathcal{T} = \frac{1}{c}(R_{AB} + \Delta\mathcal{T}_{GR} + \Delta\mathcal{T}_\Lambda) \tag{18}$$

$$\Delta\mathcal{T}_{GR} = \frac{GM}{c^2}\left[2\ln\frac{x_B + \sqrt{x_B^2 + b^2}}{x_A + \sqrt{x_A^2 + b^2}} - \left(\frac{x_B}{\sqrt{x_B^2 + b^2}} - \frac{x_A}{\sqrt{x_A^2 + b^2}}\right)\right] \tag{19}$$

$$\Delta\mathcal{T}_\Lambda = \frac{\Lambda}{18}\left[2(x_B^3 - x_A^3) + 3b^2(x_B - x_A)\right] \tag{20}$$

From Equations (18)–(20), the direction of light at the emission point A and reception point B are computed using Equations (10) and (11).

Since we are now choosing the emission point A and reception point B as being located upon the line $y = b$ and $x_A < x_B$, then $R_{AB} = |\vec{x}_B - \vec{x}_A| = x_B - x_A$, $\vec{N}_{AB} = ((x_B - x_A)/R_{AB}, 0) = (1, 0)$; the photon may travel along this straight-line with the minimum value of the coordinate $r = b$ (the impact parameter) if the light ray were un-deflected in the absence of central mass M and cosmological constant Λ. Thus, let us define the angle θ_A as that between $\vec{N}_{AB}$ and $\vec{k}_A$ and angle θ_B as that between $\vec{N}_{AB}$ and $\vec{k}_B$. Further, we suppose that θ_A and θ_B have a small value, $\theta_A \ll 1$, $\theta_B \ll 1$, then the direction vectors $\vec{k}_A$ and $\vec{k}_B$ can be expressed by the following form

$$\vec{k}_A = \begin{pmatrix} \cos\theta_A \\ \sin\theta_A \end{pmatrix} \simeq \begin{pmatrix} 1 \\ \theta_A \end{pmatrix} = \vec{N}_{AB} + \delta\vec{k}_A = \begin{pmatrix} 1 \\ 0 \end{pmatrix} + \begin{pmatrix} \delta k_{xA} \\ \delta k_{yA} \end{pmatrix} \tag{21}$$

$$\vec{k}_B = \begin{pmatrix} \cos\theta_B \\ \sin\theta_B \end{pmatrix} \simeq \begin{pmatrix} 1 \\ \theta_B \end{pmatrix} = \vec{N}_{AB} + \delta\vec{k}_B = \begin{pmatrix} 1 \\ 0 \end{pmatrix} + \begin{pmatrix} \delta k_{xB} \\ \delta k_{yB} \end{pmatrix} \tag{22}$$

Hence, we may obtain θ_A and θ_B from the y components of $\delta\vec{k}_A$ and $\delta\vec{k}_B$, namely, δk_{yA} and δk_{yB}, respectively.

Let us take the deflection angle α in such a way that $\alpha > 0$. As a consequence, the deflection angle α is given by

$$\alpha \equiv \theta_A - \theta_B = \alpha_{GR} + \alpha_\Lambda + \mathcal{O}(r_g^2, \Lambda^2) \tag{23}$$

$$\alpha_{GR} = \frac{GM}{c^2}b\left(\frac{2}{x_A\sqrt{x_A^2 + b^2} + x_A^2 + b^2} - \frac{2}{x_B\sqrt{x_B^2 + b^2} + x_B^2 + b^2}\right.$$

$$\left. + \frac{x_A}{\sqrt{x_A^2 + b^2}^{\,3}} - \frac{x_B}{\sqrt{x_B^2 + b^2}^{\,3}}\right) \tag{24}$$

$$\alpha_\Lambda = -\frac{2\Lambda}{3}b(x_B - x_A) \tag{25}$$

Again, we note $x_B > x_A$ in our case. The Equation (25) is similar and comparable with previous results, that is, the third term of Equation (13) in [20], and the fourth term of Equation (25) in [22].

The reason why $\mathcal{O}(m\Lambda)$ term disappear in our result comes from the fact that present calculation is first (linear) order with respect to $\epsilon \sim r_g \sim \Lambda$, see Equations (9) and (15). If we extend to second order $\mathcal{O}(\epsilon^2)$, $\mathcal{O}(m\Lambda)$ terms appear. See, e.g., Equation (41) in [29].

In the case of Schwarzschild spacetime, the total deflection angle α_{GR} is obtained for the limit $r \to \infty$; however, in the case of Schwarzschild–de Sitter/Kottler spacetime, we cannot impose this limit since the term $(1 - r_g/r - \Lambda r^2/3)^{-1}dr^2$ in Equation (12) diverges at $r = \sqrt{3/\Lambda}$ and the coordinate value r does not range $r > \sqrt{3/\Lambda}$ (here we assume $r_g/r \ll 1$). Then we shall define the total deflection angle due to Λ, α_Λ, in such a way that $x_A = -\sqrt{3/\Lambda}$ and $x_B = \sqrt{3/\Lambda}$. Hence, inserting these values into Equation (25), we have,

$$\alpha_\Lambda = -\frac{4\sqrt{3\Lambda}}{3}b \tag{26}$$

We note that the transformation from coordinate distance into angular distance is discussed, e.g., in [34].

It is worthwhile to show that Equation (24) can result in $\alpha_{GR} = 4GM/(c^2 b)$ when $\Lambda = 0$. Equation (24) is rewritten as

$$\alpha_{GR} = \frac{GM}{c^2 b}\left[2(\cos\phi_B - \cos\phi_A) + \sin^2\phi_B\cos\phi_B - \sin^2\phi_A\cos\phi_A\right] \tag{27}$$

where we introduced

$$\sin\phi_A = \frac{b}{\sqrt{x_A^2 + b^2}}, \quad \cos\phi_A = \frac{x_A}{\sqrt{x_A^2 + b^2}} \tag{28}$$

$$\sin\phi_B = \frac{b}{\sqrt{x_B^2 + b^2}}, \quad \cos\phi_B = \frac{x_B}{\sqrt{x_B^2 + b^2}} \tag{29}$$

For $\phi_A \to \pi$ (the emission point A is located at $-\infty$) and $\phi_B \to 0$ (the reception point B is located at $+\infty$), Equation (29) gives

$$\alpha_{GR} = \frac{4GM}{c^2 b} \tag{30}$$

thus replicating the light deflection in the Schwarzschild case.

It should be mentioned that the time transfer function Equation (9), which is used to determine the defection, is justified as long as the zeroth-order straight line that joins $\vec{x}_A$ and $\vec{x}_B$ does not intersect an event horizon such as the Schwarzschild horizon. This implies that $|\phi_B - \phi_A| < \pi$ is a necessary condition to apply the method. However, this condition may be violated if $\vec{x}_A$ and/or $\vec{x}_B$ are sufficiently far from the mass center. To avoid this difficulty, it may be a much more satisfactory procedure to introduce the periapsis $\vec{x}_P$ of the light ray, calculate the defection angle between $\vec{x}_A$ and $\vec{x}_P$ as well as that between $\vec{x}_P$ and $\vec{x}_B$, and finally add these two contributions.

4. Summary

We revisited the effect of the cosmological constant Λ on light deflection by means of the Schwarzschild–de Sitter or Kottler metric. To obtain the deflection angle α, we adopted the time transfer function approach, instead of solving the geodesic equation of photon. We showed that the cosmological constant appears in the deflection angle α, and it diminishes the light bending due to the mass of the central body M.

We list in Table 1 the expressions of bending angle due to the cosmological constant previously obtained [14–25], and estimate the numerical value using $c = 3.0 \times 10^8$ [m/s], $G = 6.674 \times 10^{-11}$ [m³ · kg⁻¹ · s⁻²], Mass of galaxy $M \approx 10^{12} M_\odot = 2.0 \times 10^{42}$ [kg], $\Lambda \approx 10^{-52}$ [m⁻²],

$b, R, r_0, B \sim 10^5$ [ly] $\sim 10^{21}$ [m] (typical radius of galaxy), $r_s, r_o, d_{OL}, d_{LS}, r_S, r_{obs}, x_B, -x_A \sim 10$ [Mpc] $\sim 10^{23}$ [m]. R_1 in [24] is calculated $1/R_1 = 2GM/c^2B^2 + 15\pi(GM)^2/8c^4B^3$. The underline indicates the leading order term.

Table 1. Comparison with previous results.. We estimate the numerical value using $c = 3.0 \times 10^8$ [m/s], $G = 6.674 \times 10^{-11}$ [m$^3 \cdot$ kg$^{-1} \cdot$ s^{-2}], Mass of galaxy $M \approx 10^{12} M_\odot = 2.0 \times 10^{42}$ [kg], $\Lambda \approx 10^{-52}$ [m^{-2}], $b, R, r_0, B \sim 10^5$ [ly] $\sim 10^{21}$ [m] (typical radius of galaxy), $r_s, r_o, d_{OL}, d_{LS}, r_S, r_{obs}, x_B, -x_A \sim 10$ [Mpc] $\sim 10^{23}$ [m]. R_1 in [24] is calculated $1/R_1 = 2GM/c^2B^2 + 15\pi(GM)^2/8c^4B^3$. The underline indicates the leading order term.

Authors	Deflection Due to Λ	Numerical Value [rad]
Rindle & Ishak [13,14]	$-\frac{c^2 \Lambda R^3}{6GM}$	-1.1×10^{-5}
Park [15]	Not contribute	-
Khriplovich & Pomeransky [16]	Not contribute	-
Sereno [17,18]	$+\frac{2GMb\Lambda}{3c^2} + \frac{b^3\Lambda}{6}\left(\frac{1}{r_s} + \frac{1}{r_o}\right)$	$+3.3 \times 10^{-13}$
Simpson *et al.* [19]	Not contribute	-
Bhadra *et al.* [20]	$\frac{2GM\Lambda b}{3c^2} - \frac{\Lambda b}{6}(d_{OL} + d_{LS}) + \frac{\Lambda b^3}{6}\left(\frac{1}{d_{OL}} + \frac{1}{d_{LS}}\right)$	-3.3×10^{-9}
Miraghaei *et al.* [21]	$-\sqrt{\frac{2\Lambda}{3}}R$	-8.2×10^{-6}
Biressa *et al.* [22]	$+\frac{2GMb\Lambda}{3c^2} - \frac{b\Lambda}{6}(r_S + r_{obs})$ $-\frac{b^3\Lambda}{12}\left(\frac{1}{r_S} + \frac{1}{r_{obs}}\right) + \frac{GMb^3\Lambda}{6c^2}\left(\frac{1}{r_S^2} + \frac{1}{r_{obs}^2}\right)$	-3.3×10^{-9}
Arakida & Kasai [23]	Not contribute	-
Hammad [24]	$-\frac{\sqrt{2}}{3}\Lambda\sqrt{\frac{GMR_1^3}{c^2}}$	-1.1×10^{-5}
Batic *et al.* [25]	$-\frac{2}{\sqrt{3}}r_0\sqrt{\Lambda} - \frac{2\sqrt{\Lambda}}{\sqrt{3}}\frac{2GM}{c^2} - \frac{\sqrt{3\Lambda}}{4}\frac{(2GM)^2}{c^4 r_0} - \frac{5\sqrt{\Lambda}}{8\sqrt{3}}\frac{(2GM)^3}{c^6 r_0^2}$	-1.1×10^{-5}
Present Paper	$-\frac{2\Lambda}{3}b(x_B - x_A)$	-1.3×10^{-8}

Our result seems to be similar and comparable with the third term of Equation (13) in [20], and the fourth term of Equation (25) in [22]. Also, as [13,14,20–22,24,25], the cosmological constant leads to diminishing the bending angle due to the mass of the central body M.

However, contrary to previous results such as [13,14,17,18,20,22,24,25], in our case the bending angle due to Λ does not couple with the mass of the central body M. As mentioned in Section 3, it comes from the fact that our calculation is first (linear) order with respect to $\epsilon \sim r_g \sim \Lambda$, (see Equations (9) and (15)), then if we extend to second order $\mathcal{O}(\epsilon^2)$, the coupling term $\mathcal{O}(m\Lambda)$ appears.

Acknowledgments: We would like to acknowledge anonymous referees for reading our manuscript carefully and for giving fruitful comments and suggestions, which significantly improved the quality of the manuscript. This work was supported by JSPS KAKENHI Grant Number 15K05089.

Conflicts of Interest: The authors declare no conflict of interest.

References

1. Weinberg, S. The cosmological constant problem. *Rev. Mod. Phys.* **1989**, *61*, 1–23.
2. Carroll, S.M. The Cosmological Constant. *Living Rev. Relativ.* **2001**, *4*, doi:10.12942/lrr-2001-1.
3. Riess, A.G.; Filippenko, A.V.; Challis, P.; Clocchiatti, A.; Diercks, A.; Garnavich, P.M.; Gilliland, R.L.; Hogan, C.J.; Jha, S.; Kirshner, R.P.; *et al.* Observational Evidence from Supernovae for an Accelerating Universe and a Cosmological Constant. *Astron. J.* **1998**, *116*, 1009–1038.
4. Schmidt, B.P.; Suntzeff, N.B.; Phillips, M.M.; Schommer, R.A.; Clocchiatti, A.; Kirshner, R.P.; Garnavich, P.; Challis, P.; Leibundgut, B.; Spyromilio, J.; *et al.* The High-Z Supernova Search: Measuring Cosmic Deceleration and Global Curvature of the Universe Using Type Ia Supernovae. *Astrophys. J.* **1998**, *507*, 46–63.
5. Perlmutter, S.; Aldering, G.; Goldhaber, G.; Knop, R.A.; Nugent, P.; Castro, P.G.; Deustua, S.; Fabbro, S.; Goobar, A.; Groom, D.E.; *et al.* The Supernova Cosmology, Measurements of Ω and Λ from 42 High-Redshift Supernovae. *Astrophys. J.* **1999**, *517*, 565–586.
6. Kerr, A.W.; Hauck, J.C.; Mashhoon, B. Standard clocks, orbital precession and the cosmological constant. *Class. Quant. Grav.* **2003**, *20*, 2727–2736.

7. Rindler, W. *Relativity: Special, General, and Cosmological.* 2nd ed.; Oxford University Press: New York, NY, USA, 2006.
8. Iorio, L. Solar System Motions and the Cosmological Constant: A New Approach. *Adv. Astron.* **2008**, *2008*, 268647.
9. Fienga, A.; Laskar, J.; Kuchynka, P.; Manche, H.; Desvignes, G.; Gastineau, M.; Cognard I.; Theureau, G. The INPOP10a planetary ephemeris and its applications in fundamental physics. *Celest. Mech. Dyn. Astron.* **2011**, *111*, 363–385.
10. Pitjeva, E.V.; Pitjev, N.P. Relativistic effects and dark matter in the Solar system from observations of planets and spacecraft. *Mon. Not. R. Astron. Soc.* **2013**, *432*, 3431–3437.
11. Iorio, L. Gravitational Anomalies in the Solar System? *Int. J. Mod. Phys. D* **2015**, *24*, 1530015.
12. Jetzer, P.; Sereno, M. Solar system tests of the cosmological constant. *Il Nuovo Cimento B* **2007**, *122*, 489–498.
13. Rindler, W.; Ishak, M. Contribution of the cosmological constant to the relativistic bending of light revisited. *Phys. Rev. D* **2007**, *76*, 043006.
14. Ishak, M.; Rindler, W. The relevance of the cosmological constant for lensing. *Gen. Relativ. Gravit.* **2010**, *42*, 2247–2268.
15. Park, M. Rigorous approach to gravitational lensing. *Phys. Rev. D* **2008**, *78*, 023014.
16. Khriplovich, I.B.; Pomeransky, A.A. Does the cosmological term influence gravitational lensing? *Int. J. Mod. Phys. D* **2008**, *17*, 2255–2259.
17. Sereno, M. Influence of the cosmological constant on gravitational lensing in small systems. *Phys. Rev. D* **2008**, *77*, 043004.
18. Sereno, M. Role of Λ in the Cosmological Lens Equation *Phys. Rev. Lett.* **2009**, *102*, 021301.
19. Simpson, F.; Peacock, J.A.; Heavens, A.F. On lensing by a cosmological constant. *Mon. Not. R. Astron. Soc.* **2010**, *402*, 2009–2016.
20. Bhadra, A.; Biswas, S.; Sarkar, K. Gravitational deflection of light in the Schwarzschild–de Sitter space-time. *Phys. Rev. D* **2010**, *82*, 063003.
21. Miraghaei, H.; Nouri-Zonoz, M. Classical tests of general relativity in the Newtonian limit of the Schwarzschild–de Sitter spacetime. *Gen. Relativ. Gravit.* **2010**, *42*, 2947–2956.
22. Biressa, T.; de Freitas Pacheco, J.A. The cosmological constant and the gravitational light bending. *Gen. Relativ. Gravit.* **2011**, *43*, 2649–2659.
23. Arakida, H.; Kasai, M. Effect of the cosmological constant on the bending of light and the cosmological lens equation. *Phys. Rev. D* **2012**, *85*, 023006.
24. Hammad, F. A Note on the Effect of the Cosmological Constant on the Bending of Light. *Mod. Phys. Lett. A* **2013**, *28*, 1350181.
25. Batic, D.; Nelson, S.; Nowakowski, M. Light on curved backgrounds. *Phys. Rev. D* **2015**, *91*, 104015.
26. Ishak, M.; Rindler, W.; Dossett, J. More on lensing by a cosmological constant. *Mon. Not. R. Astron. Soc.* **2010**, *403*, 2152–2156.
27. Ghosh, S.; Bhadra, A. Influences of dark energy and dark matter on gravitational time advancement. *Eur. Phys. J. C* **2015**, *75*, 1–6.
28. Le Poncin-Lafitte, C.; Linet, B.; Teyssandier, P. World function and time transfer: General post-Minkowskian expansions. *Class. Quant. Gravit.* **2004**, *21*, 4463–4483.
29. Teyssandier, P.; Le Poncin-Lafitte, C. General post-Minkowskian expansion of time transfer functions. *Class. Quant. Gravit.* **2008**, *25*, 145020.
30. Synge, J.L. *Relativity: The General Theory*; North-Holland: Amsterdam, The Netherlands, 1964.
31. Arakida, H. Application of time transfer function to McVittie spacetime: Gravitational time delay and secular increase in astronomical unit. *Gen. Relativ. Gravit.* **2011**, *43*, 2127–2139.
32. Kottler, F. Über die physikalischen Grundlagen der Einsteinschen Gravitationstheorie. *Ann. Phys.* **1918**, *361*, 401–462.
33. Brumberg, V.A. *Essential Relativistic Celestial Mechanics*; Adam Hilger: Bristol, UK, 1991.
34. Schücker, T. Cosmological constant and lensing. *Gen. Relativ. Gravit.* **2007**, *41*, 66–75.

universe

MDPI

Article

Charged and Electromagnetic Fields from Relativistic Quantum Geometry

Marcos R. A. Arcodía [2] and Mauricio Bellini [1,2,]*

[1] Departamento de Física, Facultad de Ciencias Exactas y Naturales, Universidad Nacional de Mar del Plata, Funes 3350, C.P. 7600 Mar del Plata, Argentina
[2] Instituto de Investigaciones Físicas de Mar del Plata (IFIMAR), Consejo Nacional de Investigaciones Científicas y Técnicas (CONICET), C.P. 7600 Mar del Plata, Argentina; marcodia@mdp.edu.ar
[*] Correspondence: mbellini@mdp.edu.ar

Academic Editors: Lorenzo Iorio and Elias C. Vagenas
Received: 3 May 2016; Accepted: 7 June 2016; Published: 21 June 2016

Abstract: In the recently introduced Relativistic Quantum Geometry (RQG) formalism, the possibility was explored that the variation of the tensor metric can be done in a Weylian integrable manifold using a geometric displacement, from a Riemannian to a Weylian integrable manifold, described by the dynamics of an auxiliary geometrical scalar field θ, in order that the Einstein tensor (and the Einstein equations) can be represented on a Weyl-like manifold. In this framework we study jointly the dynamics of electromagnetic fields produced by quantum complex vector fields, which describes charges without charges. We demonstrate that complex fields act as a source of tetra-vector fields which describe an extended Maxwell dynamics.

Keywords: Relativistic Quantum Geometry; Quantum Complex Fields

1. Introduction

The consequences of non-trivial topology for the laws of physics has been a topic of perennial interest for theoretical physicists [1], with applications to non-trivial spatial topologies [2] like Einstein-Rosen bridges, wormholes, non-orientable spacetimes, and quantum-mechanical entanglements.

Geometrodynamics [3,4] is a picture of general relativity that studies the evolution of the spacetime geometry. The key notion of the Geometrodynamics was the idea of *charge without charge*. The Maxwell field was taken to be source free, and so a non-vanishing charge could only arise from an electric flux line trapped in the topology of spacetime. With the construction of ungauged supergravity theories it was realised that the Abelian gauge fields in such theories were source-free, and so the charges arising therein were therefore central charges [5] and as consequence satisfied a BPS bound [6] where the embedding of Einstein-Maxwell theory into $N = 2$ supergravity theory was used. The significant advantages of geometrodynamics, usually come at the expense of manifest local Lorentz symmetry [7]. During the 70s and 80s decades a method of quantization was developed in order to deal with some unresolved problems of quantum field theory in curved spacetimes [8–10].

In a previous work [11] the possibility was explored that the variation of the tensor metric must be done in a Weylian integrable manifold using a geometric displacement, from a Riemannian to a Weylian integrable manifold, described by the dynamics of an auxiliary geometrical scalar field θ, in order that the Einstein tensor (and the Einstein equations) can be represented on a Weyl-like manifold. An important fact is that the Einstein tensor complies with the gauge-invariant transformations studied in a previous work [12]. This method is very useful because can be used to describe, for instance, nonperturbative back-reaction effects during inflation [13]. Furthermore, the relativistic quantum dynamics of θ was introduced by using the fact that the cosmological constant Λ is a relativistic invariant. In this letter, we extend our study to complex charged fields that act as the source of vector fields A^μ.

2. RQG Revisited

The first variation of the Einstein-Hilbert (EH) action $\mathcal{I}$ (Here, g is the determinant of the covariant background tensor metric $g_{\mu\nu}$, $R = g^{\mu\nu} R_{\mu\nu}$ is the scalar curvature, $R^\alpha_{\mu\nu\alpha} = R_{\mu\nu}$ is the covariant Ricci tensor and $\mathcal{L}_m$ is an arbitrary Lagrangian density which describes matter. If we deal with an orthogonal base, the curvature tensor will be written in terms of the connections: $R^\alpha_{\beta\gamma\delta} = \Gamma^\alpha_{\beta\delta,\gamma} - \Gamma^\alpha_{\beta\gamma,\delta} + \Gamma^\epsilon_{\beta\delta}\Gamma^\alpha_{\epsilon\gamma} - \Gamma^\epsilon_{\beta\gamma}\Gamma^\alpha_{\epsilon\delta}$).

$$\mathcal{I} = \int_V d^4x \sqrt{-g} \left[\frac{R}{2\kappa} + \mathcal{L}_m \right] \tag{1}$$

is given by

$$\delta\mathcal{I} = \int d^4x \sqrt{-g} \left[\delta g^{\alpha\beta} \left(G_{\alpha\beta} + \kappa T_{\alpha\beta} \right) + g^{\alpha\beta} \delta R_{\alpha\beta} \right] \tag{2}$$

where $\kappa = 8\pi G$, G is the gravitational constant and $g^{\alpha\beta}\delta R_{\alpha\beta} = \nabla_\alpha \delta W^\alpha$, where $\delta W^\alpha = \delta\Gamma^\alpha_{\beta\gamma}g^{\beta\gamma} - \delta\Gamma^\epsilon_{\beta\epsilon}g^{\beta\alpha} = g^{\beta\gamma}\nabla^\alpha\delta\Psi_{\beta\gamma}$. When the flux of δW^α that cross the Gaussian-like hypersurface defined in an arbitrary region of the spacetime, is nonzero, one obtains in the last term of Equation (2), that $\nabla_\alpha \delta W^\alpha = \delta\Phi(x^\alpha)$, such that $\delta\Phi(x^\alpha)$ is an arbitrary scalar field that takes into account the flux of δW^α across the Gaussian-like hypersurface. This flux becomes zero when there are no sources inside this hypersurface. Hence, in order to make $\delta\mathcal{I} = 0$ in Equation (2), we must consider the condition: $G_{\alpha\beta} + \kappa T_{\alpha\beta} = \Lambda g_{\alpha\beta}$, where Λ is the cosmological constant. Additionally, we must require the constriction $\delta g_{\alpha\beta}\Lambda = \delta\Phi g_{\alpha\beta}$. Then, we propose the existence of a tensor field $\delta\Psi_{\alpha\beta}$, such that $\delta R_{\alpha\beta} \equiv \nabla_\beta \delta W_\alpha - \delta\Phi g_{\alpha\beta} \equiv \Box\delta\Psi_{\alpha\beta} - \delta\Phi g_{\alpha\beta} = -\kappa\delta S_{\alpha\beta}$ (We have introduced the tensor $S_{\alpha\beta} = T_{\alpha\beta} - \frac{1}{2}T g_{\alpha\beta}$, which takes into account matter as a source of the Ricci tensor $R_{\alpha\beta}$), and hence $\delta W^\alpha = g^{\beta\gamma}\nabla^\alpha\delta\Psi_{\beta\gamma}$, with $\nabla^\alpha\delta\Psi_{\beta\gamma} = \delta\Gamma^\alpha_{\beta\gamma} - \delta^\alpha_\gamma\delta\Gamma^\epsilon_{\beta\epsilon}$. *Notice that the fields δW_α and $\delta\Psi_{\alpha\beta}$ are gauge-invariant under transformations:*

$$\delta\bar{W}_\alpha = \delta W_\alpha - \nabla_\alpha\delta\Phi, \qquad \delta\bar{\Psi}_{\alpha\beta} = \delta\Psi_{\alpha\beta} - \delta\Phi g_{\alpha\beta} \tag{3}$$

where the scalar field $\delta\Phi$ complies $\Box\delta\Phi = 0$. On the other hand, we can make the transformation

$$\bar{G}_{\alpha\beta} = G_{\alpha\beta} - \Lambda g_{\alpha\beta} \tag{4}$$

and the transformed Einstein equations with the equation of motion for the transformed gravitational waves, hold

$$\bar{G}_{\alpha\beta} = -\kappa T_{\alpha\beta}, \tag{5}$$

$$\Box\delta\bar{\Psi}_{\alpha\beta} = -\kappa\delta S_{\alpha\beta} \tag{6}$$

with $\Box\delta\Phi(x^\alpha) = 0$ and $\delta\Phi(x^\alpha) g_{\alpha\beta} = \Lambda \delta g_{\alpha\beta}$. The Equation (5) provides us the Einstein equations with cosmological constant included, and Equation (6) describes the exact equation of motion for gravitational waves with an arbitrary source $\delta S_{\alpha\beta}$ on a closed and curved space-time. A very important fact is that the scalar field $\delta\Phi(x^\alpha)$ appears as a scalar flux of the tetra-vector with components δW^α through the closed hypersurface $\partial\mathcal{M}$. This arbitrary hypersurface encloses the manifold by down and must be viewed as a 3D Gaussian-like hipersurface situated in any region of space-time. This scalar flux is a gravitodynamic potential related to the gauge-invariance of δW^α and the gravitational waves $\delta\bar{\Psi}_{\alpha\beta}$. Another important fact is that since $\delta\Phi(x^\alpha) g_{\alpha\beta} = \Lambda \delta g_{\alpha\beta}$, the existence of the Hubble horizon is related to the existence of the Gaussian-like hypersurface. The variation of the metric tensor must be done in a Weylian integrable manifold [11] using an auxiliary geometrical scalar field θ, in order to the Einstein tensor (and the Einstein equations) can be represented on a Weyl-like manifold, in agreement with the gauge-invariant transformations Equation (3). If we consider a zero covariant derivative of the metric tensor in the Riemannian manifold (we denote with ";" the Riemannian-covariant derivative): $\Delta g_{\alpha\beta} = g_{\alpha\beta;\gamma} \, dx^\gamma = 0$, hence the Weylian covariant derivative $g_{\alpha\beta|\gamma} = \theta_\gamma \, g_{\alpha\beta}$, described with respect to the Weylian connections (To simplify the notation we shall denote $\theta_\alpha \equiv \theta_{,\alpha}$).

$$\Gamma^\alpha_{\beta\gamma} = \left\{ \begin{matrix} \alpha \\ \beta\,\gamma \end{matrix} \right\} + g_{\beta\gamma}\theta^\alpha \tag{7}$$

will be nonzero

$$\delta g_{\alpha\beta} = g_{\alpha\beta|\gamma} \, dx^\gamma = - \left[\theta_\beta g_{\alpha\gamma} + \theta_\alpha g_{\beta\gamma} \right] dx^\gamma \tag{8}$$

2.1. Gauge-Invariance and Quantum Dynamics

From the action's point of view, the scalar field $\theta(x^\alpha)$ is a generic geometrical transformation that leaves invariant the action

$$\mathcal{I} = \int d^4x \sqrt{-\hat{g}} \left[\frac{\hat{R}}{2\kappa} + \mathcal{L} \right] = \int d^4x \left[\sqrt{-\hat{g}}e^{-2\theta} \right] \left\{ \left[\frac{\hat{R}}{2\kappa} + \mathcal{L} \right] e^{2\theta} \right\} \tag{9}$$

where we shall denote with a hat, $\hat{}$, the quantities represented on the Riemannian manifold. Hence, Weylian quantities will be varied over these quantities in a Riemannian manifold so that the dynamics of the system preserves the action: $\delta\mathcal{I} = 0$, and we obtain

$$-\frac{\delta V}{V} = \frac{\delta\left[\frac{\hat{R}}{2\kappa} + \mathcal{L} \right]}{\left[\frac{\hat{R}}{2\kappa} + \mathcal{L} \right]} = 2\,\delta\theta \tag{10}$$

where $\delta\theta = -\theta_\mu dx^\mu$ is an exact differential and $V = \sqrt{-\hat{g}}$ is the volume of the Riemannian manifold. Of course, all the variations are in the Weylian geometrical representation, and assure us gauge invariance because $\delta\mathcal{I} = 0$. Using the fact that the tetra-length is given by $S = \frac{1}{2}x_\nu \hat{U}^\nu$ and the Weylian velocities are given by

$$u^\mu = \hat{U}^\mu + \theta^\mu \left(x_\epsilon \hat{U}^\epsilon \right) \tag{11}$$

can be demonstrated that

$$u^\mu u_\mu = 1 + 4S \left(\theta_\mu \hat{U}^\mu - \frac{4}{3}\Lambda S \right) \tag{12}$$

The components u^μ are the relativistic quantum velocities, given by the geodesic equations

$$\frac{du^\mu}{dS} + \Gamma^\mu_{\alpha\beta}u^\alpha u^\beta = 0 \tag{13}$$

such that the Weylian connections $\Gamma^\mu_{\alpha\beta}$ are described by Equation (7). In other words, the quantum velocities u^μ are transported with parallelism on the Weylian manifold, meanwhile $\hat{U}^\mu$ are transported with parallelism on the Riemann manifold. The quantum velocities u^μ (given by Equation (11)), must be considered as nondeterministic because they depend on θ^μ, so that the only quantity that has classical sense is its quantum expectation value on the classical Riemannian background manifold:

$$\langle B|u^\mu|B \rangle = \hat{U}^\mu + \langle B|\theta^\mu|B \rangle \left(x_\epsilon \hat{U}^\epsilon \right) \tag{14}$$

If we require that $u^\mu u_\mu = 1$, we obtain the gauge

$$\hat{\nabla}_\mu A^\mu = 4\frac{d\Phi}{dS} = \frac{2}{3}\Lambda^2 S(x^\mu) \tag{15}$$

where A^μ is given by [11,12]

$$A^\mu = \frac{\delta W^\alpha}{\delta S} = \frac{\delta \Gamma^\alpha_{\beta\gamma}}{\delta S} g^{\beta\gamma} - \frac{\delta \Gamma^\epsilon_{\beta\epsilon}}{\delta S} g^{\beta\alpha} \tag{16}$$

Hence, we obtain the important result

$$d\Phi = \frac{1}{6}\Lambda^2 S\, dS \tag{17}$$

or, after integrating

$$\Phi(x^\mu) = \frac{\Lambda^2}{12} S^2(x^\mu) \tag{18}$$

such that $d\Phi(x^\mu) = -\frac{\Lambda}{2} d\theta(x^\mu)$. Hence, from Equation (9) we obtain that the quantum volume is given by

$$V_q = \sqrt{-\hat{g}}\, e^{-2\theta} = \sqrt{-\hat{g}}\, e^{\frac{1}{3}\Lambda S^2} \tag{19}$$

where $\Lambda S^2 > 0$. This means that $V_q \geq \sqrt{-\hat{g}}$, for $S^2 \geq 0$, $\Lambda > 0$ and $\theta < 0$. This implies a signature for the metric: $(-,+,+,+)$ in order for the cosmological constant to be positive and a signature $(+,-,-,-)$ in order to have $\Lambda \leq 0$. Finally, the action Equation (9) can be rewritten in terms of both quantum volume and the quantum Lagrangian density $\mathcal{L}_q = \left[\frac{\hat{R}}{2\kappa} + \hat{\mathcal{L}}\right] e^{2\theta}$

$$\mathcal{I} = \int d^4 x\, V_q\, \mathcal{L}_q \tag{20}$$

As was demonstrated in [11] the Einstein tensor can be written as

$$\bar{G}_{\alpha\beta} = \hat{G}_{\mu\nu} + \theta_{\alpha;\beta} + \theta_\alpha \theta_\beta + \frac{1}{2} g_{\alpha\beta} \left[(\theta^\mu)_{;\mu} + \theta_\mu \theta^\mu \right] \tag{21}$$

and we can obtain the invariant cosmological constant Λ

$$\Lambda = -\frac{3}{4} \left[\theta_\alpha \theta^\alpha + \hat{\Box}\theta \right] \tag{22}$$

so that we can define a geometrical Weylian quantum action $\mathcal{W} = \int d^4 x \sqrt{-\hat{g}}\, \Lambda$, such that the dynamics of the geometrical field, after imposing $\delta W = 0$, is described by the Euler-Lagrange equations which take the form

$$\hat{\nabla}_\alpha \Pi^\alpha = 0, \quad \text{or} \quad \hat{\Box}\theta = 0 \tag{23}$$

where the momentum components are $\Pi^\alpha \equiv -\frac{3}{4}\theta^\alpha$ and the relativistic quantum algebra is given by [11]

$$[\theta(x), \theta^\alpha(y)] = -i\Theta^\alpha\, \delta^{(4)}(x-y), \quad [\theta(x), \theta_\alpha(y)] = i\Theta_\alpha\, \delta^{(4)}(x-y) \tag{24}$$

with $\Theta^\alpha = i\hbar\, \hat{U}^\alpha$ and $\Theta^2 = \Theta_\alpha \Theta^\alpha = \hbar^2 \hat{U}_\alpha\, \hat{U}^\alpha$ for the Riemannian components of velocities $\hat{U}^\alpha$.

2.2. Charged Geometry and Vector Field Dynamics

In order to extend the previous study we shall consider that the scalar field θ is given by

$$\theta(x^\alpha) = \phi(x^\alpha)\, e^{-i\theta(x^\alpha)}, \quad \text{or} \quad \theta(x^\alpha) = \phi^*(x^\alpha)\, e^{i\theta(x^\alpha)} \tag{25}$$

where $\phi(x^\alpha)$ is a complex field and $\phi^*(x^\alpha)$ its complex conjugate. In this case, since $\theta^\alpha = e^{i\theta}\left(\hat{\nabla}^\alpha + i\theta^\alpha\right)\phi^*$, the Weylian connections hold

$$\Gamma^\alpha_{\beta\gamma} = \left\{ \begin{matrix} \alpha \\ \beta\,\gamma \end{matrix} \right\} + e^{i\theta} g_{\beta\gamma} \left(\hat{\nabla}^\alpha + i\theta^\alpha\right) \phi^* \equiv \left\{ \begin{matrix} \alpha \\ \beta\,\gamma \end{matrix} \right\} + g_{\beta\gamma}\, e^{i\theta} \left(D^\alpha \phi^*\right) \tag{26}$$

where we use the notation $D^{\alpha}\phi^* \equiv \left(\hat{\nabla}^{\alpha} + i\theta^{\alpha}\right)\phi^*$. The Weylian components of the velocity u^{μ} and the Riemannian ones U^{μ}, are related by

$$u^{\mu} = \hat{U}^{\mu} + e^{i\theta}\left(D^{\mu}\phi^*\right)\left(x_{\epsilon}\hat{U}^{\epsilon}\right) \tag{27}$$

Furthermore, using the fact that

$$\delta g_{\alpha\beta} = e^{-i\theta}\left[\left(\hat{\nabla}_{\beta} - i\theta_{\beta}\right)\hat{U}_{\alpha} + \left(\hat{\nabla}_{\alpha} - i\theta_{\alpha}\right)\hat{U}_{\beta}\right]\phi\,\delta S \tag{28}$$

we can obtain from the constriction $\Lambda\delta g_{\alpha\beta} = g_{\alpha\beta}\delta\Phi$, that

$$\delta\Phi = \frac{\Lambda}{4}g^{\alpha\beta}\,\delta g_{\alpha\beta} \tag{29}$$

so that, using Equation (28), the flux of A^{μ} across the Gaussian-like hypersurface can be expressed in terms of the quantum derivative of the complex field:

$$\frac{\delta\Phi}{\delta S} \equiv \frac{d\Phi}{dS} = \frac{\Lambda}{2}e^{i\theta}\hat{U}_{\alpha}\left(D^{\alpha}\phi^*\right) \tag{30}$$

Using the fact that $\hat{\nabla}_{\alpha}\delta W^{\alpha} = \delta\Phi$, it is easy to obtain

$$\hat{\nabla}_{\mu}A^{\mu} = \frac{\Lambda}{2}e^{i\theta}\hat{U}_{\alpha}\left(D^{\alpha}\phi^*\right) \tag{31}$$

where we have defined $A^{\mu} = \frac{\delta W^{\mu}}{\delta S}$. Notice that the velocity components $\hat{U}^{\alpha}$ of the Riemannian observer define the gauge of the system. Furthermore, due to the fact that $\delta W^{\alpha} = g^{\beta\gamma}\hat{\nabla}^{\alpha}\delta\Psi_{\beta\gamma}$, hence we obtain that

$$\frac{\delta W^{\alpha}}{\delta S} \equiv A^{\alpha} = g^{\beta\gamma}\hat{\nabla}^{\alpha}\chi_{\beta\gamma} \equiv \hat{\nabla}^{\alpha}\chi \tag{32}$$

where $\chi_{\beta\gamma}$ are the components of the gravitational waves:

$$\hat{\nabla}_{\alpha}A^{\alpha} = g^{\beta\gamma}\hat{\nabla}_{\alpha}\hat{\nabla}^{\alpha}\chi_{\beta\gamma} \equiv \hat{\Box}\chi \tag{33}$$

3. Quantum Field Dynamics

In this section we shall study the dynamics of charged and vector fields, in order to obtain their dynamical equations.

3.1. Dynamics of the Complex Fields

The cosmological constant Equation (22) can be rewritten in terms of $\phi = \theta\,e^{i\theta}$ and $\phi^* = \theta e^{-i\theta}$

$$\Lambda = -\frac{3}{4}\left[\left(\hat{\nabla}_{\nu}\phi\right)\left(\hat{\nabla}^{\nu}\phi^*\right) + \theta_{\nu}\,J^{\nu}\right] \tag{34}$$

where the current due to the charged fields is

$$J^{\nu} = i\left[\delta_{\epsilon}^{\nu}\left(\hat{\nabla}^{\epsilon}\phi\right)\phi^* - \left(\hat{\nabla}^{\nu}\phi^*\right)\phi - i\theta^{\nu}\left(\phi\phi^*\right)\right] \tag{35}$$

The important fact in Equation (34) is that the geometrical current J^{μ} interacts with the geometrical Weylian manifold. In other words, the cosmological constant can be viewed in this context as due to a purely quantum excitation (of charged fields), of the Riemannian (classical) background.

As can be demonstrated, $\hat{\nabla}_{\nu}J^{\nu} = -\frac{8}{3}\int d\theta = \left(\frac{2}{3}\Lambda\,S\right)^2$, so that we obtain the condition

$$\phi^*\,e^{i\left(\theta - \frac{\pi}{2}\right)} = \phi\,e^{-i\left(\theta - \frac{\pi}{2}\right)} \tag{36}$$

The zeroth-component of the current is

$$J^0 = i \left[\delta_\epsilon^0 \left(\hat{\nabla}^\epsilon \phi \right) \phi^* - \left(\hat{\nabla}^0 \phi^* \right) \phi - i\theta^0 \left(\phi \phi^* \right) \right] \tag{37}$$

which represents the density of electric charge, so that the charge is

$$Q = \int d^3x \sqrt{|\det[g_{ij}]|} \, J^0 \tag{38}$$

once we require that $\hat{\nabla}_i J^i = \left(\frac{2}{3} \Lambda \, S \right)^2$, and consequently $\hat{\nabla}_0 Q = 0$.

The second equation in Equation (23) results in two different equations

$$\left(\hat{\Box} + i\theta_\mu \hat{\nabla}^\mu + \frac{4}{3} \Lambda \right) \phi^* = 0 \tag{39}$$

$$\left(\hat{\Box} - i\theta^\mu \hat{\nabla}_\mu + \frac{4}{3} \Lambda \right) \phi = 0 \tag{40}$$

where the gauge equations are

$$- \left[i\theta_\mu \hat{\nabla}^\mu + \frac{3}{4} \Lambda \right] \phi^* = \frac{3}{4} \Lambda \, e^{-i\left(\theta - \frac{\pi}{2}\right)} \tag{41}$$

$$\left[i\theta^\mu \hat{\nabla}_\mu - \frac{3}{4} \Lambda \right] \phi = \frac{3}{4} \Lambda \, e^{i\left(\theta - \frac{\pi}{2}\right)} \tag{42}$$

so that finally we obtain the equations of motion for both fields

$$\hat{\Box} \phi^* = \frac{3}{4} \Lambda \, e^{-i\left(\theta - \frac{\pi}{2}\right)} \tag{43}$$

$$\hat{\Box} \phi = \frac{3}{4} \Lambda \, e^{i\left(\theta - \frac{\pi}{2}\right)} \tag{44}$$

Notice that the functions $e^{\pm i\left(\theta - \frac{\pi}{2}\right)}$ are invariant under $\theta = 2n\pi$ (n- integer) rotations, so that the complex fields are vector fields of spin 1. Using the expressions Equation (26) to find the commutators for the complex fields, we obtain that

$$[\phi^*(x), D^\mu \phi^*(y)] = \frac{4}{3} i\Theta^\mu \, \delta^{(4)}(x-y), \qquad [\phi(x), D_\mu \phi(y)] = -\frac{4}{3} i\Theta_\mu \, \delta^{(4)}(x-y) \tag{45}$$

where $D^\mu \phi^* \equiv \left(\hat{\nabla}^\mu + i\theta^\mu \right) \phi^*$ and $D_\mu \phi \equiv \left(\hat{\nabla}_\mu - i\theta_\mu \right) \phi$.

3.2. Dynamics of the Vector Fields

On the other hand, if we define $F^{\mu\nu} \equiv \hat{\nabla}^\mu A^\nu - \hat{\nabla}^\nu A^\mu$, such that A^α is given by Equation (32), we obtain the equations of motion for the components of the electromagnetic potentials A^ν: $\hat{\nabla}_\mu F^{\mu\nu} = J^\nu$

$$\hat{\Box} A^\nu - \hat{\nabla}^\nu \left(\hat{\nabla}_\mu A^\mu \right) = J^\nu \tag{46}$$

where J^ν being given by the expression Equation (35) and from Equation (15) we obtain that $\hat{\nabla}_\mu A^\mu = -\frac{\Lambda}{2} \theta_\mu \hat{U}^\mu = \frac{2}{3} \Lambda^2 \, S(x^\mu) = 4\frac{d\Phi}{dS}$ determines the gauge that depends on the Riemannian frame adopted by the relativistic observer. Notice that for massless particles the Lorentz gauge is fulfilled, but it does not work for massive particles, where $S \neq 0$.

4. Final Remarks

We have studied charged and electromagnetic fields from relativistic quantum geometry. In this formalism, the Einstein tensor complies with gauge-invariant transformations studied in a previous work [12]. The quantum dynamics of the fields is described on a Weylian manifold which comes from a geometric extension of the Riemannian manifold, on which is defined the classical geometrical background. The connection that describes the Weylian manifold is given in Equation (26) in terms of the quantum derivative of the complex vector field with a Lagrangian density described by the cosmological constant Equation (34). We have demonstrated that vector fields A^μ describe an extended Maxwell dynamics (see Equation (46)), where the source is provided by the charged fields current density J^μ, with a nonzero tetra-divergence. Furthermore, the gauge of A^μ is determined by the relativistic observer: $\hat{\nabla}_\mu A^\mu = \frac{\Lambda}{2}\theta_\mu \hat{U}^\mu$. Finally, it is important to notice that the cosmological constant appears as a Riemannian invariant, but not a Weylian one. It can be viewed in this context as due to a purely quantum excitation. In this paper these excitations of a Riemannian (classical) background, are driven by charged complex fields.

Author Contributions: Both authors contributed in the paper. Both authors have read and approved the final version.

Conflicts of Interest: The authors declare no conflict of interest.

References

1. Weyl, H. *Philosophy of Mathematics and Natural Science*; Princeton University Press: Princeton, UK, 1949.
2. Cvetic, M.; Gibbons, G.W.; Pope, C.N. Super-geometrodynamics. *J. High Energy Phys.* **2015**, *2015*, 029.
3. Wheeler, J.A. On the nature of quantum geometrodynamics. *Ann. Phys.* **1957**, *2*, 604–614.
4. Wheeler, J.A. Superspace and the Nature of Quantum Geometrodynamics. In *Battelle Rencontres, 1967 Lectures in Mathematics and Physics*; De Witt, C.M., Wheeler, J.A., Eds.; W.A. Benjamin: New York, NY, USA, 1968.
5. Gibbons, G.W. Soliton States and Central Charges in Extended Supergravity Theories. *Lect. Notes Phys.* **1982**, *160*, 145–151.
6. Gibbons, G.W.; Hull, C.M. A bogomolny bound for general relativity and solitons in N=2 supergravity. *Phys. Lett. B* **1982**, *109*, 190–194.
7. Rácz, I. Cauchy problem as a two-surface based 'geometrodynamics'. *Class. Quantum Gravity* **2015**, *32*, 015006.
8. Prugovečki, E. Liouville dynamics for optimal phase-space representations of quantum mechanics. *Ann. Phys.* **1978**, *110*, 102–121.
9. Prugovečki, E. Consistent formulation of relativistic dynamics for massive spin-zero particles in external fields. *Phys. Rev. D* **1978**, *18*, 3655–3675.
10. Prugovečki, E. A Poincaré gauge-invariant formulation of general-relativistic quantum geometry. *Nuo. Cim. A* **1987**, *97*, 597–628.
11. Ridao, L.S.; Bellini, M. Towards relativistic quantum geometry. *Phys. Lett. B* **2015**, *751*, 565–571.
12. Ridao, L.S.; Bellini, M. Discrete modes in gravitational waves from the big-bang. *Astrophys. Space Sci.* **2015**, *357*, 94.
13. Bellini, M. Inflationary back-reaction effects from Relativistic Quantum Geometry. *Phys. Dark Univ.* **2016**, *11*, 64–67.

universe

MDPI

Article

Symplectic Structure of Intrinsic Time Gravity

Eyo Eyo Ita III [1,*] and Amos S. Kubeka [2]

[1]   Physics Department, U.S. Naval Academy, Annapolis, MD 21401, USA
[2]   Mathematical Sciences Department, University of South Africa, Pretoria 0002, South Africa;
     kubekas@unisa.ac.za
*   Correspondence: ita@usna.edu

Academic Editor: Lorenzo Iorio
Received: 13 June 2016; Accepted: 5 August 2016; Published: 30 August 2016

Abstract: The Poisson structure of intrinsic time gravity is analysed. With the starting point comprising a unimodular three-metric with traceless momentum, a trace-induced anomaly results upon quantization. This leads to a revision of the choice of momentum variable to the (mixed index) traceless momentric. This latter choice unitarily implements the fundamental commutation relations, which now take on the form of an affine algebra with SU(3) Lie algebra amongst the momentric variables. The resulting relations unitarily implement tracelessness upon quantization. The associated Poisson brackets and Hamiltonian dynamics are studied.

Keywords: intrinsic time; quantum gravity; canonical; quantization; symmetry

1. Introduction

A crucially important question in the quantization of gravity in 3+1 dimensions, as for any theory, is the choice of the fundamental dynamical variables of the classical theory, which upon quantization become promoted to quantum operators. In Loop Quantum Gravity (LQG) [1] the starting point for the classical theory are the Ashtekar variables, where a SU(2) gauge connection and a densitized triad form a canonically conjugate pair. This choice of variables turns the initial value constraints of GR from intractable non polynomial phase space functions, as they appear in the Arnowitt Deser Misner (ADM) theory [2], into polynomial form at the expense of an additional set of constraints related to the SU(2) gauge symmetry inherent in the theory. It is hoped that the polynomial form of the constraints in LQG make the constraints more tractable for quantization and the construction of a physical Hilbert space. The actual configuration variable in LQG which is subject to the quantization procedure is not the connection itself, but rather the holonomy of the connection, since the latter is well-defined in the quantum theory whereas the connection fails to exist [3]. Furthermore, the transformation properties of the holonomy more aptly are representative, at the kinematical level, of the symmetry properties of the theory [4]. Consequently, upon quantization in LQG all constraints and quantities must be rewritten in terms of the holonomies and the densitized triad, which themselves no longer form a canonical pair.

In LQG there exists only a manifold structure with no metric, and the metric is no longer fundamental, but becomes a derived quantity in terms of more fundamental variables. A main difficulty in LQG is the construction of a physical Hilbert space from solution of the Hamiltonian constraint. Whether one utilizes the self-dual version of the connection or its real counterparts as in the Barbero variables [5], the solution to the Hamiltonian constraint and its subsequent delineation of the physical Hilbert space, is a long and standing unresolved problem [2]. Consequently, the quantization of LQG remains complete only at the kinematical level (which is more suitably adapted to the fundamental variables), and the physical dynamics of gravity remain to be completely encoded within this procedure [4]. LQG can be contrasted with the standard ADM approach [4], wherein the fundamental variables are the spatial three metric and its conjugate momentum, constructed from the extrinsic curvature of the spatial slice of four-dimensional spacetime upon which the quantization must be performed. The corresponding initial value constraints are intractable due to various technical issues particularly related to ultraviolet divergences associated with operator products, which in LQG are absent. The choice of fundamental variables in the ADM approach poses the problem that as canonically conjugate variables, the momentum generates translations of the spatial three metric. Since the spectrum of both variables is the real line, then the positivity of the metric in the quantum theory cannot be guaranteed while having self-adjoint variables. Positivity of the spatial three metric is a crucially important condition that any quantum theory of gravity must satisfy, since spatial distances as measured by the theory must always be positive.

The theory of Intrinsic time quantum gravity (ITQG) [6] presented in this paper is driven by the motivation to solve all of the above difficulties. The choice of the configuration space variable in ITQG will be a unimodular spatial three metric metric and a momentum variable (ultimately known as a momentric) which generates dilations (more precisely SU(3) and SL(3,R) transformations of the metric). The importance of this choice of fundamental variables is that they will be self-adjoint in the quantum theory, while preserving the positivity and the unimodularity of the spatial three metric forming the configuration space variable. A common misconception of the price for such a result is that the variables cannot be canonically related, resulting in complications in their quantization. However, in the case of ITQG we will see that it is precisely their non canonical nature that makes them perfectly suited for quantization, and admits a group-theoretical interpretation as such, which resolves all of the aforementioned difficulties in the LQG and ADM approaches in one stroke.

In [6], a new formulation for quantization of the gravitational field in ITQG, is presented. The basic idea, as introduced in [7] and [8], is the concept of a new phase space for gravity which breaks the paradigm of four-dimensional spacetime covariance, shifting the emphasis to three dimensional spatial diffeomorphism invariance combined with a physical Hamiltonian which generates evolution with respect to intrinsic time. Through the constructive interference of wavefronts, classical spacetime emerges from the formalism, with direct correlation between intrinsic time intervals and proper time intervals of spacetime. In the present paper we will take a step back to analyse the motivations and canonical structure of ITQG, and then construct the fundamental variables and their commutation relations of the theory. These relations are noncanonical, which lead to the uncovering of an inherent $SU(3)$ structure for gravity. This presents certain advantages from the standpoint of quantization. The paper is thus structures as follows: Section 2 discusses the Poisson structure of the barred classical variables, Section 3 highlights the prelude to the quantum theory, Section 4 discusses the momentric operators and the SU(3) Lie algebra, Section 5 revisits the classical theory, and then lastly, Section 6 concludes the paper with some recommendations for similar future work in this direction.

2. Poisson Structure of the Barred Classical Variables

Let $q_{ij}, \tilde{\pi}^{ij}$ denote the spatial 3-metric and its conjugate momentum defined on a spatial slice Σ of a four dimensional spacetime of topology $M = \Sigma \times R$. In the ADM metric theory, the basic variables provide a canonical one form

$$\Theta_{ADM} = \int_{\Sigma} d^3x \, \tilde{\pi}^{ij}(x)\delta q_{ij}(x). \tag{1}$$

Starting from this canonically conjugate pair, let us define as fundamental classical variables the following barred quantities $\bar{q}_{ij}$, a unimodular metric with $\det \bar{q}_{ij} = 1$, and a traceless momentum variable $\bar{\pi}^{ij}$ via the relations [7,8]

$$\bar{q}_{ij} = q^{-1/3}q_{ij}; \quad \bar{\pi}^{ij} = q^{1/3}\left(\tilde{\pi}^{ij} - \frac{1}{3}q^{ij}\tilde{\pi}\right), \tag{2}$$

where $\tilde{\pi} = q_{ij}\tilde{\pi}^{ij}$ with $\bar{q}_{ij}\bar{\pi}^{ij} = 0$. From Equation (2) we get the following cotangent space decomposition

$$\delta q_{ij} = q^{1/3}\left(\bar{q}_{ij}\delta\ln q^{1/3} + \delta\bar{q}_{ij}\right) \longrightarrow \delta\bar{q}_{ij} = \bar{P}_{ij}^{kl}\delta q_{kl} \tag{3}$$

where we have defined the traceless projector $\bar{P}^{ij}_{kl} = \frac{1}{2}(\delta^i_k \delta^j_l + \delta^j_k \delta^i_l) - \frac{1}{3}\bar{q}^{ij}\bar{q}_{kl}$, with $\bar{P}^{ij}_{kl}\bar{q}^{kl} = \bar{q}_{ij}\bar{P}^{ij}_{kl} = 0$. So we have $\bar{q}^{ij}\delta\bar{q}_{ij} = 0$, namely that the cotangent space elements $\delta\bar{q}^{ij}$ are traceless. The inverse relations

$$q_{ij} = q^{1/3}\bar{q}_{ij}; \quad \tilde{\pi}^{ij} = q^{-1/3}\left(\bar{\pi}^{ij} + \frac{1}{3}\bar{q}^{ij}\bar{\pi}\right) \tag{4}$$

take us from the barred back to the unbarred variables. Substitution of the left side of the arrow of Equation (3) into Equation (1) provides a clean separation of the barred gravitational degrees of freedom with canonical one-form [7]

$$\Theta = \int_\Sigma d^3x\, \tilde{\pi}^{ij}\delta q_{ij} = \int_\Sigma d^3x \left(\tilde{\pi}\delta\ln q^{1/3} + \bar{\pi}^{ij}\delta\bar{q}_{ij}\right), \tag{5}$$

where we have used $\bar{\pi}^{ij}\bar{q}_{ij} = \bar{q}^{ij}\delta\bar{q}_{ij} = 0$. Equation (5) yields a corresponding symplectic two-form

$$\Omega = \delta\Theta = \int_\Sigma d^3x \left(\delta\tilde{\pi} \wedge \delta\ln q^{1/3} + \delta\bar{\pi}^{ij} \wedge \delta\bar{q}_{ij}\right). \tag{6}$$

While this may be the case, as we will see, the Poisson brackets which can arise from (6) are not unique, on account of subtleties due to the implementation of tracelessness of $\bar{\pi}^{ij}$.

A necessary condition for a consistent canonical quantization of the theory is that the correct Poisson brackets comprise the starting point at the classical level. So let us directly calculate via Equation (2) barred Poisson brackets with respect to the unbarred canonical structure, which is clearly known to be unambiguous. For the metric components we have $\{\bar{q}_{ij}(x), \bar{q}_{kl}(y)\} = 0$ which is encouraging, as the unbarred metric clearly is devoid of any momentum dependence. However, using the following relations

$$\frac{\delta q^{1/3}}{\delta q_{ij}} = \frac{1}{3}q^{1/3}q^{ij}; \quad \frac{\delta q^{ij}}{\delta q_{mn}} = -q^{(im}q^{j)n}; \quad \frac{\delta\bar{q}_{kl}}{\delta q_{ij}} = q^{-1/3}\bar{P}^{ij}_{kl}; \quad \frac{\delta\tilde{\pi}^{ij}}{\delta\tilde{\pi}^{kl}} = q^{1/3}\bar{P}^{ij}_{kl}, \tag{7}$$

in conjunction with

$$\frac{\delta\tilde{\pi}^{ij}}{\delta q_{kl}} = \frac{1}{3}\left(q^{kl}\tilde{\pi}^{ij} + q^{1/3}\left(q^{(ik}q^{j)l}q_{rs}\tilde{\pi}^{rs} - q^{ij}\tilde{\pi}^{kl}\right)\right) = \frac{1}{3}q^{-1/3}\left(\bar{q}^{kl}\bar{\pi}^{ij} - \bar{q}^{ij}\bar{\pi}^{kl}\right) + \frac{1}{3}q^{1/3}q^{(ik}q^{j)l}\tilde{\pi}, \tag{8}$$

we obtain the following Poisson bracket relations between barred metric and momentum

$$\{\bar{q}_{ij}(x), \bar{\pi}^{kl}(y)\} = \int_\Sigma d^3z \left(\frac{\delta\bar{q}_{ij}(x)}{\delta q_{mn}(z)}\frac{\delta\bar{\pi}^{kl}(y)}{\delta\tilde{\pi}^{mn}(z)} - \frac{\delta\bar{\pi}^{kl}(y)}{\delta q_{mn}(z)}\frac{\delta\bar{q}_{ij}(x)}{\delta\tilde{\pi}^{mn}(z)}\right) = \bar{P}^{kl}_{ij}\delta^{(3)}(x,y). \tag{9}$$

Finally, we obtain the following relation amongst the barred momentum components

$$\{\bar{\pi}^{ij}(x), \bar{\pi}^{kl}(y)\} = \int_\Sigma d^3z \left(\frac{\delta\bar{\pi}^{ij}(x)}{\delta q_{mn}(z)}\frac{\delta\bar{\pi}^{kl}(y)}{\delta\tilde{\pi}^{mn}(z)} - \frac{\delta\bar{\pi}^{kl}(y)}{\delta q_{mn}(z)}\frac{\delta\bar{\pi}^{ij}(x)}{\delta\tilde{\pi}^{mn}(z)}\right) = \frac{1}{3}\left(\bar{q}^{kl}\bar{\pi}^{ij} - \bar{q}^{ij}\bar{\pi}^{kl}\right)\delta^{(3)}(x,y). \tag{10}$$

The Poisson brackets between barred variables are noncanonical. But we will show that they yield the same barred contribution as the symplectic two form (6) which can be seen as follows. From the calculated Poisson brackets the following Poisson matrix can be constructed

$$P^{IJ} = \begin{pmatrix} \{\bar{q}_{ij}(x), \bar{q}_{kl}(y)\} & \{\bar{q}_{ij}(x), \bar{\pi}^{kl}(y)\} \\ \{\bar{\pi}^{kl}(y), \bar{q}_{ij}(x)\} & \{\bar{\pi}^{ij}(x), \bar{\pi}^{kl}(y)\} \end{pmatrix} = \begin{pmatrix} 0 & \bar{P}^{kl}_{ij} \\ -\bar{P}^{kl}_{ij} & \frac{1}{3}\left(\bar{q}^{kl}\bar{\pi}^{ij} - \bar{q}^{ij}\bar{\pi}^{kl}\right) \end{pmatrix} \delta^{(3)}(x,y). \tag{11}$$

In Poisson geometry, a two form $\Omega = \frac{1}{2}\Omega_{IJ}\delta q^I \wedge \delta q^J$ on the phase space $q^I \equiv \bar{q}_{ij}, \bar{\pi}^{ij}$ can be constructed whose components are the inverse of the Poisson matrix. If Ω is closed ($\delta\Omega = 0$) and nondegenerate, then it is said to be a symplectic two form. Making the identifications $\{\bar{q}, \bar{\pi}\} \sim \beta$ and $\{\bar{\pi}, \bar{\pi}\} \sim \alpha$, then the inverse of the Poisson matrix for the barred variables is of the form

$$P^{-1} = \begin{pmatrix} 0 & \beta \\ -\beta & \alpha \end{pmatrix}^{-1} = \begin{pmatrix} \beta^{-1}\alpha\beta^{-1} & -\beta^{-1} \\ \beta^{-1} & 0 \end{pmatrix}, \tag{12}$$

which does not exist since the projector $\bar{P}_{ij}^{kl}$ is uninvertible. This suggests, naively, that the symplectic structure associated with the above Poisson brackets does not exist.

One method of quantization of a theory is to promote Poisson brackets directly into quantum commutators. The Poisson brackets for a generic theory can be read off directly from its symplectic two form, and which in turn is defined from the Poisson matrix by constructing the inverse of the latter. We would like to construct the symplectic two form for ITQG by inverting the Poisson matrix P^{IJ} constructed in equation Equation (11). The Poisson matrix in its present form is uninvertible since it consists of projectors $\bar{P}_{ij}^{kl}$ in its block off-diagonal positions denoted by the symbol β. In the process of inversion of P^{IJ}, as shown above with P^{-1}, it is necessary to have β^{-1}. But β^{-1} does not exist on account of the fact that projectors are not invertible, which suggests, naively, that ITQG does not have a well-defined symplectic structure.

To get around this technical difficulty we will add a trace part to the Poisson matrix, parametrized by a parameter γ which we will ultimately remove after all calculations have been performed. While this distorts the theory of ITQG to a new theory parametrized by γ, it renders the resulting Poisson matrix invertible to allow progress to the corresponding symplectic two form, parametrized by γ, since the previously offending terms β now become β_γ, which as in Equation (12) are now invertible. Thus we have

$$\beta_\gamma \equiv (P_\gamma)_{kl}^{ij} = P_{kl}^{ij} + \gamma \bar{q}^{ij}\bar{q}_{kl} \longrightarrow \beta_\gamma^{-1} = P_{mn}^{kl} + \frac{1}{9\gamma}\bar{q}^{kl}\bar{q}_{mn}. \tag{13}$$

So now, we can invert the resulting object, and we have that

$$\beta_\gamma^{-1}\alpha\beta_\gamma^{-1} = \frac{1}{3}\left(\bar{P}_{kl}^{mn} + \frac{1}{9\gamma}\bar{q}^{mn}\bar{q}_{kl}\right)\left(\bar{q}^{kl}\bar{\pi}^{ij} - \bar{q}^{ij}\bar{\pi}^{kl}\right)\left(\bar{P}_{ij}^{rs} + \frac{1}{9\gamma}\bar{q}^{rs}\bar{q}_{ij}\right) = -\frac{1}{9\gamma}\left(\bar{\pi}^{mn}\bar{q}^{rs} - \bar{q}^{mn}\bar{\pi}^{rs}\right), \tag{14}$$

where we have used $\bar{P}_{kl}^{ij}\bar{q}_{ij} = \bar{q}^{kl}\bar{\pi}_{kl} = 0$ and $\bar{P}_{kl}^{ij}\bar{\pi}^{kl} = \bar{\pi}^{ij}$, which assumes that $\bar{\pi}^{ij}$ is traceless. So the inverse of the Poisson matrix parametrized by γ is given by

$$P^{-1} = \begin{pmatrix} -\frac{1}{9\gamma}\left(\bar{q}^{kl}\bar{\pi}^{ij} - \bar{q}^{ij}\bar{\pi}^{kl}\right) & -\left(\bar{P}_{mn}^{kl} + \frac{1}{9\gamma}\bar{q}^{kl}\bar{q}_{mn}\right) \\ \bar{P}_{rs}^{ij} + \frac{1}{9\gamma}\bar{q}^{ij}\bar{q}_{rs} & 0 \end{pmatrix}\delta^{(3)}(x, y),$$

and the associated two form Ω inherits the γ dependence

$$\Omega^\gamma = \frac{1}{2}\Omega_{IJ}^\gamma\delta q^I\delta q^J$$
$$= \int_\Sigma d^3x\left[-\frac{1}{18\gamma}\left(\bar{q}^{kl}\bar{\pi}^{ij} - \bar{q}^{ij}\bar{\pi}^{kl}\right)\delta\bar{q}_{ij}\wedge\delta\bar{q}_{kl} + \left(\bar{P}_{kl}^{ij} + \frac{1}{9\gamma}\bar{q}^{ij}\bar{q}_{kl}\right)\delta\bar{q}_{ij}\wedge\delta\bar{\pi}_{kl} + \frac{1}{2}(0)_{ijkl}\delta\bar{\pi}^{ij}\wedge\delta\bar{\pi}^{kl}\right]. \tag{15}$$

But $\bar{q}^{ij}\delta\bar{q}_{ij} = 0$, causing the $\delta\bar{q}\wedge\delta\bar{q}$ term and the γ contribution to the $\delta\bar{q}\wedge\delta\bar{\pi}$ term of (15) vanish. The quantity $(0)_{ijkl}$ in Equation (15) is basically to highlight the fact that that term, while zero is nontrivially so. Rather than omit this term, we wanted to highlight the fact that it is a tensorial quantity forming the coefficient of the $\delta\bar{\pi}\wedge\delta\bar{\pi}$ two form. This facilitates the keeping track for the reader of each individual term, of which there should be of the type including $\delta\bar{q}\wedge\delta\bar{q}$ and $\delta\bar{q}\wedge\delta\bar{\pi}$. There is no $\delta\bar{\pi}\wedge\delta\bar{\pi}$ term since $\{\bar{q}_{ij}, \bar{q}_{kl}\} = 0$. All explicit γ dependence in the symplectic form has disappeared, so the $\gamma \to 0$ limit can be safely taken, yielding

$$\lim_{\gamma\to 0}\Omega^\gamma = \int_\Sigma d^3x P_{kl}^{ij}\delta\bar{q}_{ij}\wedge\delta\bar{\pi}^{kl}$$

$$= \int_\Sigma d^3x\delta\bar{q}_{ij}\wedge\delta\bar{\pi}^{ij} - \frac{1}{3}\int_\Sigma d^3x(\bar{q}^{ij}\delta\bar{q}_{ij})\wedge(\bar{q}_{kl}\delta\bar{\pi}^{kl}) = \int_\Sigma d^3x\delta\bar{q}_{ij}\wedge\delta\bar{\pi}^{ij}. \tag{16}$$

The vanishing of the $\frac{1}{3}$ term is due to $\bar{q}^{ij}\delta\bar{q}_{ij} = 0$ or alternatively by the Leibniz rule for the momentum term

$$(\bar{q}^{ij}\delta\bar{q}_{ij})\wedge(\bar{q}_{kl}\delta\bar{\pi}^{kl}) = \delta\bar{q}_{ij}\wedge\delta(\bar{q}^{ij}\bar{q}^{kl}\bar{\pi}_{kl}) - \delta q_{ij}\wedge\delta q^{ij}(\bar{q}_{kl}\bar{\pi}^{kl}) - (\bar{q}^{ij}\delta\bar{q}_{ij})\wedge(\bar{\pi}^{kl}\delta\bar{q}_{kl}) = 0 \tag{17}$$

due additionally to $\bar{q}_{ij}\bar{\pi}^{ij} = 0$. This implies that the tracelessness of $\bar{\pi}^{ij}$ must be conjugate to the fact that infinitesimal variations in $\bar{q}_{ij}$ are traceless. Hence (16) is the same as the barred contribution to (6), with the difference that the tracelessness of $\bar{\pi}^{ij}$ has been implicitly enforced due to a unimodular metric. This calculation demonstrates that extreme care must be exercised when extracting Poisson brackets from a symplectic two form, particular when the index structure of the fundamental variables has implicit symmetries. The requirement to implement the noncanonical Poisson brackets at the quantum level will pose nontrivial issues, which we will address in the next few sections. Let us display, for completeness, the fundamental Poisson brackets for the barred phase space

$$\{\bar{q}_{ij}(x),\bar{q}_{kl}(y)\} = 0; \quad \{\bar{q}_{ij}(x),\bar{\pi}^{kl}(y)\} = P_{kl}^{ij}(x,y); \quad \{\bar{\pi}^{ij}(x),\bar{\pi}^{kl}(y)\} = \frac{1}{3}(\bar{q}^{kl}\bar{\pi}^{ij} - \bar{q}^{ij}\bar{\pi}^{kl})\delta^{(3)}(x,y). \tag{18}$$

The basic Poisson brackets are noncanonical, which can be seen as the price to be paid for choosing $\bar{\pi}^{ij}$ to be traceless at the classical level, or alternatively, the price for choosing unimodular metric variables.

The original motivation was to obtain a symplectic form parametrized by γ and then to take the limit as γ approaches zero. But as one can see from the above that the wedge products in the resulting symplectic two form have coefficients proportional to γ^{-1}, which in the limit as γ approaches zero would be ill-defined. However, note form the arguments provided from Equation (15) through to Equation (18), that the individual wedge products of the fundamental variables all vanish on account of the unimodularity of the configuration space variable $\bar{q}_{ij}$ and the tracelessness of the momentum $\bar{p}^{ij}$. Hence the proper procedure is to leave γ arbitrary in the symplectic two form, which is immaterial since all terms which depend on γ automatically vanish. The result is that the symplectic two form reduces to $\delta\bar{q}\wedge\delta\bar{\pi}$ form as in Equation (17), whence γ is conspicuously absent. So the justification that the parametrization of the Poisson matrix by the parameter does not affect the results of the symplectic two form is that for all nonzero γ, we can transition from the Poisson matrix to the symplectic two form by inversion as per the standard procedure, yielding a symplectic two form which is independent of the parameter γ. It is the unique choice of unimodular and traceless variables, which makes this the case, which admits a complete quantization of these variables.

3. A Prelude into the Quantum Theory

Having determined the Poisson brackets for the barred phase space, the next step is to implement them at the quantum level. In proceeding to the quantum theory according to the Heisenberg–Dirac prescription, we must promote all classical variables A, B to operators $\hat{A}, \hat{B}$ and all Poisson brackets to commutators $\{A, B\} \to \frac{1}{(i\hbar)}[\hat{A}, \hat{B}]$. So the fundamental Poisson brackets (18) yield the following equal-time commutation relations

$$[\bar{q}_{ij}(x,t),\bar{q}_{kl}(y,t)] = 0; [\bar{q}_{ij}(x,t),\hat{\bar{\pi}}^{kl}(y,t)] = i\hbar\bar{P}_{kl}^{ij}(x,y); \quad [\hat{\bar{\pi}}^{ij}(x,t),\hat{\bar{\pi}}^{kl}(y,t)] = \frac{i\hbar}{3}(\bar{q}^{kl}\hat{\bar{\pi}}^{ij} - \bar{q}^{ij}\hat{\bar{\pi}}^{kl})\delta^{(3)}(x,y), \tag{19}$$

where we have chosen an operator ordering with the momenta to the right. Since the momentum components fail to commute, then we are restricted to wavefunctionals $\psi[\bar{q}]$ in the metric representation. A representation of the classically traceless momentum as a vector field

$$\widehat{\pi}^{ij}(x)\psi[\bar{q}] \longrightarrow \frac{\hbar}{i}\left[\overline{P}^{ij}_{kl}\frac{\delta}{\delta\bar{q}_{kl}} + \frac{1}{3}\left(\bar{q}^{ij}\overline{\pi}^{kl} - \bar{q}^{kl}\overline{\pi}^{ij}\right)\frac{\delta}{\delta\overline{\pi}^{kl}}\right]\psi[\bar{q}] = \frac{\hbar}{i}\overline{P}^{ij}_{kl}(x)\frac{\delta\psi[\bar{q}]}{\delta\bar{q}_{kl}(x)} \tag{20}$$

correctly reproduces the commutation relations (19) (The term of (20) from the $\overline{\pi}^{ij}, \overline{\pi}^{kl}$ commutation relation does not contribute for wavefunctionals $\psi[\bar{q}]$ polarized in the metric representation.). However, Equation (20) does not constitute a self-adjoint operator since

$$\frac{\hbar}{i}\frac{\delta}{\delta\bar{q}_{kl}(x)}\overline{P}^{ij}_{kl}(x) = \frac{\hbar}{i}\overline{P}^{ij}_{kl}(x)\frac{\delta}{\delta\bar{q}_{kl}(x)} - \frac{2\hbar}{3i}\bar{q}^{ij}\delta^{(3)}(0). \tag{21}$$

So $\bar{q}_{ij}\widehat{\overline{\pi}}^{ij} = 0 \neq \widehat{\overline{\pi}}^{ij}\bar{q}_{ij}$, namely that the momentum in (20) is left-traceless, but is not right-traceless. A self-adjoint operator can be constructed by averaging the left-traceless and right-traceless versions $\frac{1}{2}\left(\frac{\delta}{\delta\bar{q}_{ij}}\overline{P}^{kl}_{ij} + \overline{P}^{kl}_{ij}\frac{\delta}{\delta\bar{q}_{ij}}\right)$. However, the resulting operator, while self-adjoint, is neither traceless from the left nor from the right. So it appears that tracelessness is a property which is nontrivial to enforce at the quantum level in the $\bar{q}_{ij}, \overline{\pi}^{ij}$ variables.

The quantity $\delta^{(3)}(0)$ in Equation (21) is an ultraviolet singularity in field theory, which results from evaluating the commutation relations at the same spatial point. It is a formal expression more rigorously defined by a limiting procedure in the coincidence limit of the arguments x and y. It is necessary to perform the commutation relations at the same spatial point in order to reorder the fundamental operators in Equations (20) and (21), which are defined at the same spatial point, which is necessary in order to evaluate self-adjointness. This operator ordering induced ambiguity, parametrized by $\delta^{(3)}(0)$, highlights that the variables in their present form, while solving the aforementioned problem of the symplectic structure, are still not ideally suited for quantization. This will ultimately lead us to the choice of the momentric π^i_j, in lieu of the momentum variable $\overline{\pi}^{ij}$, which being self adjoint as we will demonstrate in the remainder of this paper, will eliminate the presence of any such $\delta^{(3)}(0)$ divergences in the quantum theory.

4. Momentric Operators and the SU(3) Lie Algebra

Let us define a mixed-index version of the momentum, namely the momentric variables $\overline{P}^i_j = \bar{q}_{jm}\overline{\pi}^{im}$. We first compute the commutator of $\overline{P}^i_j$ with the barred metric. This is given by

$$[\overline{P}^i_j(x), \bar{q}_{kl}(y)] = [\bar{q}_{jm}(x)\overline{\pi}^{mi}(x), \bar{q}_{kl}(y)] = \bar{q}_{jm}(x)[\overline{\pi}^{mi}(x), \bar{q}_{kl}(y)] = -i\hbar\bar{q}_{jm}\overline{P}^{mi}_{kl}\delta^{(3)}(x,y) \equiv \frac{\hbar}{i}\overline{E}^i_{j(kl)}\delta^{(3)}(x,y) \tag{22}$$

where we have used (19), with the "superspace vielbein" defined as $\overline{E}^i_{j(kl)} = \frac{1}{2}\left(\delta^i_k\bar{q}_{jl} + \delta^i_l\bar{q}_{jk}\right) - \frac{1}{3}\delta^i_j\bar{q}_{kl}$. So we will rather adopt the pair $\bar{q}_{ij}, \overline{P}^i_j$ as the fundamental variables, and recompute the fundamental relations (19) with respect to them.

For the commutators amongst the momentric components themselves the following identity involving commutation relations regarding generic operators $\widehat{A}, \widehat{B}, \widehat{C}, \widehat{D}$ will be useful

$$[\widehat{A}\widehat{B}, \widehat{C}\widehat{D}] = \widehat{A}[\widehat{B}, \widehat{C}]\widehat{D} + \widehat{C}[\widehat{A}, \widehat{D}]\widehat{B} + [\widehat{A}, \widehat{C}]\widehat{B}\widehat{D} + \widehat{C}\widehat{A}[\widehat{B}, \widehat{D}]. \tag{23}$$

Note that the proper operator ordering has been preserved in (23). So we have the following, suppressing the x-y dependence in the intermediate steps and suppressing the hats to avoid cluttering up the notation,

$$[\overline{P}_j^i(x), \overline{P}_l^k(y)] = [\overline{q}_{jm}(x)\overline{\pi}^{im}(x), \overline{q}_{ln}(y)\overline{\pi}^{kn}(y)]$$

$$= \overline{q}_{jm}[\overline{\pi}^{im}, \overline{q}_{ln}]\overline{\pi}^{kn} + \overline{q}_{ln}[\overline{q}_{jm}, \overline{\pi}^{kn}]\overline{\pi}^{im} + [\overline{q}_{jm}, \overline{q}_{ln}]\overline{\pi}^{im}\overline{\pi}^{kn} + \overline{q}_{ln}\overline{q}_{jm}[\overline{\pi}^{im}, \overline{\pi}^{kn}] \quad (24)$$

$$= \frac{\hbar}{i}\left[\overline{q}_{jm}\overline{P}_{ln}^{im}\overline{\pi}^{kn} - \overline{q}_{ln}\overline{P}_{jm}^{kn}\overline{\pi}^{im} + 0 + \frac{1}{3}\overline{q}_{ln}\overline{q}_{jm}(\overline{q}^{kn}\overline{\pi}^{im} - \overline{q}^{im}\overline{\pi}^{kn})\right]\delta^{(3)}(x,y).$$

In the third line of Equation (24) we have used the fundamental equal time commutation relations (19). For completeness, let us display some of the intermediate steps from Equation (24). For the first term on the right hand side we have

$$\overline{q}_{jm}\overline{P}_{ln}^{im}\overline{\pi}^{kn} = \overline{q}_{jm}\left(\frac{1}{2}(\delta_l^i\delta_n^m + \delta_n^i\delta_l^m) - \frac{1}{3}\overline{q}^{im}\overline{q}_{ln}\right)\overline{\pi}^{kn} = \frac{1}{2}(\delta_l^i\overline{P}_j^k + \overline{q}_{jl}\overline{\pi}^{ki}) - \frac{1}{3}\delta_j^i\overline{P}_l^k. \quad (25)$$

For the middle term we have

$$\overline{q}_{ln}\overline{P}_{jm}^{kn}\overline{\pi}^{im} = \overline{q}_{ln}\left(\frac{1}{2}(\delta_j^k\delta_m^n + \delta_m^k\delta_j^n) - \frac{1}{3}\overline{q}^{kn}\overline{q}_{jm}\right)\overline{\pi}^{im} = \frac{1}{2}(\delta_j^k\overline{P}_l^i + \overline{q}_{lj}\overline{\pi}^{ik}) - \frac{1}{3}\delta_l^k\overline{P}_j^i. \quad (26)$$

For the last term on the right hand side of (24) we have

$$\frac{1}{3}\overline{q}_{ln}\overline{q}_{jm}(\overline{q}^{kn}\overline{\pi}^{im} - \overline{q}^{im}\overline{\pi}^{kn}) = \frac{1}{3}(\delta_l^k\overline{P}_j^i - \delta_j^i\overline{P}_l^k). \quad (27)$$

Substitution of Equations (25)–(27) into Equation (24) yields the result that

$$[\overline{P}_j^i(x), \overline{P}_l^k(y)] = \frac{\hbar}{i}\left[\frac{1}{2}(\delta_l^i\overline{P}_j^k - \delta_j^k\overline{P}_l^i) + \frac{2}{3}(\delta_l^k\overline{P}_j^i - \delta_j^i\overline{P}_l^k)\right]\delta^{(3)}(x,y). \quad (28)$$

Note that the algebra closes (if not for the precise cancellation of terms of the form $\overline{q}_{jl}\overline{\pi}^{ki}$, this would not be the case). While the algebra (28) closes on the momentric variables $\overline{P}_j^i$, it does not enforce the vanishing of the trace $\overline{P} = \delta_i^j\overline{P}_j^i$. This can be seen by contraction of (28) with δ_j^i, wherein

$$[\overline{P}(x), \overline{P}_j^i(y)] = -\frac{2\hbar}{i}(\overline{P}_j^i - \frac{1}{3}\delta_j^i\overline{P})\delta^{(3)}(x,y) \equiv 2i\hbar\overline{\pi}_j^i\delta^{(3)}(x,y), \quad (29)$$

where $\overline{\pi}_j^i$ denotes the traceless part of the momentric. Note that $\overline{P} = 0$ in Equation (29) leads to a contradiction, whereas the relation (22) implies $[\overline{P}, \overline{q}_{ij}] = 0$ due to tracelessless of $\overline{E}_{j(kl)}^i$.

Still, it is interesting in Equation (29) that the commutator of $\overline{P}_j^i$ with its trace yields it traceless part $\overline{\pi}_j^i$. So let us evaluate the commutation relations involving the traceless part (suppressing the coordinate dependence for simplicity)

$$[\overline{\pi}_j^i(x), \overline{\pi}_l^k(y)] = [\overline{P}_j^i - \frac{1}{3}\delta_j^i\overline{P}, \overline{P}_l^k - \frac{1}{3}\delta_l^k\overline{P}] = [\overline{P}_j^i, \overline{P}_l^k] - \frac{1}{3}\delta_l^k[\overline{P}_j^i, \overline{P}] - \frac{1}{3}\delta_j^i[\overline{P}, \overline{P}_l^k] + \frac{1}{9}\delta_j^i\delta_l^k[\overline{P}, \overline{P}]$$

$$= \frac{i\hbar}{2}(\delta_l^i\overline{P}_j^k - \delta_j^k\overline{P}_l^i)\delta^{(3)}(x,y) \quad (30)$$

where we have used (28) and (29). We can now make the substitution $\overline{P}_j^i = \overline{\pi}_j^i + \frac{1}{3}\delta_j^i\overline{P}$, and the trace part cancels out to yield $[\overline{\pi}_j^i, \overline{\pi}_l^k] = \frac{i\hbar}{2}(\delta_l^i\overline{\pi}_j^k - \delta_j^k\overline{\pi}_l^i)$. The final result of our commutation relations (19), in terms of the traceless momentric variables $\overline{\pi}_j^i$ is given by

$$[\overline{q}_{ij}(x), \overline{q}_{kl}(y)] = 0; \quad [\widehat{\overline{\pi}}_j^i(x), \overline{q}_{kl}(y)] = \frac{\hbar}{i}\overline{E}_{j(kl)}^i\delta^{(3)}(x,y); \quad [\widehat{\overline{\pi}}_j^i(x), \widehat{\overline{\pi}}_l^k(y)] = \frac{i\hbar}{2}(\delta_l^i\overline{\pi}_j^k - \delta_j^k\overline{\pi}_l^i)\delta^{(3)}(x,y). \quad (31)$$

Note that Equation (31) implies a representation of the momentric as a vector field

$$\widehat{\overline{P}}^i_j = \frac{\hbar}{i} \frac{\delta}{\delta \overline{q}_{kl}} \overline{E}^i_{j(kl)} = \frac{\hbar}{i} \overline{E}^i_{j(kl)} \frac{\delta}{\delta \overline{q}_{kl}} + \frac{\hbar}{i} \left[\frac{\delta}{\delta \overline{q}_{kl}}, \overline{E}^i_{j(kl)} \right] = \frac{\hbar}{i} \overline{E}^i_{j(kl)} \frac{\delta}{\delta \overline{q}_{kl}}, \tag{32}$$

which is both self-adjoint and left-right traceless, implements the commutation relations, and is traceless in the sense that $\delta^j_i \overline{\pi}^i_j = \overline{\pi}^i_j \delta^j_i = 0$. There are a few things to note regarding (31). First, upon contraction with δ^j_i, yields consistently that the trace $\delta^j_i \overline{\pi}^i_j = 0$ vanishes as well as its comutator with all quantities. Secondly, the traceless momentric variables by themselves form a $SU(3)$ current algebra, and also generate an affine algebra with the metric, which unlike (19) preserves the positivity of the metric $\overline{q}_{ij}$. Thus, the fundamental variables $\overline{q}_{ij}, \overline{\pi}^i_j$ will be the prime choice for the quantum theory which, at the kinematical level, will involve constructing unitary, irreducible representations of the $SU(3)$ Lie algebra. Also of note is that the the the object $\Delta = \overline{\pi}^i_j \overline{\pi}^j_i$ encodes to the quadratic Casimir of $SU(3)$, which by definition must commute with all traceless momentric components $[\Delta, \overline{\pi}^i_j] = 0$.

The Gell–Mann matrices satisfy the relations

$$[\lambda_A, \lambda_B]^i_j = i f^C_{AB} (\lambda_C)^i_j; \quad \{\lambda_A, \lambda_B\}^i_j = d_{ABC} (\lambda_C)^i_j \tag{33}$$

with totally antisymmetric structure constants f_{ABC}, and totally symmetric d_{ABC}. We will exploit the aforementioned index structure by projection of the momentric onto the Gell–Mann matrices

$$T^A = (\lambda^A)^j_i \overline{\pi}^i_j \longrightarrow \overline{\pi}^i_j = 2 T^A (\lambda_A)^i_j, \tag{34}$$

where we have used the $SU(3)$ completeness relation $(\lambda^A)^i_j (\lambda^A)^k_l = \frac{1}{2} (\delta^k_j \delta^i_l - \frac{1}{3} \delta^i_j \delta^k_l)$. The $SU(3)$ Lie algebra is of rank 2, and therefore has two Casimir operators, $C^{(2)}$ and $C^{(3)}$ given by

$$C^{(2)} = (\lambda_A)^j_i (\lambda_A)^i_j = T^A T^A; \quad C^{(3)} = d_{ABC} (\lambda_A)^i_j (\lambda_B)^j_k (\lambda_C)^k_i = \epsilon^{ijk} \epsilon_{mnl} \overline{\pi}^m_i \overline{\pi}^n_j \overline{\pi}^l_k \propto 6 \det \overline{\pi}^i_j. \tag{35}$$

Note for $C^{(3)}$ that the pair of epsilon symbols is totally symmetric under interchange of any index pair $(i, m), (j, n), (k, l)$, which is consistent with the total symmetry of d_{ABC}.

5. The Classical Theory, Revisited

Having determined the ideal variables for quantization as the unimodular- traceless momentric pair $\overline{q}_{ij}, \overline{\pi}^i_j$, we will now re-evaluate the Poisson brackets of the theory. This provides a basis for correlation of quantum predictions to the classical dynamics. First, the fundamental Poisson brackets are given by

$$\{\overline{q}_{ij}(x), \overline{q}_{kl}(y)\} = 0, \{\overline{q}_{ij}(x), \overline{\pi}^k_l(y)\} = \overline{E}^i_{j(kl)} \delta^{(3)}(x, y); \quad \{\overline{\pi}^i_j(x), \overline{\pi}^k_l(y)\} = \frac{1}{2} (\delta^i_l \overline{\pi}^k_j - \delta^k_j \overline{\pi}^i_l) \delta^{(3)}(x, y). \tag{36}$$

So the Poisson brackets between phase space functions A and B is given by

$$\{A, B\} = \int_\Sigma d^3x \int_\Sigma d^3y \left[\frac{\delta A}{\delta \overline{q}_{ij}(x)} \{\overline{q}_{ij}(x), \overline{q}_{kl}(y)\} \frac{\delta B}{\delta \overline{q}_{kl}(y)} + \frac{\delta A}{\delta \overline{q}_{ij}(x)} \{\overline{q}_{ij}(x), \overline{\pi}^k_l(y)\} \frac{\delta B}{\delta \overline{\pi}^k_l(y)} \right.$$

$$\left. + \frac{\delta A}{\delta \overline{\pi}^i_j(x)} \{\overline{\pi}^i_j(x), \overline{q}_{kl}(y)\} \frac{\delta B}{\delta \overline{q}_{kl}(y)} + \frac{\delta A}{\delta \overline{\pi}^i_j(x)} \{\overline{\pi}^i_j(x), \overline{\pi}^k_l(y)\} \frac{\delta B}{\delta \overline{\pi}^k_l(y)} \right] \tag{37}$$

$$= \int_\Sigma d^3z \left[\overline{E}^k_{j(ij)} \left(\frac{\delta A}{\delta \overline{q}_{ij}} \frac{\delta B}{\delta \overline{\pi}^k_l} - \frac{\delta B}{\delta \overline{q}_{ij}} \frac{\delta A}{\delta \overline{\pi}^k_l} \right) + \frac{\delta A}{\delta \overline{\pi}^i_j} \overline{\pi}^i_l \frac{\delta B}{\delta \overline{\pi}^l_j} - \frac{\delta A}{\delta \overline{\pi}^i_j} \overline{\pi}^k_j \frac{\delta B}{\delta \overline{\pi}^i_l} \right].$$

In General relativity, we will be interested in the evolution of the basic variables with respect to T, gauge-invariant part of intrinsic time $\ln q^{1/3}$, under the action of a physical Hamiltonian

$$H_{Phys} = \int_\Sigma d^3 x \bar{H}(x) = \int_\Sigma d^3 x \sqrt{\bar{\pi}^j_i \bar{\pi}^i_j} + \mathcal{V}[q_{ij}], \tag{38}$$

where $\mathcal{V}$ is a potential term which depends on the metric. The Hamilton's equations for the basic variables with respect to the Poisson brackets (37) are given by

$$\frac{\delta \bar{q}_{ij}(x)}{\delta T} = \{\bar{q}_{ij}(x), H_{Phys}\} = \frac{1}{\bar{H}} \bar{E}^k_{l(ij)} \bar{\pi}^l_k;$$

$$\frac{\delta \bar{\pi}^i_j(x)}{\delta T} = \{\bar{\pi}^i_j(x), H_{Phys}\} = \frac{1}{\bar{H}} \left[\frac{1}{2} \bar{E}^i_{j(kl)} \frac{\delta \mathcal{V}}{\delta \bar{q}_{kl}} + \bar{\pi}^i_l \bar{\pi}^l_j - \bar{\pi}^k_j \bar{\pi}^i_k \right] = \frac{1}{2\bar{H}} \bar{E}^i_{j(kl)} \frac{\delta \mathcal{V}}{\delta \bar{q}_{kl}}. \tag{39}$$

As a quick consistency check, contraction of the first equation of (39) with $\bar{q}^{ij}$ and contraction of the second equation with δ^j_i shows that if $\bar{q}_{ij}$ is unimodular and $\bar{\pi}^i_j$ is traceless at time T_0, then these properties will be preserved under evolution in intrinsic time by the Hamilton's equations.

6. Conclusions

The consistent quantization of 3+1 gravity is one of the biggest unsolved problems in theoretical physics spanning the past 100 years of approaches which, while leading to insights into certain often complementary aspects of the problem, have so far not provided a complete solution due to various technical and conceptual difficulties and issues. The novelty of the author's approach is the claim that with ITQG, one has a complete and consistent quantization of gravity which provides a possible resolution to the long-standing problem, while solving the difficulties inherent in all of the approaches so far, in one stroke.

For future work, we aim to follow the work of this paper with a similar work by focusing on some 2+1 aspect of ITQG, with the aim of studying the thermodynamic aspects of the BTZ black hole. Also, looking at the initial wave function, one difference from the case of 3+1 gravity seems to be the observation that there is no Cotton-York tensor in two spatial dimensions. So we should expect just a Ricci curvature-squared higher derivative rendition of the theory. This then will help us to be able to exploit the SU(2) structure of the theory, which will go a long way towards learning about the physical Hilbert space.

Acknowledgments: Eyo Eyo Ita III and Amos S. Kubeka would like to thank the U.S. Naval Academy and the University of South Africa for the financial support.

Author Contributions: Both authors contributed equally to this work.

Conflicts of Interest: The authors declare no conflict of interest.

References

1. Ashtekar, A. New Variables for Classical and Quantum Gravity. *Phys. Rev. Lett.* **1986**, *57*, 2244–2247.
2. Arnowitt, R.; Deser, S.; Misner, C. Dynamical Structure and Definition of Energy in General Relativity. *Phys. Rev.* **1959**, *5*, 1322–1330.
3. Gambini, R.; Pullin, J. *A First Course in Loop Quantum Gravity*; Oxford University Press: Oxford, UK, 2011.
4. Rovelli C. *Quantum Gravity*; Cambridge University Press: Cambridge, UK, 2004.
5. Fatibene, L.; Francaviglia, M.; Rovelli, C. On a covariant formulation of the Barbero–Immirzi connection. *Class. Quantum Grav.* **2007**, *24*, 11.
6. Ita, E.; Soo, C.; Yu, H.-L. Intrinsic time quantum geometrodynamics. *Prog. Theor. Exp. Phys.* **2015**, *2015*, 083E01.
7. Soo, C.; Yu, H.L. General Relativity without paradigm of space-time covariance, and resolution of the problem of time. *Prog. Theor. Phys.* **2014**, *2014*, 013E01.
8. O'Murchada, N.; Soo, C.; Yu, H.L. Intrinsic time gravity and the Lichnerowicz-York equation. *Class. Quantum Grav.* **2013**, *30*, 095016.

![universe logo] *universe*

|MDPI|

Article

The Teleparallel Equivalent of General Relativity and the Gravitational Centre of Mass

José Wadih Maluf

Instituto de Física, Universidade de Brasília, C. P. 04385, 70.919-970 Brasília DF, Brazil; wadih@unb.br or jwmaluf@gmail.com

Academic Editor: Lorenzo Iorio
Received: 9 June 2016; Accepted: 22 August 2016; Published: 31 August 2016

Abstract: We present a brief review of the teleparallel equivalent of general relativity and analyse the expression for the centre of mass density of the gravitational field. This expression has not been sufficiently discussed in the literature. One motivation for the present analysis is the investigation of the localization of dark energy in the three-dimensional space, induced by a cosmological constant in a simple Schwarzschild-de Sitter space-time. We also investigate the gravitational centre of mass density in a particular model of dark matter, in the space-time of a point massive particle and in an arbitrary space-time with axial symmetry. The results are plausible, and lead to the notion of gravitational centre of mass (COM) distribution function.

Keywords: teleparallel gravity; gravitational centre of mass moment; dark energy; teleparallel equivalent of general relativity

PACS: 04.20.Cv; 04.20.-q; 04.70.Bw

1. Introduction

The most popular and acceptable approach to the relativistic theory of gravitation is given by Einstein's theory of general relativity. However, nowadays there are several alternative formulations of theories for the gravitational field that attempt to explain the dark energy and dark matter problems, which do not find satisfactory explanations within the framework of Einstein's general relativity. Moreover, concepts such as energy, momentum, angular momentum and centre of mass of the gravitational field are usually defined only for asymptotically flat space-times, in the context of a 3+1 type formulation. The latter are definitions for the total quantities, and suffer from at least two restrictions: the definitions are valid only for asymptotically flat space-times, and there do not exist localized expressions for the densities of the energy-momentum and 4-angular momentum of the gravitational field. The ADM definition for the gravitational energy-momentum [1] is constructed out of the metric tensor, and by means of the metric tensor it is not possible to construct suitable scalar densities in the form of total divergences. The approach via pseudo-tensors is certainly not satisfactory. The notions of energy-momentum and angular momentum of the gravitational field have been extensively discussed in the literature, but not the concept of gravitational centre of mass.

The notion of centre of mass can be made clear in flat space-time. Any relativistic field theory in flat space-time is expected to be covariant under the inhomogeneous Lorentz transformations, or Poincaré transformations: the 4-rotations and space-time translations. The generators of these transformations satisfy an algebra, the algebra of the Poincaré group. The generators of the 4-rotations are composed by the generators of the ordinary 3 dimensional rotations, and by the generators of the boosts. The latter are related to the centre of mass moment of the field. Energy, momentum and angular momentum of the field constitute seven conserved integral quantities associated to the symmetries of the theory. The integrals are carried out over the whole three-dimensional space. The three other integral quantities are associated to the centre of mass of the field, which sometimes is also called the centre of energy [2].

In the notation of Ref. [2], the centre of mass integrals read

$$J^{0i} = tP^i - \int d^3x \, x^i T^{00} \,, \tag{1}$$

where 0 and i are time and space indices, P^i is the i-th component of the momentum of the field, and T^{00} is the energy component of the energy-momentum tensor of field. It is argued [2] that the J^{0i} components have no clear physical significance since J^{0i} can be made to vanish if the coordinate system is chosen to coincide with the "centre of energy" at $t = 0$. However, in the context of general relativity, Dixon [3–5] developed a procedure for describing the dynamics of extended bodies in an arbitrary gravitational field, and for this purpose a definition of the centre of mass of such bodies (considered as quasi-rigid bodies) was proposed. A general relation between the centre of mass 4-velocity and the energy-momentum of the body was obtained [6]. One is led to the concept of centre of mass world line, whose uniqueness depends on the strength of the gravitational field [6].

In the standard metric formulation of general relativity, the centre of mass moment for the gravitational field has been first considered by Regge and Teitelboim [7,8], and reconsidered by several other authors (see References [9–12] and references therein). The centre of mass integral was obtained in the context of the Hamiltonian formulation of general relativity. The idea was to require the variation of the total Hamiltonian to be well defined in an asymptotically flat space-time, where the standard asymptotic space-time translations and 4-rotations are considered as coordinate transformations at spacelike infinity. This requirement leads to the addition of boundary (surface) terms to the primary Hamiltonian, so that the latter has well defined functional derivatives, and therefore one may obtain the field equations in the Hamiltonian framework (Hamilton's equations) by means of a consistent procedure. In this way, one arrives at the total energy, momentum, angular momentum and centre of mass moment of the gravitational field, given by surface terms of the total Hamiltonian.

In this article we address the centre of mass moment of the gravitational field in the realm of the teleparallel equivalent of general relativity (TEGR), which is an alternative and mathematically consistent formulation of general relativity [13] (see also Reference [14], and chapters 5 and 6 of Reference [15] and references therein). The geometrical structure of the TEGR was already considered by Einstein [16,17] in his attempt to unify gravity and electromagnetism, and later on by Cho [18,19], Hayashi and Shirafuji [20,21], Hehl et al. [22], Nitsch [23], Schweizer et al. [24], Nester [25] and Wiesendanger [26]). In recent years the teleparallel geometrical structure has been used in modified theories of gravity, with the purpose of constructing cosmological models that provide a consistent explanation to the dark energy problem (see the review article [27] and references therein).

The TEGR is constructed out of the tetrad fields $e^a{}_\mu$, where $a = \{(0), (i)\}$ and $\mu = \{0, i\}$ are SO(3,1) and space-time indices, respectively. The extra six components of the tetrads (compared to the 10 components of the metric tensor) yield additional geometric structure, that allows to define field quantities that cannot be constructed in the ordinary metric formulation of the theory (such as non-trivial total divergences, for instance). The tetrad fields allow to use concepts and definitions of both Riemannian and Weitzenböck geometries.

The definitions of the gravitational energy, momentum, angular momentum and centre of mass moment in the TEGR are not obtained according to the procedure described above, based on surface integrals of the total Hamiltonian. In the TEGR we first consider the Hamiltonian formulation of the theory [28,29]. The constraint equations of the theory (typically as $C = 0$) are equations that define the energy-momentum and the 4-angular momentum of the gravitational field [13] (i.e., $C = H - E = 0$). Moreover, the definitions of the energy-momentum and 4-angular momentum satisfy the algebra of the Poincaré group in the phase space of theory [13,30]. However, the energy-momentum definition, together with the gravitational energy-momentum tensor (but not the 4-angular momentum) may also be obtained directly from the Lagrangian field equations [13].

The gravitational centre of mass moment to be considered here yields the concept of gravitational centre of mass (COM) distribution function. One purpose of the present article is to show that a cosmological constant, which might be responsible for the dark energy, induces a very intense (divergent) gravitational COM distribution function in the vicinity of the cosmological horizon $r = R \simeq \sqrt{3/\Lambda}$ in a simple Schwarzschild-de Sitter space-time, in agreement with the hypothetical existence of dark energy. It seems that this result, obtained by means of tetrad fields, cannot be obtained in the context of the metric formulation of general relativity.

In Section 2 we present a brief review of the TEGR, emphasizing a recent simplified definition of the 4-angular momentum of the gravitational field, given by a total divergence. In Section 3 we investigate the gravitational COM distribution function of (i) the space-time of a massive particle in isotropic coordinates; (ii) the Schwarzschild-de Sitter space-time; (iii) a particular model of dark energy that arises from the non-local formulation of general relativity; and (iv) of an arbitrary space-time with axial symmetry. In the analysis of the first three cases above, which are spherically symmetric, we arrive at interesting results, that share similarities with the standard expressions in classical mechanics. For such space-times, the total centre of mass moment vanishes, as expected.

Notation: space-time indices $\mu, \nu, ...$ and SO(3,1) (Lorentz) indices $a, b, ...$ run from 0 to 3. The torsion tensor is given by $T_{a\mu\nu} = \partial_\mu e_{a\nu} - \partial_\nu e_{a\mu}$. The flat space-time metric tensor raises and lowers tetrad indices, and is fixed by $\eta_{ab} = e_{a\mu}e_{b\nu}g^{\mu\nu} = (-1, +1, +1, +1)$. The frame components are given by the inverse tetrads $\{e_a{}^\mu\}$. The determinant of the tetrad fields is written as $e = \det(e^a{}_\mu)$.

It is important to note that we assume that the space-time geometry is determined by the tetrad fields only, and thus the only possible non-trivial definition for the torsion tensor is given by $T^a{}_{\mu\nu}$. This tensor is related to the antisymmetric part of the Weitzenböck connection $\Gamma^\lambda_{\mu\nu} = e^{a\lambda}\partial_\mu e_{a\nu}$, which determines the Weitzenböck space-time and the distant parallelism of vector fields.

2. A Review of the Lagrangian and Hamiltonian Formulations of the TEGR

The TEGR is constructed out of the tetrad fields only. The first relevant consideration is an identity between the scalar curvature and an invariant combination of quadratic terms in the torsion tensor,

$$eR(e) \equiv -e\left(\frac{1}{4}T^{abc}T_{abc} + \frac{1}{2}T^{abc}T_{bac} - T^a T_a\right) + 2\partial_\mu(eT^\mu), \tag{2}$$

where $T_a = T^b{}_{ba}$ and $T_{abc} = e_b{}^\mu e_c{}^\nu T_{a\mu\nu}$. The Lagrangian density for the gravitational field in the TEGR is given by [31]

$$
\begin{aligned}
L(e) &= -ke\left(\frac{1}{4}T^{abc}T_{abc} + \frac{1}{2}T^{abc}T_{bac} - T^a T_a\right) - \frac{1}{c}L_M \\
&\equiv -ke\Sigma^{abc}T_{abc} - \frac{1}{c}L_M,
\end{aligned}
\tag{3}
$$

where $k = c^3/(16\pi G)$, L_M represents the Lagrangian density for the matter fields, and Σ^{abc} is defined by

$$\Sigma^{abc} = \frac{1}{4}\left(T^{abc} + T^{bac} - T^{cab}\right) + \frac{1}{2}\left(\eta^{ac}T^b - \eta^{ab}T^c\right). \tag{4}$$

Thus, the Lagrangian density is geometrically equivalent to the scalar curvature density. The variation of $L(e)$ with respect to $e^{a\mu}$ yields the fields equations

$$e_{a\lambda}e_{b\mu}\partial_\nu(e\Sigma^{b\lambda\nu}) - e(\Sigma^{b\nu}{}_a T_{b\nu\mu} - \frac{1}{4}e_{a\mu}T_{bcd}\Sigma^{bcd}) = \frac{1}{4kc}eT_{a\mu}, \tag{5}$$

where $T_{a\mu}$ is defined by $\delta L_M/\delta e^{a\mu} = eT_{a\mu}$.

The field equations are equivalent to Einstein's equations. It is possible to verify by explicit calculations that the equations above can be rewritten as

$$\frac{1}{2}[R_{a\mu}(e) - \frac{1}{2}e_{a\mu}R(e)] = \frac{1}{4kc}T_{a\mu}, \tag{6}$$

Since the Lagrangian density (3) does not contain the total divergence that arises on the right hand side of Equation (2), it is not invariant under arbitrary local SO(3,1) transformations, but the field Equation (5) are covariant under such transformations.

The equivalence between the TEGR and the standard metric formulation of general relativity is based on the equivalence of Equations (5) and (6). However, in the TEGR there are additional field quantities (like third order tensors) constructed by means of the tetrad fields, such as total divergences, for instance, that cannot be obtained in the standard metric formulation. These additional field quantities are covariant under the global Lorentz transformations, but not under local transformations. In the ordinary formulation of arbitrary field theories, energy, momentum, angular momentum and COM moment are frame dependent field quantities, that transform under the global SO(3,1) transformations. In particular, energy transforms as the zero component of the energy-momentum four-vector. This feature must hold also in the presence of the gravitational field. As an example, consider the total energy of a black hole, represented by the mass parameter m. As seen by a distant observer, the total energy of a static Schwarzschild black hole is given by $E = mc^2$. However, at great distances the black hole may be considered as a particle of mass m, and if it moves with constant velocity v, then its total energy as seen by the same distant observer is $E = \gamma mc^2$, where $\gamma = (1 - v^2/c^2)^{-1/2}$. Likewise, the gravitational momentum, angular momentum and the COM moment are also frame dependent field quantities in general, whose values are different for different frames and different observers. On physical grounds, energy, momentum, angular momentum and COM moment cannot be local Lorentz *invariant* field quantities, since these quantities depend on the frame, as we know from special relativity, which is the limit of the general theory of relativity when the gravitational field is weak or negligible.

After some rearrangements, Equation (5) may be written in the form [13]

$$\partial_\nu(e\Sigma^{a\mu\nu}) = \frac{1}{4k}ee^a{}_\nu(t^{\mu\nu} + \frac{1}{c}T^{\mu\nu}), \tag{7}$$

where

$$t^{\mu\nu} = k(4\Sigma^{bc\mu}T_{bc}{}^\nu - g^{\mu\nu}\Sigma^{bcd}T_{bcd}), \tag{8}$$

is interpreted as the gravitational energy-momentum tensor [13,32] and $T^{\mu\nu} = e_a{}^\mu T^{a\nu}$.

The Hamiltonian density of the TEGR is constructed as usual in the phase space of the theory. We first note that the Lagrangian density (3) does not depend on the time derivatives of e_{a0}. Therefore, the latter arise as Lagrange multipliers in the Hamiltonian density H. The momenta canonically conjugated to e_{a0} are denoted by Π^{a0}. The latter are primary constraints of the theory: $\Pi^{a0} \approx 0$. The momenta canonically conjugated to e_{ai} are given by $\Pi^{ai} = \delta L/\delta\dot{e}_{ai} = -4k\Sigma^{a0i}$. The Hamiltonian density is obtained by rewriting the Lagrangian density in the form $L = \Pi^{ai}\dot{e}_{ai} - H$, in terms of e_{ai}, Π^{ai} and Lagrange multipliers. After the Legendre transform is performed, we obtain the final form of the Hamiltonian density. It reads [29,30]

$$H(e,\Pi) = e_{a0}C^a + \lambda_{ab}\Gamma^{ab}. \tag{9}$$

where λ_{ab} are Lagrange multipliers. In the above equation we have omitted a surface term. $C^a = \delta H/\delta e_{a0}$ is a long expression of the field variables, and $\Gamma^{ab} = -\Gamma^{ba}$ are defined by

$$\Gamma^{ab} = 2\Pi^{[ab]} + 4ke(\Sigma^{a0b} - \Sigma^{b0a}). \tag{10}$$

After solving the field equations, the Lagrange multipliers are identified as $\lambda_{ab} = (1/4)(T_{a0b} - T_{b0a} + e_a{}^0T_{00b} - e_b{}^0T_{00a})$. The constraints C^a may be written as

$$C^a = -\partial_i \Pi^{ai} - p^a = 0, \tag{11}$$

where p^a is an intricate expression of the field quantities.

The quantities C^a and Γ^{ab} are first class constraints. They satisfy an algebra similar to the algebra of the Poincaré group [29]. The integral form of the constraint equations $C^a = 0$ yields the gravitational energy-momentum P^a [13],

$$P^a = -\int_V d^3x \, \partial_i \Pi^{ai}, \tag{12}$$

where V is an arbitrary volume of the three-dimensional space and $\Pi^{ai} = -4k\Sigma^{a0i}$. In similarity to the definition above, the definition of the gravitational 4-angular momentum follows from the constraint equations $\Gamma^{ab} = 0$ [30]. However, it has been noted [33] that the second term on the right hand side of Equation (10) can be rewritten as a total divergence, so that the constraints Γ^{ab} become

$$\Gamma^{ab} = 2\Pi^{[ab]} - 2k\partial_i[e(e^{ai}e^{b0} - e^{bi}e^{a0})] = 0. \tag{13}$$

Therefore, the definition of the total 4-angular momentum of the gravitational field L^{ab} may be given by an integral of a total divergence, in similarity to Equation (12). We have

$$L^{ab} = -\int_V d^3x \, 2\Pi^{[ab]}, \tag{14}$$

where

$$2\Pi^{[ab]} = (\Pi^{ab} - \Pi^{ba}) = 2k\partial_i[e(e^{ai}e^{b0} - e^{bi}e^{a0})]. \tag{15}$$

It is easy to show [30] that expressions (12) and (14) satisfy the algebra of the Poincaré group in the phase space of the theory,

$$\begin{aligned}
\{P^a, P^b\} &= 0, \\
\{P^a, L^{bc}\} &= \eta^{ab}P^c - \eta^{ac}P^b, \\
\{L^{ab}, L^{cd}\} &= \eta^{ad}L^{cb} + \eta^{bd}L^{ac} - \eta^{ac}L^{db} - \eta^{bc}L^{ad}.
\end{aligned} \tag{16}$$

Therefore, from a physical point of view, the interpretation of the quantities P^a and L^{ab} is consistent.

Definitions (12) and (14) are invariant under coordinate transformations of the three-dimensional, under time reparametrizations, and under global SO(3,1) transformations. The gravitational energy is the zero component of the energy-momentum four vector P^a.

3. The Centre of Mass Moment

The gravitational centre of mass (COM) moment is given by the components

$$L^{(0)(i)} = -\int d^3x \, M^{(0)(i)}, \tag{17}$$

where

$$M^{(0)(i)} = 2\Pi^{[(0)(i)]} = 2k\partial_j[e(e^{(0)j}e^{(i)0} - e^{(i)j}e^{(0)0})], \tag{18}$$

according to definition (15). The quantity $-M^{(0)(i)}$ is identified as the gravitational COM density. The evaluation of the expression above is very simple. One needs just to establish the suitable set of tetrad fields that define a frame in space-time.

The inverse tetrads $e_a{}^\mu$ are interpreted as a frame adapted to a particular class of observers in space-time. Let the curve $x^\mu(\tau)$ represent the timelike worldline C of an observer in space-time, where τ is the proper time of the observer. The velocity of the observer along C is given by $u^\mu = dx^\mu/d\tau$. A frame adapted to this observer is constructed by identifying the timelike component of the frame

$e_{(0)}{}^\mu$ with the velocity u^μ of the observer: $e_{(0)}{}^\mu = u^\mu(\tau)$. The three other components of the frame, $e_{(i)}{}^\mu$, are orthogonal to $e_{(0)}{}^\mu$, and may be oriented in the three-dimensional space according to the symmetry of the physical system. If the space-time has axial symmetry, for instance, then the $e_{(3)}{}^\mu$ components of the tetrad fields are chosen to be oriented, asymptotically, along the z axis of the coordinate system, i.e., $e_{(3)}{}^\mu(t, x, y, z) \simeq (0, 0, 0, 1)$ in the limit $r \to \infty$. A static observer in space-time is defined by the condition $u^\mu = (u^0, 0, 0, 0)$. Thus, a frame adapted to a static observer in space-time must satisfy the conditions $e_{(0)}{}^i(t, x^k) = (0, 0, 0)$.

An alternative way to characterise a frame in space-time is by means of the acceleration tensor ϕ_{ab} [34–38],

$$\phi_{ab} = \frac{1}{2}[T_{(0)ab} + T_{a(0)b} - T_{b(0)a}]. \tag{19}$$

This tensor is invariant under coordinate transformations and covariant under global SO(3,1) transformations, but not under local SO(3,1) transformations. It yields the inertial (i.e., the non-gravitational) accelerations that are necessary to impart to a frame in space-time in order to maintain the frame in a given inertial state. Three components of ϕ_{ab} yield the translational accelerations, and three other components yield the frequency of rotation of the frame. Altogether, these six components cancel the gravitational acceleration, so that the frame is kept in a particular inertial state.

In the following, we will evaluate the density of the centre of mass moment of four space-time configurations that exhibit spherical symmetry. In the four cases we will establish the frame of a static observer in space-time.

3.1. The Space-Time of a Massive Point Particle

The Schwarzschild solution in isotropic coordinates represents the space-time of a point massive particle [39,40]. It is obtained as an exact solution of Einstein's equations by writing the energy-momentum tensor in terms of a δ function of a point particle of mass M, with support at the origin of the coordinate system. The solution is described by the line element

$$ds^2 = -\alpha^2 c^2 dt^2 + \beta^2 [dr^2 + r^2(d\theta^2 + \sin^2\theta \, d\phi^2)], \tag{20}$$

where

$$\alpha^2 = \left(\frac{1 - \frac{m}{2r}}{1 + \frac{m}{2r}}\right)^2, \quad \beta^2 = \left(1 + \frac{m}{2r}\right)^4. \tag{21}$$

The parameter $m = GM/c^2$ represents the mass of the point particle that appears in the energy-momentum tensor. The line element above is clearly a solution of Equation (6), with the appropriate energy-momentum tensor $T_{a\mu}$ described in Reference [40].

By performing a coordinate transformation to (x, y, z) coordinates where

$$
\begin{aligned}
x &= r \sin\theta \cos\phi, \\
y &= r \sin\theta \sin\phi, \\
z &= r \cos\theta,
\end{aligned}
\tag{22}
$$

the line element becomes

$$ds^2 = -\alpha^2 c^2 dt^2 + \beta^2 (dx^2 + dy^2 + dz^2). \tag{23}$$

The tetrad fields adapted to static observers is given by

$$e_{a\mu}(t, x, y, z) = \begin{pmatrix} -\alpha & 0 & 0 & 0 \\ 0 & \beta & 0 & 0 \\ 0 & 0 & \beta & 0 \\ 0 & 0 & 0 & \beta \end{pmatrix}. \tag{24}$$

Taking into account Equation (18), straightforward calculations yield $M^{(0)(1)} = 2k\,\partial_1\beta^2$, $M^{(0)(2)} = 2k\,\partial_2\beta^2$ and $M^{(0)(3)} = 2k\,\partial_3\beta^2$. It is easy to obtain

$$
\begin{aligned}
-M^{(0)(1)} &= d_g\,x\,, \\
-M^{(0)(2)} &= d_g\,y\,, \\
-M^{(0)(3)} &= d_g\,z\,.
\end{aligned}
\tag{25}
$$

The quantity d_g is defined by

$$
d_g = \frac{4\,k\,m}{r^3}\left(1 + \frac{m}{2r}\right)^3.
\tag{26}
$$

Therefore,

$$
\begin{aligned}
L^{(0)(1)} &= \int d^3x\,d_g x\,, \\
L^{(0)(2)} &= \int d^3x\,d_g y\,, \\
L^{(0)(3)} &= \int d^3x\,d_g z\,,
\end{aligned}
\tag{27}
$$

where $d^3x = dx\,dy\,dz$ and $r^2 = x^2 + y^2 + z^2$. The expressions above remind the definition of centre of mass in classical mechanics. Given that $M^{(0)(i)} = 2k\,\partial_i\beta^2$, it is easy to see that all integrals given by Equation (17) vanish, namely, all components of the total centre of mass moment vanish. However, the field quantity (26) has the following properties:

$$
\begin{aligned}
r \to \infty &\;:\; d_g \to 0\,, \\
r \to 0 &\;:\; d_g \to \infty\,.
\end{aligned}
\tag{28}
$$

Thus, d_g is more intense in the vicinity of the particle, and vanishes at spatial infinity. In view of Equations (27) and (28), d_g may be interpreted as the gravitational COM distribution function. It is clearly related to the intensity of the gravitational field. The analyses of the space-time configurations below support this interpretation, as we will see.

3.2. The Schwarzschild-de Sitter Space-Time

The line element of the Schwarzschild-de Sitter space-time is given by

$$
ds^2 = -\alpha^2\,dt^2 + \frac{1}{\alpha^2}dr^2 + r^2 d\theta^2 + r^2\sin^2\theta\,d\phi^2\,,
\tag{29}
$$

where

$$
\alpha^2 = 1 - \frac{2m}{r} - \frac{r^2}{R^2}\,,
\tag{30}
$$

$R = \sqrt{3/\Lambda}$ and Λ is the cosmological constant. Here we are considering the speed of light $c = 1$. The Schwarzschild-de Sitter space-time has been considered in the TEGR in Reference [41]. The set of tetrad fields adapted to stationary observers in space-time is given by

$$
e_{a\mu} = \begin{pmatrix}
-\alpha & 0 & 0 & 0 \\
0 & \alpha^{-1}\sin\theta\,\cos\phi & r\cos\theta\,\cos\phi & -r\sin\theta\,\sin\phi \\
0 & \alpha^{-1}\sin\theta\,\sin\phi & r\cos\theta\,\sin\phi & r\sin\theta\,\cos\phi \\
0 & \alpha^{-1}\cos\theta & -r\sin\theta & 0
\end{pmatrix}.
\tag{31}
$$

After long but simple calculations we find that the components of Equation (18) read

$$-M^{(0)(1)} = 4k\sin\theta\left(\frac{1}{\alpha}-1\right)r\sin\theta\cos\phi,$$

$$-M^{(0)(2)} = 4k\sin\theta\left(\frac{1}{\alpha}-1\right)r\sin\theta\sin\phi,$$

$$-M^{(0)(3)} = 4k\sin\theta\left(\frac{1}{\alpha}-1\right)r\cos\theta. \tag{32}$$

We identify $x = r\sin\theta\cos\phi$, $y = r\sin\theta\sin\phi$, $z = r\cos\theta$ as usual, and write Equation (17) as

$$L^{(0)(1)} = \int d^3x\, 4k\sin\theta\left(\frac{1}{\alpha}-1\right)x,$$

$$L^{(0)(2)} = \int d^3x\, 4k\sin\theta\left(\frac{1}{\alpha}-1\right)y,$$

$$L^{(0)(3)} = \int d^3x\, 4k\sin\theta\left(\frac{1}{\alpha}-1\right)z, \tag{33}$$

where $d^3x = dr\, d\theta\, d\phi$. Integration in the angular variables implies the vanishing of the three integrals $L^{(0)(i)}$, i.e., the total centre of mass vanishes, as expected. The equations above may be written exactly as Equation (27) provided we identify

$$d_g = 4k\sin\theta\left(\frac{1}{\alpha}-1\right) \equiv 4k\sin\theta\, f(r). \tag{34}$$

The analysis of the expression above leads to interesting results. Let r_1 and r_2 denote the two horizons of the Schwarzschild-de Sitter space-time, $\alpha(r_1) = 0$ and $\alpha(r_2) = 0$, so that $r_1 < r_2$. The radius r_1 is close to the Schwarzschild radius, $r_1 \approx \frac{2m}{r}$, and $r_2 \approx R$. We have

$$r \to r_1 \quad : \quad f(r) \to \infty,$$

$$r \to r_2 \quad : \quad f(r) \to \infty. \tag{35}$$

The function $f(r)$ is defined by Equation (34). The minimum of $f(r)$ is given by

$$\frac{df}{dr} = -\frac{1}{\alpha^2}\frac{d\alpha}{dr} = 0,$$

and takes place at $r_{min} = (mR^2)^{1/3}$. Thus, d_g is intense close to both r_1 and r_2, i.e., close to the Schwarzschild and cosmological horizons.

The radial position r_{min} is related to the inertial accelerations of an observer. In order to understand this feature, we evaluate the translational (non-gravitational) accelerations of a frame given by Equation (19). We find

$$\phi_{(0)(1)} = \frac{d\alpha}{dr}\sin\theta\cos\phi,$$

$$\phi_{(0)(2)} = \frac{d\alpha}{dr}\sin\theta\sin\phi,$$

$$\phi_{(0)(3)} = \frac{d\alpha}{dr}\cos\theta. \tag{36}$$

We define the inertial acceleration vector Φ as

$$\Phi(r) = (\phi_{(0)(1)}, \phi_{(0)(2)}, \phi_{(0)(3)}) \equiv \phi(r)\hat{r} = \frac{d\alpha}{dr}\hat{r}, \tag{37}$$

where $\hat{r} = (\sin\theta\cos\phi, \sin\theta\sin\phi, \cos\theta)$. Since

$$\frac{d\alpha}{dr} = \frac{1}{\alpha}\left(\frac{m}{r^2} - \frac{r}{R^2}\right),$$

we see that

$$r_1 < r < r_{min}: \quad \frac{d\alpha}{dr} > 0 \quad \to \quad \phi(r) > 0,$$

$$r_{min} < r < r_2: \quad \frac{d\alpha}{dr} < 0 \quad \to \quad \phi(r) < 0. \tag{38}$$

Thus, given that the inertial acceleration $\phi(r) > 0$ is repulsive in the region $r_1 < r < r_{min}$, the gravitational acceleration is attractive in this interval, as expected. By means of a similar argument, we see that the gravitational acceleration is repulsive in the region $r_{min} < r < r_2$, as expected.

In view of the analysis above, we may interpret d_g given by Equation (34) as the gravitational COM distribution function, in similarity to Equation (26), and therefore one may understand the gravitational repulsion as attraction to a region of intense gravitational COM distribution function, which, in the present case, is the region in the vicinity of the cosmological horizon. If dark energy is indeed related to the existence of a cosmological constant, then it is natural that it is concentrated close to the radius $r_2 \approx R = \sqrt{3/\Lambda}$ in the context of a simple Schwarzschild-de Sitter model.

The function d_g plays the role of a gravitational COM density. However, mathematically it is not a density. The integrands in Equations (27) and (33) are in fact densities, but not d_g alone. In Newtonian mechanics, d_g in Equations (27) and (33) plays the role of mass density.

3.3. Dark Matter Simulated by Non-Local Gravity

A non-local formulation of general relativity, based on a geometrical framework similar to the one established by Equations (2)–(4) has been developed by Hehl, Mashhoon and collaborators [42–44]. One interesting consequence of this development is an extension of Newtonian gravity that may play a relevant role in the dynamics of galaxies, and might provide an explanation that is expected to come from dark matter models of gravity. We restrict the considerations to a simplified space-time with spherical symmetry, so that Equations (29), (31), (33) and (34) remain valid.

The Newtonian approximation is established by

$$\alpha^2 = -g_{00} \simeq 1 + \frac{2\Phi_g}{c^2}, \tag{39}$$

where Φ_g is the Newtonian potential, and $2\Phi_g/c^2 \ll 1$.

It follows that

$$f(r) = \frac{1}{\alpha} - 1 \simeq -\frac{\Phi_g}{c^2}. \tag{40}$$

The Newtonian potential that arises in the non-local formulation of gravity is given by [43,44]

$$\Phi_g \simeq -\frac{GM}{r} + \frac{GM}{\lambda}\ln\left(\frac{r}{\lambda}\right), \tag{41}$$

where λ is a constant length, and is taken to be $\lambda \approx 1kpc = 3260$ light-years. Consequently, the influence of the second term on the right hand side of Equation (41) in the solar system is negligible. Therefore, we find

$$f(r) = \frac{1}{\alpha} - 1 \simeq \frac{m}{r} - \frac{m}{\lambda}\ln\left(\frac{r}{\lambda}\right). \tag{42}$$

In the expression above, $m = GM/c^2$. For values of r within a galaxy, $r < \lambda$ and thus $-(m/\lambda)\ln(r/\lambda)$ is positive, and decreases as $1/r$ with increasing values of r, a result that shows that the gravitational field is sufficiently intense at the borders of a galaxy to explain the rotation curves of spiral galaxies. The function $d_g = 4k \sin\theta f(r)$ may again be understood as the gravitational COM distribution function of the spherically symmetric space-time.

3.4. Arbitrary Space-Time with Axial Symmetry

The analysis of a space-time that is not spherically symmetric allows to obtain the generalization of Equations (27) and (33). Let us consider an arbitrary space-time with axial symmetry. It is described by following line element,

$$ds^2 = g_{00}dt^2 + g_{11}dr^2 + g_{22}d\theta^2 + g_{33}d\phi^2 + 2g_{03}d\phi\,dt\,, \tag{43}$$

where all metric components depend on r and θ, but not on ϕ : $g_{\mu\nu} = g_{\mu\nu}(r,\theta)$. The determinant $e = \sqrt{-g}$ is $e = [g_{11}g_{22}\delta]^{1/2}$, where

$$\delta = g_{03}g_{03} - g_{00}g_{33}\,.$$

The inverse metric components are $g^{00} = -g_{33}/\delta$, $g^{03} = g_{03}/\delta$ and $g^{33} = -g_{00}/\delta$.

The set of tetrad fields in spherical coordinates that is adapted to static observers in space-time is given by

$$e_{a\mu} = \begin{pmatrix} -A & 0 & 0 & -C \\ 0 & \sqrt{g_{11}}\sin\theta\cos\phi & \sqrt{g_{22}}\cos\theta\cos\phi & -Dr\sin\theta\sin\phi \\ 0 & \sqrt{g_{11}}\sin\theta\sin\phi & \sqrt{g_{22}}\cos\theta\sin\phi & Dr\sin\theta\cos\phi \\ 0 & \sqrt{g_{11}}\cos\theta & -\sqrt{g_{22}}\sin\theta & 0 \end{pmatrix}. \tag{44}$$

The functions A, C and D are defined such that Equation (44) yields (43). They read

$$\begin{aligned} A(r,\theta) &= (-g_{00})^{1/2}, \\ C(r,\theta) &= -\frac{g_{03}}{(-g_{00})^{1/2}}, \\ D(r,\theta) &= \frac{1}{(r\sin\theta)}\left[\frac{\delta}{(-g_{00})}\right]^{1/2}. \end{aligned} \tag{45}$$

After simple calculations, we find that Equations (17) and (18) yield

$$\begin{aligned} L^{(0)(1)} &= \int d^3x\, d_{g1}\,(r\sin\theta\cos\phi), \\ L^{(0)(2)} &= \int d^3x\, d_{g2}\,(r\sin\theta\cos\phi), \\ L^{(0)(3)} &= \int d^3x\, d_{g3}\,(r\cos\theta), \end{aligned} \tag{46}$$

where now we have

$$\begin{aligned} d_{g1} = d_{g2} &= 2k\left\{-\frac{1}{r}\partial_1\left[\frac{g_{22}\delta}{(-g_{00})}\right]^{1/2} - \frac{1}{r\sin\theta}\partial_2\left[\left(\frac{g_{11}\delta}{(-g_{00})}\right)^{1/2}\cos\theta\right]\right. \\ &\left.+\frac{1}{r\sin\theta}(g_{11}g_{22})^{1/2}\right\}, \\ d_{g3} &= 2k\left\{-\frac{1}{r}\partial_1\left[\frac{g_{22}\delta}{(-g_{00})}\right]^{1/2} + \frac{1}{r\cos\theta}\partial_2\left[\left(\frac{g_{11}\delta}{(-g_{00})}\right)^{1/2}\sin\theta\right]\right\}. \end{aligned} \tag{47}$$

In the flat space-time, the quantities above vanish. It is not difficult to see that if the metric tensor components above represent the exterior gravitational field of a typical rotating source, the expressions above are not divergent. Note that $L^{(0)(1)}$ and $L^{(0)(2)}$ vanish due to integration in ϕ, as a consequence of the axial symmetry, but $L^{(0)(3)}$ is non-vanishing in general.

In the equations above we obtain $d_{g1} = d_{g2}$ because of the axial symmetry of the space-time. We see that, in general, we may have three different COM distribution functions, one for each direction in the three-dimensional space, in contrast to the situation in classical mechanics, where there is a single mass density in the definition of centre of mass.

4. Conclusions

In this article we have investigated the definition of centre of mass of the gravitational field, in the realm of the teleparallel equivalent of general relativity. The analysis of the gravitational centre of mass density leads to the concept of COM distribution function. We may understand the latter as a quantity that provides a description of the intensity of the gravitational field in space-time. The emergence of this quantity justifies the analysis of the centre of mass density of arbitrary configurations of the gravitational field, including gravitational wave configurations. We have applied this definition to the space-time endowed with a positive cosmological constant. At the speculative level, dark energy might be a consequence of the existence of a positive cosmological constant that induces a strong gravitational acceleration very far from our present location in the universe. In the simple model established by the Schwarzschild-de Sitter space-time, dark energy is roughly located in the region beyond $r = r_{min} = (mR^2)^{1/3}$, according to Equation (38).

The centre of mass moment naturally arises in the Hamiltonian formulation of the teleparallel equivalent of general relativity, and its definition is obtained from the primary constraints of the theory—Equation (13). It is given by Equations (17) and (18). The analysis led us to interpret the quantity d_g in the integrand of Equations (27), (33) and (46) as the gravitational COM distribution function. Although d_g plays the role of a density, mathematically it is not a density. It vanishes when the gravitational field is turned off. The expressions of $L^{(0)(i)}$ given by Equations (27) and (33) do remind the standard expression of centre of mass in classical mechanics. The distribution function d_g in the three-dimensional space is related to the intensity of the gravitational field. In the space-time of a point massive particle, d_g is intense (and in fact diverges) in the vicinity of the particle, and in the Schwarzschild-de Sitter space-time d_g is positive definite and diverges at both the Schwarzschild and cosmological horizons, which are precisely the regions where the gravitational field is more intense.

In relativistic field theory or in the Newtonian approximation of general relativity, energy, momentum and angular momentum are frame dependent field quantities, and so they are, in general, in the present context. In particular, the gravitational COM moment is evaluated in the frame adapted to an arbitrary observer in space-time. The gravitational centre of mass given by Equations (17) and (18) is invariant under coordinate transformations of the three-dimensional space, and under time reparametrizations. It transforms covariantly under global SO(3,1) transformations, provided the tetrad fields transform as $\tilde{e}^a{}_\mu = \Lambda^a{}_b e^b{}_\mu$, where $\Lambda^a{}_b$ are matrices of the SO(3,1) group. However, definition (17) is not covariant under local SO(3,1) transformations. In relativistic field theory, the COM definition is also not covariant under local SO(3,1) transformations.

We conclude that repulsion, in the Schwarzschild-de Sitter space-time, is in fact attraction to a region of intense gravitational COM distribution function. We have seen that in the region $r < r_{min} = (mR^2)^{1/3}$ the gravitational acceleration is attractive, and is repulsive in the dark energy region $r > r_{min}$. We expect the present analysis to be useful in the investigation of realistic cosmological models endowed with a positive cosmological constant.

Acknowledgments: I am grateful to B. Mashhoon for enlightening comments and for pointing out relevant references.

Conflicts of Interest: The authors declare no conflict of interest.

References

1. Arnowitt, R.; Deser, S.; Misner, C.W. *Gravitation: An Introduction to Current Research*; Witten, L., Ed.; Wiley: New York, NY, USA, 1962.
2. Weinberg, S. *Gravitation and Cosmology*; Wiley: New York, NY, USA, 1972.

3. Dixon, W.G. Dynamics of Extended Bodies in General Relativity. I. Momentum and Angular Momentum. *Proc. R. Soc. Lond. A* **1970**, *314*, 499–527.

4. Dixon, W.G. Dynamics of Extended Bodies in General Relativity. II. Moments of the Charge-Current Vector. *Proc. R. Soc. Lond. A* **1970**, *319*, 509–547.

5. Dixon, W.G. The definition of multipole moments for extended bodies. *Gen. Relativ. Gravit.* **1973**, *4*, 199–209.

6. Ehlers, J.; Rudolph, E. Dynamics of extended bodies in general relativity center-of-mass description and quasirigidity. *Gen. Relativ. Gravit.* **1977**, *8*, 197–217.

7. Regge, T.; Teitelboim, C. Role of surface integrals in the Hamiltonian formulation of general relativity. *Ann. Phys.* **1974**, *88*, 286–318.

8. Hanson, A.; Regge, T.; Teitelboim, C. *Constrained hamiltonian systems*; Accademia Nazionale dei Lincei: Roma, Italy, 1976.

9. Beign, R.; ÓMurchadha, N. The Poincaré group as the symmetry group of canonical general relativity. *Ann. Phys.* **1987**, *174*, 463–498.

10. Baskaran, D.; Lau, S.R.; Petrov, A.N. Center of mass integral in canonical general relativity. *Ann. Phys.* **2003**, *307*, 90–131.

11. Nester, J.M.; Ho, F.H.; Chen, C.M. Quasilocal Center-of-Mass for Teleparallel Gravity. 2004, arXiv:gr-qc/0403101.

12. Nester, J.M.; Meng, F.F.; Chen, C.M. Quasilocal Center-of-Mass. *J. Korean Phys. Soc.* **2004**, *45*, S22–S25.

13. Maluf, J.W. The teleparallel equivalent of general relativity. *Ann. Phys.* **2013**, *525*, 339–357.

14. Aldrovandi, R.; Pereira, J.G. *Teleparallel Gravity: An Introduction*; Springer: Heidelberg, Germany, 2013.

15. Blagojevic, M.; Hehl, F.W. *Gauge Theories of Gravitation*; Imperial College: London, UK, 2013.

16. Einstein, A. Riemannsche Geometrie unter Aufrechterhaltung des Begriffes des Fernparallelismus (Riemannian Geometry with Maintaining the Notion of Distant Parallelism). In *Sitzungsberichte der Preussischen Akademie der Wissenshcaften*; Verlag der Akademie der Wissenschaften: Berlin, Germany, 1928; pp. 217–221.

17. Einstein, A. Unified Field Theory based on Riemannian Metrics and Distant Parallelism. *Math. Ann.* **1930**, *102*, 685–697.

18. Cho, Y.M. Einstein Lagrangian as the translational Yang-Mills Lagrangian. *Phys. Rev. D* **1976**, *14*, 2521–2525;

19. Cho, Y.M. Gauge theory of Poincaré symmetry. *Phys. Revs. D* **1976**, *14*, 3335–3341.

20. Hayashi, K.; Shirafuji, T. New general relativity. *Phys. Rev. D* **1979**, *19*, 3524–3554.

21. Hayashi, K.; Shirafuji, T. Addendum to "New general relativity". *Phys. Rev. D* **1981**, *24*, 3312–3315.

22. Hehl, F.W. Four lectures on poincaré Gauge field theory. In *Cosmology and Gravitation: Spin, Torsion, Rotation and Supergravity*; Bergmann, P.G., de Sabbata, V., Eds.; Plenum Press: New York, NY, USA, 1980.

23. Nitsch, J. The macroscopic limit of the poincaré gauge field theory on gravitation. In *Cosmology and Gravitation: Spin, Torsion, Rotation and Supergravity*; Bergmann, P.G., de Sabbata, V., Eds.; Plenum Press: New York, NY, USA, 1980.

24. Schweizer, M.; Straumann, N.; Wipf, A. Post-Newtonian Generation of Gravitational Waves in a Theory with Torsion. *Gen. Relativ. Gravit.* **1980**, *12*, 951–961.

25. Nester, J.M. Positive energy via the teleparallel Hamiltonian. *Int. J. Mod. Phys. A* **1989**, *4*, 1755–1772.

26. Wiesendanger, C. Translational gauge invariance and classical gravitodynamics. *Class. Quantum Gravity* **1995**, *12*, 585–603.

27. Cai, Y.-F.; Capozziello, S.; de Laurentis, M.; Saridakis, E.N. $f(T)$ Teleparallel Gravity and Cosmology. 2015, arXiv:1511.07586.

28. Maluf, J.W.; da Rocha-Neto, J.F. Hamiltonian formulation of general relativity in the teleparallel geometry. *Phys. Rev D* **2001**, *64*, 084014.

29. Da Rocha-Neto, J.F.; Maluf, J.W.; Ulhoa, S.C. Hamiltonian formulation of unimodular gravity in the teleparallel geometry. *Phys. Rev. D* **2010**, *82*, 124035.

30. Maluf, J.W.; Ulhoa, S.C.; Faria, F.F.; da Rocha-Neto, J.F. The angular momentum of the gravitational field and the Poincaré group. *Class. Quantum Gravity* **2006**, *23*, 6245–6256.

31. Maluf, J.W. Hamiltonian formulation of the teleparallel description of general relativity. *J. Math. Phys.* **1994**, *35*, 335–343.

32. Maluf, J.W. The gravitational energy-momentum tensor and the gravitational pressure. *Ann. Phys. (Berlin)* **2005**, *14*, 723–732.

33. Da Rocha-Neto, J.F.; Maluf, J.W. The angular momentum of plane-fronted gravitational waves in the teleparallel equivalent of general relativity. *Gen. Relativ. Gravit.* **2014**, *46*, 1667.
34. Mashhoon, B.; Muench, U. Length measurement in accelerated systems. *Ann. Phys. (Berlin)* **2002**, *11*, 532–547.
35. Mashhoon, B. Vacuum electrodynamics of accelerated systems: Nonlocal Maxwell's equations. *Ann. Phys. (Berlin)* **2003**, *12*, 586–598.
36. Maluf, J.W.; Faria, F.F.; Ulhoa, S.C. On reference frames in spacetime and gravitational energy in freely falling frames. *Class. Quantum Gravity* **2007**, *24*, 2743–2754.
37. Maluf, J.W.; Faria, F.F. On the construction of Fermi-Walker transported frames. *Ann. Phys. (Berlin)* **2008**, *17*, 326–335.
38. Maluf, J.W. Repulsive gravity near naked singularities and point massive particles. *Gen. Relativ. Gravit.* **2014**, *46*, 1734.
39. Parker, E.P. Distributional geometry. *J. Math. Phys.* **1979**, *20*, 1423–1426.
40. Katanaev, M.O. Point massive particle in General Relativity. *Gen. Relativ. Gravit.* **2013**, *45*, 1861–1875.
41. Ulhoa, S.C.; da Rocha-Neto, J.F.; Maluf, J.W. The gravitational energy problem for cosmological models in teleparallel gravity. *Int. J. Mod. Phys. D* **2010**, *19*, 1925–1935.
42. Hehl, F.W.; Mashhoon, B. Formal framework for a nonlocal generalization of Einstein's theory of gravitation. *Phys. Rev. D* **2009**, *79*, 064028.
43. Blome, H.J.; Chicone, C.; Hehl, F.W.; Mashhoon, B. Nonlocal modification of Newtonian gravity. *Phys. Rev. D* **2010**, *81*, 065020 .
44. Hehl, F.W.; Mashhoon, B. Nonlocal gravity simulates dark matter. *Phys. Lett. B* **2009**, *673*, 279–282.

universe

MDPI

Article

Autoparallel *vs.* Geodesic Trajectories in a Model of Torsion Gravity

Luis Acedo

Instituto Universitario de Matemática Multidisciplinar, Universitat Politècnica de València, Building 8G, 2°
Floor, Camino de Vera 46022, Valencia, Spain; luiacrod@imm.upv.es

Academic Editor: Lorenzo Iorio
Received: 14 October 2015; Accepted: 13 November 2015; Published: 25 November 2015

Abstract: We consider a parametrized torsion gravity model for Riemann–Cartan geometry around a rotating axisymmetric massive body. In this model, the source of torsion is given by a circulating vector potential following the celestial parallels around the rotating object. Ours is a variant of the Mao, Tegmark, Guth and Cabi (MTGC model) in which the total angular momentum is proposed as a source of torsion. We study the motion of bodies around the rotating object in terms of autoparallel trajectories and determine the leading perturbations of the orbital elements by using standard celestial mechanics techniques. We find that this torsion model implies new gravitational physical consequences in the Solar system and, in particular, secular variations of the semi-major axis of the planetary orbits. Perturbations on the longitude of the ascending node and the perihelion of the planets are already under discussion in the astronomical community, and if confirmed as truly non-zero effects at a statistically significant level, we might be at the dawn of an era of torsion phenomenology in the Solar system.

Keywords: Solar system anomalies; Riemann–Cartan spacetime; gravitation models; autoparallel curves; geodesic curves

---

1. Introduction

After one hundred years since its proposal [1], gravitation is still understood in terms of the theory of general relativity (GR). This theory is considered as the pinnacle of classical physics, and the status of its agreement with experiments is very good, although the progress in its verification has been painfully slow [2] due to the weakness of the gravitational interaction and the technical difficulties in measuring/observing its predicted effects due to their smallness [3–6]. In the last few years, an important advance has been achieved with the confirmation of the geodetic and the frame-dragging effects upon a gyroscope mounted on an artificial satellite orbiting the Earth. This is known as the Gravity Probe B experiment [7]. With this outstanding result, most of the major deviations from Newtonian gravity, as predicted by GR in the Solar system, are already experimentally checked. Efforts to test a few other ones, such as the Lense–Thirring [8–12] and the post-Newtonian quadrupolar orbital precessions [13,14], are ongoing. Moreover, the discovery of exoplanets orbiting other stars provides an opportunity to obtain additional substantiation of GR [15–17]. In particular, some authors have claimed that the relativistic precession of periastra in exoplanets could be detectable in the near future [18,19]. Zhao and Xie have also studied the influence of parametrized post-Newtonian dynamics in their transit times and the possibility of testing GR to a 6% level [20]. Even testing a putative fifth-force have also been considered [21]. The hot exoplanet WASP-33b also constitutes an excellent natural laboratory for GR [22], because the predicted Lense–Thirring node precession is 3.25×10^5 larger than that of Mercury [23], and this is only one order of magnitude below the measurability threshold for these systems.

Universe **2015**, *1*, 422–455

On the other hand, since the development of GR in its standard form, there have been many attempts to propose modified theories of gravity capable of predicting new testable phenomena. From the 1920s to the 1950s, the main drive of this research was to find a way of unifying gravitation and electromagnetism, although this objective was slowly being abandoned, except for Einstein himself and his collaborators [24]. Many of these theories were characterized by including a nonzero torsion tensor field, as well as the curvature tensor of standard Riemannian geometry. Extended geometries including both curvature and torsion have been used in physics since the early work of Einstein and Cartan [24]. The resulting Einstein–Cartan theory, much later improved by Sciama [25] and Kibble [26], is still considered a viable alternative to standard GR. In fact, it is still actively investigated as attested from the papers, conferences and even books published on this topic [27–29]. The Einstein–Cartan–Sciama–Kibble theory (ECSK) has also an interesting structure, as it can be consistently described as a gauge theory of the Poincaré group [30]. This way, it was put in correspondence with the very successful gauge theory approach to other interactions.

As beautiful as it could be, ECSK theory has not received any experimental support yet. This is not sufficient to dismiss any gravitation theory beforehand, because, as has happened with GR, the experiments are very difficult to design and carry out. Net spin densities are very small in most substances, as alignment of individual atomic spins is random, but it can be large in some elements, such as helium three (^{3}He) or dysprosium-iron compounds (Dy_6Fe_{23}). Ni has suggested to use these elements to build gyroscopes capable of testing PGtheory [31]. Apart from small spin densities, there is also the peculiarity that ECSK predicts null torsion outside macroscopic bodies. The reason for that behavior is that the field equation for torsion relates it linearly with spin density. Consequently, in a vacuum, where spin densities are null, torsion is also zero. Another important application of ECSK theory has been recently found by Popławski, who showed that torsion in the early Universe generates repulsion, and this could solve the flatness and horizon problems without resorting to an *ad hoc* inflation scenario [32,33]. A class of Poincaré gauge theories with extended Lagrangians quadratic in curvature and torsion have also been studied in the last few decades. These theories follow closely the analogy of the gauge paradigms of Weyl and Yang–Mills and, in some cases, predict a propagating torsion [34].

However, exploring alternative theories and models to standard PG is still both viable and timely. Theories in which torsion propagates outside macroscopic bodies can also be developed consistently [35]. As early as 1979, Hojman *et al.* proposed a model in which torsion is connected with a massless scalar field [36,37]. In this theory, torsion propagates in a vacuum, and torsion waves can be generated by sources with variable spin. At the same time, Hayashi and Shirafuji discussed an alternative to GR in which they revived the notion of Einstein's teleparallelism [38]. In the Hayashi–Shirafuji theory the fundamental entities are the tetrads instead of the metric, and the action is varied with respect to them to obtain the field equations. Interestingly, Hayashi and Shirafuji found a static spherically-symmetric vacuum solution in Weitzenböck spacetime (characterized by a null curvature tensor and a nonzero torsion [39,40]), which replaces the Schwarzschild solution in standard GR. The so-called new general relativity agrees with the classical tests for light bending, the anomalous advance of the perihelion of Mercury and Shapiro's delay of radar signals [38]. However, this theory fails to predict the geodetic and frame-dragging effects already checked in the Gravity Probe B experiment [7,41].

In 2007, Mao, Tegmark, Guth and Cabi proposed a phenomenological parametrized model for torsion in the Solar system (MTGC model). In this model, the source of torsion is assumed to be the rotational angular momentum of the planets and the Sun [41]. This is not the case in standard PG theory in which only the microscopic spin of elementary particles can generate torsion. The MTGC model does not depend on any specific theoretical framework, as the authors deduce the form of the torsion tensor from symmetry principles as the invariance under rotation, the antisymmetry of the torsion tensor in its covariant indices and the behavior of the angular momentum vector under parity transformations [41]. By using autoparallel or extreme schemes for the spin four vector S^μ or spin fourth tensor $S^{\mu\nu}$, these authors calculate extra contributions to the geodetic and frame-dragging precessions of a gyroscope's spin orbiting around the Earth. They claim that a refined version of Gravity Probe B experiment could be used to determine the values of some combinations of the seven constant parameters used to parametrize torsion. For the time being, error bars in the GPBexperiment are so large that we can only give some estimates on the bounds of the torsion parameters, but compatibility with standard GR is still not excluded by observations. Applying the planetary equations of Lagrange in the Gauss form, March *et al.* calculated the secular variations of the orbital elements for the planets and the Earth's geodynamics satellites [42,43]. In particular, they have found extra precessions rates for the longitude of the ascending node, *i.e.*, an anomalous Lense–Thirring effect, and also a contribution to the precession of the perihelion. Unfortunately, the precision in the determination of the Lense–Thirring effect or the perihelion precession, even for the highly-accurate measurements of the geodynamics satellites, such as LAGEOS, is still not sufficient to evince an irrefutable discrepancy with the predictions of standard GR. It is hoped that the newly-launched LARES [44] satellite may yield an improvement in the accuracy of the ongoing and forthcoming tests of fundamental physics, although also, such a possibility is currently debated [45–48]. Although the authors of these works have given some bounds on the values of the torsion parameters, based on the known error bars from the most recent ephemerides, it is still premature to draw any firm conclusion on the need of modified theories of gravity to explain the data.

The approaches of Mao *et al.* [41] and the subsequent spin-off applications by March *et al.* [42,43] have been heavily criticized by advocates of the PG theory. Hehl *et al.* have argued the following [49,50]: (i) Postulating that structureless test bodies follow autoparallel trajectories is incorrect in a general relativistic setup. In standard torsion theories test, bodies follow extremal trajectories, as is also the case in GR. The extremal trajectory is derived from the field equations themselves, and this is a theoretical feature of general relativity that should be preserved in any future theory. (ii) The net orbital angular momentum is not an integral over a local density, and consequently, it cannot be the source of torsion in a local field theory of gravity. Concerning the first objection, Kleinert and Pelster argued that in a spacetime with torsion, we must notice that parallelograms are, in general, not closed [51], the closure failure being proportional to the torsion tensor. This implies that the variational principle for finding the extrema of the action must take into account that the variation at the final point is nonzero as a consequence of the closure failure [52]. Using this modified variational principle, Kleinert and Pelster found that the equation of motion of structureless test bodies is given by autoparallel trajectories instead of extremal trajectories [51]. Bel has also shown that an analogy can be established among geodesics in Riemannian spacetime and autoparallels of a Weitzenböck connection [40]. In the absence of torsion extremal and autoparallel trajectories coincides as happens in standard GR. Hehl and Obukhov criticized this approach, because the autoparallel trajectories were not derived from the energy-momentum conservation laws, as is done in the GR and PG theories [50]. However, the closure failure also implies that the energy-momentum tensor of spinless point particles satisfies a different conservation law, as shown by Kleinert [53]. The second objection is, however, lethal to the MTGC model and its consequences. Any consistent theory of gravity must admit only local quantities or quantities obtained as the integration of local densities as sources of the tensor fields.

For these reasons, we investigate in this paper an alternative source for torsion around a rotating sphere. In our model, torsion is related to an axial vector field following the celestial parallels, $\mathbf{A}(r,\theta,\phi) = A(r,\theta,\phi)\hat{\boldsymbol{\phi}}$. This field structure could be obtained from the solution of a local Laplacian equation relating the vector potential with the energy-momentum flux of the rotating body in analogy to the corresponding equation for the magnetic field around a charged rotating sphere. All quantities in this model are local, and the non-locality induced by considering the total angular momentum as the source of torsion is removed. We study the secular evolution of the orbital elements for a test particle orbiting around a rotating central body. Some new effects unknown in GR are found: (i) a secular variation of the semi-major axis of the orbit; and (ii) a secular variation of the orbital eccentricity. As the increase of the astronomical unit is currently being discussed and no conventional explanation has still been found, our model could provide such an explanation, and moreover, we can give estimations on the torsion parameters from the preliminary data on these anomalies [54–56]. If these anomalies are confirmed, torsion fields generated by a circulating potential vector around rotating bodies could provide a parsimonious explanation of these phenomena, and they would stimulate further research in torsion gravity.

The structure of the paper is as follows: In Section 2, we provide a brief review on Riemann–Cartan spacetime as a quick reference for the rest of the paper. Our proposal for the torsion around spherical rotating bodies is discussed in Section 3 by following the symmetry arguments of Mao *et al.* [41]; autoparallel trajectories and orbital equations for perturbation theory are derived in Section 4. Results for the secular variation of the elements and comparison with Solar system anomalies are used in Section 5 to estimate the torsion parameters of our model. The discussion and conclusions are given in Section 6. Appendix A A is also included, in which the relation among the perturbing forces in the Sun's and the orbital system of reference is derived.

2. The Torsion and Contortion Tensors in Riemann–Cartan Spacetime

In this section, we remind about the main definitions and relations among tensors and the affine connection in Riemann–Cartan spacetime [57]. This spacetime is characterized by a non-zero curvature tensor and a torsion tensor defined as follows:

$$S_{jk}{}^i = \frac{1}{2}\left(\Gamma^i_{jk} - \Gamma^i_{kj}\right). \tag{1}$$

Therefore, a nonvanishing torsion implies that the affine connection is not symmetrical in the two lower indices in contrast with the postulates of ordinary Riemannian geometry. Christoffel's symbols are defined in terms of the metric tensor by the same expression found in Riemannian geometry. However, as we will find below, they do not coincide with the affine connection. Therefore, we have for Christoffel's symbols:

$$\left\{ \begin{matrix} i \\ jk \end{matrix} \right\} = \frac{1}{2}g^{il}\left(g_{lk,j} + g_{jl,k} - g_{jk,l}\right), \tag{2}$$

where the commas denote, as usual, ordinary derivatives with respect to the coordinates. However, covariant and contravariant derivatives of a vector field must be defined in terms of the affine connection by the following relations:

$$A_{i|j} = A_{i,j} - A_k\Gamma^k_{ji}, \tag{3}$$

$$A^i_{|j} = A^i_{,j} + A^k\Gamma^i_{jk}. \tag{4}$$

The metric condition is given as usual:

$$g_{ij|k} = g_{ij,k} - g_{hj}\Gamma^h_{ki} - g_{ih}\Gamma^h_{kj} = 0, \tag{5}$$

so the non-metricity is null [41]. By adding up the equivalent equations resulting from Equation (5) by the cyclic permutation of the the the three indices, we have:

$$\left\{ \begin{matrix} i \\ jk \end{matrix} \right\} = \Gamma^i_{jk} + K_{jk}{}^i, \tag{6}$$

where $K_{jk}{}^i$ is the contortion tensor defined in terms of the torsion as follows:

$$\begin{aligned} K_{jk}{}^i &= -S_{jk}{}^i + g^{il}g_{hk}S_{jl}{}^h + g^{il}g_{jh}S_{kl}{}^h \\ K_{jk}{}^i &= -S_{jk}{}^i - S^i{}_{jk} + S_k{}^i{}_j, \end{aligned} \tag{7}$$

where we have used the metric tensor to raise and lower the indices. Similarly, in terms only of covariant indices, we have:

$$K_{ijk} = S_{jik} - S_{kij} + S_{jki} , \tag{8}$$

from which we deduce the following antisymmetry property:

$$K_{ijk} = -K_{ikj} , \tag{9}$$

which also implies $K_{ij}{}^j = 0$. In the case of the Riemann–Cartan spacetime, we have a generalization of a Ricci identity involving the torsion tensor as follows:

$$A^i{}_{|jk} - A^i{}_{|kj} = -A^h R_{kjh}{}^i - 2Q_{kj}{}^h A^i{}_{|h} , \tag{10}$$

where both the curvature tensor and the torsion tensor appear.

3. Parametrization of Torsion in Spherically-Symmetric and Axisymmetric Spacetimes

The main idea of the MTGC model consist of a parametrization of torsion for both the static, spherical and parity symmetric case and the stationary, spherically-axisymmetric spacetime by using dimensional and symmetry arguments [41].

In the first case, we expect torsion to be invariant under the group of spatial rotations, $O(3)$, and, consequently, to involve only invariant quantities, such as the radio vector, x^i, $i = 1, 2, 3$, the Kronecker δ-function and the mass of the spherical object generating the field. The most general torsion tensor with these conditions becomes:

$$S_{0i}{}^0 = t_1 \frac{m}{2r^3} x^i , \tag{11}$$

$$S_{jk}{}^i = t_2 \frac{m}{2r^3} \left(x^j \delta_{ki} - x^k \delta_{ji} \right) , \tag{12}$$

where i, j and k are spatial indices and t_1, t_2 are functions of r alone to be treated as constants for an orbit of fixed radius. For perturbation calculations, it is highly convenient to transform this result to spherical coordinates by using the identities: $\partial x^i / \partial r = \hat{e}_r^i$, $\partial x^i / \partial \theta = r \hat{e}_\theta^i$ and $\partial x^i / \partial \phi = r \sin \theta \hat{e}_\phi^i$, which yields for the nonvanishing components:

$$S_{tr}{}^t = S_{tr}{}^t \frac{\partial x^i}{\partial r} = t_1 \frac{m}{2r^2} \tag{13}$$

$$S_{r\theta}{}^\theta = S_{jk}{}^i \frac{\partial x^j}{\partial r} \frac{\partial x^k}{\partial \theta} \frac{\partial \theta}{\partial x^i} = t_2 \frac{m}{2r^2} , \tag{14}$$

$$S_{r\phi}{}^\phi = S_{r\theta}{}^\theta . \tag{15}$$

Now, we consider the stationary spherically-axisymmetric spacetime whose metric is given, to first order, as:

$$\begin{aligned} ds^2 = & -\left[1 - \frac{\rho_S}{r}\right] c^2 dt^2 + \left[1 + \gamma \frac{\rho_S}{r}\right] \\ & + r^2 \left(d\theta^2 + \sin^2\theta d\phi^2 \right) \\ & - (1 + \gamma + \alpha_1/4) \frac{\rho_S \rho_J}{r} c dt d\phi , \end{aligned} \tag{16}$$

where $\rho_S = 2GM/c^2$ is the Schwarzschild radius and $\rho_J = J/(Mc)$ is a distance given in terms of the total angular momentum of the rotating object, its mass, M, and the speed of light, c. The constant parameters γ and α_1 are the parametrized post-Newtonian mechanics (PPN) parameters whose values in the case of standard general relativity are $\gamma = 1$, $\alpha_1 = 0$ as we will assume in this paper. The non-zero first-order contributions to Christoffel's symbols in Equation (2) are listed below:

$$\left\{ \begin{matrix} t \\ rt \end{matrix} \right\} = \left\{ \begin{matrix} r \\ tt \end{matrix} \right\} = \frac{\rho_S}{2r} , \tag{17}$$

$$\left\{ \begin{matrix} t \\ r\phi \end{matrix} \right\} = -\tfrac{3}{2}\sin^2\theta \frac{\rho_s \rho_J}{r^2} , \tag{18}$$

$$\left\{ \begin{matrix} r \\ \phi t \end{matrix} \right\} = -\sin^2\theta \frac{\rho_s \rho_J}{2r^2} , \tag{19}$$

$$\left\{ \begin{matrix} r \\ rr \end{matrix} \right\} = -\frac{\rho_s}{2r^2} , \tag{20}$$

$$\left\{ \begin{matrix} r \\ \theta\theta \end{matrix} \right\} = \rho_s - r = \left\{ \begin{matrix} r \\ \phi\phi \end{matrix} \right\} / \sin^2\theta , \tag{21}$$

$$\left\{ \begin{matrix} \theta \\ t\phi \end{matrix} \right\} = \frac{\rho_s \rho_J}{r^3}\sin\theta\cos\theta , \tag{22}$$

$$\left\{ \begin{matrix} \theta \\ r\theta \end{matrix} \right\} = \left\{ \begin{matrix} \phi \\ r\phi \end{matrix} \right\} = \tfrac{1}{r} , \tag{23}$$

$$\left\{ \begin{matrix} \theta \\ \phi\phi \end{matrix} \right\} = -\sin\theta\cos\theta , \tag{24}$$

$$\left\{ \begin{matrix} \phi \\ tr \end{matrix} \right\} = \frac{\rho_s \rho_J}{2r^4} , \tag{25}$$

$$\left\{ \begin{matrix} \phi \\ t\theta \end{matrix} \right\} = -\frac{\cos\theta}{\sin\theta}\frac{\rho_s \rho_J}{r^3} , \tag{26}$$

$$\left\{ \begin{matrix} \phi \\ \theta\phi \end{matrix} \right\} = \frac{\cos\theta}{\sin\theta} , \tag{27}$$

and the corresponding symbols with permutated lower indices, which coincide with the listed ones by symmetry. First-order refers to the fact that we only consider terms proportional to ρ_s, ρ_J and $\rho_s\rho_J$ and ignore any higher-order power.

Finally, we will consider the torsion tensor for the stationary spherically-axisymmetric case. Torsion will be associated with an axial vector field A^k instead of the angular momentum, as proposed in the MTGC model [41]. This vector field reverses under time reversal and improper rotations. This requires that only those components with a single temporal index are nonvanishing. Moreover, to cancel the minus sign arising in improper rotations, we must include the Levi–Cività symbol, ϵ_{ijk}, because, being a pseudotensor, it also changes sign in improper rotations. With these conditions, it was found that:

$$\begin{aligned} S_{ij}{}^t &= \tfrac{f_1}{2r^3}\epsilon_{ijk}A^k + \tfrac{f_2}{2r^5}A^k x^l \left(\epsilon_{ikl}x^j - \epsilon_{jkl}x^i \right) \\ &+ \tfrac{f_6}{2r^5}A^k x_k \epsilon_{ijl}x^l , \\ S_{tij} &= \tfrac{f_3}{2r^3}\epsilon_{ijk}A^k + \tfrac{f_4}{2r^5}A^k x^l \epsilon_{ikl}x^j \\ &+ \tfrac{f_5}{2r^5}A^k x^l \epsilon_{jkl}x^i + \tfrac{f_7}{2r^5}A^k x_k \epsilon_{ijl}x^l , \end{aligned} \tag{28}$$

where $f_1, \ldots, f_7$ are constants by dimensional analysis. Notice that Mao *et al.* [41] identified the vector A^k with the angular momentum J^k, but in the model presented in this paper, it could be a more general vector field. Save for constant prefactors, we chose a vector field with the same structure that the vector potential of a rotating charged sphere is as follows:

$$\mathbf{A} = \frac{4GM\Omega R^2}{5c^3}\sin\theta(-\sin\phi\,\hat{\mathbf{m}}_1 + \cos\phi\,\hat{\mathbf{m}}_2) , \tag{29}$$

where Ω is the angular velocity of the rotating central body, m is the mass and R is its radius. The prefactor is, then, proportional to the modulus of the total angular momentum. Notice that we can also write it as $4GM\Omega R^2/(5c^3) = \rho_S\rho_J$, where $\rho_S = 2GM/c^2$ is the Schwarzschild radius of the central body and $\rho_J = J/(Mc) = 2/5(\Omega R/c)R$. The unit vector $\hat{m}_1$ points towards the ascending node of the Sun's axial rotation, and $\hat{m}_2$ is perpendicular to it in the equatorial plane of the Sun or the rotating body that we are considering. The nonvanishing components of the torsion tensor corresponding to the static spherically-symmetric case are derived from Equation (13) as given in [41] as follows:

$$S_{tr}{}^t = t_1\frac{m}{2r^2} , \quad S_{r\theta}{}^\theta = S_{r\phi}{}^\phi = t_2\frac{m}{2r^2} . \tag{30}$$

Additionally, those components, opposite in sign, correspond to the permutation of the covariant indices. Similarly, we find from Equations (28) and (29) that additional nonzero components of the torsion tensor for the stationary spherically-axisymmetric case are given by:

$$
\begin{aligned}
S_{r\theta}{}^t = -S_{\theta r}{}^t &= \chi_1\frac{\rho_S\rho_J}{2r^2}\sin\theta , \\
S_{t\theta}{}^r = -S_{\theta t}{}^r &= \chi_2\frac{\rho_S\rho_J}{2r^2}\sin\theta , \\
S_{tr}{}^\theta = -S_{rt}{}^\theta &= \chi_3\frac{\rho_S\rho_J}{2r^4}\sin\theta ,
\end{aligned}
\tag{31}
$$

where χ_1, χ_2 and χ_3 are constant parameters. Finally, we must tabulate the values of the nonvanishing components of the contortion tensor (up to first-order in ρ_S, ρ_J and $\rho_S\rho_J$) by using the relation with the torsion in Equation (7) and Equations (30) and (31). The results are listed below:

$$K_{01}{}^0 = K_{00}{}^1 = -\frac{\rho_S t_1}{r^2} , \tag{32}$$

$$K_{22}{}^1 = K_{33}{}^1/\sin^2\theta = -\rho_S t_2 , \tag{33}$$

$$K_{21}{}^2 = K_{31}{}^3 = \frac{\rho_S t_2}{r^2} , \tag{34}$$

$$K_{21}{}^0 = (\chi_1 + \chi_2 + \chi_3)\frac{\rho_S\rho_J}{2r^2}\sin\theta , \tag{35}$$

$$K_{12}{}^0 = (\chi_1 - \chi_2 - \chi_3)\frac{\rho_S\rho_J}{2r^2}\sin\theta , \tag{36}$$

$$K_{20}{}^1 = (\chi_1 + \chi_2 + \chi_3)\frac{\rho_S\rho_J}{2r^2}\sin\theta , \tag{37}$$

$$K_{02}{}^1 = (\chi_1 - \chi_2 + \chi_3)\frac{\rho_S\rho_J}{2r^2}\sin\theta , \tag{38}$$

$$K_{10}{}^2 = (\chi_2 + \chi_3 - \chi_1)\frac{\rho_S\rho_J}{2r^4}\sin\theta , \tag{39}$$

$$K_{01}{}^2 = (\chi_2 - \chi_1 - \chi_3)\frac{\rho_S\rho_J}{2r^4}\sin\theta . \tag{40}$$

In the next section, we will study the autoparallel trajectories in the spacetime with this contortion tensor. Notice that the torsion and contortion tensor fields are determined by five constant parameters: t_1, t_2, χ_1, χ_2 and χ_3. The effect of the first two, t_1 and t_2, has already been analyzed by March *et al.* [42,43], and we will be concerned in this paper with the bounds or estimated values of χ_1, χ_2 and χ_3.

4. Autoparallel Trajectories and Perturbation Theory

As in previous models [41–43], we will assume that structureless point particles move along autoparallel trajectories of the Riemann–Cartan spacetime. Therefore, we have that:

$$\frac{d^2x^\alpha}{d\tau^2} + \left\{ \begin{array}{c} \alpha \\ \mu\nu \end{array} \right\} \frac{dx^\mu}{d\tau}\frac{dx^\nu}{d\tau} = K_{\mu\nu}{}^\alpha\frac{dx^\mu}{d\tau}\frac{dx^\nu}{d\tau} , \tag{41}$$

where τ is proper time measured along the trajectory. Notice that in a purely Riemannian spacetime, the contortion tensor is null, and Equation (41) is also found for geodesic trajectories. It is usually claimed that only test bodies with a microstructure can couple to torsion [58] and that spinless particles should follow the geodesic trajectories defined by:

$$\frac{d^2 x^\alpha}{d\tau^2} + \left\{ \begin{array}{c} \alpha \\ \mu\nu \end{array} \right\} \frac{dx^\mu}{d\tau} \frac{dx^\nu}{d\tau} = 0 \,. \tag{42}$$

In standard general relativity, one finds this equation of motion in two ways: as the extremal of the integral of the spacetime element, ds, or as a consequence of the field equations, *i.e.*, as the condition that the covariant divergence of the stress-energy tensor has zero value [59]. Both ways are equivalent and lead to the geodesic equation of motion in Equation (42). Mathematically, this coincides with the autoparallels in Equation (41) for zero torsion. If the calculations are translated to a Riemann-Cartan spacetime using holonomic constraints at the initial and final points (with zero variation in these points) of the trajectory, we find again the geodesic trajectories, because only the Christoffel symbols enter into the analysis [53].

However, geodesics are a global concept defined, as they are, as the shortest paths between two points. On the contrary, autoparallels can be defined locally as the straightest paths in spacetime. Therefore, from the fundamental principle of locality in classical field theory, it seems more natural to ascribe physical significance to autoparallels instead of geodesics.

Kleinert and collaborators [51–53] have studied a new nonholonomic mapping principle from flat spacetime to curved spacetime with torsion in which curvature is described as a disclination and torsion as a dislocation of the spacetime fabric. This implies that a closure failure appears in parallelograms, and the endpoints of a variational trajectory are displaced by:

$$\delta^S b^\mu = \delta^S q^\mu - \delta q^\mu \,, \tag{43}$$

where δ^S denotes the nonholonomic variations and δ are the auxiliary variations vanishing at the endpoints [53]. Starting from the action:

$$\mathcal{A} = -\frac{1}{2} \int_{\sigma_1}^{\sigma_2} d\sigma \, g_{\mu\nu}(q(\sigma)) \dot{q}^\mu(\sigma) \dot{q}^\nu(\sigma) \,, \tag{44}$$

and applying the nonholonomic variations, we get:

$$\delta^S \mathcal{A} = -\int_{\sigma_1}^{\sigma_2} d\sigma \left(g_{\mu\nu} \dot{q}^\nu \delta^S \dot{q}^\mu + \frac{1}{2} \partial_\mu g_{\lambda\kappa} \delta^S q^\mu \dot{q}^\lambda \dot{q}^\kappa \right) \,, \tag{45}$$

where the dot denotes a derivative with respect the proper time, σ. In terms of the multivalued tetrads $e^i{}_\nu$, $i = 0, \ldots, 3$, $\mu = 0, \ldots, 3$, the metric tensor is defined as follows $g_{\mu\nu} = e^i{}_\mu e^i{}_\nu$, and the affine connection is given by $\Gamma_{\mu\nu}{}^\lambda = e_i{}^\lambda \partial_\mu e^i{}_\nu$. Taking also into account that $\partial_\mu g_{\nu\lambda} = \Gamma_{\mu\nu\lambda} + \Gamma_{\mu\lambda\nu}$ and integrating by parts, we obtain:

$$\begin{aligned} \delta^S \mathcal{A} &= -\int_{\sigma_1}^{\sigma_2} d\sigma \left[-g_{\mu\nu} \left(\ddot{q}^\nu + \left\{ \begin{array}{c} \nu \\ \lambda\kappa \end{array} \right\} \dot{q}^\lambda \dot{q}^\kappa \right) \delta q^\mu \right. \\ &\quad + \left. \left(g_{\mu\nu} \dot{q}^\nu \frac{d}{d\sigma} \delta^S b^\mu + \Gamma_{\mu\lambda\kappa} \delta^S b^\mu \dot{q}^\lambda \dot{q}^\kappa \right) \right] \,. \end{aligned} \tag{46}$$

By using the differential equation for the nonholonomic variation $\delta^S b^\mu$:

$$\frac{d}{d\sigma} \delta^S b^\mu = -\Gamma_{\lambda\nu}{}^\mu \delta^S b^\lambda \dot{q}^\nu + 2 S_{\lambda\nu}{}^\mu \dot{q}^\lambda \delta q^\nu \,, \tag{47}$$

Kleinert finds:

$$\delta^S \mathcal{A} = \int_{\sigma_1}^{\sigma_2} d\sigma \, g_{\mu\nu} \left(\ddot{q}^\nu + \Gamma^\nu_{\lambda\kappa} \dot{q}^\lambda \dot{q}^\kappa \right) \delta q^\mu = 0 \,. \tag{48}$$

As the auxiliary variations are arbitrary, but null at the endpoints, σ_1 and σ_2, we get from this the autoparallel equation instead of the geodesic. A similar derivation was discussed also by Kleinert and Pelster [51]. Further details are also given in several textbooks [60,61]. Autoparallels also satisfy a gauge invariance relating the autoparallel equations of motion in different Riemann–Cartan spacetimes [62]. This property could find a deeper physical meaning, apart from its mathematical interest, in future theories of gravity involving curvature and torsion.

The question of the relevance for the physics of the standard method or the method based on nonholonomic variations for the derivation of the equations of motion in Riemann–Cartan spacetime should be decided experimentally if, as we suggest, there is an additional torsion structure in spacetime. This question cannot be answered in the context of standard general relativity, because, in this case, autoparallel and geodesic equations coincide.

In Riemann–Cartan spacetime, the right-hand side of Equation (41) represents a perturbing force whose effects can be calculated by standard perturbation theory in celestial mechanics. We should calculate the first-order perturbation terms arising from the contortion tensor field in Equations (32)–(40). As the planets in the Solar system move with velocities much smaller than the speed of light, we can identify the proper time with the ephemeris time or the atomic time used by astronomers, *i.e.*, $\tau = t$. We will also consider that the torsion parameters $t_1 = t_2 = 0$, because it has been shown that t_1 can be absorbed in a redefinition of the mass of the source, and t_2 does not appear in the equations of the trajectories [42,43].

The components of the perturbing torsion terms are then given by:

$$\delta F^i = K_{\mu\nu}{}^i \frac{dx^\mu}{d\tau} \frac{dx^\nu}{d\tau} \,, \tag{49}$$

with $i = 1, 2, 3$ corresponding to the radial, polar and azimuthal coordinates, respectively. The leading term of the radial perturbing acceleration is found by direct substitution of Equations (32)–(40) into Equation (49):

$$\begin{aligned} \delta a_r &= \delta F^r = c \left(K_{02}{}^1 + K_{20}{}^1 \right) \dot\theta \\ &= c(\chi_1 + \chi_3) \frac{\rho_S \rho_l}{r^2} \sin\theta \, \dot\theta \,, \end{aligned} \tag{50}$$

ignoring corrections $\mathcal{O}\left(\dot\theta^2\right)$ and $\mathcal{O}\left(\dot\phi^2\right)$, c being the speed of light in a vacuum, $dx^0/dt = c$ and $\dot\theta = d\theta/dt$. Similarly, for the polar component of the perturbing force, we find:

$$\begin{aligned} \delta a_\theta &= r\delta F^\theta = r \left(K_{01}{}^2 + K_{10}{}^2 \right) \dot r \\ &= c(\chi_2 - \chi_1) \frac{\rho_S \rho_l}{r^3} \sin\theta \, \dot r \,. \end{aligned} \tag{51}$$

On the other hand, the azimuthal component of the perturbing torsion force is zero, as expected for an axisymmetric source, $\delta a_\phi = 0$. The perturbing accelerations in Equations (50) and (51) are corrections to the accelerations in spherical coordinates in the system of reference of the Sun (see Appendix A 6), which are obtained from the left-hand side of the equations for the autoparallel trajectories and the first-order Christoffel symbols in Equations (17)–(27) as follows:

$$a_r = \ddot r - r\dot\theta^2 - r\dot\phi^2 \sin^2\theta \,, \tag{52}$$

$$a_\theta = r\ddot\theta + 2\dot r\dot\theta - r\dot\phi^2 \sin\theta\cos\theta \,, \tag{53}$$

$$a_\phi = r\ddot\phi \sin\theta + 2\dot r\dot\phi \sin\theta + 2r\dot\theta\dot\phi \cos\theta \,. \tag{54}$$

As our objective is to apply celestial mechanics perturbation techniques, we will find it convenient to calculate the radial, $\mathcal{R}$, tangential to the orbit, $\mathcal{T}$, and normal to the orbital plane, $\mathcal{N}$, components of the perturbing torsion force per unit mass. In Equations (88)–(90) in the Appendix A 6, we have found those components in terms of the accelerations in spherical coordinates in a system of reference whose z axis is the rotation axis of the Sun. The relation is expressed in terms of a transformation matrix α_{ij}, $i,j = 1,2,3$ obtained as the set of scalar products among the vectors in the orbital system of reference and the Sun's system of reference, as given in Equation (79).

Direct substitution of Equations (50) and (51) and $\delta a_\phi = 0$ into Equations (88)–(90) yields:

$$\mathcal{R} = c(\chi_1 + \chi_3)\frac{\rho_S \rho_J}{r^2}(\alpha_{13}\sin\nu \tag{55}$$

$$- \alpha_{23}\cos\nu)\dot{\nu},$$

$$\mathcal{T} = c(\chi_2 - \chi_1)\frac{\rho_S \rho_J}{r^3}(\alpha_{13}\sin\nu \tag{56}$$

$$- \alpha_{23}\cos\nu)\dot{r},$$

$$\mathcal{N} = c(\chi_2 - \chi_1)\frac{\rho_S \rho_J}{r^3}\sin\theta$$

$$(\alpha_{31}\cos\theta\cos\phi + \alpha_{32}\cos\theta\sin\phi - \alpha_{33}\sin\theta)\dot{r},$$

$$= c\alpha_{33}(\chi_1 - \chi_2)\frac{\rho_S \rho_J}{r^3}\dot{r}. \tag{57}$$

Notice that all terms in Equations (55)–(57) are constants or can be written in terms of the true anomaly. The relations among the polar, θ, and azimuthal, ϕ, angles and the true anomaly, ν, are given in Equations (80)–(82). Using these relations and the identity in Equation (83), we have found the simplifications for $\mathcal{T}$ and $\mathcal{N}$.

For the radio vector and radial velocity, we have [63–65]:

$$r = \frac{p}{1+\epsilon\cos(\nu-\omega)},$$
$$\dot{r} = \sqrt{\frac{\mu}{p}}\epsilon\sin(\nu-\omega), \tag{58}$$

where $p = a(1-\epsilon^2)$ is the semilatus rectum, ϵ is the eccentricity, a is the semimajor axis, ω is the argument of the perihelion and $\mu = GM$ is the product of the gravitational constant and the mass of the Sun. The relation among time, t, and the true anomaly, ν, will also be useful in the following perturbation calculations:

$$dt = \frac{T}{2\pi}\frac{(1-\epsilon^2)^{3/2}}{(1+\epsilon\cos(\nu-\omega))^2}d\nu, \tag{59}$$

where the orbital period, T, is given by Kepler's third law: $T = 2\pi a^{3/2}/\mu^{1/2}$. We are now ready for calculating the perturbations in the orbital elements as a consequence of the torsion force in Equations (55)–(57). Following the classical treatment of Burns [65], we can write for the semimajor axis:

$$\frac{\dot{a}}{a} = \frac{2a}{\mu}\dot{E} = \frac{2a}{\mu}(\dot{r}\mathcal{R} + r\dot{\nu}\mathcal{T}), \tag{60}$$

where E is the total energy. After some simplifications using Equations (55), (56), (58) and (59), we have:

$$da = \frac{cT}{\pi}\frac{\rho_S \rho_J}{a^2}\frac{\epsilon}{(1-\epsilon^2)^{5/2}}(\chi_2 + \chi_3)\sin(\nu-\omega)$$
$$(1+\epsilon\cos(\nu-\omega))^2(\alpha_{13}\sin\nu - \alpha_{23}\cos\nu)d\nu. \tag{61}$$

For the instantaneous variation of the eccentricity, we start from:

$$\dot{\epsilon} = \frac{\epsilon^2 - 1}{2\epsilon}\left[-\frac{\dot{a}}{a} + \frac{2r\mathcal{T}}{H}\right], \tag{62}$$

where $H = \sqrt{\mu p} = \sqrt{GMa(1 - \epsilon^2)}$ is the angular momentum per unit mass. By using Equations (61), (56) and (59) conjointly with Equation (62), we arrive at:

$$d\epsilon = \frac{cT}{2\pi a}\frac{\rho_S \rho_J}{a^2}\frac{\sin(\nu - \omega)}{(1 - \epsilon^2)^{1/2}}(\alpha_{13}\sin\nu - \alpha_{23}\cos\nu)$$
$$\left[\frac{(\chi_2 + \chi_3)}{1 - \epsilon^2}(1 + \epsilon\cos(\nu - \omega))^2 + \chi_1 - \chi_2\right]d\nu. \tag{63}$$

Torsion also induces an extra precession of the longitude of the ascending node of the planetary orbits in addition to the Lense–Thirring effect arising in standard general relativity. This precession rate is proportional to the normal component to the planetary orbits of the perturbing force, as found in perturbation theory [63]:

$$\frac{d\Omega}{dt} = \frac{r\mathcal{N}\sin\nu}{H\sin I} = \sqrt{\frac{p}{GM}}\frac{\mathcal{N}}{\sin I}\frac{\sin\nu}{1 + \epsilon\cos(\nu - \omega)}, \tag{64}$$

where I is the orbital inclination and $p = a(1 - \epsilon^2)$ is the semi-latus rectum. From the expression of the normal component of the perturbing force in Equation (57) and the contribution to the precession of the longitude of the ascending node in Equation (64), we obtain:

$$d\Omega = \alpha_{33}(\chi_1 - \chi_2)\frac{cT}{2\pi a}\frac{\rho_S \rho_J}{a^2}\frac{\epsilon}{(1 - \epsilon^2)^{3/2}}\frac{\sin\nu}{\sin I}\sin(\nu - \omega)d\nu. \tag{65}$$

Finally, for the extra precession of the perihelion contributed by spacetime torsion, we find:

$$d\omega = (\chi_2 - \chi_1)\frac{cT}{2\pi a}\frac{\rho_S \rho_J}{a^2}\frac{1}{(1 - \epsilon^2)^{3/2}}$$
$$\left(\sin^2(\nu - \omega)\frac{2 + \epsilon\cos(\nu - \omega)}{(1 + \epsilon\cos(\nu - \omega))^3}(\alpha_{13}\sin\nu - \alpha_{23}\cos\nu)\right.$$
$$+ \epsilon\alpha_{33}\cot I\sin\nu\sin(\nu - \omega))d\nu$$
$$- c(\chi_1 + \chi_3)\sqrt{\frac{a}{GM}}\frac{\rho_S \rho_J}{a^2}\frac{1}{\epsilon(1 - \epsilon^2)^{3/2}}$$
$$\cos(\nu - \omega)(\alpha_{13}\sin\nu - \alpha_{23}\cos\nu)d\nu. \tag{66}$$

In the next section, we will discuss the predictions of Equations (61)–(66) for the variation of the orbital elements of the planet and its possible connection with certain anomalies recently found by astronomers.

5. Results

As we cannot give to the parameters χ_1, χ_2 and χ_3 definite values on a theoretical basis, it is not possible, in principle, to make predictions on the variation of the orbital elements as a consequence of torsion in our model. It is reasonable to assume that these parameters are of the order of unity, if there is a theory consistent with the torsion model, but this is not sufficient to suggest any reliable prediction. However, we can use an inductive approach by assuming that some anomalies recently found for the planetary orbits are the consequence of torsion gravity arising in a theory that includes the phenomenological model discussed in this paper as a particular case. Specifically, we refer to the anomalous secular increase of the astronomical unit (AU) first reported by Krasinsky and Brumberg in 2004 [66]. The analysis of databases of radar and laser ranging and spacecraft observations in the last few decades showed that the astronomical unit increases by 15 ± 4 meters per century. An independent study by Standish reduced this figure to 7 ± 2 meters per century [67]. This problem has even motivated the International Astronomical Union to redefine the AU as a constant and, as a consequence, to remove the Gaussian constant of mass from the list of astronomical constants [56]. However, these redefinitions of constants do not solve the problem pointed out by the aforementioned astronomers. However, we should notice that, according to this new definition, it is not rigorous to compare the rates of the semi-major axes of the planets with the rates of the astronomical unit, because this is fixed. On the other hand, we can use the recent reports on a secular decrease of the mass parameter $\mu = GM$ of the Sun [56]. The average of the rates determined with EPM2008, EPM2010 and EPM2011 ephemerides is $\dot{\mu} = (5.73 \pm 4.27) \times 10^{-14}$ yr^{-1}. It is known from the perturbation analysis of the planetary orbits in a scenario of a diminishing gravitational constant that the semi-major axis increase as [68]:

$$\frac{\dot{a}}{a} = -\frac{\dot{\mu}}{\mu} \, , \tag{67}$$

which implies a rate of 0.86 ± 0.64 meters per century. It is important to point out that the recent analyses with the INPOP13cephemerides are statistically compatible with a zero variation of the Sun's mass parameter [56]. Some conventional and unconventional attempts for an explanation of the variation rates for the semi-major axes of the planets have been suggested, but there is still no convincing solution of the problem [69–73].

It is interesting to notice that Equation (61) implies a variation of the semimajor orbital axes for $\eta = \chi_2 + \chi_3 \neq 0$. Moreover, if we assume that:

$$\chi_1 + \chi_3 = 0 \, , \tag{68}$$

it is found from Equations (61)–(66) that the variations of a, ϵ, Ω and ω depend only on the single parameter $\eta = \chi_2 - \chi_1$. We will choose the condition in Equation (68) without losing the perspective that other possibilities are compatible with our model, even the case in which no secular change is found for the semimajor planetary axes, *i.e.*, the case $\eta = 0$.

Using this condition and averaging Equations (61), (63), (65) and (66) over a whole orbit, we obtain the following results for the variation of the elements in one year:

$$\frac{\Delta a}{a} = \eta \frac{Ly}{a} \frac{\Omega R}{c} \left(\frac{R}{a}\right)^2 \frac{\epsilon}{(1-\epsilon^2)^{5/2}} \tag{69}$$

$$\left(1 + \frac{\epsilon^2}{4}\right)(\alpha_{13} \cos \omega + \alpha_{23} \sin \omega) \, ,$$

$$\Delta \epsilon = \eta \frac{5Ly}{8a} \frac{\rho_s \rho_l}{a^2} \frac{\epsilon^2}{(1-\epsilon^2)^{3/2}} \tag{70}$$

$$(\alpha_{13} \cos \omega + \alpha_{23} \sin \omega) \, ,$$

$$\Delta \Omega = -\eta \, \alpha_{33} \frac{Ly}{2a} \frac{\rho_s \rho_l}{a^2} \frac{\epsilon}{(1-\epsilon^2)^{3/2}} \frac{\cos \omega}{\sin I} \, , \tag{71}$$

$$\Delta \omega = \eta \frac{Ly}{2a} \frac{\rho_s \rho_l}{a^2} \frac{\epsilon}{(1-\epsilon^2)^{3/2}} \tag{72}$$

$$\left[\alpha_{33} \cot I \cos \omega - \frac{5}{4}(\alpha_{13} \sin \omega - \alpha_{23} \cos \omega)\right] \, ,$$

where Ly stands for a light year. By assuming that the increase of the astronomical unit is obtained as an average over the inner planets, Mercury, Venus, the Earth and Mars, and taking the value reported by Standish, $\Delta AU = 7 \pm 2$ meters per century, we find that $\eta = -0.154$. If we take the values deduced from the variation of the mass parameter GM of the Sun as reported in the EPM2008–2011 ephemerides, a smaller value $\eta = -0.019$ is found. In these calculation, the elements of the planets as given in [74,75] were used. We can also make some predictions for the variation of the other orbital elements and check if they are consistent with the presented observations. We should see that, at least, our model is not inconsistent in this sense.

For the secular variation of the eccentricity for this value of η, we find $\Delta\epsilon = 4.57 \times 10^{-13}$ in one year in the case of Mercury and even lower values for the other planets. This is below the precision threshold for present determinations of this magnitude [69].

Possible anomalous contributions to the secular node precessions are currently considered in recent ephemerides. For the moment, no statistically-significant results have been obtained with the attained precisions, but these values set an upper limit to any prediction by theoretical models. For INPOP10a, these corrections are listed in milliarcseconds per century and compared to our predictions from Equation (71) in Table 1. The most recent ephemerides, INPOP10a and EPM2011, in connection with possible Solar system anomalies have been discussed by Iorio [56].

Table 1. Corrections to the secular node precessions as obtained in the INPOP10a ephemeris and predictions of the torsion gravity model discussed in this paper in milliarcseconds per century.

Planet	$\Delta\dot{\Omega}$ (INPOP10a)	$\Delta\dot{\Omega}$(Torsion Gravity)
Mercury	1.4 ± 1.8	0.30
Venus	0.2 ± 1.5	2.14×10^{-2}
Earth	0.0 ± 0.9	-1.70×10^{-4}
Mars	-0.05 ± 0.13	2.62×10^{-3}
Jupiter	-40 ± 42	1.51×10^{-5}
Saturn	-0.1 ± 0.4	1.31×10^{-5}

The corrections to the standard secular perihelion precessions are given in Table 2 and compared to the predictions of our torsion gravity model.

Inspection of Tables 1 and 2 shows that the predictions of the model are compatible with both ephemerides at the 2σ level. Therefore, in the context of the torsion gravity model, the anomalous increase of the astronomical unit is consistent with the rest of measurements on possible corrections to other orbital elements. On the other hand, we have shown that only for Mercury, it seems that an improved ephemeris could detect a statistically-significant nonzero correction for $\Delta\dot{\Omega}$ and $\Delta\dot{\omega}$ in the foreseeable future.

Table 2. Corrections to the secular perihelion precessions as obtained in the INPOP10a and EPM2011 ephemerides and predictions of the torsion gravity model discussed in this paper in milliarcseconds per century.

Planet	$\Delta\dot{\omega}$ (INPOP10a)	$\Delta\dot{\omega}$ (EPM2011)	$\Delta\dot{\omega}$ (Torsion Gravity)
Mercury	0.4 ± 0.6	-2.0 ± 3.0	-0.622
Venus	0.2 ± 1.5	2.6 ± 1.6	-4.28×10^{-3}
Earth	-0.2 ± 0.9	0.19 ± 0.19	-1.03×10^{-4}
Mars	-0.04 ± 0.15	-0.02 ± 0.037	-4.9×10^{-3}
Jupiter	-41 ± 42	58.7 ± 28.3	-3.48×10^{-5}
Saturn	0.15 ± 0.65	-0.32 ± 0.47	-2.69×10^{-5}

6. Conclusions

The advancement of physics can only proceed by a continuous interplay between theory and experiment. This healthy interaction allows for a selection of the most promising hypotheses among the different proposals. Gravity theory has been an exception to this methodological rule for the most part of the 20th century, because experiments are very difficult to develop, as they usually imply very accurate devices, and these must be set into orbit to perform the measurements [7,8]. Gravity being the weakest of all interactions is also the one we know least, because accurate tests of all general relativity predictions are still lacking [2]. For example, Lense–Thirring precession of orbital nodes is only known with a wide error bar from the laser range monitoring of the geodynamic satellites [8–10,76–78].

A further, non-negligible source of difficulty in gravitational experiments resides in the extremely long times required either to collect data or to analyze them: the Gravity Probe B is a case-study example [7].

The experimental situation has been complicated in recent years because of the discovery of a set of anomalies that, apparently, cannot be explained conventionally (see, e.g., the recent review by Iorio [56]). Similarly, the most recent ephemerides has allowed the determination of upper bounds on the variation of the orbital elements of the planets beyond the predictions of classical perturbation theory and general relativity [56,75]. Although some of these anomalies may lose their statistical significance in the more or less near future in view of further observations and related analyses, it is nonetheless important to discuss the possibility that they may constrain theories and extensions beyond general relativity.

Many extensions of general relativity, some of them proposed by Einstein himself, have made use of the concept of torsion [24]. These ideas coalesced in the 1960s and 1970s in the so-called Einstein–Cartan–Sciama–Kibble (ECSK) theory in which torsion is connected with the microscopic spin density and does not propagate outside massive bodies [30]. This theory is still considered a viable alternative to general relativity, but it suffers from a total lack of experimental support, despite some claims that it could explain inflation [32,33]. This situation leaves room for the study of more alternatives without restricting to a given mathematical formalism. In such a spirit, Mao *et al.* proposed in 2007 the MTGC parametrized model in which torsion is connected to macroscopic angular momentum [41]. It was shown that this model can be constrained by perturbations in the orbital elements of the planets and geodynamic satellites [42,43].

Hehl *et al.* [49,50] have pointed out that it is inconsistent to use total angular momentum, *i.e.*, a quantity not obtained by integration over local densities, as the source of a local field quantity, such as torsion. To avoid this inconsistency, we have modified the approach of the original MTGC model by connecting torsion with a local circulating vector potential as the one obtained in classical electromagnetism for a rotating charged sphere. This way, we have shown that a new phenomenon, qualitatively distinct from those obtained in general relativity, is predicted. Namely, a secular increase of the semi-major axes of the planets [66,67]. This problem has attracted the attention of several authors in recent years, who have tried to find explanations in terms of nonstandard and conventional hypotheses [56,69,70,72]. This observation remains unexplained and, although it could be dismissed by more precise analyses in the future, it deserves further attention. We have shown that our torsion model is compatible with these observations for a value of the parameters of the order of unity.

Moreover, planetary ephemerides are becoming increasingly precise year after year. It is still premature to state that statistically-significant anomalies have been revealed in the secular precessions of the longitude of the ascending node and the argument of the perihelion, but the uncertainty intervals are promisingly small [56]. Anyway, it is now clear that any deviation from the predictions of general relativity (once we take into account standard perihelion precessions and the gravitomagnetic Lense–Thirring effect) is expected only in the range of a few milliarcseconds per century.

We have also shown that extra secular precessions of these elements are the consequence of torsion, but for any planet, they are very small in relation to the confidence intervals of the INPOP10a and EPM2011 ephemerides. However, for the case of Mercury, they could be detected in the foreseeable future, because they lie in the range of a few tenths of milliarcseconds per century. The important fact is that this agreement is achieved in consistency with an increase of the astronomical unit of a few meters per century. This means that testing a torsion gravity extension of general relativity as the one discussed in this paper is within the reach of modern observation techniques and data analyses in astronomy.

We conclude that further experiments and observations are required, achieving the maximum precision possible with present-day technology, to confirm or dismiss possible anomalies beyond general relativity in the secular evolution of the elements of the planets and spacecraft. From these future observations, the model proposed in this paper could receive further support. In such a case, it could serve as the basis for a consistent theory of torsion gravity obtained by the scientific method of induction from experience.

Acknowledgments: The author gratefully acknowledges Lluís Bel for many useful comments and discussions.

Conflicts of Interest: The author declares no conflict of interest.

Appendix A. Relation among the Sun's System of Reference and the Orbital Coordinates

Firstly, we define the ecliptic system of reference: $\hat{k}$ is a unit vector perpendicular to the ecliptic plane; $\hat{i}$ points towards the point of Aries; and $\hat{j}$ is perpendicular to the preceding unit vectors in such a way that we have a right-handed Cartesian coordinate system [75].

The orbital plane of a given planet is then characterized by two angles: the inclination ι with respect to the ecliptic plane and the angle Ω among the line of nodes (*i.e.*, the intersection among the two planes) and the point of Aries. Consequently, we can write the unit vector of the orbital system of reference as follows:

$$\hat{n}_1 = \cos\Omega\,\hat{i} + \sin\Omega\,\hat{j}\,, \tag{73}$$

$$\hat{n}_2 = -\cos\iota\sin\Omega\,\hat{i} + \cos\iota\cos\Omega\,\hat{j} + \sin\iota\,\hat{k}\,, \tag{74}$$

$$\hat{n}_3 = \sin\iota\sin\Omega\,\hat{i} - \sin\iota\cos\Omega\,\hat{j} + \cos\iota\,\hat{k}\,. \tag{75}$$

Similarly, we can define the inclination of the Sun's axis and the longitude of the ascending node of its equator. These two angles determine the orientation of the Sun's rotation axis on space and were obtained in the 19th century by careful observations of Carrington. Carrington's elements, as they are called, are given by [79,80]:

$$\iota_c = 7.25°\,, \tag{76}$$

$$\Omega_c = 73.67° + 0.013958°\,(t - 1850)\,, \tag{77}$$

where t is the year of observation. We define the Sun's system of reference as the Cartesian system obtained by the three unit vectors $\hat{m}_i$, $i = 1, 2, 3$ whose expression in terms of the Carrington elements is also given by Equation (73).

The planet's orbital radius vector is usually written as [63,64]:

$$\mathbf{r} = p\,\frac{\cos\nu\,\hat{n}_1 + \sin\nu\,\hat{n}_2}{1 + \epsilon\cos(\nu - \omega)}\,, \tag{78}$$

where $p = a\left(1 - \epsilon^2\right)$ is the semilatus rectum, a the semi-major axis, ϵ the orbital eccentricity, ν is the true anomaly and ω is the argument of the perihelion. Notice that we use an unconventional definition of the true anomaly as the angle among the radius vector and the ascending node of the planet instead of measuring it from the perihelion as usual.

We now introduce the transformation matrix α_{ij} as the scalar products of the unit vectors of the orbital and the Sun's system of reference:

$$\alpha_{ij} = \hat{n}_i \cdot \hat{m}_j\,, i, j = 1, 2, 3\,. \tag{79}$$

If θ, ϕ are, respectively, the polar angle and azimuthal angle in the Sun's system of reference, we find from Equation (78) and using the definition in Equation (79) the relation among them and the true anomaly, ν, in the following form:

$$\cos\theta = \tfrac{\mathbf{r}}{r}\cdot\hat{m}_3 = \alpha_{13}\cos\nu + \alpha_{23}\sin\nu\,, \tag{80}$$

$$\sin\theta\cos\phi = \tfrac{\mathbf{r}}{r}\cdot\hat{m}_1 = \alpha_{11}\cos\nu + \alpha_{21}\sin\nu\,, \tag{81}$$

$$\sin\theta\cos\phi = \tfrac{\mathbf{r}}{r}\cdot\hat{m}_2 \tag{82}$$

$$= \alpha_{12} \cos \nu + \alpha_{22} \sin \nu .$$

The coefficients of the transformation matrix satisfy some useful identities:

$$\sum_{k=1}^{3} \alpha_{ik} \alpha_{jk} = \delta_{ij} , \ i, j = 1, 2, 3 , \tag{83}$$

where δ_{ij} is Kronecker's delta. The spherical unit vectors in the Sun's system of reference are given as:

$$\begin{aligned} \hat{\mathbf{r}} &= \sin \theta \cos \phi \, \hat{\mathbf{m}}_1 \\ &+ \sin \theta \cos \phi \, \hat{\mathbf{m}}_2 + \cos \theta \, \hat{\mathbf{m}}_3 , \end{aligned} \tag{84}$$

$$\begin{aligned} \hat{\boldsymbol{\theta}} &= \cos \theta \cos \phi \, \hat{\mathbf{m}}_1 \\ &+ \sin \theta \sin \phi \, \hat{\mathbf{m}}_2 - \sin \theta \, \hat{\mathbf{m}}_3 , \end{aligned} \tag{85}$$

$$\hat{\boldsymbol{\phi}} = - \sin \phi \, \hat{\mathbf{m}}_1 + \cos \phi \, \hat{\mathbf{m}}_2 . \tag{86}$$

From Equations (80) and (84), we can also find the tangential unit vector to the planetary orbit:

$$\begin{aligned} \hat{\boldsymbol{\nu}} = \frac{d\hat{\mathbf{r}}}{d\nu} &= \left(-\alpha_{11} \sin \nu + \alpha_{21} \cos \nu \right) \hat{\mathbf{m}}_1 \\ &+ \left(-\alpha_{12} \sin \nu + \alpha_{22} \cos \nu \right) \hat{\mathbf{m}}_2 \\ &+ \left(-\alpha_{13} \sin \nu + \alpha_{23} \cos \nu \right) \hat{\mathbf{m}}_3 . \end{aligned} \tag{87}$$

Orthogonality with $\hat{\mathbf{r}}$ and normalization to the unity modulus can be shown by applying the identities in Equation (83).

Now, we can find the radial, $\mathcal{R}$, tangential to the orbit, $\mathcal{T}$, and normal to the orbital plane, $\mathcal{N}$, components of the perturbing force per unit mass in terms of the accelerations, a_r, a_θ and a_ϕ, in the Sun's spherical system of reference. It is obvious that:

$$\mathcal{R} = a_r . \tag{88}$$

For the tangential component, we have:

$$\begin{aligned} \mathcal{T} &= \left(a_\theta \, \hat{\boldsymbol{\theta}} + a_\phi \, \hat{\boldsymbol{\phi}} \right) \cdot \hat{\boldsymbol{\nu}} \\ &= \frac{a_\theta}{\sin \theta} \left(\alpha_{13} \sin \nu - \alpha_{23} \cos \nu \right) \\ &+ \frac{a_\phi}{\sin \theta} \left(\alpha_{11} \alpha_{22} - \alpha_{12} \alpha_{21} \right) , \end{aligned} \tag{89}$$

after some simplifications using Equations (80)–(87). Finally, for the normal component, we find:

$$\begin{aligned} \mathcal{N} &= \left(a_\theta \, \hat{\boldsymbol{\theta}} + a_\phi \, \hat{\boldsymbol{\phi}} \right) \cdot \hat{\mathbf{n}}_3 \\ &= a_\theta \left(\alpha_{31} \cos \theta \cos \phi + \alpha_{32} \cos \theta \sin \phi - \alpha_{33} \sin \theta \right) \\ &+ a_\phi \left(-\alpha_{31} \sin \phi + \alpha_{32} \cos \phi \right) . \end{aligned} \tag{90}$$

Notice that the angles θ and ϕ can be formally expressed in terms of the true anomaly by using Equation (80).

References

1. Iorio, L. Editorial for the Special Issue 100 Years of Chronogeometrodynamics: The Status of the Einstein's Theory of Gravitation in Its Centenial Year. *Universe* **2015**, *1*, 38–81. [CrossRef]
2. Will, C.M. The Confrontation between General Relativity and Experiment. *Living Rev. Relativ.* **2006**, *9*, 3. [CrossRef]
3. Lämmerzahl, C.; Ciufolini, I.; Dittus, H.; Iorio, L.; Müller, H.; Peters, A.; Samain, E.; Scheithauer, S.; Schiller, S. OPTIS–An Einstein Mission for Improved Tests of Special and General Relativity. *Gen. Relativ. Gravit.* **2004**, *36*, 2373–2416. [CrossRef]

Universe **2015**, *1*, 422–455

4. Iorio, L.; Ciufolini, I.; Pavlis, E.C.; Schiller, S.; Dittus, H.; Lämmerzahl, C. On the possibility of measuring the Lense-Thirring effect with a LAGEOS LAGEOS II OPTIS mission. *Classical. Quant. Grav.* **2004**, *21*, 2139–2151. [CrossRef]

5. Schiller, S.; Tino, G.M.; Gill, P.; Salomon, C.; Sterr, U.; Peik, E.; Nevsky, A.; Görlitz, A.; Svehla, D.; Ferrari, G. Einstein Gravity Explorer-a medium-class fundamental physics mission. *Exp. Astron.* **2009**, *23*, 573–610. [CrossRef]

6. Turyshev, S.G.; Turyshev, S.G.; Sazhin, M.V.; Toth, V.T. General relativistic laser interferometric observables of the GRACE-Follow-On mission. *Phys. Rev. D* **2014**, *89*, 105029. [CrossRef]

7. Everitt, C.W.F.; de Bra, D.B.; Parkinson, B.W.; Turneaure, J.P.; Conklin, J.W.; Heifetz, M.I.; Keiser, G.M.; Silbergleit, A.S.; Holmes, T.; Kolodziejczak, J.; *et al.* Gravity Probe B: Final Results of a Space Experiment to Test General Relativity. *Phys. Rev. Lett.* **2011**, *106*, 221101. [CrossRef] [PubMed]

8. Ciufolini, I.; Pavlis, E.C. A confirmation of the general relativistic prediction of the Lense-Thirring effect. *Nature* **2004**, *431*, 958–960. [CrossRef] [PubMed]

9. Ciufolini, I.; Paolozzi, A.; Pavlis, E.C.; Ries, J.C.; Koenig, R.; Matzner, R.A.; Sindoni, G.; Neumayer, H. Towards a One Percent Measurement of Frame Dragging by Spin with Satellite Laser Ranging to LAGEOS, LAGEOS 2 and LARES and GRACE Gravity Models. *Space Sci. Rev.* **2009**, *148*, 71–104. [CrossRef]

10. Iorio, L.; Lichtenegger, H.I.M.; Ruggiero, M.L.; Corda, C. Phenomenology of the Lense-Thirring effect in the solar system. *Astrophys. Space Sci.* **2011**, *331*, 351–395. [CrossRef]

11. Ciufolini, I. Frame Dragging and Lense-Thirring Effect. *Gen. Relativ. Gravit.* **2004**, *36*, 2257–2270. [CrossRef]

12. Renzetti, G. History of the attempts to measure orbital frame-dragging with artificial satellites. *Cent. Eur. J. Phys.* **2013**, *11*, 531–544. [CrossRef]

13. Soffel, M.; Wirrer, R.; Schastok, J.; Ruder, H.; Schneider, M. Relativistic effects in the motion of artificial satellites. I—The oblateness of the central body. *Celest. Mech.* **1988**, *42*, 81–89. [CrossRef]

14. Iorio, L. A possible new test of general relativity with Juno. *Classical Quant. Grav.* **2013**, *30*, 195011. [CrossRef]

15. Iorio, L. Are we far from testing general relativity with the transiting extrasolar planet HD 209458b "Osiris"? *New Astron.* **2006**, *11*, 490–494. [CrossRef]

16. Iorio, L. Classical and relativistic long-term time variations of some observables for transiting exoplanets. *Mon. Not. R. Astron. Soc.* **2011**, *411*, 167–183. [CrossRef]

17. Adams, F.C.; Laughlin, G. Relativistic Effects in Extrasolar Planetary Systems. *Int. J. Mod. Phys. D* **2006**, *15*, 2133–2140. [CrossRef]

18. Pal, A.; Kocsis, B. Periastron precession measurements in transiting extrasolar planetary systems at the level of general relativity. *Mon. Not. R. Astron. Soc.* **2008**, *389*, 191–198. [CrossRef]

19. Jordan, A.; Bakos, G. Observability of the General Relativistic Precession of Periastra in Exoplanets. *Astrophys. J.* **2008**, *685*, 543–552. [CrossRef]

20. Zhao, S.; Xie, Y. Parametrized post-Newtonian secular transit timing variations for exoplanets. *Res. Astron. Astrophys.* **2013**, *13*, 1231–1239. [CrossRef]

21. Xie, Y.; Deng, X.-M. On the (im)possibility of testing new physics in exoplanets using transit timing variations: deviation from inverse-square law of gravity. *Mon. Not. R. Astron. Soc.* **2014**, *438*, 1832–1838. [CrossRef]

22. Iorio, L. Accurate characterization of the stellar and orbital parameters of the exoplanetary system WASP-33b from orbital dynamics. *Mon. Not. R. Astron. Soc.* **2016**, in press.

23. Iorio, L. Classical and relativistic node precessional effects in WASP-33b and perspectives for detecting them. *Astrophys. Space Sci.* **2011**, *331*, 485–496. [CrossRef]

24. Goenner, H.F.M. On the History of Unified Field Theories. *Living Rev. Relativ.* **2004**, *7*, 1830–1923. [CrossRef]

25. Sciama, D.W. The Physical Structure of General Relativity. *Rev. Mod. Phys.* **1964**, *36*, 463–469. [CrossRef]

26. Kibble, T.W.B. Lorentz invariance and the gravitational field. *J. Math. Phys.* **1961**, *2*, 212–221. [CrossRef]

27. Mielke, E.W. Is Einstein-Cartan Theory Coupled to Light Fermions Asymptotically Safe? *J. Grav.* **2013**, *2013*, 5. [CrossRef]

28. Hehl, F.W. Gauge theory of gravity and spacetime. In Proceedings of the Workshop Towards a Theory of Spacetime Theories, Wuppertal, Germany, 21–23 July 2010.

29. Aldrovandi, R.; Pereira, J.G. *Teleparallel Gravity*; Springer: Berlin, Germany, 2013.

30. Hehl, F.W.; von der Heyde, P.; Kerlick, G.D.; Nester, J.M. General Relativity with Spin and Torsion: Foundations and Prospects. *Rev. Mod. Phys.* **1976**, *48*, 393–415. [CrossRef]

31. Ni, W.T. Searches for the role of spin and polarization in gravity. *Rep. Prog. Phys.* **2010**, *73*, 056901. [CrossRef]

32. Popławski, N.J. Cosmology with torsion: An alternative to cosmic inflation. *Phys. Lett. B* **2010**, *694*, 181–185. [CrossRef]
33. Popławski, N.J. Non-singular, big-bounce cosmology from spinor-torsion coupling. *Phys. Rev. D* **2012**, *85*, 107502. [CrossRef]
34. Obukhov, Y.N. *Poincaré Gauge Gravity: Selected Topics*; Cornell University: Ithaca, NY, USA, 2006.
35. Hammond, R.T. Torsion gravity. *Rep. Prog. Phys.* **2002**, *65*, 599–649. [CrossRef]
36. Hojman, S.; Rosenbaum, M.; Ryan, M.P.; Shepley, L. Gauge invariance, minimal coupling and torsion. *Phys. Rev. D* **1978**, *17*, 3141–3146. [CrossRef]
37. Hojman, S.; Rosenbaum, M.; Ryan, M.P. Propagating torsion and gravitation. *Phys. Rev. D* **1979**, *19*, 430–437. [CrossRef]
38. Hayashi, K.; Shirafuji, T. New General Relativity. *Phys. Rev. D* **1979**, *19*, 3524–3553. [CrossRef]
39. Weitzenböck, R. *Invariantentheorie*; Popko Noordhoff: Groningen, The Netherlands, 1923.
40. Bel, L. *Connecting Connections. A Bricklayer View of General Relativity*; Cornell University: Ithaca, NY, USA, 2008.
41. Mao, Y.; Tegmark, M.; Guth, A.H.; Cabi, S. Constraining torsion with Gravity Probe B. *Phys. Rev. D* **2007**, *76*, 104029. [CrossRef]
42. March, R.; Belletini, G.; Tauraso, R.; Dell'Agnello, S. Constraining spacetime torsion with the Moon and Mercury. *Phys. Rev. D* **2011**, *83*, 104008. [CrossRef]
43. March, R.; Belletini, G.; Tauraso, R.; Dell'Agnello, S. Constraining spacetime torsion with LAGEOS. *Gen. Relativ. Gravit.* **2011**, *43*, 3099–3126. [CrossRef]
44. Paolozzi, A.; Ciufolini, I. LARES successfully launched in orbit: Satellite and mission description. *Acta Astronaut.* **2013**, *91*, 313–321. [CrossRef]
45. Iorio, L. The impact of the new Earth gravity models on the measurement of the Lense-Thirring effect with a new satellite. *New Astron.* **2005**, *10*, 616–635. [CrossRef]
46. Renzetti, G. On Monte Carlo simulations of the LAser RElativity Satellite experiment. *Acta Astronaut.* **2015**, *113*, 164–168. [CrossRef]
47. Ciufolini, I.; Moreno Monge, B.; Paolozzi, A.; Koenig, R.; Sindoni, G.; Michalak, G.; Pavlis, E.C. Monte Carlo simulations of the LARES space experiment to test General Relativity and fundamental physics. *Class. Quantum Grav.* **2013**, *30*, 235009. [CrossRef]
48. Iorio, L. The impact of the orbital decay of the LAGEOS satellites on the frame-dragging tests. *Adv. Space Res.* **2016**, in press.
49. Hehl, F.W.; Obukhov, Y.N. Élie Cartan torsion in geometry and in field theory, an essay. *Annal. Found. Louis Broglie* **2007**, *32*, 157–194.
50. Hehl, F.W.; Obukhov, Y.N.; Puetzfeld, D. On Poincaré gauge theory of gravity, its equations of motion, and Gravity Probe B. *Phys. Lett. A* **2013**, *377*, 1775–1781. [CrossRef]
51. Kleinert, H.; Pelster, A. Autoparallels from a new action principle. *Gen. Relativ. Grav.* **1999**, *31*, 1439–1447. [CrossRef]
52. Kleinert, H.; Shabanov, S.V. Spaces with Torsion from Embedding and the Special Role of Autoparallel Trajectories. *Phys. Lett. B* **1998**, *428*, 315–321. [CrossRef]
53. Kleinert, H. Nonholonomic Mapping Principle for Classical and Quantum Mechanics in Spaces with Curvature and Torsion. *Gen. Rel. Grav.* **2000**, *32*, 769–839. [CrossRef]
54. Anderson, J.D.; Nieto, M.M. *Relativity in Fundamental Astronomy: Dynamics, Reference Frames, and Data Analysis*; Klioner, S.A., Seidelmann, P.K., Soffel, M.H., Eds.; Proceedings IAU Symposium No. 261; Cambridge University Press: Cambridge, UK, 2010; pp. 189–197.
55. Lämmerzahl, C.; Preuss, O.; Dittus, H. Is the physics of the Solar System really understood? *Lasers, Clocks Drag-Free Control* **2008**, *349*, 75–101.
56. Iorio, L. Gravitational anomalies in the Solar System? *Int. J. Mod. Phys. D* **2015**, *24*, 1530015. [CrossRef]
57. Laskos-Grabowski, P. The Einstein-Cartan Theory: The Meaning and Consequences of Torsion. Master's Thesis, University of Wrocław, Wrocław, Poland, 2009.
58. Puetzfeld, D.; Obukhov, Y.N. Probing non-Riemannian spacetime geometry. *Phys. Lett. A* **2008**, *372*, 6711–6716. [CrossRef]
59. Landau, L.D.; Lifshitz, E.M. *The Classical Theory of Fields. Course of Theoretical Physics*, 4th ed.; Butterworth-Heinemann: Oxford, UK, 1987; Volume 2.

60. Kleinert, H. *Path Integrals in Quantum Mechanics, Statistics, Polymer Physics and Financial Markets*, 15th ed.; World Scientific Publish. Co.: Singapore, Singapore, 2009.

61. Kleinert, H. *Multivalued Fields: in Condensed Matter, Electromagnetism, and Gravitation*; World Scientific Publ. Co.: Singapore, Singapore, 2008.

62. Kleinert, H.; Pelster, A. Novel Geometric Gauge Invariance of Autoparallels. *Acta Phys. Pol.* **1998**, *29*, 1015–1023.

63. Pollard, H. *Mathematical Introduction to Celestial Mechanics*; Prentice-Hall Inc.: Englewood Cliffs, NY, USA, 1966.

64. Danby, J.M.A. *Fundamentals of Celestial Mechanics*, 2nd ed.; Willmann-Bell, Inc.: Richmond, Virginia, USA, 1988.

65. Burns, J.A. Elementary derivation of the perturbation equations of celestial mechanics. *Am. J. Phys.* **1976**, *44*, 944–949. [CrossRef]

66. Krasinsky, G.A.; Brumberg, V.A. Secular increase of astronomical unit from analysis of the major planet motions, and its interpretation. *Celest. Mech. Dyn. Astron.* **2004**, *90*, 267–288. [CrossRef]

67. Standish, E.M. The astronomical unit now. In *Transit of Venus: New Views of the Solar System and Galaxy*; Kurtz, D.W., Ed.; Cambridge University Press: Cambridge, UK, 2005; p. 163.

68. Vinti, J.P. Classical solution of the two-body problem if the gravitational constant diminishes inversely with the age of the Universe. *Mon. Not. R. Astron. Soc.* **1974**, *169*, 417–427. [CrossRef]

69. Iorio, L. An empirical explanation of the anomalous increases in the astronomical unit and the lunar eccentricity. *Astron. J.* **2011**, *142*, 68. [CrossRef]

70. Acedo, L. Anomalous post-Newtonian terms and the secular increase of the astronomical unit. *Adv. Space Res.* **2013**, *52*, 1297–1303. [CrossRef]

71. Li, X.; Chang, Z. Kinematics in Randers-Finsler geometry and secular increase of the astronomical unit. *Chin. Phys. C* **2011**, *35*, 914–919. [CrossRef]

72. Miura, T.; Arakida, H.; Kasai, M.; Kuramata, S. Secular increase of the astronomical unit: A possible explanation in terms of the total angular momentum conservation law. *Publ. Astron. Soc. Jpn.* **2009**, *61*, 1247–1250. [CrossRef]

73. Iorio, L. Secular increase of the astronomical unit and perihelion precessions as tests of the Dvali Gabadadze Porrati multi-dimensional braneworld scenario. *J. Cosmol. Astropart. Phys.* **2005**, *2005*, 6. [CrossRef]

74. NASA Planetary Fact Sheet. Available online: http://nssdc.gsfc.nasa.gov/planetary/factsheet/ (accessed on 14 May 2015).

75. Acedo, L. Constraints on non-standard gravitomagnetism by the anomalous perihelion precession of the planets. *Galaxies* **2014**, *2*, 466–481. [CrossRef]

76. Iorio, L. A Critical Analysis of a Recent Test of the Lense-Thirring Effect with the LAGEOS Satellites. *J. Geod.* **2006**, *80*, 128–136. [CrossRef]

77. Iorio, L.; Ruggiero, M.L.; Corda, C. Novel considerations about the error budget of the LAGEOS-based tests of frame-dragging with GRACE geopotential models. *Acta Astronaut.* **2013**, *91*, 141–148. [CrossRef]

78. Renzetti, G. Some reflections on the LAGEOS frame-dragging experiment in view of recent data analyses. *New Astron.* **2014**, *29*, 25–27.

79. Giles, P. Time-Distance Measurements of Large-Scale Flows in the Solar Convection Zone. Ph.D. Thesis, Stanford University, Stanford, CA, USA, 1999.

80. Stark, D.; Wöhl, H. On the solar rotation elements as determined from sunspot observations. *Astron. Astrophys.* **1981**, *93*, 241–244.

![universe logo] *universe*

MDPI

Article
Convexity and the Euclidean Metric of Space-Time

Nikolaos Kalogeropoulos

Carnegie Mellon University in Qatar, Education City, P.O. Box 24866 Doha, Qatar; nkaloger@cmu.edu or nikos.physikos@gmail.com

Academic Editors: Stephon Alexander, Jean-Michel Alimi, Elias C. Vagenas and Lorenzo Iorio
Received: 1 November 2016; Accepted: 2 February 2017; Published: 8 February 2017

Abstract: We address the reasons why the "Wick-rotated", positive-definite, space-time metric obeys the Pythagorean theorem. An answer is proposed based on the convexity and smoothness properties of the functional spaces purporting to provide the kinematic framework of approaches to quantum gravity. We employ moduli of convexity and smoothness which are eventually extremized by Hilbert spaces. We point out the potential physical significance that functional analytical dualities play in this framework. Following the spirit of the variational principles employed in classical and quantum Physics, such Hilbert spaces dominate in a generalized functional integral approach. The metric of space-time is induced by the inner product of such Hilbert spaces.

Keywords: space-time metric; convexity; smoothness; Hilbert spaces; Banach spaces; dualities

---

1. Introduction

When one looks at the equations describing the four fundamental interactions of nature, then s/he immediately notices that the kinematic equations in the Lagrangian formalism involve second order derivatives with respect to space-time variables. It may be worthwhile to try to understand the reasons behind this phenomenon, which has been taken for granted as an empirical fact since the earliest days of Newtonian mechanics. Modelling of the fundamental interactions except gravity relies, at the classical level, on Classical Mechanics, on Electromagnetic Theory and its Yang-Mills/non-abelian gauge "generalizations". General Relativity can also be seen as a gauge theory whose gauge group is the diffeomorphism group (re-parametrization invariance) of the underlying topological space endowed with its metric structure.

In all of the above, and in the Lagrangian approach which we employ throughout this work, the Euler-Lagrange equations that describe the underlying dynamics can be seen to emerge from variational principles; such equations could use derivatives of arbitrarily high order and a formalism for accommodating this fact has already been developed. However, in practice, when dealing with fundamental interactions and not performing perturbative or approximate calculations that rely on series expansions, one rarely needs derivatives that are of higher than the second-order with respect to space-time variables.

The statements on the number of derivatives in the equations of dynamics can be seen to be essentially equivalent, upon partial integration, to the fact that the kinetic terms as well as relevant potential energy terms are at most quadratic with respect to first order derivatives of the fundamental variables/fields/order parameters. This in turn, allows one to use Euclidean/Riemannian concepts to model the evolution of such systems; for particle systems one uses the more familiar aspects of finite-dimensional Riemannian spaces [1], and for field theories one may have to resort to using aspects of infinite dimensional manifolds [2] which involve further subtleties.

One could use, equivalently in the simplest context, a Hamiltonian approach where the equations involve first order derivatives. Each of these two lines of approach Lagrangian versus Hamiltonian has its own advantages and drawbacks, as is well-known. See, for instance, the very recent review [3] for an approach to gravity based on first order actions including boundary terms. Without denying the advantages of the Hamiltonian or first order formalisms, we will adopt in this work the Lagrangian/second-order formalism, as already mentioned above. Our viewpoint is somewhat influenced by and may have common points with that of [4], even though the methods which we use and the results we reach are substantially different from those of that work.

We will adopt the convenient technical device of Wick-rotating to a "Euclidean" (i.e., positive-definite) signature metric. We would like to state at this point, that we do not consider positive-definite metrics "superior" in any sense, to these of Lorentzian (indefinite) signature. On the contrary, all experiments, so far, point to the latter as being the physical ones, so even if unstated, our underlying view is closer to [5] rather than [6], for instance. On the other hand, on purely formalistic grounds, the Euclidean (positive-definite) metrics are easier to work with, as they obey the positivity property and the triangle inequality that their indefinite metric (Lorentzian) counterparts are lacking.

A substantial amount of effort has been spent, in recent decades, into understanding the non-trivial features of space-time, as described by the General Theory of Relativity or other theories incorporating space-time diffeomorphism invariance. Even though considerable progress has been made toward such an understanding, it is probably fair to state that many important issues still remain unresolved [7]. It is not clear, for instance, why space-time is 4-dimensional, to what extent it is smooth and how such a smoothness arises, why its Wick-rotated metric obeys the Pythagorean theorem etc. Henceforth we will work only with positive definite (Wick-rotated) metrics of space-time and by "Euclidean" we will mean only the ones obeying the Pythagorean theorem. In linear algebraic and functional analytic language the metrics we call Euclidean would be called the l^2 metrics. Moreover, we will use the term space-time in order to keep in mind that our arguments actually purport to describe (indefinite/Lorentzian signature) space-time even if we use positive-definite (Wick-rotated) metrics throughout this work.

A potentially fruitful way toward answering why the space-time metric is Euclidean, which is the subject of this work, is to look at it through the eyes of convexity. Convexity plays a central role in many branches of Mathematics, but seems to be under-appreciated and under-utilized in gravitational Physics [8], and not only. This can also be considered as a partial motivation for looking for answers to the questions of our interest through convexity. In the context of linear spaces, convexity turns out to be dual to smoothness, a fact that we also use to support the case for the Euclidean form of the space-time metric.

In Section 2, we provide a general background and a physical interpretation, wherever feasible, from the theory of normed spaces with emphasis on properties pertinent to our arguments. In Section 3, we discuss the aspects of convexity, smoothness and their duality via a Legendre-Fenchel transform, employing in particular Clarkson's modulus of convexity and to the Day-Lindenstrauss modulus of smoothness. In Section 4, we put our, less than rigorous, argument together on how the previous results result in space-time metric that has the Euclidean form. Section 5, presents some conclusions and caveats.

2. Background and Physical Interpretation of Some Concepts

Space-times are assumed to be locally flat, to first order approximation of their metric. This is an outcome of the application of the Equivalence Principle, and a well-known fact in Riemannian/Lorentzian geometry. Therefore, we can analyze the ultralocal aspects of the features of a space-time by confining our attention to vector spaces. Most features of such tangent spaces can be captured by the various closely related functional spaces that can be defined on them. There are numerous classes of functional spaces that have been investigated during the last century, in the context of Functional Analysis, such as Lebesgue, Hardy, Bergman, Sobolev, Orlicz, Besov, Triebel-Lizorkin, etc. spaces. Most of the previously named spaces are or rely on, in one form or another, constructions and results of (usually infinite dimensional) Banach spaces. Such infinite dimensional Banach spaces will be the main vector spaces of interest in this work.

In this Section we provide some preliminary information on these and related mathematical constructions which are pertinent to the subsequent sections, where most of our arguments are developed. We attempt to stress their physical motivation and interpretation, wherever feasible, from the viewpoint we adopt in this work.

2.1. Norms on Linear Spaces

To make the exposition more readable and self-contained, we recall [9,10] that a norm on a vector space V defined over a field $\mathbb{F}$, usually $\mathbb{F} = \mathbb{R}$ or $\mathbb{F} = \mathbb{C}$ in most applications in Physics, is a function $\| \cdot \| : V \to \mathbb{R}_+$ where $\mathbb{R}_+ = \{\lambda \in \mathbb{R} : \lambda \geq 0\}$, such that for all $x, y \in V$

- Positive definiteness: $\|x\| = 0$, if and only if $x = 0$.
- Homogeneity: $\|kx\| = |k| \|x\|$, for all $x \in V$, and all $k \in \mathbb{F}$.
- Triangle inequality: $\|x + y\| \leq \|x\| + \|y\|$

A vector space V endowed with a norm, which moreover is complete in this norm, namely such that all Cauchy sequences have limits belonging to V, is called a Banach space. Examples of Banach spaces that are explicitly used in this work are:

- c_0 is the space of all sequences $a = (a_n), n \in \mathbb{N}$ converging to zero, with $a_n \in \mathbb{F}$, and the sup-norm

$$\|a\| = \sup_n |a_n| \tag{1}$$

- $l^p, 1 \leq p < \infty$, which is the space of all sequences $a = (a_n), n \in \mathbb{N}$ with $a_n \in \mathbb{F}$ endowed with the norm

$$\|a\|_p = \left(\sum_{i=1}^n |a_n|^p \right)^{\frac{1}{p}} < \infty \tag{2}$$

- l^∞, the space of all bounded sequences $a = (a_n), n \in \mathbb{N}$ endowed with the supremum norm

$$\|a\| = \sup_n |a_n| \tag{3}$$

- $L^p(\mathbb{R}^n), 1 \leq p < \infty$, the space of Lebesgue integrable functions $f : \mathbb{R}^n \to \mathbb{F}$ endowed with the norm

$$\|f\|_p = \left(\int_{\mathbb{R}^n} |f|^p \right)^{\frac{1}{p}} \tag{4}$$

- $L^\infty(\mathbb{R}^n)$, the space of all essentially bounded $f : \mathbb{R}^n \to \mathbb{F}$, endowed with the norm

$$\|f\|_\infty = \inf\{C : |f| < C \text{ almost everywhere}\} \tag{5}$$

The above functional spaces $L^p(\mathbb{R}^n), 1 \leq p \leq \infty$ are, strictly speaking, spaces over equivalence classes of functions, where two functions are considered equivalent ("equal") if they differ from each other, at most, in a set of measure zero. One uses the Lebesgue measure ("volume") of $\mathbb{R}^n$ in the definition of such $L^p(\mathbb{R}^n)$. In most applications in Physics, the distinction between functions and classes of equivalent functions is tacitly assumed and not explicitly stated. For completeness, we mention that a Hilbert space is a linear space endowed with an inner product $(\cdot, \cdot)$ which is, moreover, complete [9,10]. The norm which we will assume that Hilbert spaces are endowed with, is induced by their inner product by $\|x\|^2 = (x, x)$. Among the above examples of Banach spaces, l^2 and $L^2(\mathbb{R}^n)$ are Hilbert spaces. Since the spaces that we will be referring to are normed, hence metrizable, they are endowed with the topology induced by the metric. An important topological property of such topological (vector) spaces is separability: this means that such spaces contain a countable dense subset. For such metrizable spaces, being separable is equivalent to being second countable. Topological spaces that are finite or countably infinite are, obviously, separable. For Hilbert spaces separability implies the existence of a countable orthonormal basis, hence any separable infinite dimensional Hilbert space is isometric to l^2. In most cases in quantum theory, separability of the space of the Hllbert space of wave-functions is tacitly assumed. Banach spaces can also be either separable or not separable [11]. The issue of separabilty of the spaces we use is not pertinent to the arguments of this work, so it will not be encountered anywhere in the sequel.

2.2. Norm Equivalence

Naturally, one can endow a vector space $\mathcal{V}$ with many different norms. Usually the "appropriate" choice of such a norm has substantial implications for the specific predictions of the physical model built on $(\mathcal{V}, \| \cdot \|)$. In other words, the choice of two "different", in the naive sense of the word, norms will usually result in substantially different predictions of the physical quantities resulting form such calculations. Hence the choice of a norm is usually considered to be a piece of data which is initially provided by hand, in any model. In most cases such a choice of norm is not explicitly discussed, because it is assumed that a norm arising from an inner product is the physically relevant one. In this work, we would like to know why this may be the case.

Given that a vector space may be endowed with different norms, one can ask what the word "different" may actually mean. If two norms are point-wise different but still reasonably close to each other, in some particular sense, let's say with respect to a particular metric, can they still be declared as "equivalent"? Consider, for instance, the case of Hamiltonian systems of many degrees of freedom. This paragraph operates in the context of Euclidean metrics but can still be used to highlight our point of view. Suppose that one changes the metric of the phase space. If all thermodynamic quantities remain invariant under such a change of norm, is it reasonable to consider the two metrics as "equivalent" or should someone insist as treating them as "different"? Under some additional conditions, two such metrics may turn out to be equal; this is reminiscent of the (Hamburger, Stieltjes etc.) moment problem pertaining to the equality of underlying probability distributions, if the sets of moments of these distributions are equal [12]. Something similar occurs in Geometry or Analysis: quite often two metrics or norms are declared as "equivalent" even though they may be pointwise distinct.

This re-interpretation of the concept of "equivalence" can have substantial consequences for Statistical or Quantum Physics. The issue at hand can be seen as a form of "stability". Consider, for instance, a Hamiltonian system. This determines a symplectic structure on the phase space $\mathcal{M}$ of the system or, alternatively, a Poisson structure on the space of smooth functions $C^\infty(\mathcal{M})$ [13]. However, it does not generically determine a metric structure on $\mathcal{M}$. Such a metric is usually assumed to stem from the quadratic "kinetic" term of the Hamiltonian, at least for systems having a kinetic term of such form. Then we can proceed to perform an analysis of the evolution of this Hamiltonian system, quantize the system etc. based on this symplectic and metric structures. However, one should not forget that the metric structure was not "natural", or may not be unique. For this reason if it changes, especially a little bit, one would expect the physically relevant results to remain practically invariant. This can be interpreted as a form of structural stability of the underlying Hamiltonian dynamical system, even though it applies to an auxiliary piece of data, such as the metric. It is a subject of much discussion on whether such a metric structure should be assumed, and if so, to what extent it determines the statistically significant features of the system in an appropriate many-body "thermodynamic" limit [14].

Given the above considerations, one may be willing to allow variations of the assumed metric of the underlying Hamiltonian system. Then two metrics can be declared as equivalent, as also previously mentioned, if they give the same predictions of physically relevant quantities. The question is then how to find such variations of metrics, or norms. From an analytical viewpoint, one can use a central concept of Analysis, that of the limit, to determine how to answer such questions. The most crude/rough approach is to demand that two metrics/norms $\| \cdot \|_1, \| \cdot \|_2$ to be considered as equivalent if all sequences converging with respect to one of them also converge with respect to the other. This is realized when there are constants $0 < C_1 \leq C_2$ such that

$$C_1 \|\cdot\|_1 \leq \|\cdot\|_2 \leq C_2 \|\cdot\|_1 \tag{6}$$

This always holds for any two norms on finite dimensional vector spaces. However, it is not true in general for infinite dimensional functional spaces, which are at the center of our attention.

A way to interpret (6) from the viewpoint of statistical theories is as follows: physically relevant results of the microscopic or quantum dynamics should be somehow reflected or emerge in the large scale/multi-particle or thermodynamic limit. Such features should remain largely unaffected by most "small-scale" details of the system or their perturbations. This tacitly assumes that the underlying quantities characterizing geometric characteristics of the system are proportional to the effective measure(s) used in the calculations of the pertinent statistical quantities. This is clearly true for ergodic systems, but it can also be true for non-ergodic systems such as the ones whose thermodynamic behavior is conjecturally described by any of the many recently proposed entropic functionals, such as the "Tsallis entropy" [15], or the "κ-entropy" [16], for instance. So, from a geometric viewpoint, equations like (6) express this insensitivity to small-scale details. This viewpoint motivates and pervades "coarse geometry" [17] and is frequently encountered in constructions related to hyperbolic spaces [18] or groups [19]. Metrically, it is expressed by demanding invariance under quasi-isometries. In dynamical systems one can see a similar viewpoint in several occasions, an example of which is that the topological entropy of a map or flow on a metric space $(\mathfrak{X}, d)$ does not actually depend on the specific metric/distance function d, but only on the class on metrics on $\mathfrak{X}$ that induce the same topology on $\mathfrak{X}$. Then the key/desired invariance akin to (6), is the topological conjugacy [20].

2.3. The Operator Norm and the Banach-Mazur Distance

Before continuing, for completeness of the exposition, we state two definitions that will be extensively used in the sequel [9,10]. Let $(\mathcal{X}, \|\cdot\|_{\mathcal{X}})$ and $(\mathcal{Y}, \|\cdot\|_{\mathcal{Y}})$ be two normed spaces, over $\mathbb{R}, \mathbb{C}$ (or any other field, although the general case does not appear to be of any particular interest in Physics, so far) and let T be a continuous linear map $T : \mathcal{X} \to \mathcal{Y}$. If such a map exists between $\mathcal{X}, \mathcal{Y}$ which is bijective, and its inverse T^{-1} is also bijective, then $\mathcal{X}, \mathcal{Y}$ are called isomorphic. Actually less is needed: the Open Mapping Theorem guarantees that if T is bijective and bounded, so is T^{-1}. If a mapping T is an isometry, namely if

$$\|Tx\|_{\mathcal{Y}} = \|x\|_{\mathcal{X}} \tag{7}$$

then $\mathcal{X}, \mathcal{Y}$ are called isometric. Since boundedness of T is equivalent to its continuity, one can see that isometric spaces are isomorphic. Let the space of bounded linear maps from $\mathcal{X}$ to $\mathcal{Y}$ be denoted by $\mathcal{B}(\mathcal{X}, \mathcal{Y})$. This space $\mathcal{B}(\mathcal{X}, \mathcal{Y})$ can be endowed with the operator (sup-) norm which for $T \in \mathcal{B}(\mathcal{X}, \mathcal{Y})$ is given by the following equivalent definitions

$$\|T\| = \sup_{x \in \mathcal{X}}\{\|Tx\| : \|x\| \leq 1\} = \sup_{x \in \mathcal{X}}\{\|Tx\| : \|x\| = 1\} = \sup_{x \in \mathcal{X}}\left\{ \frac{\|Tx\|}{\|x\|} : \|x\| \neq 0 \right\} \tag{8}$$

It turns out that if $\mathcal{X}$ is a normed space and if $\mathcal{Y}$ is a Banach space, then $\mathcal{B}(\mathcal{X}, \mathcal{Y})$ endowed with the operator norm (8) is a Banach space too. We know that every normed space can be isometrically embedded in a Banach space. So, most of the pertinent features of general normed spaces are contained in Banach spaces, therefore we can use the latter believing that we are not losing important aspects of the flexibility or generality of the former, for applications in Physics. Given this, our question is then reduced to asking, why among all Banach spaces, the inner product (Hilbert) spaces are the ones describing most fundamental aspects of nature most accurately, so far as we know today.

In the spirit of norm-equivalence, discussed in Subsection 2.2, one can ask how close, or how far, from each other are two linear spaces. From a metric viewpoint they are identical, if they are isometric.

We want to have a "reasonable" distance function that measures how far from each other they may be, if they are not isometric. Defining such a distance function is clearly a matter of choice which ideally, for our purposes, should also reflect some desirable physical properties. It appears that the classical Banach-Mazur distance has properties fitting such requirements. It is defined as follows. Let $\mathcal{X}, \mathcal{Y}$ be two isomorphic Banach spaces. The Banach-Mazur distance between them is defined as

$$d_{BM}(\mathcal{X}, \mathcal{Y}) = \inf_{T}\{\|T\| \cdot \|T^{-1}\|\} \tag{9}$$

where $T : \mathcal{X} \to \mathcal{Y}$ is an isomorphism. If $\mathcal{X}, \mathcal{Y}$ are not isomorphic, then their Banach-Mazur distance is infinite, by definition. We can immediately see that

$$d_{BM}(\mathcal{X}, \mathcal{Y}) \geq 1 \tag{10}$$

and that for isometric spaces such a distance is exactly equal to one. The converse is also true, but only for finite-dimensional Banach spaces. Therefore, the Banach-Mazur distance is actually a distance function on the set of equivalence classes of normed spaces (where "equivalence" is defined as "isometry"), but in a multiplicative sense, namely for three isomorphic linear spaces $\mathcal{X}, \mathcal{Y}, \mathcal{Z}$, it satisfies

$$d_{BM}(\mathcal{X}, \mathcal{Z}) \leq d_{BM}(\mathcal{X}, \mathcal{Y}) \cdot d_{BM}(\mathcal{Y}, \mathcal{Z}) \tag{11}$$

To get to the usual triangle inequality instead of (11), we have to consider the logarithm of d_{BM}. The Banach-Mazur distance is invariant under invertible linear maps T, namely

$$d_{BM}(T\mathcal{X}, T\mathcal{Y}) = d_{BM}(\mathcal{X}, \mathcal{Y}) \tag{12}$$

In some sense the Banach-Mazur distance expresses the minimum distortion that any isomorphism between two linear spaces can possibly entail.

Before closing this Subsection, one cannot help but distinguish between the functional spaces arising in a theory developed on space-time and the underlying form of the metric of space-time itself. These are clearly two quite distinct classes of spaces that have to be treated independently. However, one can hope that if and when a reasonably testable model of quantum gravity is found, then its Hilbert (or more generally, functional) space of "wave-functions" will induce the observable metric of space-time. So from now on, the working assumption will be that such quantum mechanical Hilbert spaces induce the space-time metric. Therefore, we should address the question about what is so special about inner product (Hilbert) spaces among the class of all Banach spaces. Since we will be working with spaces of functions, we will focus on infinite dimensional Banach spaces in the sequel. To be more concrete, we will have in mind spaces seen frequently in applications such as the spaces of p-summable sequences (l^p), or of Lebesgue p-integrable functions on $\mathbb{R}^n$ (L^p) endowed with their cardinality or their induced Euclidean measure "volume" respectively, appearing in (1)–(5).

2.4. Reflexive and Super-Reflexive Spaces

To proceed in determining desirable properties of the Banach spaces of functions on $\mathbb{R}^n$, we consider the following. It is widely believed among many, or even most, quantum gravity practitioners that spacetime properties such as its topology, smoothness, metric etc. should be "derived" from a quantum theory of gravity rather than be put in the models by hand. This seems to be a widespread belief, regardless of the exact approach to quantum gravity that someone follows. It is based in the fact that macroscopic properties of systems can be derived from their quantum counterparts and rely, to some extent, in the great separation of scales and numbers of degrees of freedom between the microscopic and the macroscopic scales. This is the main reason, as accepted today, of why Statistical Mechanics works so well in providing accurate predictions for systems with many degrees of freedom.

Independently of such physical considerations, there has been a recent surge of activity in Geometry and Analysis purporting to better understand first order calculus properties of non-smooth spaces [21]. An influential work in this direction has been [22], where conditions were given for the existence of a differentiable structure on metric measure spaces, based on Lipschitz maps, which also have the doubling property and admit a Poincaré inequality. Among the numerous works that clarified, elaborated and generalized [22], we could point out [23–26]. In these works, the differentiable structure appears naturally as a result of the more "primitive" assumptions stated above and presented in [22]. What is pertinent to our purposes is that [22] discovered that Sobolev spaces of functions on such metric measure spaces, which seem to be the most relevant from a physical viewpoint, turn out to be reflexive. Moreover they admit a uniformly convex norm. We will elaborate on the first condition in this Subsection, and the second in the next Section.

Reflexive Spaces. Let $\mathcal{X}$ be a Banach space and let $B_\mathcal{X}$ indicate its closed unit ball, namely

$$B_\mathcal{X} = \{x \in \mathcal{X} : \|x\| \leq 1\} \tag{13}$$

The dual of $\mathcal{X}$, denoted by $\mathcal{X}'$, is the space of (real or complex valued) continuous linear functionals of $\mathcal{X}$, namely an element of $\mathcal{B}(\mathcal{X}, \mathbb{R})$ or $\mathcal{B}(\mathcal{X}, \mathbb{C})$. Examples of such dual spaces are $(c_0)' = l^1, (l^1)' = l^\infty, (L^p(\mathbb{R}^n))' = L^q(\mathbb{R}^n), 1 < p < \infty, p^{-1} + q^{-1} = 1$ etc. In case $\mathcal{X}$ is finite-dimensional, the closed unit ball of its dual $B_{\mathcal{X}'}$ is the polar body of the unit ball of $\mathcal{X}$, namely

$$B_{\mathcal{X}'} = B_\mathcal{X}^\circ \tag{14}$$

Generalizing, one can define the bi-dual of $\mathcal{X}$ as the dual of $\mathcal{X}'$. These dualities induce a natural linear mapping $F : \mathcal{X} \to (\mathcal{X}')'$ given by

$$F(f) = f(x) \tag{15}$$

for $x \in \mathcal{X}$ and $f \in \mathcal{X}'$ being its dual. There is no a priori reason why the double dual of $\mathcal{X}$ should be equal to $\mathcal{X}$. Usually $\mathcal{X} \subset (\mathcal{X}')'$. A simple example of this inclusion is that the dual of c_0 is l^1 and the dual of l^1 is l^∞. Therefore the inclusion $F : c_0 \to l^\infty$ is not surjective. If, however, it happens that under the canonical map F

$$\mathcal{X} = (\mathcal{X}')' \tag{16}$$

then the Banach space $\mathcal{X}$ is called reflexive. It is important that $\mathcal{X}$ is isometric to $(\mathcal{X}')'$ under the canonical embedding (15). It is possible for a non-reflexive space to be isometric to its bi-dual; an example is provided by James' space. Obviously every finite-dimensional Banach space is reflexive, due to the rank-nullity theorem. It should be immediately noticed that reflexivity is a topological property and not a property of the norm of a particular space. This is quite important given the fact that sometimes we will consider renormings of Banach spaces for the reasons mentioned in Section 2.2. An example of reflexive Banach spaces are the Lebesgue spaces $L^p(\mathbb{R}^n), 1 < p < \infty$. Reflexive Banach spaces have numerous desirable properties: one can mention, for instance

1. the dual of reflexive space is reflexive.
2. the closed subspaces of reflexive spaces are reflexive.
3. the quotient spaces of reflexive spaces are reflexive, etc.

The question that comes to mind is whether there are physical reasons why a linear space should be reflexive. This is unclear in our opinion, beyond the desirable mathematical properties previously mentioned. It is not clear, for instance, what would be the physical consequences if the bi-dual of a Banach space were dense, rather than surjective, under the canonical mapping (15).

Discrete spacetime and"internal" symmetries, such as parity, time-inversion and charge conjugation whose "double dual" is the identity, namely idempotent operations, have played and continue to play an important role in several branches of Physics. Despite this, it is not clear to us if the linear functional duality (16) has or reflects some deeper physical origin. One would certainly not want to preclude spaces such as $L^1(\mathbb{R}^n)$ or $L^\infty(\mathbb{R}^n)$ from being used as functional spaces in applications due to their non-reflexivity. The actual question that is relevant to our purposes is whether such functional spaces have anything to do with the determination of the Euclidean metric of space-time. The work of [22] and subsequent developments seem to point out that reflexive (Sobolev) spaces may have some special geometric significance under the assumptions of his work. For this reason when combined with the implications of reflexivity for convexity and smoothness, we will restrict our attention to reflexive Banach spaces only, in the sequel.

Finite representability. One may be able to demand a stronger property along the lines of reflexivity, from physically relevant Banach spaces, for the purposes of determining the space-time metric, First a definition: a Banach space $\mathcal{X}$ is finitely representable in a Banach space $\mathcal{Z}$ if for every $\epsilon > 0$ and for every finite-dimensional subspace $\mathcal{X}_0 \subset \mathcal{X}$ there is a subspace $\mathcal{Z}_0 \subset \mathcal{Z}$ such that $d_{BM}(\mathcal{X}_0, \mathcal{Z}_0) < 1 + \epsilon$. This essentially means that any finite-dimensional subspace of $\mathcal{X}$ can be represented, almost isometrically, in $\mathcal{Z}$. Equivalently, one controls the distortion of the embedding of every finite-dimensional subspace of $\mathcal{X}$ into $\mathcal{Z}$. From a physical viewpoint the above definition may be of interest, since it is at the confluence of two ideas: one has to do with the fact that based on quantum physics, or on the statistical interpretation of theories of many degrees of freedom, one may have to reconsider or even dispense with the concept of strict, "point-wise", equality. Instead one should think much more along the lines of probabilistic equivalence, something that of course needs further qualifications. From such a perspective though, an approximate rather than strict, demand for isometry such as required in the definition of finite representability of a finite dimensional linear space is not unreasonable. The second idea relies on the fact that in any physical measurement we have a finite number of pieces of data on which to rely. As a result, the infinite dimensional spaces are excellent mathematical models, but from a very pragmatic perspective we see only their finite subspaces and then we mentally and technically extrapolate to the infinite dimensional counterparts. From this viewpoint, properties of finite dimensional vector spaces is the most of what someone can realistically expect to have to deal with in physical applications.

Super-reflexive spaces. A Banach space $\mathcal{X}$ is called super-reflexive if every Banach space which is finitely representable in $\mathcal{X}$ is reflexive. Equivalently, a Banach space $\mathcal{X}$ is super-reflexive if no non-reflexive Banach space $\mathcal{Y}$ is finitely representable in $\mathcal{X}$. Examples of super-reflexive spaces, pertinent to our discussion, are the Lebesgue spaces $L^p(\mathbb{R}^n), 1 < p < n$. Super-reflexive spaces are reflexive, but the converse is not true. Super-reflexive spaces have numerous desirable properties, from a physical viewpoint, some of which will be encountered in the next Section as they are pertinent to convexity and smoothness properties. One property is that if a Banach space is isomorphic to a super-reflexive space then it is itself super-reflexive. Another useful property is is that a Banach space $\mathcal{X}$ is super-reflexive if and only if its dual $\mathcal{X}'$ is super-reflexive. The super-reflexivity of Banach spaces is a property which allows the structure of infinite dimensional Banach spaces to be determined by the embedding properties of its finite-dimensional subspaces. Since c_0 and l^1 are not reflexive Banach spaces, they are not super-reflexive either. For completeness, we mention there are reflexive spaces are not necessarily super-reflexive: indeed consider a Banach space such that $L^\infty(\mathbb{R}^n)$ is finitely representable in it; then it cannot be not super-reflexive. In closing, one would like to notice that super-reflexivity, very much like reflexivity, is a topological property: as it does not really depend on the specific norm with which the underlying linear space is endowed.

Super-properties. One can be more general at this point and talk about "super-properties", a term that we will occasionally use in the sequel. These were defined by R.C. James in [27]. Here, we follow the excellent "pedestrian" exposition of [28]. Consider a property P that is valid on a Banach space $\mathcal{X}$. Consider two finite-dimensional subspaces $\mathcal{Y}, \mathcal{Z} \subset \mathcal{X}$ and numbers $n_P(\mathcal{Y}), n_P(\mathcal{Z})$ respectively such that

$$n_P(\mathcal{Y}) \to n_P(\mathcal{Z}) \quad \text{as} \quad d_{BM}(\mathcal{Y}, \mathcal{Z}) \to 1 \tag{17}$$

Probably the most straightforward case of this occurrence is when a relation like

$$n_P(\mathcal{Z}) \leq d_{BM}(\mathcal{Y}, \mathcal{Z}) n_P(\mathcal{Y}) \tag{18}$$

holds. Then the Banach space $\mathcal{X}$ has the property P when

$$n_P(\mathcal{X}) = \sup_{\mathcal{Y}} n_P(\mathcal{Y}) < \infty \tag{19}$$

where the supremum is taken over all finite dimensional subspaces of $\mathcal{X}$. Whether a property P holds for the Banach space $\mathcal{X}$ evidently depends on the family $\mathfrak{F}(\mathcal{Y})$ of all finite-dimensional spaces $\mathcal{V}$ such that

$$\forall \epsilon > 0, \exists \mathcal{Y} \subset \mathcal{X} : d_{BM}(\mathcal{V}, \mathcal{Y}) < 1 + \epsilon \tag{20}$$

In this terminology, finite representability of $\mathcal{Y}$ in $\mathcal{X}$ amounts to $\mathfrak{F}(\mathcal{Y}) \subset \mathfrak{F}(\mathcal{X})$. The property P is called a super-property if whenever a Banach space $\mathcal{X}$ has P, then every Banach space $\mathcal{Y}$ finitely representable in $\mathcal{X}$ also has P.

The Radon-Nikodým property. An additional property which is quite desirable at the technical level, and which is extensively used in Statistical Mechanics, where it is usually taken for granted, is the Radon-Nikodým property. It basically provides a way to make a transition between two different measures in a measure space. From a certain viewpoint, it can be seen as a generalization of the change of variables formula employed in multivariable calculus integration. For "practical purposes", it states that one can use a function alongside the volume of a manifold as equivalent to any absolutely continuous measure. Such a density function is the micro-canonical distribution employed extensively in equilibrium Statistical Mechanics. This mathematical result extends to vector-valued measures as follows: consider a probability space (Ω, μ) with a σ- algebra of (Borel) sets Σ, $U \in \Sigma$ and a Banach space $\mathcal{X}$. Let the vector-valued measure $\nu : \Sigma \to \mathcal{X}$ be countably additive and of bounded variation. Then there is a (Bochner) integrable function $f : \Omega \to \mathcal{X}$ such that

$$\nu(U) = \int_U f d\mu \tag{21}$$

For details and the generalization of this concept, from a geometric viewpoint, see [24]. As stated previously, this is a convenient technical theorem which allows one to use continuous functions f as a form of a "derivative" $d\nu/d\mu$. Such functions play an important role in Statistical Mechanics and quantum/thermal field theories as they do allow the coarse-grained distributions of interest to be treated as continuous rather than as discrete variables, something that many times simplifies the calculations. Moreover, it is certainly true that thermodynamic potentials such as the entropy have the drawback of being coordinate-dependent for continuous distributions; the only known way around such a difficulty is the use of reference measures which invariably lead to Radon-Nikodým derivatives. Hence it may be a relief to know that reflexive Banach spaces $\mathcal{X}$ do obey the Radon-Nikodým property and that the same is true for super-reflexive spaces: the obey the super Radon-Nikodým property.

3. Convexity and Smoothness

There are numerous characterizations, singling out inner product spaces among all Banach spaces. One can consult, for instance, the book [29] for a classical, extensive exposition and numerous results. We have also found a host of pertinent information in [30,31].

In this Section, we provide some information about one modulus of convexity and one modulus of smoothness and use them to single out Hilbert spaces among all the Banach spaces of interest.

3.1. Why Convexity and Smoothness?

There is little doubt that convexity is a fundamental concept, whose origin and initial developments can be traced as far as the Greek antiquity schools of Geometry such as the Pythagoreans, having far-reaching consequences in many branches of Mathematics. We are interested in aspects of convexity, in this work, mainly in the context of vector spaces [32] with our approach oriented toward the infinite dimensional cases.

Convexity enters dynamics very early, both historically and at a stage of its development. It is present as early as Newton's equations, at least. A particle trajectory in 3-dimensional space "bends" locally in the general direction of the total force acting on the particle. In other words, the total force acts toward the "convex interior", vaguely speaking, of the curve. The resulting acceleration is associated to the curvature of the trajectory in space. Such a curvature is an extrinsic concept though, namely it depends on the way the curve has been embedded in 3-dimensional space. Incidentally, the intrinsic geometry of a line is trivial. However this hints at a strong connection between curvature and convexity of the embedded curve. One can generalize this observation for higher dimensional sub-manifold embeddings. Such embeddings become crucial in the configuration or in the phase space of a system. Consider, for instance, an isolated system; its energy os constant. Hence, as is well-known, its evolution takes place in a co-dimension one sub-manifold of its configuration space. This can be seen as a simple example of an embedding. The quantity characterizing such embeddings locally, in Riemannian geometry, is the second fundamental form or its closely associated shape operator. This topic is a classical one. For an overview and many references, see the recent thesis [33]. The statement of interest, for our purposes, is that if one considers a compact hyper-surface $\mathcal{M}$ of $\mathbb{R}^n$ endowed with the induced Euclidean metric, and the second fundamental form of this embedding is positive-definite everywhere on $\mathcal{M}$, then $\mathcal{M}$ bounds a convex subset of $\mathbb{R}^n$ [34]. This provides a local characterization of convexity for embeddings and can be seen as a generalization of the kinematical framework that originated with Newton's equations.

One also expects convexity to be related to smoothness at least in the context of finite-dimensional vector spaces $\mathcal{V}$. A hand-waving argument that may illustrate this relationship is as follows: Consider a curve $\gamma : [0,1] \rightarrow \mathcal{V}$ which is locally rectifiable and which is arc-length parametrized by $s \in [0,1]$. For simplicity, and in order to make this argument more clear, assume that γ rests on a 2-plane, at least in some neighborhood U_x around a point x corresponding to $\gamma(s_0)$. Then consider the osculating circle of γ at x: this is a circle with a common tangent to the tangent $\frac{d\gamma(s_0)}{ds}$ whose center is along the normal line to the tangent in the 2-plane around U_x. The radius of the osculating circle is equal to the radius of curvature $R(s_0)$ at x. The smaller the radius of curvature $R(s_0)$ the more "steeply" the curve turns. Now assume that $R(s_0) \rightarrow 0$. This will result in the formation of a corner at $\gamma(s_0)$ so γ will no longer be differentiable at s_0. At the same time, intuitively speaking, the "amount of convexity", a concept that really needs to be made precise, of $\gamma(s_0)$ will increase as $R(s_0) \rightarrow 0$. A word of caution at this point: using Clarkson's modulus of convexity which is a primary object of interest in this work, such a shrinking of the radius of curvature would leave that modulus unaffected. Still the statements regarding convexity in this paragraph were meant to be heuristic and suggestive, rather than precise, in order to visually illustrate our motivation for the use of convexity and smoothness and their inter-relation, in the rest of this work.

Convexity can also be seen at the level of Einstein's field equations of General Relativity. These are, in 4-dimensions,

$$R_{\mu\nu} - \frac{1}{2}R g_{\mu\nu} = \frac{8\pi G}{c^4} T_{\mu\nu} \tag{22}$$

assuming that the cosmological constant $\Lambda = 0$ and $\mu, \nu = 0, 1, 2, 3$. Consider the null energy condition [35], which is (arguably) the most fundamental of the energy conditions [36],

$$T_{\mu\nu} l^{\mu} l^{\nu} \geq 0 \tag{23}$$

where l is a null vector. (23) amounts to essentially demanding that

$$R_{\mu\nu} l^\mu l^\nu \geq 0 \qquad (24)$$

(24) can be interpreted as a (mean) convexity condition in a null direction: one way is to see that the Ricci tensor involves two derivatives of the components of the metric tensor hence the non-negativity of (24) signifies convexity in analogy with the case of functions. Such convexity can also be seen in a Riemannian context. There, the non-negativity of the Ricci curvature (24) can be shown [37] to imply a generalized Brunn-Minkowski inequality, which is essentially a statement about the concavity, the convexity of the opposite function, of the volume in Euclidean and Riemannian spaces.

3.2. A Modulus of Convexity

Convex sets. To be more precise and attempt to make the exposition somewhat self-contained, we present the following well-known definitions and statements. Consider $\mathcal{V}$ to be a vector space over $\mathbb{R}$ or $\mathbb{C}$. A subset $\mathcal{A} \subset \mathcal{V}$ is called (affinely) convex if

$$\{ta_1 + (1-t)a_2, t \in [0,1]\} \subset \mathcal{A}, \forall a_1, a_2 \in \mathcal{A} \qquad (25)$$

Equivalently, for every $n \in \mathbb{N}$ and for every $t_0, t_1, \ldots, t_n \in [0,1]$ such that $t_0 + t_1 + \ldots + t_n = 1$ and for every $a_0, a_1, \ldots, a_n \in \mathcal{A}$, we have $t_0 a_0 + t_1 a_1 + \ldots + t_n a_n \in \mathcal{A}$. Obviously $\mathbb{R}^n$ as well as its linear and affine subspaces are convex. The same conclusion holds about open and closed unit balls in normed vector spaces.

Convex functions. Convex functions can be considered as generalizations of convex sets. Let $\mathcal{A} \subset \mathcal{V}$ be a convex subset of the linear space $\mathcal{V}$ and let $f : \mathcal{A} \to \mathbb{R}$ be a function. The epigraph of f is defined to be the set

$$Ep(f) = \{(a,t) \in \mathcal{A} \times \mathbb{R} : f(a) \leq t\} \qquad (26)$$

Then, such a function f is called convex, if $Ep(f)$ is a convex subset of $\mathcal{A} \times \mathbb{R}$. Equivalently, if for all $a, b \in \mathcal{A}$ and $t \in [0,1]$, such f satisfies the inequality

$$f(ta + (1-t)b) \leq (1-t)f(a) + tf(b) \qquad (27)$$

The combination of homegeneity and the triangle inequality shows immediately that a norm $\| \cdot \|$ on $\mathcal{V}$ is a convex function. Other convex functions are the distance functions between two lines in Euclidean space ($\mathbb{R}^n$ endowed with the Euclidean metric) and, more generally, distance functions on metric spaces of negative curvature. Moreover, if in a normed (more generally: a metric) space $\mathcal{V}$ with $\mathcal{A} \subset \mathcal{V}$ and for all $a \in \mathcal{V}$, one defines

$$d_{\mathcal{A}}(a) = \inf_{b \in \mathcal{A}} \|a - b\| \qquad (28)$$

when $\mathcal{A}$ is a nonempty closed convex subset of $\mathcal{V}$, then the distance function $d_{\mathcal{A}} : \mathcal{V} \to \mathbb{R}_+$ is convex.

Convex functions have many nice properties: they are semicontinuous and almost everywhere differentiable, they possess left and right derivatives (which however need not coincide), limits of sequences of convex functions defined on convex sets are convex functions, they have a unique minimum and they obey the local-to-global property, namely a locally convex function is actually (globally) convex. It is properties like these that make convex functions so useful and widespread in Physics and, in particular, in Analytical Mechanics and Thermodynamics.

A modulus of convexity. A next logical step is to find a way to quantify the extent of convexity of a set or of a function. Naturally, determining such a "modulus of convexity" is not a unique process and it involves making certain choices. Simplicity and computability in, at least, simple cases are usually good guidelines, as far as physical applications are concerned. A relatively recent list of such moduli which is quite extensive, even if not necessarily comprehensive, can be found in [38].

The oldest and most studied among the moduli of convexity is due to J.A. Clarkson [39,40]. To define it, consider the normed linear space $(\mathcal{X}, \|\cdot\|)$, let $B_{\mathcal{X}}$ indicate its closed unit ball, as before, and let $\varepsilon \in [0,2]$. The modulus of local convexity is defined by

$$\delta(x,\varepsilon) = \inf\left\{1 - \left\|\frac{x+y}{2}\right\| : y \in B_{\mathcal{X}}, \|x-y\| \geq \varepsilon\right\} \tag{29}$$

Then Clarkson's modulus of convexity of $\mathcal{X}$ denoted by $\delta_{\mathcal{X}} : [0,2] \to [0,1]$ is defined as

$$\delta_{\mathcal{X}}(\varepsilon) = \inf\{\delta(x,\varepsilon) : x \in B_{\mathcal{X}}\} \tag{30}$$

This can be equivalently expressed as

$$\delta_{\mathcal{X}}(\varepsilon) = \inf\left\{1 - \left\|\frac{x+y}{2}\right\| : x,y \in \mathcal{X}, \|x\| = \|y\| = 1, \|x-y\| = \varepsilon\right\} \tag{31}$$

It should be immediately noticed that this modulus of convexity is really a property of the 2-plane spanned by $x,y \in \mathcal{X}$ which is then inherited by $\mathcal{X}$. This should look familiar: the curvature of an n-dimensional Riemannian manifold $\mathcal{M}$ is a genuinely 2-dimensional concept which is actually formulated on its Grassmann manifold $G_{2,n}(\mathcal{M})$. The (sectional) curvature expresses the deviation of the metric of $\mathcal{M}$ from its flat counterpart [41]. The characteristic of convexity of $(\mathcal{X}, \|\cdot\|)$ is defined as [42]

$$\varepsilon_0(\mathcal{X}) = \sup\{\varepsilon \in [0,2] : \delta_{\mathcal{X}}(\varepsilon) = 0\} \tag{32}$$

Uniformly convex spaces. A Banach space $(\mathcal{X}, \|\cdot\|)$ is uniformly convex when it has non-zero modulus of convexity, namely $\delta_{\mathcal{X}}(\varepsilon) > 0, \varepsilon \in (0,2]$ or equivalently when $\varepsilon_0(\mathcal{X}) = 0$. Geometrically, the idea of the definition is simple: uniformly convex spaces have a unit ball $B_{\mathcal{X}}$ whose boundary unit sphere $S_{\mathcal{X}}$ does not contain any (affine) line segments. Roughly speaking: the further away from containing an affine segment $S_{\mathcal{X}}$ is, the higher the modulus of convexity of $\mathcal{X}$ is. It should be noticed that according the D.P. Milman [43] - B.J. Pettis [44] theorem, uniformly convex spaces are reflexive. Actually one can see that uniformly convex spaces are actually super-reflexive. Hence, if one deems reflexivity or super-reflexivity to be a desirable, or pertinent, property in an argument about the Euclidean nature of the space-time metric, as was stated above, then confining their attention to uniformly convex Banach spaces will not miss this property.

From a different viewpoint, it is known from the work of P. Enflo [45] and R.C. James [46] that $\mathcal{X}$ is super-reflexive if and only if it has an equivalent uniformly convex norm. "Equivalence" in this theorem is meant to be understood in the sense described in Subsection 2.2. Therefore, insisting on having super-reflexivity of the underlying functional spaces that determine/induce the space-time metric usually has to allow for a change of their norm to a uniformly convex one, if this is feasible. This change may have substantial physical implications for the underlying model. If however, as mentioned above, one is only interested in large-scale "coarse" phenomena that ignore spatially "small" details and arise as a result of statistical averaging of many degrees of freedom, then one believes that such a renorming will leave the macroscopic quantities of interest unaffected.

Modulus of convexity of Lebesgue spaces. Explicitly calculating the modulus of convexity for specific Banach spaces has proved to be more difficult than one might have naively anticipated. This is one reason why so many different moduli of convexity have been defined over the decades, after Clarkson's work [38]. For completeness, we mention that for $p = 1$ and for $p = \infty$, $\delta_{L^p} = 0$ as these two Banach spaces are not uniformly convex. The fact that the other Lebesgue spaces $L^p(\mathbb{R}^n), 1 < p < \infty$ are uniformly convex was already known to [39]. For a simpler and more recent proof, see [47]. However, the explicit asymptotic form of their modulus of convexity was determined by [48] who relied on the inequalities bearing his name [48,49] to reach his result. Hanner proved that

$$
\delta_{L^p}(\varepsilon) =
\begin{cases}
(p-1)\frac{\varepsilon^2}{8} + o(\varepsilon^2), & \text{if } 1 < p \leq 2 \\[2mm]
\frac{1}{p}\left(\frac{\varepsilon}{2}\right)^p + o(\varepsilon^p), & \text{if } 2 \leq p < \infty
\end{cases}
\tag{33}
$$

One cannot fail to notice the "phase transition" in the asymptotic behavior of this modulus of convexity occurring around $p = 2$, namely around the Hilbert space case. An inevitable question is how generic such a behavior might be, among all Banach spaces. If it is, then it would be an initial indication that Hilbert spaces may be "special", from a convexity viewpoint. It turns out that this is essentially true, and this is the statement on which the main argument of the present work relies. This is the realization that among all Banach spaces, the Hilbert space is the "most" convex. To be more precise, it was proved in [50] that among all Banach spaces $\mathcal{X}$, the Hilbert spaces $\mathcal{H}$ have the largest modulus of convexity, namely

$$
\delta_{\mathcal{X}}(\varepsilon) \leq \delta_{\mathcal{H}}(\varepsilon)
\tag{34}
$$

where

$$
\delta_{\mathcal{H}}(\varepsilon) = \frac{2 - \sqrt{4 - \varepsilon^2}}{2}
\tag{35}
$$

It is also interesting to notice that the results of [50] when combined with those of [51,52] prove that if a linear space $\mathcal{X}$ is such that the equality in (34) holds, then $\mathcal{X}$ is an inner product space.

Modulus of convexity and equivalences. It should be noted that the modulus of convexity $\delta_{\mathcal{X}}(\varepsilon)$ is not necessarily itself a convex function of ε. What we know, for instance, is that for an infinite dimensional uniformly convex Banach space $\mathcal{X}$, we have that

$$
\delta_{\mathcal{X}}(\varepsilon) \leq C\varepsilon^2
\tag{36}
$$

For a Hilbert space we have (35) which is, obviously, compatible with (36). So, in a quantitative sense, the closer $\delta_{\mathcal{X}}$ is to ε^2 for a normed space $\mathcal{X}$, the closer $\mathcal{X}$ is being maximally convex, hence the closer it is to being a Hilbert space $\mathcal{H}$. T. Figiel [53] showed that one can consider as a modulus of convexity instead of $\delta_{\mathcal{X}}(\varepsilon)$ the greatest convex function $\tilde{\delta}_{\mathcal{X}}(\varepsilon)$, namely

$$
\tilde{\delta}_{\mathcal{X}}(\varepsilon) \leq \delta_{\mathcal{X}}(\varepsilon)
\tag{37}
$$

Then [53]

$$
c_1 \delta_{\mathcal{X}}(c_2 \varepsilon) \leq \tilde{\delta}(\varepsilon)
\tag{38}
$$

for constants $c_1 > 0, c_2 > 0$.

The re-definition employed in [53] is reminiscent of the equivalence of growth functions in geometric group theory when one has to decide whether such a growth function is exponential, polynomial or has an intermediate growth rate [54]. The reason for such similarity is quite clear: in geometric group theory, as in our case, one really cares about equivalences that may distort distances, but leave large-scale details of the structure invariant. The metric equivalence employed there is that of quasi-isometric maps, which distorts distances between two metric spaces $(\mathfrak{X}_1, d_1), (\mathfrak{X}_2, d_2)$ according to

$$
\frac{1}{c_1} d_1(x, y) - c_2 \leq d_2(f(x), f(y)) \leq c_1 d_1(x, y) + c_2, \qquad \forall x, y \in \mathfrak{X}_1
\tag{39}
$$

where $c_1 > 1$ and $c_2 > 0$. The difference between (6) and (39) is the presence of c_2 in (39) which completely ignores structures whose length is smaller than c_2. We have previously employed such equivalences in our work in [55,56] in our attempt to determine the dynamical basis of a power-law entropy, something that may have implications for the derivation of the metric of space-time from microscopic models of quantum gravity. The use of such maps acquires even greater significance when one considers that it is intimately related to properties of (Gromov) hyperbolic spaces. After a $3+1$ decomposition, 3-dimensional manifolds represent space-like hyper-surfaces in spacetimes. Following the results of the Thurston geometrization program (see [57] for an overview), it seems that "most" of the 3-dimensional manifolds are hyperbolic. Hence quasi-isometries and similar ideas may be relegated to a central role in determining classical/long-distance structures of the (4-dimensional) spacetimes from their microscopic/quantum foundations.

3.3. A Modulus of Smoothness and a Duality

A quantity intimately related to convexity, in the context of normed linear spaces, is smoothness. In line with convexity, one also needs a way to quantify the amount of smoothness of a linear space. Once more, there is no unique way of how to go about constructing such a modulus of smoothness and actually several such moduli of smoothness have been constructed over the years [38].

A modulus of smoothness. The oldest and most studied modulus of smoothness is due to M.M. Day [40] and J. Lindenstrauss [58]. The modulus of smoothness of a normed space $(\mathcal{X}, \|\cdot\|)$ is a function $\rho_{\mathcal{X}} : [0, \infty) \to \mathbb{R}$ which is defined by

$$\rho_{\mathcal{X}}(t) = \sup \left\{ \frac{\|x + ty\| + \|x - ty\|}{2} - 1, x, y \in B_{\mathcal{X}} \right\} \tag{40}$$

This can be alternatively defined as

$$\rho_{\mathcal{X}}(t) = \sup \left\{ \frac{\|x + y\| + \|x - y\|}{2} - 1 : x \in B_{\mathcal{X}}, \|y\| \leq t \right\} \tag{41}$$

or as

$$\rho_{\mathcal{X}}(t) = \sup \left\{ \frac{\|x + y\| + \|x - y\|}{2} - 1, x, y \in S_{\mathcal{X}} \right\} \tag{42}$$

One defines the coefficient of smoothness of $\mathcal{X}$ by

$$\rho_0(\mathcal{X}) = \lim_{t \to 0^+} \frac{\rho_{\mathcal{X}}(t)}{t} \tag{43}$$

Uniformly smooth spaces. The Banach space $(\mathcal{X}, \|\cdot\|)$ is called uniformly smooth if $\rho_0(\mathcal{X}) = 0$. One sees immediately that uniform smoothness is a point-wise property and is essentially 2-dimensional, as is the case for uniform convexity. In a pictorial sense, this modulus of smoothness captures the fact that the unit sphere $S_{\mathcal{X}}$ of the Banach space $\mathcal{X}$ is smooth, i.e., that it has not corners. More issue on this issue is discussed below.

Before continuing it may be worth developing a pictorial idea about uniform convexity and uniform smoothness. Consider, for instance, a square with rounded corners. This is not uniformly convex, since it includes line segments in its boundary, but it lacks corners, so it is uniformly smooth. Consider now the, assumed non-empty and not a point, intersection of two equal radius disks: it is uniformly convex as its boundary contains no line segments, but it is not uniformly smooth since at the two intersection points of the two circles which are the boundaries of the two disks, the figure has corners.

Modulus of smoothness of Lebesgue spaces. Unlike the modulus of convexity, the modulus of smoothness is a convex function, essentially by definition. The asymptotic behavior of the modulus of smoothness of the Lebesgue spaces $L^p(\mathbb{R}^n)$ were also calculated in [48]. O. Hanner found that

$$\rho_{L^p}(t) = \begin{cases} \frac{t^p}{p} + o(t^p), & \text{if } 1 < p \leq 2 \\ \frac{p-1}{2} t^2 + o(t^2), & \text{if } 2 \leq p < \infty \end{cases} \tag{44}$$

The same comments and thoughts regarding the phase transition around $p = 2$ apply here as in the case of the modulus of convexity as stated above. The counterpart of (34) was also established, and it states that that for any Banach space $\mathcal{X}$ one has

$$\rho_{\mathcal{X}}(t) \geq \rho_{\mathcal{H}}(t) \tag{45}$$

Therefore Hilbert spaces are the least smooth among all Banach spaces. Moreover, the exact same statement as the one following (35) applies for the modulus of smoothness: if a Banach space $\mathcal{X}$ obeys

$$\rho_{\mathcal{X}}(t) = \rho_{\mathcal{H}}(t) \tag{46}$$

then $\mathcal{X}$ is an inner product space. The modulus of smoothness, very much like for the modulus of convexity, of a Hilbert space is a result of the validity of the parallelogram equality and it is given by

$$\rho_{\mathcal{H}}(t) = \sqrt{1 + t^2} - 1 \tag{47}$$

The analogue of (36) is that for an infinite dimensional Banach space $\mathcal{X}$ one has

$$\rho_{\mathcal{X}}(t) \geq ct^2 \tag{48}$$

The Milman-Pettis theorem established, in addition to uniform convexity, that uniformly smooth spaces are reflexive. Even though the dual of a super-property may not be a super-property, using the "convexity-smoothness" duality of the next paragraph, one sees that uniformly smooth spaces are indeed super-reflexive. Analogous things can be stated about the uniform smoothability of a Banach space as were stated about their uniform convexifiability, with equivalent norms.

A duality and the Legendre-Fenchel transforms. One observes form the above that Clarkson's modulus of convexity and the Day-Lindenstrauss modulus of smoothness appear to behave very much like dual concepts. The fact is that this suspected duality is true. More precisely, [58] proved that a Banach space $\mathcal{X}$ is uniformly convex if and only if its dual $\mathcal{X}'$ is uniformly smooth. The exact relation between the corresponding moduli is given by the Legendre-Fenchel transform

$$\rho_{\mathcal{X}'}(t) = \sup\left\{\frac{\varepsilon t}{2} - \delta_{\mathcal{X}}(\varepsilon), 0 \leq \varepsilon \leq 2\right\} \tag{49}$$

or, equivalently,

$$\rho_{\mathcal{X}}(t) = \sup\left\{\frac{\varepsilon t}{2} - \delta_{\mathcal{X}'}(\varepsilon), 0 \leq \varepsilon \leq 2\right\} \tag{50}$$

Therefore, every theorem valid for convexity has a dual analogue valid for smoothness. We see, for instance, that

$$\rho_0(\mathcal{X}') = \frac{1}{2}\varepsilon_0(\mathcal{X}), \qquad \rho_0(\mathcal{X}) = \frac{1}{2}\varepsilon_0(\mathcal{X}') \tag{51}$$

which shows that a Banach space $\mathcal{X}$ is uniformly convex if and only if its dual $\mathcal{X}'$ is uniformly smooth. Continuing along the lines of the quantification of convexity and smoothness, one can state the following: consider the Banach space $\mathcal{X}$. It turns out that

$$\delta_{\mathcal{X}}(\varepsilon) \geq C\varepsilon^q, \qquad q \geq 2 \tag{52}$$

in which case $\mathcal{X}$ is called a $q-$convex space. According to a theorem of Figiel and Assouad, this is equivalent to the existence of some constant α such that

$$\frac{1}{2}\left(\|x + y\|^q + \|x - y\|^q\right) \geq \|x\|^q + \alpha\|y\|^q, \qquad \forall x, y \in \mathcal{X} \tag{53}$$

For smoothness, the corresponding statement is that for the Banach space $\mathcal{Y}$ one has

$$\rho_\mathcal{Y}(t) \leq ct^p, \qquad 1 < p \leq 2 \tag{54}$$

in which case Y is called p−smooth. Then, according to a theorem of Pisier and Assouad, this is equivalent to the existence of some constant β such that

$$\frac{1}{2}(\|x+y\|^p + \|x-y\|^p) \leq \|x\|^p + \beta\|y\|^p, \qquad \forall x,y \in \mathcal{Y} \tag{55}$$

Based on the Legendre-Fenchel duality noted above, one can also state that a Banach space $\mathcal{X}$ is q-convex if and only if its dual $\mathcal{X}'$ is p-smooth, where q qnd p are harmonic conjugates, namely

$$\frac{1}{p} + \frac{1}{q} = 1 \tag{56}$$

3.4. Smoothness, Derivatives and Equivalences

The word "smoothness" is associated, in a typical Physicist's mind, with the concept of the derivative. It may be of interest to know that the same can be stated for the cases of the Banach spaces of our interest. Using infinite dimensional spaces though introduces additional complexities and also possible counter-intuitive features that should be carefully accounted for. For one, there are several possible definitions for the (directional) derivative in a normed space. We only need the definition of the Frechét derivative in this work [31]. Let $\mathcal{X}, \mathcal{Y}$ be Banach spaces and let $f : \mathcal{X} \to \mathcal{Y}$. Then f is called Frechét differentiable at $x \in \mathcal{X}$, if there is a bounded linear operator $A_x : \mathcal{X} \to \mathcal{Y}$ such that the following limit exists uniformly for $y \in S_\mathcal{X}$

$$A_x y = \lim_{t \to 0} \frac{f(x+ty) - f(x)}{t} \tag{57}$$

If this limit indeed exists, then it is called the Frechét derivative of f at x and is indicated by $D_f(x)$. The pertinent statement is that if $\mathcal{X}$ is a uniformly smooth Banach space then its norm $f(x) = \|x\|$ is Frechét differentiable for every $x \in \mathcal{X}\backslash\{0\}$. In such a case, one can see that the linear approximation to f at x through $D_f(x)$ is valid, namely we can write

$$f(x+y) = f(x) + D_f(x)y + o(\|y\|) \tag{58}$$

If a function is Frechét differentiable at a point, it turns out that it is also continuous at that point, which is the Banach space analogue of a well-known and frequently used result of elementary calculus.

In addition to the above, it turns out [31] that for a Banach space $\mathcal{X}$ the following statements are equivalent:

- $\mathcal{X}$ is super-reflexive.
- $\mathcal{X}$ admits an equivalent, uniformly convex norm, whose modulus of convexity satisfies, for some $q \geq 2$,

$$\delta_\mathcal{X}(\varepsilon) \geq c_1 \varepsilon^q \tag{59}$$

- $\mathcal{X}$ admits an equivalent, uniformly smooth norm, whose modulus of smoothness satisfies, for some $1 < p \leq 2$

$$\rho_\mathcal{X}(t) \leq c_2 t^p \tag{60}$$

Moreover, Asplund [59] showed that a space which admits two equivalent norms one of which is uniformly convex and the other uniformly smooth, admits a third one which is equivalent to the previous two and which has both properties. As a special case, one can state that a super-reflexive space admits an equivalent norm which is both uniformly convex and uniformly smooth.

3.5. Type, Co-Type and Moduli

Rademacher functions. The powers q and p in the lower bound of the modulus of convexity (52) and in the upper bound in the modulus of smoothness (54), respectively, have a nice geometric-probabilistic interpretation. To formulate it, we need to use the Rademacher functions which are defined as $r_i : [0,1] \rightarrow \pm 1$ by

$$r_i(t) = \text{sign}[\sin(2^i \pi t)], \quad i \in \mathbb{N} \tag{61}$$

We see that the Rademacher functions can be interpreted as a sequence of identically distributed random variables on $[0,1]$ endowed with its Lebesgue measure, taking values ± 1, each with probability 0.5. It may be worth noting that for vectors $x_i, i = 1, \ldots n$ in a Banach (or more generally in a normed) space $(\mathcal{X}, \| \cdot \|)$, one has

$$\int_0^1 \left\| \sum_{i=1}^n r_i(t) x_i \right\|^p dt = \mathbb{E} \left(\left\| \sum_{i=1}^n \epsilon_i x_i \right\|^p \right) \tag{62}$$

where the expectation value $\mathbb{E}$ is taken over $\epsilon_i = \pm 1$.

Type. Suppose now that for any finite number n and any choice of vectors $x_i, i = 1, \ldots, n$ of $\mathcal{X}$ there is a constant $C_p > 0$ such that

$$\left(\int_0^1 \left\| \sum_{i=1}^n r_i(t) x_i \right\|^p dt \right)^{\frac{1}{p}} \leq C_p \left(\sum_{i=1}^n \|x_i\|^p \right)^{\frac{1}{p}} \tag{63}$$

then the Banach space $\mathcal{X}$ is said to have type p. The best constant $C_p(\mathcal{X})$ in the definition (63) is called type-p constant of $\mathcal{X}$. Consider A.I. Khintchine's inequality for $a_i, a_i \in \mathbb{R}, i = 1, \ldots, n, 0 < p < \infty$, the Rademacher functions $r_i(t)$, and constants A_p, B_p, namely

$$A_p \left(\int_0^1 \left| \sum_{i=1}^n a_i r_i(t) \right|^p dt \right)^{\frac{1}{p}} \leq \left(\int_0^1 \left| \sum_{i=1}^n a_i r_i(t) \right|^2 dt \right)^{\frac{1}{2}} \leq B_p \left(\int_0^1 \left| \sum_{i=1}^n a_i r_i(t) \right|^p dt \right)^{\frac{1}{p}} \tag{64}$$

If we assume that all vectors are equal in (63), then by using (64) we see that that $p \leq 2$. By using the triangle inequality, one can also see that $p \geq 1$. Therefore the only allowed values for a type p Banach space are $1 \leq p \leq 2$. This is equivalent to stating that the elements of the family $\mathfrak{F}(\mathcal{Z})$ of finite-dimensional subspaces $\mathcal{Z} \subset \mathcal{X}$ satisfy

$$\sup_{\mathfrak{F}(\mathcal{Z})} C_p(\mathcal{Z}) < \infty \tag{65}$$

which shows that $\mathcal{X}$ having type p is a super-property. One can also see that if $p' < p$ then type p implies type p'.

Co-type. With similar notation as for type, a Banach space $\mathcal{Y}$ has co-type q if there is a constant $C_q' > 0$ such that

$$\left(\int_0^1 \left\| \sum_{i=1}^n r_i(t) x_i \right\|^q dt \right)^{\frac{1}{q}} \geq C_q' \left(\sum_{i=1}^n \|x_i\|^q \right)^{\frac{1}{q}} \tag{66}$$

for any $n \in \mathbb{N}$ and for any set of vectors $x_i \in \mathcal{Y}$. Again, the best constant $C_q'(\mathcal{Y})$ in (66) is called the co-type q constant for $\mathcal{Y}$. This again is equivalent to stating that the elements of the family $\mathfrak{F}(\mathcal{W})$, for all finite-dimensional subspaces $\mathcal{W} \subset \mathcal{Y}$ satisfy

$$\sup_{\mathfrak{F}(\mathcal{W})} C_q(\mathcal{W}) < \infty \tag{67}$$

which also shows that $\mathcal{Y}$ having co-type q is a super-property. By using Khintchine's inequality again, one can see that the co-type of any Banach space that has one has to be $q \geq 2$. One can also see that if $q' > q$, then co-type q implies co-type q'.

Type and co-type properties. As an explicit example of type and co-type we know [30] that the Lebesgue spaces $L^p(\mathbb{R}^n)$ have

- Type p and co-type 2, if $1 \leq p \leq 2$
- Type 2 and co-type p, if $2 \leq p < \infty$

One again cannot fail to notice the "phase transition" in the behavior of type and co-type of $L^p(\mathbb{R}^n)$ taking place around $p = 2$. Hilbert spaces are singled out in a stronger sense: indeed, using the parallelogram equality (38) below, one can see that a Hilbert space has type 2 and co-type 2. That the converse is also true, is a non-trivial result due to S. Kwapień [60].

The type and co-type of a Banach space $\mathcal{X}$ can be seen in a variety of ways: one way is to state that they are a way of quantifying how far $\mathcal{X}$ is from being a Hilbert space. The comparison of the definitions of type and co-type with the parallelogram equality that only Hilbert spaces satisfy

$$\|x + y\|^2 + \|x - y\|^2 = 2(\|x\|^2 + \|y\|^2) \tag{68}$$

is quite suggestive. In view of the last paragraph of this Subsection (see below) the same can be stated by comparing (68) to (53) and (55) which alternatively quantify the moduli of convexity and smoothness of $\mathcal{X}$.

Now, one can inquire about the robustness of the concepts of type and co-type. J.P. Kahane's inequality [61] is a vector-valued extension of A.I. Khintchine's inequality (64) and states, with the above notation, that for every $0 < p < q < \infty$ there is a constant $C(p, q) > 0$ such that

$$\left(\int_0^1 \left\| \sum_{i=1}^n r_i(t) x_i \right\|^p dt \right)^{\frac{1}{p}} \leq \left(\int_0^1 \left\| \sum_{i=1}^n r_i(t) x_i \right\|^q dt \right)^{\frac{1}{q}} \leq C(p, q) \left(\int_0^1 \left\| \sum_{i=1}^n r_i(t) x_i \right\|^p dt \right)^{\frac{1}{p}} \tag{69}$$

This shows that the type p and co-type q are properties that are maintained under an equivalent norm. Moreover, instead of using Rademacher functions in the definitions, something which is technically convenient, one could use centered Bernoulli random variable, Gaussian random variables etc. with just a change in the values of the constants in the definitions [62].

Type, co-type and moduli. The relation of the type and co-type of a Banach space with its moduli of convexity and smoothness is contained in the following theorem due to T. Figiel, G. Pisier [63]: Let $\mathcal{X}$ be a uniformly convex Banach space with modulus of convexity satisfying $\delta_{\mathcal{X}}(\varepsilon) \geq C\varepsilon^q$ for some $q \geq 2$. Then $\mathcal{X}$ has co-type q. Let $\mathcal{Y}$ be a uniformly smooth Banach space whose modulus of smoothness satisfies $\rho_{\mathcal{Y}}(t) \leq ct^p$ for some $1 < p \leq 2$. Then $\mathcal{Y}$ has type p. Therefore the moduli of smoothness and convexity of a Banach space are bounded by the type and co-type of that Banach space, assuming that the latter exist. It may be worth noticing at this point the behavior of type and co-type under duality: It is known that when a Banach space $\mathcal{X}$ has type p, then its dual $\mathcal{X}'$ has co-type q where p and q are harmonic conjugates of each other. However the converse is not true without one additional assumption. The accurate statement is that if a Banach space $\mathcal{X}$ has non-trivial type, and co-type q, then its dual $\mathcal{X}'$ has type p, where p and q are harmonic conjugates of each other. For excellent expositions of the type and co-type of normed spaces, including proofs of all of the above statements, one may consult [30,31,62].

4. The Space-Time Metric from Variational Principles

One cannot fail to notice from the content of the previous Sections, the unique role that inner product (Hilbert) spaces $\mathcal{H}$ play among all Banach spaces $\mathcal{X}$ and, in particular, the distinct role of $L^2(\mathbb{R}^n)$ among all $L^p(\mathbb{R}^n)$. Such Hilbert spaces are, at the same time, the most convex and least smooth among all Banach spaces. They are the only Lebesgue spaces that have the same type and co-type 2. In addition, they are super-reflexive. Moreover such Hilbert spaces are the only self-dual Lebesgue spaces under harmonic conjugation, a basic fact reflecting properties of the polarity of convex bodies and of the Legendre-Fenchel transformations [64].

Functional integrals and variational principles. Using extremal (more accurately: stationary) properties of functionals under infinitesimal variations subject to appropriate boundary conditions has been a fundamental aspect of Classical and Quantum Physics since the time of Maupertuis, D'Alembert and Lagrange at least [65], if not earlier. In particular, a large number of works in Quantum Physics have used and continue to use as starting point, especially for calculational purposes, the stationary phase or saddle point approximation which rely on the vanishing variation under small perturbations of a judiciously chosen functional (the "classical action" S) [66], an approach that can be traced back to an original idea of P.A.M. Dirac [67]. In this path-integral approach, as is very well-known, one starts with the path-integral/canonical partition function as the primary object encoding the statistically significant properties of the system

$$\mathcal{Z} = \int e^{-S} [\mathcal{D}\phi] \tag{70}$$

where ϕ corresponds to the variables ("fields") in the action S to be integrated over and $[\mathcal{D}\phi]$ represents an appropriate integration measure, which may not rigorously be known on whether it even exists, but whose ad hoc choice (usually being of Gaussian form) allows concrete calculations to be performed and eventually the results derived from it to be compared with experimental data. In the case of quantum gravity the most immediate choice for S is usually taken to be the Einstein-Hilbert or the Palatini actions, a Chern-Simons action, of their discretized counterparts etc., each of which may be augmented with boundary terms and/or topological terms etc.

Other entropies and robustness. Before proceeding, we would like to have a short digression. During the last three decades, there have been several functionals that have been proposed purporting to capture the collective/thermodynamic properties of systems of many degrees of freedom. One motivation for the formulation of such functionals, such as the "Tsallis entropy" [15] or the "κ-entropy" [16] is to determine the thermodynamic properties of systems with long-range interactions. From the Newtonian viewpoint, gravity clearly falls in this category, as well as Maxwell's theory of electrodynamics etc. Assuming that such functionals may prove to be applicable to a path-integral formulation of aspects of semi-classical or even quantum gravity, the arguments of the present work will still hold without any major modifications. The minor modification needed in case such functionals are pertinent, is to use in (70) another convex function instead of the exponential one, something akin to the aptly named "q-exponential" [15], whose form will have to be determined. One could possibly use the maximum entropy principle subject to appropriate constraints, for such a purpose. A second minor, for our purposes, modification may be to substitute some other measure in the place of the often used Gaussian measure in (70). Beyond these points, we expect the above analysis to still be valid. Another point that will most likely change is the rate of convergence to the limit of the saddle-point approximation, which is intimately related to the probability of dealing with space-time geometries that may be non-Euclidean. This in the spirit of theories having a statistical interpretation, where even when the "classical" limit is known, it is the form of the "semi-classical" contributions/corrections that is used to distinguish between several competing models purporting to describe the same physical phenomenon.

Generalized path integrals. Going back to our argument, in the spirit of the path-integral, an often discussed but still unsolved question is whether one should extend (70) by considering additional contributions by summing over more "primitive", than the metric, structures such over all topological, piecewise-linear, differentiable etc. structures. Most of the treatments to quantum gravity that we are aware of, address the issue of a possible sum in the right-hand-side of (70) over all topologies. Then the modification of (70) states that the partition function of quantum gravity should be

$$\mathcal{Z}_{top} = \sum_{topologies} \int e^{-S}[\mathcal{D}\phi] \tag{71}$$

Such possible summation over all topologies has presented insurmountable difficulties, which can be credited in large part for the eventual demise of the dynamical triangulation approach to quantum gravity [68], where one is faced with hard problems of Gödelian type indecisive propositions.

On the other hand, (71) can be used as as starting point for a similar question, where the summation is not over all topologies of 4-dimensional manifolds, which are of course locally spaces endowed with a Euclidean metric. One could instead ask for summation over all metrics that can be placed on the underlying topological or uniform structure of the space whose classical limit will eventually be a space-time. This however is would be a very broad set, hence a very difficult to analyze class of metrics. To be closer to something manageable, and also be in accordance with the equivalence principle demanding the local approximation of the underlying structure with linear spaces, one may wish to consider locally, only p-integrable Banach metrics, namely metrics/norms induced from $L^p(\mathbb{R}^n)$. This effectively generalizes the underlying space-time structure from that of a Riemannian to something akin to a Finslerian space. Therefore one could write a modified path-integral/canonical partition function, instead of (71), as

$$\mathcal{Z}_B = \sum_{p \geq 1} \int e^{-S}[\mathcal{D}\phi] \tag{72}$$

with the summation being over all metrics of a space(-time) which are locally induced by the $L^p(\mathbb{R}^n)$. In conventional path-integral approaches to quantum gravity such a summation does not arise, since one has already determined to only use Euclidean metrics on the space-time underlying manifold.

One could use the lack of renormalizability of the path-integral expressions like (70) around their saddle points, namely around a fixed background metric, to argue against such an approach [69]. After all, if (70) gives rise to non-renormalizable interactions around Minkowski space, this should force us to believe that it precludes (72) from being more successful to that end. Such a criticism would be misguided though. Lack of renormalizability of (70) takes into account only metrics of Riemannian (quadratic) form on the underlying space and the saddle point is calculated within the set of such metrics. What we propose is to enlarge such a set to the more general locally p-integrable Banach metrics. Since our argument is quite empirical/"phenomenological" rather than fundamental/dynamical, the issue of renormalizability of the underlying path-integral does not even enter our considerations.

Demanding a summation over such p-integrable metrics as in (72) may be reasonable, or not, but only after someone can properly write a finite classical action $\mathcal{A}$ for them. How exactly to do this is not clear to us at this stage. There are synthetic definitions of the Ricci and even the scalar curvature for metric measure spaces with very little regularity [70]. One could use them and alongside a general minimal cost transportation to possibly argue in this direction. What is sorely lacking in such cases though is the formalism that could accommodate expressions for the non-gravitational fields capturing the essence of the stress-energy tensor and recasting it in such a synthetic framework. Therefore a dynamical argument, which would be the most desirable, in favor of Hilbert spaces and their induced Euclidean metrics does not seem to be feasible, in any obvious way, at this stage.

A "kinematic" approach via smoothness. Given the above difficulties, we have to resort to ad hoc decisions in order to proceed. To the extent of our knowledge, there has never been a variational principle of "maximum convexity" or a principle of "minimal smoothness" that would single out Hilbert spaces among all Banach spaces. The path-integral/partition function approach to quantum Physics can be interpreted as suggesting that all allowed possibilities in quantum evolution should be considered in calculating quantities of interest, each possibility however being assigned with a different weight factor. Following this viewpoint one can extend/stretch the domain of this interpretation to allow not only for a set of Riemannian metrics to contribute to the evolution of a gravitational system but also consider a broader class of possible metrics. To keep things close to the familiar territory of Riemannian/Lorentzian metrics we have used induced metrics on space-time locally induced by $L^p(\mathbb{R}^n)$, as was mentioned before. The familiar picture of space-time appears then as the classical limit of a theory of quantum gravity, and so are its associated properties like smoothness, etc.

Given the assumed irregularity/granularity of space-time at a fundamental level (expressed through spin networks in loop quantum gravity, partially ordered sets with discrete measures in the causal set approach, simplicial approximations and Regge calculus in causal dynamical triangulations etc.) it may not be out of place to assume that nature chooses the least smooth class of metrics among such induced metrics from $L^p(\mathbb{R}^n)$, $1 \leq p < \infty$. Therefore it is the Hilbert space metric/norm $L^2(\mathbb{R}^n)$ that provides the only extremal, hence dominant, contribution in an "extended" path-integral approach (72) which, in turn, induces its properties to the classical space-time limit of the quantum gravitational theory. In short, the induced metric of space-time inherits its Euclidean character from that of $L^2(\mathbb{R}^n)$ which dominates the path-integral (72), by being the least smooth.

Convexity and predictability. A somewhat complementary argument for Hilbert spaces and the induced Euclidean metric form on space-time, can be made based on convexity and predictability. As stated in the previous sections, the Hilbert spaces $\mathcal{H} = L^2(\mathbb{R}^n)$ are the most convex among all $L^p(\mathbb{R}^n)$. Contrast the behavior of the norm/metric of $\mathcal{H}$ to those of the family of $L^p(\mathbb{R}^n)$ that are the least convex. These are $L^1(\mathbb{R}^n)$ and $L^\infty(\mathbb{R}^n)$ which are neither uniformly convex nor uniformly smooth, nor are they reflexive, so they have been largely ignored in most part of this work. Nevertheless, use of these two spaces can help make this argument more transparent.

Consider, for concreteness, the $L^1(\mathbb{R}^n)$ space, or to be more intuitive, the metric induced by the related l^1 norm on $\mathbb{R}^2$. This metric, for $\vec{x} = (x_1, x_2), \vec{y} = (y_1, y_2)$ where the coordinates are considered with respect to a Cartesian system, is given by

$$d(\vec{x}, \vec{y}) = |x_1 - x_2| + |y_1 - y_2| \tag{73}$$

Then one can see that there is an infinity of geodesics connecting $\vec{x}$ and $\vec{y}$. This by itself is not a drawback: after all, the north and south poles of a sphere with the induced metric from the Euclidean space are also connected by infinite geodesics (the meridians). The definition of geodesics is not a problem either: in metric geometry [71] they can be defined to be the isometric images of the unit interval. The problem exists because many of the geodesics between $\vec{x}$ and $\vec{y}$ in the (73) are branching: geodesics that have initially a common segment can separate after a while. If one assumes strong locality in a theory of gravity, whose metric is even "Euclidianized" (made positive-definite) after a Wick rotation, then this presents a problem with predictability. Since the theory does not possess any "memory" in its formulation, how then one can make any prediction based on the behavior of geodesics which largely encode the underlying geometry, if the theory has branching geodesics? If, for instance, the action $\mathcal{S}$ or the resulting kinematic equations possessed some form "memory" as in the case of systems being modelled by fractional derivatives [72–74], then the use of geometric structures with branching geodesics might not pose a serious problem to predictability. Knowing this, one may wish to stay as far away as possible from using metrics that may allow for the possibility of branching geodesics. It turns out that the Hilbert space $L^2(\mathbb{R}^n)$ is the furthest away from resembling $L^1(\mathbb{R}^n)$ which has branching geodesics, at least when one uses the modulus of convexity to quantify such a difference.

The counter-argument to the above is that $L^1(\mathbb{R}^n)$ has branching geodesics exactly because it is not uniformly convex. If someone chose any other $L^p(\mathbb{R}^n)$, $p \neq 1$ then this problem would not exist. This is largely correct. However it assumes very much like many occasions in classical Physics that some objects can be well-approximated by point particles, which in the absence of (non-gravitational) forces move along causal (in the Lorentzian signature framework) geodesics. When the quantum nature of such an object comes into consideration though, even in a fixed, classical background space-time, this statement would not be accurate. The uncertainty principle would prevent such point-like structures from existing; this introduces substantial technical complications for any operator in the assumed Fock space of a quantum gravity theory following the canonical approach, such as loop gravity for instance: the operator has to be "smeared", namely to act on test not at a particular point but on an appropriately chosen neighborhood of it before applying the canonical commutation relations [75].

The result is that a wave-function will sweep out a tube, for short times, rather than a line, in such a space-time as it evolves. For predictability purposes, since the Schrödinger, the Klein-Gordon, the Dirac etc. equations involve usual derivatives, as opposed to being integro-differential equations that may signify that memory effects are taken into account, it is quite important for such tubes not to have a branching property. Naively speaking and without getting into any details, we believe that this goal has the best chances of being realized, if the underlying space has a metric which is as far away from having branching geodesics as possible, which again brings us back to favoring the use of a Hilbert space.

The space-time metric from Hilbert spaces. Going from $L^2(\mathbb{R}^n)$ to the metric of space-time itself is quite straightforward, in principle. Consider as the linear spaces of interest to be appropriate $A \subset \mathbb{R}^n$. For the case of point particles this will be the tangent to the particles's space-time evolution trajectory. We can then confine ourselves to the analysis of a subspace of $L^2(\mathbb{R}^n)$ which is comprised of the characteristic functions χ_A of such subspaces. Then the quadratic metric on $L^2(\mathbb{R}^n)$ gets induced on such A which acquires itself a quadratic metric. Use the equivalence principle and "patch together" such A endowed with their Euclidean metrics to form the space-time of interest. This is a kinematic construction. The dynamics is provided by Einstein's equations, after a Wick rotation back to indefinite signature metrics. This transition between metrics of different signatures may involve several subtleties which may have to be addressed at that stage, but this is outside the scope of the present work.

5. Discussion and Outlook

In this work we have presented a non-rigorous, conjectural argument that aims to explain why the metric of space-time, after being suitably Wick-rotated, obeys the Pythagorean theorem. Such a metric can be seen as descending from the metrics of appropriate Banach spaces which provide a reasonable kinematical framework of a mesoscopic description/quasi-classical limit of models of quantum gravity. The advantage of this approach is also its disadvantage: it is kinematical with ad hoc aspects. It does not attempt to delve into the actual realm of the approaches to quantum gravity, however it is motivated by and uses some of the common points of such approaches. We relied on aspects of General Relativity and the Einstein equations for some motivation, used the spirit of the variational principles and path-integrals to formulate our approach in the spirit of classical and quantum Physics and then for our proposal we used standard results from functional analysis, convexity and the theory of normed spaces.

As in any, partly ad hoc and conjectural, work the virtues of the current work may be seen as out-weighting their shortcomings, or vice versa. The value of work like the present, which may appear to purport to justify the "obvious", may lie in the ideas involved in it and in the methods used to reach the conclusions. Much more importantly from a physical viewpoint, it may also point out to reasons about the inapplicability of its conclusions should pertinent experiments refute our currently held ideas about space-time properties, or should one observe such exceptions at, or beyond, the galaxy cluster or the Planck scales.

We would like to add that there should be some skepticism regarding the role of the Wick rotation of the space-time signature to a positive-definite one. There is little doubt that the Wick rotation has been very successful in obtaining results in perturbative quantum field theories, or more generally when employing saddle point approximations, which may otherwise be ill-defined or inaccessible in a covariant approach to such relativistic theories. It is not clear, to us at least, to what extent employing such Euclidean (positive-definite) metrics is equivalent to purely Lorentzian arguments and results, in particular in regimes outside the possible domain of validity of such saddle-point approximations. It is well-known now that many fundamental results of Riemannian geometry (such as the Hopf-Rinow theorem etc.) are either not valid in Lorentzian geometry or may become valid when appropriate modifications are made. Such discrepancies become even more pronounced when one considers the topological and causal formalism of space-times and other such statements of Lorentzian geometry which have no obvious analogue in the Riemannian case [35,76,77]. Since our results rely on positive-definite metrics of the underlying (linear) spaces, we can only be skeptical about their applicability in the physical, indefinite (Lorentzian) signature, case.

A technical point that may be worth addressing before closing, is the fact that all the analysis in this work uses in a very essential way the particular moduli of convexity and smoothness whose definitions and properties have been stated above. Naturally, these are mathematical constructions and cannot conceivably be unique or automatically be considered the "most useful" among their peers. And actually they are not. As mentioned in the opening paragraphs of this work, there are several, generally inequivalent, moduli encoding convexity and smoothness [38]. If history is any guide, new such moduli will keep being defined, their properties being examined and their values will be calculated in concrete cases in the future. We have chosen the above two moduli because they appear to us to be the simplest, the most intuitive and also the most developed. It is theoretically possible that other moduli may single out other spaces, as opposed to Hilbert ones, but we have not been able to find any such results in the existing literature, nor can we see any overwhelming physical reason for their implementation.

Acknowledgments: This work was, largely, performed when the author was a faculty member at the Weill Cornell Medicine-Qatar.

Conflicts of Interest: The author declares no conflict of interest.

References

1. Pettini, M. Geometry and Topology in Hamiltonian Dynamics and Statistical Mechanics. In *Interdisciplinary Applied Mathematics*; Springer: New York, NY, USA, 2007; Volume 33.
2. Kriegl, A.; Michor, P.W. The Convenient Setting of Global Analysis. In *Mathematical Surveys and Monographs*; American Mathematical Society: Providence, RI, USA, 1997; Volume 53.
3. Corichi, A.; Rubalcava-García, I.; Vukašinac, T. Actions, topological terms and boundaries in first order gravity: A review. *Int. J. Mod. Phys. D* **2016**, *25*, 1630011.
4. Kar, A.; Rajeev, S.G. A Non-Riemannian Metric on Space-Time Emergent from Scalar Quantum Field Theory. *Phys. Rev. D* **2012**, *86*, 065022.
5. Sorkin, R.D. Is the spacetime metric Euclidean rather than Lorentzian? In *Recent Research in Quantum Gravity*; Dasgupta, A., Ed.; Nova Science Publishers Inc.: New York, NY, USA, 2013.
6. Bojowald, M. Canonical Relativity and the Dimensionality of the World. In *Relativity and the Dimensionality of the World*; Petkov, V., Ed.; Springer: New York, NY, USA, 2007; pp. 137–152.
7. Nawarajan, D.; Visser, M. Global properties of physically interesting Lorentzian spacetimes. *arXiv* **2016**, arXiv:1601.03355.
8. Gibbons, G.W.; Ishibashi, A. Convex Functions and Spacetime Geometry. *Class. Quant. Grav.* **2001**, *18*, 4607–4628.
9. Rudin, W. *Functional Analysis*, 2nd ed.; McGraw-Hill Inc.: New York, NY, USA, 1991.
10. Lieb, E.H.; Loss, M. *Analysis*, 2nd ed.; American Mathematical Society: Providence, RI, USA, 2001.
11. Hájek, P.; Santalucía, V.M.; Vanderwerff, J.; Zizler, V. *Biorthogonal Systems in Banach Spaces*; Springer: New York, NY, USA, 2008.
12. Simon, B. The classical moment problem as a self-adjoint finite difference operator. *Adv. Math.* **1998**, *137*, 82–203.
13. Weinstein, A. The local structure of Poisson manifolds. *J. Diff. Geom.* **1983**, *18*, 523–557.
14. Klauder, J. Enhanced quantum procedures that resolve difficult problems. *Rev. Math. Phys.* **2015**, *27*, 1530002.
15. Tsallis, C. *Introduction to Nonextensive Statistical Mechanics: Approaching a Complex World*; Springer: New York, NY, USA, 2009.

16. Kaniadakis, G. Theoretical Foundations and Mathematical Formalism of the Power-Law Tailed Statistical Distributions. *Entropy* **2013**, *15*, 3983–4010.

17. Roe, J. *Lectures on Coarse Geometry*; American Mathematical Society: Providence, RI, USA, 2003.

18. Bridson, M.R.; Haefliger, A. *Metric Spaces of Non-Positive Curvature*; Springer: Berlin, Germany, 1999.

19. Gromov, M. Hyperbolic Groups. In *Essays in Group Theory*; Gesten, S.M., Ed.; Springer: New York, NY, USA, 1987; pp. 75–264.

20. Katok, A.; Hasselblatt, B. *Introduction to the Modern Theory of Dynamical Systems*; Cambridge University Press: Cambridge, UK, 1995.

21. Heinonen, J. Nonsmooth calculus. *Bull. Amer. Math. Soc.* **2007**, *44*, 163–232.

22. Cheeger, J. Differentiability of Lipschitz functions on metric measure spaces. *Geom. Funct. Anal.* **1999**, *9*, 428–517.

23. Keith, S. A differentiable structure for metric measure spaces. *Adv. Math.* **2004**, *183*, 271–315.

24. Cheeger, J.; Kleiner, B. Differentiability of Lipschitz maps from metric measure spaces to Banach spaces with the Radon-Nikodým property. *Geom. Funct. Anal.* **2009**, *19*, 1017–1028.

25. Bate, D. Structure of measures in Lipschitz differentiability spaces. *J. Amer. Math. Soc.* **2015**, *28*, 412–482.

26. Cheeger, J.; Kleiner, B.; Schioppa, A. Infinitesimal structure of differentiability spaces, and metric differentiation. *arXiv* **2016**, arXiv:1503.07348.

27. James, R.C. Some self-dual properties of normed linear spaces. In *Symposium on Infinite Dimensional Topology, 1967*; Princeton University Press: Princeton, NJ, USA, 1972; pp. 159–175.

28. Maurey, B. Type, Cotype and *K*-convexity. In *Handbook of the Geometry of Banach Spaces*; Johnson, W.B., Lindenstrauss, J., Eds.; Elsevier Science B.V.: Amsterdam, The Netherlands, 2003; pp. 1299–1332.

29. Amir, D. *Characterizations of Inner Product Spaces*; Springer: Basel, Switzerland, 1986.

30. Lindenstrauss, J.; Tzafriri, L. *Classical Banach Spaces I and II*; Reprint of the 1977 and 1979 Editions; Springer: Berlin, Germany, 1996.

31. Benyamini, Y.; Lindenstrauss, J. *Geometric Nonlinear Functional Analysis*; American Mathematical Society: Providence, RI, USA, 2000.

32. Rockafellar, R.T. *Convex Analysis*; Princeton University Press: Princeton, NJ, USA, 1970.

33. Verpoort, S. The Geometry of the Second Fundamental form: Curvature Properties and Variational Aspects. Ph.D. Thesis, Katholieke Universiteit Leuven, Leuven, Belgium, May 2008.

34. Bishop, R. Infinitesimal convexity implies local convexity. *Ind. Univ. Math. J.* **1974/1975**, *24*, 168–172.

35. Chruściel, P.T.; Galloway, G.J.; Pollack, D. Mathematical general relativity: A sampler. *Bull. Amer. Math. Soc.* **2010**, *47*, 567–638.

36. Parikh, M. Two Roads to the Null Energy Condition. *Int. J. Mod. Phys. D* **2015**, *24*, 1544030.

37. Cordero-Erausquin, D.; McCann, R.J.; Schmuckenschläger, M. A Riemannian interpolation inequality à la Borell, Brascamp and Lieb. *Invent. Math.* **2001**, *146*, 219–257.

38. Fuster, E.L. Moduli and Constants: ... what a show! Available online: http://www.uv.es/llorens/Documento. pdf (accessed on 5 November 2016).

39. Clarkson, J.A. Uniformly convex spaces. *Trans. Amer. Math. Soc.* **1936**, *40*, 394–414.

40. Day, M.M. Uniform convexity in factor and conjugate spaces. *Ann. Math.* **1944**, *45*, 375–385.

41. Gromov, M. Sign and geometric meaning of curvature. *Rend. Mat. Sem. Mat. Fis. Milano Milan J. Math.* **1991**, *61*, 9–123.

42. Goebel, K. Convexity of balls and fixed point theorems for mapping with nonexpansive maps. *Comp. Math.* **1970**, *22*, 269–274.

43. Milman, D.P. On some criteria for the regularity of spaces of the type (B). *Dokl. Acad. Nauk. SSSR* **1938**, *20*, 243–246.

44. Pettis, B.J. A proof that every uniformly convex space is reflexive. *Duke Math. J.* **1939**, *5*, 249–253.

45. Enflo, P. Banach spaces which can be given an equivalent uniformly convex norm. *Israel J. Math.* **1972**, *13*, 281–288.

46. James, R.C. Super-reflexive Banach spaces. *Can. J. Math.* **1972**, *24*, 896–904.

47. Hanche-Olsen, H. On the uniform convexity of L^p. *Proc. Amer. Math. Soc.* **2006**, *134*, 2359–2362.

48. Hanner, O. On the uniform convexity of L^p and l^p. *Arkiv. Mat.* **1956**, *3*, 239–244.

49. Ball, K.; Carlen, E.A.; Lieb, E.H. Sharp uniform convexity and smoothness inequalities for trace norms. *Invent. Math.* **1994**, *115*, 463–482.

50. Nordlander, G. The modulus of convexity in normed linear spaces. *Arkiv. Mat.* **1960**, *4*, 15–17.
51. Day, M.M. Some characterizations of inner product spaces. *Trans. Amer. Math. Soc.* **1947**, *62*, 320–337.
52. Schoenberg, I.J. A remark on M.M. Day's characterization of inner product spaces and a conjecture of L.M. Blumenthal. *Proc. Amer. Math. Soc.* **1952**, *3*, 961–964.
53. Figiel, T. On the moduli of convexity and smoothness. *Stud. Math.* **1976**, *56*, 121–155.
54. Grigorchuk, R.; Pak, I. Groups of intermediate growth, an introduction. *L'Enseign. Math.* **2008**, *54*, 251–272.
55. Kalogeropoulos, N. Tsallis entropy composition and the Heisenberg group. *Int. J. Geom. Methods Mod. Phys.* **2013**, *10*, 1350032.
56. Kalogeropoulos, N. Groups, non-additive entropy and phase transitions. *Int. J. Mod. Phys. B* **2014**, *28*, 1450162.
57. Scott, P. The geometries of 3-manifolds. *Bull. London Math. Soc.* **1983**, *15*, 401–487.
58. Lindenstrauss, J. On the modulus of smoothness and divergent series in Banach spaces. *Mich. Math. J.* **1963**, *10*, 241–252.
59. Asplund, E. Averaged norms. *Israel J. Math.* **1967**, *5*, 227–233.
60. Kwapień, S. Isomorphic characterizations of inner product spaces by orthogonal series with vector valued coefficients. *Stud. Math.* **1972**, *44*, 583–595.
61. Kahane, J.P. *Some Random Series of Functions*, 2nd ed.; Cambridge University Press: Cambridge, UK, 1985.
62. Milman, V.D.; Schechtman, G. *Asymptotic Theory of Finite Dimensional Normed Spaces*; Springer: Berlin, Germany, 1986.
63. Figiel, T.; Pisier, G. Séries aléatoires dans les espaces uniformément convexes ou uniformément lisses. *C. R. Acad. Sci. Paris Ser. A* **1974**, *279*, 611–614.
64. Artstein-Avidan, S.; Milman, V. The concept of duality in convex analysis, and the characterization of the Legendre transform. *Ann. Math.* **2009**, *169*, 661–674.
65. Lanczos, C. *The Variational Principles of Mechanics*, 4th ed.; University Toronto Press: Toronto, ON, Canada, 1970.
66. Zinn-Justin, J. *Quantum Field Theory and Critical Phenomena*, 3rd ed.; Oxford University Press: Oxford, UK, 1996.
67. Dirac, P.A.M. The Lagrangian in Quantum Mechanics. In *Physikalische Zeitschrift der Sowjetunion*; Charkow, Technischer staatsverlag: Charkow, Ukraine, 1933; pp. 64–72.
68. Ambjørn, J.; Görlich, A.; Jurkiewicz, J.; Loll, R. CDT and the Search for a Theory of Quantum Gravity. *arXiv* **2013**, arXiv:1305.6680.
69. Veltman, M.J.G. Quantum Theory of Gravitation. In *Methods in Field Theory, Proceedings of the Les Houches Summer School of Theoretical Physics, Session 28, 28 July–6 September 1975*; Balian, R., Zinn-Justin, J., Eds.; North Holland Publisher: Amsterdam, The Netherlands, 1981.
70. Villani, C. Synthetic Theory of Ricci Curvature Bounds, Takagi Lectures 2015. *Jpn. J. Math.* **2016**, *11*, 219–263.
71. Burago, D.; Burago, Y.; Ivanov, S. *A Course in Metric Geometry*; American Mathematical Society: Providence, RI, USA, 2001.
72. Laskin, N. Fractional Quantum Mechanics. *Phys. Rev. E* **2000**, *62*, 3135–3145.
73. Tarasov, V.E. *Fractional Dynamics: Applications of Fractional Calculus to Dynamics of Particles, Fields and Media*; Springer: New York, NY, USA, 2011.
74. Calgagni, G. Geometry and field theory in multi-fractional spacetime. *J. High Energy Phys.* **2012**, *2012*, 65.
75. Ashtekar, A.; Lewandowski, J. Background Independent Quantum Gravity: A Status Report. *Class. Quant. Grav.* **2004**, *21*, R53.
76. Beem, J.K.; Ehrlich, P.E.; Easley, K.L. *Global Lorentzian Geometry*, 2nd ed.; Marcel Dekker Inc.: New York, NY, USA, 1996.
77. Minguzzi, E.; Sánchez, M. The causal hierarchy of spacetimes. In *Recent Developments in Pseudo-Riemannian Geometry*; Alekseevsky, D.V., Baum, H., Eds.; ESI Lectures in Mathematics and Physics; European Mathematical Society: Zürich, Switzerland, 2008; pp. 299–358.

MDPI AG

St. Alban-Anlage 66

4052 Basel, Switzerland

Tel. +41 61 683 77 34

Fax +41 61 302 89 18

http://www.mdpi.com

Universe Editorial Office

E-mail: universe@mdpi.com

http://www.mdpi.com/journal/universe

www.ingramcontent.com/pod-product-compliance
Lightning Source LLC
Chambersburg PA
CBHW051703210326
41597CB00032B/5350